Introduction to Biomimetics and Bioinspiration

Bharat Bhushan

Introduction to Biomimetics and Bioinspiration

Materials and Surfaces for Green Science and Technology

Springer

Bharat Bhushan
San Jose, CA, USA

ISBN 978-3-031-62343-1 ISBN 978-3-031-62344-8 (eBook)
https://doi.org/10.1007/978-3-031-62344-8

This Springer imprint is published by the registered company Springer Nature Switzerland AG
The registered company address is: Gewerbestrasse 11, 6330 Cham, Switzerland

To my grandkids Sahana, Ashwin, Joya, and Nivaan

Preface

Biomimetics is derived from the Greek word biomimesis. It means **mimicking biology or living nature, or living organisms**, and is also called biomimicry. **Biologically inspired** research allows derivation of optimal designs benefiting from improvements made during the evolution of living nature and efficient use of natural resources in a more sustainable and environmentally friendly (green) manner. Biological materials are highly organized from the molecular to the nano-, micro-, and macroscales, often in a hierarchical manner with intricate nanoarchitecture that ultimately makes up a myriad of different functional elements. The field of biomimetics or bioinspired materials and surfaces started to take off in the early 2000s with a major emphasis on nanoscience and nanotechnology. Nanofabrication techniques made it possible to create natural surfaces with features ranging from the molecular scale to the macroscale. The bioinspired materials and surfaces can be eco-friendly and are used for various commercial applications. They have generated significant interest and are helping to shape what we call "green science and technology," a term used for the first time in the first edition of author's biomimetics monogram published in 2012. The interest in green science and technology has also provided impetus for advancement. The emphasis should be on durable and self-healing materials and surfaces. The field is highly interdisciplinary, spanning from biology, physics, chemistry, materials science, and engineering. As of 2023, the field contains only a handful of visionaries and leaders. This author is fortunate to be one of the early explorers.

Since the early 2000s, there have been significant advances in research, and many ideas have been commercialized. With a continued understanding of the mechanisms relevant in species of living nature and the development of new materials and nanofabrication techniques, rapid advancements are expected in the next decade and beyond. It is expected that new inventions will play a major role in human life.

The first edition of the biomimetics research monogram (Biomimetics: Bioinspired Hierarchical-Structured Surfaces for Green Science and Technology) was published in 2012. Given rapid advancements in the author's lab, there was a need to publish an updated second edition in 2016 and a third edition in 2018. These monograms provide a state of the art of the biomimetics field primarily related to interface science.

Given the significant interest in the field of biomimetics, several universities offer undergraduate and graduate courses in this field. A textbook is needed. There is also a need for a simpler and shorter introductory book for researchers and practitioners. By taking some chapters from his biomimetics 2018 monogram and supplementing with new chapters, the author has prepared this textbook on biomimetics and bioinspiration. The book starts with a fundamental understanding of how various living species provide functionality, followed by modeling to develop optimum structures and then fabrication of optimum structures using smart materials and fabrication techniques to provide functionality of commercial interest. The emphasis is on durable materials and surfaces. The book contains overviews of several living species and examples of various structures fabricated in the lab.

The book should serve as an excellent textbook in biomimetics/bioinspiration for a one-semester graduate or a two-semester undergraduate course. The book can also serve as a textbook for an applied nanotechnology course. The book is also intended for use by novices as well as experts in the field, practitioners, solution seekers, and the curious. The book should help in the advancement of the field.

The book is based on the work of outstanding contributions by past and present collaborators. These include:

Past and present students:

Zach Burton, Yong Chae Jung, James Hunt, Robert A. Sayer, Eun Kyu Her (Seoul National University, Korea), Brian Dean, Daniel Ebert, Gregory D. Bixler, Yunlu Pan and Dalei Jing (Harbin Institute of Technology, China), Srimala Perara (University of Moratuwa, Sri Lanka), Andrew Theiss, Shan Peng (Wuhan University of Science and Technology, China), Samuel G. Martin, Wenjing Ma (Nanjing Forestry University, China), Feiran Li (Harbin Institute of Technology, China), Wenting Kong (Harbin Institute of Technology, China), Wei Feng (Chongqing University, China), and Dev Gurera.

Past and present postdoctoral fellows and visitors:

Dr. Michael Nosonovsky (University of Wisconsin, Milwaukee), Dr. Andrei G. Peressadko (Russia), Dr. Tae-Wan Kim (Pukyong National University, Korea), Prof. Kerstin Koch (University of Bonn, Germany), Dr. Manuel L. B. Palacio (Western Digital Corp., San Jose, California), Dr. Hyungoo Lee, Dr. P. K. Muthiah, Dr. Yongxin Wang, Dr. Philip S. Brown, Dr. Victor Multanen, Dr. Dong Song, and Dr. Joseph C. Cremaldi.

External Collaborators:

Prof. Wilhelm Barthlott (University of Bonn, Germany), Prof. Kerstin Koch (Rhine-Waal University of Applied Science, Kleve, Germany, Dr. Eui-Sung Yoon (KIST,

Korea), Dr. Patrick Hoffman (EPFL, Lausanne, Switzerland), Dr. Andre Meister (CSEM, Switzerland), Prof. Scott. C. Schricker (The Ohio State University, Columbus, Ohio), Prof. Joao F. Mano (University of Minho, Guimaraes, Portugal), Prof. Jiyu Sun (Jilin University, China), Prof. Shunsuke Nishimoto (Okayama University, Japan), Prof. Stanislav N. Gorb (University of Kiel, Germany), Prof. Eduard Arzt (Max Planck Institute for Metals Research, Stuttgart, Germany), Prof. Stephen C. Lee (The Ohio State University, Columbus, Ohio), and Prof. Aditya Kumar (Indian Institute of Technology, Dhanbad, India).

Their contributions have shaped the emerging field of biomimetics and are gratefully acknowledged. Particularly, pioneering and sustained contributions by Y. C. Jung and D. Gurera (Ph.D. Students), Drs. M. Nosonovsky and P. S. Brown (postdoctoral fellows), and Prof. K. Koch (external collaborator) are noteworthy. Finally, the author would like to thank his family for their constant support and encouragement.

March 2024

Bharat Bhushan
The Ohio State University
San Jose, USA

Nelumbo nucifera (Lotus leaf)

Blue Morpho didius (Butterfly)

Isurus oxyrinchus (Shark)

Rynchops (Skimmer bird)

Gekko gecko (Tokay gecko)

Opuntia microdasys (Cactus)

Aedes vexans (Mosquito)

Contents

About the Author

Dr. Bharat Bhushan is Academy Professor (San Jose, CA) and has served as an Ohio Eminent Scholar and the Howard D. Winbigler Professor in the College of Engineering, Director of the Nanoprobe Laboratory for Bio- and Nanotechnology and Biomimetics (NLB^2), and affiliated faculty in John Glenn College of Public Affairs at the Ohio State University, Columbus, Ohio. He received an M.S. in mechanical engineering from the Massachusetts Institute of Technology, an M.S. in mechanics and a Ph.D. in mechanical engineering from the University of Colorado at Boulder, an MBA from Rensselaer Polytechnic Institute at Troy, NY, Doctor Technicae from the University of Trondheim at Trondheim, Norway in 1990, a Doctor of Technical Sciences from the Warsaw University of Technology at Warsaw, Poland, in 1996, Honorary Doctor of Science from the National Academy of Sciences, Gomel, Belarus, in 2000, University of Kragujevac, Serbia, in 2011, and University of Tyumen, Russia, in 2019. He is a registered professional engineer. In 2013–14, he served as ASME/AAAS Science and Technology Policy Fellow, House Committee on Science, Space and Technology, United States Congress, Washington, DC. He has served as Expert Investigator on IP-related issues in the US and International Courts. His research interests are in fundamental studies in the interdisciplinary areas of bio/nanotribology/nanomechanics, nanomaterials characterization, scanning probe techniques, magnetic storage, bio/nanotechnology, nanomanufacturing, bioinspired liquid repellency, self-cleaning, anti-icing, anti-fouling, and water harvesting, science and technology policy. He is an internationally recognized expert of bio/nanotribology and bio/nanomechanics using scanning probe microscopy and biomimetics. He is considered by some of the pioneers of tribology and mechanics of magnetic storage devices, nanotribology, green tribology, and biomimetics. He had introduced the word nanotribology in the title of a Nature paper in 1995 and in the title of a first book on green tribology in 2010. He is one of the most prolific authors. He has authored 10 scientific books, 100+ handbook chapters, and 900+ scientific papers. He is Google Scholar's one of 1248 Highly Cited Researchers in All Fields, h-index—145+ with 100 k+ citations; Scopus's one of 401 Scientists for Career-long Citation Impact Across All Fields out of over 8 million scientists from around the world; 4th Highly Cited Researcher in Mechanical Engineering; 149th Most Cited in Materials Science in the World; and ISI Highly Cited Researcher in Materials Science and in Cross-field Category. His research was listed as one of the Top Ten Science Stories of 2015. He has also edited more than 60 books and holds more than 25 US and foreign patents. He is a co-editor of Springer NanoScience and Technology Series and a co-editor of Microsystem Technologies. He has given more than 400 invited presentations including 300+ keynote/plenary addresses at major international conferences on six continents. He delivered a *TEDx 2019 lecture* on Lessons from Nature.

His biography has been listed in over two dozen Who's Who books including Who's Who in the World. He has received more than two dozen awards for his contributions to science and technology from professional societies, industry, and US government agencies including International Tribology Gold Medal and Institution of Chemical Engineers (UK) Global Award for bioinspired surfaces. He received NASA's Certificate of Appreciation to recognize the critical tasks performed in support of President Reagan's Commission investigating the Space Shuttle Challenger Accident. He is also the recipient of various international fellowships including the Alexander von Humboldt Research Prize for Senior Scientists, Max Planck Foundation Research Award for Outstanding Foreign Scientists, and Fulbright Senior Scholar Award. He is foreign member of the International Academy of Engineering (Russia), Byelorussian Academy of Engineering and Technology, and the Academy of Triboengineering of Ukraine, an honorary member of the Society of Tribologists of Belarus and STLE, a fellow of ASME, IEEE, and the New York Academy of Sciences, and a member of ASEE, Sigma Xi, and Tau Beta Pi.

He is an accomplished organizer. He organized the First Symposium on Tribology and Mechanics of Magnetic Storage Systems in 1984 and the First Int. Symposium on Advances in Information Storage Systems in 1990, both of which are now held annually. He organized two international NATO institutes in Europe. He is the founder of an ASME Information Storage and Processing Systems Division founded in 1992 and served as founding chair during 1993–1998.

He has previously worked for Mechanical Technology Inc., Latham, NY; SKF Industries Inc., King of Prussia, PA; IBM, Tucson, AZ; and IBM Almaden Research Center, San Jose, CA. He has held visiting professorship at University of California at Berkeley, University of Cambridge, UK, Vienna University of Technology, Austria, University of Paris, Orsay, ETH Zurich, EPFL Lausanne, University of Southampton, UK, University of Kragujevac, Serbia, Tsinghua University, China, Harbin Institute, China, Indian Institute of Science, Bengaluru, BITS Pilani and Hyderabad, and KFUPM, Saudi Arabia.

Introduction 1

Look deep into nature and you will understand everything—Albert Einstein

Nature always tends to act in the simplest way—Bernoulli.

(In nature,) Nothing is lacking and nothing is superfluous—Leonardo da Vinci.

Biomimetics research allows derivation of optimal designs benefitting from improvements made during evolution of living nature and efficient use of natural resources in a more sustainable and environmentally friendly (green) manner, we call it "green science and technology"—Bharat Bhushan (2012)

1.1 Biomimetics/Bioinspiration and Green Science and Technology

Biomimetics is derived from the Greek word biomimesis. It means **mimicking biology or living nature or living organisms**. It is also called biomimicry. The word biomimetics was coined by polymath Otto Schmitt in 1957, who, in his doctoral research, developed a physical device that mimicked the electrical action of a nerve. Another word often used in Europe and Asia is bionics, coined by Jack Steele of Wright-Patterson Air Force Base in Dayton, Ohio in 1960. Bionics combines two words, biology, and technology/electronics. It translates to the creation of products, devices, and processes by using materials and processes found in living nature. Bionics is referred to as robotics and replacement or enhancement of living matter, tissue, body parts, and organs. An example of bionics

B. Bhushan, *Introduction to Biomimetics and Bioinspiration*,
https://doi.org/10.1007/978-3-031-62344-8_1

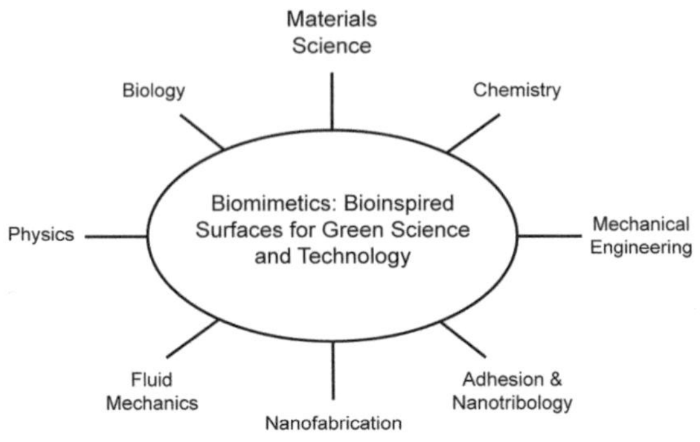

Fig. 1.1 Interdisciplinary nature of the field of biomimetics (Bhushan 2018)

would be the "bionic man". Finally, another word used is biognosis, which is defined as the scientific investigation of life (Bhushan 2009, 2012a, 2016a, 2018, 2019a, 2020a).

The word biomimetics is commonly used in the English-speaking world. It first appeared in Webster's dictionary in 1974 and is defined as "the study of the formation, structure or function of biologically produced substances and materials (as enzymes or silk) and biological mechanisms and processes (as protein synthesis or photosynthesis) especially for the purpose of synthesizing similar products by artificial mechanisms which mimic natural ones."

The field of biomimetics is highly interdisciplinary, as shown in Fig. 1.1. It involves the understanding of biological functions, structures, and principles of various objects found in nature by biologists, physicists, chemists, material scientists, fluid mechanics experts, engineers, nanotribologists, and nanotechnologists (Bhushan 2018). It can lead to biologically inspired design, adaptation, or derivation from living nature. Bioinspired or biodesigns can be fabricated using smart materials and modern manufacturing processes in a more sustainable and environmentally friendly manner. An appropriate term for the field would be **Bioinspiration** or **biodesign**, as it applies to engineering applications (Bhushan 2009, 2012a, 2016a, 2018, 2020a).

Albert Einstein stated, "Look deep into nature and you will understand everything". The reason behind that was that nature has gone through evolution over the 3.8 billion years since life is estimated to have appeared on the Earth (Darwin's theory of biological evolution; Darwin 1859). It has evolved species with desired functionality using commonly found materials and routine fabrication methods materials (Gordon 1976; Bhushan 2009, 2018). Given that nature has gone through evolution, Bernoulli stated that "Nature always tends to act in the simplest way," and Leonardo da Vinci stated that "(in nature,) nothing is lacking and nothing is superfluous." The author literally believes in these quotes (Bhushan 2012a, 2016a, 2018).

Nature has a limited toolbox. To provide functionality, biological materials generally have laminated and/or hierarchical structure with dimensions of features ranging from the macroscale to the nanoscale (Alberts et al. 2008; Nosonovsky and Bhushan 2008; Bhushan 2009, 2018, 2020a). Properties of materials and surfaces result from a complex interplay between surface structure and morphology and physical and chemical properties. Many materials, surfaces, and objects provide multi-functional properties.

Various materials, structures, and devices have been fabricated for commercial interest by engineers, material scientists, chemists, and biologists, and for beauty, structure, and design by artists and architects (Bhushan 2009, 2012a, 2016a, 2018, 2020a). The economic impact of bioinspired materials and surfaces is significant, on the order of several hundred billion dollars per year worldwide.

1.1.1 Green Science and Technology

The major benefit of biomimetics research is that it allows derivation of optimal designs benefitting from improvements made during evolution of living nature and efficient use of natural resources in a more sustainable and environmentally friendly (green) manner. Bioinspired designs can also provide strategies to recover from environmental impact and for survival, referred to as self-healing or adaptation. Biologically inspired materials and surfaces have generated significant interest and are helping to shape what we call "green science and technology" (Bhushan 2012a).

1.1.2 Climate Change and Lack of Recycling Impact on Sustainable Environment

Climate change is a global issue. The amount of carbon dioxide and other greenhouse gases in the atmosphere significantly affects Earth's surface temperature. As plants photosynthesize and grow, they inhale carbon dioxide from the atmosphere into leaves and other parts and they exhale oxygen through their openings called stomata and this helps the climate. Decomposition of plant matter releases carbon dioxide into the air. Burning of fossil fuels also releases carbon dioxide locked in hundreds of millions of years in nature. In addition, animals (including humans) contribute to carbon dioxide as they inhale oxygen and exhale carbon dioxide. Ice core studies have shown that atmospheric carbon dioxide has been on the order of 300 parts per million (ppm) for some 800,000 years. However, it has started to increase since the 1960's. In 2017 it was recorded over 410 ppm. It is believed that more than 90% comes from burning fossil fuels.

Global warming interferes with the environmental cycle. That has an impact on survival of natural species that is interdependent. For example, every year, as seasons change trees in the northern hemisphere start to produce leaves as frost recedes. Caterpillars hatch to gorge on leaves. Insects emerge to pollinate flowers. Birds migrate from southern to northern hemisphere to lay eggs and feast on insects in the north. All these species stay in sync with one another by relying on environmental cues, like the ballet dancers moving with an orchestra. Global warming is constantly changing environmental cues

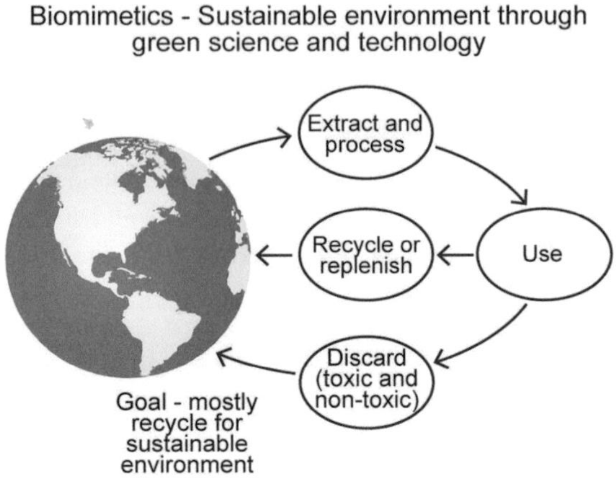

Fig. 1.2 Biomimetics to achieve a sustainable environment through green science and technology (Bhushan 2018)

and some species cannot adjust and are falling out of step; referred to as phenological mismatch. This lack of adjustment is having significant negative impact such as lack of food supply for animals and the rise of sea level. Carbon dioxide emissions need to be cut. Furthermore, negative emission approaches need to be developed such as "carbon capture" (Bhushan 2018).

Another global environmental issue is that today humans extract, process and use more than they recycle or replenish which is unsustainable, Fig. 1.2 (Bhushan 2018). Humans need to maintain a sustainable environment for their survival. Otherwise, planet Earth will head towards disastrous consequences.

1.2 Biodiversity

It is estimated that about 1.7 million different living organisms are known to exist. According to some estimates, Earth is home to some 10 million different living species (Mora et al. 2011). About 80% of these species are yet to be discovered. Given the loss of biodiversity, which began with the industrial era, some of the undiscovered organisms may be lost forever. Major groups of organisms account for very different proportions of biodiversity. Barthlott et al. (2014) have reported that arthropods make up the largest category, with roughly 5 million species, followed by bacteria (~2 million), other animals (~1.8 million), fungi (~600,000), and terrestrial plants (~370,000). The number of species per area varies considerably in different geographical regions. The highest diversity occurs in subtropical areas, and especially in the tropics.

1.3 Lessons from Living Nature

The understanding of the functions provided by species and processes found in living nature can guide us to design and produce nanomaterials, nanodevices, and processes (Bhushan 2009, 2012a, 2016a, 2018, 2020a). In nature, many flora and fauna exist. Many lessons can be learnt from these species, which then can be used to solve many problems. Molecular scale devices, superhydrophobicity, self-cleaning, drag reduction in fluid flow, antifouling, superhydrophilicity, energy conversion and conservation, reversible adhesion, aerodynamic lift, materials and fibers with high mechanical strength, biological self-assembly, water harvesting, painless piercing, anti-reflection, structural coloration, thermal insulation, self-healing, and sensory aid mechanisms are examples of functionalities found in living nature that are of commercial interest (Bhushan 2009, 2018, 2020a; Bar-Cohen 2011).

Figure 1.3 provides an overview of various species and objects from living nature and their selected functions (Bhushan 2009, 2012a, 2016a, 2018, 2019a, 2020a). These include bacteria (Jones and Aizawa 1991), plants (Koch et al. 2008, 2009; Barthlott et al. 2017), insects/spiders/lizards/frogs (Autumn et al. 2000; Gorb 2001; Bhushan 2007), aquatic animals (Bechert et al. 1997, 2000; Dean and Bhushan 2010; Bixler and Bhushan 2013a, 2015; Martin and Bhushan 2014), birds (Jakab 1990), seashells/bones/teeth (Lowenstam and Weiner 1989; Sarikaya and Aksay 1995; Mann 2001; Alexander and Diskin 2004; Meyers et al. 2006, 2008; Sun and Bhushan 2012a), spider web (Jin and Kaplan 2003; Bar-Cohen 2011), desert species (Bhushan 2020a, 2020b; Gurera and Bhushan 2020), mosquitoes (Gurera et al. 2018), and biological systems with self-healing capability (Fratzl and Weinkamer 2007; Nosonovsky and Bhushan 2009; Cremaldi and Bhushan 2018), and sensory aid devices (Barth et al. 2003; Bar-Cohen 2011).

Figure 1.4 shows a montage of selected examples from living nature (Bhushan 2009, 2012a, 2016a, 2018, 2019a, 2020a). Relevant details follow.

1.3.1 Bacteria

Many bacteria propel themselves by a type of extremely miniature motor called a flagellum motor. These motors rotate up to 100,000 rpm (Jones and Aizawa 1991). These are similar to an electrical motor and have a stator, rotor, and drive shaft with a bushing. The diameter of the bushing is approximately 20–30 nm with a clearance of approximately 1 nm. The flagellum motor is driven by the proton flow caused by the electrochemical potential difference across the cell membrane. It is an example of a biological motor (Bhushan 2009, 2018).

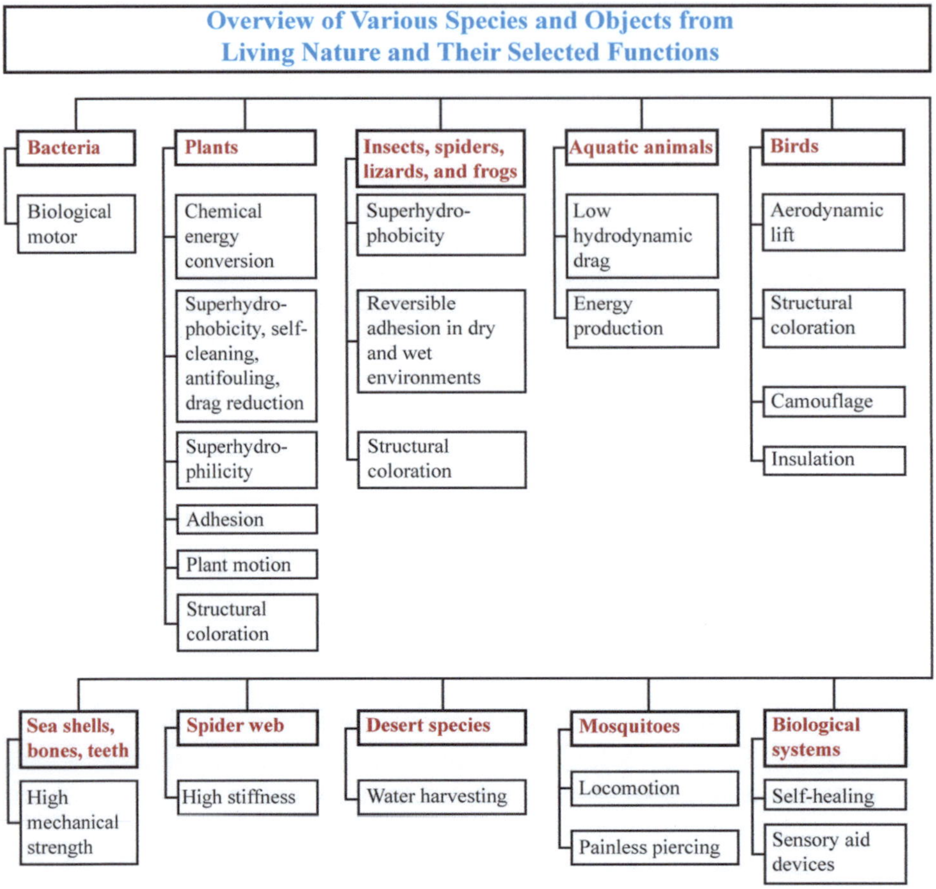

Fig. 1.3 An overview of various species and objects from living nature and their selected functions. (Adapted from Bhushan 2009, 2012a, 2016a, 2018, 2019a, 2020a)

1.3.2 Plants

The diversity of the structure and morphology of plant leaf surfaces provides multifunctional properties (Koch et al. 2008, 2009; Bhushan 2009, 2018; Barthlott et al. 2017). Plants use photosynthesis to harness solar energy to support plant life. Plant leaves can exhibit superhydrophobicity, self-cleaning, antifouling, drag reduction, superhydrophilicity, adhesion, and plant motion.

Leaves of water-repellent plants, such as *Nelumbo nucifera* (lotus), are known to be superhydrophobic, self-cleaning, and antifouling. The leaf surface consists of microbumps formed by convex papilla epidermal cells, covered with a 3-D epicuticular wax which self-assembles as nanotubules (Neinhuis and Barthlott 1997; Barthlott and Neinhuis 1997;

5 nm
flagellum motor
2 μm
lotus leaf
lid
peristome
pitcher plant
20 μm
water strider
butterfly wings
BR
SP
5 μm
gecko feet
0.5 mm
shark skin
bird & peacock
0.3 mm
nacre of abalone shell
spider web
cactus
mosquito
self-healing
sensory aid devices: ear and eye

◂**Fig. 1.4** Montage of some examples from living nature: schematic of bacteria and flagellum motor; lotus leaf with water repellant and self-cleaning properties (Bhushan et al. 2009); The slippery rim (peristome) of the pitcher plant captures insects by insect aquaplaning (Koch et al. 2009); water strider walking on water (Gao and Jiang 2004); butterfly wing with water repellant and self-cleaning properties and structural coloration (Bixler and Bhushan 2013b); gecko foot exhibiting reversible adhesion (Gao et al. 2005); scale structure of shark reducing drag (Jung and Bhushan 2010); wings of a bird in landing approach; brilliant iridescent colors in bird and peacock feathers produced using structural coloration; nacre of the inner abalone shell showing mesostructure with mesolayers (Meyers et al. 2006); spider web made of silk material (Bar-Cohen 2011); cactus with water harvesting from fog and condensation (Bhushan 2020a, b; Gurera and Bhushan 2020); mosquito with painless-piercing using fascicle mouthpart (Gurera et al. 2018); bruised or skinned knee which heals itself; and ears and eyes as examples of sensory aid devices. (Adapted from Bhushan 2009, 2012a, 2016a, 2018, 2019b, 2020b)

Wagner et al. 2003; Burton and Bhushan 2006; Bhushan and Jung 2006, 2011; Bhushan 2009, 2011, 2012b, 2018; Bhushan et al. 2009). The presence of wax makes the surface hydrophobic and multi-scale structure of the rough surface makes it superhydrophobic. Water droplets on these surfaces readily sit on the apex of nanostructures because air bubbles fill in the valleys of the structure under the droplet. Water droplets roll off the surface; rolling droplets clean any dirt present. Therefore, these leaves exhibit considerable superhydrophobicity, self-cleaning, and low adhesion (Bhushan et al. 2009).

Rice leaves survive even when submerged in water. Hierarchical structures consisting of micropapillae covered by nanobumps of epicuticular wax provide superhydrophobicity and self-cleaning with low adhesion (Bixler and Bhushan 2012b, 2013b, 2015; Bhushan 2018). Micropatterns consist of dual-height cylindrical pillars having sinusoidal pattern in the transverse direction, which provide anisotropic flow with low drag.

Unlike the lotus leaf, various species of rose petals exhibit superhydrophilicity with high or low adhesion. For example, rose petals species, rosea Rehd and Rosa Hybrid Tea, cv. Bairage, exhibit superhydrophobicity with high adhesion and Rosa Hybrid Tea cv. Showtime exhibit superhydrophobicity with low adhesion (Bhushan and Her 2010; Bhushan 2012b, 2016a, 2018). The different behaviour of wetting was explained by petals having different surface microstructures. The microstructure of a rose petal with high adhesion consists of a larger pitch value and lower height than of the one with low adhesion. Therefore, the liquid is allowed to impregnate between the microstructure and partially into the nanostructure, which increases the wetted area and leads to high adhesion.

The floating water ferns of genus *Salvinia* can trap and hold an air film underwater for up to several months (Koch et al. 2009; Bhushan 2012a, 2016a, 2018). Within the floating water ferns of the genus *Salvinia*, morphologically different kinds of water repellent (superhydrophobic) hairs exist. Depending on the species, the hair size varies in the order of several hundreds of micrometers, and hairs are visible in the order of several hundred micrometers, and hairs are visible with the naked eye. These multi-cellular hairs on the upper (adaxial) side of the leaves form complex hierarchical surface structures with hydrophilic patches at the tip of each hair can retain an air layer at the surface, even when leaves were fixed under water for several days. The hairs are eggbeater-shaped with tiny crowns. The ability to retain air prevents wetting and submersion.

Some plants provide defense mechanisms to prevent feeding by insects whereas some feed on insects (Koch et al. 2008; Bhushan 2009, 2012a, 2018, 2019a). As an example for defense, the surface of common Horsetail (*Equisetum arvensis*) consists of subcuticular inserts of silica crystals (not shown in Fig. 1.4). Silicon is a bioactive element and increases the resistance of plants to pathogenic fungi (Koch et al. 2008). The two strategies used by plants for catching insects for digestion are having either sticky or slippery surfaces. As an example of plants using sticky surfaces to catch insects, the glands of carnivorous plants of genus *Pinguicula* (butterworts) and *Drosera* (sundew) secrete adhesives and enzymes to trap and digest small insects (not shown in Fig. 1.4). In another example, pitcher plants (carnivorous, genus *Nepenthes*) create slippery surfaces using surface wetting, which causes insects to aquaplane into the digestion area (Bohn and Federle 2004; Bhushan 2009, 2018; Koch et al. 2009). The pitcher rim (peristome) consists of microstructure with radial ridges of smooth overlapping epidermal cells, which form a series of steps facing the inside of the pitcher. This anisotropic surface is completely wettable by nectar secreted at the inner margin of the peristome and by rainwater, so a water film covers the surface under humid conditions. The wet surface is slippery and causes the aquaplaning effect, forcing insects to slip inside the pitcher plants.

1.3.3 Insects, Spiders, Lizards, and Frogs

Insects, spiders, lizards, and frogs can exhibit superhydrophobicity, reversible adhesion in dry and wet environments, and some exhibit structural coloration. Water striders (*Gerris remigis*) can stand and walk upon a water surface without getting wet. Even the impact of rain droplets with a size greater than the water strider's size does not immerse it in the water. The water strider's legs consist of a hierarchical structure with a large number of oriented tiny hairs (microsetae) with fine nanogrooves that are covered with cuticle wax that makes the leg surfaces superhydrophobic and is responsible for their water resistance. Because of these structures, water striders can stand and walk quickly on the water surface (Gao and Jiang 2004; Bhushan 2009, 2012a, 2012b, 2016a, 2018).

Wings of butterflies, a family of insects, exhibit superhydrophobicity, self-cleaning, antifouling, and low drag (Bixler and Bhushan 2012b, 2015; Bhushan 2018). The wings have a hierarchical structure consisting of shingle-like scales covered by microgrooves oriented radially outward from the main body of the butterfly. The hierarchical scales covered with wax create a superhydrophobic and self-cleaning surface, like the lotus effect. Superhydrophobicity and anisotropic grooves, like those in shark skin (to be described later), lead to low drag. Thus, the wing surface combines both the lotus and shark skin effects.

The beautiful colors of the wings are produced by structural coloration, not by color pigments which are not environmentally friendly and degrade with time (Sun et al. 2013). In color pigments, colors are produced by selective absorption of light. Whereas, in structural coloration, colors are produced by selective reflection of light from nanostructured surfaces.

Leg attachment pads of several animals including many insects (e.g. beetles and flies), spiders, lizards (e.g. Geckos), and amphibians (e.g. tree and torrent frogs) can attach to a variety of surfaces and are used for locomotion (Bhushan 2009, 2012b, 2018). They consist of a leg attachment system to provide them with dynamic attachment ability, referred to as reversible adhesion or smart adhesion. A gecko is the largest animal that can attach, detach, and climb on a variety of surfaces, ranging from smooth glass windows to rough terrain. Gecko feet produce high (dry) adhesion to support its weight with a high factor of safety. The attachment pads on two feet of the Tokay gecko have an area of approximately 220 mm^2. They develop high adhesion by cumulative van der Waals attraction by splitting contact area into a large number of contacts. Gecko skin is comprised of a complex hierarchical structure of lamellae, setae, branches, and spatula (Autumn et al. 2000; Gao et al. 2005; Bhushan 2007). Approximately 3 million setae on their toes that branch off into about three billion spatulas on two feet with a three-level hierarchical structure which produces a strong adhesive force by van der Waals adhesion. An adhesive force of approximately 20 N (vertical force required to pull a lizard down a nearly vertical surface) is produced which allows them to climb vertical surfaces at speeds of over 1 m/s. They can detach their toes from the attachment surface at will by peeling action in milliseconds (Bhushan 2007, 2009, 2018).

Some amphibians, such as trees and torrent frogs and arboreal salamanders, can attach to and move over wet or even flooded surfaces without falling by wet adhesion (not shown in Fig. 1.4) (Federle et al. 2006; Bhushan 2009, 2018). In a tree frog, toe attachment pads consist of channels and are made of an extremely soft, inhomogeneous material, the epithelium. It has an effective modulus of about 15 MPa, equivalent to silicone rubber (Bhushan 2009, 2018). The pads are permanently wetted by mucous secreted from glands that open into the channels between epidermal cells. They attach to mating surfaces by meniscus contribution. They conform to rough surfaces and form meniscus around the edges. The toe pads maintain a very thin fluid film at the interface, responsible for locomotion and maneuverability. Frogs can climb on wet rocks even when water is flowing over the surface.

1.3.4 Aquatic Animals

Shark skin, dolphins, and humpback whales are models of aquatic animals from nature for low drag surfaces (Bhushan 2009, 2012a, 2012b, 2016a, 2018). Sharks can swim at high speeds in turbulent flow with low hydrodynamic drag. At high speeds in turbulent flow, vortices form on the surface, which causes high shear stresses across the entire surface. Their skin is covered by tiny tooth-like scales called dermal denticles (little skin teeth), shaped like riblets with longitudinal grooves aligned parallel to the local flow direction of the water. The grooved scales lift vortices to the tips of the scales, resulting in water moving efficiently over their surface (Bechert et al. 1997, 2000; Dean and Bhushan 2010; Jung and Bhushan 2010; Bixler and Bhushan 2013a; Martin and Bhushan 2014). The spacing between these dermal denticles is such that microscopic aquatic organisms have difficulty adhering to the surface, making the skin surface antifouling (Genzer and Efimenkox 2006; Bixler and Bhushan 2012a, 2013a).

The compliant skin of dolphins allows the high speed of dolphins (not shown in Fig. 1.4) (Bhushan 2009). The compliant skin, interacting with the water flowing over the body's surface, stabilizes the flow and delays transition to turbulence. The delay in the onset of turbulence is believed to reduce skin friction and drag. Dolphins also possess an optimum shape for drag reduction of submerged bodies. Submarines and swimsuits use the shape of dolphins.

The boxfish (*Ostracion Meleagris*) has a streamlined form with squared-off contours (not shown in Fig. 1.4) (Mueller 2008). The body provides a low drag, and the fish can swim up to six body lengths per second. Its shape inspired Mercedes-Benz's bionic concept car with a low aerodynamic drag, providing high fuel economy.

Humpback whales on their top body have scalloped edges with series of bumps or teeth called tubercles on their flippers (not shown in Fig. 1.4) (Fish and Battle 1995). The tubercles enable these huge whales to make tight turns in maneuvering. They also decrease drag, which allows whales to use less energy to turn. Inspired by this, the wind turbine blades with tubercles are designed to reduce drag performance at low wind speeds without stalling by increasing the angle of attack (Bhushan 2009, 2018).

1.3.5 Birds

Bird wings are covered by several consecutive rows of covering feathers that are flexible. These movable flaps develop the aerodynamic lift necessary for flight. When a bird lands, a few feathers are deployed in front of the leading edges of the wings, which help to reduce the drag on the wings (Jakab 1990). Various tiny insects such as beetles and dragonflies can fly at high speeds (Sun and Bhushan 2012b; Sun et al. 2014, 2016). Insect wings consist of complex mechanical structure and provide an inspiration for Micro Air Vehicles (MAVs).

The skimmer bird is the only bird that does fly-fishing. These birds fly close to water while searching for fish. When they spot a fish, their beak moves quickly in the water to catch it. Their long and thin beak has riblets, like that of sharkskin, and are responsible for low drag (not shown in Fig. 1.4) (Martin and Bhushan 2016; Bhushan 2018).

The brilliant iridescent colors in various plants, butterfly wings, and peacock feathers are created by structural coloration rather than the use of color pigments (Sun et al. 2013). Structural coloration occurs by selective light reflection by nanostructured surfaces. Structural coloration does not fade with time unlike color pigments. Structural colors often appear iridescent, meaning that color changes when the angle of view or the angle of illumination changes. The special functions of color changes in some animals serve as camouflage, predation, signal communication, sex selection, and reproductive behavior (Bhushan 2009, 2018).

1.3.6 Moth Eyes

For nocturnal insects, it is important not to reflect the light since the reflection makes them vulnerable to predators. The moth eyes are antireflective to visible light as they consist of hundreds of hexagonally organized nano-pillars, each approximately 200 nm in diameter and height, which result in the very low reflectance for visible light (Genzer and Efimenko 2006; Mueller 2008; Bhushan 2009, 2018).

These nanostructures' optical surfaces make the eye surface nearly antireflective in any direction. The front part of the eyes, cornea, is lubricated with a liquid tear film. It also protects the eye from contaminants including bacteria and dirt. The cornea consists of folds and ridges to store the liquid tear (not shown in Fig. 1.4) (Braun 2012; Bhushan 2009, 2018).

1.3.7 Fur and Skin of Polar Bear

The polar bear, one of the biggest predators on Earth, has to survive in arctic cold, down to −60 °C. Owing to its black skin up to 100 mm thick, the polar bear can capture incident sunlight (not shown in Fig. 1.4). The yellowish white hollow hairs reflect the sunlight along their length until it is transferred into heat in the black skin. Hollow hairs in the fur; together with fat skin, provide high thermal insulation (Bhushan 2009, 2018; Stegmaier et al. 2009).

1.3.8 Seashells, Bones, and Teeth

Seashells, bones, and teeth are biomaterials that are nanocomposites with a laminated, hierarchical structure that exhibit superior mechanical properties with a combination of high strength and toughness (Lowenstam and Weiner 1989; Sarikaya and Aksay 1995; Mann 2001; Alexander and Diskin 2004; Meyers et al. 2006, 2008; Sun and Bhushan 2012a). The nacre of the inner abalone shell (*Haliotis*) contains both inorganic mineral components and organic components in which the mineral components provide high hardness, and the organic components hinder crack protection responsible for high fracture toughness and durability. Monolithic $CaCO_3$ with a high hardness consists of a hierarchically organized structure starting with single crystals of the aragonite polymorph of $CaCO_3$, consisting of 0.5–10 μm thick bricks (microstructure) and finishing with 0.3 μm thick layers (nanostructure). Between the layers, there exists a 20–30 nm thick organic protein substance that inhibits transverse crack propagation and makes the material strong and resilient (Bhushan 2009, 2018).

It should be noted that bone joints are covered with a thin layer of strong, smooth, articular cartilage (not shown in Fig. 1.4). It is composed of a porous and permeable collagen network impregnated with synovial fluid. The thin layer of slippery viscous synovial fluid with egg white consistency, also called synovia, lubricates the cartilage—covered bone surfaces and reduces friction between them during movement (Bhushan 2009, 2018).

1.3.9 Spider Web

A spider generates silk fiber and has a sufficient supply of raw material for its silk to span great distances (Jin and Kaplan 2003; Bar-Cohen 2011; Bhushan 2009, 2018). A spider web is a structure built of continuous fibers made of 99% protein with hierarchical structure. They have high toughness (area under stress–strain curve) on the order of 150 MJ m^{-3}. To put it into perspective, the toughness of carbon fibers is on the order of 20 MJ m^{-3}. They are insoluble in water. The web can hold a significant amount of water droplets, and it is resistant to rain, wind, and sunlight (Sarikaya and Aksay 1995).

1.3.10 Desert Species

Various flora and fauna can survive in desert environment by water collection from fog in the night. However, collected water evaporates as soon as the sun comes out in the early mornings. Desert species have mechanisms to transport collected water to a location where, either it is stored or consumed (Bhushan 2018, 2020a, 2020b).

As an example, in the case of cactus, water droplets grow on tips of small barbs then move down onto a conical spine due to Laplace pressure gradient, and they are absorbed in the trunk. Conical geometry (or spines in the shape of cones) creates the Laplace pressure gradient. The pressure gradient is large enough that water droplets can defy gravity and climb up.

1.3.11 Arthropods

Arthropods are a diverse group of invertebrates and include insects, arachnids (such as spiders and scorpions), crustaceans (like crabs and shrimp), and myriapods (such as centipedes and millipedes). Arthropods feed on various kinds of food including plants, nectar, blood, and flesh (Das et al. 2018; Gurera et al. 2018; Bhushan 2018). Among various medically important arthropods, which feed on living hosts, mosquitoes have been reported to cause the highest number of deaths annually. The primary cause of the deaths is malaria, which is spread by infected mosquitoes piercing. To survive, mosquitoes need to hunt an animal for blood. Mosquitoes are mobile since they possess unusual locomotory attributes: standing on water, sticking to any surface, and flying in rain, because of their superhydrophobic legs, dry adhesion in the foot, hydrophobic wings, and antifogging eyes.

Female mosquitoes pray on host for blood. It takes them some five minutes to suck the blood and the host does not even know they are there. They painlessly pierce into the host's skin using a needle-like fascicle in their mouth. They pierce using a combination of numbing by releasing saliva, the fascicle's serrated design with sharp teeth and frequency-dependent graded mechanical properties and a soft tip, and vibratory actuation. They first numb the surface by spraying saliva and then pierce. Fascicle pierces with sharp teeth with a soft tip in the vibratory motion to minimize pain. Once they pierce the blood vessel, they suck the blood until repletion. While sucking, they release saliva again which now works as an anti-blood-clotting agent. After the repletion, the fascicle is retracted.

1.3.12 Anti-Freeze Proteins (AFPs)

Some species use anti-freeze proteins (AFPs) or ice structuring proteins (ISPs) to prevent freezing. These refer to a class of polypeptides, found in serum, produced by many species including polar fish, insects, bacteria, fungi, and plants. AFPs bind to small ice crystals to inhibit growth and recrystallization of ice. AFPs are responsible for the freezing-point depression of serum. The interfacial curvature results in a reduction of local freezing point due to the Kelvin effect (Nutt and Smith 2008; Meister et al. 2013; Bhushan 2018).

An AFP containing carbohydrate (glycoprotein) has been reported in serum in *P. borchgrevinki* in Antarctic fishes that do not freeze at seawater temperature (DeVries and Wohlschlag 1969). An AFP called *choristoneura fumiferana* has been reported in spruce budworm insect (Graether et al. 2000). AFPs have also been reported in the darkling beetle, *Mercantha contracta* (Duman 1977), bacteria and fungi (Duman and Olsen 1993; Gilbert et al. 2004), and over wintering perennials such as rye grass, *Lolium perenne* which are frost tolerant (Sidebottom et al. 2000) and various other plants (Urrutia et al. 1992).

1.3.13 Biological Systems—Self-healing Properties

Biological systems have self-healing properties. There are several biological mechanisms of self-repair or self-healing (Bhushan 2009; Cremaldi and Bhushan 2018). At the molecular level, they are dynamically breaking and repairing 'sacrificial' bonds, which allow material to deform in a quasi-plastic manner without fracture (Cremaldi and Bhushan 2018; Multanen and Bhushan 2019). In bones, there is a cyclic replacement of material by specialized cells, which allows a bone to adapt to changing conditions and to repair damage. Many fractured or critically damaged living tissues can heal themselves by the formation of an intermediate tissue based on the response to inflammation followed by the scar tissue (Fratzl and Weinkamer 2007; Cremaldi and Bhushan 2018). Tissue damage triggers bleeding, which is followed by the formation of a fibrin clot. Fibroblast cells migrate to the wound site, enabling the creation of granulation tissue to fill the wound (Singer and Clark 1999; Cremaldi and Bhushan 2018). As an example, a bruised or skinned knee heals itself.

1.3.14 Biological Systems—Sensory Aid Devices

Biological systems consist of sensory-aid devices. Examples include ears, eyes, and nose (Bhushan 2009, 2018). Animals and humans use their ears to hear or detect sound or noise by detecting frequency domains. They use a basilar membrane that separates sound according to their frequency, which is then conducted to the sensory cells that transfer the vibration of the basilar membrane into a neural code. In an eye, the light rays enter through the cornea, the clear, curved layer in front of the iris and pupil. The cornea helps focus light on the retina at the back of the eye. The retina functions like a camera. It receives the image that the cornea focuses through the eye's internal lens and transforms the image into electrical impulses through millions of tiny nerve fibers. These impulses are sent to the optic nerve and to the brain. A nose is a chemosensory system which consists of senses of smell and taste. Objects emit tiny air molecules that can enter through either the nostrils or the back of the throat from eating food. Once inside the nostrils, these air molecules land on the olfactory epithelium—a tissue covered in mucus that lines the nasal cavity. Epithelium contains millions of olfactory receptors or neurons that are capable of binding with specifically ordered molecules that help identify certain smells. Once the olfactory receptors bind with a specific odor, they send their electrical pulses to a certain microregion (called glomerulus) which passes it along to other parts of the brain. The so-called odorant patterns that are released from the glomerulus are interpreted in the brain as smell.

To conclude, these examples show the bounty of properties that living nature has to offer scientists and engineers. These properties, and others yet undiscovered, could hold the keys to solving science and engineering problems in a more sustainable and environmentally friendly manner.

1.4 Locomotion in Living Nature

1.4.1 Walking

Locomotion in animals in living nature occurs by walking, running, or slithering like in a snake. Animals do not use wheels, although they are technologically important for humans (Gould 1981). Records show the use of a rolling wheel by humans about 3500 BC to reduce friction in translatory motion and to be able to move at high speeds (Bhushan 2018).

Hard wheels are more energy efficient on hard, flat ground. However, in living nature, most surfaces are rough and may encounter soft surfaces as well which deform and become less energy efficient. Furthermore, to survive and thrive, an animal needs to catch its prey as well as escape its predator. To catch its prey requires fast linear acceleration from rest. However, bicycles using wheels with no gear do not accelerate very quickly from rest. Human runners can achieve the maximum speed much faster than a cyclist can. It is well understood that running is more efficient for acceleration from rest at short distances under 100 m or so. To escape from a predator, the prey generally does not try to escape from the predator linearly but will turn at an angle to the direction of travel of the predator. Bicycles cannot change direction as quickly as compared to human beings or animals. Therefore, wheels are less useful than other ways of moving, like walking, running, or slithering of a snake (Bhushan 2018).

It should also be noted that a wheel is driven by the application of torque on it. In animals, movement is usually caused by skeletal muscles which are not useful for generating torque.

Finally, for moving, there are two ways that objects can use rotation, either the entire object rolls or part of it turns like a wheel or propeller with a stationary axle. Some species move around by rolling in which the whole organism rotates, without any part staying still. Some animals can roll by forming their bodies into a circle that includes caterpillars, mantis shrimp and salamanders (Full et al. 1993; Scholtz 2008) or into a ball including pangolins, the armadillo lizards and wheel spiders. However, none of these uses wheels (Bhushan 2018).

1.4.2 Gear Systems for Precise Movement

Juvenile hopping insects from the *Issus coleoptratus* that are commonly called planthoppers use gears located on the top segment of each of their hind legs for coordination,

Fig. 1.5 (Burrows and Sutton 2013). To jump, both insect's hind legs must push forward at the exact same time. Because both swings laterally, if one were extended a fraction of a second earlier than the other, it would push the insect off course to the right or left. The gear is nature's solution which locks their back legs together, allowing both appendages to rotate at the exact same instant with each moving within 30 ms of the other (Bhushan 2018).

The gears appear to mechanically function like man-made gears. They consist of 10 to 12 tapered teeth, each about 80 μm wide. The teeth even have filleted curves at the base, similar to the man-made gears designed to reduce wear over time (Bhushan 2013a, b). See magnified view of gear teeth in Fig. 1.5 (Burrows and Sutton 2013; Bhushan 2018).

It appears that living nature designed gears before human beings did in about 300 BC.

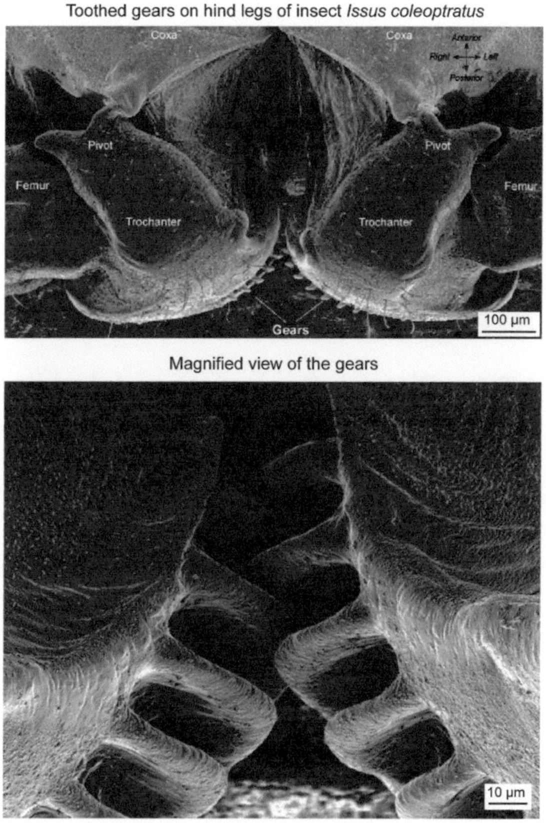

Fig. 1.5 SEM micrographs of the toothed gears located on the top segment of each of the hind legs of the juvenile hopping insect *Issus coleoptratus* to precisely synchronize the kicks of the hind legs as it jumps and the magnified view of meshing gear teeth showing tooth profiles (Adapted from Burrows and Sutton 2013)

1.5 Biomimetics and Bioinspiration in Art and Architecture—Bioarchitecture

Scientists and engineers take inspiration from living nature for functionality and commercial applications, while artists and architects take inspiration from nature for the added purpose of beauty and function (Ripley and Bhushan 2016). Various patterns found in nature are used by artists for aesthetics and design. Further, structural coloration from nanostructured surfaces is being developed for use by artists to produce vivid, durable colors that may not be possible by color pigments. Architects develop design strategies by incorporating patterns and organization found in natural forms (Gruber 2011; Imhof and Gruber 2013; Bhushan 2016a, 2018, 2019a). Moreover, as nanostructured surfaces or materials with functional hierarchy are developed, these, too, are being incorporated into design practice ranging from nanoscale to the macroscale. The field of bioinspired art and architecture has been called "bioarchitecture" by Bhushan (2016a).

Bioarchitecture attempts to draw into the practice of art and architecture not merely biomimetics, but bioinspired design as well. Bioarchitecture incorporates bioinspired design from the outset in all parts of the work at all scales. This is not simply copying natural shapes or forms in plan, section, or elevation, or in ornamentation (on the structure, components, or in discrete art objects). This is not simply incorporating bioinspired materials in the construction of a structure or the making of art. It is also not simply scaling up natural principles. Instead, bioarchitecture is a multiscale approach to the incorporation into the work of solutions and opportunities that nature presents in solving universal human problems. It is an approach that is evident in the theory behind the overall design, the careful choice of materials for construction or creation, the extraction and adaptation of principles, and the cohesion of the parts in the whole (Ripley and Bhushan 2016).

1.5.1 Biomimetics in Arts and Architecture

At its most basic, biomimetics is the mimicking of living nature. At a macroscale, art accomplishes this through direct copying of natural figures. Figure 1.6 shows examples of biomimetics in art from living nature. Pictured are four examples of antique jewelry pieces, crafted by Van Cleef and Arpels in the early 1900s, and three paintings: "Mount Corcoran" by Bierstadt (1877), "Water Lilies" by Monet (1906), and "The Afternoon Meal" by Melendez (1771) (Bhushan 2012a, 2016a, 2018, 2019a; Ripley and Bhushan 2016). The brooches are inspired by snowflakes, butterflies, orchids, and sycamore leaves. The forms are copied exactly—or as exactly as the medium allows—and do not extract or adapt any function or forms. Similarly, the paintings demonstrate a fascination with representing landscape and living nature, but without adaptation or use of functions derived from the natural world.

Fig. 1.6 Examples of biomimetics in art: antique jewelry pieces, crafted by Van Cleef and Arpels in the early 1900s; "Mount Corcoran" by Bierstadt (1877); "Water Lilies" by Monet (1906); and "The Afternoon Meal" by Melendez (1771). (Adapted from Bhushan 2012a, 2016a, 2018, 2019a; Ripley and Bhushan 2016)

Figure 1.7 shows examples of biomimetics in architectures from living nature with direct representation thereof (Ripley and Bhushan 2016). First, morning glory vines are compared to the enclosure panels of the art nouveau at Porte Dauphine for the Paris Metro by Hector Guimard (1899). Art nouveau aspired to incorporate nature into architecture using new material discoveries (cast iron) and decorative motifs. Next, an acanthus leaf incorporated into the capitals of the columns adorning the neoclassical U.S. Capitol building by William Thornton (1800) is shown. Neoclassical architecture discarded the divine proportions of the golden ratio used so prominently in Greek architecture as too limiting, instead relying solely on the material (marble) and the shapes to reference classical architecture. Finally, a dahlia head is compared to the rose window of the gothic Sainte

Chapelle Cathedral in Paris, France, designed for King Louis IX in 1248 (Watkin 2005). Gothic architecture drew inspiration from the divine: light, space, and, as here, nature.

Fig. 1.7 Examples of biomimetics in architecture: enclosure panels at Porte Dauphine in Paris by Hector Guimard (1899), inspired by vine plants (left); columns on the U.S. Capitol building by William Thornton (1800), inspired by the acanthus leaf (left); rose window of Sainte Chapelle Cathedral in Paris (Watkin 2005) for Louis IX by an unknown architect (1248), inspired by the flower petals (left) (Adapted from Ripley and Bhushan 2016)

1.5.2 Bioinspiration in Arts and Architecture

Direct representation of nature is a facile way to incorporate nature into design without actually delving into the organization or properties nature provides. Artists and architects alike strive to derive these properties and incorporate this organization into their work in this manner; not as merely direct representation or ornament to the design, but as an integral part of the making of the work (Ripley and Bhushan 2016).

Figure 1.8 shows the work of three artists that have made strides in this regard: Bone Chair and Bone Armchair by Joris Laarman in 2006 and 2008, Quadversal (fashion collection) by Iris van Herpen with Warren Du Preez and Nick Thornton Jones (Spring/Summer 2016), and Biocouture Skirt by Suzanne Lee (c. 2013) (Ripley and Bhushan 2016). For the Bone Chair and Bone Armchair, Joris Laarman (2006 and 2008) studied the way that the interior structure of bones grows, and, using sophisticated software, created the structure for his chair in much the same way (Ripley and Bhushan 2016). The process involved reducing material where the structure could be light and increasing it where the structure needed to be strong (Fairs 2006). This created a chair that is structurally very strong and sturdy, yet light, both in actual weight and in appearance, and, since the chairs were 3D printed, they were "grown" in the studio, deriving not only composition and structure but also construction from nature's processes.

Quadversal by Iris van Herpen with Warren Du Preez and Nick Thornton Jones (Spring/Summer 2016), capitalizes on the idea of growth and intermixing inherent in living nature, particularly with trees (Ripley and Bhushan 2016). Central to the Quadversal collection is a dress that interprets growth, subverting the "premade" notion of fashion, and creating a dress in real time to a particular setting and model on which it is made. Fashion construction techniques such as weaving, cutting, and folding are translated into a means of growth. Further, the form taken by the dress suggests the way tree canopies interact—branches weaving and intermixing above—and the way the eye travels through a densely wooded area.

Inspired by how nature creates, Suzanne Lee began to explore how microbes, bacteria, fungi, and yeast all create fabric as their means of survival (Grushkin 2015). In harnessing that growth potential, Lee created fabric and garments from bacteria and other materials that are environmentally friendly in their creation. They result in zero waste and are compostable at the end of their life cycle. In her exploratory practice, Lee has inspired others to incorporate nature's processes, resulting in some prototypes of future products, such as self-healing concrete (Ripley and Bhushan 2016).

Figure 1.9 shows examples of architecture drawing inspiration from plants, animals, and from replication at the molecular level in nature (Ripley and Bhushan 2016). Pictured are La Sagrada Familia in Barcelona, Spain, by Antoni Gaudí (1882); the Nakagin Capsule Tower in Tokyo, Japan, by Kisho Kurakawa (1972); the Milwaukee Art Museum in Milwaukee, Wisconsin, USA, by Santiago Calatrava (2001); Exeter Cathedral in Exeter,

Fig. 1.8 Examples of bioinspiration in art: Bone Chair and Bone Armchair by Joris Laarman (2006, 2008), inspired by bone structure (left); Quadversal (fashion collection) by Iris van Herpen with Warren Du Preez and Nick Thornton Jones (Spring/Summer 2016), inspired by a tree bridge in India (left); and Biocouture Skirt by Suzanne Lee (c. 2013), inspired by the growth of bacteria in a kombucha mother culture (Adapted from Ripley and Bhushan 2016)

England, by an unknown architect (completed c. 1400); and Habitat 67 in Montreal, Quebec, Canada, by Moshe Safdie (1967). Inspired by much in the natural world, Antoni Gaudí began construction on the cathedral of La Sagrada Familia (The Holy Family) in 1882. It is wrought in a southern style of art nouveau influenced by the political discord in Spain of the Catalan people, subjugated by the Castilians several centuries prior (Watkin 2005). Gaudí eschewed the rigidity of style dictated by writers of northern architectural and artistic theory. Instead, Gaudí solved the classical problems of height and weight in his cathedral in novel ways using nature's solutions. He noted that very tall, thin trees support the load of their tops and of external forces by tilting, rather than remaining rigidly upright. Thus, for the load-bearing columns of the interior of his cathedral, Gaudí chose tall, thin columns that branched to draw load from various points, distributing the weight over a greater area, yet drawing it down one, thin column. This allowed Gaudí to build a tall nave with more dematerialized external walls (Ripley and Bhushan 2016).

Nakagin Capsule Tower, produced out of the Metabolist movement in post-war Japan, used cellular discoveries of the time to reduce the essential living requirements to discrete, interconnected blocks suitable for rapid growth and expansion (Kurokawa 1977; von Richthofen 2007). The Metabolists were fascinated by the new scanning electron microscope (SEM) images of metabolic and cellular organizations. They extracted from these images' ideas of cellular replication, organization, and efficiency of function. From these ideas, they scaled up and created projects that addressed rapid, post-war urban growth (von Richthofen 2007; Ripley and Bhushan 2016). Thus, Nakagin Capsule Tower was constructed of nearly identical individual pods, or cells. These cells would cycle with the building framework: growth of the framework, growth (addition) of the cells, use during a cellular lifecycle, removal of cells that had reached the end of their useful life, and replacement with new cells. The idea of this building was to allow the framework to remain whole but allow a life cycle for the individual cells.

The Milwaukee Art Museum by Santiago Calatrava (2001) is another example of bioinspiration derived from animals (Ripley and Bhushan 2016). Here, Calatrava designed a solar-sensitive wing that, like a bird, can open and close. While taking the principle of the mechanical operation of the bird wing, it does not attempt to make the building fly, of course. Instead, the wings open and close to provide solar protection for the museum and for attendees, allowing shade to take over the roof structure of the museum building.

Exeter Cathedral in Exeter, England, by an unknown architect (completed c. 1400), approached the issue of the weight of the ceiling and roof structure not from the position of support for increasingly high loads, but from the idea of reducing the actual load of that roof structure (Ripley and Bhushan 2016). To accomplish this, the designers turned to the structure of the human body, specifically the ribs and intercostal spaces. Architects found that by replacing the ceiling with vaults of strong ribs and filling the spaces between these ribs with much thinner, lighter materials, the weight of the ceiling was reduced. This allowed for taller ceilings without concomitant significant increases in the load (Ripley and Bhushan 2016).

Through these projects, a progression can be seen from mimesis to inspiration; biomimetics transcends mimicry and enters bioinspiration. In bioinspiration, no longer are the works mere copies of living nature, as in biomimetics, but instead, they extract a principle (or principles) from nature, seek to understand that principle, and then incorporate the principle into a new medium, design, or other applications. Incorporation of bioinspired parts into art and architectural practice has evolved and advanced those disciplines. There is a balance to achieve between form and function; between ornament/

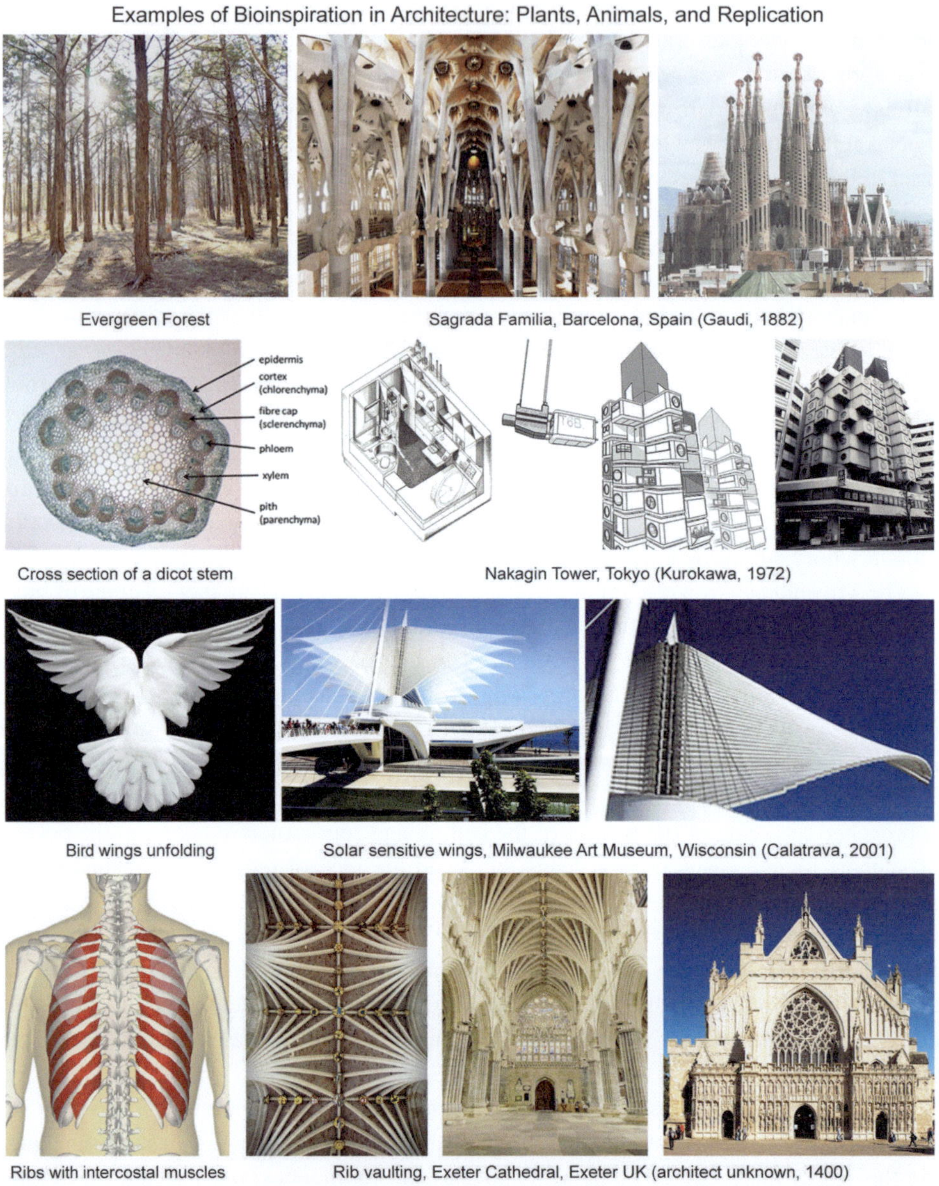

Fig. 1.9 Examples of bioinspiration in architecture: support columns in La Sagrada Familia in Barcelona, Spain, by Antoni Gaudí (1882), inspired by groups of trees (left); Nakagin Capsule Tower in Tokyo, Japan, by Kisho Kurakawa (1972), inspired by dicot structure (left); the Milwaukee Art Museum in Milwaukee, Wisconsin, USA, by Santiago Calatrava (2001), inspired by bird wings (left); and rib vaulting in Exeter Cathedral in Exeter, England, by an unknown architect (completed c. 1400), inspired by ribs with intercostal muscles (Adapted from Ripley and Bhushan 2016)

aesthetics, and in adaptation of principles. Architects have found a solution to this balance through previous practice; in architectural parlance, a balance is struck between "ducks and decorated sheds" (Venturi et al. 1977; Ripley and Bhushan 2016).

Figure 1.10 provides a schematic offering of bioarchitecture, blending properties and functionalities from various species (Ripley and Bhushan 2016). Looking at the lotus leaf, its shape is engaged not only for aesthetic purposes, but also for functional purposes, as it is ideal for rainwater collection. Using nature-inspired superhydrophobic materials for the surfaces of the roof structure, efficient collection of that rainwater is provided, and a self-cleaning and antifouling surface is ensured as well, reducing cleaning and maintenance costs. Filtration of the collected water by selective porosity can provide potable water for building uses. By employing the principles behind plant circulation, the efficient transport of people en masse through the structure can be achieved. Finally, energy production and self-regulation of temperature and light can be achieved using principles from photosynthesis and the operation of the human iris.

Bioarchitecture, as a practice, can provide a means to cross disciplines and take advantage of scientific opportunities presented without pastiche or amalgamation, but instead through cohesive, thoughtful design practice (Ripley and Bhushan 2016).

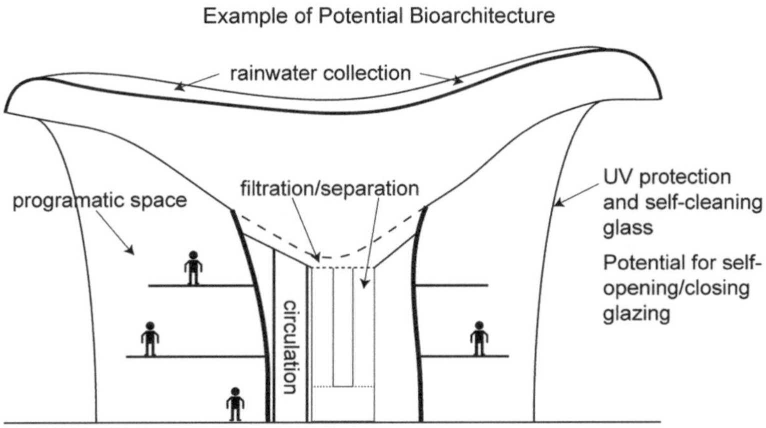

Fig. 1.10 Schematic example of a potential bioarchitecture building by blending functionalities from various species in nature: the form of the lotus leaf can be used for both aesthetic and functional purposes to collect rainwater, as well as form the roof structure; superhydrophobic materials for surfaces can provide efficient collection of rainwater and ensure self-cleaning and antifouling surfaces; filtration provided by selective porosity can provide water for building uses; the principles of plant circulation can provide efficient transport of people through the structure; and principles from the operation of the human iris can provide self-regulation of heat and light by opening and closing the glazing (windows) (Adapted from Ripley and Bhushan 2016)

1.6 Unique Patterns Used by Nature: Golden Ratio and Fibonacci Numbers

Unique patterns are used by nature throughout the universe. Various species and objects found in nature follow the so-called "Golden Ratio" and "Fibonacci Number" (Vajda 1989; Livio 2002; Posamentier and Lehmann 2007). To understand the secrets of the order that nature practices, researchers have studied them. They describe unique patterns found throughout not only our terrestrial environment, but the universe as well (Vajda 1989; Livio 2002; Posamentier and Lehmann 2007). Scientists and mathematicians began to study the Golden Ratio around 450 B.C., though it was used for some time before by the Greeks, and Egyptians even before them, in art and architecture (Wittkower 1960). Leonardo da Pisa, nicknamed Fibonacci (1175–1230), discovered that a certain sequence of numbers, the Fibonacci numbers, approached the Golden Ratio, and he began studying the preponderance of examples of those numbers in nature around 1200 A.D. (Wittkower 1960; Bixler and Bhushan 2013b, 2015; Bhushan 2016a, 2018, 2019a).

The Golden Ratio is a unique ratio whose proportions are considered to represent a functional and aesthetic ideal. It is defined as the ratio of length to width of a rectangle ≈ 1.618. A rectangle produced using the Golden Ratio is called the Golden Rectangle, consisting of a square combined with another Golden Rectangle, as shown in Fig. 1.11. A Golden Rectangle is produced by drawing a line from the midpoint of one side of the square to an opposite corner, then using that line as the radius to draw an arc that defines the height of the rectangle. Each Golden Rectangle contains an infinite sequence of squares and adjacent Golden Rectangles. The figure also shows how the Golden Spiral can be produced from the Golden Rectangle by drawing an arc tangent to the square, from the bottom corner of the square to the opposite corner, where it meets the square encompassed in the adjacent encompassed Golden Rectangle (Wittkower 1960; Bhushan 2016a, 2018, 2019a).

The Golden Ratio is found throughout nature, with examples ranging from the proportions of the human body to the spiral shapes of the Milky Way galaxy and seashells. Inspired by beauty found in nature, famous artists have included the Golden Ratio in their works, such as Leonardo da Vinci's paintings "A Head of an Old Man" and the "Mona Lisa." Architects have also relied on the Golden Ratio for functional aesthetics since its discovery (Wittkower 1960). The "mystical" beauty found using the golden section is exemplified in the proportions of the Parthenon (Zeising 1854), the construction of medieval churches (Zeising 1854; Wittkower 1960), and, more recently in the mid-twentieth century, in the work of Le Corbusier (Wittkower 1960), who used the Golden Ratio to define his body of work (Bhushanx 2016a, 2018, 2019a).

Fibonacci numbers describe geometries of various structures found in nature. While these whole numbers do not produce the Golden Ratio precisely, the ratios of the numbers quickly approach the Golden Ratio, as the sequence continues into infinity (Wittkower 1960). The Fibonacci numbers are described by the mathematical expression $F_n = F_{n-1} + F_{n-2}$. The sequence begins with the seed values of $F_0 = 0$ and $F_1 = 1$, , then, the

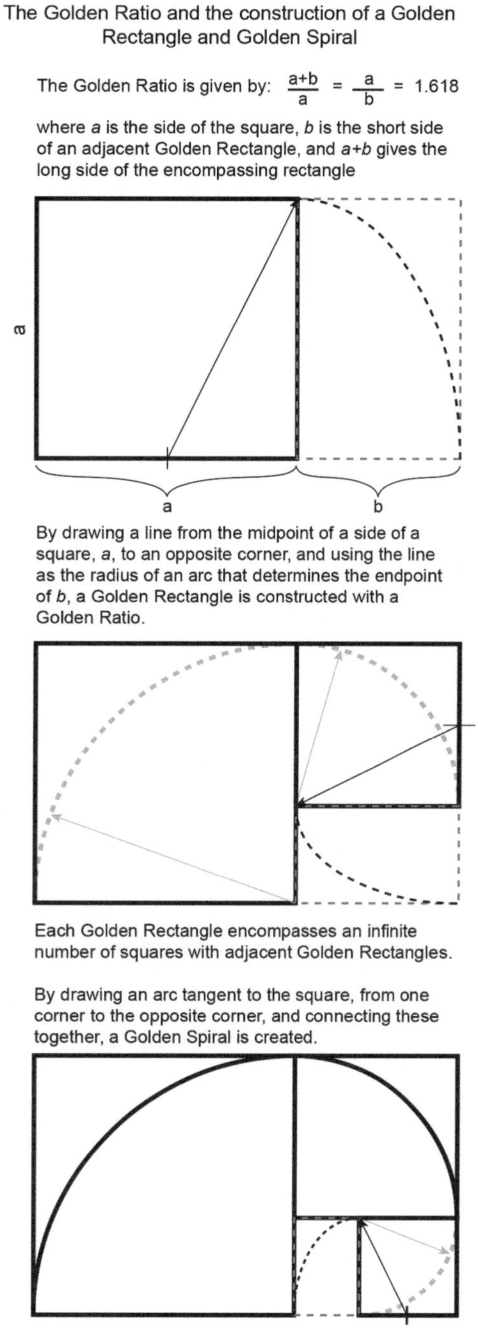

Fig. 1.11 The Golden Ratio, and the construction of the Golden Rectangle and the Golden Spiral, as derivations of the Golden Ratio (Bhushan 2016a, 2018)

recursive expression provides the values following, resulting in 0, 1, 1, 2, 3, 5, 8, 13, 21, 34,... and so on (Wittkower 1960; Vajda 1989; Posamentier and Lehmann 2007). Figure 1.12 shows how the sequence is created out of these whole numbers, and how their ratios approach the Golden Ratio. In the figure, the Fibonacci sequence, F_n, is placed to the left; in the center, the sequence is displaced by one number, F_{n-1}, and to the right, displaced by 2 numbers, F_{n-2} (Bhushan 2018, 2019a). This shows how the sequence is calculated. Examples of the Fibonacci numbers in nature include the number and sequence of patterns on pineapples and sunflower heads, as well as the number of petals on flowers.

Figure 1.13 illustrates examples that demonstrate the prevalence of the Fibonacci numbers and the Golden Ratio in nature as well as in art (Bixler and Bhushan 2013b; 2015; Bhushan 2016a, 2018, 2019a). Shown are the Milky Way galaxy, a sunflower head, flower petals, a nautilus seashell, the Mona Lisa, and pineapples. The Mona Lisa highlights the Golden Rectangle, as well as the Golden spiral, which draws the eye to the focus of the art. The Golden Spiral is demonstrated in the images of the Milky Way galaxy, the seed

Fibonacci sequence

The Fibonacci sequence is defined by:

$$F_n = F_{n-1} + F_{n-2}$$

		F_n	F_{n-1}	F_{n-2}
Seed Values	F_0	0	0	0
	F_1	1	0	0
Recursive Values	F_2	1	1	0
	F_3	2	1	1
	F_4	3	2	1
	F_5	5	3	2
	F_6	8	5	3
	F_7	13	8	5
	F_8	21	13	8

The ratio of each number in the sequence to its preceding number rapidly approaches the Golden Ratio:

$\frac{1}{1}, \frac{2}{1}, \frac{3}{2}, \frac{5}{3}, \frac{8}{5}, \frac{13}{8}, \frac{21}{13}$ gives:

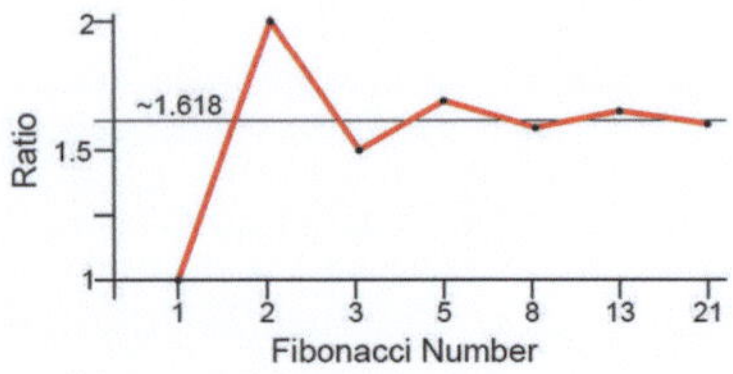

Fig. 1.12 Fibonacci sequence shows how the sequence is calculated from whole numbers, and how the ratios of these numbers in sequence approach the Golden Ratio (Adapted from Wittkower 1960 and Bhushan 2016a, 2018)

growth in the sunflower head, and a nautilus seashell. When the Golden Rectangle height is divided by its width, the Golden Ratio of 1.618 is calculated, as demonstrated with the galaxy, sunflower head, seashell, and Mona Lisa painting. Furthermore, Fibonacci numbers are present in the arrangement of petals in a dahlia flower, and interlocking helices of 0, 5, 8, and 13 wrapping around the skin of a pineapple (Vajda 1989; Livio 2002; Posamentier and Lehmann 2007).

The Golden Ratio and Fibonacci numbers also are present in shark skin, a rice leaf, and butterfly wings (Bhushan 2016a, 2018, 2019a). It has been found that riblets from

Fig. 1.13 Montage of images showing examples of the Golden Ratio, Golden Rectangle, and Fibonacci numbers found in nature and art. Examples include galaxy, sunflower heads, dahlia flower petals, seashells, Mona Lisa painting, and pineapple fruit (from top left to bottom right). The Golden Rectangle and Fibonacci or Golden spiral are superimposed on images. Such proportions and patterns are considered aesthetically pleasing and are believed to enhance performance of systems found throughout nature. (Adapted from Bixler and Bhushan 2013b; Bhushan 2016a, 2018, 2019a)

the mako (*Isurus oxyrinchus*) and the dogfish (*Sqaulus acanthias*) shark species contain five and three riblets per dermal denticle, respectively. These values represent Fibonacci numbers (Bixler and Bhushan 2013b). Furthermore, it has been reported that butterfly wing scales are approximately 75 μm wide by 125 μm long (Bixler and Bhushan 2013b). Such scales exhibit a ratio value approximately equating to Fibonacci ratios, $\frac{125\,\mu\text{m}}{75\,\mu\text{m}} \approx \frac{13}{8}$, which approach the Golden Ratio, 1.618 (Bhushan 2016a, 2018, 2019a).

To help explain the significance of these numbers, possible explanations have been developed (Bixler and Bhushan 2013b, 2015). For shark skin, it is believed that each riblet must be large enough to properly lift and pin the turbulent vortices to lower drag yet be small enough to easily flex with the shark. Furthermore, it is suspected that dermal denticles with fewer than three riblets would be too disjointed and not effectively control the vortices. Conversely, it is suspected that dermal denticles with more than five riblets would be too large to flex with the shark. As for the butterfly wings, it is believed that the scales exhibit the Golden Ratio to create effective overlapping shingle-like structures, with the longer edge in the fluid flow direction. This arrangement is suspected to help encourage the desired anisotropic flow away from the butterfly body, which leads to self-cleaning. It is believed that scales veering from the Golden Ratio may not effectively direct the fluid flow (Bhushan 2016a, 2018, 2019a).

Since Fibonacci numbers and the Golden Ratio are found throughout nature, it is believed such findings may hold clues for designing low drag and self-cleaning surfaces (Bixler and Bhushan 2013b, 2015; Bhushan 2016a, 2018, 2019a).

1.7 Industrial Applications

The understanding of the functions provided by species and processes found in living nature can guide us to produce nanomaterials, nanodevices, and processes with desirable functionality. Bioinspired designs, adaptation, or derivation from living nature are expected to play a major role in industrial adaptations. This may be referred to as biodesign.

Although the word biomimetics is relatively new, our ancestors looked to living nature for inspiration and development of various materials and devices many centuries ago (Ball 2002; Anonymous 2007; Meyers et al. 2008; Bhushan 2009, 2018, 2019a; Bar-Cohen 2011). For example, the Chinese tried to make artificial silk some 3000 years ago. Leonardo da Vinci, a genius of his time, studied how birds fly and proposed designs for flying machines. In the twentieth century, various products, including the design of aircraft, have been inspired by nature.

In 1948, Swiss engineer George de Mestral discovered that burrs of the cocklebur plant get attached to clothes readily. Their spines are tipped with tiny hooks that stick out from the seeds and provide a hook-and-loop construction which grips instantly but can be ungripped with a light force (Bhushan 2009, 2018). This led to the invention of the zip, called Velcro, which works as a zipper. One side has stiff hooks like burrs and the other has loops in the fabric. The word Velcro was derived from the French words velour (velvet) and crochet (hooks).

Since the 1980s, artificial intelligence (AI) and neural networks in information technology have been inspired by the desire to mimic the human brain. The existence of biocells and deoxyribonucleic acid (DNA) serves as a source of inspiration for nanotechnologists who hope to one day build self-assembled molecular-scale devices. In molecular biomimetics, proteins are being utilized in controlling materials formation in practical engineering towards self-assembled, hybrid, functional materials structure (Grunwald et al. 2009; Tamerler and Sarikaya 2009).

Liquid repellent and self-cleaning surfaces are being considered for various applications which require self-cleaning/anti-smudge/fingerprint resistance, anti-icing, antifogging, optical transparency, low adhesion, drag reduction in fluid flow, and anti-fouling properties (Bhushan 2009, 2012a, 2016a, 2018, 2019a). Self-cleaning applications include self-cleaning windows, windshields, smart screens, camera lenses, exterior paint for buildings, utensils, roof tiles, textiles, solar panels, self-cleaning shampoo and detergent bottles, and exterior surfaces in transportation. Anti-smudge and fingerprint resistance properties are important in applications which require finger touch such as in smart devices and automotive and aerospace interior and exterior surfaces. Anti-icing properties are of interest in applications such as automobiles and aircraft surfaces which can be exposed in winter interest to temperatures as low as −60 °C. Antifogging properties are of interest in optical applications such as glass windows and eyeglasses. Optical transparency is of interest in electronic mobile devices and many other optical surfaces.

Low drag surfaces are of interest in both internal and external fluid flow applications. Major application is in transportation such as boat hulls, automobile skin and aircraft skin. Low adhesion/low drag surfaces also exhibit anti-fouling of interest in applications such as biomedical and food packaging as well as membranes for desalination and water purification. These surfaces can also be used for energy conservation because of low adhesion/drag, and energy conversion by using switchable superliquiphobic/philic surfaces (Bhushan 2018, 2019a; Li et al. 2019; Liu et al. 2019).

Surfaces with repellency of water or oil and affinity of the other can be used for oil–water separation for use in oil spill cleanup and water purification (Bhushan 2018, 2019b, 2020a).

For industrial applications, surfaces and selected materials used in the construction of structures should be durable. It is desired that the surfaces have self-healing properties for a long life. Species found in nature self-heal (Cremaldi and Bhushan 2018).

Since the mid-1990s, the lotus effect has provided inspiration to develop a variety of surfaces for superliquiphobicity and self-cleaning. Self-cleaning products include hydrophobic façade paints Lotusan®, introduced by Sto in 1999, and Erlus Lotus roof tiles and NanoSphere® fabric by Schoeller Technologies. Self-cleaning hydrophilic SunClean® glass by PPG Industries contains a titanium dioxide coating, which uses the sheeting action of water to efficiently gather and remove contaminants.

Table 1.1 shows selected examples of objects and desired properties in some superliquiphilic/phobic applications (Bhushan 2018).

Common man-made adhesives such as tape or glue involve the use of wet adhesives that permanently attach to surfaces. However, replication of the characteristics of gecko feet would enable the development of a superadhesive surface capable of clean dry

Table 1.1 Selected examples of objects, typical materials used, and desired properties including mechanical durability for superliquiphilic/phobic applications (Adapted from Bhushan 2018)

(a) *Biomedical applications*		
Objects	Typical materials	Desired properties
Biomedical devices, catheters, drainage tubing, micropumps, microvalves, electronic circuits	PDMS, Polystyrene, PMMA	Superoleophobic, self-cleaning, anti-biofouling
(b) Automotive applications		
Windshield, window glass, side/rear view mirrors, camera lens (back of car)	Glass	Superoleophobic, self-cleaning, anti-finger touch, antifouling, optically transparent/ antireflective
Display, headlights	PMMA for smart screens for capacitive touch material, Polycarbonate (head lights)	Superoleophobic, self-cleaning, anti-finger touch, antifouling, optically transparent/ antireflective
Dashboard, door liners, seats, carpet, leather	Polyurethane, nylon (dashboard), Polypropylene (door liner), Polyurethane (seats), Leather (natural, synthetic polycarbonate)	Superoleophobic, self-cleaning, anti-finger touch, antifouling
Car body	Painted steel	Self-cleaning, anti-smudge, low drag
(c) Smart screens for electronic display		
Screens	Soda-lime glass, polyethylene terephthalate, polycarbonate	Superoleophobic, self-cleaning, anti-finger touch, antifouling, optically transparent
(d) Plastic bottles and caps and plastic sheets		
Bottles	High-density polyethylene, polyethylene terephthalate, polypropylene	Superoleophobic, superhydrophobic
Bottle caps	Polypropylene	Superoleophobic, superhydrophobic
Plastic packaging sheets	Polypropylene, acrylic	Superoleophobic, superhydrophobic, anti-fouling

adhesion (Bhushan 2007, 2018, 2019a). They have the potential for use in everyday objects such as adhesive tapes, fasteners, and toys and in high technology, such as microelectronic and space applications. These structures also can be used in prosthetics and surgical masks for reversible adhesion (Bhushan and Lee, 2012; Lee and Bhushan 2012; Bhushan 2016a, 2018). These can bind to components in microelectronics without the high heat associated with various soldering processes. These structures will never dry out in a vacuum—a common problem in aerospace applications. Replication of the dynamic climbing and peeling ability of geckos could find use in the treads of wall climbing robots (Cutkosky and Kim 2009; Bhushan 2018).

Low drag is of interest in both internal and external fluid flow. Examples of internal flow include pipelines, catheters, and micro/nano channels such as in micro/nanofluidic based sensors (Bhushan 2016b, 2017, 2018). A major example of external flow is transportation—ships, airplanes, and automobiles. According to the U.N. International Maritime Organization, about 90% of worldwide goods are carried by sea-faring vessels, and about 35–70% of energy consumed in transportation is estimated to be lost due to drag. In airplanes and automobiles, energy losses due to drag are also important. In the oil industry, oil flow in the pipeline must overcome high drag due to viscous effects and sludge fouling. Low drag surfaces often equate to energy conservation and less fouling. Shark skin-inspired surfaces have been created to reduce drag (Dean and Bhushan, 2010; Bixler and Bhushan 2013a; Bhushan 2016a, 2018). Shark skin-inspired experimental vinyl sawtooth riblets have been developed by 3 M Corp. (Marentic and Morris 1992). They have been reported to reduce drag by 3% when 70% of the Airbus plane surface was covered by these sheets. Their major commercial application is in the Speedo FastSkin® racing swimsuit. It was worn by American Swimmer Michael Phelps first in the 2008 in Beijing Olympics, who won 8 gold medals, the most of any athlete in any single Olympic showing, and broke 8 Olympic and 7 World records. However, full-body swimwear was banned in international sports in January 2010.

The shape of the Kingfisher beak reduces noise when moving in air and this helps to avoid detection by prey. A train makes a lot of noise when exiting the tunnel. To minimize noise, inspired by Kingfisher beak, bullet trains have this shape (Bhushan 2009, 2016a, 2018).

There is not enough fresh water for everyone on earth. Desert species survive in arid region regions by collecting water from fog and condensation at night. The current supply of freshwater needs to be supplemented to meet future needs. Many plants and species in arid regions exhibit efficient solutions for water harvesting from fog and condensation of water vapor (Bhushan 2020a, 2020b). Based on cactus-inspiration, water harvesting towers have been designed to supplement water supply in arid regions.

Mosquitoes can pierce painlessly in the human body to suck the blood. Mosquito-inspired painless microneedles designs have been developed which can be used for flu shots, drug delivery, and other biomedical applications (Bhushan 2018).

In the field of biomimetic materials, bioinspired ceramics have been fabricated based on seashells and other biomimetic materials. Two-component biological materials such as bones, teeth, and abalone shells exhibit high mechanical strength, as well as fracture toughness, responsible for durability. Nanoscale architecture used in nature for optical reflection and anti-reflection has been used to develop reflecting and anti-reflecting surfaces, as well as structural coloration. Inspired by the fur of the polar bear, artificial furs and textiles have been developed. Self-healing of biological systems found in nature is of interest for self-repair. Biomimetics is also guiding in the development of sensory-aid devices (Bhushan 2009, 2016a, 2018).

The major emphasis on nanoscience and nanotechnology since the early 1990s has provided a significant impetus in mimicking nature using nanofabrication techniques for commercial applications (Bhushan 2016b, 2017). Biomimetics has spurred interest across many disciplines. Several thousand research papers are published each year. Revenues from biomimetic products in 2017 were estimated to be several billion dollars annually. Annual sales are expected to continue to increase dramatically.

1.8 Economic Impact

The derivation of bioinspired optimal designs allows development of multifunctional materials with unique properties and can provide efficient operation in applications. This improves quality of life. Selected properties, applications, and expected benefits of some high economic–impact examples are presented in Table 1.2 (Bhushan 2018). The expected benefits include improved operating efficiency, low maintenance, low drag, and energy conservation. A major application of low drag and energy conservation is transportation including ships, airplanes and automobiles. Economic impact of bioinspired materials and surfaces on the U.S. economy is expected to be significant; couple of hundred billion U.S. dollars per year or about one percent of its gross national product (GDP) in 2019.

1.9 Research Objective and Approach

The objective of biomimetics research is to develop biologically inspired materials and surfaces of commercial interest. The approach is fivefold:

(1) Species are selected from nature, which provide functionality of commercial interest.
(2) The species are characterized to understand how a natural species provides functionality.
(3) Often, structures are fabricated in the lab using nature's route to verify one's understanding. Functional properties of lab structures should be comparable to that of natural species.

Table 1.2 Selected properties, applications, and expected benefits of some high economic-impact examples (Adapted from Bhushan 2018)

Selected properties	Applications	Expected benefits
Superliquiphobicity	Self-cleaning windows, windshields, smart screens, camera lenses, exterior paint for buildings, roof tiles, textiles, exterior surfaces for transportation, anti-icing for transportation, anti-biofouling for biomedical and food packaging	Improved operating efficiency, low maintenance, anti-fouling, low drag, energy conservation
Superliquiphilicity/phobicity	Anti-fogging for optical applications, oil–water separation for oil spill cleanup and water purification	Improved operating efficiency, low maintenance, environmental control
Gecko feet-inspired reversible adhesion	Organic-free surfaces with reversible adhesion for climbing robots and surgical masks	New applications, better designs
Shark skin inspired low drag surfaces	Pipelines, catheters, micro/nanofluidic based sensors, transportation—ships, airplanes, and automobiles	Low drag, emission reduction, energy conservation
Cactus-inspired water harvesting	Safe drinking water in arid regions	Economical water supply
Mosquito-inspired painless piercing	Painless microneedles for shots, drug delivery and other biomedical applications	Painless treatment

(4) Biology works on "the principle of good enough". For example, in living nature, if tree branch breaks, it replaces. However, in many commercial applications, bioinspired products need to have an optimum design, for example, we cannot afford for aircraft to crash. Structures are modeled to develop optimal designs.

(5) Once one understands how nature does it, one can then fabricate durable and optimum structures using smart materials and modern fabrication techniques to provide functionality of interest in a more sustainable and environmentally friendly manner.

(6) For commercial applications, structures should be durable and would be desirable to be self-healing.

1.10 Organization of the Book

After a broad introduction, the book starts with roughness-induced superliquiphobic surfaces and the lotus effect which exhibits superhydrophobicity, self-cleaning, low adhesion/drag reduction, self-healing, and antifouling. It follows with shark skin which exhibits low drag and antifouling; gecko adhesion which exhibits reversible adhesion; bio- and inorganic fouling, water harvesting from fog and condensation and purification, and mosquitos' painless piercing.

References

Alberts, B., Johnson, A., Lewis, J., Raff, M. Roberts, K., Walter, P. (eds.) (2008) *Molecular Biology of the Cell*, Garland Science, New York.

Alexander, R. M. and Diskin, A. (2004), *Human Bones: A Scientific and Pictorial Investigation,* Pi Press, New York.

Anonymous (2007), *Biomimetics: Strategies for Product Design Inspired by Nature*, Dept. of Trade and Industry, U. K.

Autumn, K., Liang, Y. A., Hsieh, S. T., Zesch, W., Chan, W. P., Kenny, T. W., Fearing, R. and Full, R. J. (2000), "Adhesive Force of a Single Gecko Foot-Hair," *Nature* **405**, 681–685.

Ball, P. (2002), "Natural Strategies for the Molecular Engineer," *Nanotechnology* **13**, R15–R28.

Bar-Cohen, Y. (2011), *Biomimetics: Nature-Based Innovation*, CRC Press, Boca Raton, FL.

Barth, F. G., Humphrey, J. A. C., and Secomb, T. W. (2003), *Sensors and Sensing in Biology and Engineering*, Springer-Verlag, New York.

Barthlott, W. and Neinhuis, C. (1997), "Purity of the Sacred Lotus, or Escape from Contamination in Biological Surfaces," *Planta* **202**, 1–8.

Barthlott, W., Erdelen, W. R., and Rafiqpoor, M. D. (2014), "Biodiversity and Technical Innovations: Bionics," in *Concepts and Values in Biodiversity* (Lanzerath, D. and Friele, M. (eds.), Routledge, London, U. K.

Barthlott, W., Mail, M., Bhushan, B., and Koch, K. (2017), "Plant Surfaces: Structures and Functions for Biomimetic Innovations," *Nano-Micro Lett.* **9**, #23.

Bechert, D. W., Bruse, M., Hage W., Van Der Hoeven, J. G. T., and Hoppe, G. (1997), "Experiments on Drag-Reducing Surfaces and Their Optimization with an Adjustable Geometry," *J. Fluid Mech.* **338**, 59–87.

Bechert, D. W., Bruse, M., and Hage, W. (2000), "Experiments with Three-Dimensional Riblets as an Idealized Model of Shark Skin," *Exp. Fluids* **28**, 403–412.

Bhushan, B. (2007), "Adhesion of Multi-Level Hierarchical Attachment Systems in Gecko Feet," *J. Adhes. Sci. Technol.* **21**, 1213–1258.

Bhushan, B. (2009), "Biomimetics: Lessons from Nature – An Overview," *Phil. Trans. R. Soc. A* **367**, 1445–1486.

Bhushan, B. (2011), "Biomimetics Inspired Surfaces for Drag Reduction and Oleophobicity/philicity," *Beilstein J. Nanotechnol.* **2**, 66–84.

Bhushan, B. (2012a), *Biomimetics: Bioinspired Hierarchical-structured Surfaces for Green Science and Technology*, Springer International, Cham, Switzerland.

Bhushan, B. (2012b), "Bioinspired Structured Surfaces," *Langmuir* **28**, 1698–1714.

Bhushan, B. (2013a), *Introduction to Tribology*, second ed., Wiley, New York.

Bhushan, B. (2013b), *Principles and Applications of Tribology*, second ed., Wiley, New York.

Bhushan, B. (2016a), *Biomimetics: Bioinspired Hierarchical-structured Surfaces for Green Science and Technology*, second ed., Springer International, Cham, Switzerland.

Bhushan, B. (2016b), *Encyclopedia of Nanotechnology*, Vols. 1–6, second ed., Springer International, Cham, Switzerland.

Bhushan, B. (2017), *Springer Handbook of Nanotechnology,* fourth ed., Springer International, Cham, Switzerland.

Bhushan, B. (2018), *Biomimetics: Bioinspired Hierarchical-structured Surfaces for Green Science and Technology*, third ed., Springer International, Cham, Switzerland.

Bhushan, B. (2019a), "Lessons from Nature for Green Science and Technology: An Overview and Bioinspired Superliquiphobic/philic Surfaces," *Phil. Trans. R. Soc. A* **377**, 20180274.

Bhushan, B.(2019b), "Bioinspired Oil-water Separation Approaches for Oil Spill Clean-up and Water Purification," *Phil. Trans. R. Soc. A* **377**, 20190120.

Bhushan, B. (2020a), *Bioinspired Water Harvesting, Purification, and Oil-Water Sepa*ration, Springer International, Cham, Switzerland.

Bhushan, B. (2020b), "Design of Water Harvesting Towers and Projections for Water Collection from Fog and Condensation," *Phil. Trans. R. Soc. A* **378**, 20190440.

Bhushan, B. and Her, E. K. (2010), "Fabrication of Superhydrophobic Surfaces with High and Low and Low Adhesion Inspired by Rose Petal," *Langmuir* **26**, 8207–8217

Bhushan, B. and Jung, Y. C. (2006), "Micro and Nanoscale Characterization of Hydrophobic and Hydrophilic Leaf Surface," *Nanotechnology* **17**, 2758–2772.

Bhushan, B. and Jung, Y. C. (2011), "Natural and Biomimetic Artificial Surfaces for Superhydrophobicity, Self-Cleaning, Low Adhesion, and Drag Reduction," *Prog. Mater. Sci.* **56**, 1–108.

Bhushan, B. and Lee, H. (2012), "Fabrication and Characterization of Multi-Level Hierarchical Surfaces," *Faraday Discussion* **156**, 235–241.

Bhushan, B., Jung, Y. C., and Koch, K. (2009), "Micro-, Nano- and Hierarchical Structures for Superhydrophobicity, Self-Cleaning and Low Adhesion," *Phil. Trans. R. Soc. A* **367**, 1631–1672.

Bixler, G. D. and Bhushan, B. (2012a), "Biofouling: Lessons from Nature," *Phil. Trans. R. Soc. A* **370**, 2381–2417.

Bixler, G. D. and Bhushan, B. (2012b), "Bioinspired Rice Leaf and Butterfly Wing Surface Structures Combining Shark Skin and Lotus Effects," *Soft Matter* **9**, 1620–1635.

Bixler, G. D. and Bhushan, B. (2013a), "Fluid Drag Reduction with Shark-skin Riblet Inspired Microstructured Surfaces," *Adv. Func. Mater.* **23**, 4507–4528.

Bixler, G. D. and Bhushan, B. (2013b), "Fluid Drag Reduction and Efficient Self-Cleaning with Rice Leaf and Butterfly Wing Bioinspired Surfaces," *Nanoscale* **5**, 7685–7710.

Bixler, G. D. and Bhushan, B. (2015), "Rice- and Butterfly-Wing Effect Inspired Low Drag and Antifouling Surfaces: A Review," *Crit. Rev. in Solid State Mater. Sci.* **40**, 1-37.

Bohn, H. F. and Federle, W. (2004), "Insect Aquaplaning: Nepenthes Pitcher Plants Capture Prey with the Peristome, A Fully Wettable Water-lubricated Anisotropic Surface," *PNAS* **101**, 14138–14143.

Braun, R. J. (2012), "Dynamics of the Tear Film," *Annual Rev. Fluid Mech.* **44**, 267–297.

Brown, P. S. and Bhushan, B. (2015), "Bioinspired, Roughness-Induced, Water and Oil Super-philic and Super-phobic Coatings Prepared by Adaptable Layer-by-Layer Technique," *Sci. Rep.–Nature* **5**, 14030.

Brown, P. S. and Bhushan, B. (2016), "Bioinspired Materials for Water Supply and Management: Water Collection, Water Purification and Separation of Water from Oil," *Phil. Trans. R. Soc. A* **374**, 20160135.

Burrows, M. and Sutton, G. (2013), "Interacting Gears Synchronize Propulsive Leg Movements in a Jumping Insect," *Science* **341,** 1254–1256.

Burton, Z. and Bhushan, B. (2006), "Surface Characterization and Adhesion and Friction Properties of Hydrophobic Leaf Surfaces," *Ultramicroscopy* **106**, 709–719.

Cremaldi, J. C. and Bhushan, B. (2018), "Bioinspired Self-healing Materials: Lessons from Nature," *Beilstein J. Nanotech.* **9**, 907–935.

Cutkosky, M. R. and Kim, S. (2009), "Design and Fabrication of Multi-Materials Structures for Bio-Inspired Robots," *Philos. Trans. R. Soc. A* **367**, 1799–1813.

Darwin, C. R. (1859), *On the Origin of Species,* John Murray, London, U. K.

Das, R., Yadav, R., Sihota, P., Uniyal, P., Kumar, N., and Bhushan, B. (2018), "Biomechanical Evaluation of Wasp and Honeybee Stingers," *Sci. Rep. – Nature* **8**, 14945.

Dean, B. and Bhushan, B. (2010), "Shark-Skin Surfaces for Fluid-Drag Reduction in Turbulent Flow: A Review," *Phil. Trans. R. Soc. A.* **368**, 4775–4806; **368**, 5737.

DeVries, A. L. and Wohlschlag, D. E. (1969), "Freezing Resistance in Some Antarctic Fishes," *Science* **163**, 1073–1075.

Duman, J. G. (1977), "The Role of Macromolecular Antifreeze in the Darkling Beetle *Meracantha contracta*," *J. Comp. Physiol. B* **115**, 279–286.

Duman, J. G. and Olsen, T. M. (1993), "Thermal Hysteresis Protein Activity in Bacteria, Fungi, and Phylogenetically Diverse Plants," *Cryobiology* **30**, 322–328.

Fairs, M. (2006), "Joris Laarman Works with Opel," *Dezeen Magazine* (http://www.dezeen.com/2006/12/27/joris-laarman-works-with-open/)

Federle, W., Barnes, W. J. P., Baumgartner, W., Drechsler, P., and Smith, J. M. (2006), "Wet but Not Slippery: Boundary Friction in Tree Frog Adhesive Toe Pads," *J. R. Soc. Interface* **3,** 689–697.

Fish, F. E. and Battle, J. M. (1995), "Hydrodynamic Design of the Humpback Whale Flipper." *J. Morphol.* **225**, 51–60.

Fratzl, P. and Weinkamer, R. (2007), "Nature's Hierarchical Materials," *Prog. Mat. Sci.* **52**, 1263–1334.

Full, R., Earls, K., Wong, M., and Caldwell, R. (1993), "Locomotion Like a Wheel," *Nature* **365**, 495.

Gao, X. F. and Jiang, L. (2004), "Biophysics: Water-repellent Legs of Water Striders," *Nature* **432**, 36.

Gao, H., Wang, X., Yao, H., Gorb, S. and Arzt, E. (2005), "Mechanics of Hierarchical Adhesion Structures of Geckos," *Mech. Mater.* **37**, 275–285

Genzer, J. and Efimenko, K. (2006), "Recent Developments in Superhydrophobic Surfaces and Their Relevance to Marine Fouling: A Review," *Biofouling* **22**, 339–360.

Gilbert, J. A., Hill, P. J., Dodd, C. E., and Laybourn-Parry, J. (2004), "Demonstration of Antifreeze Protein Activity in Antarctic Lake Bacteria," *Microbiology* **150**, 171–180.

Gorb, S. (2001), *Attachment Devices of Insect Cuticle*, Kluwer Academic, Dordrecht, Netherlands.

Gordon, J. E. (1976), *The New Science of Strong Materials, or Why You Don't Fall Through the Floor*, second ed., Pelican–Penguin, London, U. K.

Gould, S. J. (1981), "Kingdoms Without Wheels," *Nat. Hist.* **90**, 42–48.

Graether, S. P., Kuiper, M. J., Gagné, S. M., Walker, V. K., Jia, Z., Sykes, B. D., and Davies, P. L. (2000), "β-Helix Structure and Ice- Binding Properties of a Hyperactive Antifreeze Protein from an Insect," *Nature* **406**, 325–328.

Gruber, P. (2011), *Biomimetics in Architecture*, Springer-Verlag, Germany.

Grunwald, I., Rischka, K., Kast, S. M., Scheibel, T., and Bargel, H. (2009), "Mimicking Biopolymers on a Molecular Scale: Nano(bio)technology Based on Engineering Proteins," *Phil. Trans. R. Soc. A* **367**, 1727–1726.

Grushkin, D. (2015), "Meet the Woman Who Wants to Grow Clothing in a Lab," *Popular Science*, www.popsci.com/meet-woman-who-wants-growing-clothing-lab, accessed 1/6/2016.

Gurera, D. and Bhushan, B. (2020), "Passive Water Harvesting by Desert Plants and Animals: Lessons from Nature," *Phil Trans. R. Soc. A* **378**, 20190444.

Gurera, D., Bhushan, B., and Kumar, N. (2018), "Lessons from Mosquitoes' Painless Piercing," *J. Mech. Behav. Biomed.* **84**, 178–187.

Imhof, B. and Gruber, P. (2013), *What is the Architect Doing in the Jungle?*, Springer Wein, New York.

Jakab, P. L. (1990), *Vision of a Flying Machine*, Smithsonian Institution Press, Washington D.C.

Jin, H. -J. and Kaplan, D. L. (2003), "Mechanism of Silk Processing in Insects and Spiders," *Nature* **424**, 1057–1061

Jones, C. J. and Aizawa, S. (1991), "The Bacterial Flagellum and Flagellar Motor: Structure, Assembly, and Functions," *Adv. Microb. Physiol.* **32**, 109–172.

Jung, Y. C. and Bhushan, B. (2010), "Biomimetic Structures for Fluid Drag Reduction in Laminar and Turbulent Flows," *J. Phys.: Condens. Matter* **22**, 035104.

Koch, K., Bhushan, B., and Barthlott, W. (2008), "Diversity of Structure, Morphology, and Wetting of Plant Surfaces," *Soft Matter* **4**, 1943–1963.

Koch, K., Bhushan, B., and Barthlott, W. (2009), "Multifunctional Surface Structures of Plants: An Inspiration for Biomimetics," *Prog. Mater. Sci.* **54**, 137–178.

Kurokawa, K. (1977), *Metabolism in Architecture*, Westview Press, Boulder, Colorado.

Lee, H. and Bhushan, B. (2012), "Fabrication of and Characterization of Hierarchical Nanostructured Smart Adhesion Surfaces," *J. Colloid Interface Sci.* **372**, 231–238.

Li, F., Kong, W., Bhushan, B., Zhao, X., and Pan, Y. (2019), "Ultraviolet-Driven Switchable Superliquiphobic/superliquiphilic Coating for Separation of Oil-water Mixtures and Emulsions and Water Purification," *J. Colloid and Interface Sci.* **557**, 395–407.

Liu, L., Pan, Y., Bhushan, B., and Zhao, Z. (2019), "Mechanochemical Robust, Magnetic-Driven, Superhydrophobic 3D Porous Materials for Contaminated Oil Recovery," *J. Colloid and Interface Sci.* **538**, 25–33.

Livio, M. (2002), *The Golden Ratio: The Story of Phi, the World's Most Astonishing Number*, Broadway Books, New York.

Lowenstam, H. A. and Weiner, S. (1989), *On Biomineralization*, Oxford University Press, Oxford, U. K.

Mann, S. (2001), *Biomineralization*, Oxford Univ. Press, Oxford, U. K.

Marentic, F. J. and Morris, T. L. (1992), "Drag Reducing Article," U.S. Patent No. 5,133,516.

Martin, S. and Bhushan, B. (2014), "Fluid Flow Analysis of a Shark-Inspired Microstructure," J. Fluid Mech. 756, 5–29.

Martin, S. and Bhushan, B. (2016), "Discovery of Riblets in a Bird Beak (Rynchops) for Low Fluid Drag," *Phil Trans. R. Soc. A* **374**, 20160134.

Meister, K., Ebbinghaus, S., Xu, Y, Duman, J. G., DeVries, A., Gruebele, M., Leitner, D. M., and Havenith, M. (2013), "Long-Range Protein-Water Dynamics in Hyperactive Insect Antifreeze Proteins," *Proc. Natl. Acad. Sci. U. S. A.* **110**, 1617–1622.

Meyers, M. A., Lin, A. Y. M., Seki, Y., Chen, P. Y., Kad, B. K., and Bodde, S. (2006), "Structural Biological Composites: An Overview," *JOM* **58**, 35–41.

Meyers, M. A., Chen, P. Y., Lin, A. Y. M., and Seki, Y. (2008), "Biological Materials: Structure and Mechanical Properties," *Prog. Mater. Sci.* **53**, 1–206.

Mora, C., Tittensor, D. P., Adl, S., Simpson, A. G. B., and Worm, B. (2011), "How Many Species Are There on Earth and in the Ocean?," *PLoS Biology* **9(8)**, e1001127.

Mueller, T. (2008), "Biomimetics Design by Natures," *National Geographic* April 2008, 68–90.

Multanen, V. and Bhushan, B. (2019), "Bioinspired Self-healing, Superliquiphobic and Self-cleaning Hydrogel- coated Surfaces with High Durability," *Phil. Trans. R. Soc. A* **377**, 20190117.

Neinhuis, C., and Barthlott, W. (1997), "Characterization and Distribution of Water-Repellent, Self-Cleaning Plant Surfaces," *Annals of Botany* **79**, 667–677.

Nosonovsky, M and Bhushan, B. (2008), *Multiscale Dissipative Mechanisms and Hierarchical Surfaces*, Springer, Heidelberg, Germany.

Nosonovsky, M. and Bhushan, B. (2009), "Thermodynamics of Surface Degradation, Self-Organization, and Self-Healing for Biomimetic Surfaces," *Phil. Trans. R. Soc. A* **367**, 1607–1627.

Nutt, D. R. and Smith, J. C. (2008), "Dual Function of the Hydration Layer around an Antifreeze Protein Revealed by Atomistic Molecular Dynamics Simulations," *J. Am. Chem. Soc.* **130**, 13066–13073.

Posamentier, A. S. and Lehmann, I. (2007), *The Fabulous Fibonacci Numbers*, Prometheus Books, New York.

Ripley, R. and Bhushan, B. (2016), "Bioarchitecture: Bioinspired Art and Architecture – A Perspective," *Phil. Trans. Roy. Soc. A.* **374**, 20160192.

Sarikaya, M. and Aksay, I. A. (1995), *Biomimetic Design and Processing of Materials*, American Institute of Physics, Woodbury, NY.

Scholtz, G. (2008), "Scarab Beetles at the Interface of Wheel Invention in Nature and Culture?," *Contr. Zool.* **77**,139–148.

Sidebottom, C., Buckley, S., Pudney, P., Twigg, S., Jarman, C., Holt, C., Telford, J., McArthur, A., Worrall, D., Hubbard, R., and Lillford, P. (2000), "Heat-Stable Antifreeze Protein from Grass," *Nature* **406**, 256.

Singer, A. J. and Clark, R. A. F. (1999), "Cutaneous Wound Healing," *N. Engl. J. Med.* **341**, 738–746.

Stegmaier, T., Linke, M., and Planck, H. (2009), "Bionics in Textiles: Flexible and Translucent Thermal Insulations for Solar Thermal Applications," *Phil. Trans. R. Soc. A* **367**, 1749–1758.

Sun, J. and Bhushan, B. (2012a), "Hierarchical Structure and Mechanical Properties of Nacre: A Review," *RSC Adv.* **2**, 7617–7632.

Sun, J. and Bhushan, B. (2012b), "The Structure and Mechanical Properties of Dragonfly Wings and Their Role in Flyability," *C. R. Mecanique* **340**, 3–17.

Sun, J., Bhushan, B., and Tong, J. (2013), "Structural Coloration in Nature," *RSC Adv.* **3**, 14862–14899.

Sun, J., Ling, M., Wu, W., Bhushan, B., and Tong, J. (2014), "The Hydraulic Mechanism of the Unfolding of Hind Wings in *Dorcus titanus platymelus* (Order: Coleoptera)," *Int. J. Mol. Sci.* **15**, 6009–6018.

Sun, J., Wu, W., Ling, M., Bhushan, B., and Tong, J. (2016), "The Hydraulic Mechanism in the of Hind Wing Veins of *Cybister japonicas* Sharp (Order: Coleoptera)," *Beilstein J. Nanotechnol.* **7**, 904–913.

Tamerler, C. and Sarikaya, M. (2009), "Molecular Biomimetics: Nanotechnology and Molecular Medicine Utilizing Genetically Engineered Peptides," *Phil. Trans. R. Soc. A* **367**, 1705–1726.

Urrutia, M. E., Duman, J. G., and Knight, C. A. (1992), "Plant Thermal Hysteresis Proteins," *Biochim. Biophys. Acta, Protein Struct. Mol. Enzymol.* **1121**, 199–206.

Vajda, S. (1989), *Fibonacci & Lucas Numbers, and the Golden Section: Theory and Applications*, Halsted Press, New York.

Venturi, R., Brown, D. S., and Izenour, S. (1977), *Learning from Las Vegas: The Forgotten Symbolism of Architectural Form*, MIT Press, Cambridge, MA.

von Richthofen, A. (2007), "3m's of Metabolism," *Pidgin* **3,** 146–163.

Wagner, P., Fürstner, R., Barthlott, W. and Neinhuis, C. (2003), "Quantitative Assessment to the Structural Basis of Water Repellency in Natural and Technical Surfaces," *J. Exper. Botany* **54**, 1295–1303.

Watkin, D. (2005) *A History of Western Architecture*, fourth ed., Watson-Guptill Publications, New York.

Wittkower, R. (1960), "The Changing Concept of Proportion," *Daedalus* **89**, 199–215.

Zeising, A. (1854), *Neue Lehre von den Proportionen des menschlichen Korpers*, R. Weigel, Leipzig, Germany.

2 Roughness-Induced Superliquiphilic/Phobic Surfaces: Wetting States and Lessons from Living Nature

God created solids, but the devil made their surfaces—Wolfgang Pauli, Nobel Laureate in Physics

2.1 Introduction

A solid surface, or more exactly a solid-gas or solid-liquid interface, has a complex structure and properties dependent upon nature of solids, the method of preparation, and the interaction between the surface and the environment. Physical and chemical properties of surfaces affect their interaction with other surfaces.

The wetting state of a surface (i.e., whether it is -philic or -phobic to a liquid) depends on a variety of surface characteristics, including surface roughness. Surface roughness affects the angle at which a liquid contacts a surface, called the contact angle. If there is a high contact angle, the liquid is less likely to wet the surface. If the contact angle is low, the liquid is more likely to wet the surface. Contact angle depends not only on roughness, but also surface energy and cleanliness of the surface (Adamson 1990; Israelachvili 1992; Nosonovsky and Bhushan 2008a). Contact angle is the parameter that is most important for characterizing the wetting state. Various flora and fauna found in living nature exhibit a variety of wetting states that can be of interest in commercial applications.

2.2 Wetting States

To define various wetting states of surfaces, we start with the definition of a few Greek words of interest for liquids: hydro- = water, oleo- = oil, amphi- = both (water and oil in this context), omni- = all or everything, and liquid- = an unspecified liquid. Within oils,

B. Bhushan, *Introduction to Biomimetics and Bioinspiration*,
https://doi.org/10.1007/978-3-031-62344-8_2

edible oils have a surface tension larger than 30 mN/m and alkanes-based oils have a surface tension 20–30 mN/m. Greek suffixes of interest are: -philic = friendly or attracting, and -phobic = afraid of or repelling.

Figure 2.1 shows schematics of various wetting states. If a liquid wets a surface, it is referred to as a wetting liquid and the value of the static contact angle is $0 \leq \theta \leq 90°$. A surface that is wetted by a wetting liquid is referred to as hydrophilic if that wetting liquid is water, oleophilic if it is oil, amphiphilic if it can be wetted by water and oil, omniphilic if all liquids wet the surface, and liquiphilic if the liquid is unspecified. If the liquid does not wet the surface, it is referred to as a non-wetting liquid, and the value of the contact angle is $90° < \theta \leq 180°$. A surface that repels a liquid is referred to as hydrophobic if it repels water, oleophobic for oil, amphiphobic if it repels both water and oil, omniphobic if it repels all liquids, and liquiphobic if the liquid is unspecified. Surfaces with a contact angle of less than 10° are called superliquiphilic, while surfaces with a contact angle between 150° and 180° are called superliquiphobic. The words superliquiphilic and superliquiphobic were coined by Bhushan (2016a).

Surfaces with high energy with polar molecules, tend to be -philic to liquids with low surface tension. Conversely, those with low energy with non-polar molecules tend to be -phobic to those with high surface tension. It should be noted that by using low surface energy, generally fluorinated materials, the highest water contact angles that can be achieved are on the order of 120°. The contact angle is lower for liquids with lower surface energy, such as oils. In order to achieve a contact angle higher than 120°, and certainly superliquiphobicity with a contact angle over 150°, a suitable roughness distribution needs to be introduced. This design approach is referred to as "roughness-induced" superliquiphobicity.

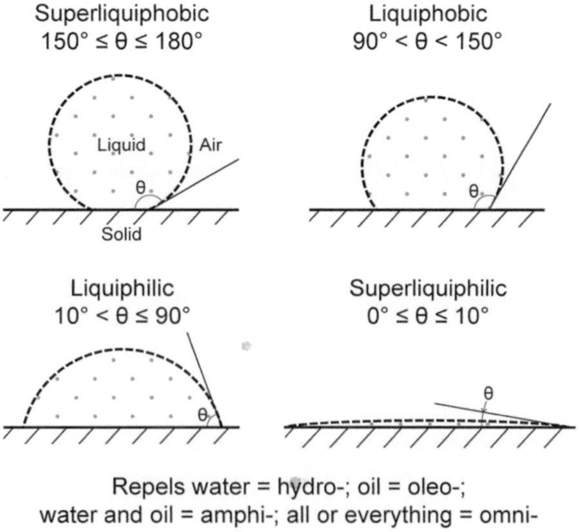

Fig. 2.1 Schematics of liquid droplets in contact with superliquiphobic, liquiphobic, liquiphilic, and superliquiphilic solid surfaces (Bhushan 2016a)

For liquid repellant and self-cleaning surfaces, high contact angle is required. In fluid flow, in order to have low drag and for applications requiring the self-cleaning feature, in addition to the high contact angle, superliquiphobic surfaces should also have very low liquid contact angle hysteresis. The contact angle at the front of the droplet (advancing contact angle) is greater than that at the back of the droplet (receding contact angle), resulting in contact angle hysteresis (CAH), which is the difference between the advancing and receding contact angles, representing two stable values. It occurs due to surface roughness and surface heterogeneity. CAH reflects the irreversibility of the wetting/dewetting cycle. In a surface with high CAH, the receding contact angle is low and may pin the liquid and provide resistance to flow. CAH is a measure of energy dissipation during the flow of a droplet along a solid surface. At a low value of CAH, the droplets may roll, in addition to slide, and take contaminants from the surface with them, providing a self-cleaning ability known as the "lotus effect." Surfaces with low CAH have a low liquid roll-off (tilt) angle, which denotes the angle to which a surface must be tilted for liquid drops to roll off. The tangent of the tilt angle is equal to the coefficient of friction of a droplet sliding/rolling on a surface. Therefore, surfaces with CAH or a tilting angle of less than $<10°$ are self-cleaning surfaces and with low adhesion/drag and antifouling properties.

2.3 Applications

Superliquiphobic surfaces can be used for liquid repellency, self-cleaning, low adhesion, drag reduction in fluid flow, energy conservation, and energy conversion (Nosonovsky and Bhushan 2008a, d, 2009a, b, 2010; Bhushan 2009, 2011; Bhushan and Jung 2011). Liquid repellent and self-cleaning surfaces are of interest in various applications, including self-cleaning windows, windshields, smart screens, camera lenses, exterior paints for buildings, utensils, roof tiles, textiles, solar panels, and exterior surfaces in transportation (Bhushan and Jung 2011). They also exhibit antifouling of interest in various applications including biomedical and membranes used in desalination and water purification (Bixler and Bhushan 2012a). Recent advances in superliquiphobic surfaces make such applications possible.

Selection of a proper superliquiphobic surface allows the reduction of energy dissipation, and, therefore, energy conservation. Reversible superliquiphobicity provides potential for new means of energy conversion. Reversible superliquiphobic surfaces have the ability to switch between liquiphobic and liquiphilic properties using an external parameter such as heat, electric potential or ultraviolet irradiation. Reversible surfaces can be used for movement of liquid droplets for fluid flow or to develop a microcapillary engine. A microcapillary engine converts the change of surface energy into mechanical energy. In Fig. 2.2 a tip held down by a spring moves vertically back and forth, as the surface is switched between hydrophilicity and hydrophobicity. For hydrophilic state of the surface, the surface energy is high enough to form a meniscus capillary bridge between the tip and the surface. Whereas in the hydrophobic state, the surface energy is lower and the meniscus is destroyed and the tip is returned to its original position by the spring. As soon as the capillary bridge is formed again, the tip snaps onto the surface (Nosonovsky and Bhushan 2008b, 2009a).

Wetting may lead to the formation of concave shaped menisci at the interface between hydrophilic solid bodies during static or sliding contact, shown in Fig. 2.3. These menisci develop a negative pressure leading to an intrinsic attractive force, which increases adhesion and friction. In some cases, the wet friction force can be greater than the dry friction force, which is usually undesirable (Bhushan 1996, 1998, 2001, 2003, 2013, 2016b, 2017a, b). In some cases, high wet adhesion/friction may lead to sticking of two surfaces, referred to as stiction. On the other hand, high adhesion is desirable in some applications, such as adhesive tapes and adhesion of cells to biomaterial surfaces. Enhanced wetting would be desirable in these applications.

Numerous applications, such as magnetic storage devices and micro/nanoelectromechanical systems (MEMS/NEMS), require surfaces with low adhesion

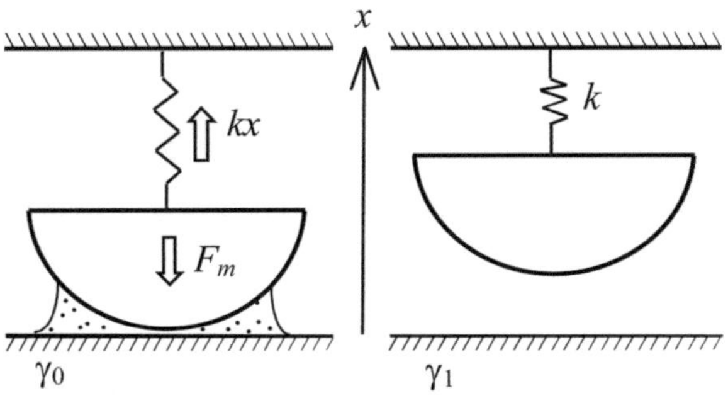

Fig. 2.2 A microcapillary engine that converts the change of the surface energy into the mechanical energy. In the first state (left), the surface energy (γ_0) is high enough to form a meniscus bridge, and the meniscus force deflects the tip. In the second state (right), the surface energy is lower, the meniscus is destroyed and the tip is returned by the elastic force to its equilibrium position. k represents spring constant, x represents the vertical displacement, and F_m represents the meniscus force (adapted from Nosonovsky and Bhushan 2008b)

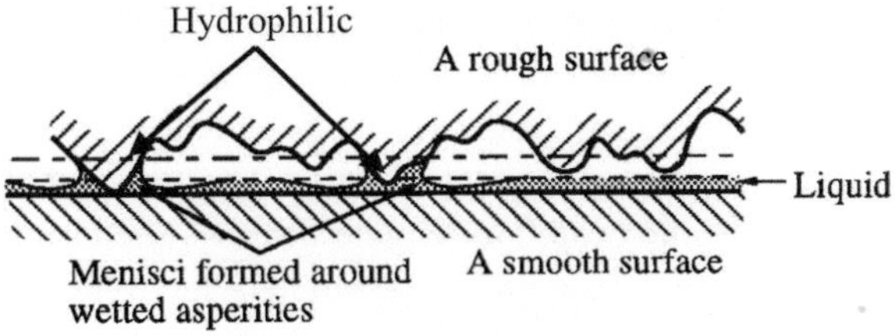

Fig. 2.3 A schematic diagram of condensed water vapor from the environment forming meniscus bridges at asperity contacts, which lead to an intrinsic attractive force

and stiction (Bhushan et al. 1995; Bhushan 1996, 1998, 2001, 2003, 2016b, 2017a, b). As the size of these devices decreases, surface forces tend to dominate over the volume forces, and adhesion and stiction constitute a challenging problem for proper operation of these devices. This makes the development of superhydrophobic surfaces with non-adhesive characteristics crucial for many of these emerging applications.

A related wetting problem is icing. Icing may occur due to sticking of supercooled water droplets onto a solid surface, also known as freezing rain or atmospheric icing. It is undesirable, as it leads to glazed roadways, broken tree limbs and power lines, and stalled aircraft airfoils. Icephobicity is defined as a surface's ability to prevent ice formation or to have very low adhesion to form ice. Anti-icing surfaces are typically superhydrophobic, as super-cooled droplets of water are able to roll off the surface before freezing, and any ice formed is weakly adhered compared to a hydrophilic surface, due to an air cushion (Cao et al. 2009; Wang et al. 2013; Brown and Bhushan 2015). Freezing time scale is about 15 ms so time of contact of droplet during bouncing should be about 15 ms or larger.

Another related wetting problem is fogging. Fogging occurs due to condensation of water in the form of small droplets on a surface. Antifogging is of interest in transparent glass or plastic surfaces used in optical applications. Hydrophilic surfaces are useful for anti-fogging applications, as they spread out any water into a thin film, thereby maintaining transparency. This is more favorable than hydrophobic surfaces for antifogging, as these require a surface to be tilted for the droplets to roll off, in order to maintain transparency. Various superhydrophilic surfaces have been fabricated using polymer-nanoparticle coatings (Grosu et al. 2004; Cebeci et al. 2006; Brown and Bhushan 2015).

2.4 Natural Superhydrophobic, Self-cleaning, Low Adhesion/Drag Reduction Surfaces with Antifouling

In the 1990s, biologists and materials scientists started to study natural superhydrophobic surfaces. It has been reported that all superhydrophobic and self-cleaning leaves consist of an intrinsic hierarchical structure (Koch et al. 2008, 2009a). Hierarchical structure provides air pocket formation, leading to a low contact area for an applied water droplet, shown in Fig. 2.4. Air pocket formation results in a reduction of the CAH, tilt angle, and adhesive force responsible for self-cleaning, low adhesion, and antifouling (Nosonovsky and Bhushan 2008a; Bhushan et al. 2009; Bhushan and Jung 2011; Bixler and Bhushan 2012a).

The leaves of water-repellent plants, such as *Nelumbo nucifera*, (lotus), are the most studied plant leaves (Barthlott and Neinhuis 1997; Neinhuis and Barthlott 1997; Wagner et al. 2003; Bhushan and Jung 2006, 2011; Burton and Bhushan 2006; Koch et al. 2008, 2009a; Bhushan 2009). The lotus leaf was known as sacred for some 2000 years

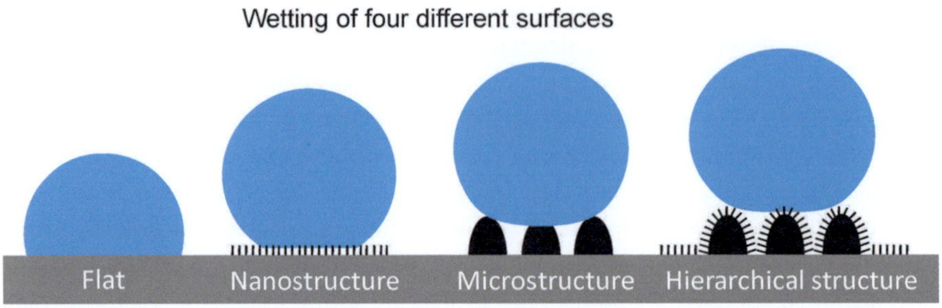

Fig. 2.4 Schematic and wetting of the four different surfaces. The largest contact area between the droplet and the surface is given in flat and microstructured surfaces, but is reduced in nanostructured surfaces and is minimized in hierarchical structured surfaces

by Hindus and Buddhists due to its self-cleaning ability. It was scientifically examined in 1997. The leaf surface is rough due to so-called papillose epidermal cells, which form papillae or microasperities. In addition to the microscale roughness, the surface of the papillae is also rough, with nanoscale asperities composed of three-dimensional epicuticular waxes which are long chain hydrocarbons and are hydrophobic, Fig. 2.5 (adapted from Bhushan et al. 2009). The wax on the lotus leaf self-assembles as tubules, but on other leaves, waxes exist also in the form of platelets or other morphologies (Koch et al. 2008, 2009a). The hierarchical structure of these leaves has been studied by Burton and Bhushan (2006) and Bhushan and Jung (2006). The water droplets on these surfaces readily sit on the apex of the nanostructures because air bubbles fill in the valleys of the structure under the droplet. Therefore, these leaves exhibit considerable superhydrophobicity and extremely low CAH. Static contact angle and CAH of a lotus leaf are about 164° and 3°, respectively (Fig. 2.5) (Bhushan et al. 2009; Koch et al. 2009b). Simply, wax makes the surface hydrophobic, and hierarchical structure makes it superhydrophobic with low CAH. Because of the latter, the water droplets on the leaves remove any contaminant particles from their surfaces when they roll off, leading to self-cleaning (Neinhuis and Barthlott 1997).

Other examples of superhydrophobicity in biological objects include rice leaves and butterfly wings (Bixler and Bhushan 2012b), water striders (*Gerris remigis*) (Gao and Jiang 2004), and mosquito (*Culex pipiens*) eyes (Gao et al. 2007). Their superhydrophobicity results from hierarchical structures. Duck feathers also provide superhydrophobicity (Bhushan 2009). Their corrugated surfaces provide air pockets that prevent water from completely touching the surface.

Fig. 2.5 SEM micrograph of a lotus (*Nelumbo nucifera*) leaf surface, consisting of microstructure formed by papillose epidermal cells covered with 3-D epicuticular wax tubules on surface, which create nanostructure, and an image of a water droplet sitting on a lotus leaf (adapted from Bhushan et al. 2009)

2.5 Natural Superhydrophobic and High Adhesion Surfaces

Unlike the lotus leaf, some rose petals (*rosea Rehd*), scallions, and garlic exhibit superhydrophobicity with high CAH (Feng et al. 2008; Chang et al. 2009; Bhushan and Her 2010; Ebert and Bhushan 2012). While a water droplet can easily roll off the surface of a lotus leaf, it stays pinned to the surface of these leaves. The different behavior of wetting between the lotus leaf and the rose petal can be explained by different designs in the surface hierarchical micro- and nanostructure. Since the rose petal's microstructures have a larger pitch value than those of the lotus leaf, the liquid is allowed to impregnate between the microstructure, and partially penetrates into the nanostructure. This is referred to as the Cassie impregnating wetting regime, in which the wetted surface area is less than that in the Wenzel regime but greater than that in the Cassie-Baxter regime, to be described later. Such an explanation implies that the extent of CAH increases with increasing wetted surface area, which is governed by surface micro- and nanostructure. These surfaces exhibit high adhesion.

2.6 Natural Superoleophobic Self-cleaning and Low Drag Surfaces with Antifouling

A model surface for superoleophobicity in water and self-cleaning is provided by sea animals such as fish (including carp and trout) and sharks, which are known to be protected from contamination by oil pollution although they are wetted by water (Jung and Bhushan 2009; Liu et al. 2009; Bixler and Bhushan 2012b). Fish are covered with oriented scales with concentric rings overlapping and hinged that promote anisotropic flow from head to tail (Bixler and Bhushan 2012b). Shark skin, which is a model from nature for a low drag surface, is covered by very small individual tooth-like scales called dermal denticles (little skin teeth), shaped like small riblets with longitudinal grooves (aligned parallel to the local flow direction of the water). When a shark swims fast during turbulent flow, vortices form on the surface that cause high stresses across the entire surface, leading to high drag. The riblets on the skin lift the vortices to the tips of the grooves and constrain them, resulting in water moving efficiently over their surface (Bechert et al. 2000; Bhushan 2009, 2011; Jung and Bhushan 2009, 2010; Dean and Bhushan 2010; Bixler and Bhushan 2013a). The space between these dermal denticles is such that microscopic aquatic organisms have difficulty adhering to and colonizing the surface (Genzer and Efimenko 2006; Jung and Bhushan 2009, 2010; Dean and Bhushan 2010; Bixler and Bhushan 2012a, 2013a). If oil is present on the surfaces of sea animals, the scales repel the oil and it is superoleophobic (Bixler and Bhushan 2013b). Hydrophilic surfaces exhibit superoleophobicity underwater as it is difficult for water to dewet the surface.

The lower side of the lotus leaf also provides a model surface for superoleophobicity in water. The lower side does not contain three-dimensional wax crystals (Neinhuis and Barthlott 1997) and consists of rather flat, tubular, and slightly convex papillae (Koch et al. 2009a). Therefore, the bottom surface is superhydrophilic but superoleophobic in water (Cheng et al. 2011).

Model surfaces in living nature with superoleophobicity in air are rare. Examples include certain species of insects such as some leafhoppers (Rakitov and Gorb 2013) and springtails (Hensel et al. 2013). Superoleophobic surfaces are of interest for drag reduction in oil pipelines and antifouling.

2.7 Closure

The term omniphobic/philic is used with regard to wetting by all liquids of scientific interest. The term amphiphobic/philic is used with regard to wetting by water and oil. If liquid is not specified, the surface is simply referred to as liquiphilic or liquiphobic, dependent upon the contact angle. A surface is superliquiphobic if it has a liquid contact angle above 150°. These surfaces are liquid repellent. These surfaces with low contact angle hysteresis (less than 10°) also have a self-cleaning effect, called the "lotus effect." Liquid droplets roll off the surface and take contaminants with them.

The wetting state of a surface depends on a variety of factors, including surface roughness, surface energy, and cleanliness of the surface. Various flora and fauna found in living nature exhibit a variety of wetting states. A model surface for superhydrophobicity is the lotus leaf. Water repellant and self-cleaning surfaces with low adhesion/drag are of interest in various applications. Low adhesion and drag reduction in fluid flow is also of interest in many applications. In addition, superhydrophobic surfaces can be used for energy conservation and energy conversion. Surfaces with low adhesion/drag also exhibit antifouling properties. Superhydrophobic surfaces are also desirable in the ambient environment to minimize stiction.

Unlike lotus leaves, some rose petals, scallions, and garlic exhibit superhydrophobicity with high contact angle hysteresis. These surfaces exhibit high adhesion.

Oleophobic/philic is the term used with regard to wetting by oil and other organic liquids. A model surface for superoleophobicity in water and self-cleaning is provided by many sea animals including fish and sharks, which are known to be well protected from contaminants from oil pollution although they are wetted by water. Microstructured surfaces of sea animals also exhibit antifouling. Another model surface for superoleophobicity in water is the lower side of the lotus leaf. Model surfaces for superoleophobicity in air is rare and include certain species of insects such as some leafhoppers and springtails.

References

Adamson, A. V. (1990), *Physical Chemistry of Surfaces*, Wiley, New York.

Barthlott, W. and Neinhuis, C. (1997), "Purity of the Sacred Lotus, or Escape from Contamination in Biological Surfaces," *Planta* **202**, 1–8.

Bechert, D. W., Bruse, M., and Hage, W. (2000), "Experiments with Three-Dimensional Riblets as an Idealized Model of Shark Skin," *Exp. Fluids* **28**, 403–412.

Bhushan, B. (1996), *Tribology and Mechanics of Magnetic Storage Systems*, second ed., Springer-Verlag, New York.

Bhushan, B. (1998), *Tribology Issues and Opportunities in MEMS*, Kluwer Academic Publishers, Dordrecht, Netherlands.

Bhushan, B. (2001), *Modern Tribology Handbook, Vol. 1—Principles of Tribology; Vol. 2—Materials, Coatings, and Industrial Applications*, CRC Press Inc., Boca Raton, Florida.

Bhushan, B. (2003), "Adhesion and Stiction: Mechanisms, Measurement Techniques and Methods for Reduction," *J. Vac. Sci. Technol. B* **21**, 2262–2296.

Bhushan, B. (2009), "Biomimetics: Lessons from Nature—An Overview," *Phil. Trans. R. Soc. A* **367**, 1445–1486.

Bhushan, B. (2011), "Biomimetics Inspired Surfaces for Drag Reduction and Oleophobicity/philicity," *Beilstein J. Nanotechnol.* **2**, 66–84.

Bhushan, B. (2013), *Introduction to Tribology*, second ed., Wiley, New York.

Bhushan, B. (2016a), *Biomimetics—Bioinspired Hierarchical-Structured Surfaces for Green Science and Technology*, second ed., Springer International, Switzerland.

Bhushan, B. (2016b), *Encyclopedia of Nanotechnology, Vols. 1–6*, second ed., Springer International, Switzerland.

Bhushan B. (2017a) *Nanotribology and Nanomechanics—An Introduction*, fourth ed., Springer International, Switzerland.

Bhushan, B. (2017b), *Springer Handbook of Nanotechnology*, fourth ed., Springer International, Switzerland.

Bhushan, B. and Her, E. K. (2010), "Fabrication of Superhydrophobic Surfaces with High and Low Adhesion Inspired from Rose Petal," *Langmuir* **26**, 8207–8217.

Bhushan, B. and Jung, Y. C. (2006), "Micro and Nanoscale Characterization of Hydrophobic and Hydrophilic Leaf Surface," *Nanotechnology* **17**, 2758–2772.

Bhushan, B. and Jung, Y. C. (2011), "Natural and Biomimetic Artificial Surfaces for Superhydrophobicity, Self-Cleaning, Low Adhesion, and Drag Reduction," *Prog. Mater. Sci.* **56**, 1–108.

Bhushan, B., Israelachvili, J. N., and Landman, U. (1995), "Nanotribology: Friction, Wear and Lubrication at the Atomic Scale," *Nature* **374**, 607–616.

Bhushan, B., Jung, Y. C., and Koch, K. (2009), "Micro-, Nano- and Hierarchical Structures for Superhydrophobicity, Self-Cleaning and Low Adhesion," *Phil. Trans. R. Soc. A* **367**, 1631–1672.

Bixler, G. D. and Bhushan, B. (2012a), "Biofouling: Lessons from Nature," *Phil. Trans. R. Soc. A* **370**, 2381–2417.

Bixler, G. D. and Bhushan, B. (2012b), "Bioinspired Rice Leaf and Butterfly Wing Surface Structures Combining Shark Skin and Lotus Effects," *Soft Matter* **9**, 1620–1635.

Bixler, G. D. and Bhushan, B. (2013a), "Fluid Drag Reduction with Shark-Skin Riblet Inspired Microstructured Surfaces," *Adv. Func. Mater.* **23**, 4507–4528.

Bixler, G. D. and Bhushan, B. (2013b), "Fluid Drag Reduction and Efficient Self-Cleaning with Rice Leaf and Butterfly Wing Bioinspired Surfaces," *Nanoscale* **5**, 7685–7710.

Brown, P. S. and Bhushan, B. (2015), "Bioinspired, Roughness-induced, Water and Oil Super-philic and Super-phobic Coatings Prepared by Adaptable Layer by Layer Technique," *Sci. Rep.* **5**, 14030.

Burton, Z. and Bhushan, B. (2006), "Surface Characterization and Adhesion and Friction Properties of Hydrophobic Leaf Surfaces," *Ultramicroscopy* **106**, 709–719.

Cao, L., Jones, A. K., Sikka, V. K., Wu, J., and Gao, D. (2009), "Anti-Icing Superhydrophobic Surfaces," *Langmuir* **25**, 12444–12448.

Cebeci, F. C., Zhizhang, W., Zhai, L., Cohen, R. E., and Rubner, M. F. (2006), "Nanoporisity-Driven Superhydrophilicity: A Means to Create Multifunctional Antifogging Coatings," *Langmuir* **22**, 2856–2862.

Chang, F. M., Hong, S. J., Sheng, Y. J., and Tsao, H. K. (2009), "High Contact Angle Hysteresis of Superhydrophobic Surfaces: Hydrophobic Defects," *Appl. Phys. Lett.* **95**, 064102.

Cheng, Q., Li, M., Zheng, Y., Su, B., Wang, S., and Jiang, L. (2011), "Janus Interface Materials: Superhydrophobic Air/Solid Interface and Superoleophobic Water/Solid Interface Inspired by a Lotus Leaf," *Soft Matter* **7**, 5948–5951.

Dean, B. and Bhushan, B. (2010), "Shark-Skin Surfaces for Fluid-Drag Reduction in Turbulent Flow: A Review," *Phil. Trans. R. Soc. A.* **368**, 4775–4806; **368**, 5737.

Ebert, D. and Bhushan, B. (2012), "Wear-Resistant Rose Petal-Effect Surfaces with Superhydrophobicity and High Droplet Adhesion Using Hydrophobic and Hydrophilic Nanoparticles," *J. Colloid Interf. Sci.* **384**, 183–188.

Feng, L., Zhang, Y., Xi, J., Zhu, Y., Wang, N., Xia, F., and Jiang, L. (2008), "Petal Effect: A Superhydrophobic State with High Adhesive Force," *Langmuir* **24**, 4114–4114.

Gao, X. F. and Jiang, L. (2004), "Biophysics: Water-Repellent Legs of Water Striders," *Nature* **432**, 36.

Gao, X., Yan, X., Yao, X., Xu, L., Zhang, K., Zhang, J., Yang, B., and Jiang, L. (2007), "The Dry-Style Antifogging Properties of Mosquito Compound Eyes and Artificial Analogues Prepared by Soft Lithography," *Adv. Mater.* **19**, 2213–2217.

Genzer, J. and Efimenko, K. (2006), "Recent Developments in Superhydrophobic Surfaces and Their Relevance to Marine Fouling: A Review," *Biofouling* **22**, 339–360.

Grosu, G., Andrzejewski, L., Veilleux, G., and Ross, G. G. (2004), "Relation between the Size of Fog Droplets and their Contact Angles with CR39 Surfaces," *J. Phys. D* **37**, 3350–3355.

Hensel, R., Helbig, R., Aland, S., Braun, H. G., Voigt, A., Neinhuis, C., and Werner, C. (2013) "Wetting Resistance at Its Topographical Limit: The Benefit of Mushroom and Serif T Structures," *Langmuir* **29**, 1100–1112.

Israelachvili, J. N. (1992), *Intermolecular and Surface Forces*, second edition, Academic Press, London, UK.

Jung, Y. C. and Bhushan, B. (2009), "Wetting Behavior of Water and Oil Droplets in Three Phase Interfaces for Hydrophobicity/philicity and Oleophobicity/philicity," *Langmuir* **25**, 14165–14173.

Jung, Y. C. and Bhushan, B. (2010), "Biomimetic Structures for Fluid Drag Reduction in Laminar and Turbulent Flows," *J. Phys.: Condens. Matter* **22**, 035104.

Koch, K., Bhushan, B., and Barthlott, W. (2008), "Diversity of Structure, Morphology, and Wetting of Plant Surfaces," *Soft Matter* **4**, 1943–1963.

Koch, K., Bhushan, B., and Barthlott, W. (2009a), "Multifunctional Surface Structures of Plants: An Inspiration for Biomimetics," *Prog. Mater. Sci.* **54**, 137–178.

Koch, K., Bhushan, B., Jung, Y. C., and Barthlott, W. (2009b), "Fabrication of Artificial Lotus Leaves and Significance of Hierarchical Structure for Superhydrophobicity and Low Adhesion," *Soft Matter* **5**, 1386–1393.

Liu, M., Wang, S., Wei, Z., Song, Y., and Jiang, L. (2009), "Bioinspired Design of a Superoleophobic and Low Adhesive Water/Solid Interface," *Adv. Mater.* **21**, 665–669.

Neinhuis, C. and Barthlott, W. (1997), "Characterization and Distribution of Water-Repellent, Self-Cleaning Plant Surfaces," *Ann. Bot.* **79**, 667–677.

Nosonovsky, M. and Bhushan, B. (2008a), *Multiscale Dissipative Mechanisms and Hierarchical Surfaces: Friction, Superhydrophobicity, and Biomimetics*, Springer-Verlag, Heidelberg, Germany.

Nosonovsky, M. and Bhushan, B. (2008b), "Superhydrophobicity for Energy Conversion and Conservation Applications," *J. Adhesion Sci. Tech.* **22**, 2105–2115.

Nosonovsky, M. and Bhushan, B. (2009a), "Multiscale Effects and Capillary Interactions in Functional Biomimetic Surfaces for Energy Conversion and Green Engineering," *Phil. Trans. R. Soc. A* **367**, 1511–1539.

Nosonovsky, M. and Bhushan, B. (2009b), "Superhydrophobic Surfaces and Emerging Applications: Non-Adhesion, Energy, Green Engineering," *Curr. Opin. Colloid Interface Sci.* **14**, 270–280.

Nosonovsky, M. and Bhushan, B. (2010), "Green Tribology: Principles, Research Areas and Challenges," *Phil. Trans. R. Soc. A* **368**, 4677–4694.

Rakitov, R. and Gorb, S. N. (2013), "Brochosomal Coats Turn Leafhopper (*Insecta, Hemiptera, Cicadellidae*) Integument to Superhydrophobic State," *Proc. R. Soc. B.* **280**, 20122391.

Wagner, P., Fürstner, R., Barthlott, W., and Neinhuis, C. (2003), "Quantitative Assessment to the Structural Basis of Water Repellency in Natural and Technical Surfaces," *J. Exp. Bot.* **54**, 1295–1303.

Wang, Y. Xue, J., Wang, Q., Chen, Q., and Ding, J. (2013), "Verification of Icephobic/Anti-icing Properties of a Superhydrophobic Surface," *ACS Appl. Mater. Inter.* **5**, 3370–3381.

3 Modeling of Contact Angle for a Liquid in Contact with a Rough Surface for Various Wetting Regimes

3.1 Introduction

The roughness distribution affects contact angle and surface wetting. Three surface wetting regimes are discussed, which include Wenzel, Cassie-Baxter, and Cassie regimes. In the Wenzel regime, a liquid droplet completely wets the rough surface with a homogeneous interface. In the Cassie-Baxter regime, a heterogeneous or composite interface with air pockets trapped between the asperities is formed. In a Cassie regime, a liquid film impregnates some of the cavities in an area surrounding the droplet as well.

3.2 Contact Angle Definition

The surface atoms or molecules of liquids or solids have fewer bonds with neighboring atoms. Therefore, they have higher energy than similar atoms and molecules in the interior. This additional energy is characterized quantitatively by the surface tension or free surface energy γ, which is equal to the work that is required to create a unit area of the surface at a constant pressure and temperature. The unit of γ is J/m^2 or N/m, and it can be interpreted as either energy per unit surface area or as tension force per unit length of a line at the surface. When a solid is in contact with liquid, the molecular attraction will reduce the energy of the system below that for the two separate surfaces. This is expressed by the Dupré equation,

$$W_{SL} = \gamma_{SA} + \gamma_{LA} - \gamma_{SL} \tag{3.1}$$

where W_{SL} is the (thermodynamic) work of adhesion per unit area, γ_{SA}, γ_{SL}, and γ_{LA} are the solid-air, solid-liquid, and liquid-air surface tensions, respectively.

B. Bhushan, *Introduction to Biomimetics and Bioinspiration*,
https://doi.org/10.1007/978-3-031-62344-8_3

If a liquid droplet is placed on a smooth solid surface, the liquid and solid surfaces come together under equilibrium at a static contact angle θ_0 (Fig. 3.1). This contact angle can be determined by minimizing the net surface free energy of the system (Adamson 1990; Israelachvili 1992; Nosonovsky and Bhushan 2008a; Bhushan 2013a, b). The total energy E_{tot} is given by,

$$E_{tot} = \gamma_{LA}(A_{LA} + A_{SL}) - W_{SL}A_{SL} \tag{3.2}$$

where A_{SL} and A_{LA} are the contact areas of the liquid with the solid and air, respectively. It is assumed that the droplet of density ρ is smaller than the capillary length, $(\gamma_{LA}/\rho g)^{1/2}$, so that the gravitational potential energy can be neglected. It is also assumed that the volume and pressure are constant, so that the volumetric energy does not change. At the equilibrium $\mathrm{d}E_{tot} = 0$, which yields,

$$\gamma_{LA}(\mathrm{d}A_{LA} + \mathrm{d}A_{SL}) - W_{SL}\mathrm{d}A_{SL} = 0 \tag{3.3}$$

For a droplet of constant volume, using geometrical considerations,

$$\mathrm{d}A_{LA}/\mathrm{d}A_{SL} = \cos\theta_0 \tag{3.4}$$

Combining (3.1), (3.3) and (3.4), the well-known Young equation for the static contact angle is obtained,

$$\cos\theta_0 = \frac{\gamma_{SA} - \gamma_{SL}}{\gamma_{LA}} \tag{3.5}$$

Equation (3.5) provides an expression for the static contact angle for given surface tensions. Using (3.1) and (3.5), an expression for the work of adhesion is obtained

$$W_{SL} = \gamma_{LA}(1 + \cos\theta_0) \tag{3.6}$$

This is called the Young-Dupre equation. W_{SL} should be reduced to reduce interfacial adhesion which implies a high value of θ_0.

As discussed in Chap. 2, the term "wetting" implies a value of contact angle, θ_0, between 0° and 90°. If θ_0 is greater than 90°, the liquid does not wet the surface. If θ_0 is equal to 0°, the liquid spreads over the surface. Thus, spreading occurs when the work of adhesion is greater than the fluid adhesion,

$$W_{SL} \geq 2\gamma_{LA} \tag{3.7}$$

Note that, although the term "air" is used, the analysis does not change in the case of another gas, such as water vapor.

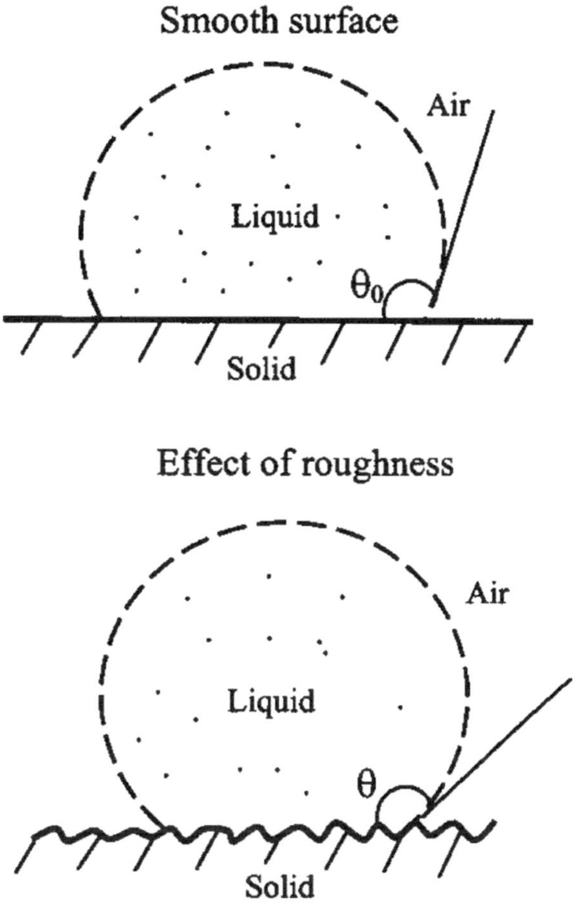

Fig. 3.1 Schematic of a liquid droplet in contact with a smooth solid surface (contact angle θ_0) and with a rough solid surface (contact angle θ)

3.3 Homogeneous and Heterogeneous Interfaces and the Wenzel, Cassie-Baxter and Cassie Equations

In this section, the equations that govern the contact angle of liquid with a rough surface for homogeneous and heterogeneous interfaces are introduced and discussed. First, we consider a liquid droplet on a rough surface with a homogeneous interface. The interface area increases with respect to that for a smooth surface. Using the surface force balance and empirical considerations, the contact angle of a liquid droplet upon a rough solid surface, θ, is related to that upon a smooth surface, θ_0 for a homogeneous interface (Fig. 3.1), through the non-dimensional surface roughness factor, $R_f > 1$, which is equal to the ratio of the rough surface area, A_{SL}, to its flat projected area, A_F (Wenzel 1936),

$$\cos\theta = \frac{dA_{LA}}{dA_F} = \frac{A_{SL}}{A_F}\frac{dA_{LA}}{dA_{SL}} = R_f\cos\theta_0 \tag{3.8}$$

where,

$$R_f = \frac{A_{SL}}{A_F} \tag{3.9}$$

This is called the Wenzel equation.

The dependence of the contact angle on the roughness factor is presented in Fig. 3.2a for different values of θ_0 (Nosonovsky and Bhushan 2005). The Wenzel model predicts that a liquiphobic surface ($\theta_0 > 90°$) becomes more liquiphobic with an increase in R_f, while a liquiphilic surface ($\theta_0 < 90°$) becomes more liquiphilic with an increase in R_f (Nosonovsky and Bhushan 2005; Jung and Bhushan 2006). As an example, Fig. 3.2b shows a geometry with square-based hemispherically-topped pyramidal asperities with a rounded top, which has complete packing. The size and shape of the asperities can be optimized for a desired roughness factor.

Next, we consider a liquid droplet on a rough surface with a heterogeneous interface. For a heterogeneous interface composed of two fractions, one with the fractional area f_1 and the contact angle θ_1 and the other with f_2 and θ_2, respectively (so that $f_1 + f_2 = 1$), the contact angle is given by the Cassie equation (Cassie 1948),

$$\cos\theta = f_1 \cos\theta_1 + f_2 \cos\theta_2 \tag{3.10}$$

A composite interface shown in Fig. 3.3, consists of a fractional geometrical area of the solid-liquid interface under the droplet ($f_1 = f_{SL}$, $\theta_1 = \theta$) and of the liquid-air interface ($f_2 = f_{LA} = 1 - f_{SL}$, $\cos\theta_2 = -1$). For this case, combining (3.9) and (3.10) yields the so-called Cassie-Baxter equation (Cassie and Baxter 1944),

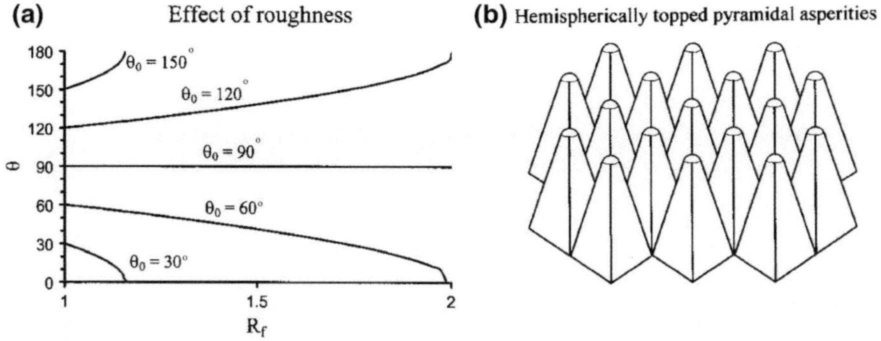

Fig. 3.2 **a** Contact angle for a rough surface (θ) as a function of the roughness factor (R_f) for a liquid wetting the surface with various contact angles of the smooth surface (θ_0), and **b** schematic of square-based hemispherically-topped pyramidal asperities with complete packing (Nosonovsky and Bhushan 2005)

Fig. 3.3 Schematic of the formation of a composite solid-liquid-air interface for a rough surface

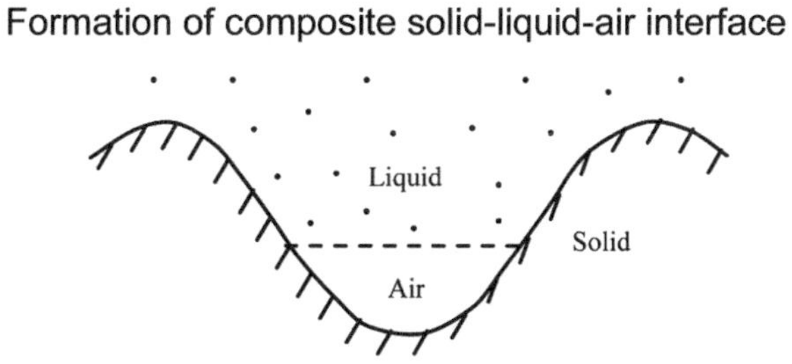

$$\cos\theta = R_f f_{SL} \cos\theta_0 - 1 + f_{SL}$$
$$\text{or, } \cos\theta = R_f \cos\theta_0 - f_{LA}(R_f \cos\theta_0 + 1) \tag{3.11}$$

Another interface in which liquid film impregnates some of the cavities in an area surrounding the liquid droplet, while the top of the rough surface remains dry (Cassie 1948). It consists of two fractions—solid-liquid homogeneous interface (Wenzel interface) and some of the neighboring cavities filled with liquid with dry tops. The energy of such a rough surface with filled cavities is different from the energy of the surface with empty cavities. Filled cavities correspond to the liquid-liquid interface with cos $\theta_2 = 1$ ($\theta_2 = 0°$). Using (3.8) and (3.10) for this case, one gets,

$$\cos\theta = 1 + f_{SL}(R_f \cos\theta_0 - 1) \tag{3.12}$$

A liquid film filling the cavities surrounding the droplet is possible when $\theta_0 < \theta_c$, where θ_c is the critical contact angle given by (de Gennes et al. 2003),

$$\cos\theta_c = \frac{1 - f_{SL}}{R_f - f_{SL}} \tag{3.13}$$

This wetting interface is referred to as the Cassie-impregnated interface or simply the Cassie interface.

Pure Cassie-Baxter and Wenzel wetting situations rarely occur. An intermediate state between the Wenzel and Cassie-Baxter interfaces is more common. In this interface, underneath the droplet, liquid penetrates and the cavities are partially impregnated or filled, and is referred to as the mixed/impregnated state (Marmur 2003; Bhushan and Nosonovsky 2010). This state is metastable.

For easy comparison, Fig. 3.4 schematically shows liquid sitting on rough surfaces showing the Wenzel, Cassie-Baxter, Cassie-impregnated, and mixed/impregnated interfaces. An extension of these three regimes to nine different regimes for surfaces with hierarchical roughness has been presented by Bhushan and Nosonovsky (2010).

Equation (3.11) for the composite interface was derived using (3.8) and (3.9), and it could also be obtained independently. For this purpose, two sets of interfaces are considered: a liquid-air interface, and a composite interface under the droplet, which itself

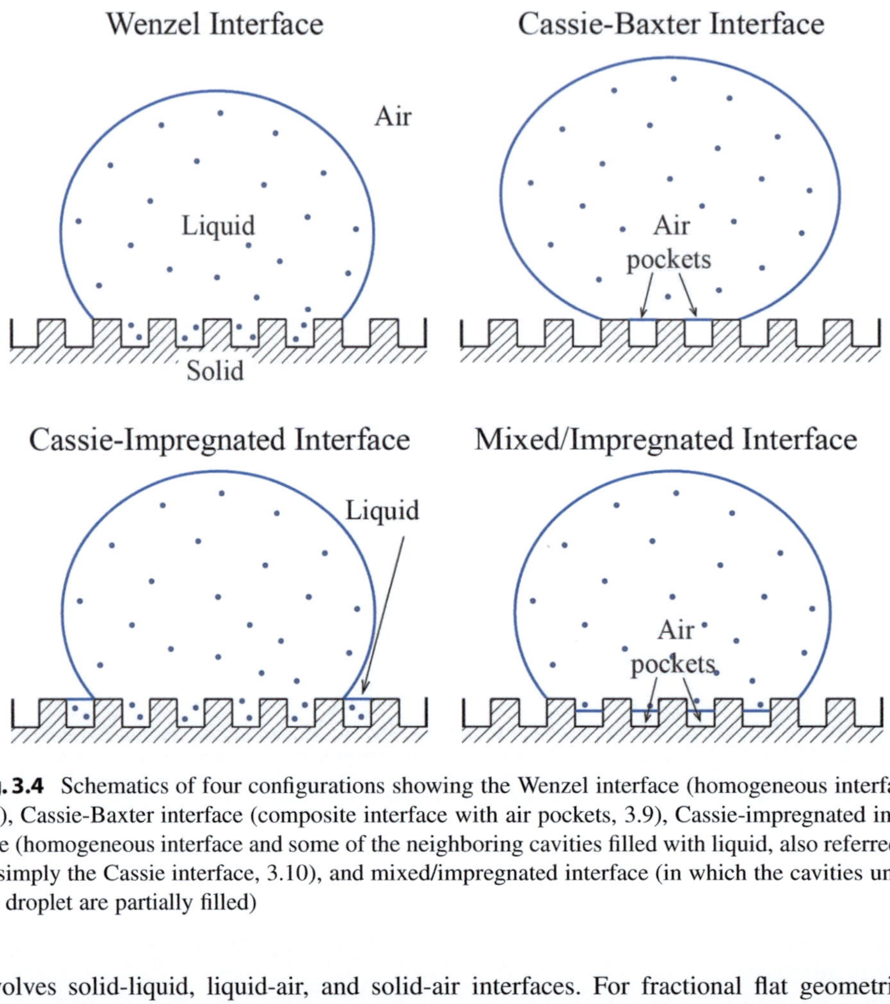

Fig. 3.4 Schematics of four configurations showing the Wenzel interface (homogeneous interface, 3.6), Cassie-Baxter interface (composite interface with air pockets, 3.9), Cassie-impregnated interface (homogeneous interface and some of the neighboring cavities filled with liquid, also referred to as simply the Cassie interface, 3.10), and mixed/impregnated interface (in which the cavities under the droplet are partially filled)

involves solid-liquid, liquid-air, and solid-air interfaces. For fractional flat geometrical areas of the solid-liquid and liquid-air interfaces under the droplet, f_{SL} and f_{LA} ($f_{SL} = 1 - f_{LA}$), the flat area of the composite interface is (Nosonovsky and Bhushan 2008a)

$$A_F = f_{SL}A_F + f_{LA}A_f = R_f A_{SL} + f_{LA}A_F \tag{3.14}$$

In order to calculate the contact angle in a manner similar to the derivation of (3.6), the differential area of the liquid-air interface under the droplet, $f_{LA}\mathrm{d}A_F$, should be subtracted from the differential of the total liquid-air area $\mathrm{d}A_{LA}$, which yields the Cassie-Baxter equation (3.11),

$$\begin{aligned}\cos\theta &= \frac{\mathrm{d}A_{LA} - f_{LA}\mathrm{d}A_F}{\mathrm{d}A_F} = \frac{A_{SL}}{A_F}\frac{\mathrm{d}A_{LA}}{\mathrm{d}A_{SL}} - f_{LA} = R_f f_{SL}\cos\theta_0 - f_{LA} \\ &= R_f\cos\theta_0 - f_{LA}\left(R_f\cos\theta_0 + 1\right)\end{aligned}$$

The dependence of the contact angle on the roughness factor and fractional liquid-air area for hydrophilic and hydrophobic surfaces with a composite interface is presented in Fig. 3.5a (Jung and Bhushan 2006). According to (3.11), even for a liquiphilic surface, the contact angle increases with an increase of f_{LA}. At a high value of f_{LA}, a surface can become liquiphobic; however, the value required may be unachievable, or the formation of air pockets may become unstable. Using the Cassie-Baxter equation, the value of f_{LA} at which a liquiphilic surface could turn into a liquiphobic one is given as (Jung and Bhushan 2006)

$$f_{LA} \geq \frac{R_f \cos\theta_0}{R_f \cos\theta_0 + 1} \quad \text{for } \theta_0 < 90^\circ \tag{3.15}$$

Figure 3.5b shows the value of f_{LA} required as a function of R_f for four surfaces with different contact angles, θ_0. Hydrophobic surfaces can be achieved above a certain f_{LA} value, as predicted by (3.15). The upper part of each contact angle line is the hydrophobic region. For the hydrophobic surface, contact angle increases with an increase in f_{LA}, both for smooth and rough surfaces.

For high f_{LA}, a nanopattern is desirable because whether a liquid-air interface is generated depends upon the ratio of distance between two adjacent asperities and droplet radius. Furthermore, asperities can pin liquid droplets and thus prevent liquid from filling the valleys between asperities. High R_f can be achieved by both micropatterns and nanopatterns.

Based on Nosonovsky and Bhushan (2008a, 2009), spreading of a liquid over a rough solid surface continues until (3.5) (Young equation) is satisfied locally at the triple line (the line of contact of the solid, liquid, and air interfaces) and, simultaneously, the surface area is minimized over the entire liquid-air interface. The minimal surface area condition states that the sum of the inverse of the principal radii of curvature R_1 and R_2 of the liquid surface along the two mutually orthogonal planes (mean curvature) is constant at any point, which governs the shape of the liquid-air interface. The same condition is also the consequence of the Laplace equation, which relates pressure change through an interface, ΔP, with its mean curvature (Nosonovsky and Bhushan 2008a),

$$\frac{1}{R_1} + \frac{1}{R_2} = \frac{\Delta P}{\gamma_{LA}} \tag{3.16}$$

Note that at the thermodynamic equilibrium (when condensation and evaporation occurs at the same speed), ΔP is dependent on the partial vapor pressure. For contact with saturated vapor the mean curvature of the liquid-air interface is zero at equilibrium. A convex interface ($1/R_1 + 1/R_2 > 0$) results in evaporation prevailing over condensation. This is why small droplets tend to evaporate. However, a concave interface ($1/R_1 + 1/R_2 < 0$) results in condensation of saturated vapor prevailing over evaporation.

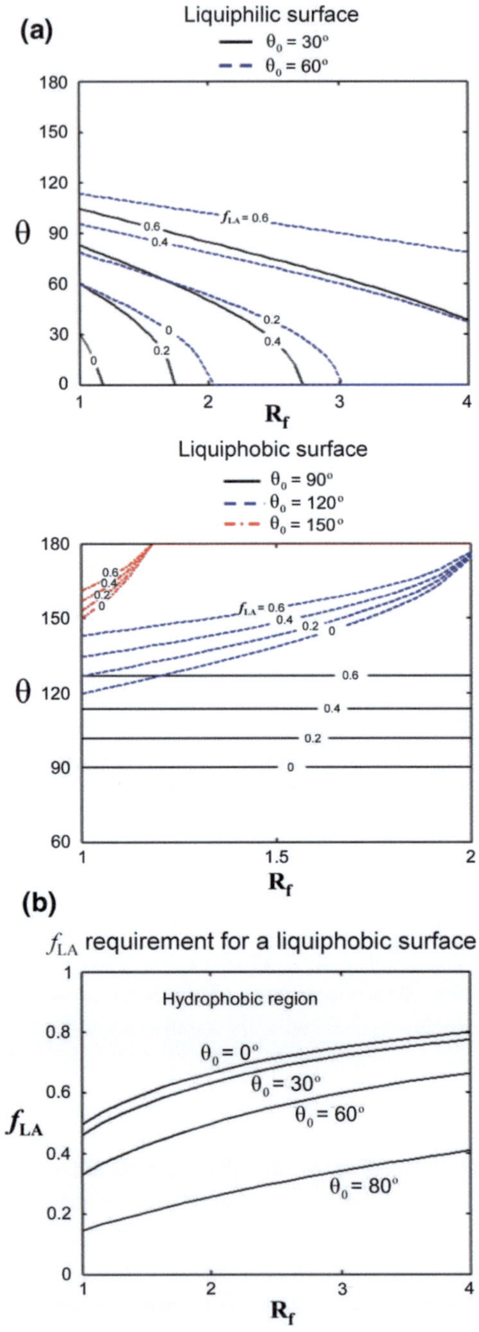

Fig. 3.5 **a** Contact angle of a rough surface (θ) for a liquid with composite interface as a function of the roughness factor (R_f) for various f_{LA} values on a liquiphilic surface and a liquiphobic surface, and **b** f_{LA} requirement for a liquiphilic surface to be liquiphobic as a function of the roughness factor (R_f) and θ_0 (Jung and Bhushan 2006)

Since the condensation prevails, a concave interface may be in thermodynamic equilibrium with undersaturated vapor. This is why concave menisci tend to condense even when the relative humidity is less than 100% (Nosonovsky and Bhushan 2008a, 2009).

3.3.1 Limitations of the Wenzel and Cassie-Baxter Equations

Cassie-Baxter equation (3.10) is based on the assumption that the heterogeneous surface is composed of well-separated distinct patches of different material, so that the free surface energy can be averaged (Nosonovsky and Bhushan 2008a, d). It has also been suggested that when the size of the chemical heterogeneities is very small of atomic or molecular dimensions, the quantity that should be averaged is not the energy but the dipole moment of a macromolecule (Israelachvili and Gee 1989). Furthermore, (3.10) should be replaced by the following equation (Nosonovsky and Bhushan 2008d),

$$(1+\cos\theta)^2 = f_1(1+\cos\theta_1)^2 + f_2(1+\cos\theta_2)^2 \tag{3.17}$$

Experimental studies of polymers with different functional groups showed a good agreement with (3.17) (Tretinnikov 2000).

The contact angle provided by (3.5), (3.8), (3.10), and (3.12) is a macroscale parameter, so it is sometimes called "the apparent contact angle." Based on Nosonovsky and Bhushan (2008a, c), the actual angle under which the liquid-air interface comes in contact with the solid surface at the micro- and nanoscale can be different. There are several reasons that this is the case. First, liquid molecules tend to form a thin layer upon the surfaces of many materials. This is because of a long-distance van der Waals adhesion force that creates the so-called disjoining pressure (Derjaguin and Churaev 1974). This pressure is dependent upon the liquid layer thickness and may lead to the formation of stable thin films. In this case, the shape of the droplet near the triple line (line of contact of the solid, liquid and air, to be shown later in Fig. 3.8) transforms gradually from the spherical surface into a precursor layer, and thus the nanoscale contact angle is much smaller than the apparent contact angle. In addition, adsorbed liquid monolayers and multilayers are common for many materials. Second, even carefully prepared atomically smooth surfaces exhibit a certain roughness and chemical heterogeneity. Liquid tends first to cover the liquiphilic spots with high surface energy and low contact angle (Checco et al. 2003). The tilt angle due to roughness can also contribute to the apparent contact angle. Third, the very concept of the static contact angle is not well defined. For practical purposes, the contact angle, which is formed after a droplet is gently placed upon a surface and stops propagating, is considered the static contact angle. However, depositing the droplet involves adding liquid while leaving it involves evaporation, so it is difficult to avoid dynamic effects. Fourth, for small droplets and curved triple lines, the effect of the contact line tension may be significant. Molecules at the surface of a liquid or solid phase have higher energy because they are bonded to fewer molecules than those in the bulk. This is responsible for surface

Table 3.1 Wetting of a superhydrophobic surface as a multiscale process (Nosonovsky and Bhushan 2007c, 2008c)

Scale level	Characteristic length	Parameters	Phenomena	Interface
Macroscale	Droplet radius (mm)	Contact angle, droplet radius	Contact angle hysteresis	2-D
Microscale	Roughness detail (μm)	Shape of the droplet, position of the liquid-air interface (h)	Kinetic effects	3-D solid surface, 2-D liquid surface
Nanoscale	Molecular heterogeneity (nm)	Molecular description	Thermodynamic and dynamic effects	3-D

tension and surface energy. In a similar manner, molecules at the concave surface and, especially, at the edge have fewer bonds than those at the surface, which leads to line tension and curvature dependence of the surface energy. This effect becomes important when the radius of curvature is comparable with the so-called Tolman's length, normally of the molecular size (Anisimov 2007). However, the triple line at the nanoscale can be curved so that the line tension effects become important (Pompe et al. 2000).

The contact angle, accounting for the contact line effect, for a droplet with radius R is given by $\cos\theta = \cos\theta_0 + 2\tau/(R\gamma_{LA})$, where τ is the contact line tension, and θ_0 is the value given by Young's equation (Boruvka and Neumann 1977). Thus, while the contact angle is a convenient macroscale parameter, wetting is governed by interactions at the micro- and nanoscale, which determine the contact angle hysteresis and other wetting properties. Table 3.1 shows various scale levels, which affect wetting of a superhydrophobic surface.

3.3.2 Range of Applicability of the Wenzel and Cassie-Baxter Equations

Gao and McCarthy (2007) showed experimentally that the contact angle of a droplet is defined by the triple line and does not depend upon the roughness under the bulk of the droplet. A similar result for chemically heterogeneous surfaces was obtained by Extrand (2003). Gao and McCarthy (2007) concluded that the Wenzel and Cassie-Baxter equations "should be used with the knowledge of their fault". However, the question remains, under what circumstances the Wenzel and Cassie-Baxter equations can be used safely and under what circumstances they become irrelevant?

Based on Nosonovsky (2007c) and Nosonovsky and Bhushan (2008a), for a liquid front propagating along a rough two-dimensional profile (Fig. 3.6a–b), the derivative of the free surface energy (per liquid front length), W, by the profile length, t, yields the surface tension force $\sigma = dW/dt = \gamma_{SL} - \gamma_{SA}$ (Nosonovsky 2007c). The quantity of practical interest is the component of the tension force that corresponds to the advancing of the liquid front in the horizontal direction for dx. This component is given by $dW/dx = (dW/dt)(dt/dx) = (\gamma_{SL} - \gamma_{SA})dt/dx$. It is noted that the derivative dt/dx is equal to Wenzel's roughness factor (R_f), in the case when the roughness factor is constant throughout the surface. Therefore, Young's equation (3.5), which relates the contact angle with solid, liquid, and air interface tensions, is modified as (Nosonovsky 2007c)

$$\gamma_{LA} \cos\theta = R_f(\gamma_{SA} - \gamma_{SL}) \tag{3.18}$$

The empirical Wenzel equation (3.8) is a consequence of (3.18) combined with Young's equation.

Nosonovsky (2007c) and Nosonovsky and Bhushan (2008a) showed that for a more complicated case of a non-uniform roughness given by the profile $z(x)$, the local value of derivative $r(x) = dt/dx = (1 + (dz/dx)^2)^{1/2}$ matters. In the cases that were studied experimentally by Gao and McCarthy (2007) and Extrand (2003), the roughness was present ($r > 1$) under the bulk of the droplet, but there was no roughness ($r = 0$) at the triple line, and the contact angle was given by (3.8) (Fig. 3.6c). In the general case of a 3-D rough surface $z(x, y)$, the roughness factor can be defined as a function of the coordinates $r(x, y) = (1 + (dz/dx)^2 + (dz/dy)^2)^{1/2}$.

Equation (3.8) is valid for uniformly rough surfaces, that is, surfaces with r = constant, and for non-uniformly rough surfaces, the generalized Wenzel equation is formulated to determine the local contact angle (a function of x and y) with rough surfaces at the triple line (Nosonovsky 2007c; Nosonovsky and Bhushan 2008a, b)

$$\cos\theta = r(x, y)\cos\theta_0 \tag{3.19}$$

The difference between the Wenzel equation (3.8) and the Nosonovsky-Bhushan equation (3.19) is that the latter is valid for a non-uniform roughness with the roughness factor as the function of the coordinates. Equation (3.19) is consistent with the experimental results of the researchers who showed that roughness beneath the droplet does not affect the contact angle, since it predicts that only roughness at the triple line matters. It is also consistent with the results of the researchers who confirmed the Wenzel equation (for the case of the uniform roughness) and of those who reported that only the triple line matters (for non-uniform roughness). A summary of experimental results for uniform and non-uniform rough and chemically heterogeneous surfaces is shown in Table 3.2.

The Cassie equation for the composite surface can be generalized in a similar manner introducing the spatial dependence of the local densities, f_1 and f_2 of the solid-liquid interface with the contact angle, as a function of x and y. It is given as (Nosonovsky 2007c),

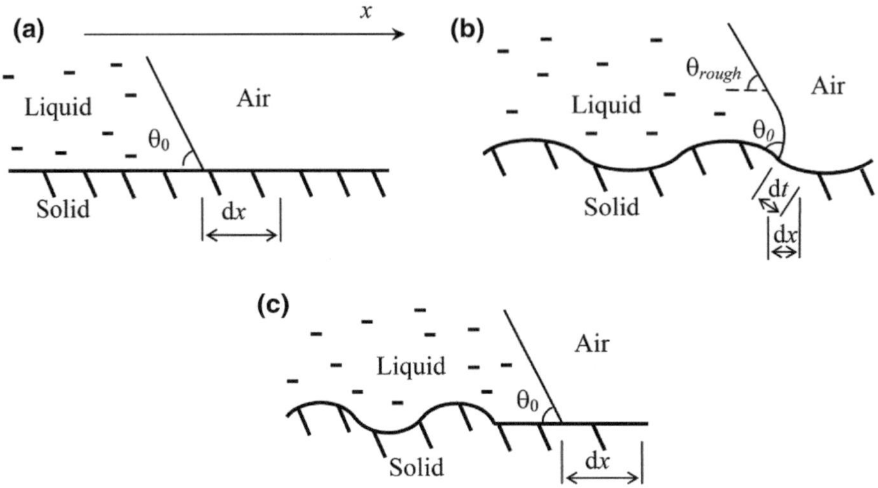

Fig. 3.6 Liquid front in contact with **a** a smooth solid surface and **b** a rough solid surface. Propagation for a distance dt along the curved surface corresponds to the distance dx along the horizontal surface. **c** Surface roughness under the bulk of the droplet does not affect the contact angle (Nosonovsky 2007c)

Table 3.2 Summary of experimental results for uniform and non-uniform rough and chemically heterogeneous surfaces

Experiment	Roughness/ hydrophobicity at the triple line and at the rest of the surface	Roughness at the bulk (under the droplet)	Experimental contact angle (compared with that at the rest of the surface)	Theoretical contact angle, Wenzel/ Cassie equations	Theoretical contact angle, generalized Wenzel-Cassie equations (3.17)–(3.18)
Gao and McCarthy (2007)	Hydrophobic	Hydrophilic	Not changed	Decreased	Not changed
	Rough	Smooth	Not changed	Decreased	Not changed
	Smooth	Rough	Not changed	Increased	Not changed
Extrand (2003)	Hydrophilic	Hydrophobic	Not changed	Increased	Not changed
	Hydrophobic	Hydrophilic	Not changed	Decreased	Not changed
Bhushan et al. (2007)	Rough	Rough	Increased	Increased	Increased
Barbieri et al. (2007)	Rough	Rough	Increased	Increased	Increased

For non-uniform surfaces, the results are shown for droplets larger than the islands of non-uniformity. Detailed quantitative values of the contact angle in various sets of experiments can be found in the referred sources (Nosonovsky 2007c)

$$\cos\theta_{composite} = f_1(x, y)\cos\theta_1 + f_2(x, y)\cos\theta_2 \tag{3.20}$$

where $f_1 + f_2 = 1$, and θ_1 and θ_2 are contact angles of the two components.

According to Nosonovsky (2007c) and Nosonovsky and Bhushan (2008a), the important question remains, what should be the typical size of roughness/heterogeneity details in order for the generalized Wenzel and Cassie equations (3.19) and (3.20) to be valid? Some scholars have suggested that roughness/heterogeneity details should be comparable with the thickness of the liquid-air interface and thus "the roughness would have to be of molecular dimensions to alter the equilibrium conditions" (Bartell and Shepard 1953). Conversely, others have claimed that roughness/heterogeneity details should be small compared with the linear size of the droplet (Johnson and Dettre 1964; Jung and Bhushan 2006, 2007, 2008a, b; Li and Amirfazli 2006; Bhushan and Jung 2007, 2008; Barbieri et al. 2007; Bhushan et al. 2007). The interface in the analysis proposed earlier is an idealized 2-D object, which has no thickness. In reality, the triple line zone has two characteristic dimensions: the thickness (of the order of molecular dimensions) and the length (of the order of the droplet size).

According to Nosonovsky and Bhushan (2008a, b), the apparent contact angle, given by (3.19) and (3.20), may be viewed as the result of averaging of the local contact angle at the triple line over its length, and thus the size of the roughness/heterogeneity details should be small compared to the length (and not the thickness) of the triple line (Nosonovsky and Bhushan 2008b). A rigorous definition of the generalized equation requires the consideration of several length scales. The length dx needed for the averaging of the energy gives the length over which the averaging is performed to obtain $r(x, y)$. This length should be larger than the roughness details. However, it is still smaller than the droplet size, and the length scale at which the apparent contact angle is observed (at which local variations of the contact angle level out). Since of these lengths (the roughness size, dx, the droplet size), the first and the last are of practical importance, it is concluded that the roughness details should be smaller than the droplet size. When the liquid-air interface is studied at the length scale of roughness/heterogeneity details, the local contact angle, θ_0, is given by (3.8)–(3.12). The liquid-air interface at that scale has perturbations caused by the roughness/heterogeneity, and the scale of the perturbations is the same as the scale of the roughness/heterogeneity details. However, when the same interface is studied at a larger scale, the effect of the perturbation vanishes, and apparent contact angle is given by (3.19) and (3.20) (Fig. 3.6c). This apparent contact angle is defined at the length scale for which the small perturbations of the liquid-air interface vanish, and the interface can be treated as a smooth surface. The values of $r(x, y)$, $f_1(x, y)$, $f_2(x, y)$ in (3.19) and (3.20) are average values over an area (x, y) with a size larger than a typical roughness/heterogeneity detail size. Therefore, the generalized Wenzel and Cassie equations can be used at the scale at which the effect of the interface perturbations vanish, or in other words, when the size of the solid surface roughness/heterogeneity details is small compared with the size of the liquid-air interface, which is of the same order as the size of the droplet (Nosonovsky and Bhushan 2008b).

Nosonovsky and Bhushan (2008a, d) used the surface energy approach to find the domain of validity of the Wenzel and Cassie equations (uniformly rough surfaces) and generalized it for a more complicated case of non-uniform surfaces. The generalized equations explain a wide range of existing experimental data, which could not be explained by the original Wenzel and Cassie equations.

3.4 Contact Angle Hysteresis, Tilt Angle, and Energy Dissipation

There is no one single value of the contact angle for a rough or heterogeneous surface (Johnson and Dettre 1964; Marmur 2003; Li and Amirfazli 2006; Nosonovsky and Bhushan 2008a, d). The contact angle can be in a range of values between the receding contact angle, θ_{rec}, and the advancing contact angle, θ_{adv}. The system tends to achieve the receding contact angle when liquid is removed (for example, at the rear end of a moving droplet), whereas the advancing contact angle is achieved when the liquid is added (for example, at the front end of a moving droplet) (Fig. 3.7a). When the liquid is neither added nor removed, the system tends to have a static or "most stable" contact angle, which is given approximately by (3.5), (3.8), (3.10), and (3.12). Another way to define θ_{adv} and θ_{rec} is to consider a moving liquid droplet over a tilted surface as shown in Fig. 3.7b. The liquid is pumped at the leading edge, and the contact angle at this edge corresponds to the advancing contact angle. The liquid is pumped away from the trailing edge, and the angle at the trailing edge corresponds to the receding contact angle. It is noted that the advancing and receding contact angles of a droplet just before sliding need not necessarily be identical to the apparent advancing and receding contact angles during sliding (Krasovitski and Marmur 2005).

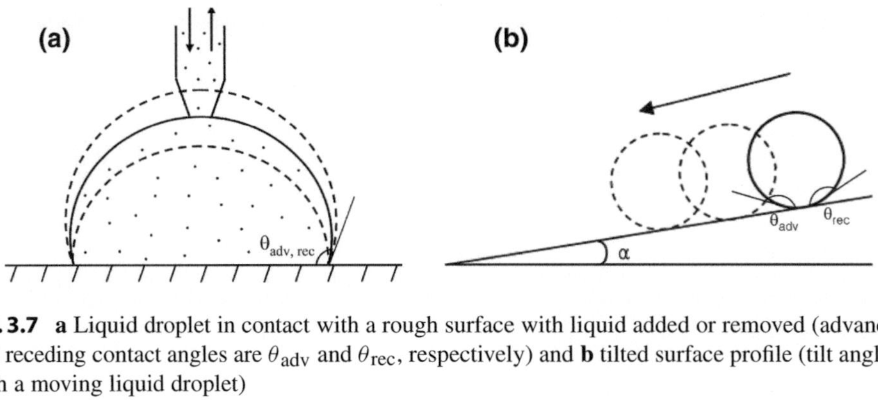

Fig. 3.7 **a** Liquid droplet in contact with a rough surface with liquid added or removed (advancing and receding contact angles are θ_{adv} and θ_{rec}, respectively) and **b** tilted surface profile (tilt angle, α with a moving liquid droplet)

The difference between the advancing and receding contact angles is called contact angle hysteresis (CAH). It is another important characteristic of a solid-liquid interface, of interest in fluid flow. Contact angle hysteresis occurs due to surface roughness and surface heterogeneity. Although for surfaces with roughness carefully controlled on the molecular scale it is possible to achieve contact angle hysteresis as low as $<1°$ (Gupta et al. 2005), hysteresis cannot be eliminated completely, since even atomically smooth surfaces have a certain roughness and heterogeneity. Contact angle hysteresis reflects the irreversibility of the wetting/dewetting cycle. In a surface with high contact angle hysteresis, the receding contact angle is low and pins the liquid, and provides resistance to flow.

Low contact angle hysteresis results in a low tilt angle, the angle to which a surface may be tilted for roll-off of liquid drops (i.e., very low drag) (Extrand 2002; Kijlstra et al. 2002; Bhushan and Jung 2007, 2008, 2011; Jung and Bhushan 2007, 2008a) (Fig. 3.7b). A relationship between the contact angle hysteresis and tilt angle can be derived by force balance between gravitation pulling the droplet downhill and the capillary force (integral of surface tension around the contact line of the drop). A droplet rolls off an inclined surface when the tilt angle, α exceeds a value given by (Yoshimitsu et al. 2002; Antonini et al. 2009),

$$\sin\alpha = \frac{k\gamma_{LA}w}{V\rho g}(\cos\theta_{rec} - \cos\theta_{adv}) \tag{3.21}$$

where w is the width of the apparent contact area measured perpendicular to the sliding direction, V is the volume of the droplet, ρ is the liquid density, g is the gravitational acceleration ($= 9.81$ m/s^2) and k is a geometrical factor (~ 1). For a superhydrophobic surface of $\theta_{adv} \sim 180°$, the (3.21) reduces to

$$\sin\alpha = \frac{\gamma_{LA}w}{V\rho g}(\cos\theta_{rec} + 1) \tag{3.22}$$

From (3.21) and (3.22), it is observed that high values of $(\cos\theta_{rec} - \cos\theta_{adv})$ or high value of θ_{rec} results in low tilt angle. Low tilt angle is important in liquid flow applications for low drag and surfaces with self-cleaning ability.

For completeness, for one solid body sitting on an inclined plane, the tangent of the tilt angle at which the body starts to slide is defined as the Coulomb coefficient of friction (Bhushan 2013a, b).

The motion of a liquid droplet along a solid surface is a dissipative process. The contact angle hysteresis or tilt angle is a measure of energy dissipation during the flow of a droplet along a solid surface. At a lower CAH, the droplet can roll, in addition to slide along the surface and take contaminants from the surface with them. Liquid-repellant surfaces such as Lotus leaves, have low CAH or tilt angle of about 2–4°.

For rough surfaces, Nosonovsky and Bhushan (2007b, d) assumed that the factors that affect CAH include adhesion hysteresis (AH), surface roughness, and surface heterogeneity. It is known that the energy required to separate the two solid surfaces is always greater than the energy gained by bringing them together (work of adhesion), due to the so called adhesion hysteresis (Bhushan 2013a, b, 2017).

Based on (3.6), for a smooth surface, energies gained by bringing liquid and solid surfaces together and separating them are related to $\gamma_{LA}(1+\cos\theta_{adv})$ and $\gamma_{LA}(1+\cos\theta_{rec})$, respectively. Thus,

$$AH_0 = \gamma_{LA}(\cos\theta_{adv0} - \cos\theta_{rec0}) \tag{3.23}$$

Energy dissipation is proportional to $(\cos\theta_{adv0} - \cos\theta_{rec0})$. For a rough surface, adhesion hysteresis AH is given as

$$AH = \gamma_{LA}(\cos\theta_{adv} - \cos\theta_{rec}) \tag{3.24}$$

For a composite interface, the fraction of the solid-liquid area is given by $R_f\, f_{SL}$ or R_f $(1 - f_{LA})$. Therefore

$$AH = R_f(1 - f_{LA})AH_0 \tag{3.25}$$

Combining (3.23) and (3.25) and incorporating the term H_r corresponding to the effect of roughness, we get

$$\cos\theta_{adv} - \cos\theta_{rec} = R_f(1 - f_{LA})(\cos\theta_{adv0} - \cos\theta_{rec0}) + H_r \tag{3.26}$$

The first term in the right-hand part of the equation, which corresponds to the inherent contact angle hysteresis of a smooth surface, is proportional to the fraction of the solid-liquid contact area, $1 - f_{LA}$. The second term, H_r, is the effect of surface roughness, which is equal to the total perimeter of the asperity per unit area, or in other words, to the length density of the triple line (Bhushan et al. 2007). Thus (3.26) involves both the term proportional to the solid-liquid interface area and to the triple line length. It is observed from (3.11) and (3.26) that increasing $f_{LA} \rightarrow 1$ results in increasing the contact angle ($\cos\theta \rightarrow -1$, $\theta \rightarrow \pi$) and decreasing the contact angle hysteresis ($\cos\theta_{adv} - \cos\theta_{rec} \rightarrow 0$). In the limiting case of very small solid-liquid fractional contact area under the droplet, when the contact angle is large ($\cos\theta \approx -1 + (\pi - \theta)^2/2$, $\sin\theta \approx \theta - \pi$) and the contact angle hysteresis is small ($\theta_{adv} \approx \theta \approx \theta_{rec}$), based on (3.11) and (3.26) (Nosonovsky and Bhushan 2007b, d),

$$\pi - \theta = \sqrt{2(1 - f_{LA})(R_f\cos\theta_0 + 1)} \tag{3.27}$$

$$\theta_{adv} - \theta_{rec} = R_f(1 - f_{LA})\frac{\cos\theta_{adv0} - \cos\theta_{rec0}}{-\sin\theta} = R_f\left(\sqrt{1 - f_{LA}}\right)\frac{\cos\theta_{rec0} - \cos\theta_{adv0}}{\sqrt{2(R_f\cos\theta_0 + 1)}} \tag{3.28}$$

This is called the Nosonovsky-Bhushan equation.

For the homogeneous interface, $f_{LA} = 0$, whereas for the composite interface f_{LA} is a non-zero number. It is observed from (3.27) and (3.28) that for a homogeneous interface, increasing roughness (high R_f) leads to increasing the contact angle hysteresis (high values of $\theta_{adv} - \theta_{rec}$), while for the composite interface, an approach to unity of f_{LA} provides both high contact angle and small contact angle hysteresis (Jung and Bhushan 2006; Bhushan et al. 2007; Nosonovsky and Bhushan 2007b, d). Therefore, the composite interface is desirable for low contact angle hysteresis.

A sharp edge can pin the line of contact of the solid, liquid, and air (or the "triple line") at a position far from stable equilibrium, i.e., at contact angles different from θ_0 (Eustathopoulos et al. 1999). This effect is illustrated in Fig. 3.8, which shows a droplet propagating along a solid surface with grooves (Nosonovsky and Bhushan 2005). Based on Nosonovsky and Bhushan (2005, 2008a), at the edge point, the contact angle is not defined and can have any value between the values corresponding to contact with the horizontal and inclined surfaces. For a droplet moving from left to right, the triple line will be pinned at the edge point until it will be able to proceed to the inclined plane. As it is observed from Fig. 3.8, the change of the surface slope (α) at the edge is the cause of the pinning. Because of the pinning, the value of the contact angle at the front of the droplet (dynamic maximum advancing contact angle or $\theta_{adv} = \theta_0 + \alpha$) is greater than θ_0, whereas the value of the contact angle at the back of the droplet (dynamic minimum receding contact angle or $\theta_{rec} = \theta_0 - \alpha$) is smaller than θ_0. A hysteresis domain of the dynamic contact angle is thus defined by the difference $\theta_{adv} - \theta_{rec}$. The liquid can travel easily along the surface if the contact angle hysteresis is small. It is noted that the static contact angle lies within the hysteresis domain. Therefore, increasing the static contact angle up to the values of a superliquiphobic surface (approaching 180°) will also result in a reduction of the contact angle hysteresis. In a similar manner, contact angle hysteresis can also exist even if the surface slope changes smoothly, without sharp edges. There is an analogy between the two mechanisms leading to contact angle hysteresis (energy dissipation at the solid-liquid interface and pinning of the triple line) and dissipation mechanisms of dry friction (adhesion and deformation) (Nosonovsky 2007b, Nosonovsky and Bhushan 2008a).

3.5 Stability of a Composite Interface and Role of Hierarchical Structure with Convex Surfaces

Stability of the composite interface is an important issue. Even though it may be geometrically possible for the system to become composite, it may be energetically profitable for the liquid to penetrate into the valleys between asperities and form a homogeneous interface. Marmur (2003) formulated geometrical conditions for a surface under which the energy of the system has a local minimum and the composite interface may exist.

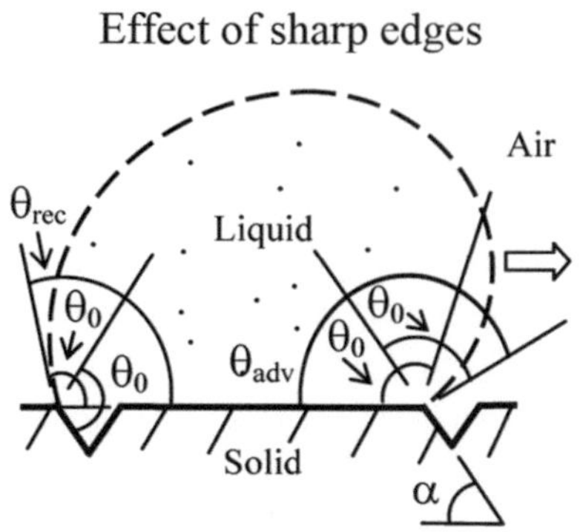

Fig. 3.8 A liquid droplet in contact with a surface with sharp edges. For a droplet moving from left to right on a sharp edge (shown by arrow), the contact angle at a sharp edge may take any value between the contact angle with the horizontal and inclined planes. This effect results in the difference between advancing ($\theta_{adv} = \theta_0 + \alpha$) and receding ($\theta_{rec} = \theta_0 - \alpha$) contact angles (Nosonovsky and Bhushan 2005)

Patankar (2004) pointed out that whether the homogeneous or composite interface exists depends on the system's history, i.e., on whether the droplet was formed at the surface or deposited.

Based on Nosonovsky and Bhushan (2007b, 2008a), formation of a composite interface is a multiscale phenomenon that depends upon the relative sizes of the liquid droplet and roughness details. The composite interface is fragile and can be irreversibly transformed into the homogeneous interface, thus damaging superhydrophobicity. In order to form a stable composite interface with air pockets between solid and liquid, the destabilizing factors such as capillary waves, nanodroplet condensation, surface inhomogeneity, and liquid pressure should be avoided. A description of these destabilizing factors follows.

- First, the capillary waves at the liquid-air interface may destabilize the composite interface. Due to an external perturbation, a standing capillary wave can form at the liquid-air interface. If the amplitude of the capillary wave is greater than the height of the asperity, the liquid can touch the valley between the asperities, and if the angle under which the liquid comes in contact with the solid surface is greater than θ_0, it is energetically profitable for the liquid to fill the valley (Nosonovsky and Bhushan 2005, 2006a). When the composite interface is destroyed and space between the asperities is filled with liquid, it is highly unlikely that the composite interface will be formed again because the transition from the non-composite solid-liquid interface to a composite interface would require a large activation energy. Such a transition has never been observed. The effect of capillary waves is more pronounced for small asperities with height comparable with wave amplitude.

- Second, nanodroplets may condense and accumulate in the valleys between asperities and eventually destroy the composite interface. Cheng et al. (2005) observed condensation of submicron sized droplets on a lotus leaf surface, and found that droplets tend to condense at areas adjacent to bumps (i.e., in the valleys) and have a contact angle of less than 90°, whereas larger droplets have higher contact angles, thus demonstrating that the contact angle is scale dependent. The scale effect is observed for small droplets or at small distances near the triple line (Nosonovsky and Bhushan 2007b, 2008a). At nanoscale distances from the triple line, the liquid touches the solid under a much lower contact angle.
- Third, even hydrophobic surfaces are usually not chemically homogeneous and can have hydrophilic spots. It is known from experiments that for droplets of submicron size, the value of the contact angle is usually smaller than for droplets at the macroscale (Lafuma and Quéré 2003). Checco et al. (2003) suggested that surface inhomogeneity is responsible for this scale effect, since nanodroplets tend to sit at the highest free surface energy (most hydrophilic) spots and thus have lower contact angles. Their phenomenological numerical simulations showed good agreement with experimental data.

Nosonovsky and Bhushan (2007a, b, c, d, 2008a, b, c, e) have demonstrated that a combination of microroughness and nanoroughness (multiscale roughness) with convex surfaces can help resist the destabilization by pinning the interface. It also helps in preventing the gaps between the asperities from filling with liquid, even in the case of a hydrophilic material. The effect of roughness on wetting is scale-dependent, and the mechanisms that lead to destabilization of a composite interface are also scale-dependent. To avoid these scale-dependent mechanisms, it is expected that a multiscale roughness is optimum for superhydrophobicity.

Nosonovsky and Bhushan (2008a) considered a two-dimensional structure with rectangular pillars of height h separated by distance b, each pillar covered with small semi-circular bumps (convex) and grooves (concave surface) of radius r (Fig. 3.9a). It is referred to as a geometry with re-entrant curvature. If the distance between the pillars is small in comparison with the capillary length, the effect of gravity is negligible, it can be assumed that the liquid-air interface is a horizontal plane, and its position is characterized by the vertical coordinate z. The free energy is given by the following equation (Nosonovsky 2007a; Nosonovsky and Bhushan 2008a)

$$W = A_{SL}\gamma_{SL} + A_{SA}\gamma_{SA} + A_{LA}\gamma_{LA} = rL\gamma_{LA}(\sin\alpha - \alpha\cos\theta_0), \quad 0 < z < h \tag{3.29}$$

where $\alpha = a\cos((r - z)/r) + 2\pi N$ is the angle corresponding to vertical position of the interface z, N is the number of ridges or grooves, and L is length of the grooves in the y-direction. The dependence is presented in Fig. 3.9b, for the cases of -phobic ($\theta_0 = 150°$) and -philic ($\theta_0 = 30°$) materials for both the convex surface (with bumps) and concave surface (with grooves). It is seen that for the convex surface, there are many states of stable equilibrium (shown in Fig. 3.9a with dotted lines), separated by energy barriers which correspond to every ridge, whereas for the concave surface, equilibrium states are unstable. Therefore, the ridges can pin the triple line and thus lead to a composite interface. In the case of a -philic surface, each lower position of the equilibrium state corresponds to a lower value of W. Therefore, when the liquid advances from one equilibrium state to the next, the total energy decreases, and the liquid's advance is energetically profitable. When the liquid reaches the bottom of the valley and completely fills the space between the pillars forming a homogeneous interface, the total energy decreases dramatically by the following value (Nosonovsky 2007a; Nosonovsky and Bhushan 2008a)

$$\Delta W = bL(\gamma_{SA} + \gamma_{LA} - \gamma_{SL}) = bL\gamma_{LA}(1 + \cos\theta_0) \tag{3.30}$$

The opposite transition from a homogeneous interface to a composite interface requires high activation energy ΔW, and is thus unlikely, making the transition from composite interface to homogeneous interface irreversible. If the distance between the pillars b is much greater than r, the energy barriers that separate the equilibrium states, $2\pi rL\gamma_{LA}\cos\theta_0$, will be relatively small compared to ΔW, and low activation energy will be required for the liquid to spread and propagate from one equilibrium state to the other (Nosonovsky and Bhushan 2007b, d).

For the interface to be stable, the value of the contact angle should decrease when the liquid-air interface advances, whereas for receding liquid the contact angle should increase. For a two-dimensional surface, the change of angle is equal to the change of the

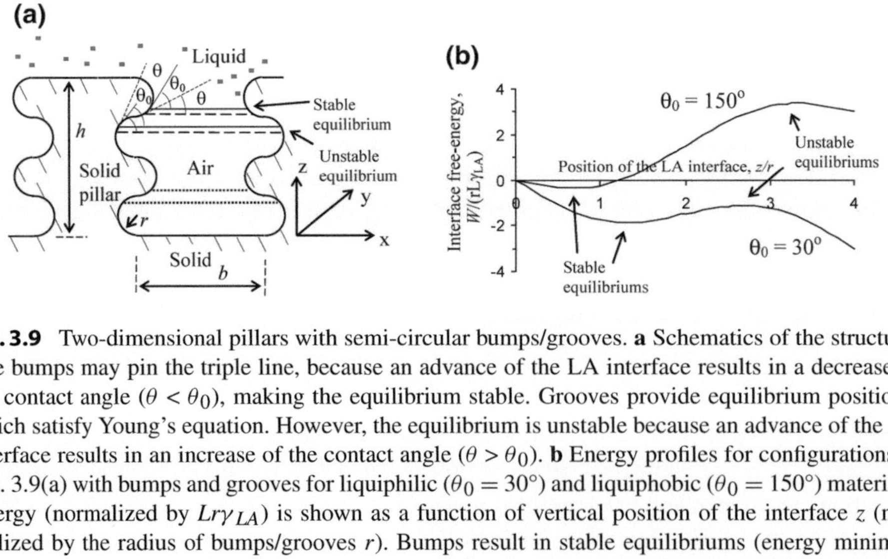

Fig. 3.9 Two-dimensional pillars with semi-circular bumps/grooves. **a** Schematics of the structure. The bumps may pin the triple line, because an advance of the LA interface results in a decrease of the contact angle ($\theta < \theta_0$), making the equilibrium stable. Grooves provide equilibrium positions, which satisfy Young's equation. However, the equilibrium is unstable because an advance of the LA interface results in an increase of the contact angle ($\theta > \theta_0$). **b** Energy profiles for configurations in Fig. 3.9(a) with bumps and grooves for liquiphilic ($\theta_0 = 30°$) and liquiphobic ($\theta_0 = 150°$) materials. Energy (normalized by $Lr\gamma_{LA}$) is shown as a function of vertical position of the interface z (normalized by the radius of bumps/grooves r). Bumps result in stable equilibriums (energy minima), whereas grooves result in unstable equilibriums (energy maxima) (Nosonovsky and Bhushan 2008a)

slope of the surface, and whether the configuration is stable or not depends on the sign of curvature of the surface. As indicated earlier, the convex surface (with bumps) leads to a stable interface, whereas a concave surface (with grooves) leads to an unstable interface. A structure consisting of bumps and grooves, a re-entrant surface structure, initially proposed by Nosonovsky and Bhushan (2008a), has been used by various researchers to create superoleophobic surfaces. Since the surface tension of oil and other organic liquids is much lower than that of water, such structures can repel oils with low surface tension as low as close to zero. For a review, see Brown and Bhushan (2016).

Sun et al. (2005) conducted an experiment to verify that the sign of curvature is indeed important for hydrophobicity. They produced both a positive and a negative replica of a lotus leaf surface by nanocasting using poly(dimethylsiloxane) , which has a contact angle with water of about 105°. This value is close to the contact angle of the wax, which covers lotus leaves (about 103° as reported by Kamusewitz et al. 1999). The positive and negative replicas have the same roughness factor and thus should produce the same contact angle in the case of a homogeneous interface, according to (3.6). However, the values of surface curvature are opposite. The value of contact angle for the positive replica was found to be 160° (same as for lotus leaf), while for the negative replica it was only 110°. This result suggests that the high contact angle for lotus leaf is due to the composite, rather than homogeneous interface, and that the sign of surface curvature indeed plays a critical role in the formation of the composite interface.

3.6 The Cassie-Baxter and Wenzel Wetting Regime Transition

Since superhydrophobicity requires a stable composite interface, it is important to understand the destabilization mechanisms for the Cassie-Baxter and Wenzel wetting transition. Based on Nosonovsky and Bhushan (2008a, d), it is known from experimental observations that the transition from the Cassie-Baxter to the Wenzel regime can be an irreversible event. Such a transition can be induced, for example, by applying pressure or force to the droplet (Jung and Bhushan 2008b; Nosonovsky and Bhushan 2008f), electric voltage (Krupenkin et al. 2004; Bahadur and Garimella 2007), light for a photocatalytic texture (Feng et al. 2004), and vibration (Bormashenko et al. 2007). However, the opposite transition is never observed.

Several approaches have been proposed for investigation of the transition between the metastable Cassie-Baxter and Wenzel regimes, referred to as "the Cassie-Wenzel transition." It has been suggested that the transition takes place when the net surface energy of the Wenzel regime becomes equal to that of the Cassie-Baxter regime, in other words, when the contact angle predicted by the Cassie-Baxter equation is equal to that predicted by the Wenzel equation. Lafuma and Quéré (2003) noticed that in certain cases the transition does not occur even when it is energetically profitable, and considered such a Cassie-Baxter state metastable. Extrand (2003) suggested that the weight of the droplet is responsible for the transition and proposed the contact line density model, according to which the transition takes place when the weight exceeds the surface tension force at the triple line. Patankar (2004) proposed a transition criterion based on energy balance. There is an energy barrier in going from a higher energy Cassie-Baxter droplet to a lower energy Wenzel droplet. The most probable mechanism is that the decrease in the gravitational potential energy during the transition helps in overcoming the energy barrier. This energy barrier was estimated by considering an intermediate state in which the liquid fills the grooves below the contact area of a Cassie-Baxter droplet but the liquid-solid contact is yet to be formed at the bottom of the valleys. Quéré (2005) also suggested that the droplet curvature (which depends upon the pressure difference between the inside and the outside of the droplet) governs the transition. Nosonovsky and Bhushan (2005, 2006a, b) proposed a probabilistic model for wetting of rough surfaces with a certain probability associated with every equilibrium state. According to their model, the overall contact angle with a two-dimensional rough profile is calculated by assuming that the overall configuration of a droplet occurs as a result of superposition of numerous metastable states. Numerous experimental results support many of these approaches, however, it is not clear which particular mechanism prevails.

Based on Nosonovsky and Bhushan (2008a, d), there is an asymmetry between the wetting and dewetting processes, since less energy is released during wetting than the amount required for dewetting due to adhesion hysteresis. Adhesion hysteresis is one of the reasons that leads to contact angle hysteresis, and it also results in the hysteresis of the Wenzel and Cassie-Baxter state transition. Figure 3.10 shows the contact angle of a rough surface as a function of surface roughness parameter, given by (3.9) (Nosonovsky and Bhushan 2008d). Here it is assumed that $R_f \sim 1$ for a Cassie-Baxter regime with a stable composite interface, and the liquid droplet sits flat over the surface. It is noted that, at a certain point, the contact angles given by the Wenzel and Cassie-Baxter equations are the same, and $R_f = (1 - f_{LA}) - f_{LA}/\cos\theta_0$. At this point, the lines corresponding to the Wenzel and Cassie-Baxter regimes intersect. This point corresponds to an equal net energy of the Cassie-Baxter and Wenzel regimes. For a lower roughness (e.g., larger pitch between the asperities) the Wenzel state is more energetically profitable, whereas for a higher roughness, the Cassie-Baxter regime is more energetically profitable.

It is observed from Fig. 3.10 that an increase of roughness may lead to the transition between the Wenzel and Cassie-Baxter regimes at the intersection point. With decreasing roughness, the system is expected to transition to the Wenzel state. However, experiments show that, despite the energy in the Wenzel regime being lower than that in the Cassie-Baxter regime, the transition does not necessarily occur, and the droplet may remain in the metastable Cassie-Baxter regime (Bhushan and Jung 2007, 2008; Barbieri et al. 2007; Bhushan et al. 2007; Jung and Bhushan 2007, 2008a, b). This is because there are energy barriers associated with the transition, which occurs due to destabilization by dynamic effects (such as waves and vibration) (Nosonovsky and Bhushan 2008a, d).

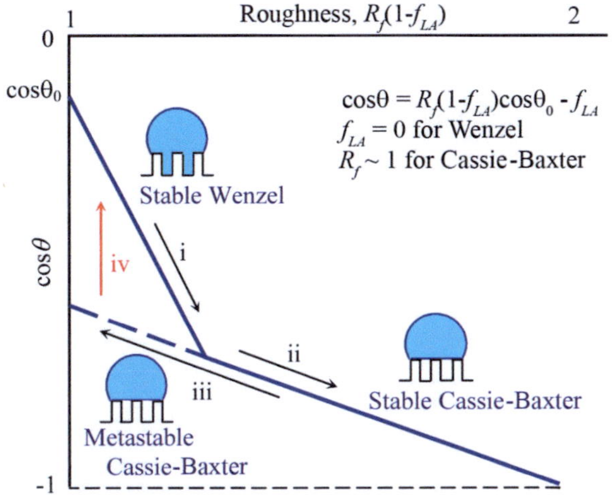

Fig. 3.10 Wetting hysteresis for a superliquiphobic surface. Contact angle as a function of roughness. The stable Wenzel state (i) can transform into the stable Cassie-Baxter state with increasing roughness (ii). The metastable Cassie-Baxter state (iii) can abruptly transform (iv) into the stable Wenzel state. The transition i–ii corresponds to equal Wenzel and Cassie-Baxter states free energies, whereas the transition iii–iv corresponds to a significant energy dissipation, and thus it is irreversible (adapted from Nosonovsky and Bhushan 2008d)

In order to understand contact angle hysteresis and the transition between the Cassie-Baxter and Wenzel regimes, Nosonovsky and Bhushan (2008a, d) analyzed the shape of the free surface energy profile. The free surface energy of a droplet upon a smooth surface as a function of the contact angle has a distinct minimum which corresponds to the most stable contact angle. As shown in Fig. 3.11a, the macroscale profile of the net surface energy allows us to find the contact angle (corresponding to energy minimums). However, it fails to predict contact angle hysteresis and the Cassie-Baxter and Wenzel transition, which are governed by micro- and nanoscale effects (Nosonovsky and Bhushan 2008d). As soon as microscale substrate roughness is introduced, the droplet shape cannot be considered as an ideal truncated sphere anymore, and energy profiles have multiple energy minimums, corresponding to the location of the asperities (Fig. 3.11b). The microscale energy profile (solid line) has numerous energy maxima and minima due to surface asperities. While exact calculation of the energy profile for a 3-D droplet is complicated, a qualitative shape may be obtained by assuming a periodic sinusoidal dependence (Johnson and Dettre 1964), superimposed upon the macroscale profile, as shown in Fig. 3.11b (Nosonovsky and Bhushan 2008a, d). Thus, the advancing and receding contact angles can be identified as the maximum and minimum possible contact angles corresponding to energy minimum points. However, the transition between the Wenzel and Cassie-Baxter branches still cannot be explained. Note also that Fig. 3.11b explains qualitatively the

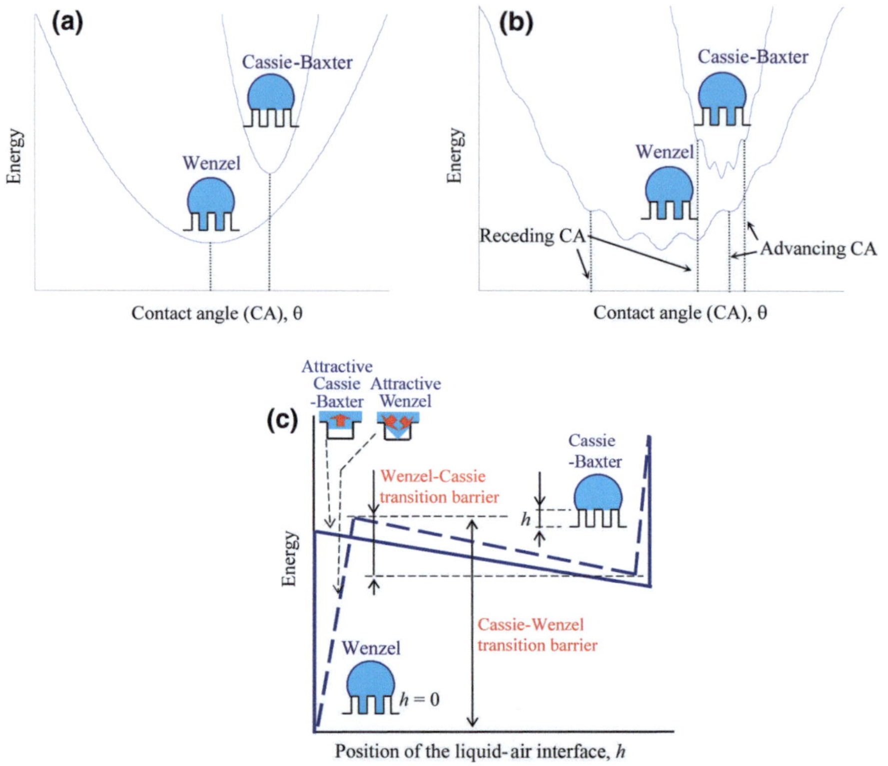

Fig. 3.11 Schematics of net free energy profiles. **a** Macroscale description; energy minimums correspond to the Wenzel and Cassie states. **b** Microscale description with multiple energy minimums due to surface texture. Largest and smallest values of the energy minimum correspond to the advancing and receding contact angles, respectively. **c** Origin of the two branches (Wenzel and Cassie-Baxter) is found when a dependence of energy upon h (air layer thickness or vertical position of the liquid-air interface) is considered for the microscale description (solid line) and nanoscale imperfectness (dashed line) (adapted from Nosonovsky and Bhushan 2008d). When the nanoscale imperfectness is introduced, it is observed that the Wenzel state corresponds to an energy minimum and the energy barrier for the Wenzel-Cassie transition is much smaller than for the opposite transition

hysteresis due to the kinetic effect of the asperities, but not the inherited adhesion hysteresis, which is characterized by the molecular length scale and cannot be captured by the microscale model.

Based on Nosonovsky and Bhushan (2008a, d), the energy profile as a function of the contact angle does not provide information on how the transition between the Cassie-Baxter and Wenzel regimes occurs, because their two states correspond to completely isolated branches of the energy profile in Fig. 3.11a–b. However, the energy may depend not only upon the contact angle, but also upon micro/nanoscale parameters such as, for example, the vertical position of the liquid-air interface under the droplet, h (assuming that the interface is a horizontal plane) or similar geometrical parameters (assuming a more complicated shape of the interface). In order to investigate the Wenzel and Cassie-Baxter transition, Nosonovsky and Bhushan (2008a, d) studied the dependence of the energy upon these parameters. They assumed that the liquid-air interface under the droplet is a flat horizontal plane. When such air layer thickness or the vertical position of the liquid-air interface, h, is introduced, the energy can be studied as a function of the droplet's shape, the contact angle, and h (Fig. 3.11c). For an ideal situation, the energy profile has an abrupt minimum at the point corresponding to the Wenzel state, which corresponds to the sudden net energy change due to the destruction of the solid-air and liquid-air interfaces ($\gamma_{SL} - \gamma_{SA} - \gamma_{LA} = -\gamma_{LA}(\cos\theta + 1)$ times the interface area) (Fig. 3.11c).

In a more realistic case, the liquid-air interface cannot be considered horizontal due to nanoscale imperfectness or dynamic effects such as the capillary waves (Nosonovsky and Bhushan 2006a). A typical size of the imperfectness is much smaller than the size of details of the surface texture and thus belongs to the molecular scale level. The height of the interface, h, can now be treated as an average height. The energy dependence upon h is now not as abrupt as in the idealized case. For example, for the "triangular" shape as shown in Fig. 3.11c, the Wenzel state may become the second attractor for the system. It is seen that there are two equilibriums that correspond to the Wenzel and Cassie-Baxter regimes, with the Wenzel state corresponding to a much lower energy level. The energy dependence upon h governs the transition between the two states, and it is observed that a much larger energy barrier exists for the transition from Wenzel to Cassie-Baxter regime than for the opposite transition. This is why the first transition has never been observed experimentally (Nosonovsky and Bhushan 2007c).

3.7 Closure

The modeling of wetting of rough surfaces is presented. There are two primary wetting regimes—Wenzel and Cassie-Baxter regimes, summarized in Fig. 3.12. In the Wenzel regime, a liquid droplet completely wets the rough surface with a homogeneous interface. The contact angle of a rough surface is altered by the roughness details—roughness factor R_f, which is a ratio of the surface area to its flat projected area. The Wenzel equation predicts that a liquiphobic surface becomes more liquiphobic, while a -philic surface becomes more liquiphilic with an increase in roughness. In the Cassie-Baxter regime, a heterogeneous or composite interface with air pockets trapped between the asperities is formed. The contact angle is altered by the roughness details and fractional liquid-air area (f_{LA}). The Cassie-Baxter equation predicts that, even for a liquiphilic surface, the surface can become liquiphobic with an increase of f_{LA}. However, the f_{LA} value required may be very high, or the formation of air pockets may become unstable. Whether a liquid-air

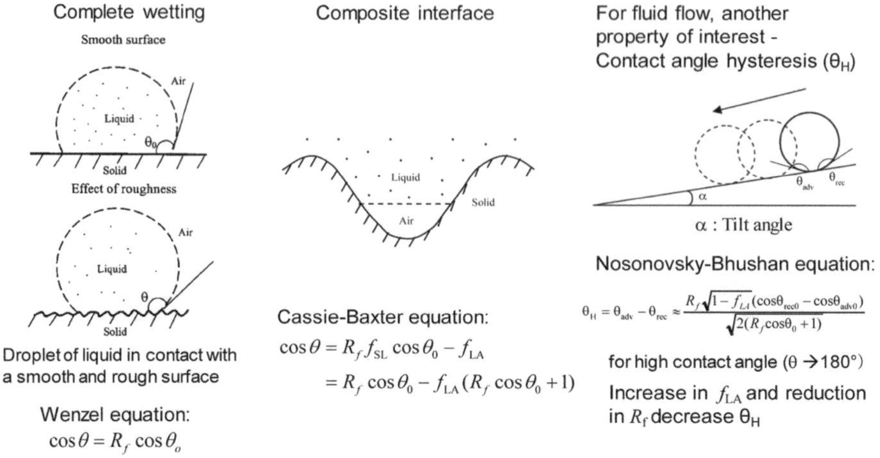

Fig. 3.12 Schematic of a liquid droplet in contact with smooth and rough solid surfaces (left), schematic of the formation of composite solid-liquid-air interface for a rough surface (middle), and schematic of a tilted surface (with tilt angle α) with a moving liquid droplet (θ_{adv} and θ_{rec} are advancing and receding contact angles, respectively), showing the definition of contact angle hysteresis (Bhushan 2012)

interface is generated depends upon the ratio of distance between two adjacent asperities and droplet radius.

In an intermediate wetting regime, referred to as the Cassie-impregnated regime, or simply the Cassie regime, a liquid film impregnates some of the cavities in an area surrounding the droplet as well. It consists of two interfaces—a Wenzel interface and some of the neighboring cavities filled with liquid. Pure Cassie-Baxter and Wenzel wetting situations rarely occur. An intermediate state between the Wenzel and Cassie-Baxter interfaces is more common. In this interface, underneath the droplet, liquid penetrates and the cavities are partially impregnated or filled, and is referred to as the mixed/impregnated state.

Another property of interest in fluid flow is contact angle hysteresis, which is the difference between the advancing and receding contact angles, also summarized in Fig. 3.12. It occurs due to surface roughness and heterogeneity. Hysteresis cannot be eliminated completely, since even atomically smooth surfaces have a certain roughness and heterogeneity. Contact angle hysteresis is a measure of energy dissipation during the flow of a droplet along a solid surface. Low contact angle hysteresis results in a very low tilt or roll-off angle, implying low drag and self-cleaning ability. It turns out that it is efficient to increase both contact angle and reduce contact angle hysteresis by having a moderate roughness and large fractional liquid-air area.

A composite interface is metastable, and its stability is an important issue. The formation of a composite interface is a multiscale phenomenon that depends on the relative sizes of the liquid droplet and the roughness details. Destabilizing factors include capillary wave, condensation and accumulation of nanodroplets, and surface inhomogeneity (with hydrophilic spots). A microstructure resists capillary waves present at the liquid-air interface. A nanostructure prevents nanodroplets from filling the valleys between asperities and pin the droplet. A combination of microstructure and nanostructure with convex surfaces can help resist the destabilization by pinning the interface. Based on the modeling and observation of natural objects (present in the next chapter), it is widely understood that a hierarchical surface structure is needed to develop a composite interface with high stability. A so-called re-entrant geometry is useful in creating superoleophobic surfaces.

References

Adamson A. V. (1990), *Physical Chemistry of Surfaces*, Wiley, New York.

Anisimov, M. A. (2007), "Divergence of Tolman's Length for a Droplet Near the Critical Point," *Phys. Rev. Lett.* **98**, 035702.

Antonini, C., Carmona, F. J., Pierce, E., Marengo, M., and Amirfazli, A. (2009), "General Methodology for Evaluating the Adhesion Force of Drops and Bubbles on Solid Surfaces," *Langmuir* **25**, 6143–6154.

Bahadur, V. and Garimella, S. V. (2007), "Electrowetting-Based Control of Static Droplet States on Rough Surfaces," *Langmuir* **23**, 4918–4924.

Barbieri, L., Wagner, E., and Hoffmann, P. (2007), "Water Wetting Transition Parameters of Perfluorinated Substrates with Periodically Distributed Flat-Top Microscale Obstacles," *Langmuir* **23**, 1723–1734.

Bartell, F. E. and Shepard, J. W. (1953), "Surface Roughness as Related to Hysteresis of Contact Angles" *J. Phys. Chem.* **57**, 455–458.

Bhushan, B. (2012), "Bioinspired Structured Surfaces," *Langmuir* **28**, 1698–1714.

Bhushan, B. (2013a), *Principles and Applications of Tribology*, second ed., Wiley, New York.

Bhushan, B. (2013b), *Introduction to Tribology*, second ed., Wiley, New York.

Bhushan, B. (2017), *Nanotribology and Nanomechanics—An Introduction*, fourth ed., Springer International, Switzerland.

Bhushan, B. and Jung, Y. C. (2007), "Wetting Study of Patterned Surfaces for Superhydrophobicity," *Ultramicroscopy* **107**, 1033–1041.

Bhushan, B. and Jung, Y. C. (2008), "Wetting, Adhesion and Friction of Superhydrophobic and Hydrophilic Leaves and Fabricated Micro/nanopatterned Surfaces," *J. Phys.: Condens. Matter* **20**, 225010.

Bhushan, B. and Jung, Y. C. (2011), "Natural and Biomimetic Artificial Surfaces for Superhydrophobicity, Self-Cleaning, Low Adhesion, and Drag Reduction," *Prog. Mater. Sci.* **56**, 1–108.

Bhushan, B. and Nosonovsky, M. (2010), "The Rose Petal Effect and the Modes of Superhydrophobicity," *Phil. Trans. R. Soc. A* **368**, 4713–4728.

Bhushan, B., Nosonovsky, M., and Jung, Y. C. (2007), "Towards Optimization of Patterned Superhydrophobic Surfaces" *J. R. Soc. Interface* **4**, 643–648.

Bormashenko, E., Pogreb, R., Whyman, G., and Erlich, M. (2007), "Cassie-Wenzel Wetting Transition in Vibrated Drops Deposited on the Rough Surfaces: Is Dynamic Cassie-Wenzel Transition 2D or 1D Affair?" *Langmuir* **23**, 6501–6503.

Boruvka, L. and Neumann, A. W. (1977), "Generalization of the Classical Theory of Capillarity," *J. Chem. Phys.* **66**, 5464–5476.

Brown, P. S. and Bhushan, B. (2016), "Designing Bioinspired Superoleophobic Surfaces," *APL Mater.* **4**, 015703.

Cassie, A. B. D. (1948), "Contact Angles," *Discuss. Faraday Soc.* **3**, 11–16.

Cassie, A. B. D. and Baxter, S. (1944), "Wettability of Porous Surfaces," *Trans. Faraday Soc.* **40**, 546–551.

Checco, A., Guenoun, P., and Daillant, J. (2003), "Nonlinear Dependence of the Contact Angle of Nanodroplets on Contact Line Curvatures," *Phys. Rev. Lett.* **91**, 186101.

Cheng, Y. T., Rodak, D. E., Angelopoulos, A., and Gacek, T. (2005), "Microscopic Observations of Condensation of Water on Lotus Leaves," *Appl. Phys. Lett.* **87**, 194112.

de Gennes, P. G., Brochard-Wyart, F., and Quėrė, D. (2003) *Capillarity and Wetting Phenomena*, Springer, Berlin.

Derjaguin, B. V. and Churaev, N. V. (1974), "Structural Component of Disjoining Pressure," *J. Colloid Interface Sci.* **49**, 249–255.

Eustathopoulos, N., Nicholas, M. G., and Drevet, B. (1999), *Wettability at High Temperatures*, Pergamon, Amsterdam.

Extrand, C. W. (2002), "Model for Contact Angle and Hysteresis on Rough and Ultraphobic Surfaces," *Langmuir* **18**, 7991–7999.

Extrand, C. W. (2003), "Contact Angle Hysteresis on Surfaces with Chemically Heterogeneous Islands," *Langmuir* **19**, 3793–3796.

Feng, X. J., Feng, L., Jin, M. H., Zhai, J., Jiang, L., and Zhu, D. B. (2004), "Reversible Super-hydrophobicity to Super-hydrophilicity Transition of Aligned ZnO Nanorod Films," *J. Am. Chem. Soc.* **126**, 62–63.

Gao, L. and McCarthy, T. J. (2007). "How Wenzel and Cassie Were Wrong," *Langmuir* **23**, 3762–3765.

Gupta, P., Ulman, A., Fanfan, F., Korniakov, A., and Loos, K. (2005), "Mixed Self-Assembled Monolayer of Alkanethiolates on Ultrasmooth Gold do not Exhibit Contact Angle Hysteresis," *J. Am. Chem. Soc.* **127**, 4–5.

Israelachvili, J. N. (1992), *Intermolecular and Surface Forces*, second ed., Academic Press, London, UK.

Israelachvili, J. N. and Gee, M. L. (1989), "Contact Angles on Chemically Heterogeneous Surfaces," *Langmuir* **5**, 288–289.

Johnson, R. E. and Dettre, R. H. (1964), "Contact Angle Hysteresis," in *Contact Angle, Wettability, and Adhesion, Adv. Chem. Ser.* (ed. F. M. Fowkes), Vol. 43, pp. 112–135, American Chemical Society, Washington, D. C.

Jung, Y. C. and Bhushan, B. (2006), "Contact Angle, Adhesion, and Friction Properties of Micro- and Nanopatterned Polymers for Superhydrophobicity," *Nanotechnology* **17**, 4970–4980.

Jung, Y. C. and Bhushan, B. (2007), "Wetting Transition of Water Droplets on Superhydrophobic Patterned Surfaces," *Scripta Mater.* **57**, 1057–1060.

Jung, Y. C. and Bhushan, B. (2008a), "Wetting Behavior during Evaporation and Condensation of Water Microdroplets on Superhydrophobic Patterned Surfaces" *J. Micros.* **229**, 127–140.

Jung, Y. C. and Bhushan, B. (2008b), "Dynamic Effects of Bouncing Water Droplets on Superhydrophobic Surfaces," *Langmuir* **24**, 6262–6269.

Kamusewitz, H., Possart, W., and Paul, D. (1999), "The Relation between Young's Equilibrium Contact Angle and the Hysteresis on Rough Paraffin Wax Surfaces," *Colloid Surf. A-Physicochem. Eng. Asp.* **156**, 271–279.

Kijlstra, J., Reihs, K., and Klami, A. (2002), "Roughness and Topology of Ultra-Hydrophobic surfaces," *Colloid Surf. A-Physicochem. Eng. Asp.* **206**, 521–529.

Krasovitski, B. and Marmur, A. (2005) "Drops Down the Hill: Theoretical Study of Limiting Contact Angles and the Hysteresis Range on a Tilted Plate," *Langmuir* **21**, 3881–3885.

Krupenkin, T. N., Taylor, J. A., Schneider, T. M., and Yang, S. (2004), "From Rolling Ball to Complete Wetting: The Dynamic Tuning of Liquids on Nanostructured Surfaces," *Langmuir* **20**, 3824–3827.

Lafuma, A. and Quéré, D. (2003), "Superhydrophobic States," *Nat. Mater.* **2**, 457–460.

Li, W. and Amirfazli, A. (2006), "A Thermodynamic Approach for Determining the Contact Angle Hysteresis for Superhydrophobic Surfaces," *J. Colloid. Interface Sci.* **292**, 195–201.

Marmur, A. (2003), "Wetting on Hydrophobic Rough Surfaces: To be Heterogeneous or Not to be?" *Langmuir* **19**, 8343–8348.

Nosonovsky, M. (2007a), "Multiscale Roughness and Stability of Superhydrophobic Biomimetic Interfaces," *Langmuir* **23**, 3157–3161.

Nosonovsky, M. (2007b), "Model for Solid-Liquid and Solid-Solid Friction for Rough Surfaces with Adhesion Hysteresis," *J. Chem. Phys.* **126**, 224701.

Nosonovsky, M. (2007c), "On the Range of Applicability of the Wenzel and Cassie Equations" *Langmuir* **23**, 9919–9920.

Nosonovsky, M. and Bhushan, B. (2005), "Roughness Optimization for Biomimetic Superhydrophobic Surfaces," *Microsyst. Technol.* **11**, 535–549.

Nosonovsky, M. and Bhushan, B. (2006a), "Stochastic Model for Metastable Wetting of Roughness-Induced Superhydrophobic Surfaces," *Microsyst. Technol.* **12**, 231–237.

Nosonovsky, M. and Bhushan, B. (2006b), "Wetting of Rough Three-Dimensional Superhydrophobic Surfaces," *Microsyst. Technol.* **12**, 273–281.

Nosonovsky, M. and Bhushan, B. (2007a), "Multiscale Friction Mechanisms and Hierarchical Surfaces in Nano- and Bio-Tribology," *Mater. Sci. Eng.:R* **58**, 162–193.

Nosonovsky, M. and Bhushan, B. (2007b), "Hierarchical Roughness Makes Superhydrophobic Surfaces Stable," *Microelectronic Eng.* **84**, 382–386.

Nosonovsky, M. and Bhushan, B. (2007c), "Biomimetic Superhydrophobic Surfaces: Multiscale Approach," *Nano Lett.* **7**, 2633–2637.

Nosonovsky, M. and Bhushan, B. (2007d), "Hierarchical Roughness Optimization for Biomimetic Superhydrophobic Surfaces," *Ultramicroscopy* **107**, 969–979.

Nosonovsky, M. and Bhushan, B. (2008a), *Multiscale Dissipative Mechanisms and Hierarchical Surfaces: Friction, Superhydrophobicity, and Biomimetics*, Springer-Verlag, Heidelberg, Germany.

Nosonovsky, M. and Bhushan, B. (2008b), "Roughness-Induced Superhydrophobicity: A Way to Design Non-Adhesive Surfaces," *J. Phys.: Condens. Matter* **20**, 225009.

Nosonovsky, M. and Bhushan, B. (2008c), "Biologically-Inspired Surfaces: Broadening the Scope of Roughness," *Adv. Func. Mater.* **18**, 843–855.

Nosonovsky, M. and Bhushan, B. (2008d), "Patterned Non-Adhesive Surfaces: Superhydrophobicity and Wetting Regime Transitions," *Langmuir* **24**, 1525–1533.

Nosonovsky, M. and Bhushan, B. (2008e), "Do Hierarchical Mechanisms of Superhydrophobicity Lead to Self-Organized Criticality?" *Scripta Mater.* **59**, 941–944.

Nosonovsky, M. and Bhushan, B. (2008f), "Energy Transitions in Superhydrophobicity: Low Adhesion, Easy Flow and Bouncing," *J. Phys.: Condens. Matter* **20**, 395005.

Nosonovsky, M. and Bhushan, B. (2009), "Superhydrophobic Surfaces and Emerging Applications: Non-Adhesion, Energy, Green Engineering," *Curr. Opin. Colloid Interface Sci.* **14**, 270–280.

Patankar, N. A. (2004), "Transition between Superhydrophobic States on Rough Surfaces" *Langmuir* **20**, 7097–7102.

Pompe, T., Fery, A., and Herminghaus, S. (2000), "Measurement of Contact Line Tension by Analysis of the Three-Phase Boundary with Nanometer Resolution," in *Apparent and Microscopic*

Contact Angles (eds. J. Drelich, J. S. Laskowski, and K. L. Mittal), pp. 3–12, VSP Publishing, Utrecht, Netherlands.

Quéré, D. (2005), "Non-Sticking Drops," *Rep. Prog. Phys.* **68**, 2495–2535.

Sun, M., Luo, C., Xu, L., Ji, H., Ouyang, Q., Yu, D., and Chen, Y. (2005), "Artificial Lotus Leaf by Nanocasting," *Langmuir* **21**, 8978–8981.

Tretinnikov, O. N. (2000), "Wettability and Microstructure of Polymer Surfaces: Stereochemical and Conformational Aspects" in *Apparent and Microscopic Contact Angles* (eds. J. Drelich, J. S. Laskowski, and K. L. Mittal), pp. 111–128, VSP Publishing, Utrecht, Netherlands.

Wenzel, R. N. (1936), "Resistance of Solid Surfaces to Wetting by Water," *Indust. Eng. Chem.* **28**, 988–994.

Yoshimitsu, Z., Nakajima, A., Watanabe, T., and Hashimoto, K. (2002), "Effects of Surface Structure on the Hydrophobicity and Sliding Behavior of Water Droplets," *Langmuir* **18**, 5818–5122.

Plant Leaf Surfaces in Living Nature

4

brahmany adhaya karmani

sangam tyaktva karoti yah

lipyate na sa papena

padma-patram ivambhasa

Translation—*One who performs his duty without attachment, surrendering the results unto the Supreme God, is not affected by sinful action, as the lotus leaf is untouched by water*

—Bhagwat Gita, Chapter 5, Text 10.

4.1 Introduction

Nature has evolved species with multifunctional properties by using diversity of the structure and morphology in surfaces. Hierarchical structured surfaces are common in nature (Koch et al. 2008, 2009a; Bhushan 2009). The complexity of hierarchical structures and their functionality in biological organisms surpass all abiotic natural surfaces. As an example, superhydrophobicity is restricted in nature to living organisms and it evolved with the invasion of terrestrial habitats some 400 million years ago in plants and insects. Plants are one of the dominating groups of organisms on our planet (Barthlott et al. 2017). Today, plant biodiversity contains approximately 270,000 different species, but recent calculations suggest that most species on earth have not yet been described and will not be because of high rates of species extinction (50 per day) (Koch et al. 2008). Superhydrophobicity is one of the most remarkable features of many plant surfaces. Plant surfaces exposed to air are very often superhydrophobic. Differences exist between land-living vascular and aquatic non-vascular plants. It is estimated that superhydrophobic land-living

B. Bhushan, *Introduction to Biomimetics and Bioinspiration*,
https://doi.org/10.1007/978-3-031-62344-8_4

plant leaves (e.g. around 12,000 species grasses) comprise a total area of 250 million km^2, which is about half of the total surface of earth (Barthlott et al. 2017).

The outermost layer of the primary plant surface is known as cuticle, and it stabilizes the plant tissue and has several protective properties. Its most prominent functions for a hydrophobic plant surface are presented in Fig. 4.1. One of the most important attributes of the cuticle is its hydrophobicity, which enables plants to overcome the physical and physiological problems connected to an ambient environment, such as desiccation. The cuticle also stabilizes the plant tissue and has several protective properties. It provides transpiration barrier function. This property is based on the material, made of a polymer called cutin and integrated and superimposed lipids called "waxes," which are hydrophobic. In addition to the reduction of water loss, the cuticle prevents leaching of ions from inside the cells to the environment. In plants, a wide spectra of surface structures exist which modify surface wettability and also have a significant influence on particle adhesion. Evolutionary optimized wettable or non-wettable surfaces can be found in water- and wetland plants, as for example, the water repellent leaves of Lotus (*Nelumbo nucifera*). In submerged water growing plants and some tropical and subtropical plants, cuticles provide hydrophilicity by providing a permanently wet surface or by water spreading, respectively. The plant cuticle also plays an important role for insect and microorganism interaction. In some cases, it protects the plants against overheating by reflecting radiation and/or by heat transfer via turbulent air flow and convection (Koch et al. 2008, 2009a; Bhushan 2009).

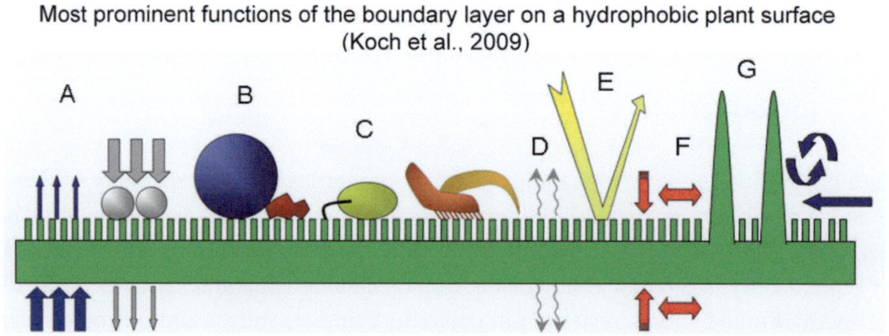

Fig. 4.1 Schematic of the most prominent functions of the boundary layer on a hydrophobic microstructured plant surface. (A) Transport barrier limitation of uncontrolled water loss/leaching from interior and foliar uptake, (B) surface wettability, (C) anti-adhesive, self-cleaning properties: reduction of contamination, pathogen attack and reduction of attachment/locomotion of insects, (D) signaling: cues for host–pathogens/insect recognition and epidermal cell development, (E) optical properties: protection against harmful radiation, (F) mechanical properties: resistance against mechanical stress and maintenance of physiological integrity, and (G) reduction of surface temperature by increasing turbulent air flow over the boundary air later (adapted from Koch et al. 2009a)

The polymer cuticle consists of papillose epidermal cells forming the papillae or microbumps on the surface covered with an epicuticular wax layer which generally self-assembles into nanostructure. The epicuticular waxes present on the cuticle form thin two-dimensional films and/or three-dimensional structures. If three-dimensional waxes exist, these are combined with a thin two-dimensional wax film, which covers the cuticle surface. The dimensions of wax structures range from a few nanometers to several micrometers. Waxes exist in different morphologies (Barthlott et al. 1998). Most common are tubules and platelets. The three-dimensional waxes and the wax films are crystals, which originate by self-assembly. Three-dimensional epicuticular waxes are responsible for non-wetting and self-cleaning properties, sliding of insects, reflection of visible light, adsorption of UV radiation, and reduction of particle adhesion (Koch et al. 2008, 2009a). The chemistry of waxes is a mixture of long chain hydrocarbons and, in some waxes, cyclic hydrocarbons. The chemical composition of plant waxes is highly variable amongst plant species as well as the organs on one specie (e.g. different leaves) and varies during organ ontogeny (Koch et al. 2008, 2009a; Barthlott et al. 2017).

The micro- and nanostructures of plant surfaces have a great influence on their attributes as interfaces. Even in a cursory look at different plant surfaces, they show different optical appearances, which arise from the surface structures in the micro- and nanoscale dimension (Koch et al. 2009a). The optical appearance of selected plant surfaces and their surface microstructures are shown in Fig. 4.2. Based on their microscopically smooth surface, the leaves of *Magnolia grandiflora* appear glossy (Fig. 4.2a and b), whereas the rougher surfaces of the flower leaves (petals) of *Dalia* appear velvet and soft (Fig. 4.2c and d). The leaves of *Leucadendron argenteum* appear white because of a dense layer of air-filled hollow hairs (Fig. 4.2e and f). The leaves of *Eucalyptus Macrocarpa* appear bluish due to a coverage of three-dimensional waxes (Fig. 4.2g and h). The leaves in (a–d) are hydrophilic and (g–h) are superhydrophobic. The great diversity in surface structures originates from the diversity of species combined with different structures in a simple plant.

Hierarchical structure present on a large number of plant surfaces is found and is of much interest to provide a large range of properties which are desirable for a given specie (Koch et al. 2008). Hierarchical structures are built by a combination of micro- and nanostructures. In plants, there are two frequently occurring combinations: a convex cell with cuticular folding or verrucose cuticle structure, and a convex cell with three-dimensional wax crystals. Both three-dimensional epicuticular waxes and cuticular folding are able to create the double-structure. In the top row of Fig. 4.3 examples are characterized by papilla and convex cells with three-dimensional waxes on top. In *Colocasia esculenta* (Fig. 4.3a) the wax crystal are wax platelets, whereas the waxes in *Nelumbo nucifera* (Fig. 4.3b) are nonacosanol tubules. The bottom row of Fig. 4.3 shows convex cells with cuticular folding. In a flower leaf of *Rosa Montana* (adaxial side), a rippled folded cuticle forms the central field of the cells and parallel folds occur at the anticline field (Fig. 4.3c). Another example is the hair-papilla cells of the inner side of the tube-like leaf of the

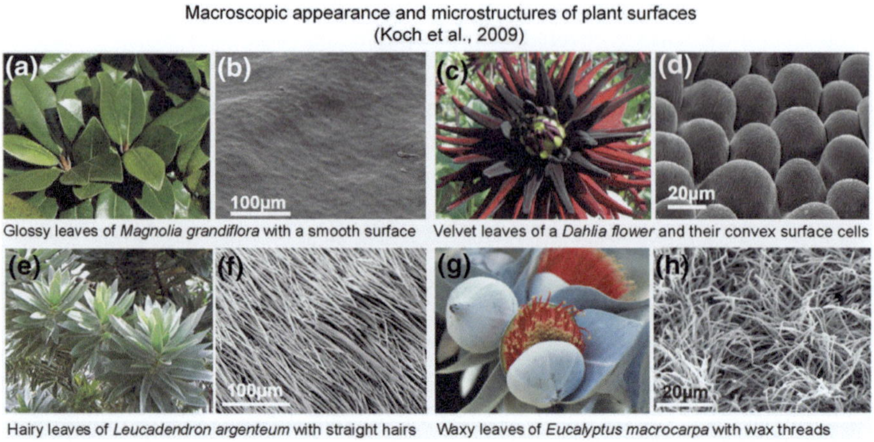

Fig. 4.2 Macroscopically visible optical appearance of plant surfaces and their surface microstructures, shown in scanning electron microscope (SEM) micrographs. In (**a**) leaves (*Magnolia grandiflora*) appear glossy because of the flat surface structure of the surface, shown in (**b**). In (**c**) the flower leaves (*Dahlia*) appear velvety, because of the microstructure of the epidermal cells shown in (**d**). In (**e**) the white appearance of the leaves (*Leucadendron argenteum*) is caused by a dense layer of hairs, shown in (**f**). **g** A white or bluish leaf surface (*Eucalyptus macrocarpa*) that is densely covered with three-dimensional waxes shown in (**h**) (adapted from Koch et al. 2009a)

carnivorous plant *Sarracenia Leucophilla* (Fig. 4.3d). The cells are downward trending hair-papilla, with a parallel cuticle folding and with larger distances at the bases and denser arrangement at the cell tip.

Based on the large variations of surface chemistry and structures, plant surfaces provide multifunctional properties. Some surfaces, like the leaves of the Lotus plant (*Nelumbo nucifera),* are extremely water repellent (superhydrophobic) and self-cleaning. However, others, such as the air-roots of epiphytic orchids, some lichens and mosses, show opposite behavior, and are constructed for the most efficient water absorption (superhydrophilic) through their surfaces. Such wetting phenomena are based on physicochemical factors.

In this chapter, various plant leaves, their roughness, and wax coatings in relation to their hydrophobic/hydrophilic and self-cleaning properties (Bhushan and Jung 2011; Bhushan 2012) will be discussed. Surface characterization of hydrophobic and hydrophilic leaves on the micro- and nanoscale is presented to understand the role of microbumps and nanobumps. In addition, the contact angle and adhesion and friction properties of these leaves are considered. The knowledge gained by examining these properties of the leaves and by quantitatively analyzing the surface structure helps in the design of superhydrophobic and self-cleaning surfaces with low adhesion/drag. Finally, various self-cleaning approaches are presented.

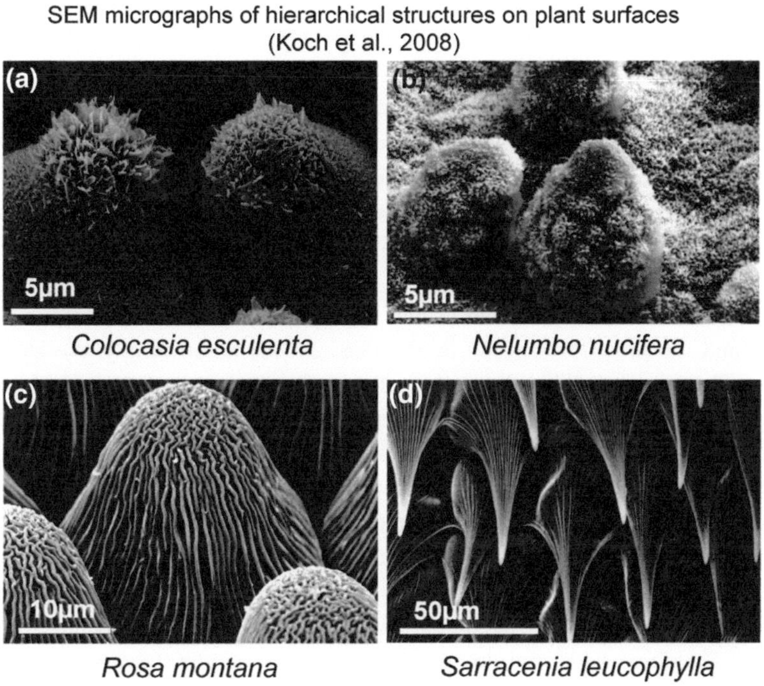

Fig. 4.3 SEM micrographs of hierarchical structures on plant surfaces. **a**, **b** Double structured plant surfaces with convex cell shapes and superimposed three-dimensional epicuticular waxes on the upper (adaxial) leaf sides of **a** *Colocasia esculenta* and **b** *N. nucifera*. **c**, **d** Convex cells with cuticular folding. **c** The flower leaf of *Rosa montana* (adaxial side) with a rippled folded cuticle in the central field of the cells and parallel folding. **d** The cells of the inner side of a tube-like leaf of the carnivorous plant *Sarracenia leucophylla*. These cells are downward trending hair papilla with a parallel cuticle folding (adapted from Koch et al. 2008)

4.2 Plant Leaves

Hydrophobic and water-repellent abilities of many plant leaves have been long known. Scanning electron microscope (SEM) studies since the 1970s have revealed that the hydrophobicity of the leaf surface is related to its microstructure. The outer cells covering a plant, especially the leaf, are called epidermis cells. The epidermis in all plant surfaces is covered by a thin extracellular membrane, called a cuticle. The plant cuticle is a composite material mainly built up of a cutin network and hydrophobic waxes (Barthlott and Neinhuis 1997; Koch et al. 2008, 2009a; Barthlott et al. 2017). The chemical structure of the epicuticular waxes has been studied extensively by plant scientists and lipid chemists (Baker 1982; Jetter et al. 2006). The epicuticular waxes can be thin with a 2-D structure, thicker with a 3-D structure, or a combination thereof. It is believed that waxes diffuse through the cuticle via a lipidic pathway, and after diffusion of the wax, the tubular or platelet type wax morphologies grow by crystallization or self-assembly (Koch et al. 2009c). The plants are able to repair the wax layer by self-assembly.

Neinhuis and Barthlott (1997) studied the surfaces and wetting properties of about 200 water-repellent plants; for comprehensive reviews, see Koch et al. (2008, 2009a). Among the epidermal relief features are the papillose epidermal cells with either every epidermal cell forming a single papilla or cells being divided into papillae. The scale of the epidermal relief features ranges from 5 μm in multipapillate cells to 100 μm in large epidermal cells. Some cells also are convex (rather than having real papillae) and/or had hairs (trichomes). Various types and shapes of wax crystals are also found at the surface (Neinhuis and Barthlott 1997; Koch et al. 2008, 2009a). Interestingly, the hairy surfaces with a thin film of wax exhibited water-repellency for short periods (minutes), after which water penetrated between the hairs. Conversely, hairs with a thick film led to strong water-repellency. The wax crystal creates nanoroughness, in addition to the microroughness created by the papillae. Hierarchical roughness with microroughness superimposed with nanoroughness plays the dominant role in self-cleaning leaves.

The hydrophobicity of the leaves is related to another important effect: the ability to remain clean after being immersed in dirty water, known as self-cleaning. This ability is best known from the lotus (*Nelumbo nucifera*) leaf, which has been considered by Hindus and Buddhists as sacred due to its purity for some 2000 years. Not surprisingly, the self-cleaning and water repellent abilities of lotus-like surfaces was dubbed the "lotus effect." As far as the biological implications of the lotus effect, self-cleaning plays an important role in the defense against pathogens binding to the leaf surface. Many spores and conidia of pathogenic organisms—most fungi—require water for germination and can infect leaves only in the presence of water. The lotus effect also provides antifouling for leaf surfaces.

SEM micrographs of a lotus leaf are shown in Fig. 4.4. The lotus leaf surface is covered by "bumps," more exactly called papillae (papillose epidermal cells), which, in turn, are covered by an additional layer of three-dimensional epicuticular waxes (Barthlott and Neinhuis 1997). The wax is present in crystalline tubules, composed of a mixture of long-chain aliphatic compounds, principally nonacosanol and nonacosanediols (Koch et al. 2006, 2008, 2009a). The wax on the lotus leaf self-assembles as tubules, but on other leaves, waxes exist also in the form of platelets or other morphologies (Koch et al. 2008, 2009a). The wax is hydrophobic with a water contact angle of about 95°–110°, and the hierarchical structure present results in a high contact angle, making the surface superhydrophobic with a low contact angle hysteresis, based on the Wenzel and Cassie-Baxter models discussed in the preceding chapter. The experimental value of the static water contact angle with the lotus leaf was reported to be about 164° and a contact angle hysteresis of 3° (Bhushan et al. 2009a; Koch et al. 2009b). Taking the papillae density of 3400 per mm^2, the average radius of the hemispherical asperities is 10 μm, and the aspect ratio (height/radius) is 1, provides the value of the roughness factor $R_f \approx 4$ based on

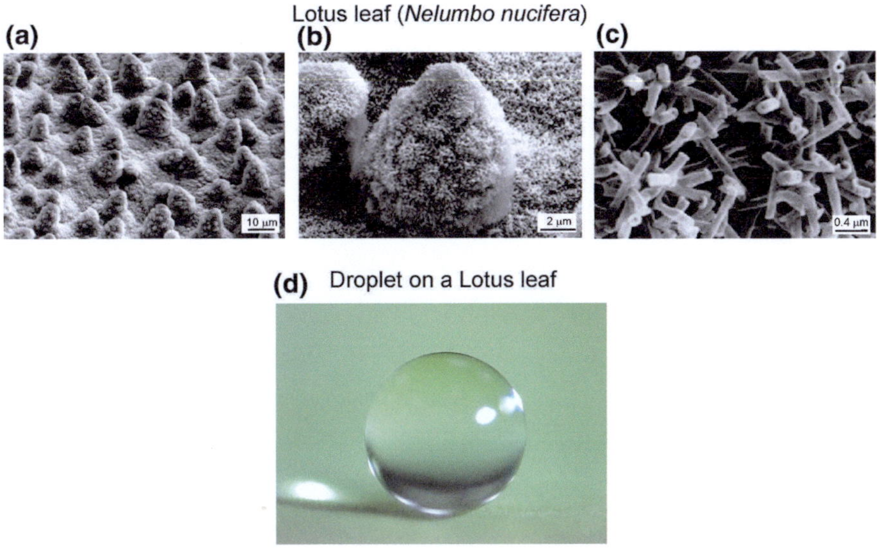

Fig. 4.4 SEM micrographs (shown at three magnifications) of a lotus (*Nelumbo nucifera*) leaf surface, consisting of microstructure formed by papillose epidermal cells covered with 3-D epicuticular wax tubules on surface, which create nanostructure, and an image of a water droplet sitting on a lotus leaf with a contact angle of about 164° (Bhushan et al. 2009a)

Wenzel's equation (Nosonovsky and Bhushan 2005). Taking the value of the contact angle for wax, $\theta_0 = 104°$ (Kamusewitz et al. 1999), the calculation with the Wenzel equation yields $\theta = 165°$, which is close to the experimentally observed values (Nosonovsky and Bhushan 2005). However, the simple Wenzel model is not sufficient to explain the lotus effect, as the roughness structure forms a composite interface, represented by the Cassie-Baxter equation. Moreover, its structure has hierarchical roughness and models have been developed to study the role of hierarchical roughness on contact angle (Nosonovsky and Bhushan 2008).

The self-cleaning ability of the lotus leaf is derived from the low contact angle hysteresis or a tilt angle of about 3° with low adhesion/drag. An explanation for self-cleaning is that on a smooth surface, the droplet slides on the surface and contaminant particles are mainly redistributed by the sliding droplet. Conversely, on a rough, lotus-like surface, contaminant particles adhere to the droplet and are removed from the surface when the droplet rolls off because of low adhesion/drag, Fig. 4.5a. Figure 4.5b shows optical micrographs of a water droplet cleaning the lotus leaf and a mercury droplet cleaning a taro leaf (Barthlott and Neinhuis 1997).

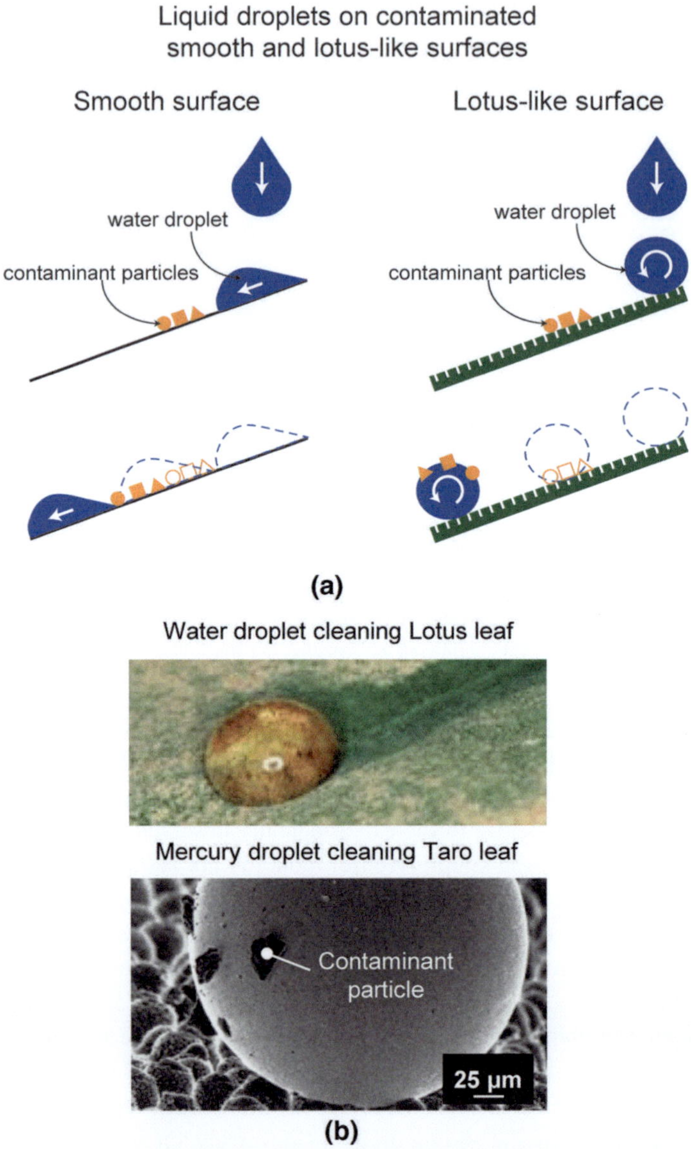

Fig. 4.5 **a** Schematic showing a droplet sliding on a smooth surface, leading to redistribution of contaminant particles, and a droplet rolling on a lotus-like surface with low contact angle hysteresis and low adhesion/drag, removing contaminant particles from the leaf surface, **b** optical micrographs showing a water droplet cleaning the lotus leaf and a mercury droplet cleaning the taro leaf (adapted from Barthlott and Neinhuis 1997)

4.3 Characterization of Superhydrophobic and Hydrophilic Leaf Surfaces

In order to understand the mechanisms of hydrophobicity in plant leaves, a comprehensive comparative study of various superhydrophobic and hydrophilic leaf surfaces and their properties was carried out by Bhushan and Jung (2006) and Burton and Bhushan (2006). Below is a discussion of the findings of the study.

4.3.1 Experimental Techniques

The static contact angles were measured using a Rame-Hart model 100 contact angle goniometer with droplets of deionized (DI) water (Bhushan and Jung 2006; Burton and Bhushan 2006). Droplets of about 5 μL in volume (with diameter of a spherical droplet about 2.1 mm) were gently deposited on the substrate using a microsyringe for the static contact angle. All measurements were made at five different points for each sample at 22 ± 1 °C and 50 ± 5% relative humidity (RH). The measurement results were reproducible within ±3°.

An optical profiler was used to measure surface roughness for different surface structures (Bhushan and Jung 2006; Burton and Bhushan 2006). A greater Z-range of the optical profiler of 2 mm is a distinct advantage over the surface roughness measurements with an atomic force microscope (AFM), which has a Z-range on the order of 7 μm, but an AFM has only a maximum lateral resolution of approximately 0.6 μm (Bhushan 2013a, b). A commercial AFM was used for additional surface roughness measurements with a high lateral resolution (sub nm) and for adhesion and friction measurements (Burton and Bhushan 2005; Bhushan and Jung 2006). The measurements for surface roughness were performed in the tapping mode with a square pyramidal Si(100) tip with a native oxide layer, which had a nominal radius of 20 nm, on a rectangular Si(100) cantilever with a spring constant of 3 N m^{-1}.

Adhesion and friction force at various relative RH were measured using a 15 μm radius borosilicate ball (Bhushan 2013a, b, 2017). A large tip radius was used to measure contributions from several microbumps and a large number of nanobumps. Friction force was measured under a normal load ranging from 20 to 250 nN using a 90° scan angle. It was measured at a velocity of 100 μm/s in a 50 μm scan and at a velocity of 4 μm/s in a 2 μm scan. The quantitative measurement of friction force was calibrated by the method described by Bhushan (2017). The normal load was varied (20–250 nN), and a friction force measurement was taken at each increment. By plotting the friction force as a function of normal load, an average coefficient of friction was obtained from the slope of the fit line of the data. The adhesive force was measured using the force distance curve approach. In this technique, the AFM tip is brought into contact with the sample by extending the piezo vertically, then retracting the piezo and calculating the force required to separate the tip from the sample. The adhesive force is obtained by multiplying the cantilever spring constant with the cantilever deflection during the retraction between zero value and the maximum negative value. The method is described in detail by Bhushan (2013a, b, 2017).

4.3.2 SEM Micrographs

Figure 4.6 shows the SEM micrographs of two superhydrophobic leaves and two hydrophilic leaves. The two superhydrophobic leaves are the lotus (*Nelumbo nucifera*) and the elephant ear or taro plant (*Colocasia esculenta*), referred to as lotus and colocasia, respectively. The two hydrophilic leaves are the beech (*Fagus sylvatica*) and Magnolia (*Magnolia grandiflora*), referred to as fagus and magnolia, respectively (Bhushan and Jung 2006). Lotus and colocasia are characterized by papillose epidermal cells responsible for the creation of papillae or microbumps on the surfaces, and an additional layer of 3-D epicuticular waxes. These waxes are a mixture of very long chain fatty acids molecules (compounds with chains > 20 carbon atoms) and create nanostructure on the entire surface. Fagus and magnolia are characterized by rather flat, tabular cells with a thin wax film with a 2-D structure (Barthlott and Neinhuis 1997). The leaves don't repel water, are not self-cleaning, and contaminant particles from ambient conditions are accumulated.

4.3.3 Contact Angle Measurements

Figure 4.7a shows the contact angles for the superhydrophobic and hydrophilic leaves before and after applying acetone. The acetone was applied in order to remove any wax present on the surface. As a result, for the superhydrophobic leaves, the contact angle was reduced, whereas for the hydrophilic leaves, the contact angle was almost unchanged. It is known that there is a very thin 2-D wax layer on the hydrophilic leaves, which introduces little roughness. In contrast, superhydrophobic leaves are known to have a 3-D wax layer on their surface consisting of nanoscale roughness over microroughness created by the papillae, which results in a hierarchical roughness. The combination of this wax and the roughness of the leaf creates a superhydrophobic surface.

Bhushan and Jung (2006) calculated the contact angles for leaves with smooth surfaces using the Wenzel equation and the calculated R_f and the contact angle of the four leaves. The results are presented in Fig. 4.7a. The approximate values of R_f for lotus and colocasia are 5.6 and 8.4 and for fagus and magnolia are 3.4 and 3.8, respectively. Based on the calculations, the contact angles on smooth surfaces were approximately 99° for lotus and 96° for colocasia. For both fagus and magnolia, the contact angles for the smooth surfaces were found to be approximately 86° and 88°. A further discussion on the effect of R_f on the contact angle will be presented later.

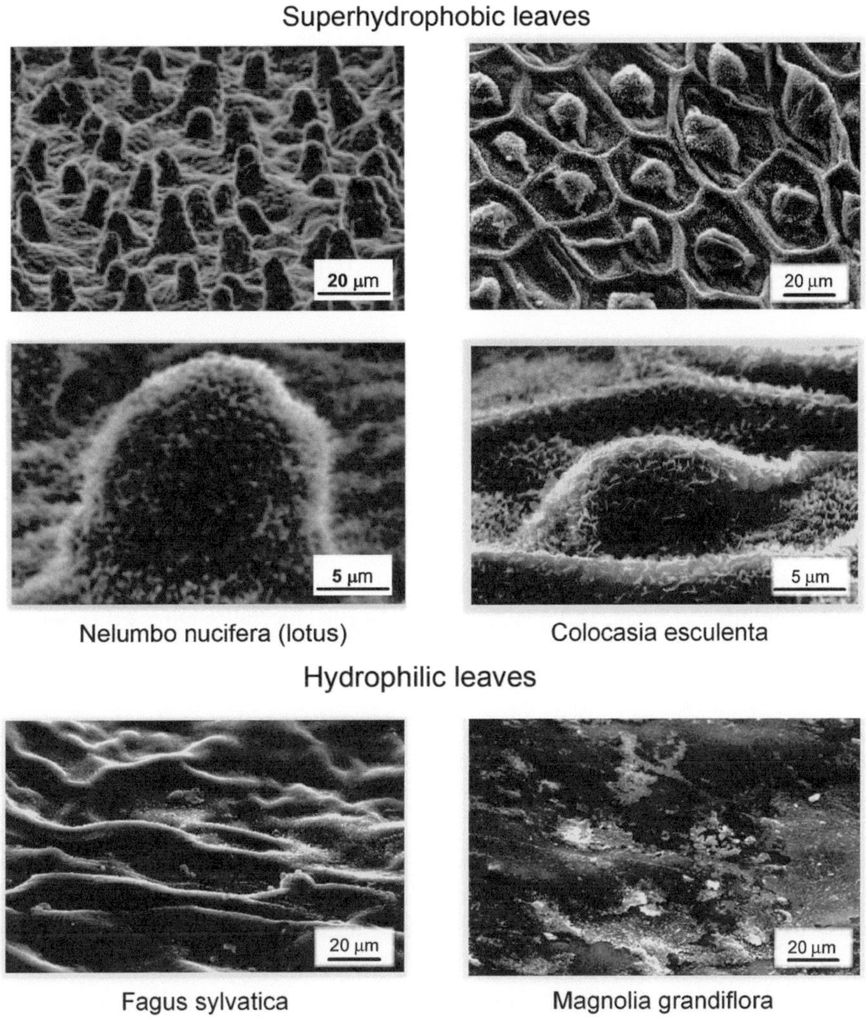

Fig. 4.6 Scanning electron micrographs of the relatively rough, water-repellent leaf surfaces of *Nelumbo nucifera* (lotus) and *Colocasia esculenta* and the relatively smooth, wettable leaf surfaces of *Fagus sylvatica* and *Magnolia grandiflora* (Bhushan and Jung 2006)

Figure 4.7b shows the contact angles for both fresh and dried states for the four leaves. There is a decrease in the contact angle for all four leaves when they are dried. For lotus and colocasia, this decrease is present because it is found that a fresh leaf has taller bumps than a dried leaf (data to be presented later), which will give a larger contact angle according to the Wenzel equation. When the surface area is at a maximum compared to the footprint area, as with a fresh leaf, the roughness factor will be at a maximum and will

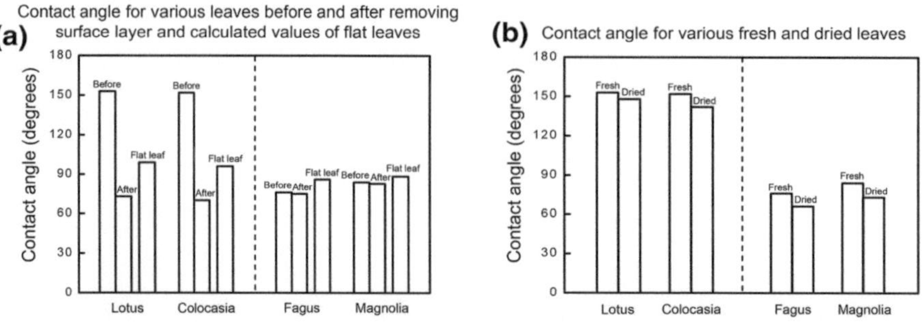

Fig. 4.7 Contact angle measurements and calculations for the leaf surfaces, **a** before and after removing the surface layer as well as calculated values, and **b** fresh and dried leaves. The contact angle on a smooth surface for the four leaves was obtained using the roughness factor calculated (Bhushan and Jung 2006)

only reduce when shrinking has occurred after drying. To understand the reason for the decrease of contact angle after drying of hydrophilic leaves, dried Magnolia leaves were also measured using an AFM. It is found that the dried leaf (peak-valley (P-V) height = 7 μm, mid-width = 15 μm, and peak radius = 18 μm) has taller bumps than a fresh leaf (P-V height = 3 μm, mid-width = 12 μm, and peak radius = 15 μm), which increases the roughness, and the contact angle decreases, leading to a more hydrophilic surface. The mid-width is defined as the width of the bump at a height equal to half of the peak to mean line value.

4.3.4 Surface Characterization Using an Optical Profiler

The use of an optical profiler allows measurements to be made on fresh leaves, which have a large P-V distance. Three different surface height maps for superhydrophobic and hydrophilic leaves are shown in Figs. 4.8 and 4.9 (Bhushan and Jung 2006). In each figure, a 3-D map and a flat map along with a 2-D profile in a given location of the flat 3-D map are shown. A scan size of 60 μm × 50 μm was used to obtain a sufficient amount of bumps to characterize the surface and also to maintain enough resolution to get an accurate measurement.

The structures found with the optical profiler correlate well with the SEM images shown in Fig. 4.6. The bumps on the lotus leaf are distributed on the entire surface, but the colocasia leaf shows a very different structure from that of the lotus. The surface structure for colocasia not only has bumps similar to lotus, but also surrounding each bump, there is a ridge that keeps the bumps separated. With these ridges, the bumps have a hexagonal (honeycomb) packing geometry that allows for the maximum number of bumps in a given area. The bumps of the lotus and both the bumps and ridges of colocasia contribute to the

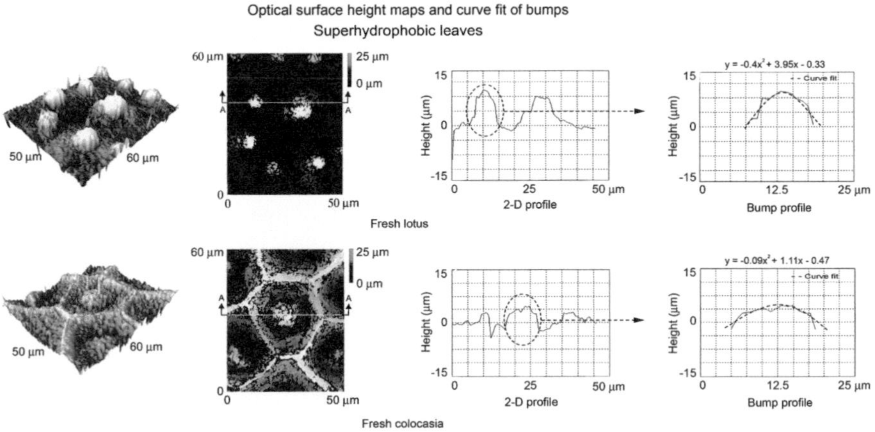

Fig. 4.8 Surface height maps and 2-D profiles of superhydrophobic leaves using an optical profiler. For lotus leaf, a microbump is defined as a single, independent microstructure protruding from the surface. For colocasia leaf, a microbump is defined as the single, independent protrusion from the leaf surface, whereas a ridge is defined as the structure that surrounds each bump and is completely interconnected on the leaf. A curve has been fitted to each profile to calculate dimensions of the bump. The radius of curvature is calculated from the parabolic curve fit of the bump (Bhushan and Jung 2006)

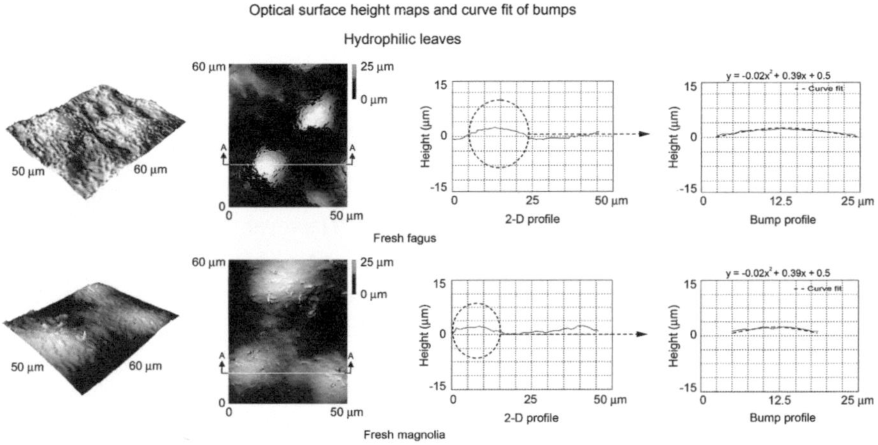

Fig. 4.9 Surface height maps and 2-D profiles of hydrophilic leaves using an optical profiler. For fagus and magnolia leaves, a microbump is defined as a single, independent microstructure protruding from the surface. A curve has been fitted to each profile to calculate dimensions of the bump. The radius of curvature is calculated from the parabolic curve fit of the bump (Bhushan and Jung 2006)

superhydrophobic nature, since they both increase the R_f factor and result in air pockets between the droplet of water and the surface. In fagus and magnolia height maps, short bumps can be seen on the surface. This means that with decreased bump height, the probability of air pocket formation decreases, and bumps have a less beneficial effect on the contact angle.

As shown in 2-D profiles of superhydrophobic and hydrophilic leaves in Figs. 4.8 and 4.9, a curve has been fitted to each profile to calculate dimensions of the bump. For each leaf, a second order curve fit has been given to the profiles to show how closely the profile is followed. By using the second order curve fitting of the profiles, the radius of curvature can be found (Bhushan and Jung 2006; Burton and Bhushan 2006).

Using these optical surface height maps, different statistical parameters of bumps and ridges can be found to characterize the surface: P-V height, mid-width, and peak radius (Bhushan 2013a, b). Table 4.1 shows these quantities found in the optical height maps for the four leaves. Comparing the superhydrophobic and hydrophilic leaves, it can be seen that the P-V height for the bumps of lotus and colocasia is much taller than that for the bumps of fagus and magnolia. The peak radius for the bumps of lotus and colocasia is also smaller than that for the bumps of fagus and magnolia. However, the values of mid-width for the bumps of the four leaves are similar.

4.3.5 Surface Characterization, Adhesion, and Friction Using an AFM

4.3.5.1 Comparison of Two AFM Measurement Techniques

To measure topographic images of the leaf surfaces, both the contact and tapping modes were first used (Bhushan and Jung 2006). Figure 4.10 shows surface height maps of dried lotus obtained by using the two techniques. In the contact mode, local height variation for a lotus leaf was observed in a 50 μm scan size. However, little height variation was obtained in a 2 μm scan, even at loads as low as 2 nN. This could be due to the substantial frictional force generated as the probe scanned over the sample. Frictional force can damage the sample. The tapping mode technique allows high-resolution topographic imaging of sample surfaces that are easily damaged, loosely held to their substrate, or difficult to image by other AFM techniques (Bhushan 2013a, b). As shown in Fig. 4.10, with the tapping mode technique, the soft and fragile leaves were imaged successfully. Therefore, the tapping mode technique was used to examine the surface roughness of the superhydrophobic and hydrophilic leaves using an AFM.

4.3.5.2 Surface Characterization

The AFM has a Z-range on the order of 7 μm and cannot be used for measurements in a conventional way because of the high P-V distances of a lotus leaf. Burton and Bhushan (2006) developed a new method to fully determine the bump profiles. In order

Table 4.1 Microbump and nanobump map statistics for superhydrophobic and hydrophilic leaves, measured both fresh and dried using an optical profiler and AFM (Bhushan and Jung 2006)

Leaf		Microbump Scan size (50 × 50 μm)			Nanobump Scan size (2 × 2 μm)		
		P-V height (μm)	Mid-width (μm)	Peak radius (μm)	P-V height (μm)	Mid-width (μm)	Peak radius (μm)
Lotus							
Fresh		13[a]	10[a]	3[a]	0.78[b]	0.40[b]	0.15[b]
Dried		9[b]	10[b]	4[b]	0.67[b]	0.25[b]	0.10[b]
Colocasia							
Fresh	Bump	9[a]	15[a]	5[a]	0.53[b]	0.25[b]	0.07[b]
	Ridge	8[a]	7[a]	4[a]	0.68[b]	0.30[b]	0.12[b]
Dried	Bump	5[b]	15[b]	7[b]	0.48[b]	0.20[b]	0.06[b]
	Ridge	4[b]	8[b]	4[b]	0.57[b]	0.25[b]	0.11[b]
Fagus							
Fresh		5[a]	10[a]	15[a]	0.18[b]	0.04[b]	0.01[b]
		4[b]	5[b]	10[b]			
Magnolia							
Fresh		4[a]	13[a]	17[a]	0.07[b]	0.05[b]	0.04[b]
		3[b]	12[b]	15[b]			

[a]Data measured using optical profiler
[b]Data measured using AFM

to compensate for the large P-V distance, two scans were made for each height: one measurement that scans the tops of the bumps and another measurement that scans the bottom or valleys of the bumps. The total height of the bumps is embedded within the two scans. Figure 4.11 shows the 50 μm surface height maps obtained using this method (Bhushan and Jung 2006). The 2-D profiles in the right column take the profiles from the top scan and the bottom scan for each scan size and splice them together to get the total profile of the leaf. The 2 μm surface height maps for both fresh and dried lotus can also be seen in Fig. 4.11. This scan area was selected on the top of a microbump obtained in the 50 μm surface height map. It can be seen that nanobumps are randomly and densely distributed on the entire surface of the lotus leaf.

Bhushan and Jung (2006) also measured the surface height maps for hydrophilic leaves in both 50 and 2 μm scan sizes as shown in Fig. 4.12. For fagus and magnolia, microbumps were found on the surface, and the P-V distance of these leaves is lower than that of lotus and colocasia. It can be seen in the 2 μm surface height maps that nanobumps selected on the peak of the microbump have an extremely low P-V distance.

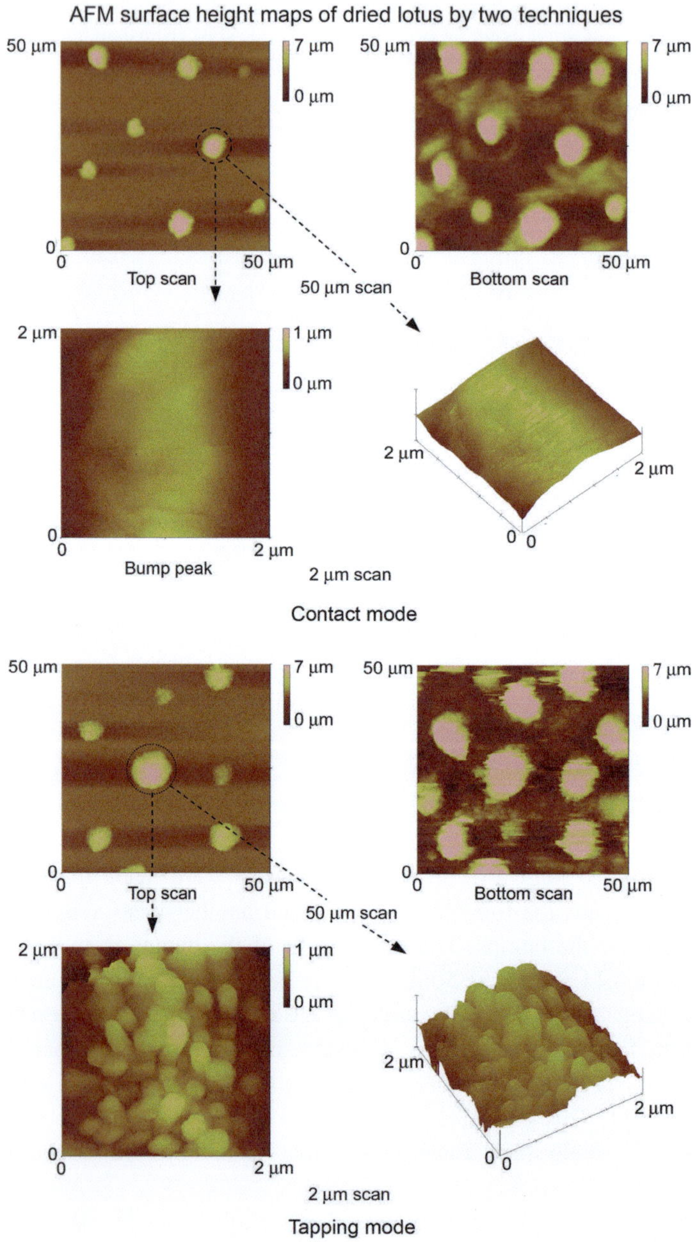

Fig. 4.10 Surface height maps showing the top scan and bottom scan in a 50 μm scan size and the bump peak scan selected in a 2 μm scan size for a lotus leaf using an AFM in contact mode and tapping mode. Two methods were used to determine a suitable method to obtain high resolution of nanotopography for a lotus leaf (Bhushan and Jung 2006)

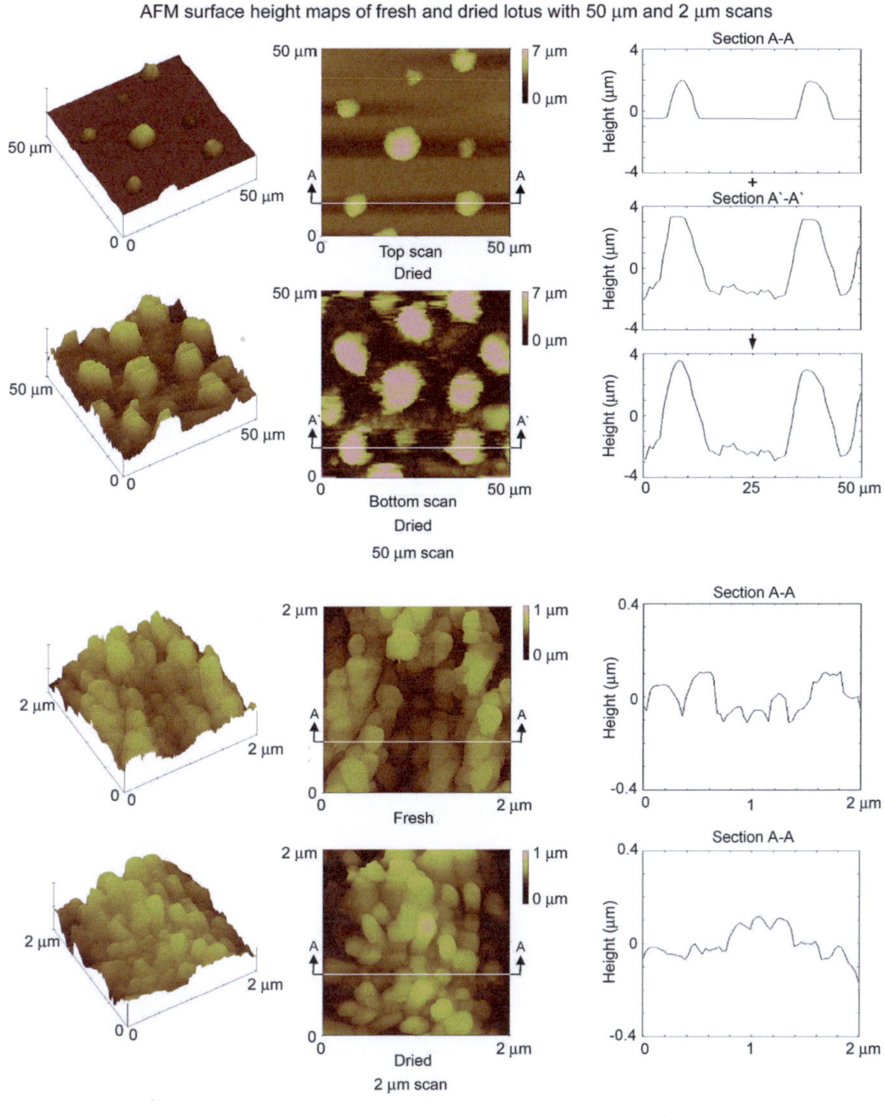

Fig. 4.11 Surface height maps and 2-D profiles showing the top scan and bottom scan of a dried lotus leaf in 50 μm scan (because the P-V distance of a dried lotus leaf is greater than the Z-range of an AFM), and the top scan of both fresh and dried lotus in a 2 μm scan (Bhushan and Jung 2006). A splicing technique was used to determine the bump profiles. In order to compensate for the large P-V distance, the total height of the bumps is embedded within the top scan and bottom scan

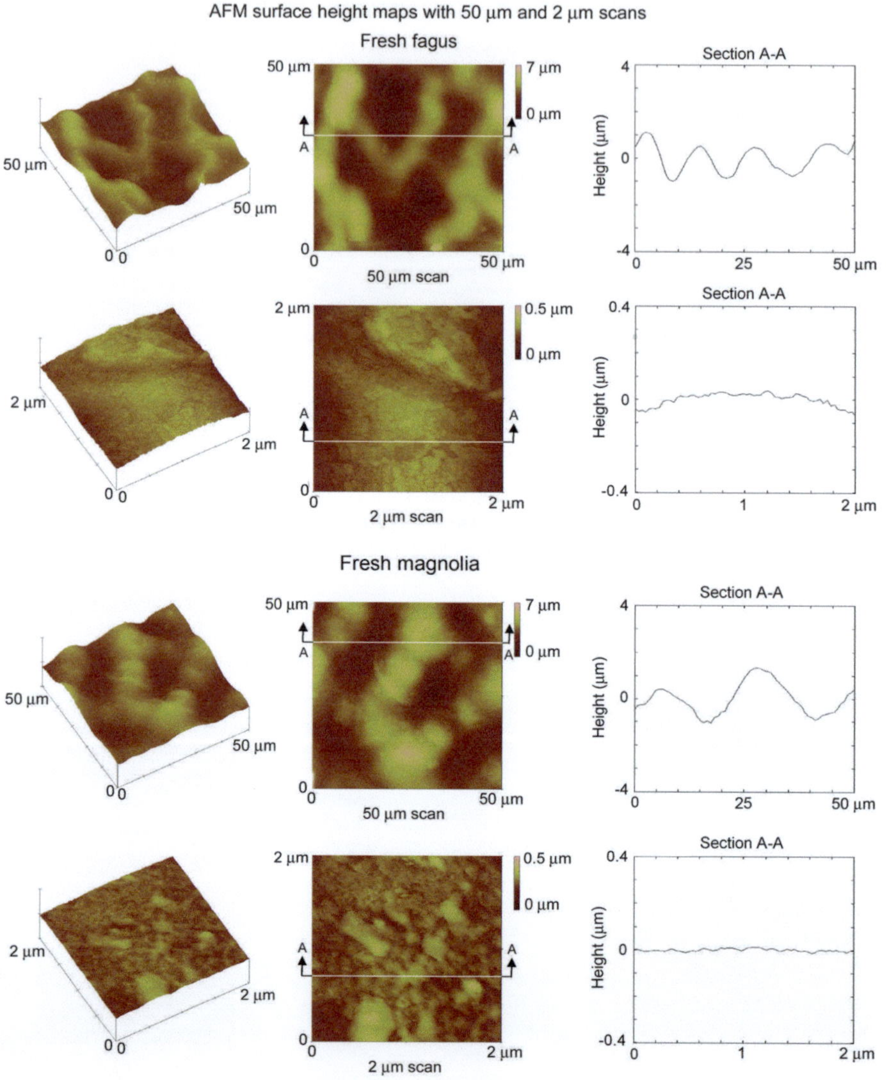

Fig. 4.12 Surface height maps and 2-D profiles of fagus and magnolia using an AFM in both 50 and 2 μm scans (Bhushan and Jung 2006)

Using the AFM surface height maps, different statistical parameters of bumps and ridges can be obtained: P-V height, mid-width, and peak radius. These quantities for the four leaves are listed in Table 4.1. It can be seen that the values correlate well with the values obtained from optical profiler scans except for the bump height, which decreased by more than half because of leaf shrinkage.

4.3.5.3 Adhesive Force and Friction

Adhesive force and the coefficients of friction of superhydrophobic and hydrophilic leaves obtained using an AFM are presented in Fig. 4.13 (Bhushan and Jung 2006). For each type of leaf, adhesive force measurements were made for both fresh and dried leaves using an AFM tip with 15 μm radius. It was found that the dried leaves had a lower adhesive force than the fresh leaves. Adhesive force arises from several sources in changing the presence of a thin liquid film, such as an adsorbed water layer that causes meniscus bridges to build up around the contacting and near contacting bumps as a result of surface energy effects (Bhushan 2013a, b). When the leaves are fresh, there is moisture within the plant material that causes the leaf to be soft. When the tip comes into contact with the fresh leaf sample, the sample will deform, a larger real area of contact between the tip and sample will occur, and the adhesive force will increase. After the leaf has dried, the moisture that was in the plant material is gone, and there is not as much deformation of the leaf when the tip comes into contact with the leaf sample. Hence, the adhesive force is decreased because the real area of contact has decreased.

The adhesive force of fagus and magnolia is higher than that of lotus and colocasia because the real area of contact between the tip and leaf surface is expected to be higher in hydrophilic leaves than in superhydrophobic leaves. Fagus and magnolia are hydrophilic, and thus, have a high affinity to water. The combination of a large real area of contact and an affinity to water are responsible for higher meniscus forces (Bhushan 2013a, b). The coefficient of friction was only measured on a dried plant surface using the same sliding velocity (10 μm/s) in different scan sizes. Fresh surfaces were not included because their P-V was too large to scan back and forth with the AFM to obtain friction force. As expected, the coefficient of friction for superhydrophobic leaves is lower than that for hydrophilic leaves due to the real area of contact between the tip and leaf sample, similar to the adhesive force results. When the scan size decreases from microscale to nanoscale, the coefficient of friction also decreases in each leaf. The reason for such dependence is the scale dependent nature of the roughness of the leaf surface. Figures 4.11 and 4.12 show AFM topography images and 2-D profiles of the surfaces for different scan sizes. The scan size dependence of the coefficient of friction has been reported previously (Poon and Bhushan 1995; Koinkar and Bhushan 1997; Tambe and Bhushan 2004).

4.3.6 Role of the Hierarchical Roughness

The approximation of the roughness factor for the leaves on the micro- and nanoscale was made using AFM scan data (Bhushan and Jung 2006). Roughness factors for various leaves are presented in Table 4.2. As mentioned earlier, the open space between asperities on a surface has the potential to collect air, and its probability appears to be higher in nanobumps as the distance between bumps in the nanoscale is smaller than those in the

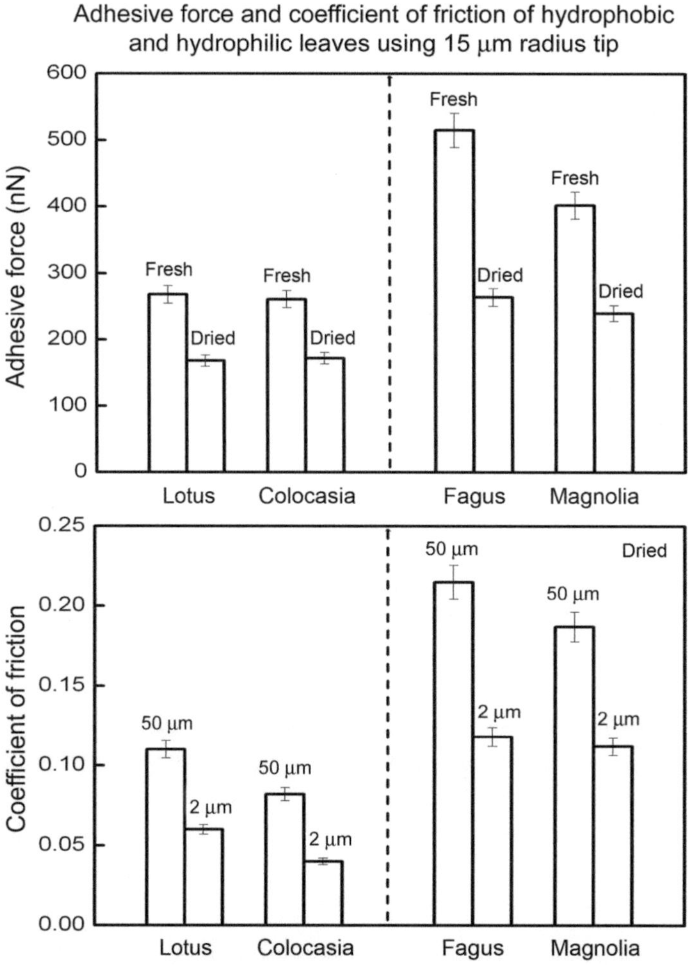

Fig. 4.13 Adhesive force for fresh and dried leaves, and the coefficient of friction for dried leaves for 50 and 2 μm scan sizes for hydrophobic and hydrophilic leaves. All measurements were made using an AFM with 15 μm radius borosilicate tip. Reproducibility for both adhesive force and coefficient of friction is ±5% for all measurements (Bhushan and Jung 2006)

microscale. The contact angle (θ_0) for smooth surfaces can be calculated using roughness factor values and contact angles (θ) from both superhydrophobic and hydrophilic surfaces, 153° and 152° in lotus and colocasia, and 76° and 84° in fagus and magnolia, respectively. The Wenzel equation (Chap. 3) is used for microbumps and the Cassie-Baxter equation (Chap. 3) for nanobumps. The contact angle ($\Delta\theta$) calculated using R_f on the smooth surface can be found in Table 4.2. It can be seen that the roughness factors and the differences ($\Delta\theta$) between θ and θ_0 on the nanoscale are higher than those on the

Table 4.2 Roughness factor and contact angle ($\Delta\theta = \theta - \theta_0$) calculated using R_f on the smooth surface for hydrophobic and hydrophilic leaves measured using an AFM, both microscale and nanoscale (Bhushan and Jung 2006)

Leaf (contact angle)	Scan size	State	R_f	$\Delta\theta$ (°)
Lotus (153°)	50 μm	Dried	5.6	54[a]
	2 μm	Fresh	20	61[b]
		Dried	16	60[b]
Colocasia (152°)	50 μm	Dried	8.4	56[a]
	2 μm bump	Fresh	18	60[b]
		Dried	14	59[b]
	2 μm ridge	Fresh	18	60[b]
		Dried	15	59[b]
Fagus (76°)	50 μm	Fresh	3.4	−10[a]
	2 μm	Fresh	5.3	2[b]
Magnolia (84°)	50 μm	Fresh	3.8	−4[a]
	2 μm	Fresh	3.6	14[b]

[a]Calculations made using Wenzel equation
[b]Calculations made using Cassie-Baxter equation. It is assumed that the contact area between the droplet and air is the half of the whole area of the rough surface

microscale. This means that nanobumps on the top of a microbump increase contact angle more effectively than microbumps alone. In the case of hydrophilic leaves, the values of R_f and $\Delta\theta$ change very little on both scales.

Based on the data in Fig. 4.13, the coefficient of friction values on the nanoscale are much lower than those on the microscale. It is observed that friction values are scale dependent. The height of a bump and the distance between bumps on the microscale is much larger than those on the nanoscale, which may be responsible for larger values of friction force on the microscale.

One difference between microbumps and nanobumps for surface enhancement of water repellency is the effect on contact angle hysteresis, which is the ease with which a droplet of water can roll on the surface. It has been stated earlier that contact angle hysteresis decreases and contact angle increases due to the decreased contact with the solid surface caused by the air pockets beneath the droplet. The surface with nanobumps has a high roughness factor compared with that of microbumps. With large distances between microbumps, the probability of air pocket formation decreases and is responsible for high contact angle hysteresis. Therefore, on the surface with nanobumps, the contact angle is high and contact angle hysteresis is low. Drops rebound easily and can begin to roll with a small tilt angle (Bhushan and Jung 2006).

Natural water-repellent and self-cleaning surfaces, such as the lotus leaf (Koch et al. 2008, 2009a) or water strider leg (Gao and Jiang 2004), have a hierarchical structure. However, the functionality of this hierarchical roughness remains a subject of discussion, and several explanations have been suggested. Nosonovsky and Bhushan (2008) showed that the mechanisms involved in superhydrophobicity are scale dependent, and thus, the roughness must be hierarchical in order to respond to these mechanisms. The surface must be able to repel both macroscopic and microscopic droplets. Fürstner et al. (2005) pointed out that artificial surfaces with one level of roughness can well repel large "artificial rain" droplets, but they cannot repel small "artificial fog" droplets trapped in the valleys between the bumps. Thus, the hierarchy may have to do with the ability to repel droplets of various size ranges. According to Gao and McCarthy (2006), large bumps allow a surface to maintain a composite interface, while small ones enhance the contact angle in accordance with the Wenzel model. Jung and Bhushan (2008) showed that a droplet with a radius of about 100–400 μm on a micropatterned surface goes through transition from the composite interface to the solid-liquid interface as the pitch increases. Bhushan et al. (2008, 2009b, c) showed that hierarchical structure can prevent the gaps between the pillars from filling with liquid until the droplet evaporates completely.

4.3.7 Summary

Various leaf surfaces on the micro- and nanoscale have been characterized, and attempts are made to separate out the effect of micro- and nanobumps and epicuticular wax on hydrophobicity. For the superhydrophobic lotus and colocasia leaves, the leaf surface consists of microbumps formed by convex papilla epidermal cells covered with a 3-D epicuticular wax (crystalline tubules composed of a mixture of secondary alcohol nonacoscan-10-01 and nonacosanedoils), which self-assemble as nanotubules. A hierarchical surface and the presence of wax creates a superhydrophobic and self-cleaning surface with low adhesion and antifouling. Hydrophilic fagus and magnolia have rather flat tabular cells with a 2-D thin wax film (not continuous) on the surface.

4.4 Various Self-cleaning Approaches

In addition to the lotus effect approach for self-cleaning, other self-cleaning approaches have been developed (Nishimoto and Bhushan 2013). The self-cleaning surfaces can be divided into three wettability regimes: superhydrophobicity, superoleophobicity, and superhydrophilicity. An overview of nine examples from living nature is provided in Fig. 4.14: lotus leaves, rice leaves, cicada wings, butterfly wings, snail shell, fish scale, shark skin, pitcher plant, and photosynthesis. An example of a photosynthesis-inspired self-cleaning surface is a TiO_2 photocatalyst coated surface, which has been developed by mimicking the photosynthesis process of green leaves.

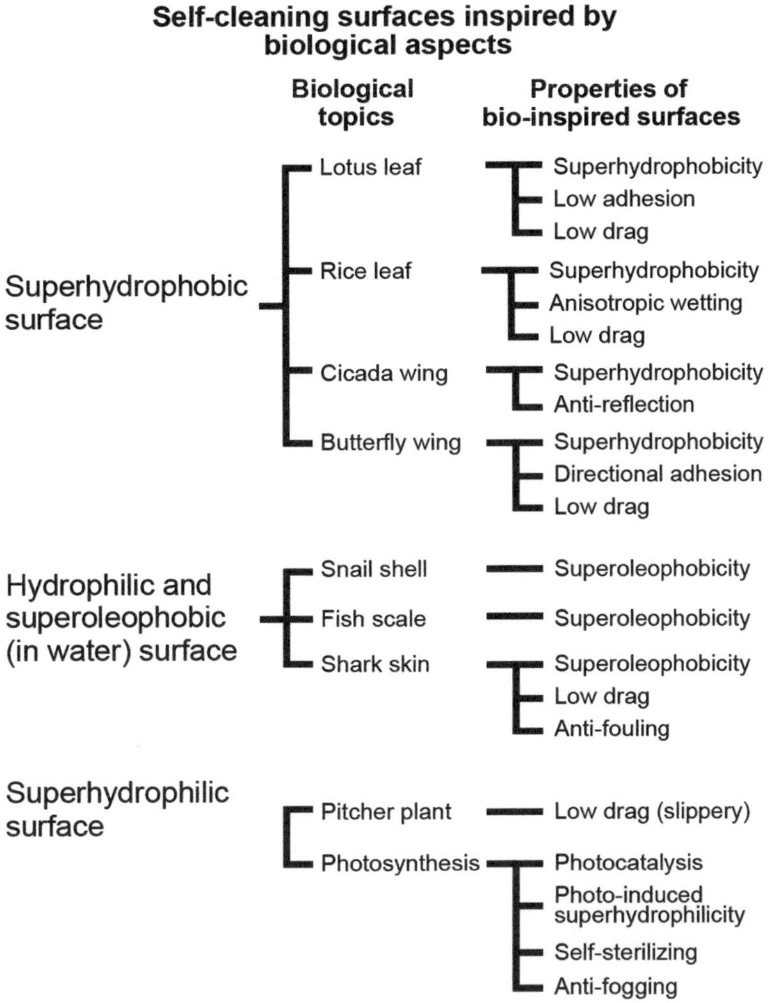

Fig. 4.14 Overview of self-cleaning surfaces inspired by biological objects (adapted from Nishimoto and Bhushan 2013)

4.4.1 Comparison Between Superhydrophobic and Superhydrophilic Surface Approaches for Self-cleaning

Several types of self-cleaning surfaces have been developed for practical applications. In particular, two types of surfaces, i.e. superhydrophobic surface (lotus) and superhydrophilic surfaces (TiO_2 photocatalyst), are notably important (Nishimoto and Bhushan 2013). We briefly describe the comparison between superhydrophobic and superhydrophilic surface approaches.

Nishimoto and Bhushan (2013) presented a comparison of the self-cleaning processes on superhydrophilic, superhydrophobic, and hydrophilic and superoleophobic surfaces, as shown in Fig. 4.15. On the superhydrophilic surface shown in Fig. 4.15a, the water droplets go beneath the contaminant and wash it away. Water on this surface spreads quickly creating no streaks. The thin film of water produced on the surface may evaporate rapidly. Water film moves the contaminants out during spreading. However, the surface may remain wet for a longer period in this case, as compared to that for superhydrophobic surfaces, which may be undesirable in some applications.

On the superhydrophobic surface shown in Fig. 4.15b, water is repelled from the surface and water droplets may roll off the surface taking contaminant particles with them. In this case, considering low adhesion of superhydrophobic surfaces, it would seem that higher self-cleaning efficiency would be obtained with a superhydrophobic surface than a superhydrophilic one.

On the superhydrophobic surface with oil-based contaminant shown in Fig. 4.15c, because both the contaminants and the superhydrophobic surface repel the water droplet, the contaminants cannot be removed by the water droplet. On the other hand, as shown in Fig. 4.15a, the oil-based contaminants, such as printing ink, can be washed away on a superhydrophilic surface. Nishimoto et al. (2009) observed that the oil-based ink that was adhered to the superhydrophilic area on the superhydrophobic-superhydrophilic pattern could be easily removed by water supply, while the ink was stable on the superhydrophobic area. The water droplets can go beneath an oil-based contaminant on the superhydrophilic surfaces, raise it up, and wash it away. It should be noted that most hydrophilic surfaces exhibit oleophobicity in water.

On the hydrophilic and superoleophobic (in water) surfaces shown in Fig. 4.15d, an oil-based contaminant could be efficiently washed away. Due to the superoleophobicity in water, the water droplets efficiently go beneath an oil-based contaminant and raise it up. However, the hydrophilic and superoleophobic (in water) surfaces may suffer from hydrophilic contaminants, because a certain amount of water would be necessary to wash the contaminants away.

It should be noted that the superhydrophobic and superhydrophilic surfaces have different surface roughnesses. For a surface to be superhydrophobic, the surface should have some roughness. On the other hand, a TiO_2 photocatalyst surface becomes superhydrophilic even if the surface is quite smooth. A superhydrophilic flat surface has the advantage of protecting the surface from contamination, and may compensate in some degree for its high adhesion as compared with a superhydrophobic surface.

Because of the differences in the self-cleaning mechanisms between the two surfaces, the self-cleaning effect for each would depend on the environment in which the surface is used, e.g. rainfall level, contaminant species, and amounts. There are some parameters to be compared for practical applications. For example, in the use of the TiO_2 photocatalyst, UV light is necessary, which would be the most significant drawback of this material. Furthermore, since the TiO_2 photocatalyst decomposes most organic materials, special

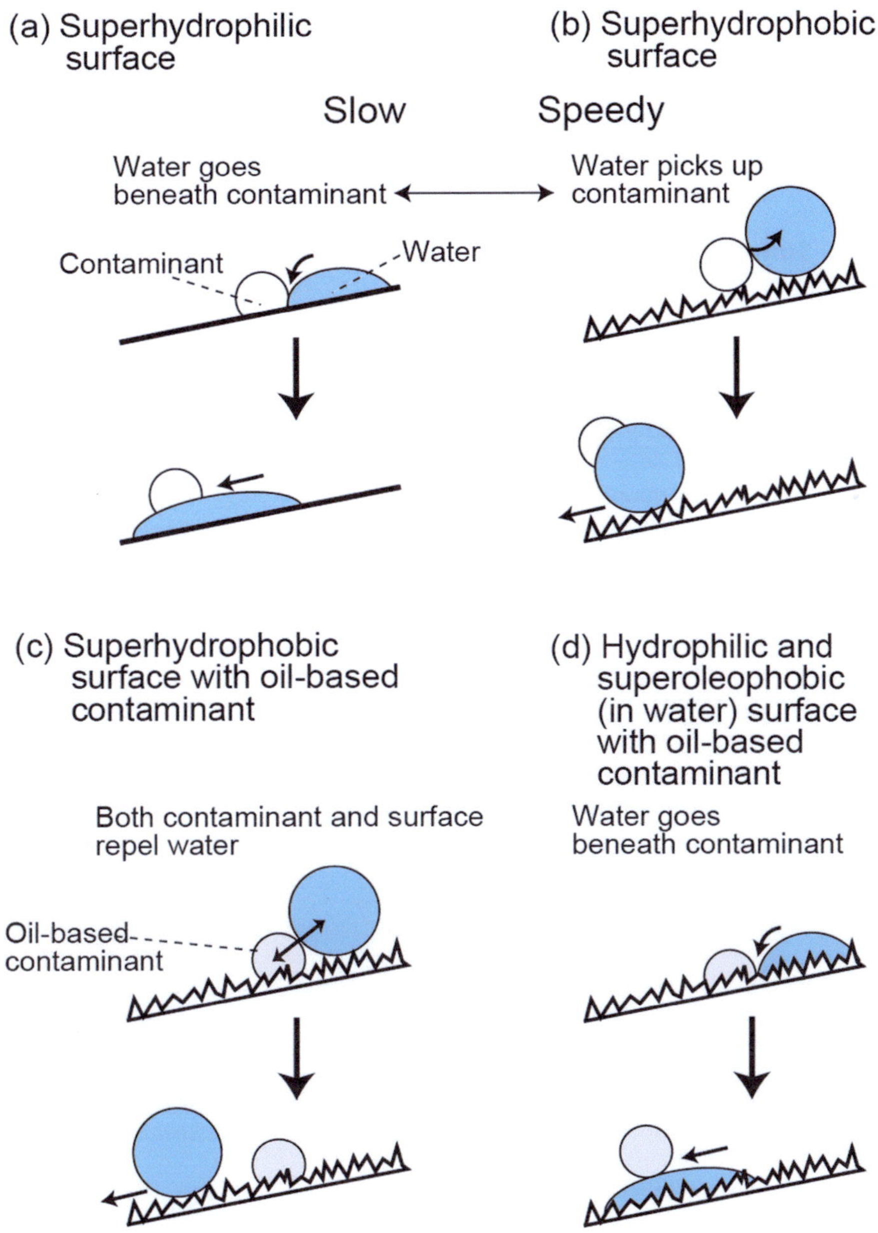

Fig. 4.15 Schematic illustrations of self-cleaning processes on **a** superhydrophilic surface, **b** superhydrophobic surface, **c** superhydrophobic surface with oil-based contaminant, and **d** hydrophilic and superoleophobic (in water) surface with oil-based contaminant (adapted from Nishimoto and Bhushan 2013)

care is necessary for the coating if it is used with the organic materials. However, the photocatalytic oxidation activity offers additional benefits such as self-sterilizing property. The durability of the smooth TiO_2 photocatalyst coating may be superior to that of the rough superhydrophobic coatings. However, recent studies show that superhydrophobic coatings also can be fabricated with high durability.

4.4.2 Summary

Superhydrophobic and superhydrophilic surfaces are commonly considered for self-cleaning properties. Both surfaces exhibit efficient self-cleaning properties in their own manner, and have been developed for practical applications. In addition to superhydrophobicity and superhydrophilicity, superoleophobicity is also a promising characteristic for self-cleaning surfaces, which can be easily achieved in hydrophilic surfaces in water. Moreover, anisotropic wetting, low drag, anti-fouling, slippery surfaces, and photocatalytic oxidation activity play an important role for the self-cleaning effect.

4.5 Closure

The lotus effect results in surfaces that are superhydrophobic and self-cleaning with low adhesion/drag. Various biological surfaces found in living nature that exhibit this effect have been presented, with a focus on the lotus leaf. Finally, various self-cleaning approaches using superhydrophobic and superhydrophilic surfaces are presented.

References

Baker, E. A. (1982), "Chemistry and Morphology of Plant Epicuticular Waxes," in *The Plant Cuticle* (eds. D. F. Cutler, K. L. Alvin, and C. E. Price), pp. 139–165, Academic Press, London, UK.

Barthlott, W. and Neinhuis, C. (1997), "Purity of the Sacred Lotus, or Escape from Contamination in Biological Surfaces," *Planta* **202**, 1–8.

Barthlott, W., Neinhuis, C., Cutler, D., Ditsch, F., Meusel, I., Theisen, I., and Wilhelm, H. (1998), "Classification and Terminology of Plant Epicuticular Waxes," *Bot. J. Linn. Soc.* **126**, 237–260.

Barthlott, W., Mail, M., Bhushan, B., and Koch, K. (2017), "Plant Surfaces: Structures and Functions for Biomimetic Innovations," *Nano-Micro Lett.* **9**, #23.

Bhushan, B. (2009) "Biomimetics: Lessons from Nature—An Overview," *Phil. Trans. R. Soc. A* **367**, 1145–1486.

Bhushan, B. (2012) "Bioinspired Structured Surfaces," *Langmuir* **28**, 1698–1714.

Bhushan, B. (2013a), *Principles and Applications of Tribology*, second ed., Wiley, New York.

Bhushan, B. (2013b), *Introduction to Tribology*, second ed., Wiley, New York.

Bhushan, B. (2017), *Nanotribology and Nanomechanics—An Introduction*, fourth ed., Springer International, Cham, Switzerland.

Bhushan, B. and Jung, Y. C. (2006), "Micro and Nanoscale Characterization of Hydrophobic and Hydrophilic Leaf Surface," *Nanotechnology* **17**, 2758–2772.

Bhushan, B. and Jung, Y. C. (2011), "Natural and Biomimetic Artificial Surfaces for Superhydrophobicity, Self-Cleaning, Low Adhesion, and Drag Reduction," *Prog. Mater. Sci.* **56**, 1–108.

Bhushan, B., Koch, K., and Jung, Y. C. (2008), "Nanostructures for Superhydrophobicity and Low Adhesion," *Soft Matter* **4**, 1799–1804.

Bhushan, B., Jung, Y. C., and Koch, K. (2009a), "Micro-, Nano- and Hierarchical Structures for Superhydrophobicity, Self-Cleaning and Low Adhesion," *Phil. Trans. R. Soc. A* **367**, 1631–1672.

Bhushan, B., Jung, Y. C., Niemietz, A., and Koch, K. (2009b), "Lotus-Like Biomimetic Hierarchical Structures Developed by the Self-Assembly of Tubular Plant Waxes," *Langmuir* **25**, 1659–1666.

Bhushan, B., Koch, K., and Jung, Y. C. (2009c), "Fabrication and Characterization of the Hierarchical Structure for Superhydrophobicity," *Ultramicroscopy* **109**, 1029–1034.

Burton, Z. and Bhushan, B. (2005), "Hydrophobicity, Adhesion, and Friction Properties of Nanopatterned Polymers and Scale Dependence for Micro- and Nanoelectromechanical Systems," *Nano Lett.* **5**, 1607–1613.

Burton, Z. and Bhushan, B. (2006), "Surface Characterization and Adhesion and Friction Properties of Hydrophobic Leaf Surfaces," *Ultramicroscopy* **106**, 709–719.

Fürstner, R., Barthlott, W., Neinhuis, C., and Walzel, P. (2005), "Wetting and Self-Cleaning Properties of Artificial Superhydrophobic Surfaces," *Langmuir* **21**, 956–961.

Gao, X. F. and Jiang, L. (2004), "Biophysics: Water-Repellent Legs of Water Striders," *Nature* **432**, 36.

Gao, L. and McCarthy, T. J. (2006), "The Lotus Effect Explained: Two Reasons Why Two Length Scales of Topography are Important," *Langmuir* **22**, 2966–2967.

Jetter, R., Kunst, L., and Samuels, A. L. (2006), "Composition of Plant Cuticular Waxes," in *Biology of the Plant Cuticle* (eds. M. Riederer and C. Müller), pp. 145–181, Blackwell Publishing, Oxford, UK.

Jung, Y. C. and Bhushan, B. (2008), "Wetting Behavior during Evaporation and Condensation of Water Microdroplets on Superhydrophobic Patterned Surfaces," *J. Micros.* **229**, 127–140.

Kamusewitz, H., Possart, W., and Paul, D. (1999), "The Relation between Young's Equilibrium Contact Angle and the Hysteresis on Rough Paraffin Wax Surfaces," *Colloid Surf. A-Physicochem. Eng. Asp.* **156**, 271–279.

Koch, K., Dommisse, A., and Barthlott, W. (2006), "Chemistry and Crystal Growth of Plant Wax Tubules of Lotus (*Nelumbo nucifera*) and Nasturtium (*Tropaeolum majus*) Leaves on Technical Substrates," *Crystl. Growth Des.* **6**, 2571–2578.

Koch, K., Bhushan, B., and Barthlott, W. (2008), "Diversity of Structure, Morphology, and Wetting of Plant Surfaces," *Soft Matter* **4**, 1943–1963.

Koch, K., Bhushan, B., and Barthlott, W. (2009a), "Multifunctional Surface Structures of Plants: An Inspiration for Biomimetics," *Prog. Mater. Sci.* **54**, 137–178.

Koch, K., Bhushan, B., Jung, Y. C., and Barthlott, W. (2009b), "Fabrication of Artificial Lotus Leaves and Significance of Hierarchical Structure for Superhydrophobicity and Low Adhesion," *Soft Matter* **5**, 1386–1393.

Koch, K., Bhushan, B., Eniskat, H. -J., and Barthlott, W. (2009c), "Self-Healing of Voids in the Wax Coating on Plant Surfaces," *Phil. Trans. R. Soc. A* **367**, 1673–1688.

Koinkar, V. N. and Bhushan, B. (1997) "Effect of Scan Size and Surface Roughness on Microscale Friction Measurements," *J. Appl. Phys.* **81**, 2472–2479.

Neinhuis, C. and Barthlott, W. (1997), "Characterization and Distribution of Water-Repellent, Self-Cleaning Plant Surfaces," *Ann. Bot.* **79**, 667–677.

Nishimoto, S. and Bhushan, B. (2013), "Bioinspired Self-Cleaning Surfaces with Superhydrophobicity, Superoleophobicity, and Superhydrophilicity," *RSC Adv.* **3**, 671–690.

Nishimoto, S., Kubo, A., Nohara, K., Zhang, X., Taneichi, N., Okui, T., Liu, Z., Nakata, K., Sakai, H., Murakami, T., Abe, M., Komine, T., and Fujishima, A. (2009), "TiO_2-Based Superhydrophobic-Superhydrophilic Patterns: Fabrication via an Ink-Jet Technique and Application in Offset Printing," *Appl. Surf. Sci.* **255**, 6221–6225.

Nosonovsky, M. and Bhushan, B. (2005), "Roughness Optimization for Biomimetic Superhydrophobic Surfaces," *Microsyst. Technol.* **11**, 535–549.

Nosonovsky, M. and Bhushan, B. (2008), *Multiscale Dissipative Mechanisms and Hierarchical Surfaces: Friction, Superhydrophobicity, and Biomimetics*, Springer-Verlag, Heidelberg, Germany.

Poon, C. Y. and Bhushan, B. (1995), "Comparison of Surface Roughness Measurements by Stylus Profiler, AFM and Non-Contact Optical Profiler," *Wear* **190**, 76–88.

Tambe, N. S. and Bhushan, B. (2004), "Scale Dependence of Micro/Nano-Friction and Adhesion of MEMS/NEMS Materials, Coatings and Lubricants," *Nanotechnology* **15**, 1561–1570.

Nanofabrication Techniques Used for Superhydrophobic Surfaces

5

5.1 Introduction

Nanofabrication of Lotus-like superhydrophobic surfaces has been an area of active research since the mid-1990s. Various nanofabrication techniques are used for micro- and nanostructure fabrication which include lithography, soft lithography, etching, deposition, and others, shown in Fig. 5.1 (Bhushan and Gupta 1991; Madou 2011; Bhushan 2017, 2018). The pros and cons of these techniques are summarized in Table 5.1. To various degrees of success, these techniques have been used in the fabrication of superhydrophobic surfaces (Bhushan 2009, 2018; Celia et al. 2013). Selected materials and nanofabrication techniques used to produce superhydrophobic surfaces are summarized in Table 5.2.

Two main requirements for a superhydrophobic surface are that the surface should be rough and hydrophobic (low surface energy). These two requirements can be met either by starting with a hydrophobic surface and making it rough or, by starting with a rough surface that may be hydrophilic and making it hydrophobic by modifying its surface chemistry or depositing a hydrophobic coating on the surface.

Transparent and superhydrophobic surfaces are of interest in optical applications and self-cleaning glasses. For the surface to be transparent, roughness details should be smaller than a quarter of the wavelength of visible light (about 100 nm or less) (Nakajima et al. 1999; Ebert and Bhushan 2012b). In a composite coating, any nanoparticles and other additives should be optically transparent. Among various nanoparticles, ITO (Indium Tin Oxide) nanoparticles exhibit high optical transmittance (Ebert and Bhushan 2012b).

Among interesting developments is the creation of switchable surfaces that can be turned from hydrophobic to hydrophilic and vice versa. This is done by surface energy modification through electrowetting, light and X-ray irradiation, dynamic effects, optical effects (Krupenkin et al. 2007; Li et al. 2019; Pan et al. 2019).

B. Bhushan, *Introduction to Biomimetics and Bioinspiration*,
https://doi.org/10.1007/978-3-031-62344-8_5

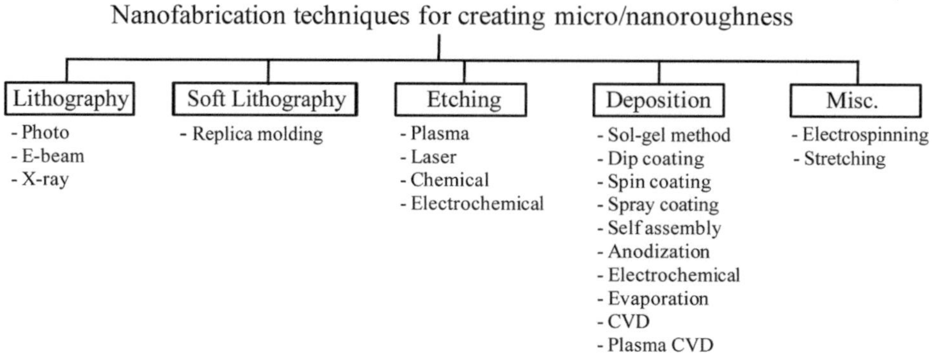

Fig. 5.1 Various nanofabrication techniques for creating micro-/nanoroughened surfaces (adapted from Bhushan 2018)

Table 5.1 Pros and cons of various nanofabrication techniques (adapted from Bhushan 2018)

Techniques	Pros	Cons
Lithography	Accuracy, large area	Slow process, high cost
Replica molding (templating)	Large area, fast, flexible, cheap	Limited to microfeatures
Etching/deposition	Flexible, cheap	Less control
Miscellaneous	Flexible, cheap	Limited applications

For commercial applications, superhydrophobic surfaces should be durable (Bhushan 2018, 2019) as well as self-healing, if possible (Multanen and Bhushan 2019).

5.2 Roughening to Create One-Level Structure

Lithography is a well-established technique and is used for creating micro/nanopatterns over a large area (Madou 2011; Bhushan 2017). Lithography uses photos, electron beam (E-beam), or X-ray beams. Bhushan and Jung (2007) and Bixler and Bhushan (2014, 2015) fabricated micropatterned Si using photolithography. To obtain a sample that is superhydrophobic, a fluorosilane coating was deposited on the micropatterned sample surface. They obtained a superhydrophobic surface with a contact angle of about 170°. Martines et al. (2005) fabricated ordered arrays of nanopits and nanopillars by using E-beam lithography. They obtained a superhydrophobic surface after hydrophobization with octadecyltrichlorosilane (OTS) on a surface consisting of tall pillars with cusped tops. Fürstner et al. (2005) created silicon wafers with regular patterns of spikes by X-ray lithography. The wafer was hydrophobized by sputtering a layer of gold and subsequent immersion in a hexadecanethiol solution. AFM can be used in nanolithography to produce

Table 5.2 Typical materials and nanofabrication techniques to produce superhydrophobic surfaces (adapted from Bhushan 2018)

Material	Technique	Notes	Source
Teflon	Plasma		Shiu et al. (2004)
PDMS	Deep reactive ion etching		Ebert and Bhushan (2016)
PDMS	Replica molding		Martin and Bhushan (2017)
PS-PDMS block copolymer	Electrospinning		Ma et al. (2005)
PS, PC, PMMA	Evaporation		Bormashenko et al. (2006)
Polyaniline nanofiber	Chemical polymerization		Chiou et al. (2007)
Polypropylene nanofibers	Porous nanomembrane patterning technique	Hierarchical	Lee and Bhushan (2012)
PET	Oxygen plasma etching		Teshima et al. (2005)
Organo-triethoxysilanes	Sol–gel	Reversible (temperature)	Shirtcliffe et al. (2005)
Al	Chemical etching		Qian and Shen (2005)
Copper	Electrodeposition	Hierarchical	Shirtcliffe et al. (2004)
Si	Replica molding		Bhushan and Jung (2007)
Si	E-beam lithography		Martines et al. (2005)
Si	X-ray lithography		Fürstner et al. (2005)
PS, PMMA	AFM nanolithography		Martin et al. (2005); Cappella and Bonaccurso (2007)
Si (Black Si)	Plasma etching	For liquid flow	Jansen et al. (1995)
Nonpolar rapeseed soot	Spray coating flame of burning rapeseed oil		Geraldi et al. (2018); Esmeryan et al. (2022); Esmeryan and Fedchenko (2023)
Silica	Sol–gel		Hikita et al. (2005)
Silica	Replica molding and spray coating	Hierarchical	Ebert and Bhushan (2012a)
Epoxy resin with synthetic and plant waxes	Replica molding and self-assembly	Hierarchical	Bhushan et al. (2008a, 2008b; 2009a, 2009b); Koch et al. (2009)

(continued)

Table 5.2 (continued)

Material	Technique	Notes	Source
Silica colloidal particles in PDMS	Spin coated	Hierarchical	Ming et al. (2005)
Au clusters	Electrochemical deposition		Zhang et al. (2004)
Carbon nanotubes	Chemical vapor deposition		Lau et al. (2003)
Carbon nanotubes	Chemical vapor deposition	Hierarchical	Huang et al. (2005)
Carbon nanotubes	Replica molding and spray coating	Hierarchical	Jung and Bhushan (2009)
ZnO, TiO_2 Nanorods	Sol–gel	Reversible (UV irradiation)	Feng et al. (2004)
SiO_2, ZnO and ITO nanoparticles	Spray coating	Hierarchical	Jung and Bhushan (2009); Ebert and Bhushan (2012b); Brown and Bhushan (2015); Martin et al. (2017); Bhushan and Martin (2018)
SiO_2	Particle encapsulation	Hierarchical	Brown and Bhushan (2016)
Liquid	Liquid impregnation	Slippery surface	Brown and Bhushan (2017)

a nanostructure with the aid of a solvent on polystyrene (PS) (Cappella and Bonaccurso 2007) or an electric field on polymethylmethacrylate (PMMA) (Martin et al. 2005).

Soft lithography or replica molding, also referred to as templating, has been commonly used to produce micropatterned surfaces of various polymeric materials. Jung and Bhushan (2006) created low aspect ratio asperities (LAR, 1:1 height-to-diameter ratio), high aspect ratio asperities (HAR, 3:1 height-to-diameter ratio), and a lotus pattern (replica from the lotus leaf), all on a PMMA surface using replica molding. A SAM of perfluorodecyltriethoxysilane (PFDTES) was deposited on the patterned surfaces using a vapor phase deposition technique. Martin and Bhushan (2017) fabricated micropatterned polydimethylsiloxane (PDMS) by a four-step replica molding using a micropatterned silicon master (Bhushan 2018).

One well-known and effective way to make rough surfaces is etching using either plasma, laser, chemical, or electrochemical techniques (Bhushan 2017). Jansen et al. (1995) etched a silicon wafer using a fluorine-based plasma by using the black silicon method to obtain isotropic, positively and negatively tapered, and vertical walls with smooth surfaces. Coulson et al. (2000) described an approach in plasma chemical roughening of poly(tetrafluoroethylene) (PTFE) (Teflon) substrates followed by the deposition of low surface energy plasma polymer layers, which give rise to high repellency towards polar and nonpolar probe liquids. Shiu et al. (2004) treated a Teflon film with oxygen plasma and obtained a superhydrophobic surface with a contact angle of 168°. Fluorinated materials have a limited solubility, which makes it difficult to roughen them. However, they may be linked or blended with other materials, which are often easier to roughen, in order to make superhydrophobic surfaces. Teshima et al. (2005) obtained a transparent superhydrophobic surface from a poly(ethylene terephthalate) (PET) substrate via selective oxygen plasma etching followed by plasma-enhanced chemical vapor deposition using tetramethylsilane (TMS) as the precursor. Qian and Shen (2005) described a simple surface roughening method by chemical etching on polycrystalline metals such as aluminum. After treatment with fluoroalkylsilane, the etched metallic surfaces exhibited superhydrophobicity.

Ebert and Bhushan (2016) fabricated transparent, superhydrophobic PDMS surfaces by deep reactive ion etching. O_2/CF_4 plasma was used to create rough surfaces. Surfaces then underwent an additional treatment of C_4F_8 plasma or vapor deposition of perfluorooctyltrichlorosilane following surface activation with O_2 plasma.

Deposition methods can also be used to make a substrate rough, including the sol–gel method, dip coating, spin coating, spray coating, self-assembly, anodization, electrochemical, evaporation, chemical vapor deposition (CVD), and plasma enhanced CVD. Solidification of wax can be used to produce a superhydrophobic surface.

Shiu et al. (2004) produced self-organized, close-packed superhydrophobic surfaces by spin coating a monodispersed polystyrene beads solution on a substrate surface. Spray coating technique is commonly used to produce superhydrophobic surfaces (Brown and Bhushan 2015; Martin et al. 2017; Bhushan and Martin 2018).

Abdelsalam et al. (2005) fabricated structured gold surfaces formed by electrodeposition through a template of submicrometer spheres. Bormashenko et al. (2006) used evaporated polymer solutions of polystyrene (PS), polycarbonate (PC) and polymethylmethacrylate (PMMA) dissolved in chlorinated solvents, dichloromethane (CH_2Cl_2) and chloroform ($CHCl_3$), to obtain a self-assembled structure with hydrophobic properties.

Chemical/physical vapor deposition (CVD/PVD) has been used for the modification of surface chemistry as well. Lau et al. (2003) created superhydrophobic carbon nanotube forests by modifying the surface of vertically aligned nanotubes with plasma enhanced chemical vapor deposition (PECVD). Superhydrophobicity was achieved down to the microscopic level where essentially spherical, micrometer-sized water droplets could be suspended on top of the nanotube forest. Zhu et al. (2005) and Huang et al. (2005) prepared surfaces with two-scale roughness by the controlled growth of carbon nanotube (CNT) arrays by CVD. Zhao et al. (2006) also synthesized vertically aligned multi-walled carbon nanotube (MWCNT) arrays by chemical vapor deposition on Si substrates using a thin film of iron (Fe) as a catalyst layer.

Other methods include electrospinning and deformation or stretching. Ma et al. (2005) produced block copolymer poly(styrene-b-dimethylsiloxane) fibers with submicrometer diameters in the range of 150–400 nm by electrospinning from a solution in tetrahydrofuran and dimethylformamide. They obtained superhydrophobic, nonwoven fibrous mats.

5.3 Coatings to Create One-Level Structures

The rapeseed soot, among the family of carbon soot materials, is nonpolar and intrinsically hydrophobic. Geraldi et al. (2018) used a spray mixture consisting of toluene, polydimethylsiloxane (PDMS) and rapeseed soot (average size of 115 ± 13 nm) to produce superhydrophobic coatings. Rapeseed soot was produced using a flame synthesis process, where rapeseed oil is combusted in an oxygen-starved environment. The soot material has also been attached to a variety of paper and textile products (Esmeryan and Fedchenko 2023) and concretes (Esmeryan et al. 2022). The soot was deposited from burning rapeseed oil by soaking a cotton-thread made lamp wick in the oil and subsequent exposure of the flame of burning oil to the substrate.

Shirtcliffe et al. (2005) prepared porous sol–gel foams from organo-triethoxysilanes which exhibited switching between superhydrophobicity and superhydrophilicity when exposed to different temperatures. Hikita et al. (2005) used colloidal silica particles and fluoroalkylsilane as the starting materials and prepared a film with super-liquid-repellency by hydrolysis and condensation of alkoxysilane compounds. Feng et al. (2004) produced superhydrophobic surfaces by sol–gel method using ZnO nanorods. They showed that superhydrophobic surfaces can be switched into hydrophilic surfaces by alternation of ultraviolet (UV) irradiation. Wu et al. (2005) made a microstructured, ZnO-based surface via a wet chemical process, and obtained superhydrophobicity after coating the surface with long-chain alkanoic acids. Chiou et al. (2007) fabricated polyaniline nanofibers using chemical oxidative polymerization to produce uniform, aligned nanofibers, and treated with CF_4 plasma treatment to create superhydrophobic surfaces with a contact angle of 175°.

Han et al. (2005) described the fabrication of lotus leaf-like superhydrophobic metal surfaces by using electrochemical reaction of Cu or Cu–Sn alloy plated on steel sheets with sulfur gas, and subsequent perfluorosilane treatment. Chemical bath deposition (CBD) has also been used to make nanostructured surfaces. Hosono et al. (2005) fabricated a nanopin film of brucite-type cobalt hydroxide (BCH) and achieved the contact angle of 178° after further modification of lauric acid (LA). Shi et al. (2006) described the use of galvanic cell reaction as a facile method to chemically deposit Ag nanostructures on a *p*-silicon wafer on a large scale. When the Ag covered silicon wafer was further modified with a self-assembled monolayer of *n*-dodecanethiol, a superhydrophobic surface was obtained with a contact angle of about 154° and a tilt angle lower than 5°.

Ebert and Bhushan (2012a, 2012b), Brown and Bhushan (2015), Martin et al. (2017), and Bhushan and Martin (2018) produced mechanically durable structures using various nanoparticles including SiO_2, ZnO and ITO (Indium Tin Oxide) nanoparticles, bonded with methylphenyl silicone binder. Nanoparticle-binder mixture was applied using a spray coating technique (Bhushan 2018). SiO_2 nanoparticles are commonly used for mechanical and chemical durability because of their high hardness and chemical inertness as well as their easy availability.

Jung and Bhushan (2009) produced mechanically durable carbon nanotube composite hierarchical structures with a static contact angle of 170° and a contact angle hysteresis of 2° by replication of a micropatterned silicon surface using an epoxy resin and by deposition of the carbon nanotube composite using a spray method. They showed that carbon nanotube composite structure had high mechanical strength and wear resistance because of the uniform distribution and strong bonding of the carbon nanotube on the substrates.

A so-called "layer-by-layer technique" was used to produce hierarchical structure by Brown and Bhushan (2015). Individual layers are used to provide specific functionality which provides flexibility in material selection and fabrication process as well as high durability. They used two polyelectrolyte layers with silica nanoparticles in between to provide roughness. The final top layer in the coating system used is referred to as functional layer, containing desired surface chemistry.

A particle-encapsulation technique was developed by Brown and Bhushan (2016) to produce hierarchical structure without a coating interface present in a coating deposition technique. In their technique, polycarbonate substrate was treated with an acetone-nanoparticles mixture. Polycarbonate dissolves readily in acetone and nanoparticles become incorporated into the polymer during swelling and subsequent polymer recrystallization. Since nanoparticles are incorporated in the polymer and no interface exists, the treated surfaces are expected to be more durable.

A liquid impregnation technique has been developed by Brown and Bhushan (2017) to produce a liquid-impregnated slippery surface. A porous surface is impregnated with a liquid that is immiscible with the liquid to be repelled. One liquid slide over another liquid surface resulting in a slippery surface.

5.4 Methods to Create Two-Level (Hierarchical) Structures

Two-level (hierarchical) roughness structures are typical for superhydrophobic surfaces in nature, as discussed earlier. Many efforts have been devoted to fabricating these hierarchical structures in various ways. Shirtcliffe et al. (2004) prepared a hierarchical copper surface by treated surface is expected to be more durable electrodeposition from acidic copper sulfate solution onto flat copper with a patterning technique of coating with a fluorocarbon hydrophobic layer.

Another way to obtain a rough surface for superhydrophobicity is by assembly of colloidal systems. Ming et al. (2005) prepared a hierarchical surface consisting of silica-based raspberry-like particles. Northen and Turner (2005) fabricated arrays of flexible silicon dioxide platforms supported by single high aspect ratio silicon pillars down to 1 μm in diameter and with heights up to about 50 μm. When these platforms were coated with polymeric organorods of approximately 2 μm tall and 50–200 nm in diameter, it showed that the surface is highly hydrophobic with a water contact angle of 145°.

Chong et al. (2006) fabricated hierarchically ordered nanowire arrays with periodic voids at the microscale and hexagonally packed nanowires at the nanoscale. This hierarchical surface was created by selective electrodeposition using nanoporous anodic alumina as a template and a porous gold film as a working electrode that was patterned by microsphere monolayers. Wang et al. (2006) also developed a novel precursor hydrothermal redox method with $Ni(OH)_2$ as the precursor to fabricate a hierarchical structure consisting of nickel hollow microspheres with nickel nanoparticles in situ.

Kim et al. (2007) fabricated a hierarchical structure by first generating nanoscale porosity by anodic aluminum oxidation. Then, the anodized porous alumina surface was replicated by PTFE. The polymer sticking phenomenon during the replication created the sub-microstructures on the negative PTFE nanostructure replica. Del Campo and Greiner (2007) reported that SU-8 hierarchical patterns comprised of features with lateral dimensions ranging from 5 to 2 mm and heights from 10 to 500 um were obtained by photolithography, which is comprised of a step of layer-by-layer exposure in soft contact printed shadow masks embedded into the SU-8 multilayer.

Bhushan et al. (2008a, 2008b, 2009a, 2009b) and Koch et al. (2009) produced hierarchical structures by replication of a micropatterned silicon surface using an epoxy resin and by self-assembly of synthetic and plant waxes as thin hydrophobic three-dimensional crystals to create hydrophobic nanostructures. The fabrication technique used is a low cost two-step process, which provides flexibility in the fabrication of a variety of hierarchical structures. They showed that a hierarchical structure has a high propensity for air pocket formation and leads to a static contact angle of 173° and contact angle hysteresis and tilt angle of about 2°.

Cortese et al. (2008) used plasma CF_4 treatment on micropatterned PDMS and obtained contact angle of 170°. Kuan et al. (2009) produced a hierarchical structure by imprinting ZnO precursor films using gratings with 830 nm and 50 μm dimensions. They achieved a contact angle of 141° by nanostructures deposited on sawtooth patterns without modifying the surface chemistry. Bhushan and Lee (2012) and Lee and Bhushan (2012) fabricated a hierarchical structure made of polypropylene fibers using two or three stacked porous membranes as a template.

5.5 Closure

Various nanofabrication techniques have been used to create superhydrophobic surfaces. These include roughening at one or two levels and/or chemical modification to reduce their surface energy. These techniques have been used to create lotus-like structures for a variety of applications.

References

Abdelsalam, M. E., Bartlett, P. N., Kelf, T., and Baumberg, J. (2005), "Wetting of Regularly Structured Gold Surfaces," *Langmuir* **21**, 1753–1757.

Bhushan, B. (2009), "Biomimetics: Lessons from Nature-An Overview," *Phil. Trans. R. Soc. A* **367**, 1445–1486.

Bhushan, B. (2017), *Springer Handbook of Nanotechnology*, fourth ed., Springer International, Cham, Switzerland.

Bhushan, B. (2018), *Biomimetics: Bioinspired Hierarchical-Structured Surfaces for Green Science and Technology*, third ed., Springer International, Cham, Switzerland.

Bhushan, B. (2019), "Lessons from Nature for Green Science and Technology: An Overview and Bioinspired Superliquiphobic/philic Surfaces," *Phil. Trans. R. Soc. A* **377**, 20180274.

Bhushan, B. and Gupta, B. K. (1991), *Handbook of Tribology: Materials, Coatings and Surface Treatments*, McGraw-Hill, New York.

Bhushan, B. and Jung, Y. C. (2007), "Wetting Study of Patterned Surfaces for Superhydrophobicity," *Ultramicroscopy* **107**, 1033–1041.

Bhushan, B. and Lee, H. (2012), "Fabrication and Characterization of Multi-level Hierarchical Surfaces," *Faraday Discuss.* **156**, 235–241.

Bhushan, B. and Martin, S. (2018), "Substrate-independent Superliquiphobic Coatings for Water, Oil, and Surfactant Repellency: An Overview," *J. Colloid Interface Sci.* **526**, 90–105.

Bhushan, B., Koch, K., and Jung, Y. C. (2008a), "Nanostructures for Superhydrophobicity and Low Adhesion," *Soft Matter* **4**, 1799–1804.

Bhushan, B., Koch, K., and Jung, Y. C. (2008b), "Biomimetic Hierarchical Structure for Self-Cleaning," *Appl. Phys. Lett.* **93,** 093101.

Bhushan, B., Jung, Y. C., Niemietz, A., and Koch, K. (2009a), "Lotus-like Biomimetic Hierarchical Structures Developed by the Self-assembly of Tubular Plant Waxes," *Langmuir* **25**, 1659–1666.

Bhushan, B., Koch, K., and Jung, Y. C. (2009b), "Fabrication and Characterization of the Hierarchical Structure for Superhydrophobicity," *Ultramicroscopy* **109**, 1029–1034.

Bixler, G. D. and Bhushan, B. (2014), "Rice- and Butterfly-Wing Effect Inspired Self-Cleaning and Low Drag Micro/nanopatterned Surfaces in Water, Oil, and Air Flow," *Nanoscale*, **6**, 76–96.

Bixler, G. D. and Bhushan, B. (2015), "Rice- and Butterfly-Wing Effect Inspired Low Drag and Antifouling Surfaces: A Review," *Crit. Rev. in Solid State Mater. Sci.* **40**, 1–37.

Bormashenko, E., Stein, T., Whyman, G., Bormashenko, Y., and Pogreb, E. (2006), "Wetting Properties of the Multiscaled Nanostructured Polymer and Metallic Superhydrophobic Surfaces," *Langmuir* **22**, 9982–9985.

Brown, P. S. and Bhushan, B. (2015), "Mechanically Durable, Superoleophobic Coatings Prepared by Layer-by-Layer Technique for Anti-smudge and Oil-water Separation," *Sci. Rep. – Nature* **5**, 8701.

Brown, P. S. and Bhushan, B. (2016), "Durable, Superoleophobic Polymer-nanoparticle Composite Surfaces with Re-entrant Geometry via Solvent-induced Phase Transformation," *Sci. Rep. – Nature* **6**, 21048.

Brown, P. S. and Bhushan, B. (2017), "Liquid-impregnated Porous Polypropylene Surfaces for Liquid Repellency," *J. Colloid Interface Sci.* **487**, 437–443.

Cappella, B. and Bonaccurso, E. (2007), "Solvent-Assisted Nanolithography on Polystyrene Surfaces using the Atomic Force Microscope," *Nanotechnology* **18**, 155307.

Celia, E., Darmanin, T., de Givenchy, E. T., Amigoni, S., and Guittard, F. (2013), "Recent Advances in Designing Superhydrophobic Surfaces," *J. Colloid Interface Sci.* **402**, 1–18.

Chiou, N., Lu, C., Guan, J., Lee, L. J., and Epstein, A. J. (2007), "Growth and Alignment of Polyaniline Nanofibres with Superhydrophobic, Superhydrophilic and Other Properties," *Nature Nanotechnol.* **2**, 354–357.

Chong, M. A. S., Zheng, Y. B., Gao, H., and Tan, L. K. (2006), "Combinational Template-Assisted Fabrication of Hierarchically Ordered Nanowire Arrays on Substrates for Device Applications," *Appl. Phys. Lett.* **89**, 233104.

Cortese, B., Amone, S. D., Manca, M., Viola, I., Cingolani, R., and Gigli, G. (2008), "Superhydrophobicity Due to the Hierarchical Scale Roughness of PDMS Surfaces," *Langmuir* **24**, 2712–2718.

Coulson, S. R., Woodward, I., Badyal, J. P. S., Brewer, S. A., and Willis, C. (2000), "Super-Repellent Composite Fluoropolymer Surfaces," *J. Phys. Chem. B* **104**, 8836–8840.

Del Campo, A. and Greiner, C. (2007), "SU-8: a Photoresist for High-Aspect-Ratio and 3D Submicron Lithography," *J. Micromech. Microeng.* **17**, R81–R95.

Ebert, D. and Bhushan, B. (2012a), "Durable Lotus-Effect Surfaces with Hierarchical Structure Using Micro- and Nanosized Hydrophobic Silica Particles," *J. Colloid Interface Sci.* **368**, 584–591.

Ebert, D. and Bhushan, B. (2012b), "Transparent, Superhydrophobic, and Wear-Resistant Coatings on Glass and Polymer Substrates using SiO_2, ZnO, and ITO Nanoparticles," *Langmuir* **28**, 11391–11399.

Ebert, D. and Bhushan, B. (2016), "Transparent, Superhydrophobic, and Wear-Resistant Surfaces Using Deep Reactive Ion Etching on PDMS Substrates," *J. Colloid Interface. Sci.* **481,** 82–90.

Esmeryan, K. D. and Fedchenko, Y. I. (2023), "Superhydrophobic Soot as a Mechanically Robust, Heat Insulating and Anti-corrosion Protective Coating for Concrete Surfaces," *Colloids Surf. A: Physicochem. Eng. Asp.* **672**, 131723.

Esmeryan, K. D., Fedchenko, Y. I., Gyoshev, S. D., Lazarov, Y., Chaushev, T. A., and Grakov, T. (2022), "On the Development of Ultradurable Extremely Water-Repellant and Oleophobic Soot-based Fabrics with Direct Relevance to Sperm Cryopreservation," *ACS Appl. Bio Mater.* **5**, 3519–3529.

Feng, X. J., Feng, L., Jin, M. H., Zhai, J., Jiang, L., Zhu, D. B. (2004), "Reversible Superhydrophobicity to Super-hydrophilicity Transition of Aligned ZnO Nanorod Films," *J. Am. Chem. Soc.* **126**, 62–63.

Fürstner, R., Barthlott, W., Neinhuis, C., and Walzel, P. (2005), "Wetting and Self-Cleaning Properties of Artificial Superhydrophobic Surfaces," *Langmuir* **21**, 956–961.

Geraldi, N. R., Dodd, L. E., Xu, B. B., Wood, D., Wells, G. G., McHale, G., and Newton, M. I. (2018), "Bioinspired Nanoparticles Spray-Coating for Superhydrophobic Flexible Materials with Oil/Water Separation Capabilities," *Bioinspir. Biomim.* **13**, 024001.

Han, J. T., Jang, Y., Lee, D. Y., Park, J. H., Song, S. H., Ban, D. Y., and Cho, K. (2005), "Fabrication of a Bionic Superhydrophobic Metal Surface by Sulfur-Induced Morphological Development," *J. Mater. Chem.* **15**, 3089–3092.

Hikita, M., Tanaka, K., Nakamura, T., Kajiyama, T., and Takahara, A. (2005), "Superliquid-Repellent Surfaces Prepared by Colloidal Silica Nanoparticles Covered with Fluoroalkyl Groups," *Langmuir* **21**, 7299–7302.

Hosono, E., Fujihara, S., Honma, I., and Zhou, H. (2005), "Superhydrophobic Perpendicular Nanopin Film by the Bottom-Up Process," *J. Am. Chem. Soc.* **127**, 13458–13459.

Huang, L., Lau, S. P., Yang, H. Y., Leong, E. S. P., and Yu, S. F. (2005), "Stable Superhydrophobic Surface via Carbon Nanotubes Coated with a ZnO Thin Film," *J. Phys. Chem.* **109**, 7746–7748.

Jansen, H., de Boer, M., Legtenberg, R., and Elwenspoek, M. (1995), "The Black Silicon Method: a Universal Method for Determining the Parameter Setting of a Fluorine-Based Reactive Ion Etcher in Deep Silicon Trench Etching with Profile Control," *J. Micromech. Microeng.* **5**, 115–120.

Jung, Y. C. and Bhushan, B. (2006), "Contact Angle, Adhesion, and Friction Properties of Micro- and Nanopatterned Polymers for Superhydrophobicity," *Nanotechnology* **17**, 4970–4980.

Jung, Y. C. and Bhushan, B. (2009), "Mechanically Durable CNT-Composite Hierarchical Structures with Superhydrophobicity, Self-Cleaning, and Low-Drag," *ACS Nano* **3**, 4155–4163.

Kim, D., Hwang, W., Park, H. C., and Lee, K. H. (2007), "Superhydrophobic Micro- and Nanostructures Based on Polymer Sticking," *Key Eng. Mat.* **334-335**, 897–900.

Koch, K., Bhushan, B., Jung, Y. C., and Barthlott, W. (2009), "Fabrication of Artificial Lotus Leaves and Significance of Hierarchical Structure for Superhydrophobicity and Low Adhesion," *Soft Matter* **5**, 1386–1393.

Krupenkin, T. N., Taylor, J. A., Wang, E. N., Kolodner, P., Hodes, M., and Salamon, T. R. (2007), "Reversible Wetting-Dewetting Transitions on Delectrically Tunable Superhydrophobic Nanostructured Surfaces," *Langmuir* **23**, 9128–9133.

Kuan, C. Y., Hon, M. H., Chou, J. M., and Leu, I. C. (2009), "Wetting Characteristics on Micro/Nanostructured Zinc Oxide Coatings," *J. Electrochem. Soc.* **156**, J32–J36.

Lau, K. K. S., Bico, J., Teo, K. B. K., Chhowalla, M., Amaratunga, G. A. J., Milne, W. I., McKinley, G. H., and Gleason, K. K. (2003), "Superhydrophobic Carbon Nanotube Forests," *Nano Lett.* **3**, 1701–1705.

Lee, H. and Bhushan, B. (2012), "Fabrication and Characterization of Hierarchical Nanostructured Smart Adhesion Surfaces," *J. Colloid Interface Sci.* **372**, 231–238.

Li, F., Kong, W., Bhushan, B., Zhao, X., and Pan, Y. (2019), "Ultraviolet-Driven Switchable Superliquiphobic/superliquiphilic Coating for Separation of Oil-water Mixtures and Emulsions and Water Purification," *J Colloid and Interface Sci.* **557**, 395–407.

Ma, M., Hill, R. M., Lowery, J. L., Fridrikh, S. V., and Rutledge, G. C. (2005), "Electrospun Poly(styrene-block-dimethylsiloxane) Block Copolymer Fibers Exhibiting Superhydrophobicity," *Langmuir* **21**, 5549–5554.

Madou, M. J. (2011), *Fundamentals of Microfabrication and Technology: Manufacturing Techniques for Microfabrication and Nanotechnology*, Vol. 2, third ed., CRC Press, Boca Raton, FL.

Martin, C., Rius, G., Borrise, X., and Perez-Murano, F. (2005), "Nanolithography on Thin Layers of PMMA using Atomic Force Microscopy," *Nanotechnology* **16**, 1016–1022.

Martin, S. and Bhushan, B. (2017), "Transparent, Wear-Resistant, Superhydrophobic and Superoleophobic Poly(dimethylsiloxane) (PDMS) Surfaces," *J. Colloid Interface Sci.* **488**, 118–126.

Martin, S., Brown, P. S., and Bhushan, B. (2017), "Fabrication Techniques for Bioinspired, Mechanically-durable, Superliquiphobic Surfaces for Water, Oil, and Surfactant Repellency," *Adv. Colloid Interf. Sci.* **241**, 1–23.

Martines, E., Seunarine, K., Morgan, H., Gadegaard, N., Wilkinson, C. D. W., and Riehle, M. O. (2005), "Superhydrophobicity and Superhydrophilicity of Regular Nanopatterns," *Nano Lett.* **5**, 2097–2103.

Ming, W., Wu, D., van Benthem, R., and de With, G. (2005), "Superhydrophobic Films from Raspberry-Like Particles," *Nano Lett.* **5**, 2298–2301.

Multanen, V. and Bhushan, B. (2019), "Bioinspired Self-healing, Superliquiphobic and Self-cleaning Hydrogel- coated Surfaces with High Durability," *Phil. Trans. R. Soc. A* **377**, 20190117.

Nakajima, A., Fujishima, A., Hashimoto, K., and Watanabe, T. (1999), "Preparation of Transparent Superhydrophobic Boehmite and Silica Films by Sublimation of Aluminum Acetylacetonate," *Adv. Mater.* **11**, 1365–1368.

Northen, M. T. and Turner, K. L. (2005), "A Batch Fabricated Biomimetic Dry Adhesive," *Nanotechnology* **16**, 1159–1166.

Pan, Y., Kong, W., Bhushan, B., and Zhao, X. (2019), "Rapid, Ultraviolet-induced, Reversibly Switchable Wettability of Superhydrophobic/Superhydrophilic Surfaces," *Beilstein J. Nanotechnology* **10**, 866–873.

Qian, B. and Shen, Z. (2005), "Fabrication of Superhydrophobic Surfaces by Dislocation-Selective Chemical Etching on Aluminum, Copper, and Zinc Substrates," *Langmuir* **21**, 9007–9009.

Shi, F., Song, Y., Niu, J., Xia, X., Wang, Z., and Zhang, X. (2006), "Facile Method To Fabricate a Large-Scale Superhydrophobic Surface by Galvanic Cell Reaction," *Chem. Mater.* **18**, 1365–1368.

Shirtcliffe, N. J., McHale, G., Newton, M. I., Chabrol, G., Perry, C. C. (2004). "Dual-Scale Roughness Produces Unusually Water-Repellent Surfaces," *Adv. Mater.* **16**, 1929–1932.

Shirtcliffe, N. J., McHale, G., Newton, M. I., Perry, C. C., and Roach, P. (2005), "Porous Materials Show Superhydrophobic to Superhydrophilic Switching," *Chem. Commun.* 3135–3137.

Shiu, J., Kuo, C., Chen, P., and Mou, C. (2004), "Fabrication of Tunable Superhydrophobic Surfaces by Nanosphere Lithography," *Chem. Mater.* **16**, 561–564.

Teshima, K., Sugimura, H., Inoue, Y., Takai, O., and Takano, A. (2005), "Transparent Ultra Water-repellent Poly(ethylene terephthalate) Substrates Fabricated by Oxygen Plasma Treatment and Subsequent Hydrophobic Coating," *Appl. Surface Sci.* **244**, 619–622.

Wang, Y., Zhu, Q., and Zhang, H. (2006), "Fabrication and Magnetic Properties of Hierarchical Porous Hollow Nickel Microspheres," *J. Mater. Chem.* **16**, 1212–1214.

Wu, X., Zheng, L., and Wu, D. (2005), "Fabrication of Superhydrophobic Surfaces from Microstructured ZnO-Based Surfaces via a Wet-Chemical Route," *Langmuir* **21**, 2665–2667.

Zhang, X., Feng, S., Yu, X., Liu, H., Fu, Y., Wang, Z., Jiang, L., and Li, X. (2004), "Polyelectrolyte Multilayer as Matrix for Electrochemical Deposition of Gold Clusters: Toward Super-Hydrophobic Surface," *J. Am. Chem. Soc.* **126**, 3064–3065.

Zhao, Y., Tong, T., Delzeit, L., Kashani, A., Meyyappan, M., and Majumdar, A. (2006), "Interfacial Energy and Strength of Multiwalled-Carbon-Nanotube-Based Dry Adhesive," *J. Vac. Sci. Technol. B* **24**, 331–335.

Zhu, L., Xiu, Y., Xu, J., Tamirisa, P. A., Hess, D. W., and Wong C. (2005), "Superhydrophobicity on Two-Tier Rough Surfaces Fabricated by Controlled Growth of Aligned Carbon Nanotube Arrays Coated with Fluorocarbon," *Langmuir* **21**, 11208–11212.

Strategies for Micropatterned, Nanopatterned, and Hierarchically Structured Lotus-like Surfaces

6

6.1 Introduction

It has been demonstrated experimentally that roughness changes contact angle (CA) in accordance with the Wenzel model or the Cassie-Baxter model, depending upon whether the surface is hydrophilic or hydrophobic (Bhushan and Jung 2011). Yost et al. (1995) found that roughness enhances wetting of a copper surface with Sn–Pb eutectic solder, which has a contact angle of 15°–20° for a smooth surface. Shibuichi et al. (1996) measured the contact angle of various liquids (mixtures of water and 1,4-dioxane) on alkylketene dimmer (AKD) substrate (contact angle not larger than 109° for a smooth surface). They found that with increasing roughness, the contact angle decreases for wetting liquids and increases for non-wetting liquids. Semal et al. (1999) investigated the effect of surface roughness on contact angle hysteresis (CAH) by studying a sessile droplet of squalane spreading dynamically on multilayer substrates (behenic acid on glass). They found that an increase in microroughness slows the rate of droplet spread. Erbil et al. (2003) measured the contact angle of polypropylene (contact angle of 104° for smooth surface) and found that the contact angle increases with increasing roughness. Burton and Bhushan (2005) and Jung and Bhushan (2006) measured the contact angle of various micro- and nanopatterned polymer surfaces with hydrophilicity and hydrophobicity. They found that for an increasing roughness factor, contact angle decreases for hydrophilic surfaces and increases for hydrophobic surfaces. Nanopatterned surfaces benefit from air pocket formation. Jung and Bhushan (2008c) studied the effect of submicron droplets on contact angle.

The contact angle on selected patterned surfaces has been measured to understand the role of pitch value on contact angle, as well as on the transition between the Cassie-Baxter and Wenzel regimes (Bhushan and Jung 2007, 2008; Jung and Bhushan 2007, 2008a, b). Evaporation studies are useful in characterizing the role of droplet size. During

B. Bhushan, *Introduction to Biomimetics and Bioinspiration*,
https://doi.org/10.1007/978-3-031-62344-8_6

evaporation, droplets with decreasing sizes exist and are used to evaluate the transition criterion on a given patterned surface (Bourges-Monnier and Shanahan 1995; Rowan et al. 1995; Erbil et al. 2002; McHale et al. 2005; Jung and Bhushan 2007, 2008a; Nosonovsky and Bhushan 2007a, c, 2008b, c, d; Bhushan et al. 2008a, 2009b, c, d). It was found that the wetting state changes from the Cassie-Baxter to the Wenzel state as the droplet becomes smaller than a critical value, and the pitch value becomes larger than a critical value on patterned surfaces.

Another important phenomenon related to wetting behavior is the bouncing of droplets. When a droplet hits a surface, it can bounce, spread, or stick. In practical applications of superhydrophobic surfaces, surfaces should maintain their ability to repel penetrating droplets under dynamic conditions. Whether or not a droplet transitions between bouncing or remaining on the surface can be determined from the impact of a droplet at a critical velocity on a patterned surface with a given geometric parameter (Richard et al. 2002; Lafuma and Quéré 2003; Bartolo et al. 2006; Reyssat et al. 2006; Jung and Bhushan 2008b, 2009; Nosonovsky and Bhushan 2008e).

An environmental scanning electron microscope (ESEM) can be used to condense or evaporate water droplets on surfaces by adjusting the pressure of the water vapor in the specimen chamber and the temperature of the cooling stage. Transfer of the water droplet has been achieved by using a specially designed micro-injector device on wool fibers and then imaging them at room temperature in the ESEM (Danilatos and Brancik 1986). Images of water droplets show strong topographic contrast in ESEM such that reliable contact angle measurements can be made on the surfaces (Stelmashenko et al. 2001). Water condensation and evaporation studies on patterned surfaces have been carried out by Jung and Bhushan (2008a). They measured changes in static contact angle and contact angle hysteresis as related to surface roughness.

Bhushan and Jung (2007, 2008), Jung and Bhushan (2006, 2007, 2008a, b), Bhushan et al. (2007, 2008a, b, 2009b, c, d), and Koch et al. (2009) used nature's route to fabricate hierarchical structures. These were created by replication of micropatterns and by self-assembly of hydrophobic alkanes and plant waxes. They studied the static contact angle, contact angle hysteresis, tilt angle, air pocket formation, and adhesive force, as well as efficiency of self-cleaning. Hierarchical structures exhibited superhydrophobicity with static contact angles of about 171°, low contact angle hysteresis of about 2°, and self-cleaning properties comparable to that of the lotus (*Nelumbo nucifera*) leaf. This verified their understanding of the wetting mechanisms of the lotus leaf. Jung and Bhushan (2008b, 2009) performed bouncing and vibrating droplet experiments to study the effect of impact velocity and vibration amplitude on the transition from the composite solid-air-liquid interface to the homogeneous solid-liquid interface.

In this chapter, characterization of various micropatterned, nanopatterned, and hierarchically structured surfaces is provided to verify theoretical models, to understand the transition between the Wenzel and Cassie-Baxter wetting regimes, and to understand the role of contact angle hysteresis and hierarchical roughness. Details on fabricated surfaces using nature's route are presented which was used to verify understanding of the lotus effect.

6.2 Experimental Techniques

6.2.1 Contact Angle, Surface Roughness, and Adhesion

Static and dynamic (advancing and receding) contact angles were measured using a Rame-Hart model 100 contact angle goniometer and droplets of DI water (Burton and Bhushan 2005; Bhushan and Jung 2007, 2008; Jung and Bhushan 2006, 2007; Bhushan et al. 2008a, b, 2009c, d; Koch et al. 2009). For measurement of the static contact angle, the droplet size should be smaller than the capillary length but larger than the dimension of the structures present on the surfaces. Droplets of about 5 μL in volume (with the diameter of a spherical droplet about 2.1 mm) were deposited on the substrate using a microsyringe for measurement of the static contact angle. The advancing and receding contact angles were measured by the addition and removal of water from a DI water sessile droplet using a microsyringe. The contact angle hysteresis was calculated as the difference between the measured advancing and receding contact angles, and the tilt angle was measured by using a tilting stage (Bhushan and Jung 2007, 2008; Bhushan et al. 2008a, b, 2009c, d; Koch et al. 2009). All measurements were made at five different points for each sample at 22 ± 1 °C and $50 \pm 5\%$ RH. The measurements were reproducible to within $\pm 3°$.

For surface roughness measurement, an optical profiler was used for different surface structures (Burton and Bhushan 2006; Bhushan and Jung 2006, 2007, 2008). The optical profiler has an advantage due to its greater Z-range (2 mm) over an AFM (Z-range about 7 μm), but it has a maximum lateral resolution of only approximately 0.6 μm (Bhushan 2013a, b, 2017a, b).

Adhesive force was measured with an AFM using a force distance curve approach (Bhushan 2017b). Experiments were performed using three different radii tips to study the effect of scale dependence. A borosilicate ball with 15 μm radius and a silica ball with 3.8 μm radius were mounted on a gold-coated triangular Si_3N_4 cantilever with a nominal spring constant of 0.58 N m^{-1}. A square pyramidal Si_3N_4 tip with a nominal radius of 30–50 nm on a triangular Si_3N_4 cantilever with a nominal spring constant of 0.58 N m^{-1} was used for the smaller radius tip.

6.2.2 Droplet Evaporation Studies

Droplet evaporation was observed and recorded by a digital camcorder with a 10× optical and 120× digital zoom for every run of the experiment. Then the decrease in the diameter of the droplets with time was determined (Jung and Bhushan 2007, 2008a; Bhushan et al. 2008a, 2009c). The frame speed of the camcorder was 0.03 s per frame. An objective lens placed in front of the camcorder during recording gave a total magnification of 10–20×. Droplet diameter as small as a few hundred microns could be measured with this method. Droplets were deposited on the substrate using a microsyringe, and the whole process of evaporation was recorded. Images obtained were analyzed using Imagetool® software (University of Texas Health Science Center) for the contact angle. To find the dust trace remaining after droplet evaporation, an optical microscope with a charge-coupled device (CCD) camera was used. All measurements were made in a controlled environment at 22 ± 1 °C and 45 ± 5% RH (Jung and Bhushan 2007, 2008a; Bhushan et al. 2008a, 2009c, d).

6.2.3 Bouncing Droplet Studies

The process of dynamic impact was recorded by a high speed camera (Kodak Ektapro HS Motion Analyzer, Model 4540) operated at 500 frames/s for each experimental run and then measuring the dynamic impact behavior of the droplet as a function of time. The impact velocity was calculated by varying the droplet release height. Droplets of about 5 μL in volume with a diameter of a spherical droplet of about 2.1 mm were used. All measurements were made in a controlled environment at 22 ± 1 °C and 45 ± 5% RH (Jung and Bhushan 2008b).

6.2.4 Vibrating Droplet Studies

The process of dynamic behavior was obtained by a system producing vertical vibrations (Jung and Bhushan 2009). The system consists of an electrodynamic shaker (Labworks Inc. Model ET-126A/B) connected with a signal generator and power amplifier, a digital camcorder with an objective lens, and a lamp for a light source. The specimen was placed on the top of the shaker, and a droplet was deposited using a microsyringe. Droplets of about 5 μL in volume with a diameter of a spherical droplet of about 2.1 mm were used. The vibration frequency was controlled between 0 and 300 Hz at 0.4 mm amplitude for measurement of the resonance frequency of a droplet. For wetting behavior of a droplet on the surface, a frequency of 30 Hz, which was less than the resonance frequency, was chosen, and the vibration amplitude was controlled between 0 and 3 mm. The vibration time applied to the droplet was 1 min for each experiment. All measurements were made in a controlled environment at 22 ± 1 °C and 45 ± 5% RH (Jung and Bhushan 2009).

6.2.5 Microdroplet Condensation and Evaporation Studies Using ESEM

A Philips XL30 ESEM equipped with a Peltier cooling stage was used to study smaller droplets (Jung and Bhushan 2008a). ESEM uses a gaseous secondary electron detector (GSED) for imaging. The ESEM column is equipped with a multistage differential pressure-pumping unit. The pressure in the upper part is about 10^{-6}–10^{-7} Torr, but a pressure of about 1–15 Torr can be maintained in the observation chamber. When the electron beam (primary electrons) ejects secondary electrons from the surface of the sample, the secondary electrons collide with gas molecules in the ESEM chamber, which in turn acts as a cascade amplifier, delivering the secondary electron signal to the positively biased GSED. The positively charged ions are attracted to the specimen to neutralize the negative charge produced by the electron beam. Therefore, the ESEM can be used to examine electrically isolated specimens in their natural state. In the ESEM, adjusting the pressure of the water vapor in the specimen chamber and the temperature of the cooling stage allows the water to condense on the sample in the chamber. For the measurement of the static and dynamic contact angles on patterned surfaces, video images were recorded. The voltage of the electron beam was 15 kV, and the distance of the specimen from the final aperture was about 8 mm. If the angle of observation is not parallel to the surface, the electron beam is not parallel to the surface but inclined at an angle, this will produce a distortion in the projection of the droplet profile. A mathematical model to calculate the real contact angle from the ESEM images was used to correct the tilting of the surfaces during imaging (Brugnara et al. 2006; Jung and Bhushan 2008a).

6.2.6 Generation of Submicron Droplets

In order to generate submicron droplets, Jung and Bhushan (2008c) developed an AFM-based technique using a modified nanoscale dispensing (NADIS) probe as shown in Fig. 6.1. The NADIS probe was fabricated by modifying a commercially available silicon nitride (Si_3N_4) cantilever (Olympus OMCL-RC800) with lengths of 100 and 200 μm, spring constants of 0.8 and 0.1 N/m, and resonance frequencies of 68.94 and 122.02 kHz, respectively (Swiss Center for Electronics and Microtechnology). The probe consisted of a loading area (30 μm diameter) for the liquid on the upper side of the cantilever. The loading area was produced by removing the material locally in and around the tip with a reflective gold layer using focused ion beam milling. The remaining gold was made hydrophobic using hexadecanethiol (in liquid phase) whereas the bare silicon nitride in the milling area remained hydrophilic. The hydrophilic–hydrophobic transition prevents spreading of the loaded liquid over the entire cantilever.

A droplet of volume V is deposited on the surface. Figure 6.2 shows an idealized spherical capped droplet. Based on the thickness of the droplet h and contact diameter

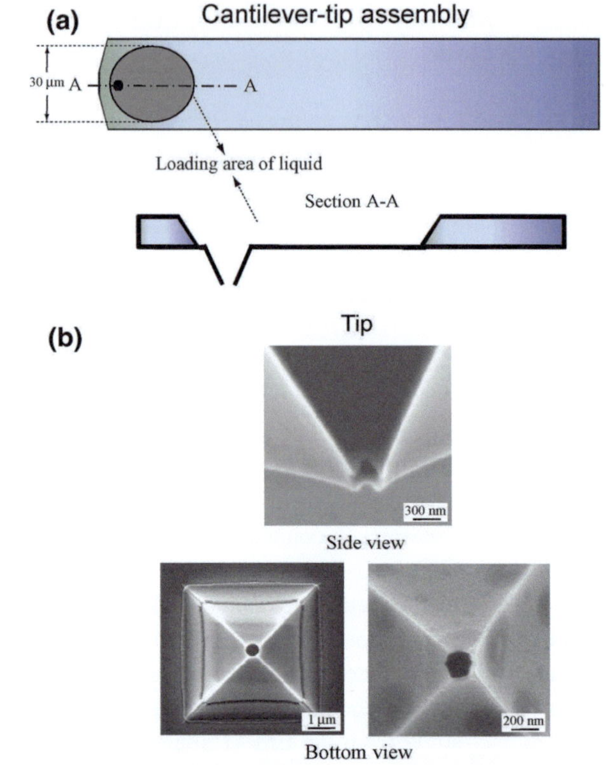

Fig. 6.1 **a** Schematic of modified nanoscale dispensing (NADIS) probe for generation of submicron size droplets. The loaded liquid is limited to the loading area (30 μm diameter circle), **b** scanning electron micrograph of tip in side view and bottom view with different aperture sizes (500 and 200 nm) at its apex (Jung and Bhushan 2008c)

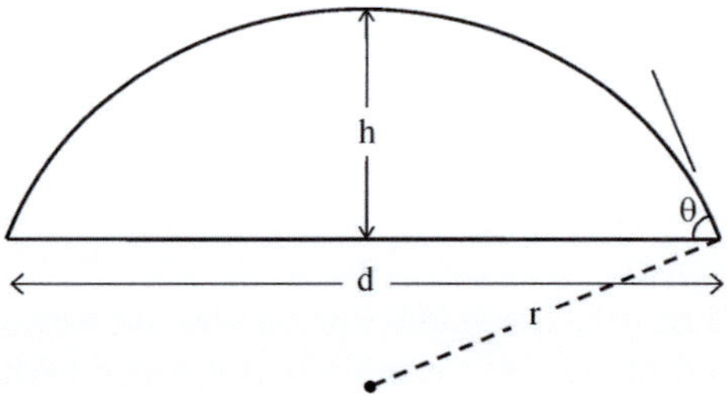

Fig. 6.2 Droplet of liquid in contact with a surface, contact angle θ. The thickness of droplet is h. The contact diameter between droplet and surface is d. The radius of curvature of droplet is r (Jung and Bhushan 2008c)

d, the contact angle is obtained by the following equation for a simple spherical capped geometry of droplet (Jung and Bhushan 2008c).

$$\theta = \sin^{-1}\left(\frac{3d}{2\left(\frac{3V}{\pi h^2} + h\right)}\right) \tag{6.1}$$

For calculation of the contact angle of the droplet, Jung and Bhushan (2008c) use the following three steps:

(1) For the measurement of the volume of a droplet deposited on the surface, the change in resonance frequency of the cantilever before and after depositing the droplet on the surface is measured. The resonance frequency of the cantilever is measured by performing a frequency sweep of the voltage driven oscillations by a thermal tune method (Palacio and Bhushan 2010).
(2) For the measurement of the thickness of a droplet deposited on the surface, the distance between the tip snap-in and the position where the tip is in contact with the surface is measured in the force calibration mode.
(3) For the measurement of the contact diameter between a droplet and surface, the image of the droplet after evaporation is measured using an Si tip.

For measurement of the thickness of a droplet deposited on the surface, a force distance curve was used (Bhushan 2013a, b, 2017b). The droplet was deposited in the first approach. The force distance curve was obtained during a second approach to measure the thickness of the droplet (Bhushan and Blackman 1991; Chen and Bhushan 2006; Lodge and Bhushan 2006). The cantilever deflection is plotted on the vertical axis against the Z-position of the piezo scanner in a force distance curve as shown in Fig. 6.3. The measurement starts at a large separation (point A), where there is no deflection of the cantilever. As the piezo moves to the sample, a sudden mechanical instability occurs between point B and point C, and the droplet jumps into contact with the tip and wicks up around the tip to form a meniscus. The cantilever bends downward because of the attractive meniscus force acting on the tip. As the piezo further approaches the surface, the deflection of the cantilever increases while the tip travels the thickness of the droplet and eventually contacts the underlying surface at point D, and the cantilever starts to bend upward. Once the piezo reaches the end of its designated ramp size at point E, it is retracted to its starting position. The tip goes beyond zero deflection and enters the adhesion region. At point F, the elastic force of the cantilever becomes equivalent to the adhesive force, causing the cantilever to snap back to point G. As the tip travels in the liquid, it is deflected as well. The tip deflection occurs in the same direction as the piezo travels for the AFM used in this study. The liquid film thickness (h) is the sum of the travel distance of the piezo (described as h_1 in Fig. 6.3) and the deflection of the cantilever (described as h_2 in Fig. 6.3). Though previous studies show that h overestimates the actual liquid film thickness, it still provides a good measurement of the thickness of the droplet (Bhushan and Blackman 1991; Chen and Bhushan 2006; Lodge and Bhushan 2006).

The resolution of volume (mass) was about 1×10^{-4} μm^3 (0.12 fg). This resolution of volume was calculated with the measured data of the shift in the resonance frequency of the cantilever from 122.01 to 122.02 kHz during the evaporation time of 10 min.

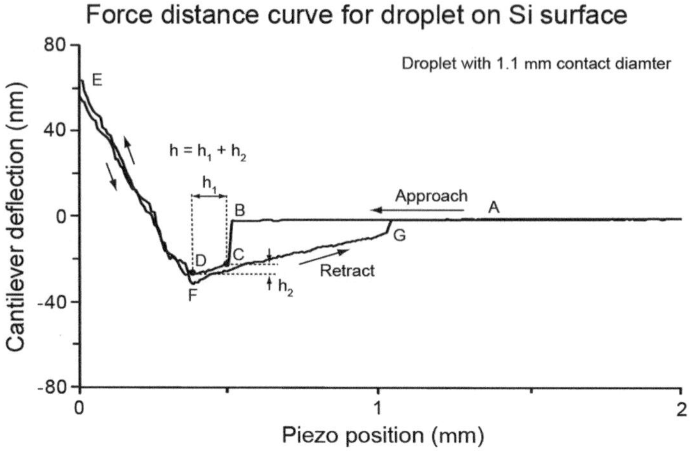

Fig. 6.3 Force distance curve for a droplet with 1.1 μm contact diameter on Si surface. The *h* is a measure of droplet thickness on the surface

The thickness resolutions was about 0.1 nm. This was measured from the calibration data of the *z* piezo by an AFM. The resolution of the contact diameter was<1 nm. This was measured from the calibration data of the x–y piezo by an AFM. The accuracy of volume, thickness, and contact diameter measurements was about ±10%, 10% and<1 nm, respectively.

6.2.7 Self-cleaning Studies

The self-cleaning characteristics were examined by first contaminating the sample by using silicon carbide particles and then comparing the removal of particles by water droplets before and after contamination (Bhushan et al. 2009a). Silicon carbide (SiC) particles in two different size ranges of 1–10 μm and 10–15 μm were used as the contaminants. The SiC particles were chosen because of their similarity in shape, sizes, and hydrophilicity to natural dirt contaminants. A divided glass contamination chamber shown in Fig. 6.4a was used for contamination studies. Samples were mounted in the bottom chamber. The particles were made airborne in the top chamber. After dispersion, the separator plate was removed to contaminate the samples mounted on a 45° stage for 30 min.

Next, the contaminated specimens were cleaned by using water droplets of approximately 2 mm diameter, using two microsyringes, Fig. 6.4b. The tilt angle at which samples were mounted was chosen to be slightly larger than the tilt angle for the samples. The water cleaning test was carried out for 2 min (water quantity, 10 mL) with nearly zero kinetic energy of droplets. For cleaning with nearly zero kinetic energy, the distance

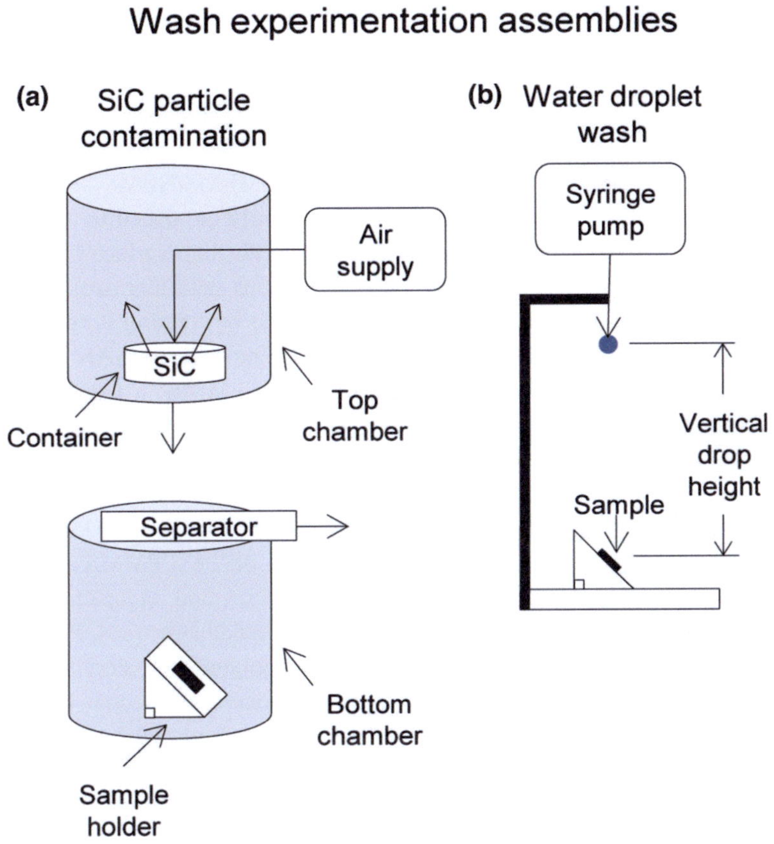

Fig. 6.4 **a** Schematic of particle contamination chamber to uniformly contaminate samples for wash experimentation. Silicon carbide (SiC) particles are dispersed with high pressure air flow into the top chamber. The separator door is removed and particles settle below onto the replica samples. **b** Schematic of wash experimentation apparatus. Water droplets at known velocities and flow rates impact the contaminated sample. Particle analysis is conducted to quantify self-cleaning efficiency

between the microsyringes and surface was set to 0.005 m (nearly zero impact velocity). The chosen impact velocity represents a low value compared to a natural rain shower, where a water droplet of 2 mm diameter can reach an impact velocity of about 6 m/s (measured under controlled conditions) (van Dijk et al. 2002). The number of particles per area was determined using automated image analysis of photographs taken by an optical microscope before and after water cleaning.

6.3 Micro- and Nanopatterned Polymers

To investigate the effects of microstructure and nanostructure on contact angle and adhesion, Jung and Bhushan (2006) studied micro- and nanopatterned surfaces made with two types of polymers: poly(methyl methacrylate) (PMMA) and polystyrene (PS). PMMA and PS were chosen because they are widely used in MEMS/NEMS devices (Bhushan 2017a). Both hydrophilic and hydrophobic surfaces can be produced by using these two polymers, as PMMA has polar (hydrophilic) groups with high surface energy while PS has electrically neutral and nonpolar(hydrophobic) groups with low surface energy. Furthermore, a PMMA structure can be made hydrophobic by treating it appropriately, for example by coating with a hydrophobic self-assembled monolayer (SAM).

Four types of surface patterns were fabricated from PMMA: a flat film, low aspect ratio asperities (LAR, 1:1 height-to-diameter ratio), high aspect ratio asperities (HAR, 3:1 height-to-diameter ratio), and a replica of the lotus leaf (the lotus pattern). Two types of surface patterns were fabricated from PS: a flat film and the lotus pattern. Figure 6.5 shows SEM images of the two types of nanopatterned structures, LAR and HAR, and the one type of micropatterned structure, lotus pattern, all on a PMMA surface (Burton and Bhushan 2005; Jung and Bhushan 2006). Both micro- and nanopatterned structures were manufactured using soft lithography. For nanopatterned structures, PMMA film was spin-coated on a silicon wafer. A UV-cured mold of polyurethane acrylate (PUA) resin with nanopatterns of interest was made which enables one to create sub-100-nm patterns with a high aspect ratio (Choi et al. 2004). The mold was placed on the PMMA film, and a slight pressure of ~10 g/cm^2 (~1 kPa) was applied and annealed at 120 °C. Finally, the PUA mold was removed from the PMMA film. For micropatterned structures, a polydimethylsiloxane (PDMS) mold was first made by casting PDMS against a lotus leaf after the wax had been removed (left with only microstructure), followed by heating. Then, the mold was placed on the PMMA and PS film to create a positive replica of a lotus leaf. As shown in Fig. 6.5, it can be seen that only microstructures exist on the surface of the lotus pattern (Jung and Bhushan 2006).

Since PMMA by itself is hydrophilic, in order to obtain a hydrophobic sample, a self-assembled monolayer (SAM) of perfluorodecyltriethoxysilane (PFDTES) was deposited on the sample surfaces using a vapor phase deposition technique. PFDTES was chosen because of its hydrophobic nature. The deposition conditions for PFDTES were 100 °C temperature, 400 Torr pressure, 20 min deposition time, and 20 min annealing time. The polymer surface was exposed to an oxygen plasma treatment (40 W, O_2 187 Torr, 10 s) prior to coating (Bhushan et al. 2006). The oxygen plasma treatment is necessary to oxidize any organic contaminants on the polymer surface and to alter the surface chemistry to allow for enhanced bonding between the SAM and the polymer surface.

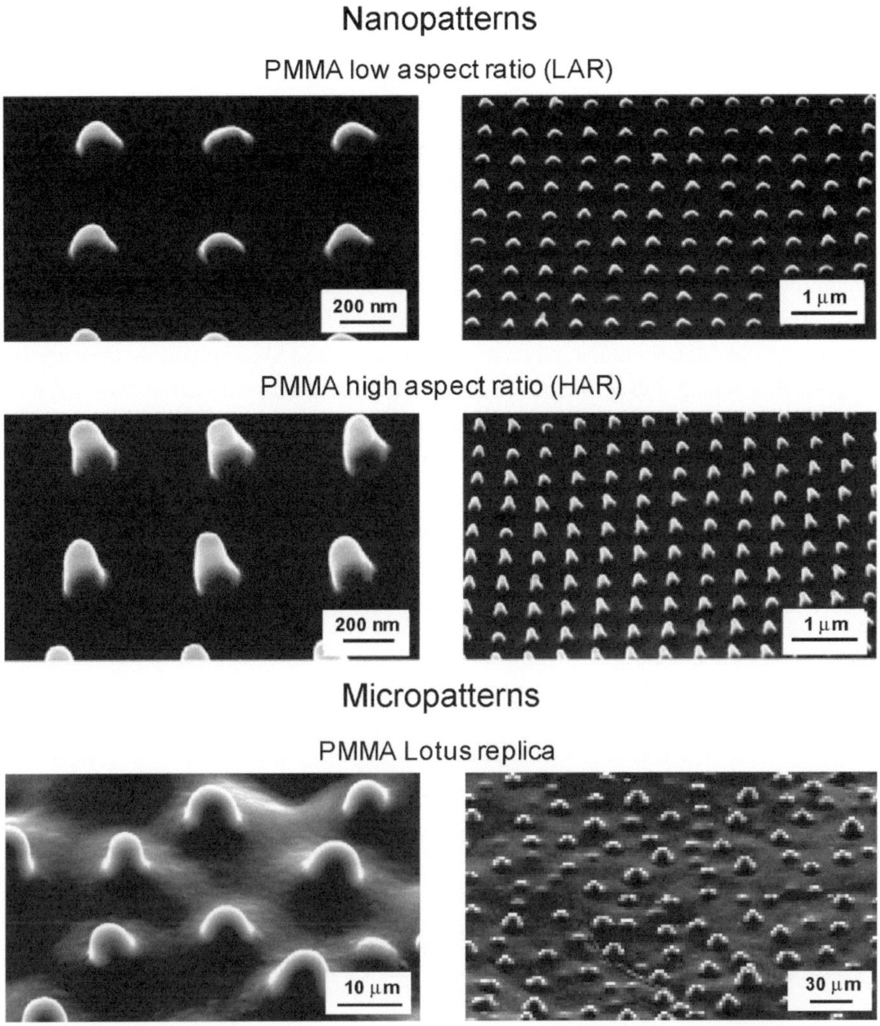

Fig. 6.5 Scanning electron micrographs of the two nanopatterned polymer surfaces (shown using two magnifications to see both the asperity shape and the asperity pattern on the surface) and the micropatterned polymer surface (lotus pattern, which has only microstructures on the surface) (Burton and Bhushan 2005; Jung and Bhushan 2006)

6.3.1 Contact Angle

Jung and Bhushan (2006) measured the static contact angle of water with the micro- and nanopatterned PMMA and PS structures; see Fig. 6.6. Since the Wenzel roughness factor is the parameter that often determines wetting behavior, the roughness factor was calculated, and it is presented in Table 6.1 for various samples. The data show that the contact angle of the hydrophilic materials decreases with an increase in the roughness factor, as predicted by the Wenzel model. When the polymers were coated with PFDTES, the film surface became hydrophobic. Figure 6.6 also shows the contact angle for various PMMA samples coated with PFDTES. For a hydrophobic surface, the standard Wenzel model predicts an increase of contact angle with roughness factor, which is what happens in the case of patterned samples. The calculated values of the contact angle for various micro- and nanopatterned samples based on the contact angle of the smooth film and Wenzel equation are also presented. The measured contact angle values for the lotus pattern were comparable with the calculated values, whereas for the LAR and HAR patterns, they are higher. It suggests that nanopatterns benefit from air pocket formation. Furthermore, pinning at the top of the nanopatterns may stabilize the droplet. For the PS material (hydrophobic), the contact angle of the lotus pattern also increased with increased roughness factor.

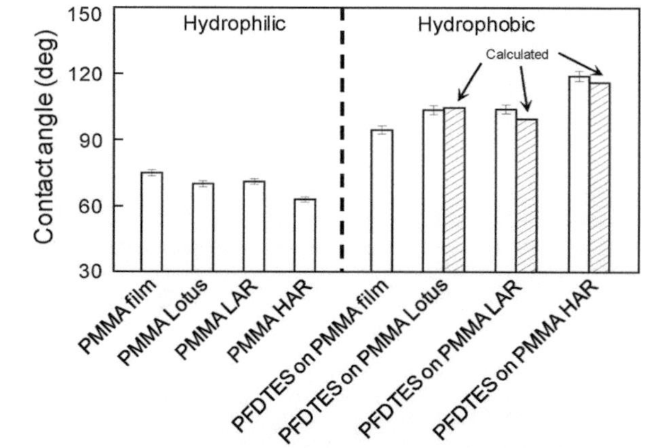

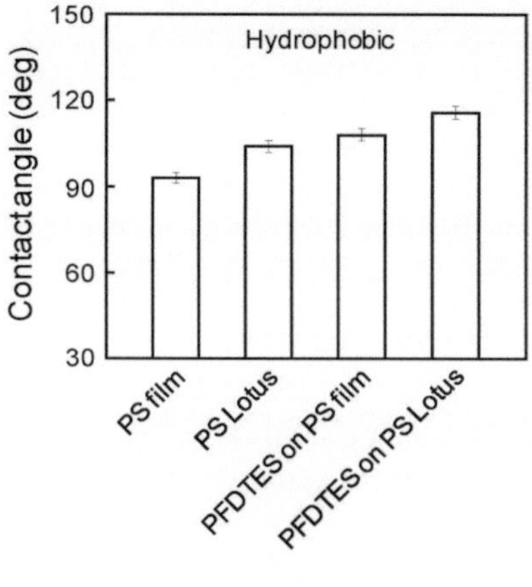

Fig. 6.6 Contact angles for various micro- and nanopatterned surfaces on PMMA and PS polymers and calculated values using Wenzel equation (Jung and Bhushan 2006)

Table 6.1 Roughness factor for micro- and nanopatterned polymers (Jung and Bhushan 2006)

	LAR	HAR	Lotus
R_f	2.1	5.6	3.2

6.3.2 Effect of Submicron Droplet on Contact Angle

Wetting phenomena have been studied and understood at the macroscale. However, micro- and nanoscale wetting mechanisms require further investigation. The actual contact angle under which the liquid-vapor interface comes in contact with the solid surface at the micro- and nanoscale is expected to be a function of the droplet size. Jung and Bhushan (2008c) measured the contact angle of micro- and nanodroplets on various surfaces using an AFM-based technique. The contact angle for different droplet sizes on various hydrophilic and hydrophobic, nanopatterned surfaces is summarized in Fig. 6.7. The data for the microdroplets with 2.4–8.1 μm diameter and nanodroplets with 0.22–1.1 μm diameter were compared with conventional contact angle measurements obtained with a droplet with 2.1 mm diameter (5 μL volume). The measured values of micro- and nanodroplets using an AFM were found to be lower than those of the macrodroplet (Pompe and Herminghaus 2000; Checco et al. 2003). There are several reasons for the scale dependence, such as the effect of contact line tension of a three-phase system (solid-liquid-vapor), which is the excess free energy of a solid-liquid-vapor system per unit length of the contact line (Pompe and Herminghaus 2000; Checco et al. 2003; Quere 2004; Nosonovsky and Bhushan 2007d). Another reason can be surface heterogeneity (Checco et al. 2003). For a thin fluid film present on a surface, the disjoining pressure of a film is repulsive, analogous to the repulsive van der Waals force across a film, and it causes a film to spread on surfaces. It decreases with the liquid layer thickness (Israelachvili 1992). This pressure may lead to a smaller contact angle at the nanoscale.

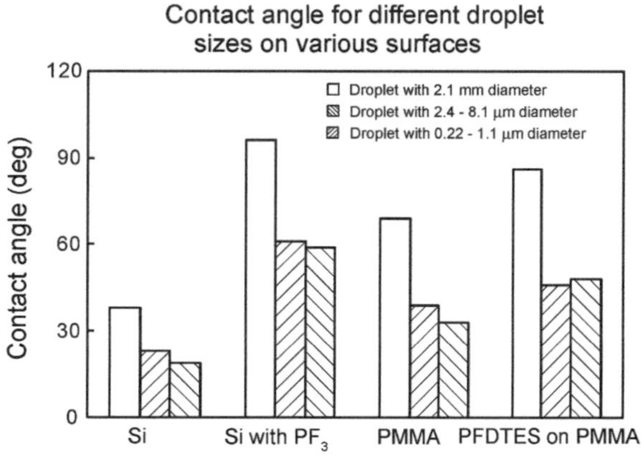

Fig. 6.7 Contact angle measurements for different droplet sizes on various nanopatterned surfaces (Jung and Bhushan 2008c)

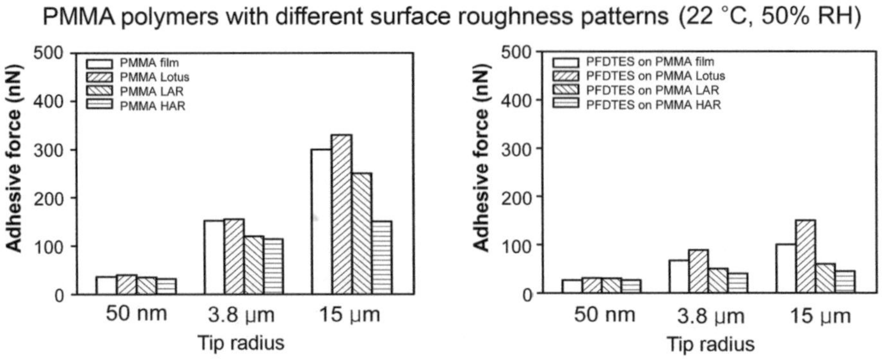

Fig. 6.8 Scale dependence of adhesive force for various micro- and nanopatterned surfaces measured using AFM tips of various radii (Jung and Bhushan 2006)

6.3.3 Adhesive Force

Adhesion force depends, among other factors, on the hydrophobic/hydrophilic nature of the surfaces, surface structure, and AFM tip radii. Figure 6.8 shows the scale dependence of adhesive force on tip radius for PMMA (hydrophilic) and PFDTES coated on PMMA (hydrophobic) surfaces with various micro- and nanopatterns (Jung and Bhushan 2006). The left bar chart in Fig. 6.8 is for hydrophilic PMMA film and lotus, LAR, and HAR patterns. For increasing radius, the adhesive force increases for each material. With a larger radius, the real area of contact and the meniscus contribution increase, resulting in increased adhesion. Adhesive force is the lowest for a nanopattern with the highest bump height. The right bar chart in Fig. 6.8 shows the results for hydrophobic PFDTES coated on each material. These samples show the same trends as the film samples, but the change in adhesion is not as dramatic. The hydrophobicity of PFDTES on material reduces meniscus forces, which in turn reduces adhesion from the surface. The dominant mechanism for the hydrophobic material is real area of contact and not meniscus force, whereas with hydrophilic material there is a combination of real area of contact and meniscus forces (Jung and Bhushan 2006).

6.3.4 Summary

The contact angle data on micro- and nanopatterned polymers show that on hydrophilic surfaces, contact angle decreases with roughness. On hydrophobic surfaces, contact angle increases with roughness, consistent with the Wenzel and Cassie-Baxter equations. The measured contact angles of nanopatterned samples are higher than the calculated values using the Wenzel equation. It suggests that nanopatterns benefit from air pocket formation.

The contact angle for micro- and nanodroplets on various hydrophilic and hydrophobic nanopatterned surfaces was found to be scale dependent.

Adhesive force decreases with an increase of the contact angle. The AFM tip radius affects the adhesion because the tip/sample interfaces area changes with tip size. The adhesive force increases with tip size for all samples because of an increased number of contacting asperities (real area of contact) and increased meniscus contribution.

6.4 Micropatterned Si Surfaces

It was reported earlier that the Cassie-Baxter regime is desirable to achieve high contact angle and low contact angle hysteresis. The relevant regime is dependent upon the micro/nanostructure and droplet radius. A criterion transition from the Cassie-Baxter to the Wenzel regime has been proposed. Using micropatterned surfaces, Jung and Bhushan (2007, 2008a) investigated the role of distance between micropillars (pitch) and droplet radius on the contact angle and contact angle hysteresis. The effect of droplet radius has been studied using evaporation studies. To provide insight into the formation of microdroplets on micropatterned surfaces, condensation and evaporation studies of microdroplets over patterned surfaces have been carried out using an ESEM. Jung and Bhushan (2008b) also studied the transition during bouncing droplet studies.

Micropatterned surfaces produced from single-crystal silicon (Si) by photolithography and coated with a SAM were used by Jung and Bhushan (2007, 2008a, b) in their study. Silicon has traditionally been the most commonly used structural material for micro/nanocomponents. A Si surface can be made hydrophobic by coating it with a SAM. One purpose of this investigation was to study the transition from the Cassie-Baxter to the Wenzel regime by changing the distance between the pillars. To create micropatterned Si, two series of nine samples each were fabricated using photolithography. Series 1 had 5-μm diameter and 10-μm height flat-top, cylindrical pillars with different pitch values (7, 7.5, 10, 12.5, 25, 37.5, 45, 60, and 75 μm), and series 2 had 14-μm diameter and 30-μm height flat-top, cylindrical pillars with different pitch values (21, 23, 26, 35, 70, 105, 126, 168, and 210 μm). The pitch is the spacing between the centers of two adjacent pillars. The SAM of 1, 1,-2, 2,-tetrahydroperfluorodecyltrichlorosilane (PF_3) was deposited on the Si sample surfaces using a vapor phase deposition technique. PF_3 was chosen because of the hydrophobic nature of the surface. The thickness and RMS roughness of the SAM of PF_3 were 1.8 and 0.14 nm, respectively (Kasai et al. 2005).

An optical profiler was used to measure the surface topography of the micropatterned surfaces (Bhushan and Jung 2007, 2008; Jung and Bhushan 2008a, b). One sample each from the two series was chosen to characterize the surfaces. Two different surface height maps can be seen for the micropatterned Si in Fig. 6.9. In each case, a 3-D map and a flat map along with a 2-D profile in a given location of the flat 3-D map are shown. A scan

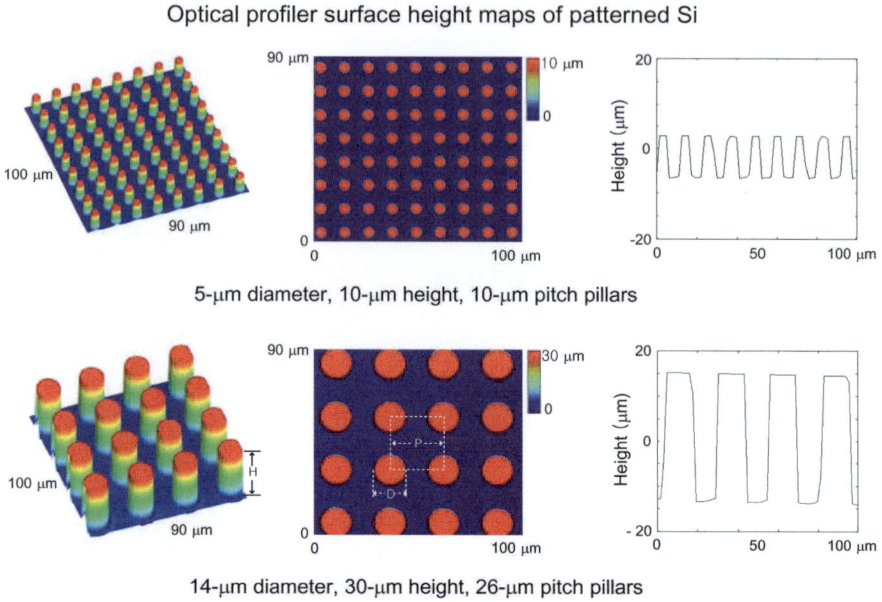

Fig. 6.9 Surface height maps and 2-D profiles of the micropatterned surfaces using an optical profiler (Bhushan and Jung 2007)

size of 100 μm × 90 μm was used to obtain a sufficient number of pillars to characterize the surface but also to maintain sufficient resolution to get an accurate measurement.

Consider the geometry of flat-top, cylindrical pillars of diameter D, height H, and pitch P, distributed in a regular square array as shown in Fig. 6.9. For the special case of a droplet size much larger than P (of interest in this study), a droplet contacts the flat-top of the pillars forming the composite interface, and the cavities are filled with air. For this case, $f_{LA} = 1 - \frac{\pi D^2}{4P^2} = 1 - f_{SL}$. Further assume that the flat tops are smooth with $R_f = 1$. The contact angles are given by the Wenzel and Cassie-Baxter equations (Bhushan and Jung 2007),

$$\text{Wenzel}: \quad \cos\theta = \left(1 + \frac{\pi DH}{P^2}\right)\cos\theta_0 \tag{6.2}$$

$$\text{Cassie-Baxter}: \quad \cos\theta = \frac{\pi D^2}{4P^2}(\cos\theta_0 + 1) - 1 \tag{6.3}$$

Geometrical parameters of the flat-top, cylindrical pillars in series 1 and 2 are used for calculating the contact angle for the above-mentioned two cases. Figure 6.10 shows the plot of the predicted values of the contact angle as a function of pitch between the pillars for the two cases. The Wenzel and Cassie-Baxter equations present two possible equilibrium states for a water droplet on the surface. This indicates that there is a critical

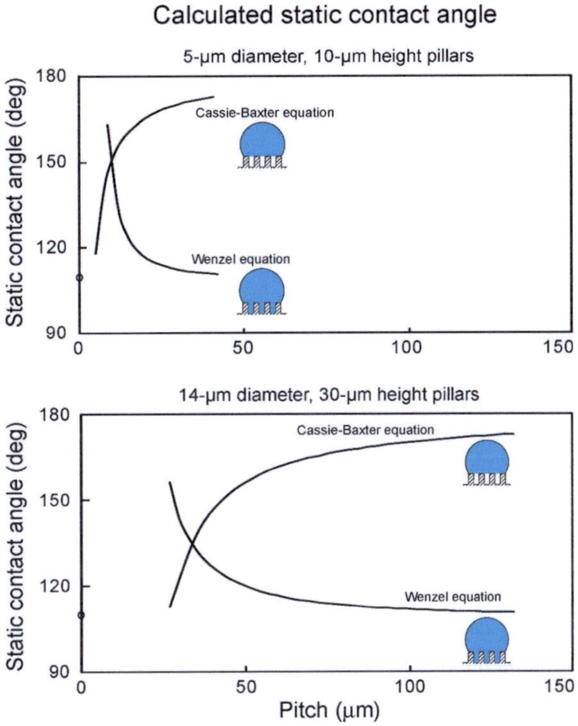

Fig. 6.10 Calculated static contact angle as a function of geometric parameters for a given value of θ_0 using the Wenzel and Cassie-Baxter equations for two series of the micropatterned surfaces with different pitch values (Bhushan and Jung 2007)

pitch below which the composite interface dominates and above which the homogeneous interface dominates the wetting behavior. Therefore, one needs to find the critical point that can be used to design superhydrophobic surfaces. Furthermore, even in cases where the liquid droplet does not contact the bottom of the cavities, the water droplet can be in a metastable state and can become unstable, with the transition from the Cassie-Baxter to Wenzel regime occurring if the pitch is large.

6.4.1 Cassie-Baxter and Wenzel Transition Criteria

A stable composite interface is essential for the successful design of superhydrophobic surfaces. However, the composite interface is fragile, and it may transform into the homogeneous interface. What triggers the transition between the regimes remains a subject of argument, although a number of explanations have been suggested. Nosonovsky and Bhushan (2007b) have studied destabilizing factors for the composite interface and found that a convex surface (with bumps) leads to a stable interface and high contact angle. In addition, they have suggested the effects of a droplet's weight and curvature among the factors which affect the transition.

Bhushan and Jung (2007, 2008) and Jung and Bhushan (2007, 2008a, b) investigated the effect of droplet curvature on the Cassie-Baxter and Wenzel regime transition. First, they considered a small water droplet suspended on a superhydrophobic surface consisting of a regular array of circular pillars with diameter D, height H, and pitch P, as shown in Fig. 6.11. The local deformation for small droplets is governed by surface effects, rather than gravity. The curvature of a droplet is governed by the Laplace equation, which relates the pressure inside the droplet to its curvature (Adamson 1990). Therefore, the curvature is the same at the top and at the bottom of the droplet (Nosonovsky and Bhushan 2007d). For the micropatterned surface considered here, the maximum droop of the droplet occurs in the center of the square formed by the four pillars as shown in Fig. 6.11a. Therefore, the maximum droop of the droplet (δ) in the recessed region can be found in the middle of two pillars which are diagonally across as shown in Fig. 6.11b, which is $(\sqrt{2}P - D)^2/(8R)$. If the droop is greater than the depth of the cavity, then the droplet will just contact the bottom of the cavities between pillars. If it is much greater, transition from the Cassie-Baxter to Wenzel regime occurs,

$$\left(\sqrt{2}P - D\right)^2/R \geq H \tag{6.4}$$

Equation (6.4) shows that geometry (P and H) and droplet radius R govern the transition. A droplet with a large radius, lower pitch, or larger height is desirable for the Cassie-Baxter regime.

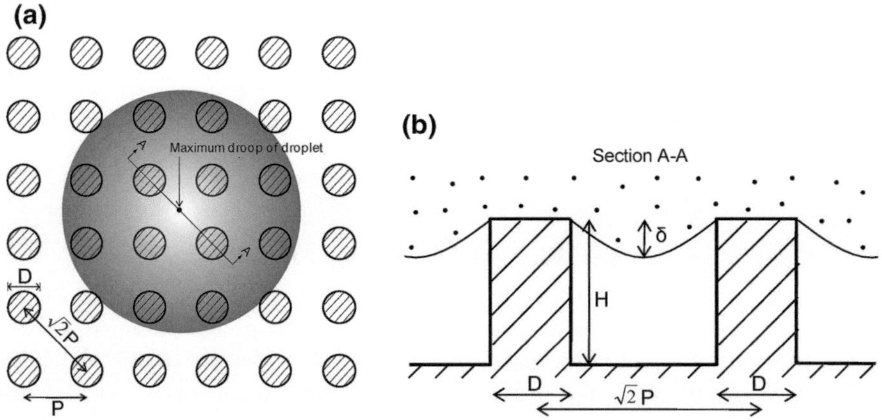

Fig. 6.11 A liquid droplet suspended on a superhydrophobic surface consisting of a regular array of circular pillars. **a** Plan view. The maximum droop of droplet occurs in the center of square formed by four pillars. **b** Side view in section A–A. The maximum droop of droplet (δ) can be found in the middle of two pillars that are diagonally across (Jung and Bhushan 2007)

To investigate the dynamic effect of a bouncing water droplet on the Cassie-Baxter and Wenzel regime transition, Jung and Bhushan (2008b) considered a water droplet hitting a superhydrophobic surface as shown in Fig. 6.11. As the droplet hits the surface at velocity V, a liquid-air interface below the droplet is formed when the dynamic pressure is less than the Laplace pressure. The Laplace pressure can be written as,

$$p_L = 2\gamma/R = 16\gamma\delta/\left(\sqrt{2}P - D\right)^2 \tag{6.5}$$

where γ is the surface tension of the liquid-air interface, and the dynamic pressure of the droplet is equal to,

$$p_d = \frac{1}{2}\rho V^2 \tag{6.6}$$

where ρ is the mass density of the liquid droplet. If the maximum droop of the droplet (δ) is larger than the height of the pillar (H), the droplet contacts the bottom of the cavities between pillars. Determination of the critical velocity at which the droplet touches the bottom is obtained by equating the Laplace pressure to the dynamic pressure. To develop a composite interface, velocity should be smaller than the critical velocity given as,

$$V < \sqrt{\frac{32\gamma H}{\rho\left(\sqrt{2}P - D\right)^2}} \tag{6.7}$$

Furthermore, in the case of large distances between the pillars, the liquid-air interface can easily be destabilized due to dynamic effects. This leads to the formation of the homogeneous solid-liquid interface (Nosonovsky and Bhushan 2007b). Equation (6.7) shows that critical velocity increases with a decrease of pitch or an increase of pillar height.

Nosonovsky and Bhushan (2008e) used the energy barrier approach to study the Cassie-Baxter and Wenzel transition. The energy barrier is given by the product of the height of the pillars (H), pillar perimeter (πD), pillar density ($1/P^2$), and the area (A_0) required to initiate the transition, and the corresponding change in the surface energy,

$$\Delta E = A_0 \frac{\pi HD}{P^2}(\gamma_{SL} - \gamma_{SA}) = -A_0 \frac{\pi HD}{P^2}\cos\theta_0 \tag{6.8}$$

where $A_0 = \pi(\mathrm{R}\sin\theta)^2$.

For a short pitch, the net energy of the Cassie-Baxter state is lower than that of the Wenzel state, whereas for larger pitch values, the energy of the Wenzel state is lower (Fig. 3.11c). However, due to the energy barriers, a metastable Cassie-Baxter state with a higher energy than the Wenzel state may be found.

The energy barrier of the Cassie-Baxter and Wenzel transition can be estimated as the kinetic energy of the droplet. The kinetic energy of a droplet of radius R_0, mass m, density

ρ, and velocity V is given by,

$$E_{kin} = \frac{(4/3)\pi\rho R^3 V^2}{2} \tag{6.9}$$

6.4.2 Effect of Pitch Value on the Transition

In order to study the effect of pitch value on the transition from the Cassie-Baxter to the Wenzel regime, the static contact angles were measured on the micropatterned Si coated with PF_3, and the data are plotted as a function of pitch between the pillars in Fig. 6.12a (Bhushan and Jung 2007, 2008; Jung and Bhushan 2007, 2008a, b). A dotted line represents the transition criteria range obtained using (6.4). The flat Si coated with PF_3 showed a static contact angle of 109°. The contact angle of selected micropatterned surfaces is much higher than that of the flat surfaces. It first increases with an increase in the pitch values, then drops rapidly to a value slightly higher than that of the flat surfaces. In the first portion, it jumps to a high value of 152° corresponding to a superhydrophobic surface and continues to increase to 170° at a pitch of 45 μm in series 1 and 126 μm in series 2. Open air space increases with an increase in pitch, which is responsible for the propensity of air pocket formation. The sudden drop at about a pitch value of 50 μm in series 1, and 150 μm in series 2, corresponds to the transition from the Cassie-Baxter to the Wenzel regime. In series 1, the value predicted from the curvature transition criteria (6.4) is a little higher than the experimental observations. However, in series 2, there is an agreement between the experimental data and the values theoretically predicted by Jung and Bhushan (2007, 2008a, b) for the Cassie-Baxter and Wenzel transition.

Figure 6.12b shows contact angle hysteresis and tilt angle as a function of pitch between the pillars (Bhushan and Jung 2007, 2008). Both angles are comparable. The flat Si coated with PF_3 showed a contact angle hysteresis of 34° and tilt angle of 37°. The angle first increases with an increase of pitch value, which has to do with pinning of the droplet at the sharp edges of the micropillars. Figure 6.13 shows droplets on micropatterned Si with 5 μm diameter and 10 μm height pillars with different pitch values. The asymmetrical shape of the droplet signifies pinning. The pinning on the micropatterned surfaces can be observed as compared to the flat surface. The micropatterned surface with low pitch (7 μm) has more pinning than the micropatterned surface with high pitch (37.5 μm), because the micropatterned surface with low pitch has more sharp edges in contact with a droplet. As the pitch increases, there is a higher propensity of air pocket formation and fewer numbers of sharp edges per unit area, which is responsible for the sudden drop in the angle. The lowest contact angle hysteresis and tilt angle are 5° and 3°, respectively, which were observed on the micropatterned Si with pitch values of 45 μm in series 1 and 126 μm in series 2. Above a pitch value of 50 μm in series 1 and 150 μm in series 2, the angle increases very rapidly because of transition to the Wenzel regime.

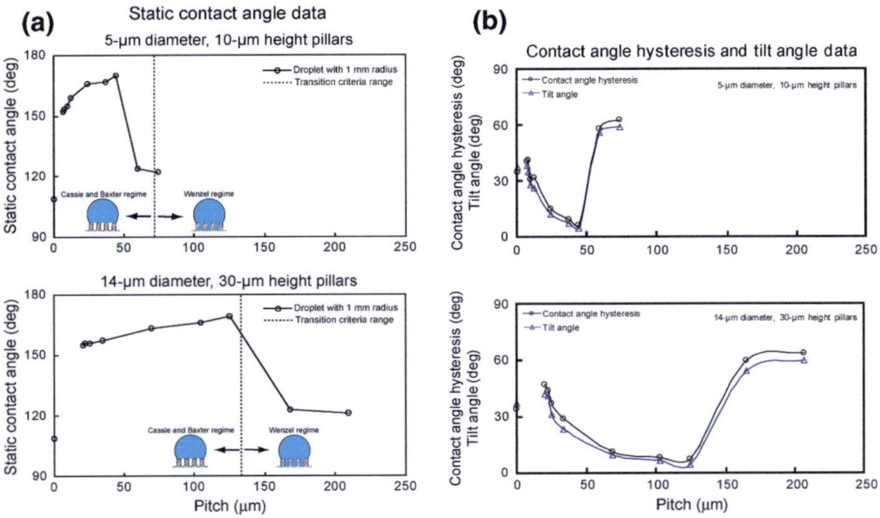

Fig. 6.12 **a** Static contact angle [a dotted line represents the transition criteria range obtained using (6.4)] and **b** contact angle hysteresis and tilt angle as a function of geometric parameters for two series of the micropatterned surfaces with different pitch values for a droplet with 1 mm in radius (5 μL volume). Data at zero pitch correspond to a flat sample (Bhushan and Jung 2007; Jung and Bhushan 2007)

These results suggest that air pocket formation and the reduction of pinning in the micropatterned surface play an important role for a surface with both low contact angle hysteresis and tilt angle (Bhushan and Jung 2007, 2008). Hence, to create superhydrophobic surfaces, it is important that they are able to form a stable composite interface with air pockets between solid and liquid.

6.4.3 Observation of Transition During the Droplet Evaporation

In order to study the effect of droplet size on the transition from a composite state to a wetted state, Jung and Bhushan (2007, 2008a) performed droplet evaporation experiments to observe the Cassie-Baxter and Wenzel transition on two different micropatterned Si surfaces coated with PF_3. The series of four images in Fig. 6.14 shows the successive photos of a droplet evaporating on the two micropatterned surfaces. The initial radius of the droplet was about 700 μm. The time interval between first two photos was 180 s, and between the latter two was 60 s. In the first three photos, the droplet is shown in a Cassie-Baxter state, and its size gradually decreases with time. However, the transition from the Cassie-Baxter to Wenzel regime occurred as the radius of the droplet reached 360 μm on the surface with 5 μm diameter, 10 μm height, and 37.5 μm pitch pillars, and

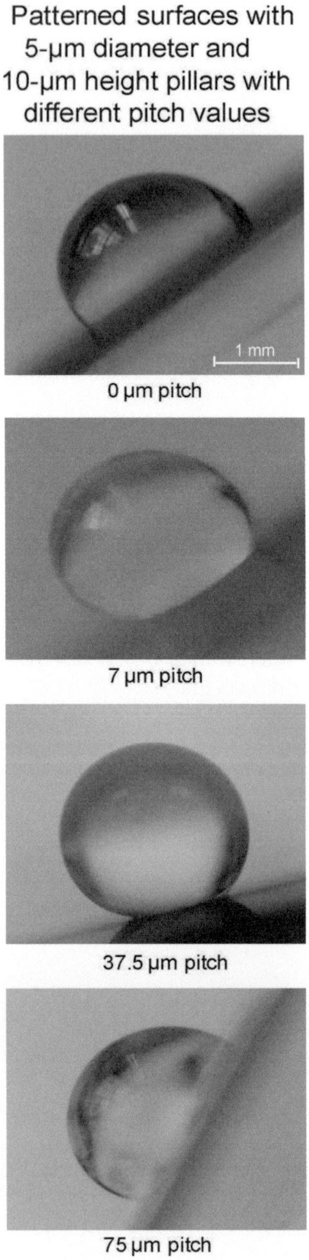

Fig. 6.13 Optical micrographs of droplets on the inclined micropatterned surfaces with different pitch values. The images were taken when the droplet started to move down. Data at zero pitch correspond to a flat sample (Bhushan and Jung 2007)

423 μm on the surface with 14 μm diameter, 30 μm height, and 105 μm pitch pillars, as indicated by the arrow. The light passes below the first droplet, indicating that air pockets exist, so that the droplet is in the Cassie-Baxter state. However, an air pocket is not visible below the last droplet, so it is in the Wenzel state. This could result from an impalement of the droplet in the micropatterned surface, characterized by a smaller contact angle.

To find the contact angle before and after the transition, the values of the contact angle are plotted against the theoretically predicted value, based on the Wenzel and Cassie-Baxter (Chap. 3) models. Figure 6.15 shows the static contact angle as a function of geometric parameters for the experimental contact angles before (circle) and after (triangle) the transition compared to the Wenzel and Cassie-Baxter equations (solid lines) with a given value of θ_0 for two series of micropatterned Si with different pitch values coated with PF_3 (Jung and Bhushan 2008a). The fit is good between the experimental data and the theoretically predicted values for the contact angles before and after transition.

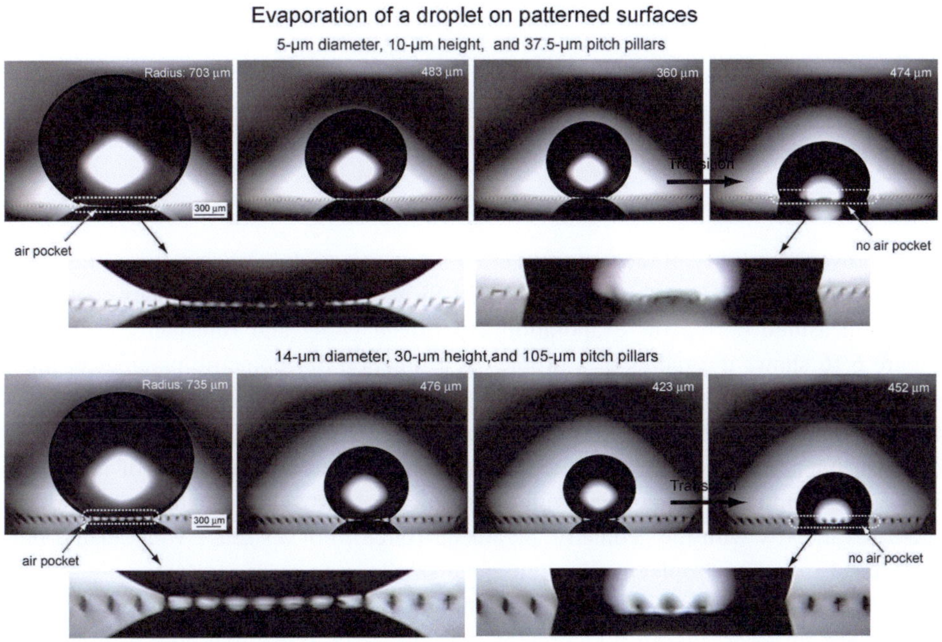

Fig. 6.14 Evaporation of a droplet on two different micropatterned surfaces. The initial radius of the droplet is about 700 μm, and the time interval between the first two photos was 180 s and between the latter two was 60 s. As the radius of the droplet reaches 360 μm on the surface with 5-μm diameter, 10-μm height, and 37.5-μm pitch pillars, and 420 μm on the surface with 14-μm diameter, 30-μm height, and 105-μm pitch pillars, the transition from the Cassie-Baxter regime to Wenzel regime occurs, as indicated by the arrow. Before the transition, an air pocket is clearly visible at the bottom area of the droplet, but after the transition, an air pocket is not found at the bottom area of the droplet (Jung and Bhushan 2008a)

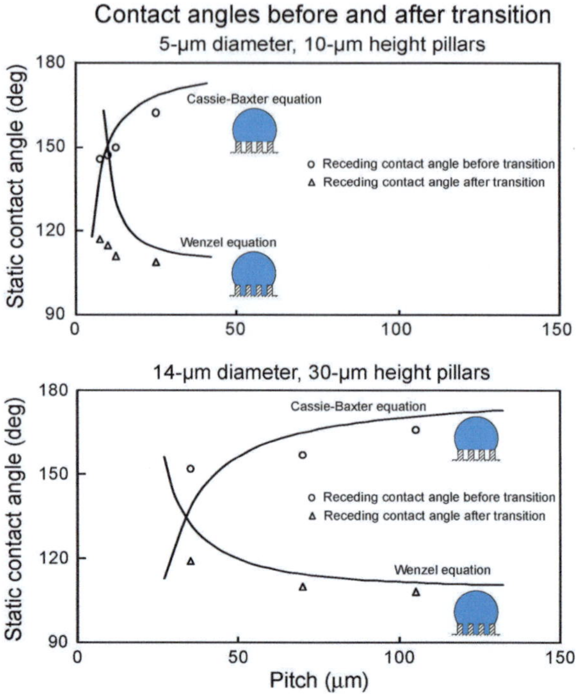

Fig. 6.15 Receding contact angle as a function of geometric parameters before (circle) and after (triangle) transition compared with predicted static contact angle values obtained using the Wenzel and Cassie-Baxter equations (solid lines) with a given value of θ_0 for two series of the micropatterned surfaces with different pitch values (Jung and Bhushan 2008a)

To prove the validity of the transition criteria in terms of droplet size, the critical radius of a droplet deposited on the micropatterned Si with different pitch values coated with PF_3 is measured during the evaporation experiment (Jung and Bhushan 2007, 2008a). Figure 6.16 shows the radius of a droplet as a function of geometric parameters for the experimental results (circle) compared with the transition criterion (6.4) from the Cassie-Baxter regime to Wenzel regime (solid lines) for two series of micropatterned Si with different pitch values coated with PF_3. It is found that the critical radius of impalement is in good quantitative agreement with the predictions. The critical radius of the droplet increases linearly with the geometric parameter (pitch). For a surface with small a pitch, the critical radius of droplet can become quite small.

To verify the transition, Jung and Bhushan (2007, 2008a) used another approach using dust mixed in water. Figure 6.17 presents the dust trace remaining after a droplet with 1 mm radius (5 μL volume) has evaporated on the two micropatterned Si surfaces. As shown in the top image, after the transition from the Cassie-Baxter regime to Wenzel regime, the dust particles remained not only at the top, but also at the bottom of the pillars with a footprint size of about 450 μm. However, as shown in the bottom image, the dust particles remained on only a few pillars with a footprint size of about 25 μm until the end of the evaporation process. From Fig. 6.16, it is observed that the transition occurs at about 300 μm radius of droplet on the 5 μm diameter and 10 μm height pillars with

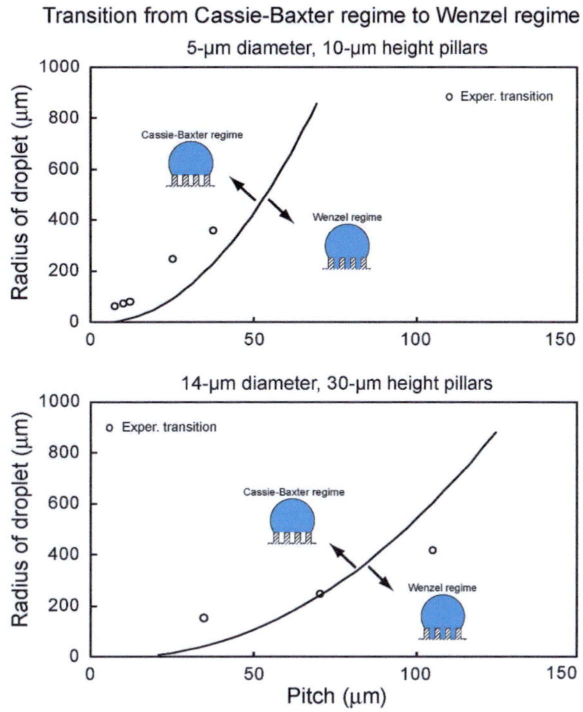

Fig. 6.16 Radius of droplet as a function of geometric parameters for the experimental results (circle) compared with the transition criteria from the Cassie-Baxter regime to Wenzel regime (solid lines) for two series of the micropatterned surfaces with different pitch values (Jung and Bhushan 2008a)

37.5 μm pitch. The transition does not occur on the patterned Si surface with pitch of less than about 5 μm. These experimental observations are consistent with model predictions. In the literature, it has been shown that on a superhydrophobic natural lotus, the droplet remains in the Cassie-Baxter regime during the evaporation process (Zhang et al. 2006). This indicates that the distance between the pillars should be minimized to improve the ability of the droplet to resist sinking.

6.4.4 Another Cassie-Baxter and Wenzel Transition for Different Series

Nosonovsky and Bhushan (2007a, c, d, 2008b, c) studied the data for the Cassie-Baxter and Wenzel transition with the two series of surfaces using the non-dimensional spacing factor

$$S_f = \frac{D}{P} \tag{6.10}$$

The values of the droplet radius at which the transition occurs during evaporation plotted against the spacing factor scale well for the two series of the experimental results,

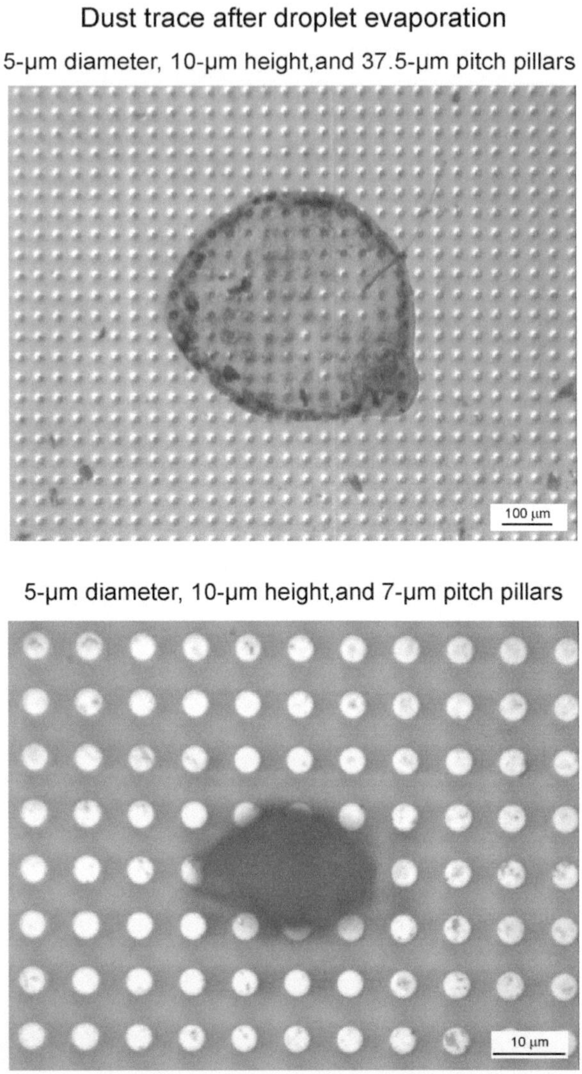

Fig. 6.17 Dust trace remained after droplet evaporation for the micropatterned surface. In the top image, the transition occurred at 360 μm radius of droplet, and in the bottom image, the transition occurred at about 20 μm radius of droplet during the process of droplet evaporation. The footprint size is about 450 and 25 μm for the top and bottom images, respectively (Jung and Bhushan 2008a)

yielding virtually the same straight line. Thus, the two series of micropatterned surfaces scale well with each other, and the transition occurs at the same value of the spacing factor multiplied by the droplet radius (Fig. 6.18a). The physical mechanism leading to this observation remains to be determined. However, it is noted that this mechanism is different from the one suggested by (6.4). The observation suggests that the transition is a linear one-dimensional phenomenon and that neither droplet droop (that would involve P^2/H), nor droplet weight (that would involve R^3) are responsible for the transition, but rather linear geometric relations are involved. Note that the experimental values approximately

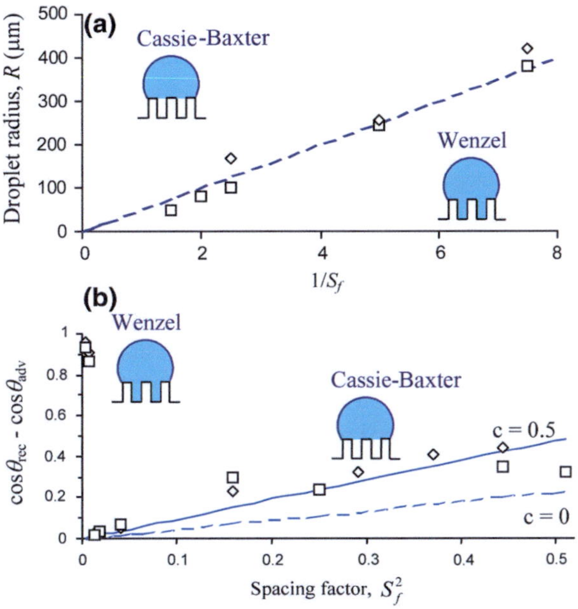

Fig. 6.18 **a** Droplet radius, R, for the Cassie-Baxter and Wenzel transition as a function of $P/D = 1/S_f$. It is observed that the transition for two series of the micropatterned surfaces takes place at a constant value of $RD/P \sim 50\ \mu$m (dashed line). This shows that the transition is a linear phenomenon. **b** The difference $\cos\theta_{rec} - \cos\theta_{adv}$ as a function of S_f for the 1st (squares) and 2nd (diamonds) series of the experiments compared with the theoretically predicted values of $\cos\theta_{adv} - \cos\theta_{rec} = (D/P)^2(\pi/4)(\cos\theta_{adv0} - \cos\theta_{rec0}) + c(D/P)^2$, where c is a proportionality constant. It is observed that when only the adhesion hysteresis/interface energy term is considered ($c = 0$), the theoretical values are underestimated by about a half, whereas $c = 0.5$ provides a good fit. Therefore, the contribution of the adhesion hysteresis is of the same order of magnitude as the contribution kinetic effects (adapted from Nosonovsky and Bhushan 2007c)

correspond to the values of the ratio $RD/P = 50\ \mu$m, or the total area of the pillar tops under the droplet $(\pi D^2/4)\pi R^2/P^2 = 6200\ \mu\text{m}^2$.

6.4.5 Contact Angle Hysteresis and Wetting/Dewetting Asymmetry

Contact angle hysteresis can be viewed as a result of two factors that act simultaneously. First, the changing contact area affects the contact angle hysteresis, since a certain value of contact angle hysteresis is inherent for even a nominally flat surface. Decreasing the contact area by increasing the pitch between the pillars leads to a proportional decrease of the contact angle hysteresis. This effect is clearly proportional to the contact area between the solid surface and the liquid droplet. Second, the edges of the pillar tops prevent the motion of the triple line. This roughness effect is proportional to the contact line density, and its contribution was, in the experiment, comparable with the contact area effect. Interestingly, the effect of the edges is much more significant for the advancing than for the receding contact angle.

Nosonovsky and Bhushan (2007a, c, d, 2008b, c) studied the wetting of two series of micropatterned Si surfaces with different pitch values coated with PF_3 based on the spacing factor (6.10) (Fig. 6.18b). They found that the contact angle hysteresis involves two terms: the term $S_f^2(\pi/4)$ ($\cos\theta_{adv0} - \cos\theta_{rec0}$) corresponding to the adhesion hysteresis (which is found even in a nominally flat surface and is a result of molecular-scale imperfectness), and the term $H_r \propto D/P^2$ corresponding to microscale roughness and proportional to the edge line density. Thus the contact angle hysteresis is given, based on the equation for rough surfaces (Chap. 3) and (6.10), by using $R_f = 1 + \frac{\pi DH}{P^2}$ and $f_{LA} = 1 - \frac{\pi D^2}{4P^2} = 1 - f_{SL}$ (Bhushan et al. 2007; Nosonovsky and Bhushan 2007b)

$$\cos\theta_{adv} - \cos\theta_{rec} = \frac{\pi}{4} S_f^2 (\cos\theta_{adv0} - \cos\theta_{rec0}) + H_r \tag{6.11}$$

Besides the contact angle hysteresis, the asymmetry of the Wenzel and Cassie-Baxter states is the result of the wetting/dewetting asymmetry. While the fragile metastable Cassie-Baxter state is often observed, as well as its transition to the Wenzel state, the opposite transition never happens. Using (6.2) and (6.3), the contact angle with micropatterned surfaces is given by Bhushan et al. Bhushan et al. (2007), Nosonovsky and Bhushan (2007b)

$$\cos\theta = \left(1 + 2\pi S_f^2\right) \cos\theta_0 \quad \text{(Wenzel state)} \tag{6.12}$$

$$\cos\theta = \frac{\pi}{4} S_f^2 (\cos\theta_0 + 1) - 1 \quad \text{(Cassie-Baxter state)} \tag{6.13}$$

For a perfect macroscale system, the transition between the Wenzel and Cassie-Baxter states should occur only at the intersection of the two regimes (the point at which the contact angle and net energies of the two regimes are equal, corresponding to $S_f = 0.51$). However, it is observed that the transition from the metastable Cassie-Baxter to stable Wenzel occurs at much lower values of the spacing factor $0.083 < S_f < 0.111$.

As shown in Fig. 6.19a, the stable Wenzel state (i) can transform into the stable Cassie-Baxter state with increasing S_f (ii). The metastable Cassie-Baxter state (iii) can abruptly transform (iv) into the stable Wenzel state (i). The transition point (i–ii) corresponds to equal free energies in the Wenzel and Cassie-Baxter states, whereas the transition (iv) corresponds to the Wenzel energy being much lower than the Cassie-Baxter energy. Thus, the transition (iv) involves significant energy dissipation and is irreversible (Nosonovsky and Bhushan 2007c). The solid and dashed straight lines correspond to the values of the contact angle, calculated from (6.12) to (6.13) using the contact angle for a nominally flat

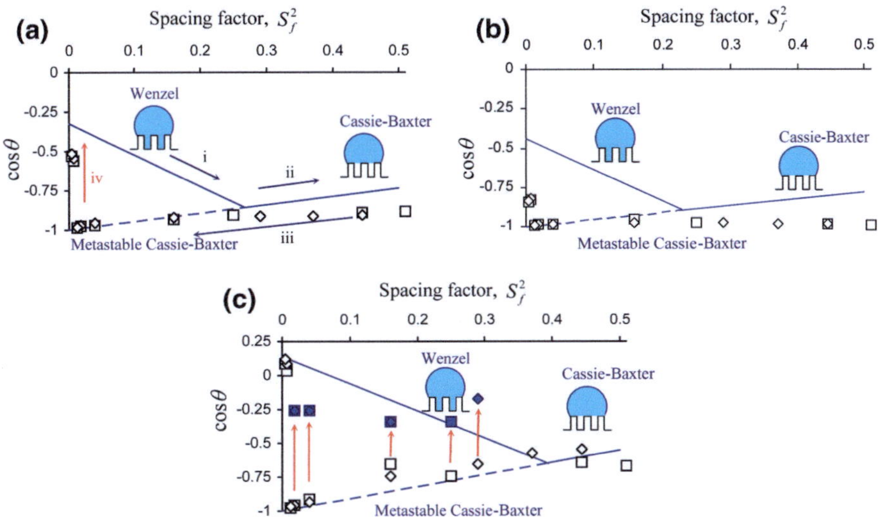

Fig. 6.19 Theoretical (solid and dashed) and experimental (squares for the 1st series, diamonds for the 2nd series) **a** contact angle as a function of the spacing factor, **b** advancing contact angle, and **c** receding contact angle and values of the contact angle observed after the transition during evaporation (blue) (adapted from Nosonovsky and Bhushan 2007c)

surface, $\theta_0 = 109°$. The two series' of the experimental data are shown with squares and diamonds.

Figure 6.19b shows the values of the advancing contact angle plotted against the spacing factor (6.10). The solid and dashed straight lines correspond to the values of the contact angle for the Wenzel and Cassie-Baxter states, calculated from (6.12) to (6.13) using the advancing contact angle for a nominally flat surface, $\theta_{adv0} = 116°$ (Nosonovsky and Bhushan 2007c). It is observed that the calculated values underestimate the advancing contact angle, especially for a large S_f (small distance between the pillars or pitch P). This is understandable, because the calculation takes into account only the effect of the contact area and ignores the effect of roughness and edge line density [it corresponds to $H_r = 0$ in (6.11)], while this effect is more pronounced for high pillar density (big S_f). In a similar manner, the contact angle is underestimated for the Wenzel state, since the pillars constitute a barrier for the advancing droplet.

Figure 6.19c shows the values of the contact angle after the transition took place (squares and diamonds), as it was observed during evaporation (Nosonovsky and Bhushan 2007c). For both series, the values almost coincide. For comparison, the values of the receding contact angle measured for millimeter-sized water droplets are also shown (squares and diamonds), since evaporation constitutes removing liquid, and thus the contact angle during evaporation should be compared with the receding contact angle. The solid and dashed straight lines correspond to the values of the contact angle, calculated from (6.12) to (6.13) using the receding contact angle for a nominally flat surface, $\theta_{rec0} = 82°$. Figure 6.19c demonstrates a good agreement between the experimental data and (6.12) and (6.13).

In the analysis of the evaporation data of micropatterned surfaces, Nosonovsky and Bhushan (2008b) found several effects specific to the multiscale character of this process. First, they discussed the applicability of the Wenzel and Cassie-Baxter equations for average surface roughness and heterogeneity. These equations relate the local contact angle with the apparent contact angle of a rough/heterogeneous surface. However, it is not obvious what should be the size of roughness/heterogeneity averaging, since the triple line at which the contact angle is defined has two very different length scales: its width is of the molecular scale while its length is on the order of the size of the droplet (that is, microns or millimeters). They presented an argument that in order for the averaging to be valid, the roughness details should be small compared to the size of the droplet (and not the molecular size). They showed that while for uniform roughness/heterogeneity the Wenzel and Cassie-Baxter equations can be applied, for a more complicated case of non-uniform heterogeneity, the generalized equations should be used. The proposed generalized Cassie-Baxter and Wenzel equations are consistent with a broad range of available experimental data. The generalized equations are valid both in the cases when the classical Wenzel and Cassie-Baxter equations can be applied as well as in the cases when the latter fail.

The macroscale contact angle hysteresis and Cassie-Baxter and Wenzel transition cannot be determined from the macroscale equations and are governed by micro- and nanoscale effects, so wetting is a multiscale phenomenon (Nosonovsky and Bhushan 2007a, c, d, 2008b, c). The kinetic effects associated with contact angle hysteresis should be studied at the microscale, whereas the effects of adhesion hysteresis and the Cassie-Baxter and Wenzel transition involve processes at the nanoscale. Their theoretical arguments are supported by the experimental data on micropatterned surfaces. The experimental study of the contact angle hysteresis demonstrates that two different processes are involved: the changing solid-liquid area of contact and pinning of the triple line. The latter effect is more significant for the advancing than for the receding contact angle. The transition between wetting states was observed for evaporating microdroplets, and droplet radius scales with the geometric parameters of the micropattern.

6.4.6 Contact Angle Measurements During Condensation and Evaporation of Microdroplets on Micropatterned Surfaces

To provide insight into the formation of microdroplets and detailed information about the contact angle on the micropatterned surfaces, ESEM experiments on micropatterned surfaces have been performed during condensation and evaporation. Figure 6.20 shows how water droplets grow and merge in an ESEM (Jung and Bhushan 2008a) that was

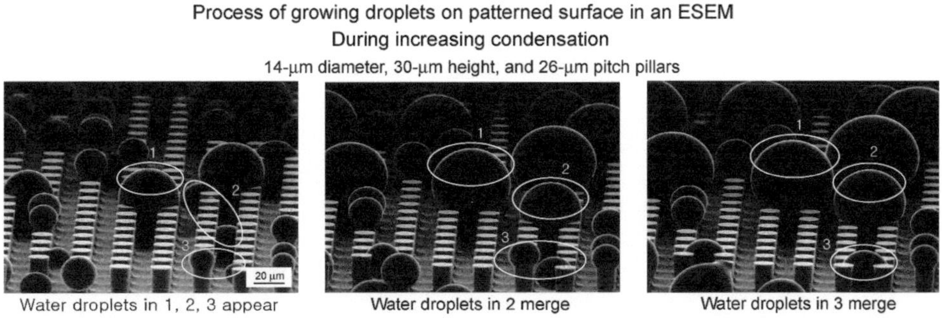

Fig. 6.20 Microdroplet (in dimensions of less than 1 mm diameter) growing and merging process under ESEM during increasing condensation by decreasing temperature. Left image: Some small water droplets appear at the beginning, i.e. water droplets 1, 2, 3. Middle image: Water droplets at locations 1 and 3 increase in size and water droplets at location 2 merge together to form one big droplet. Right image: Water droplets at locations 1 and 2 increase in size and water droplets at location 3 merge together to form one big droplet (Jung and Bhushan 2008a)

used as a contact angle analysis tool. Microdroplets (with a diameter less than 1 mm) were distributed on a micropatterned surface coated with PF_3 using condensation by decreasing temperature. At the beginning, some small water droplets appeared, i.e., the water droplets at locations 1, 2, and 3 in the left image. During further condensation with decreasing temperature, the droplets at locations 1 and 3 gradually grew, while the droplets at location 2 merged together. With further condensation, the droplets at locations 1 and 2 gradually grew, while the droplets at location 3 merged together into one large droplet in the right image. In all cases, condensation was initiated at the bottom, therefore, the droplets were in the Wenzel regime.

Compared with the conventional contact angle measurement, ESEM is able to provide detailed information about the contact angle of microdroplets on micropatterned surfaces. The diameter of the water droplets used for the contact angle measurement was 10 μm, so that the size limit pointed out by Stelmashenko et al. (2001) was avoided. For a droplet size smaller than 1 μm, substrate backscattering can distort the intensity profile, such that the images are inaccurate.

As shown in Fig. 6.21, the static contact angle and contact angle hysteresis of the microdroplets on flat and micropatterned surfaces were obtained from the images using the methodology described earlier. Once the microdroplet's condensation and evaporation reached a dynamic equilibrium, static contact angles were determined. The flat Si coated with PF_3 showed a static contact angle of 98°. The roughness of the micropatterned surfaces coated with PF_3 increased the static contact angle compared to the flat surface coated with PF_3. Advancing and receding contact angles were measured during condensation/evaporation with decreasing/increasing the temperature of the cooling stage, and the contact angle hysteresis was then calculated (Jung and Bhushan 2008a).

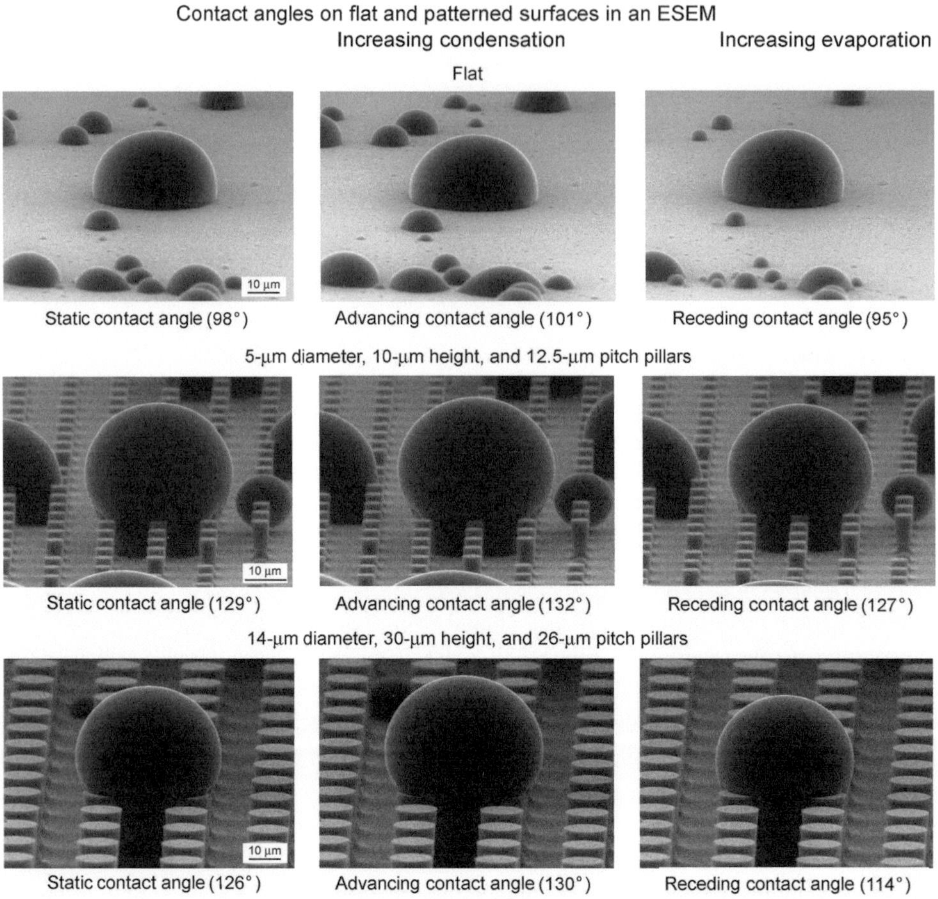

Fig. 6.21 Microdroplets on flat and two micropatterned surfaces using ESEM. Second set of images were taken during increasing condensation, and the third set of images were taken during increasing evaporation. Static contact angle was measured when the droplet was stable. Advancing contact angle was measured after increasing condensation by decreasing the temperature of the cooling stage. Receding contact angle was measured after decreasing evaporation by increasing the temperature of the cooling stage (Jung and Bhushan 2008a)

Figure 6.22 shows contact angle hysteresis as a function of pitch value for the microdroplets formed in the ESEM (triangles) for two series of micropatterned Si with different pitch values coated with PF_3. Data at zero pitch correspond to a flat Si sample. The droplets with about 20 μm radii, which are larger than the pitch, were selected in order to look at the effect of pillars in contact with the droplet. These data were compared with conventional contact angle measurements with a droplet with 1 mm radius (5 μL volume) (Bhushan and Jung 2007). When the distance between pillars increased above a certain

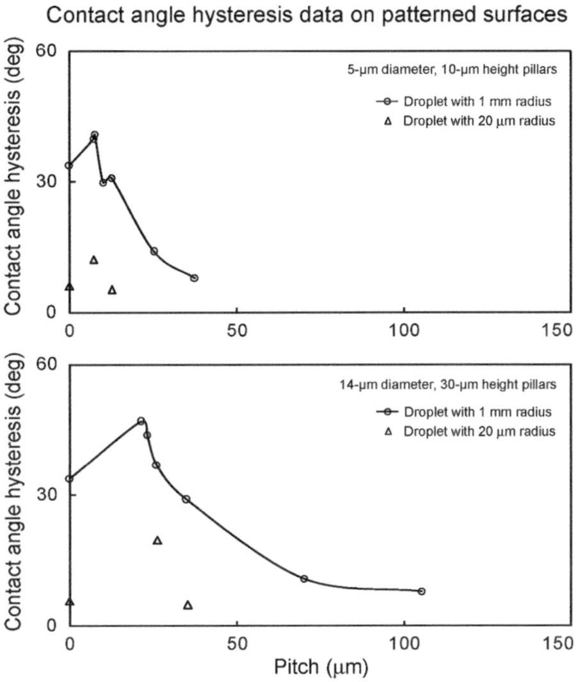

Fig. 6.22 Contact angle hysteresis as a function of pitch value for the microdroplet with about 20 μm radius from ESEM (triangle) compared with the droplet with 1 mm radius (5 μL volume) (circle and solid lines) for two series of the micropatterned surfaces with different pitch values. Data at zero pitch correspond to a flat sample (Jung and Bhushan 2008a)

value, the contact area between the micropatterned surface and the droplet decreased, resulting in a decrease of the contact angle hysteresis. Both droplets with 1 mm and 20 μm radii showed the same trend. The contact angle hysteresis for the micropatterned surfaces with low pitch were higher compared to the flat surface, due to the effect of sharp edges on the pillars, resulting in pinning (Nosonovsky and Bhushan 2005). Contact angle hysteresis for a flat surface can arise from roughness and surface heterogeneity. For a droplet advancing forward on the micropatterned surfaces, the line of contact of the solid, liquid, and air will be pinned at the edge point until it is able to move, resulting in increasing contact angle hysteresis. The contact angle hysteresis for the microdroplet from ESEM is lower as compared to that for the droplet with 1 mm radius. The difference of contact angle hysteresis between a microdroplet and a droplet with 1 mm radius could come from the different pinning effects because the latter has more sharp edges in contact with a droplet compared to the former. The results show how droplet size can affect the wetting properties of micropatterned Si surfaces (Jung and Bhushan 2008a).

6.4.7 Observation of Transition During the Bouncing Droplet

Dynamic effects, such as bouncing of a droplet (e.g. rain droplets hitting the window glass), can destroy the composite interface. Based on (6.7), the relationship between the impact velocity of a droplet and geometric parameters affects the transition from the Cassie-Baxter to the Wenzel regime. Therefore, it is necessary to study the dynamic effect of droplets under various impact velocities.

Jung and Bhushan (2008b) performed bouncing droplet experiments to observe how impact velocity influences the Cassie-Baxter and Wenzel transition during the droplet hitting the surface on two different micropatterned Si surfaces with PF_3. Figure 6.23 shows photos of a droplet with 1 mm radius hitting the surfaces. The impact velocity was obtained just prior to the droplet hitting the surface. As shown in the images in the first row for the two sets of surfaces, the droplet hitting the surface under an impact velocity of 0.44 m/s first deformed and then retracted, and bounced off the surface. Finally, the droplet sat on the surface and had a high contact angle, which suggests the formation of a solid-air-liquid interface. Next, they repeated the impact experiment by increasing the impact velocity. Bounce off did not occur, and the wetting of the surface (and possibly pinning of droplet) occurred at an impact velocity of 0.88 and 0.76 m/s, respectively. These are referred to as the critical velocity (described earlier). The second row of the two sets of images shows the droplet at the critical velocity. After the droplet hit the surface, it wetted the surface (possibly the droplet was also pinned) after the deformation of the droplet. This is because air pockets do not exist below the droplet because of droplet impalement by the pillars, characterized by a smaller contact angle. These observations indicate the transition from a Cassie-Baxter to a Wenzel regime.

To identify whether an interface is in a Wenzel regime or a Cassie-Baxter regime, the contact angle data in the static condition and after bounce off were plotted (Jung and Bhushan 2008b). Figure 6.24 shows the measured static contact angle as a function of pitch value for the droplet with 1 mm radius placed on the surface and for the droplet with 1 mm radius after hitting the surface at 0.44 m/s. The data are compared with predicted static contact angle values obtained using the Wenzel and the Cassie-Baxter equations with a given value of θ_0 (109°) for a smooth surface for two series of the micropatterned surfaces. In the case of the droplet placed on the surface, as the pitch increases up to 45 μm in series 1 and 126 μm in series 2, the static contact angle first increases gradually from 152° to 170°. Then, the contact angle starts decreasing sharply. The increase in the contact angle occurs because of an increase in the roughness factor and the formation of a composite surface (Bhushan and Jung 2007). The decrease in contact angle at pitch values higher than 60 μm for series 1 and 168 μm for series 2 occurs due to the transition from the composite interface to the solid-liquid interface. In the case of the droplet hitting the surface at 0.44 m/s, it is shown that the liquid-air interface can easily be destabilized due to dynamic impact on the surface with a pitch value higher than 12.5 μm for series 1 and 70 μm for series 2, even though the droplet was in the Cassie-Baxter regime. The static contact angle of the droplet after hitting at 0.44 m/s is lower than that of the droplet when placed. It can be interpreted that after hitting, the droplet contacts the bottom of the cavities between pillars and pushes out the entrapped air under the droplet, resulting in an abrupt increase of the solid-liquid surface area by dynamic impact. It will be shown in

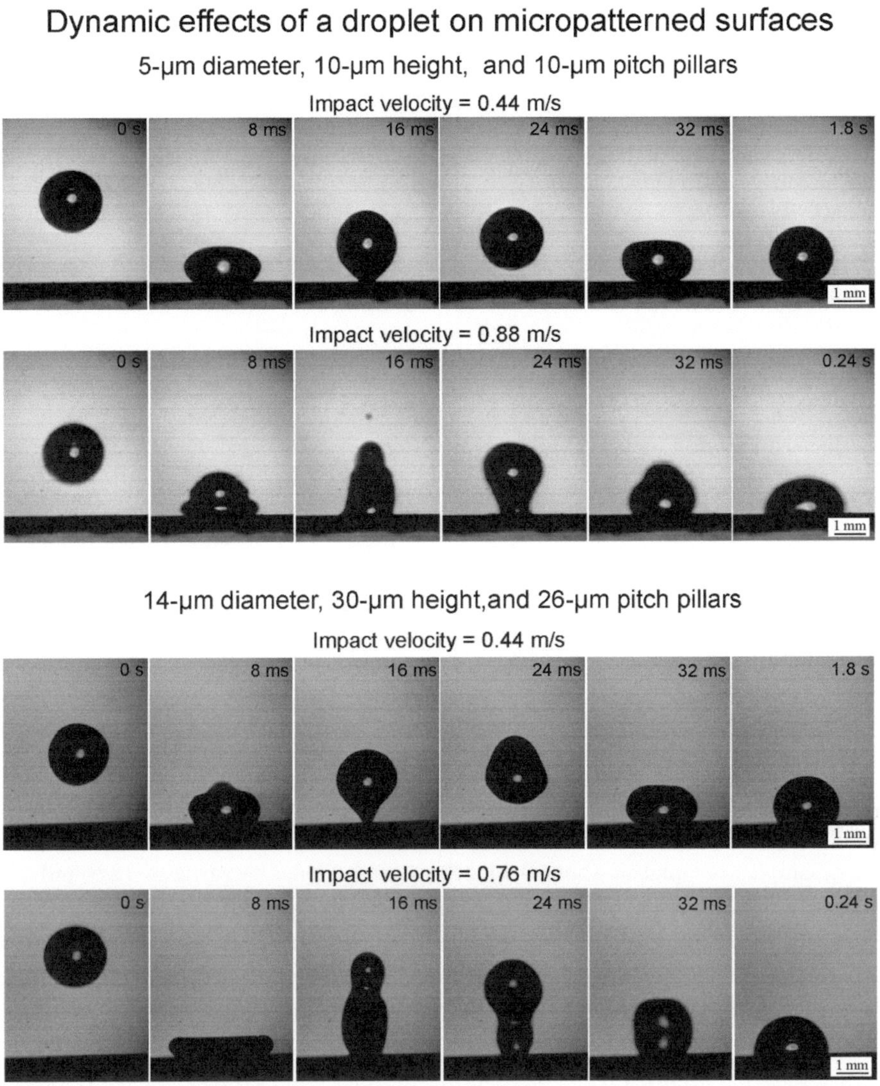

Fig. 6.23 Snapshots of a droplet with 1 mm radius hitting on two different micropatterned surfaces. The impact velocity was obtained just prior to the droplet hitting the surface. The pinning of the droplet on the surface with 5-μm diameter, 10-μm height and 10-μm pitch pillars, and on the surface with 14-μm diameter, 30-μm height and 26-μm pitch pillars occurred at impact velocity of 0.88 and 0.76 m/s, respectively (Jung and Bhushan 2008b)

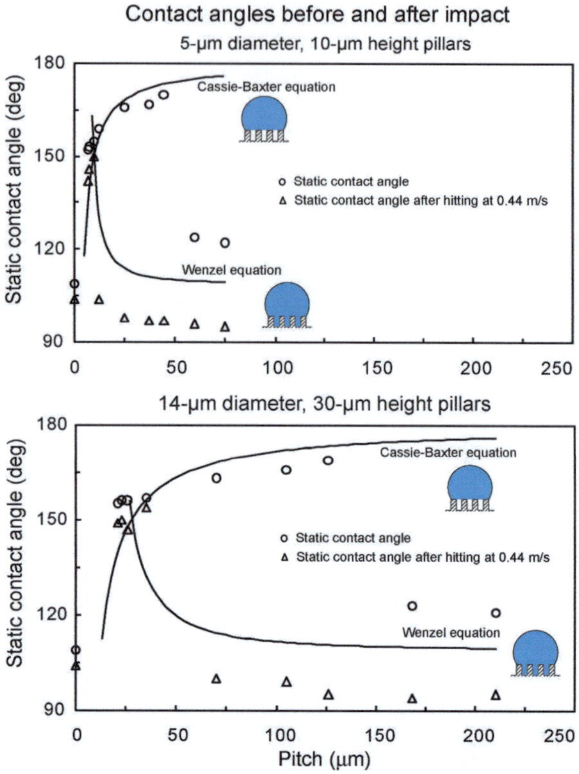

Fig. 6.24 Measured static contact angle as a function of pitch value for the droplet with 1 mm radius placed on the surface (circles) and for the droplet with 1 mm radius after hitting the surface at 0.44 m/s (triangles). The data are compared with predicted static contact angle values obtained using Wenzel and Cassie-Baxter equations (solid lines) with a given value of θ_0 (109°) for a smooth surface for two series of the micropatterned Si with different pitch values (Jung and Bhushan 2008b)

the following paragraph that the critical velocity at which wetting occurs for series 1 and series 2 samples is equal to about 0.44 m/s at pitch values larger than 12.5 and 70 μm respectively. Thus, wetting at the velocity used here is expected.

To study the validity of the transition criterion (6.7), the critical impact velocity at which wetting of the surface (possibly pinning of droplet) occurs was measured (Jung and Bhushan 2008b). For calculations, the surface tension of the water-air interface (γ) was taken at 0.073 N/m, the mass density (ρ) is 1000 kg/m^3 for water, and 1 kg m/s^2 = 1 N (Adamson 1990). Figure 6.25 shows the measured critical impact velocity of a droplet with 1 mm radius as a function of pitch value. The trends were compared with the predicted curve. It was found that the critical impact velocity at which wetting occurs was in good quantitative agreement with the predictions. The critical impact velocity of the droplet decreased with the geometric parameter (pitch). For the surface with a small pitch, the critical impact velocity of a droplet can be large.

The energy barrier of the Cassie-Baxter and Wenzel transition can be estimated as the kinetic energy of the droplets (Nosonovsky and Bhushan 2008e). Figure 6.26 shows the dependence of the kinetic energy corresponding to the transition, E_{kin}, on $\Delta E/(A_0 \cos\theta_0)$ calculated from (6.8). It was observed that the dependence was close to linear. However,

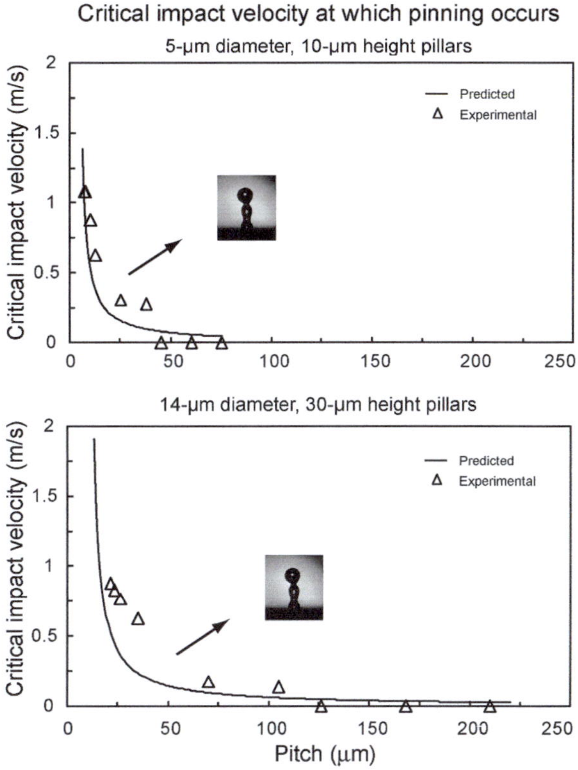

Fig. 6.25 Measured critical impact velocity of droplet with 1 mm radius as a function of pitch value (triangles). The data are compared with the criterion of impact velocity for the pinning of droplet (solid lines) for two series of the micropatterned Si with different pitch values (Jung and Bhushan 2008b)

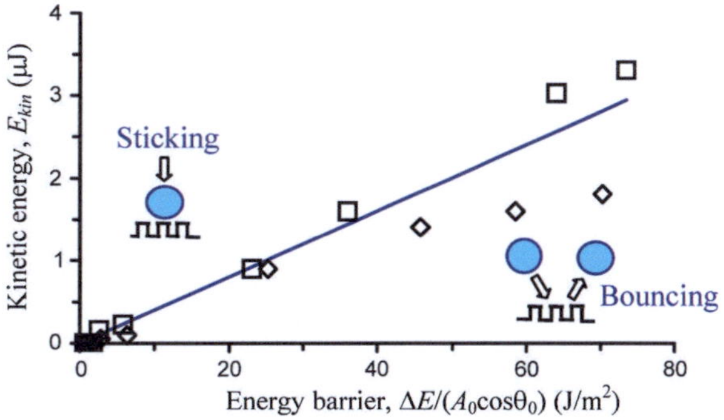

Fig. 6.26 Bouncing droplets dependency of the kinetic energy of a droplet corresponding to the regime transition upon the energy barrier calculated from (6.8) (squares for series 1 and diamonds for series 2). The fit (solid line) is shown for $A_0 = 0.12\ \text{mm}^2$ (Nosonovsky and Bhushan 2008e)

the series of smaller pillars had larger energies of transition. The value of A_0 is in the range $0.11\ \text{mm}^2 < A_0 < 0.18\ \text{mm}^2$ for series 1 and $0.05\ \text{mm}^2 < A_0 < 0.11\ \text{mm}^2$ for series 2, which is of the same order as the actual area under the droplet.

These results suggest that the energy barrier for the Cassie-Baxter and Wenzel transition is given by (6.8) and is proportional to the area under the droplet (Nosonovsky and Bhushan 2008e). For droplets sitting on the surface or evaporating, the transition takes place when the size of the barrier decreases to the value of the vibrational energy, U. The vibrational energy of the droplet is the energy associated with the vibration of the droplet due to surface waves, thermal vibration, etc. Assuming U = *const*, the proportionality of *P/D* and *R* suggests that the energy barrier is proportional to *RD/P*. This is indeed true, since the area under the droplet $A_0 = \pi(R \sin \theta)^2$. Substituting $\sin^2\theta = 0.1$, $\cos \theta_0 = \cos 109° = -0.33$, $\gamma_{LA} = 0.072$ J/m^2 in (6.8) and taking the observed value $RD/P = 50$ μm yields an estimated value of the vibrational energy $U = 1.2 \times 10^{-10}$ J. The transition happens because the size of the droplet is decreased or because the pitch between the pillars that cover the surface is increased. A different way to overcome the barrier is to hit the surface with a droplet with a certain kinetic energy.

Based on Nosonovsky and Bhushan (2008e), the vibrational energy U also plays a role in overcoming energy barriers that lead to contact angle hysteresis during liquid flow (Johnson and Dettre 1964). To estimate the effect of the energy barriers on contact angle hysteresis, equations presented in Chap. 3 are used in which the difference between the advancing and receding contact angle is given by,

$$\cos\theta_{rec} - \cos\theta_{adv} = \Delta W/\gamma_{LA} \tag{6.14}$$

where ΔW corresponds to the energy barrier associated with the wetting-dewetting cycle. Assuming that this energy barrier is of the same order as the vibrational energy per contact area, $\Delta W = U/A_0$, and taking $A_0 = 0.1$ mm^2, $\Delta W = 10^{-3}$ J/m^2 is obtained. For water ($\gamma_{LA} = 0.072$ J/m^2), Eq. (6.14) leads to a realistic value of hysteresis on a superhydrophobic surface $\cos\theta_{rec} - \cos\theta_{adv} = 0.014$. This number provides an estimate for contact angle hysteresis in the limit of small energy barriers comparable with U. The actual values for a micropatterned surface are dependent upon the solid-liquid contact area (that provides energy barriers due to so-called adhesion hysteresis) and the density of the solid-air-liquid contact line (that provides additional pinning), and were found to be between 0.0144 and 0.440 (Bhushan et al. 2007). Thus, these showed good agreement with the value calculated based on U as the lower limit. This indicates that the value of U is relevant both for the Cassie-Baxter and Wenzel regime transition and contact angle hysteresis.

6.4.8 Summary

The presence of the Wenzel and Cassie-Baxter regimes is dependent upon the micro/nanostructure and droplet radius. Cassie-Baxter and Wenzel transition criteria have been proposed. Bhushan and Jung (2007, 2008) and Jung and Bhushan (2007, 2008a, b) proposed a transition criterion based on the pitch distance between the pillars and the curvature of the droplet governed by the Laplace equation, which relates the pressure inside the droplet to its curvature. In addition, the transition can occur by applying external pressure to the droplet, or by the impact of a droplet on the patterned surfaces (Jung and

Bhushan 2008b; Nosonovsky and Bhushan 2008e). Alternatively, Bhushan et al. (2007) and Nosonovsky and Bhushan (2007c, d) found that the transition occurs at a critical value of the spacing factor, a non-dimensional parameter that is defined as the diameter of the pillars divided by the pitch distance between them for patterned surfaces and its ratio to the droplet size.

The transition criteria were validated using micropatterned surfaces with various pitch values of micropillars and droplet radii. Droplet radii were varied by performing measurements during droplet evaporation. Contact angle measurements have also been made during condensation and evaporation of droplets.

To study the dynamic effect of droplets under various impact velocities, bouncing of a droplet experiments over micropatterned surfaces have been performed. For a given micropattern geometry, as the droplet hits the surface, the droplet bounced off below a certain critical velocity. Above the critical velocity, after the droplet hits the surface, it does not bounce off and the liquid-air interface changes to the solid-liquid interface due to dynamic impact. A transition model is verified with measurement data.

6.5 Ideal Surfaces with Hierarchical Structure

It was reported earlier that a hierarchical surface is needed to develop a composite interface with high stability. The structure of an ideal hierarchical surface is shown in Fig. 6.27. The asperities should be high enough so that the droplet does not touch the valleys. As an example, for a structure with circular pillars, the following relationship should hold for a composite interface; $\left(\sqrt{2}P - D\right)^2 / R = H$, (6.4). As an example, for a droplet with a radius on the order of 1 mm or larger, H on the order of 30 μm, and D on the order of 15 μm, a value of P on the order of 130 μm (Fig. 6.12) is optimum. Nanoasperities can pin the liquid-air interface and thereby prevent liquid from filling the valleys between asperities. They are also required to support nanodroplets, which may condense in the valleys between large asperities. Therefore, nanoasperities should have a small pitch to handle nanodroplets, less than 1 mm down to few nm radius. The values of H on the order of 10 nm and D on the order of 100 nm can be fabricated.

The structures were first fabricated using nature's route to verify that properties comparable to those found in natural objects can be realized (Sect. 6.6). Next, mechanically durable structures were fabricated guided by models and by using smart materials and fabrication techniques (Sect. 6.7).

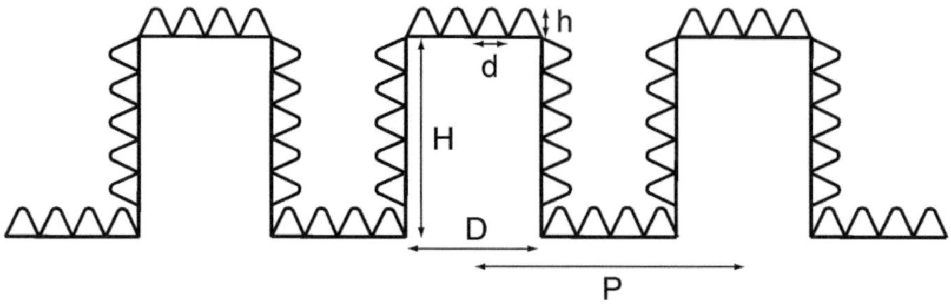

Fig. 6.27 Schematic of structure of an ideal hierarchical surface. Microasperities consist of the circular pillars with diameter *D*, height *H*, and pitch *P*. Nanoasperities consist of pyramidal nanoasperities of height *h* and diameter *d* with rounded tops

6.6 Hierarchically Structured Surfaces with Wax Platelets and Tubules Using Nature's Route

A hierarchical structure is composed of at least two levels of structuring in different length scales. Bhushan et al. (2008a, b, 2009c, d, 2012) and Koch et al. (2009) used nature's route to create various structures. They measured contact angle, contact angle hysteresis, adhesion, and self-cleaning efficiency and compared these with that of the lotus leaf. They also performed experiments to study the dynamic effects of bouncing and vibrating droplets in order to observe their influence on transition from the composite interface to the homogeneous interface.

They fabricated surfaces with a hierarchical structure with micropatterned epoxy replicas and lotus leaf microstructure, and created a second level of structuring with wax tubules and wax platelets. Tubules and platelets are the most common wax morphologies found in plant surfaces, and exist on superhydrophobic leaves. For example, lotus and *Colocasia esculenta* leaves consist of tubule and platelet morphologies, respectively. The structures developed mimic the hierarchical structures of superhydrophobic leaves. Two steps of the fabrication process include the production of microstructured surfaces by soft lithography and the subsequent development of nanostructures on top by self-assembly of plant waxes and artificial wax components.

A two-step molding process was used to fabricate several structurally identical copies of micropatterned Si surface and lotus leaves. The technique used is a fast, precise, and low cost molding process for biological and artificial surfaces (Koch et al. 2007, 2008). The technique was used to mold a microstructured Si surface with pillars 14 μm in diameter and 30 μm in height with a 23 μm pitch (Bhushan et al. 2008a, b, 2009c, d; Koch et al. 2009), fabricated by photolithography. This pitch value of 23 μm was selected to simulate lotus leaf pattern although an optimum pitch value of 45 μm was obtained in a study reported in Sect. 6.4.2. Before replication of the lotus leaf, the epicuticular wax tubules were removed in areas of approximately 6 cm^2. For this purpose, a two-component fast hardening glue was applied on the upper side of the leaves and was carefully pressed onto the leaf. After hardening, the glue with the embedded waxes was removed from the leaf, and the procedure was repeated (Koch et al. 2009).

The replication is a two-step molding process, in which first a negative replica of a template is generated and then a positive replica is generated, as shown in Fig. 6.28a (Bhushan et al. 2009d). A polyvinylsiloxane dental impression material (President Light Body® Gel, ISO 4823, Polyvinylsiloxan (PLB), Coltene Whaledent, Hamburg, Germany) was applied via a dispenser on the surface and immediately pressed down with a glass plate. After complete hardening of the molding mass (at room temperature for approximately 5 min), the silicon master surface and the mold (negative) were separated. After a relaxation time of 30 min for the molding material, the negative replicas were filled with a liquid epoxy resin (Epoxydharz L®, No. 236349, Conrad Electronics, Hirschau, Germany) with hardener (Harter S, Nr. 236365, Conrad Electronics, Hirschau, Germany). Specimens with microstructures were immediately transferred to a vacuum chamber at a pressure of 750 mTorr (100 Pa) for 10 s to remove trapped air and to increase resin infiltration through the structures. After hardening at room temperature (24 h at 22 °C), the positive replica was separated from the negative replica. The second step can be repeated to generate a number of replicas. The pillars of the master surface have been replicated without any morphological changes as shown in Fig. 6.28b (Bhushan et al. 2009d). The nanogrooves of a couple of hundred nm in lateral dimension present on the pillars of the master surface are shown to reproduce faithfully in the replica.

Nanostructures have been created by self-assembly of synthetic and plant waxes deposited by thermal evaporation. The alkane *n*-hexatriacontane ($C_{36}H_{74}$) has been used for the development of platelet nanostructures (Bhushan et al. 2008a, b, 2009d). Tubule forming waxes that were isolated from leaves of *Tropaeolum majus* (L.) and *Nelumbo nucifera*, referred to in the following as *T. majus*, and lotus, respectively, were used to create tubule structures (Bhushan et al. 2009c; Koch et al. 2009). The chemical structure of the major components of the wax forming tubule and alkane *n*-hexatriacontane are shown in Table 6.2. The complete chemistry of the plant waxes used is presented in Koch et al. (2006a). For a homogeneous deposition of the waxes and alkane, a thermal evaporation system, as shown in Fig. 6.29, has been used (Bhushan et al. 2009d). Specimens of smooth surfaces (flat silicon replicas) and microstructured replicas were placed in a vacuum chamber at 30 mTorr (4 kPa), 20 mm above a heating plate loaded with waxes of *n*-hexatriacontane (300, 500, or 1000 μg), *T. majus* wax (500, 1000, 1500, or 2000 μg), and lotus wax (2000 μg) (Bhushan et al. 2008a, b, 2009c, d; Koch et al. 2009). The wax was evaporated by heating it up to 120 °C. In a vacuum chamber the evaporation from the point source to the substrate occurs in a straight line; thus, the amount of sublimated material is equal in a hemispherical region over the point of source (Bunshah 1994). In order to estimate the amount of sublimated mass, the surface area of the half sphere was calculated by using the formula $2\pi r^2$, whereby the radius (r) represents the distance between the specimen to be covered and the heating plate with the substance to

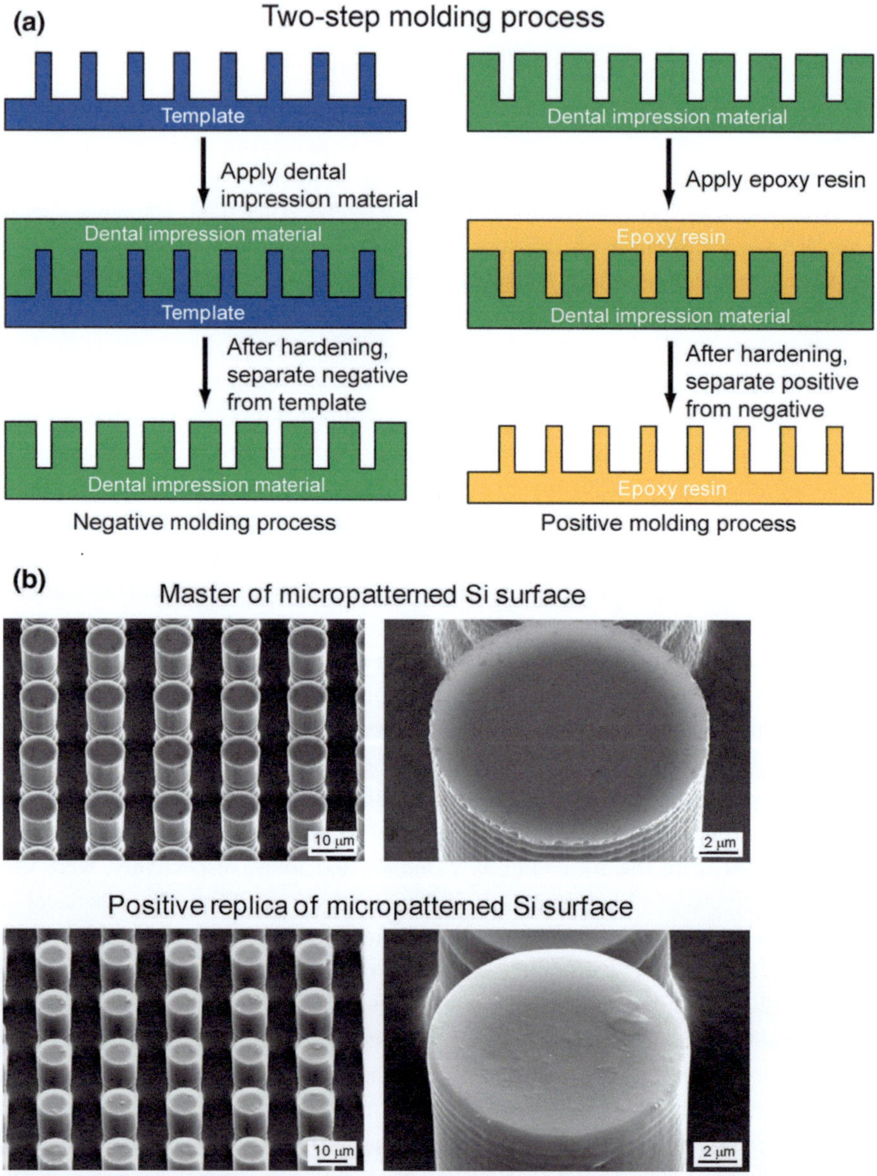

Fig. 6.28 **a** Schematic of two-step molding process used to fabricate microstructure, in which at first a negative is generated and then a positive, and **b** SEM micrographs of the master of micropatterned Si surface and positive replica fabricated from the master surface measured at 45° tilt angle (shown using two magnifications) (Bhushan et al. 2009d)

Table 6.2 Chemical structure of the major components of *n*-hexatriacontane, *T. majus*, and lotus waxes

n-hexatriacontane	$C_{36}H_{74}$	
Tropaeolum majus	Nonacosan-10-ol	$CH_3{-}(CH_2)_8{-}CH(OH){-}(CH_2)_{18}{-}CH_3$
	Nonacosane-4,10-diol	$CH_3{-}(CH_2)_2{-}CH(OH){-}(CH_2)_5{-}CH(OH){-}(CH_2)_{18}{-}CH_3$
Lotus	Nonacosane-10,15-diol	$CH_3{-}(CH_2)_8{-}CH(OH){-}(CH_2)_4{-}CH(OH){-}(CH_2)_{13}{-}CH_3$
	Nonacosan-10-ol	$CH_3{-}(CH_2)_8{-}CH(OH){-}(CH_2)_{18}{-}CH_3$

The major component is shown first (Bhushan et al. 2008a, 2009c; Koch et al. 2009)

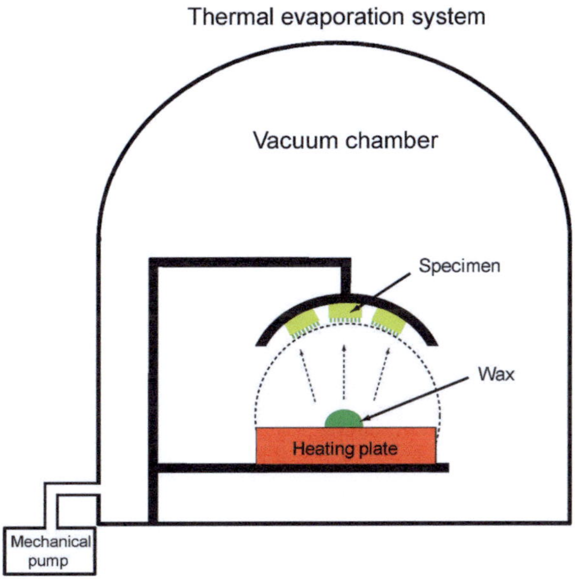

Fig. 6.29 Schematic of thermal evaporation system for self-assembly of a wax. Evaporation from the point source to the substrate occurs over hemispherical region (dotted line) (Bhushan et al. 2009d)

be evaporated. The amounts of wax deposited on the specimen surfaces were 0.12, 0.2 and 0.4 μg/mm^2 for *n*-hexatriacontane, and 0.2, 0.4, 0.6 and 0.8 μg/mm^2 for *T. majus*, and 0.8 μg/mm^2 for lotus waxes.

After coating, the specimens with *n*-hexatriacontane were placed in a desiccator at room temperature for three days for crystallization of the alkanes. A stable stage was indicated by no further increase of crystal sizes.

For the plant waxes that are a mixture of aliphatic components, different crystallization conditions were chosen. It was reported by Niemietz et al. (2009) that an increase of temperature from 21 °C (room temperature) to 50 °C had a positive effect on the mobilization and diffusion of wax molecules, required for separation of the tubule forming molecules. It is also known that chemical ambient condition has an influence on the propensity of wax crystallization, thus the specimens with evaporated plant waxes (*T. majus* and lotus) were stored for three days at 50 °C in a crystallization chamber, where they were exposed to a solvent (ethanol) in vapor phase (Fig. 6.30). Specimens were placed on metal posts, and a filter paper wetted with 20 mL of the solvent was placed below the specimens. Slow diffusive loss of the solvent in the chamber was provided by placing a thin filter paper between the glass body and the lid. After evaporation of the solvent, specimens were left in the oven at 50 °C for a total of seven days. Figure 6.31 shows the nanostructures formed by lotus wax, seven days after wax deposition on flat surfaces. Figure 6.31a shows the nanostructure after storage at 21 °C; in these no tubules were grown. Figure 6.31b shows that lotus waxes exposed to ethanol vapor for three days at 50 °C formed wax tubules. A detailed description of the nanostructure sizes and nanoroughness is given in the following.

Flat films of *n*-hexatriacontane and wax tubules were made by heating the substances above their melting point and rapidly cooling them down. This procedure interrupts the crystallization and leads to smooth films (Bhushan et al. 2008a, b, 2009c, d; Koch et al. 2009).

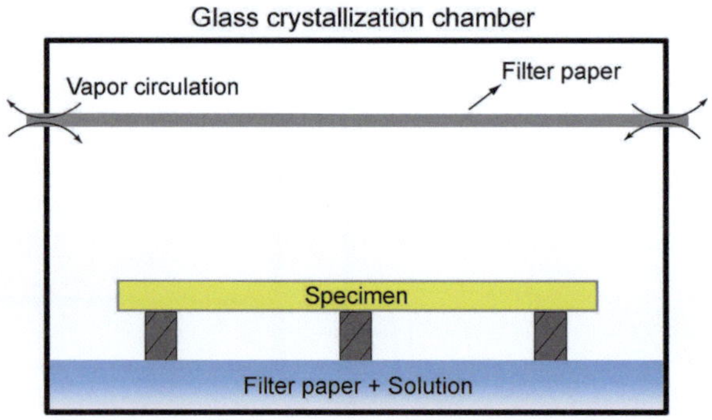

Fig. 6.30 Schematic of a glass recrystallization chamber used for tubules formation. The filter paper placed at the bottom of the chamber was wetted with 20 mL of the solvent, and slow evaporation of the solvent was provided by placing a thin filter paper between the glass body and the cap placed above. The total volume of the chamber is about 200 cm^3 (Bhushan et al. 2009b)

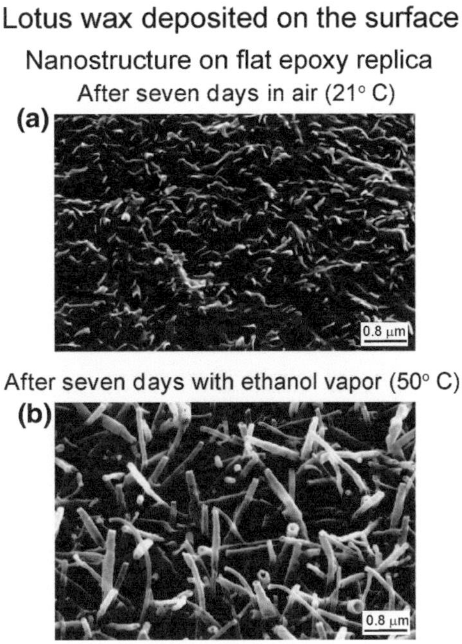

Fig. 6.31 SEM micrographs of morphology of the lotus wax deposited on the flat epoxy replica surface after two treatments of specimens measured at 45° tilt angle, **a** after seven days at 21 °C (top), nanostructure on flat epoxy replica was found with no tubules, and **b** after seven days at 50 °C with ethanol vapor (bottom), tubular nanostructures with random orientation were found on the surface (Koch et al. 2009)

6.6.1 Effect of Nanostructures with Various Wax Platelet Crystal Densities on Superhydrophobicity

Nanostructures with various wax platelet crystal densities were fabricated to study the effect of crystal density on superhydrophobicity. Figure 6.32a shows the SEM micrographs of a flat surface and nanostructures fabricated with various masses of *n*-hexatriacontane (Bhushan et al. 2008a). The nanostructure is formed by three-dimensional platelets of *n*-hexatriacontane, as shown in detail in Fig. 6.32b. Platelets are flat crystals, grown perpendicular to the substrate surface. They are randomly distributed on the surface, and their shapes and sizes show some variations. Some of the single platelets are connected to their neighboring crystals at their lateral ends. This arrangement leads to a kind of cross-linking of the single platelets. As shown in Fig. 6.32b, and based on additional specimens, the platelet thickness varied between 50 and 100 nm, and their length varied between 500 and 1000 nm. The self-assembly of *n*-hexatriacontane and most other long chain hydrocarbons leads to layered structures with a lamellae order. In these structures, the molecular axis is orientated parallel to the substrate surface. The growth of these layers results in an ordered, crystalline 3-D structure (Dorset et al. 1983). The created nanostructures are comparable to the wax crystal morphology found on superhydrophobic leaves, e.g., *Colocasia esculenta* (Neinhuis and Barthlott 1997) and *Triticum aestivum* (wheat) (Koch et al. 2006b). SEM micrographs of the nanostructures fabricated with three different masses of *n*-hexatriacontane show different densities of crystals.

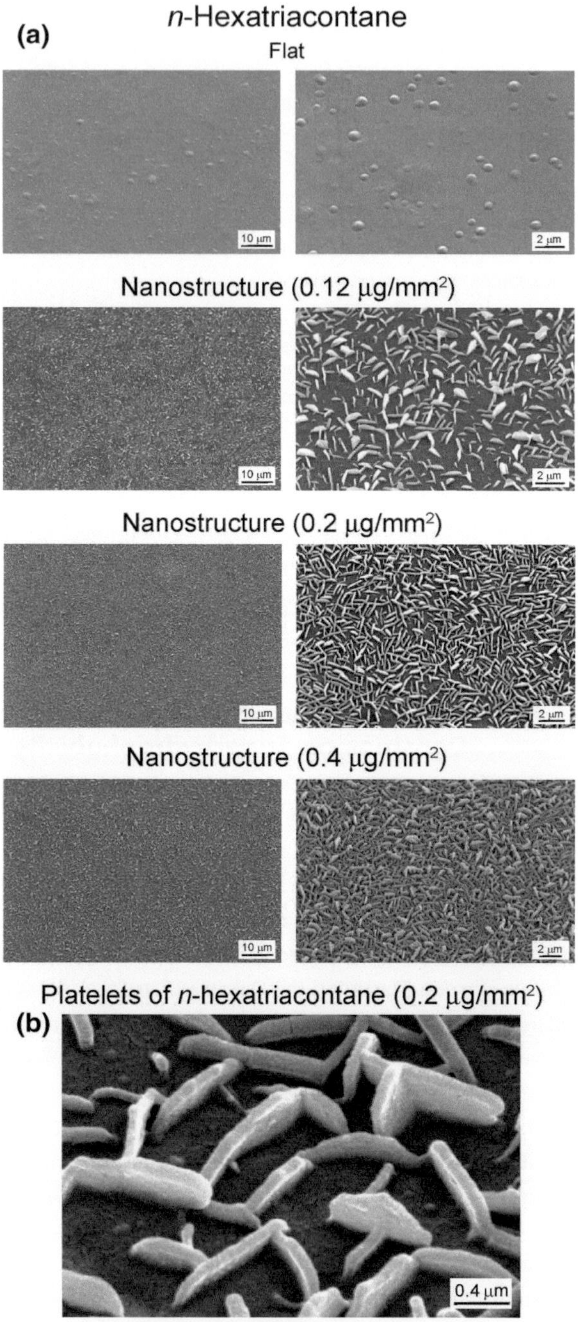

Fig. 6.32 SEM micrographs taken at 45° tilt angle (shown using two magnifications) of **a** the flat surface and nanostructures fabricated with various mass of *n*-hexatriacontane and **b** three-dimensional platelets forming nanostructures on the surface fabricated with 0.2 μg/mm^2 mass of *n*-hexatriacontane. All samples are fabricated with epoxy resin coated with *n*-hexatriacontane (Bhushan et al. 2008a)

Table 6.3 Roughness statistics for a nanostructured surface measured using an AFM (Scan size 10 μm × 10 μm). Nanostructures were fabricated with *n*-hexatriacontane, *T. majus*, and lotus waxes (Bhushan et al. 2008a, 2009c; Koch et al. 2009)

	RMS height (nm)	Peak to valley height (nm)	η (/sq. μm)
n-hexatriacontane			
Nanostructure (0.12 μg/mm^2)	46	522	0.78
Nanostructure (0.2 μg/mm^2)	65	663	1.39
Nanostructure (0.4 μg/mm^2)	82	856	1.73
Tropaeolum majus wax			
Nanostructure (0.8 μg/mm^2)	180	1570	0.57
Lotus wax			
Nanostructure (0.8 μg/mm^2)	187	1550	1.47

RMS–root mean square
η–summit density

AFM was used to characterize height distributions of the nanostructures. Statistical parameters of nanostructures [RMS height, peak to valley height, and summit density (η)] (Bhushan 2013a, b) were calculated and are presented in Table 6.3. A summit is defined as a point whose height is greater than its four nearest neighboring points above a threshold value of 10% of RMS height to avoid measurement errors. The measurement results were reproducible within ±5%.

To study the effect of nanostructures with different crystal density on superhydrophobicity, static contact angle, contact angle hysteresis, tilt angle, and adhesive forces were measured on nanostructures fabricated with various masses of *n*-hexatriacontane (Bhushan et al. 2008a). For contact angle hysteresis, the advancing and receding contact angles were measured at the front and back of the droplet moving along the tilted surface, respectively. The data are shown in Fig. 6.33. The static contact angle of a flat surface coated with a film of *n*-hexatriacontane was 91°. It showed a contact angle hysteresis of 87°, and the droplet still adhered at a tilt angle of 90°. Nanostructuring of flat surfaces with *n*-hexatriacontane platelets creates superhydrophobic surfaces with a high static contact angle and a reduction of contact angle hysteresis and tilt angle. The values are a function of crystal density. Figure 6.34 shows a plot of static contact angle and contact angle hysteresis as a function of the mass of *n*-hexatriacontane deposited. As the mass of *n*-hexatriacontane increased, the static contact angle first increased, and the contact angle hysteresis decreased. Then, above a mass of 0.2 μg/mm^2, static contact angle gradually decreased and contact angle hysteresis gradually increased with increasing mass. The highest static contact angle and lowest contact angle hysteresis are 158° and 23° at a mass of 0.2 μg/mm^2.

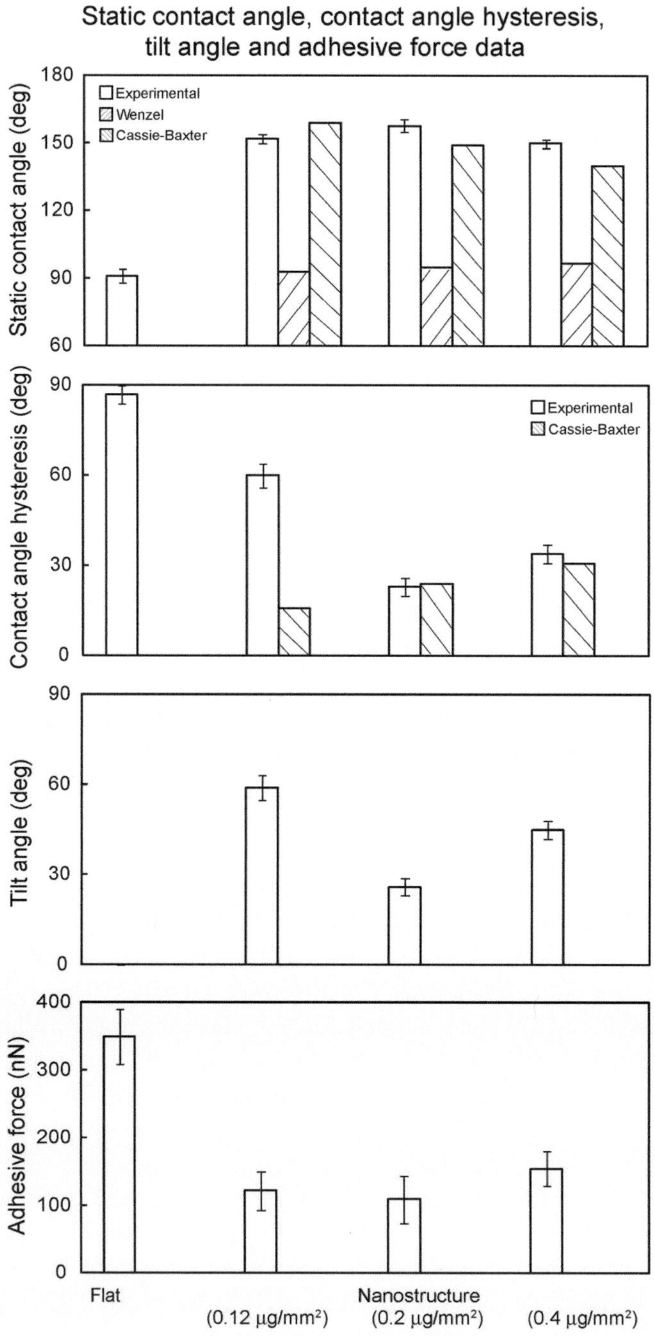
Static contact angle, contact angle hysteresis, tilt angle and adhesive force data
Static contact angle (deg)
180
150
120
90
60
Experimental
Wenzel
Cassie-Baxter
Contact angle hysteresis (deg)
90
60
30
0
Experimental
Cassie-Baxter
Tilt angle (deg)
90
60
30
0
Adhesive force (nN)
400
300
200
100
0
Flat
Nanostructure
(0.12 μg/mm²)
(0.2 μg/mm²)
(0.4 μg/mm²)

◀**Fig. 6.33** Bar chart showing the measured static contact angle, contact angle hysteresis and tilt angle on flat surface and nanostructures fabricated with various masses of *n*-hexatriacontane. Also shown are calculated static contact angles obtained using Wenzel and Cassie-Baxter equations with a given value of θ_0, and calculated contact angle hysteresis using Cassie-Baxter equation The droplet on flat surface does not move along the surface even at tilt angle of 90°. The bar chart also shows adhesive forces for various structures, measured using a 15 μm radius borosilicate tip. The error bar represents ±1 standard deviation (Bhushan et al. 2008a)

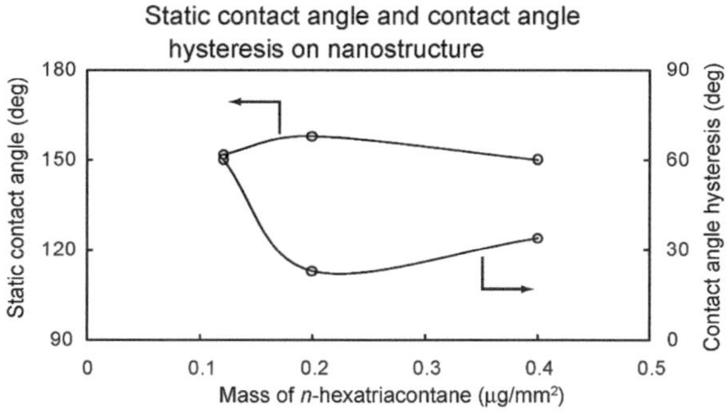

Fig. 6.34 Static contact angle and contact angle hysteresis as a function of mass of *n*-hexatriacontane deposited on nanostructures (Bhushan et al. 2008a)

The adhesive force data presented in Fig. 6.33 shows that adhesive forces of coated surfaces are lower than that of uncoated flat surface because of superhydrophobic properties. In addition, adhesive forces of the nanostructured surfaces were lower than that of the flat surface because of the lower contact area between the tip and the structured surface (Bhushan 2013a, b).

To identify wetting regimes (Wenzel or Cassie-Baxter), as well as to understand the effect of crystal density on the propensity of air pocket formation for the nanostructured surfaces, roughness factor (R_f) and fractional liquid-air interface (f_{LA}) are needed. The R_f for the nanostructures was calculated using the AFM map (Burton and Bhushan 2006; Bhushan and Jung 2006). The calculated results were reproducible within ±5%. The R_f for the nanostructured surfaces with masses of 0.12, 0.2, and 0.4 μg/mm^2 were found to be 3.4, 4.9, and 6.8, respectively. For calculation of f_{LA} of the nanostructures, only the higher crystals are assumed to come in contact with a water droplet. The fractional geometrical area of the top surface for the nanostructures was calculated from SEM micrographs with top view (0° tilt angle). The SEM images were converted to high contrast black and white images using Adobe Photoshop. The increase of contrast in the SEM image eliminates the smaller platelet structures, which were visible in the original SEM image. The higher crystals led to white signals in the SEM figure. The fractional geometrical area of the top nanostructured surfaces with masses of 0.12, 0.2, and 0.4 μg/mm^2 was found to be 0.07, 0.15, and 0.24, leading to f_{LA} of 0.93, 0.85, and 0.76, respectively. The calculated results were reproducible within ±5%. The values of static contact angle in the Wenzel and Cassie-Baxter regimes for the nanostructured surfaces were calculated using the values of R_f and f_{LA} and presented in Fig. 6.33. The values of contact angle hysteresis in Cassie-Baxter regimes for various surfaces were calculated using equations from Chap. 3. The data are presented in Fig. 6.33.

As shown in Fig. 6.33, the experimental static contact angle and contact angle hysteresis values for the two nanostructured surfaces with 0.2 and 0.4 μg/mm^2 were comparable to the calculated values in the Cassie-Baxter regime. The results suggest that a droplet on two nanostructured surfaces should exist in the Cassie-Baxter regime. However, for the nanostructured surface with 0.12 μg/mm^2, the experimental static contact angle was lower and contact angle hysteresis was higher than the calculated values in the Cassie-Baxter regime. It is believed that neighboring crystals are separated at lower crystal density and any trapped air can be squeezed out. Conversely, neighboring crystals are interconnected at higher densities and air remains trapped. At highest crystal density at a mass of 0.4 μg/mm^2, there is less open volume compared to that at 0.2 μg/mm^2, which explains a droplet static contact angle going from 158° to 150° (Bhushan et al. 2008a).

6.6.2 Effect of Hierarchical Structure with Wax Platelets on the Superhydrophobicity

To study the influence of hierarchical structure on superhydrophobicity, Bhushan et al. (2008b, 2009d) created surfaces with *n*-hexatriacontane of 0.2 μg/mm^2. Figure 6.35 shows the SEM micrographs of a flat surface and nano-, micro-, and hierarchical structures. To study the effect of structure on superhydrophobicity, static contact angle, contact angle hysteresis, tilt angle, and adhesive forces of the four structures were measured. The data are shown in Fig. 6.36. The static contact angle of a flat surface coated with a film of *n*-hexatriacontane was 91°, and increased to 158° when *n*-hexatriacontane formed a nanostructure of platelets on it. The static contact angle on a flat specimen with a microstructure was 154°, but it increased to 169° for the hierarchical surface structure. Contact angle hysteresis and tilt angle for flat, micro-, and nanostructured surfaces show similar trends. The flat surface showed a contact angle hysteresis of 87°, and the droplet still adhered at a tilt angle of 90°. The superhydrophobic micro- and nanostructured surfaces showed a reduction of contact angle hysteresis and tilt angle, but a water droplet still needs a tilt angle of 26° and 51°, respectively, before sliding. Only the hierarchical surface structure with static contact angle of 169° and low contact angle hysteresis of 2° exceeds the basic criteria for superhydrophobic and self-cleaning surfaces (Bhushan and Jung 2008).

Adhesive force data are also presented in Fig. 6.36. As observed earlier, the data shows that adhesive forces of coated surfaces are lower than that of the uncoated flat surface because of superhydrophobic properties. Adhesion force of the hierarchical surface structure was lower than that of both micro- and nanostructured surfaces because the contact between the tip and surface was lower (Bhushan 2013a, b).

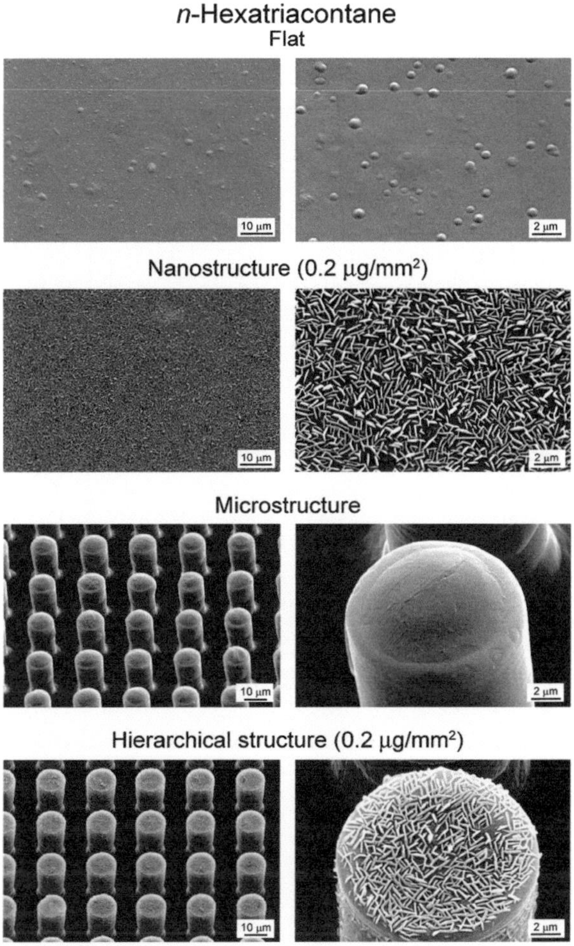

Fig. 6.35 SEM micrographs of the flat surface, nanostructure, microstructure, and hierarchical structure measured at 45° tilt angle (shown using two magnifications). All samples are fabricated with epoxy resin coated with *n*-hexatriacontane (Bhushan et al. 2008b)

In order to identify wetting regimes (Wenzel or Cassie-Baxter) for the various surfaces, roughness factor (R_f) and fractional liquid- air interface (f_{LA}) are needed. The R_f for the nanostructure was described earlier. The R_f for the microstructure was calculated for the geometry of flat-top, cylindrical pillars of diameter *D*, height *H*, and pitch *P* distributed in a regular square array. For this case, the roughness factor for the microstructure is given by $\left(R_f\right)_{micro} = \left(1 + \frac{\pi DH}{P^2}\right)$. The roughness factor for the hierarchical structure is the sum of $(R_f)_{micro}$ and $(R_f)_{nano}$. The values calculated for various surfaces are summarized in Table 6.4.

For the calculation of f_{LA}, the following assumptions are made. For the microstructure, consider that a droplet in size much larger than the pitch *P* contacts only the flat-top of the pillars in the composite interface, and the cavities are filled with air. For microstructure, the fractional flat geometrical area of the liquid-air interface under the droplet is

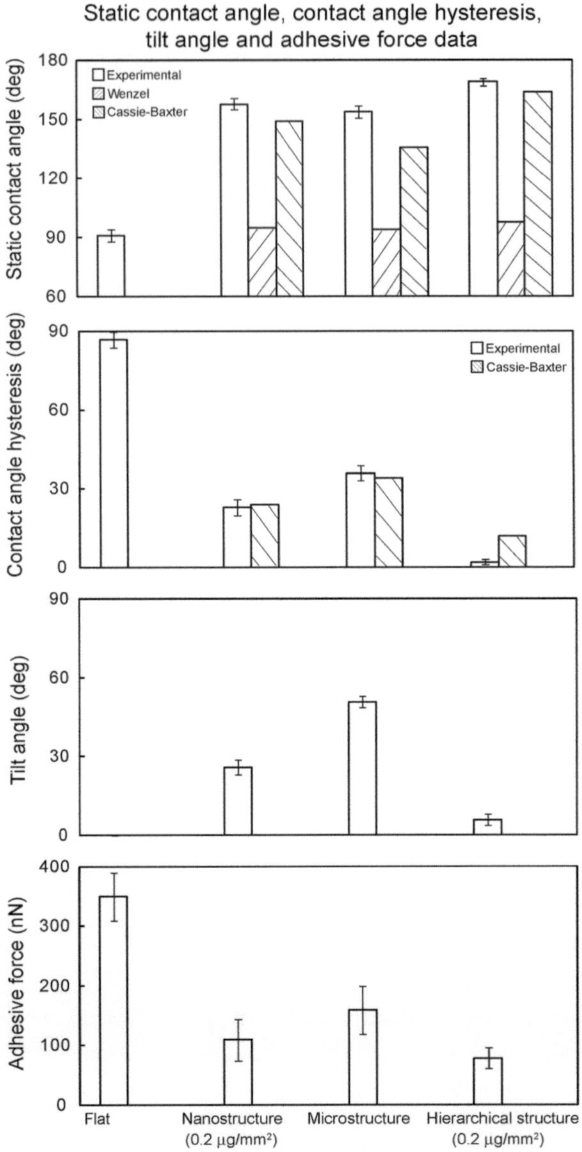

Fig. 6.36 Bar chart showing the measured static contact angle, contact angle hysteresis and tilt angle; also shown are calculated static contact angles obtained using Wenzel and Cassie-Baxter equations with a given value of θ_0, and calculated contact angle hysteresis using Cassie-Baxter equation on various structures. The droplet on a flat surface does not move along the surface even at a tilt angle of 90°. The bar chart also shows adhesive forces for various structures, measured using a 15 μm radius borosilicate tip. The error bar represents ±1 standard deviation (Bhushan et al. 2008b)

given by $(f_{LA})_{micro} = \left(1 - \frac{\pi D^2}{4P^2}\right)$ (Bhushan and Jung 2008). The fractional geometrical area of the top surface for the nanostructure was described earlier. For the hierarchical structure, the fractional flat geometrical area of the liquid-air interface is given by $(f_{LA})_{hierarchical} = 1 - \left(\frac{\pi D^2}{4P^2}\right)\left[1 - (f_{LA})_{nano}\right]$. The values of contact angle hysteresis in Cassie-Baxter regimes for various surfaces were calculated using equations from Chap. 3. The values are summarized in Table 6.4.

Table 6.4 Summary of static contact angles and contact angle hysteresis measured and calculated for droplets in the Wenzel regime and the Cassie-Baxter regime on the various surfaces with *n*-hexatriacontane using the calculated values of R_f and f_{LA} (Bhushan et al. 2008b, 2009d)

			Static contact angle (°)			Contact angle hysteresis (°)	
	R_f	f_{LA}	Measured	Calculated in Wenzel regime	Calculated in Cassie-Baxter regime	Measured	Calculated in Cassie-Baxter regime
Flat			91			87[a]	
Nanostructure	4.9	0.85	158	95	149	23	24
Microstructure	3.5	0.71	154	94	136	36	34
Hierarchical structure	8.4	0.96	169	98	164	2	12

[a]Advancing and receding contact angles are 141° and 54°, respectively

The values of static contact angle in the Wenzel and Cassie-Baxter regimes for various surfaces were calculated using the values of R_f and f_{LA} (Table 6.4). As shown in Fig. 6.36, the experimental static contact angle values for the three structured surfaces were larger than the calculated values in the Cassie-Baxter regime. The results suggest that the droplets on three of the structured surfaces were in the Cassie-Baxter regime. This indicates that the microstructure and nanostructure surface induce air pocket formation. For the contact angle hysteresis, there is good agreement between the experimental data and the theoretically predicted values for the Cassie-Baxter regime. These results show that air pocket formation in the micro- and nanostructure decreases the solid-liquid contact (Koch et al. 2009). In hierarchically structured surfaces, the air pocket formation further decreases the solid-liquid contact, and thereby reduces contact angle hysteresis and tilt angle (Bhushan et al. 2008b, 2009d).

6.6.3 Effect of Hierarchical Structure with Wax Tubules on Superhydrophobicity

6.6.3.1 *T. Majus* Tubules

Figure 6.37a shows the SEM micrographs of the nanostructure and hierarchical structure fabricated with two different masses (0.6 and 0.8 μg/mm²) of *T. majus* wax (Bhushan et al. 2009c). SEM micrographs show an increase in the tubule amount on the flat and microstructure surfaces after deposition of higher masses of wax. The tubules of *T. majus* wax grown in an ethanol atmosphere are comparable to the wax morphology found on the leaves of *T. majus*. Surfaces show a homogeneous distribution of the wax mass on the specimen surfaces, and tubules provide the desired nanostructure of three-dimensional tubules on the flat and microstructure surfaces. The tubule morphology of *T. majus* wax is shown in detail in Fig. 6.37b. The tubular crystals are hollow structures, randomly oriented on the surface and embedded into an amorphous wax layer. They are randomly distributed on the surface, and their shapes and sizes show some variations. As shown in Fig. 6.37b, and based on additional specimens, the tubular diameter varied between 100 and 300 nm, and their length varied between 300 and 1200 nm.

AFM was used to characterize the nanostructure fabricated using *T. majus* wax of 0.8 μg/mm^2 after storage at 50 °C with ethanol vapor (Bhushan et al. 2009c). Statistical parameters of the nanostructure [RMS height, peak to valley height, and summit density (η)] were calculated and are presented in Table 6.3 (Bhushan 2013a, b).

To study the effect of structures on superhydrophobicity, static contact angle, contact angle hysteresis, tilt angle, and adhesive forces of the four structures produced using *T. majus* wax were measured (Bhushan et al. 2009c). Nanostructures formed on the flat and microstructured surfaces were fabricated using *T. majus* wax of 0.8 μg/mm^2 after

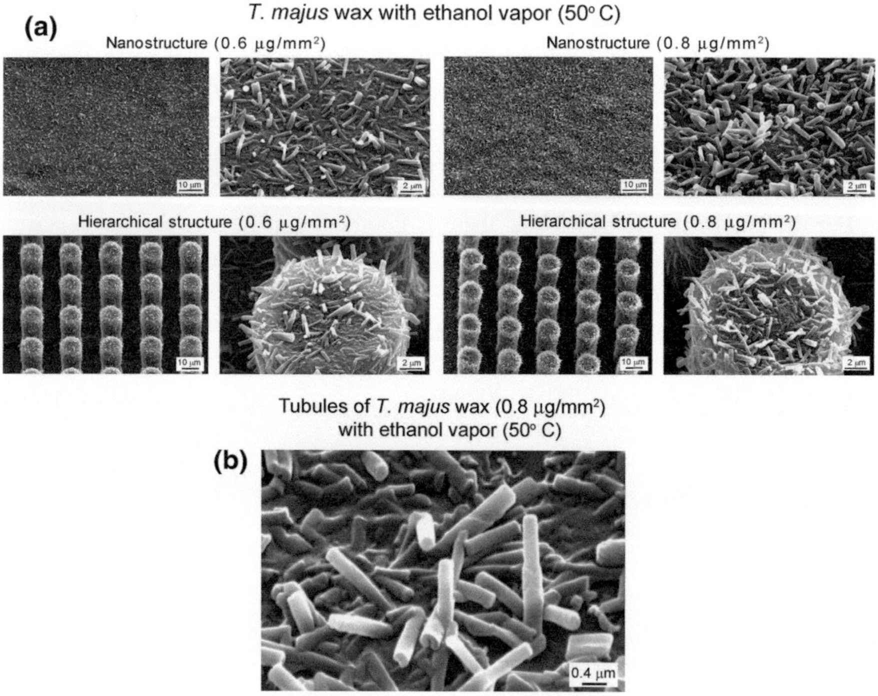

Fig. 6.37 SEM micrographs taken at 45° tilt angle (shown using two magnifications) of **a** the nanostructure and hierarchical structure fabricated with two different mass (0.6 and 0.8 μg/mm^2) of *T. majus* wax after storage at 50 °C with ethanol vapor and **b** three-dimensional tubules forming nanostructures on the surface fabricated with 0.8 μg/mm^2 mass of *T. majus* wax after storage at 50 °C with ethanol vapor (Bhushan et al. 2009c)

storage at 50 °C with ethanol vapor. The data are shown in Fig. 6.38. The static contact angle of a flat surface coated with a film of *T. majus* wax was 112°, and increased to 164° when *T. majus* wax formed a nanostructure of tubules on it. On the flat specimen with a microstructure on it, the static contact angle was 154°, but increased to 171° for the hierarchical surface structure. Contact angle hysteresis and tilt angle for flat, micro-, and nanostructured surfaces show similar trends. The flat surface showed a contact angle hysteresis of 61° and a tilt angle of 86°. The microstructured surface shows a reduction of contact angle hysteresis and tilt angle, however a water droplet still needs a tilt angle of 31° before sliding. As tubules were formed on the flat and microstructured surfaces, the nanostructured and hierarchically structured surfaces had low contact angle hysteresis of 5° and 3°, respectively. These properties are superior to plant leaves, including lotus leaves.

Adhesive force data are also presented in Fig. 6.38. As observed earlier, adhesive forces of coated surfaces are lower than that of the uncoated flat surfaces because of superhydrophobic properties. Adhesion force of the hierarchical surface structure was lower than that of micro- and nanostructured surfaces because the contact between the tip and surface was lower (Bhushan 2013a, b).

To further verify the effect of hierarchical structure on the propensity of air pocket formation, Bhushan et al. (2009c) performed evaporation experiments with a droplet on a microstructure and a hierarchical structure fabricated with 0.8 $\mu m/mm^2$ mass of *T. majus* wax with ethanol vapor at 50 °C. Figure 6.39 shows the successive photos of a droplet evaporating on the two structured surfaces. On the microstructured surface, the light passes below the droplet and air pockets can be seen. So, to begin, the droplet was in the Cassie-Baxter regime. When the radius of the droplet decreased to 425 μm, the air pockets were not visible anymore, and the droplet was in the Wenzel regime. This transition resulted from an impalement of the droplet in the patterned surface, characterized by a smaller contact angle. For the hierarchical structure, an air pocket was clearly visible at the bottom area of the droplet throughout, and the droplet was in a hydrophobic state until the droplet evaporated completely. This suggests that a hierarchical structure with nanostructures prevents liquid from filling the gaps between the pillars.

6.6.3.2 Lotus Tubules

Lotus wax was used for the development of nanostructures by tubule formation (Koch et al. 2009). Figure 6.40a shows the SEM micrographs of flat surfaces with the tubule nanostructure. The microstructures shown in Fig. 6.40b are the lotus leaf and the micropatterned Si replica covered with a lotus wax film. Hierarchical structures were fabricated with microstructured lotus leaf replicas and micropatterned Si replicas covered with a nanostructure of lotus wax tubules, as shown in Fig. 6.40c. SEM micrographs show an overview (left column), a detail in higher magnification (middle column), and a large magnification of the created flat wax layers and tubule nanostructures (right column). The grown tubules provide the desired nanostructure on flat and microstructured surfaces. The

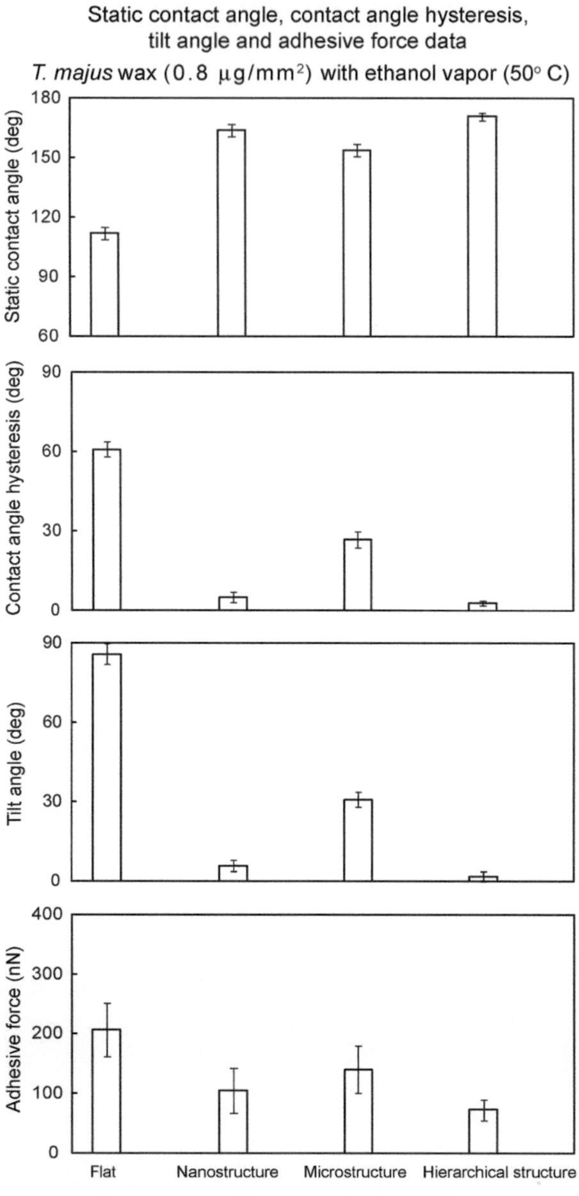

Fig. 6.38 Bar chart showing the measured static contact angle, contact angle hysteresis and tilt angle on various structures fabricated with 0.8 μm/mm^2 mass of *T. majus* wax after storage at 50 °C with ethanol vapor. The bar chart also shows adhesive forces for various structures, measured using a 15 μm radius borosilicate tip. The error bar represents ±1 standard deviation (Bhushan et al. 2009c)

recrystallized lotus wax shows tubular hollow structures, with random orientation on the surfaces. Their shapes and sizes show only a few variations. The tubular diameter varied between 100 and 150 nm, and their length varied between 1500 and 2000 nm. AFM was used to characterize the nanostructure of the lotus wax tubules. The statistical parameters

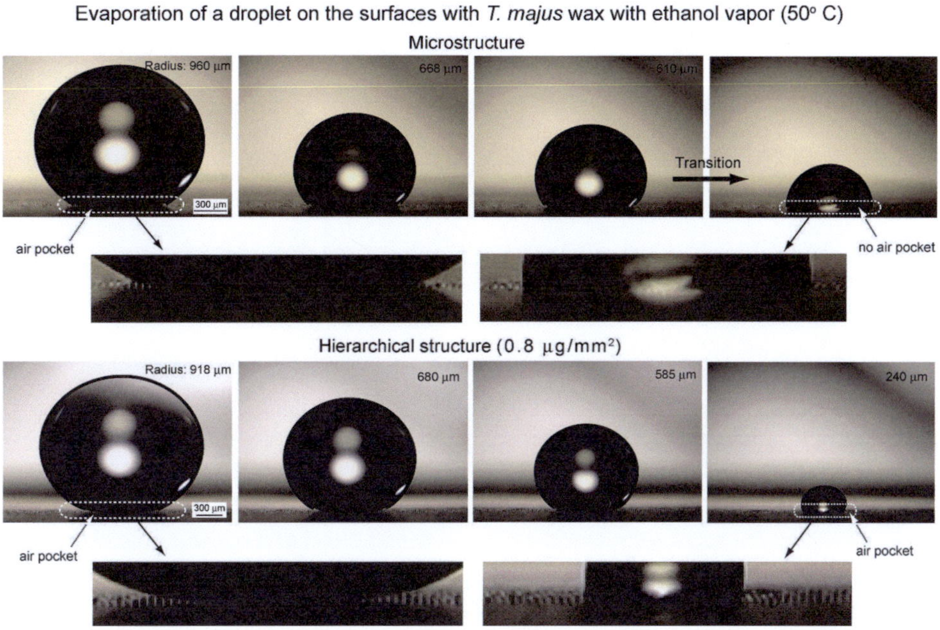

Fig. 6.39 Evaporation of a droplet on microstructured and hierarchically structured surfaces fabricated with 0.8 μm/mm^2 mass of *T. majus* wax after storage at 50 °C with ethanol vapor. The initial radius of the droplet was about 950 μm, and the time interval between first two photos was 180 s and between the latter was 60 s. As the radius of droplet reached 425 μm (foot print = 836 μm) on the microstructured surface, the transition from Cassie-Baxter regime to Wenzel regime occurred, as indicated by the arrow. On the hierarchically structured surface, air pockets are visible at the bottom area of the droplet, and exist until the droplet evaporated completely (Bhushan et al. 2009b)

of the nanostructure (RMS height, peak to valley height, and summit density (η)) were calculated and are presented in Table 6.3.

To study the effect of lotus wax tubule nanostructures on superhydrophobicity, static contact angle, contact angle hysteresis, and tilt angle were measured on flat, microstructured lotus replica, micropatterned Si replica, and hierarchical surfaces. Hierarchical surfaces were made of the lotus leaf replica and micropatterned Si replica with a nanostructure of wax tubules on top. Additionally, fresh lotus leaves were investigated to compare the properties of the fabricated structures with the original biological model.

Figure 6.41 shows that the highest static contact angles of 173°, lowest contact angle hysteresis of 1°, and tilt angle varying between 1° and 2° were found for the hierarchically structured Si replica. The hierarchically structured lotus leaf replica showed a static contact angle of 171°, and the same contact angle hysteresis (2°) and tilt angles of 1°–2° as the hierarchical Si replica. The fresh lotus leaf surface investigated here showed a static contact angle of 164°, contact angle hysteresis of 3°, and a tilt angle of 3° which

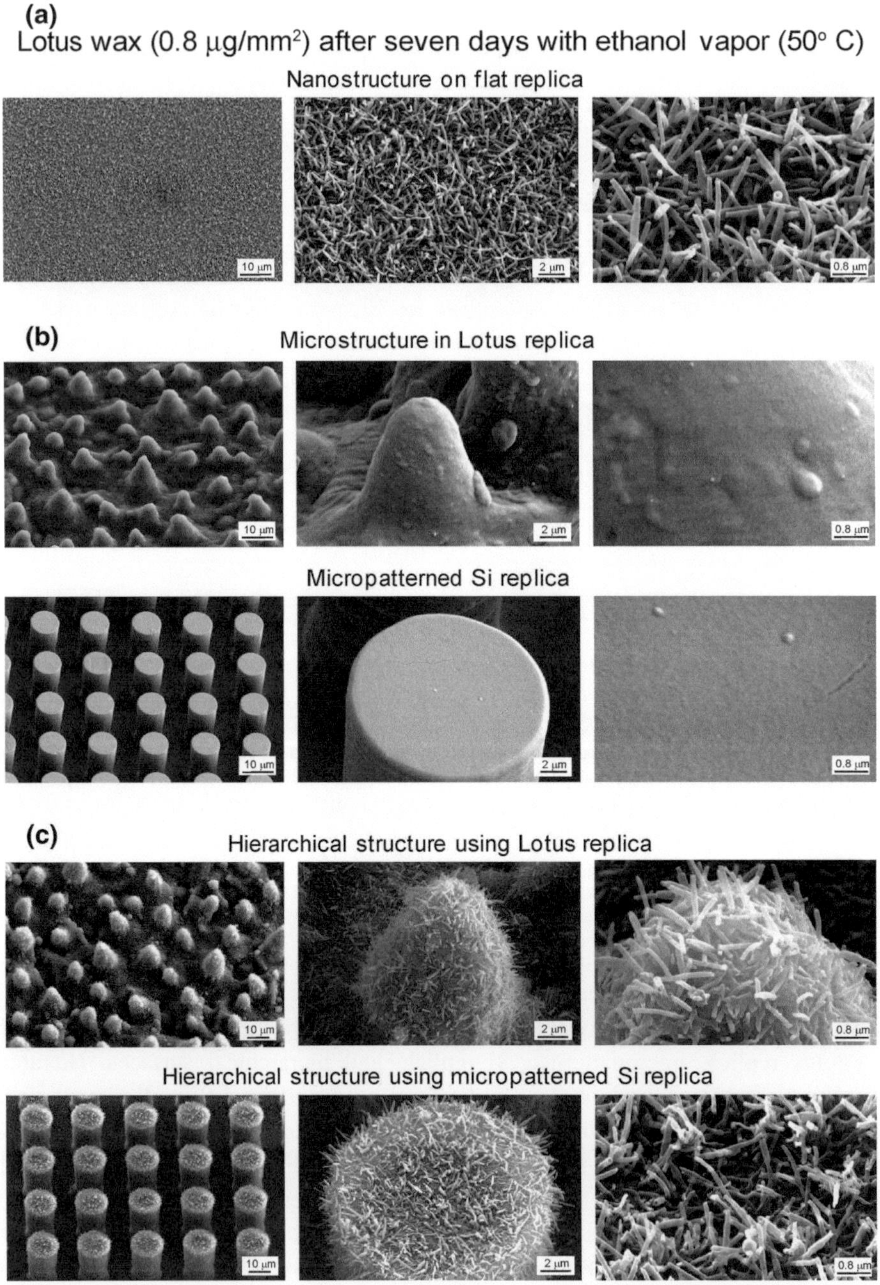

Fig. 6.40 SEM micrographs taken at 45° tilt angle (shown using three magnifications) of **a** nanostructure on flat replica, **b** microstructures in lotus replica and micropatterned Si replica, and **c** hierarchical structure using lotus and micropatterned Si replicas. Nano- and hierarchical structures were fabricated with mass 0.8 $\mu g/mm^2$ of lotus wax after storage for seven days at 50 °C with ethanol vapor. Flat epoxy resin and microstructure were covered with flat lotus wax (Koch et al. 2009)

suggests that the artificial hierarchical surfaces showed higher static contact angle and lower contact angle hysteresis. Structural differences between the original lotus leaf and the artificial lotus leaf produced here are limited to a difference in wax tubules length, which are 0.5–1 μm longer in the artificial lotus leaf (Koch et al. 2009).

The melting of the wax led to a flat surface with a flat wax film with a much lower static contact angle (119°), a higher contact angle hysteresis (71°), and a high tilt angle of 66°. The data of a flat lotus wax film on a flat replica show that the lotus wax by itself is hydrophobic. The data demonstrate that the native, flat wax of lotus leaves, with a static contact angle of 119°, is hydrophobic and can turn superhydrophobic (167°) by increasing the surface roughness after self-assembly into three-dimensional wax tubules. The static contact angle of the lotus wax film is 119°, which is higher than that of the wax film made of *T. majus* of 112°. However, films made of *n*-hexatriacontane showed static contact angles of only 91°. SEM investigations, made directly after contact angle measurements, revealed no morphological differences between these films. Based on the chemical composition, it should be assumed that the non-polar *n*-hexatriacontane molecules are more hydrophobic than the plant waxes, which contain high amounts of oxygen atoms. At this point these differences cannot be explained by structural or chemical differences of the films, but will be of interest for further studies.

Adhesive force data are also shown in Fig. 6.41 (Koch et al. 2009). As observed earlier, the adhesive force data is correlated to the wetting properties. Adhesion force of the hierarchical surface structure was lower than that of micro- and nanostructured and flat surfaces because the contact between the tip and surface was lower. However, for the fresh lotus leaf, there is moisture within the plant material, which causes softening of the leaf, and so when the tip comes into contact with the leaf sample, the sample deforms, and a larger area of contact between the tip and sample causes an increase in the adhesive force (Bhushan 2013a, b).

6.6.4 Self-cleaning Efficiency of Hierarchically Structured Surfaces

For deposition of contamination on artificial surfaces, various structures were placed in a glass contamination chamber (Bhushan et al. 2009a). In order to obtain a relative measure of the self-cleaning ability of hierarchical structures which exhibit the lowest contact angle hysteresis and tilt angle as compared to other structures (flat, nanostructures, and microstructures), the tilt angle chosen for the cleaning tests was slightly above the tilt angle for the hierarchical structures. Thus experiments were performed with 10° for surfaces covered with *n*-hexatriacontane and 3° for surfaces with lotus wax.

The cleaning tests were performed on various surfaces with n-hexatriacontane and lotus wax (Bhushan et al. 2009a). The data represent the average of five different investigated areas for each experiment. Figure 6.42 shows that none of the investigated surfaces were fully cleaned by rinsing with water. For lotus wax, which forms tubule nanostructures,

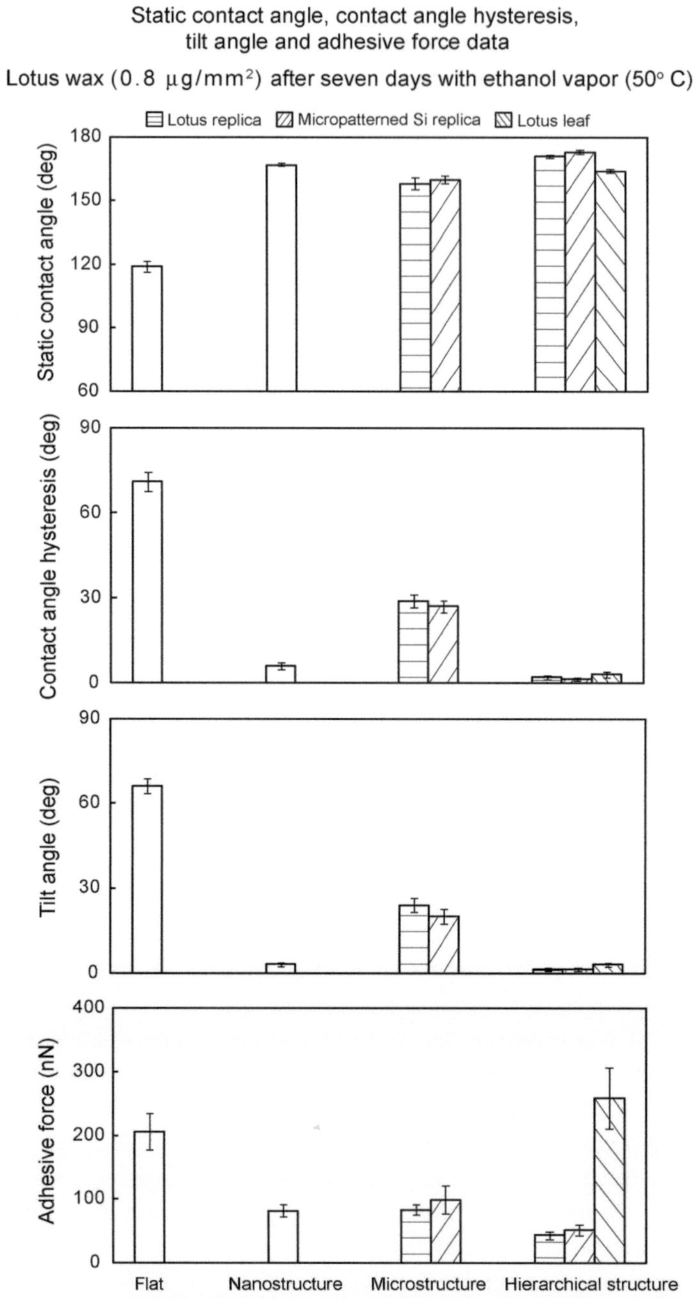

Fig. 6.41 Bar chart showing the measured static contact angle, contact angle hysteresis and tilt angle on various structures fabricated with 0.8 μg/mm^2 of lotus wax after storage for seven days at 50 °C with ethanol vapor. The bar chart also shows adhesive forces for various structures, measured using a 15 μm radius borosilicate tip. The error bar represents ± 1 standard deviation (Koch et al. 2009)

and *n*-hexatriacontane, which forms platelet nanostructures, the same tendency of particle removal was found. With the exception of hierarchical structure on all surfaces, larger particles were removed more than small ones. Most particles (70–80%) remained on smooth surfaces, and 50–70% of particles were found on microstructured surfaces. Most particles were removed from the hierarchically structured surfaces, but approximately 30% of particles remained. A clear difference in particle removal, independent of particle sizes, was only found in flat and nanostructured surfaces where larger particles were removed with higher efficiency. Observations of the droplet behavior during the movement on the surfaces showed that droplets were rolling only on the hierarchically structured surfaces. On flat, micro-, and nanostructured surfaces, the droplets first applied were not moving, but the continuous application of water droplets increased the droplet volumes and led to a sliding of these large droplets. During this, some of the particles had been removed from the surfaces. However, the rolling droplets on hierarchical structures did not collect the dirt particles trapped in the cavities of the microstructures. The data shows that hierarchical structures have superior cleaning efficiency.

6.6.5 Observation of Transition During the Bouncing Droplet

To observe how the impact velocity influences the transition from the composite solid-air-liquid interface to the homogeneous solid-liquid interface during droplet impact, Jung and Bhushan (2008b, 2009) performed bouncing droplet experiments on various surfaces with *n*-hexatriacontane and *T. majus* and lotus waxes. Figure 6.43 shows photos of a droplet of 1 mm radius hitting various surfaces fabricated with 0.2 and 0.4 μm/mm^2 of *n*-hexatriacontane after storage at room temperature. The impact velocity was obtained just prior to the droplet hitting the surface. First, on the flat surface, it was found that the droplet did not bounce off even though the impact velocity applied was up to 1.5 m/s. As shown in the images in the first row for each nano-, micro-, and hierarchically structured surface, the droplet hitting the surface under an impact velocity of 0.44 m/s first deformed and then retracted, and bounced off the surface. Finally, the droplet sat on the surface and had a high contact angle, which suggests the formation of a solid-air-liquid interface. Next, the impact experiment was repeated by increasing the impact velocity. As shown in the second row of images for the nanostructure (0.4 μg/mm^2) and microstructure, bounce off did not occur, and the wetting of the surface (and possibly pinning of droplet) occurred at impact velocities of 1.2 and 0.77 m/s, respectively, referred to as the critical impact velocity. This is because air pockets do not exist below the droplet as a result of droplet impalement by the structures, characterized by a smaller contact angle. These observations indicate the transition from the composite interface to the homogeneous interface. However, as shown in the second row of images for the nanostructure (0.2 μg/mm^2) and hierarchical structure, during application of impact velocity up to 1.5 m/s, the bounce off always occurred and the wetting of the surface did not occur.

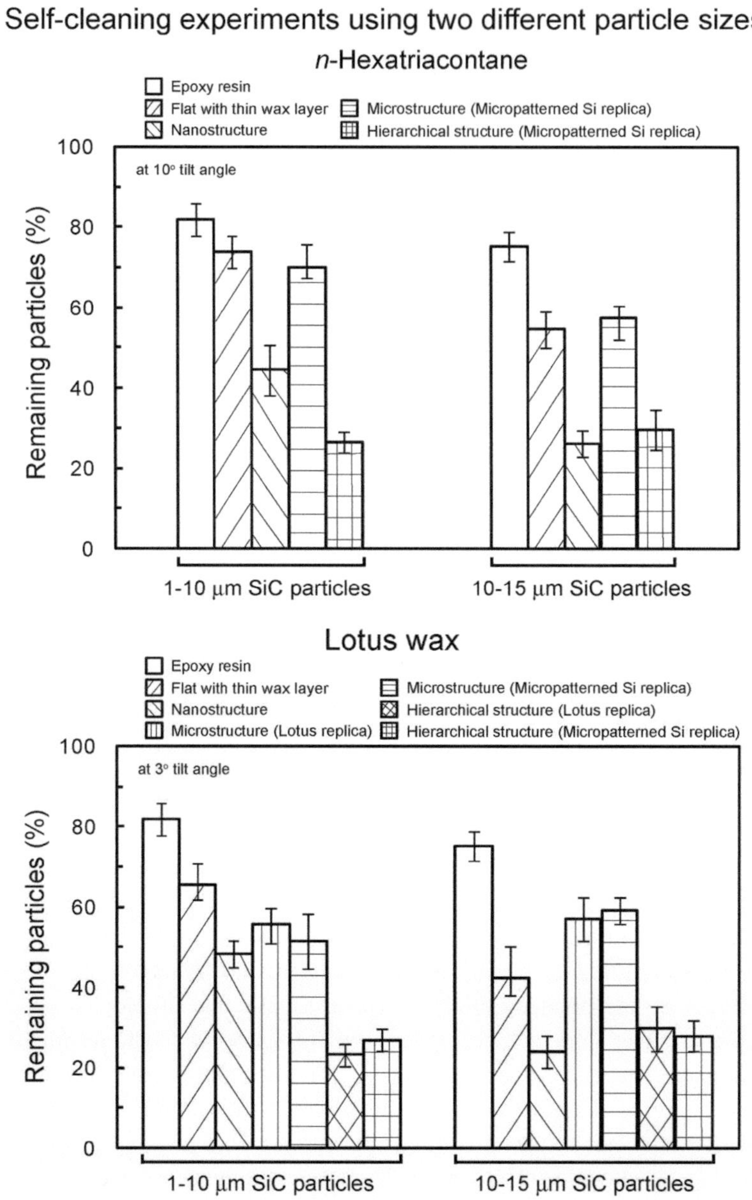

Fig. 6.42 Bar charts showing the remaining particles after applying droplets with nearly zero kinetic energy on various structures fabricated using *n*-hexatriacontane and lotus wax using 1–10 μm and 10–15 μm SiC particles. The experiments on the surfaces with *n*-hexatriacontane and lotus wax were carried out on tilted stages with 10° and 3°, respectively. The error bars represent ±1 standard deviation (Bhushan et al. 2009a)

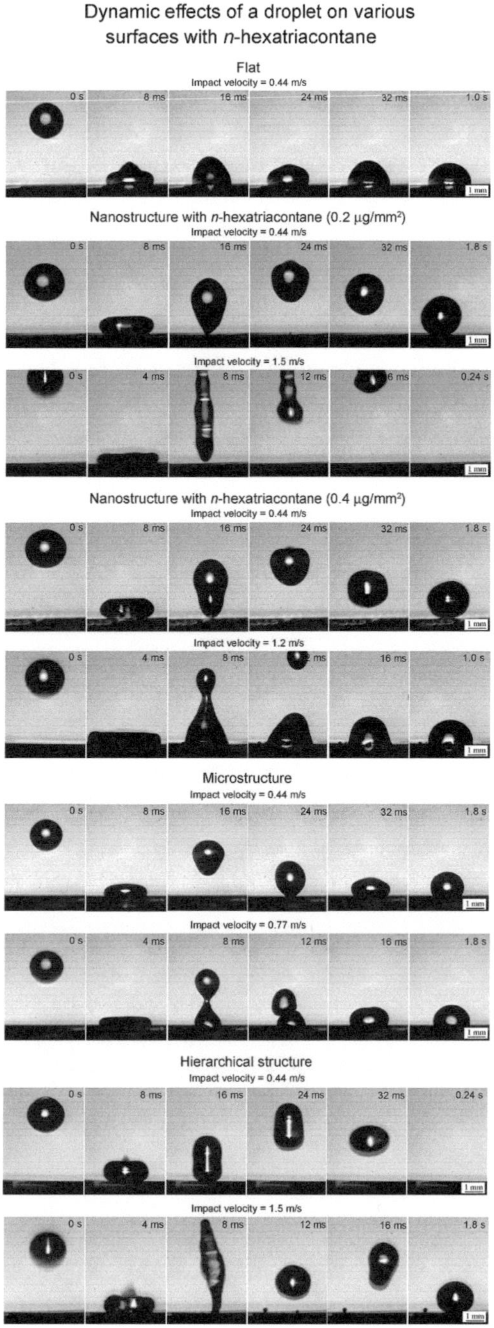

Fig. 6.43 Snapshots of a droplet with 1 mm radius hitting various surfaces fabricated with 0.2 and 0.4 μg/mm^2 of *n*-hexatriacontane after storage at room temperature. The impact velocity was obtained just prior to the droplet hitting the surface. The pinning of droplet on the nanostructure with 0.4 μg/mm^2 mass, and on the microstructure occurred at impact velocities of 1.2 and 0.77 m/s, respectively (Jung and Bhushan 2009)

As shown in Fig. 6.44, the critical impact velocity of a droplet with 1 mm radius was measured on various structures on which wetting of the surface (possibly pinning of droplet) occurs (Jung and Bhushan 2009). The arrow indicates that the critical impact velocity is more than the impact velocity used in experiments of up to 1.5 m/s. As mentioned earlier, when the mass of *n*-hexatriacontane increased, the static contact angle of the nanostructure first increased followed by a decrease at a mass of 0.2 μg/mm^2. For the critical impact velocity, the same trends were found. It is believed that if neighboring crystals are separated on a sample with lower crystal density, any trapped air can be squeezed out whereas if the neighboring crystals are interconnected on a sample with higher density, air remains trapped. At the highest crystal density at a mass of 0.4 μg/mm^2, there is less open volume compared to that at 0.2 μg/mm^2. For all microstructures with *n*-hexatriacontane and lotus wax, the critical impact velocities of the droplet are lower than those on nano- and hierarchical structures due to the larger distance between the pillars. Further, the solid-air-liquid interface can be destabilized easily from dynamic impact on

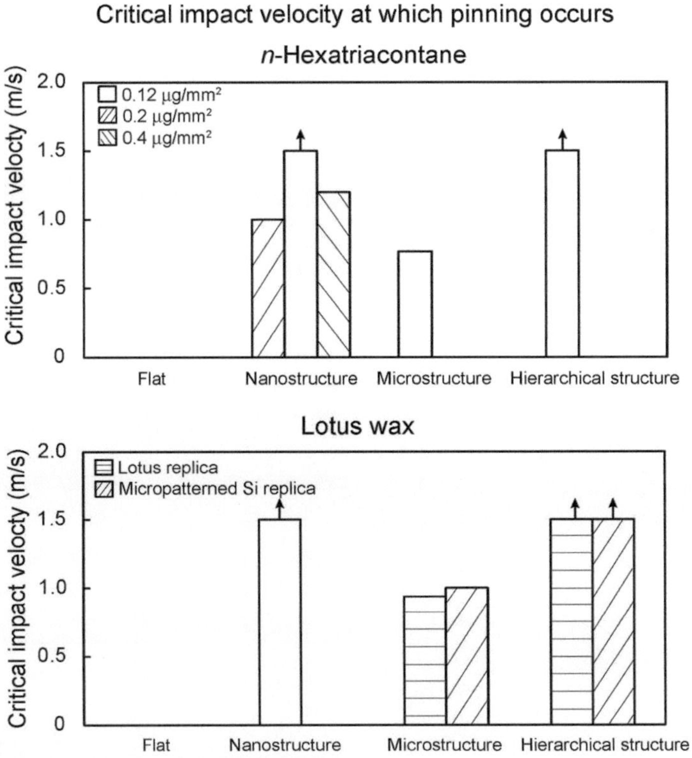

Fig. 6.44 Bar chart showing the measured critical impact velocity of a droplet with 1 mm radius at which transition occurs on various structures. The arrow indicates that the critical impact velocity can possibly be more than 1.5 m/s or the transition is not occurred (Jung and Bhushan 2009)

the surface. Based on (6.7), the critical impact velocity of the droplet decreases with the geometric parameter (pitch). The theoretical critical impact velocity for a microstructure using (6.7) is 0.5 m/s. This value is lower than the experimental values of critical impact velocity for microstructures by about 30–50% depending on the structured surfaces. In these experiments, while applying an impact velocity of up to 1.5 m/s on nano- and hierarchical structures with *n*-hexatriacontane (0.2 $\mu g/mm^2$ for nanostructure) and lotus wax, wetting of the surface did not occur. The data shows that nano- and hierarchical structures are superior to the microstructure in maintaining a stable composite solid-air-liquid interface.

To identify whether the interface is a homogeneous solid-liquid interface or a composite solid-air-liquid interface, the contact angle data in the static condition and after bounce off were measured on various surfaces, as shown in Fig. 6.45 (Jung and Bhushan 2009). The contact angles of the left bar for each sample were measured using the droplet with 1 mm radius placed on the surface. The contact angles of the middle and right bars for each sample were measured using a droplet after hitting the surface at 0.44 m/s and critical or highest impact velocity. Missing bars mean that the droplet, after hitting the surface, bounced off without coming to sit on the surface. The static contact angle of the droplet after impact at 0.44 m/s is lower than that of the droplet when placed for all of the surfaces with *n*-hexatriacontane and lotus wax. It can be interpreted that, after hitting, the droplet pushes out the entrapped air of the cavities between the pillars under the droplet, resulting in an abrupt increase of the solid-liquid surface area by dynamic impact. As mentioned earlier, after hitting the surface at a critical impact velocity, the contact angles were close to 90° and much lower than that of the droplet placed on the surface for the nanostructures with *n*-hexatriacontane (0.12 and 0.4 $\mu g/mm^2$) and the microstructures with *n*-hexatriacontane and lotus wax. Even though the droplet is in the composite interface when it is placed on the surface, these observations indicate that the composite solid-air-liquid interface was destroyed due to dynamic impact on the surface.

6.6.6 Observation of Transition During the Vibrating Droplet

To observe how a vibrating droplet influences the transition from the composite solid-air-liquid interface to the homogeneous solid-liquid interface, Jung and Bhushan (2009) performed vibrating droplet experiments on various surfaces with n-hexatriacontane and *T. majus* and lotus waxes.

6.6.6.1 Model for the Adhesion and Inertia Forces of the Vibrating Droplet

Jung and Bhushan (2009) presented a model for vibration. In this model, they calculated expressions for adhesion force and inertia force as a function of droplet properties and operating conditions. Consider a small droplet of liquid deposited on a surface. The liquid

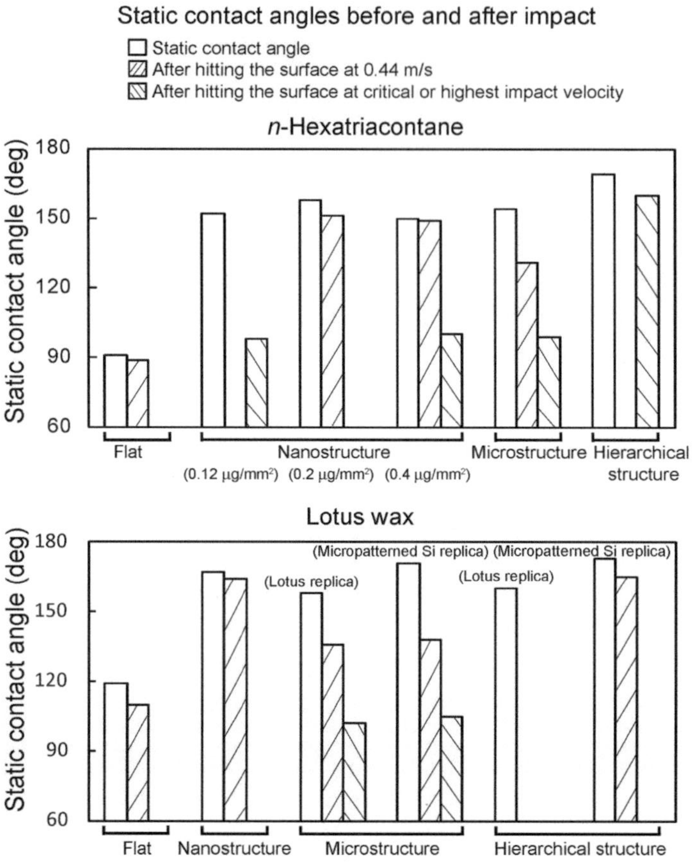

Fig. 6.45 Bar chart showing the measured static contact angle of a droplet on various surfaces. The contact angles of the left bar for each sample were measured using a droplet with 1 mm radius placed on the surface. The contact angles of the middle and right bars for each sample were measured using the droplet after hitting the surface at 0.44 m/s, and critical or highest impact velocity (Jung and Bhushan 2009)

and surface come together under equilibrium at a characteristic angle called the static contact angle θ, as shown in Fig. 6.46. When a droplet on a surface is vibrated, based on Lamb (1932), a general expression for the resonance frequency f_r of a free oscillating liquid droplet is given as,

$$f_r = \sqrt{\frac{n(n-1)(n+2)\gamma}{3\pi\rho v}} \tag{6.15}$$

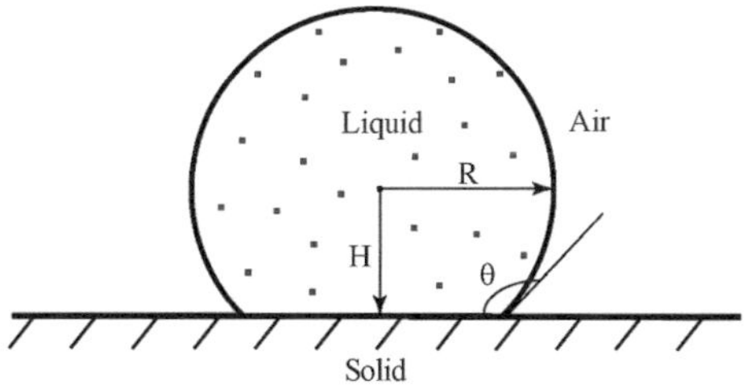

Fig. 6.46 Droplet of liquid in contact with a solid surface—contact angle θ; radius of droplet R; distance between the center of droplet and the solid H (Jung and Bhushan 2009)

where v is the volume of the droplet, and n represents the number of modes, with 2 being the 1st mode, etc. For a spherical droplet with a radius R with the distance between the center of droplet and the solid H, the volume of droplet v is given by,

$$v = \frac{1}{3}\pi(R+H)^2(2R-H) = \frac{1}{3}\pi R^3(1-\cos\theta)^2(2+\cos\theta) \tag{6.16}$$

Experimental studies were carried out by Noblin et al. (2004) and Celestini and Kofman (2006). They showed that the frequency decreases with the volume of the droplet, and the trends are compared with a theoretical model.

When liquid comes in contact with a surface, the energy gained for surfaces coming into contact is greater than the energy required for their separation (or the work of adhesion) by the quantity ΔW, which constitutes the adhesion hysteresis (Bhushan 2003). For a surface, the difference between the two values of the interface energy (measured during loading and unloading) is given by ΔW. These two values are related to the advancing contact angle, θ_a, and receding contact angle, θ_r, of the surface. For example, a model based on Young's equation has been used to calculate the work of adhesion by using contact angle hysteresis (Good 1952; Chen et al. 1991; Nosonovsky and Bhushan 2007a, 2008a),

$$\cos\theta_a - \cos\theta_r = \frac{\Delta W}{\gamma} \tag{6.17}$$

From (6.17), the dominant force (adhesion force) responsible for the separation between the droplet and surface is given by Joanny and de Gennes (1984, Lee and Laibinis (2000),

$$F_A = L\gamma(\cos\theta_a - \cos\theta_r) = 2R\sin\theta(\gamma)(\cos\theta_a - \cos\theta_r) \tag{6.18}$$

where L is the length of the triple contact line, referred to as line of contact of the solid, liquid, and air. The radius of a spherical droplet, R, depends on the contact angle and can be obtained from (6.16).

From applying vertical vibration of a droplet on a surface, the inertia force of droplet F_I is given by,

$$F_I = \rho v A \omega^2 \tag{6.19}$$

where A and ω are the amplitude and frequency of vibration, respectively. Bormashenko et al. (2007) showed that the transition occurred at a critical value of inertia force acting on the length of a triple line. However, if the inertia force of the droplet vibrated on the surface can overcome the adhesion force between the droplet and surface, then ΔF is positive value (Jung and Bhushan 2009),

$$\Delta F = F_I - F_A \tag{6.20}$$

The droplet can be vertically separated from the surface (bouncing off) before the composite solid-liquid-air interface is destroyed.

6.6.6.2 Vibration Study Results

A droplet on various surfaces was vibrated by varying the frequency with a sinusoidal excitation of relatively low amplitude (0.4 mm) (Jung and Bhushan 2009). Based on (6.15), the experiments were performed in the 1st and 2nd modes (n = 2 and 3). Figure 6.47 shows the optical micrographs of droplets on the microstructured surface with n-hexatriacontane before and after vibration at an amplitude of 0.4 mm for 1st mode (n = 2) and 2nd mode (n = 3) of vibration. The frequencies for the 1st mode and 2nd mode were measured to be 86 and 208 Hz, respectively. The resonance frequencies on various surfaces were measured, and the data are summarized in Table 6.5. For comparison, the theoretical values for the two modes were calculated using (6.15). For calculations, the surface tension of the water-air interface (γ) was taken to be 0.073 N/m, the mass density (ρ) was 1000 kg/m^3 for water, and the volume of the droplet (V) was 5 μL. The theoretical values for the 1st mode (n = 2) and 2nd mode (n = 3) from (6.15) are 110 and 214 Hz, respectively. These values are similar to those of flat surfaces with n-hexatriacontane and lotus wax. However, with the same volume, the resonance frequencies of the structured surfaces are lower than those of flat surfaces. Celestini and Kofman (2006) showed that the resonance frequency depends on the contact angle of the structured surfaces, and it decreases with increasing contact angle. The hierarchical structures with highest contact angle have the lowest resonance frequency, consistent with the results by Celestini and Kofman (2006).

To observe how the vibration of the droplet influences the transition from the composite solid-air-liquid interface to the homogeneous solid-liquid interface, Jung and Bhushan (2009) performed vibrating droplet experiments on various surfaces with n-hexatriacontane and lotus wax. Figure 6.48 shows optical micrographs of droplets on various surfaces with n-hexatriacontane before and after vibrating at a frequency of 30 Hz, which is less than the resonance frequency for the 1st mode (n = 2). The vibration amplitude was increased until transition or until the droplet bounced off. First, on the flat surface, it was found that the droplet did not change much after applying vibration at amplitudes ranging from 0 to 3 mm. As shown in the images for the nanostructure

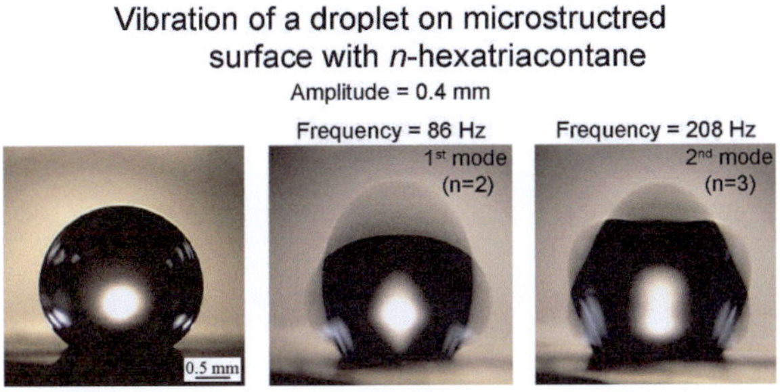

Fig. 6.47 Optical micrographs of droplets on the microstructured surface with *n*-hexatriacontane before and after vibration at amplitude of 0.4 mm for 1st mode ($n = 2$) and 2nd mode ($n = 3$) of vibration. The frequencies for 1st mode and 2nd mode were measured to be 86 and 208 Hz, respectively (Jung and Bhushan 2009)

Table 6.5 Summary of the measured resonance frequency at the fixed amplitude value of 0.4 mm, the calculated inertia force of a vibrated droplet, and adhesive force between the droplet and surface on the various surfaces fabricated with *n*-hexatriacontane and lotus wax. Positive value of ΔF means that the droplet bounced off before the transition occurred. The variation represents ± 1 standard deviation (Jung and Bhushan 2009)

	Resonance frequency (Hz)		F_A (μN)	F_I (μN) at transition or bouncing off	$\Delta F = F_I - F_A$ (μN)
	1st mode ($n = 2$)	2nd mode ($n = 3$)			
n-Hexatriacontane					
Flat	103	271	260 ± 3.7		
Nanostructure (0.12 μg/mm^2)	63	169	45 ± 2.2	11	−34 ± 2.2 (T)
Nanostructure (0.2 μg/mm^2)	50	147	10 ± 1.6	10	0 ± 1.6 (B)
Nanostructure (0.4 μg/mm^2)	54	149	13 ± 2.0	11	−2 ± 2.0 (B)
Microstructure	86	208	21 ± 1.7	9	−12 ± 1.7 (T)
Hierarchical structure	49	144	0.2 ± 0.06	3.6	3.4 ± 0.06 (B)
Lotus wax					
Flat	92	243	133 ± 3.3		
Nanostructure	47	138	0.8 ± 0.08	3.2	2.4 ± 0.08 (B)
Microstructure (lotus replica)	63	179	14 ± 1.8	11	−3 ± 1.8 (T and B)
Microstructure (micropatterned Si replica)	66	183	10 ± 1.6	10	0 ± 1.6 (B)
Hierarchical structure (lotus replica)	40	139	0.1 ± 0.04	3.6	3.5 ± 0.04 (B)
Hierarchical structure (Micropatterned Si replica)	38	137	0.04 ± 0.01	2.7	2.6 ± 0.01 (B)

B Bouncing off
T Transition

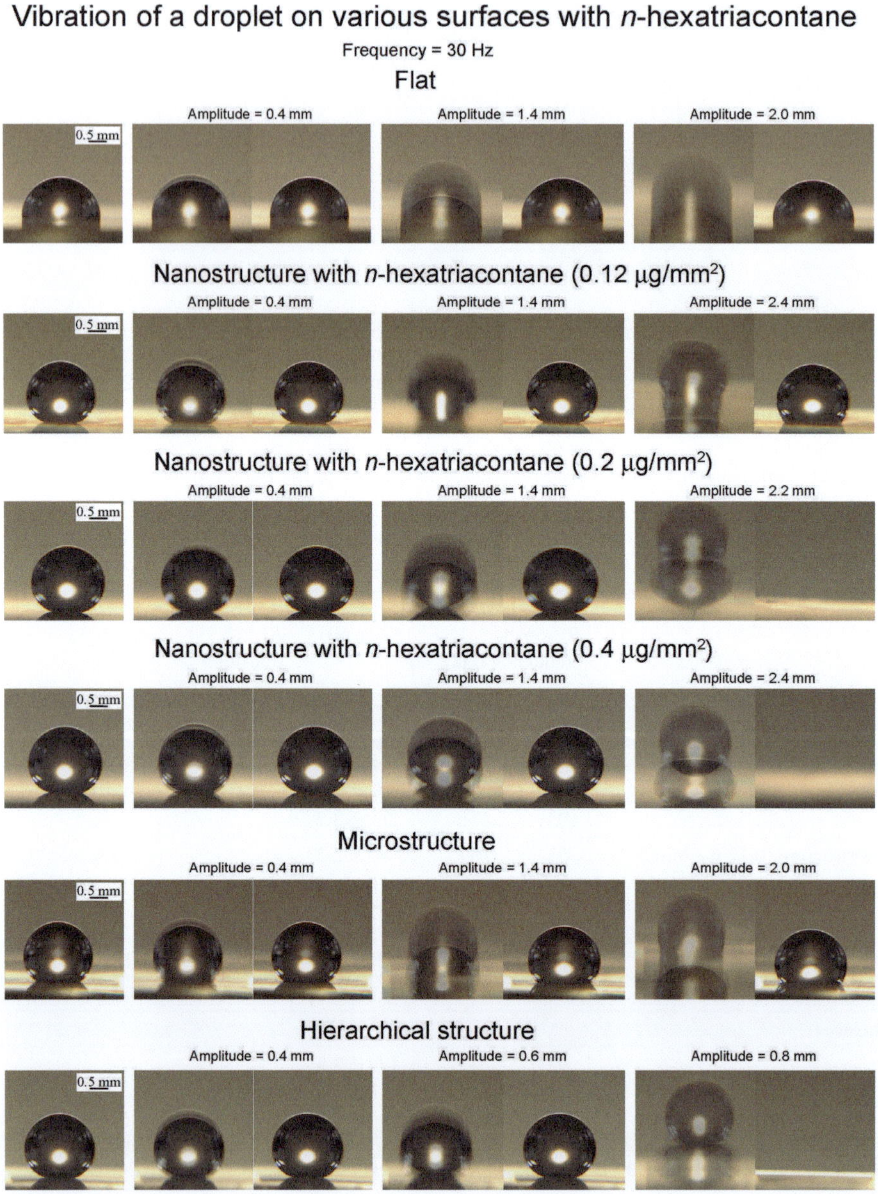

Fig. 6.48 Optical micrographs of droplets on various surfaces with *n*-hexatriacontane before and after vibrating at frequency of 30 Hz. The transition of a droplet on the nanostructure with 0.12 μg/mm^2 mass of *n*-hexatriacontane and on the microstructure occurred at amplitudes of 2.4 and 2.0 mm, respectively (Jung and Bhushan 2009)

(0.12 μg/mm^2) and microstructure, the static contact angles of the droplet before vibrating were 152° and 154°, respectively. After vibrating at amplitudes of 0.4 and 1.4 mm, the contact angles still had similar values (151° for nanostructure and 149° for microstructure), which suggests the formation of a solid-air-liquid interface. However, after vibrating at amplitudes of 2.4 mm for the nanostructure (0.12 μg/mm^2) and 2.0 mm for the microstructure, the static contact angles became 125° and 121°, respectively. As discussed earlier, this is because air pockets do not exist below the droplet as a result of droplet impalement by the structures, characterized by a smaller contact angle. These observations indicate the transition from the composite interface to the homogeneous interface. Observations of vibration on two nanostructures (0.2 and 0.4 μg/mm^2) and a hierarchical structure showed that the transition did not occur, but the droplet started to bounce off the surface from amplitudes of 2.2, 2.4, and 0.8 mm, respectively.

The adhesive force responsible for the separation between the droplet and surface and the inertia force of a droplet vibrated on various structures were calculated to study the validity of the proposed model (6.20) (Jung and Bhushan 2009). The adhesive force was obtained from (6.18) using static contact angle and contact angle hysteresis. The inertia forces were obtained from (6.19) using the amplitude and frequency of vibration as the transition or droplet bounce off occurred. The data are presented in Fig. 6.49 and summarized in Table 6.5. As shown in Fig. 6.48, it was observed that the transition occurred as a result of droplet impalement by the structures by increasing the inertia force of droplet on the surfaces. However, if the inertia force of the droplet vibrated on the surface can overcome the adhesion force between the droplet and surface (ΔF is positive), the droplet can be vertically separated from the surface (bouncing off) before the composite solid-liquid-air interface is destroyed. The experimental results for bouncing of the droplet appear to have the same trend as the proposed model (6.20). It is shown that hierarchical structures have the positive value of the difference between the inertia force and adhesive force for droplet bounce off responsible for superior resistance to the dynamic effects and maintain a stable composite solid-air-liquid interface.

6.6.7 Measurement of Fluid Drag Reduction

Superhydrophobicity reduces fluid slip and drag in fluid flow (Jung and Bhushan 2010). The drag data will be presented in Chap. 8. It will be shown that fluid drag decreases with an increase in the degree of hydrophobicity.

6.6.8 Summary

Hierarchically structured surfaces using nature's route were fabricated to verify that properties comparable to those of natural objects can be obtained. Various structures with

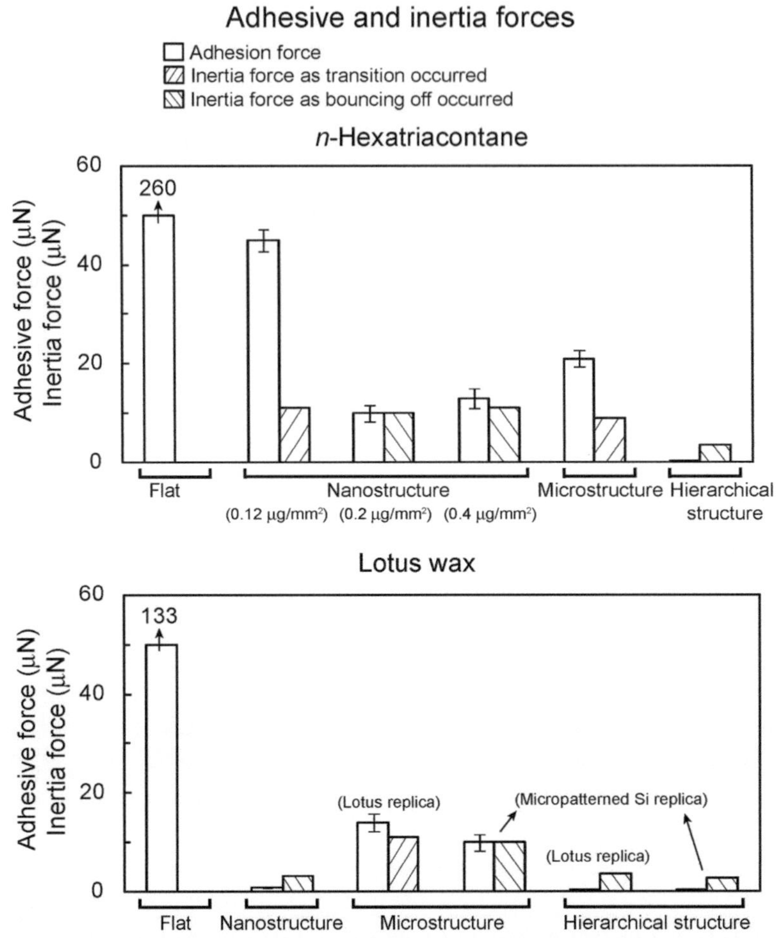

Fig. 6.49 Bar chart showing the calculated adhesive force responsible for the separation between the droplet and surface and inertia force of a droplet vibrated on various structures. The inertia forces were obtained as the transition or droplet bounce off occurred. If the inertia force of the droplet vibrated on the surface can overcome the adhesion force between the droplet and surface (ΔF is positive), the droplet can be vertically separated from the surface (bouncing off) before the composite solid-liquid-air interface is destroyed (Jung and Bhushan 2009)

two types of wax morphology with various crystal densities were fabricated for detailed analyses.

Nanostructures with different wax platelet crystal densities using n-hexatriacontane were deposited on a flat surface and a microstructure, and the effect of crystal density on the static contact angle, contact angle hysteresis, and air pocket formation, as well as adhesive force was studied. High density of crystals with sufficient spacing between them leads to a propensity to form air pockets resulting in a high static contact angle. In hierarchically structured surfaces, as compared to nanostructured and microstructured surfaces, the air pocket formation further increases contact angle and reduces both contact angle hysteresis and adhesive force.

The nanostructures with *T. majus* and lotus wax tubules were deposited on flat surfaces. Nanostructures were formed by wax tubules. Tubules were hollow structures and randomly oriented on the surface. The tubular diameter varied between 100 and 150 nm, and their length varied between 1500 and 2000 nm. The created nanostructures were comparable to the wax crystal morphology found on the lotus leaf. The influence of these nanostructures deposited on a flat surface and a microstructure was studied on the static contact angle, contact angle hysteresis, air pocket formation, and adhesive force as well as self-cleaning efficiency. The data were compared with that of a lotus plant leaf. Hierarchically structured, followed by nanostructured, surfaces have high static contact angle, low contact angle hysteresis, and low adhesive force. These properties are superior to the lotus leaf. Evidence of air pocket formation has been reported. Self-cleaning experiments performed by deposition of contaminant particles followed by cleaning with water droplets show that hierarchical structures have superior cleaning efficiency, comparable to that of the lotus leaf. To sum up, hierarchically structured surfaces using nature's route have been created and their contact angles, adhesive force, and self-cleaning efficiency are comparable to that of the lotus leaf, which verifies the understanding of the mechanisms relevant to superhydrophobicity and self-cleaning in the lotus leaf.

To investigate the transition from the composite interface to the homogeneous interface on nanostructures with *T. majus* and lotus wax tubules deposited on flat surfaces and a microstructure, bouncing and vibrating droplet studies were performed. Bouncing and vibrating droplet experiments show that in hierarchically structured and nanostructured surfaces with certain crystal densities, wetting did not occur up to 1.5 m/s. In microstructured surfaces due to the larger distance between the pillars, composite interface was destroyed above a certain critical velocity in bouncing droplet experiments and above a certain inertia force of the droplet on the surfaces in vibrating droplet experiments.

6.7 Closure

In this chapter, micropatterned, nanopatterned, and hierarchically structured surfaces were fabricated using soft lithography, photolithography, and techniques that involve the replication of micropatterns and self-assembly of hydrophobic alkanes and plant waxes. They have been characterized to validate the models and to provide design guidelines for superhydrophobic and self-cleaning surfaces. To further examine the effect of meniscus force and real area of contact, scale dependence is considered with the use of AFM tips of various radii. To investigate how the effects of spacing between pillars, droplet size, and impact velocity influence the transition, wetting, evaporation, and bouncing studies were conducted on silicon surfaces patterned with pillars of two different diameters and heights and with varying pitch values and deposited with a hydrophobic coating. In order to generate submicron droplets, an atomic force microscopy-based technique using a modified nanoscale dispensing probe was used. An ESEM study on the wetting behavior for a microdroplet with about 20 μm radius on the micropatterned Si surfaces was carried out. Hierarchical structures using nature's route were fabricated to verify the understanding of the mechanisms relevant to superhydrophobicity and self-cleaning in the lotus leaf by comparing the superhydrophobicity and self-cleaning of fabricated surfaces and the lotus leaf.

References

Adamson, A. V. (1990), *Physical Chemistry of Surfaces*, Wiley, New York.

Bartolo, D., Bouamrirene, F., Verneuil, E., Buguin, A., Silberzan, P., and Moulinet, S. (2006), "Bouncing or Sticky Droplets: Impalement Transitions on Superhydrophobic Micropatterned Surfaces," *Europhys. Lett.* **74**, 299–305.

Bhushan, B. (2003), "Adhesion and Stiction: Mechanisms, Measurement Techniques and Methods for Reduction," *J. Vac. Sci. Technol. B* **21**, 2262–2296.

Bhushan, B. (2013a), *Principles and Applications of Tribology,* second ed., Wiley, New York.

Bhushan, B. (2013b), *Introduction to Tribology*, second ed., Wiley, New York.

Bhushan, B. (2017a), *Springer Handbook of Nanotechnology,* fourth ed., Springer International, Switzerland.

Bhushan, B. (2017b), *Nanotribology and Nanomechanics – An Introduction*, fourth ed., Springer International, Switzerland.

Bhushan, B. and Blackman, G. S. (1991), "Atomic Force Microscopy of Magnetic Rigid Disks and Sliders and Its Applications to Tribology," *ASME J. Tribol.* **113**, 452–457.

Bhushan, B. and Jung, Y. C. (2006), "Micro and Nanoscale Characterization of Hydrophobic and Hydrophilic Leaf Surface," *Nanotechnology* **17**, 2758–2772.

Bhushan, B. and Jung, Y. C. (2007), "Wetting Study of Patterned Surfaces for Superhydrophobicity," *Ultramicroscopy* **107**, 1033–1041.

Bhushan, B., Nosonovsky M., and Jung, Y. C. (2007), "Towards Optimization of Patterned Superhydrophobic Surfaces" *J. R. Soc. Interface* **4**, 643–648.

Bhushan, B. and Jung, Y. C. (2008), "Wetting, Adhesion and Friction of Superhydrophobic and Hydrophilic Leaves and Fabricated Micro/nanopatterned Surfaces," *J. Phys.: Condens. Matter* **20**, 225010.

Bhushan, B. and Jung, Y. C. (2011), "Natural and Biomimetic Artificial Surfaces for Superhydrophobicity, Self-Cleaning, Low Adhesion, and Drag Reduction," *Prog. Mater. Sci.* **56**, 1–108.

Bhushan, B., Hansford, D., and Lee, K. K. (2006), "Surface Modification of Silicon and Polydimethylsiloxane Surfaces with Vapor-Phase-Deposited Ultrathin Fluorosilane Films for Biomedical Nanodevices," *J. Vac. Sci. Technol. A* **24**, 1197–1202.

Bhushan, B., Koch, K., and Jung, Y. C. (2008a), "Nanostructures for Superhydrophobicity and Low Adhesion," *Soft Matter* **4**, 1799–1804.

Bhushan, B., Koch, K., and Jung, Y. C. (2008b), "Biomimetic Hierarchical Structure for Self-Cleaning," *Appl. Phys. Lett.* **93**, 093101.

Bhushan, B., Jung, Y. C., and Koch, K. (2009a), "Self-Cleaning Efficiency of Artificial Superhydrophobic Surfaces," *Langmuir* **25**, 3240–3248.

Bhushan, B., Jung, Y. C., and Koch, K. (2009b), "Micro-, Nano- and Hierarchical Structures for Superhydrophobicity, Self-Cleaning and Low Adhesion," *Phil. Trans. R. Soc. A* **367**, 1631–1672.

Bhushan, B., Jung, Y. C., Niemietz, A., and Koch, K. (2009c), "Lotus-like Biomimetic Hierarchical Structures Developed by the Self-assembly of Tubular Plant Waxes," *Langmuir* **25**, 1659–1666.

Bhushan, B., Koch, K., and Jung, Y. C. (2009d), "Fabrication and Characterization of the Hierarchical Structure for Superhydrophobicity," *Ultramicroscopy* **109**, 1029–1034.

Bhushan, B., Jung, Y. C., and Nosonovsky, M. (2012), "Hierarchical Structures for Superhydrophobic Surfaces and Methods of Making," U.S. Patent No. 8,137,751 B2, March 20.

Bormashenko, E., Pogreb, R., Whyman, G., and Erlich, M. (2007), "Cassie-Wenzel Wetting Transition in Vibrated Drops Deposited on the Rough Surfaces: Is Dynamic Cassie-Wenzel Transition 2D or 1D Affair?" *Langmuir* **23**, 6501–6503.

Bourges-Monnier, C. and Shanahan, M. E. R. (1995), "Influence of Evaporation on Contact Angle," *Langmuir* **11**, 2820–2829.

Brugnara, M., Della Volpe, C., Siboni, S., and Zeni, D, (2006), "Contact Angle Analysis on Polymethylmethacrylate and Commercial Wax by Using an Environmental Scanning Electron Microscope," *Scanning* **28**, 267–273.

Bunshah, R. F. (1994), *Handbook of Deposition Technologies for Films and Coatings: Science, Technology and Applications*, Applied Science Publishers, Westwood, New Jersey.

Burton, Z. and Bhushan, B. (2005), "Hydrophobicity, Adhesion, and Friction Properties of Nanopatterned Polymers and Scale Dependence for Micro- and Nanoelectromechanical Systems," *Nano Lett.* **5**, 1607–1613.

Burton, Z. and Bhushan, B. (2006), "Surface Characterization and Adhesion and Friction Properties of Hydrophobic Leaf Surfaces," *Ultramicroscopy* **106**, 709–719.

Celestini, F. and Kofman, R. (2006), "Vibration of Submillimeter-size Supported Droplets," *Phys. Rev. E* **73**, 041602.

Checco, A., Guenoun, P., and Daillant, J. (2003), "Nonlinear Dependence of the Contact Angle of Nanodroplets on Contact Line Curvatures," *Phys. Rev. Lett.* **91**, 186101.

Chen, N. and Bhushan, B. (2006), "Atomic Force Microscopy Studies of Conditioner Thickness Distribution and Binding Interactions on the Hair Surface," *J. Microsc.* **221**, 203–215.

Chen, Y. L., Helm, C. A., and Israelachvili, J. N. (1991), "Molecular Mechanisms Associated with Adhesion and Contact Angle Hysteresis of Monolayer Surfaces," *J. Phys. Chem.* **95**, 10736–10747.

Choi, S. E., Yoo, P. J., Baek, S. J., Kim, T. W., and Lee, H. H. (2004), "An ultraviolet-curable mold for sub-100-nm lithography," *J. Am. Chem. Soc.* **126**, 7744–7745.

Danilatos, G. D. and Brancik, J. V. (1986), "Observation of Liquid Transport in the ESEM," *Proc. 44th Annual Meeting Electron Microscopy Soc. of America* (ed. G. W. Bailey), pp. 678–679, San Francisco Press, San Francisco, CA.

Dorset, D. L., Pangborn, W. A., and Hancock, A. J., (1983), "Epitaxial Crystallization of Alkane Chain Lipids for Electron Diffraction Analysis," *J. Biochem. Biophys. Meth.* **8**, 29–40.

Erbil, H. Y., McHale, G., and Newton, M. I. (2002), "Drop Evaporation on Solid Surfaces: Constant Contact Angle Mode," *Langmuir* **18**, 2636–2641.

Erbil, H. Y., Demirel, A. L., and Avci, Y. (2003), "Transformation of a Simple Plastic into a Superhydrophobic Surface," *Science* **299**, 1377–1380.

Good, R. J. (1952), "A Thermodynamic Derivation of Wenzel's Modification of Young's Equation for Contact Angles; Together with a Theory of Hysteresis," *J. Am. Chem. Soc.* **74**, 5041–5042.

Israelachvili, J. N. (1992), *Intermolecular and Surface Forces*, second edition, Academic Press, London, U. K.

Joanny, J. F. and de Gennes, P. G. (1984), "A Model for Contact Angle Hysteresis," *J. Chem. Phys.* **81**, 552–562.

Johnson, R. E. and Dettre, R. H. (1964), "Contact Angle Hysteresis," *Contact Angle, Wettability, and Adhesion, Adv. Chem. Ser.* (ed. F. M. Fowkes), Vol. 43, pp. 112–135, American Chemical Society, Washington, D. C.

Jung, Y. C. and Bhushan, B. (2006), "Contact Angle, Adhesion, and Friction Properties of Micro- and Nanopatterned Polymers for Superhydrophobicity," *Nanotechnology* **17**, 4970–4980.

Jung, Y. C. and Bhushan, B. (2007), "Wetting Transition of Water Droplets on Superhydrophobic Patterned Surfaces," *Scripta Mater.* **57**, 1057–1060.

Jung, Y. C. and Bhushan, B. (2008a), "Wetting Behavior during Evaporation and Condensation of Water Microdroplets on Superhydrophobic Patterned Surfaces" *J. Micros.* **229**, 127–140.

Jung, Y. C. and Bhushan, B. (2008b), "Dynamic Effects of Bouncing Water Droplets on Superhydrophobic Surfaces," *Langmuir* **24**, 6262–6269.

Jung, Y. C. and Bhushan, B. (2008c), "Technique to Measure Contact Angle of Micro/Nanodroplets using Atomic Force Microscopy," *J. Vac. Sci. Technol. A* **26**, 777–782.

Jung, Y. C. and Bhushan, B. (2009), "Dynamic Effects Induced Transition of Droplets on Biomimetic Superhydrophobic Surfaces," *Langmuir* **25**, 9208–9218.

Jung, Y. C. and Bhushan, B. (2010), "Biomimetic Structures for Fluid Drag Reduction in Laminar and Turbulent Flows," *J. Phys.: Condens. Matter* **22**, 035104.

Kasai, T., Bhushan, B., Kulik, G., Barbieri, L., and Hoffmann, P. (2005), "Micro/Nanotribological Study of Perfluorosilane SAMs for Antistiction and Low Wear," *J. Vac. Sci. Technol. B* **23**, 995–1003.

Koch, K., Dommisse, A., and Barthlott, W. (2006a), "Chemistry and Crystal Growth of Plant Wax Tubules of Lotus (*Nelumbo nucifera*) and Nasturtium (Tropaeolum majus) Leaves on Technical Substrates," *Crystl. Growth Des.* **6**, 2571–2578.

Koch, K., Barthlott, W., Koch, S., Hommes, A., Wandelt, K., Mamdouh, W., De-Feyter, S. and Broekmann, P. (2006b), "Structural Analysis of Wheat Wax (*Triticum aestivum, c.v.* 'Naturastar' L.): from the Molecular Level to Three Dimensional Crystals," *Planta* **223**, 258–270.

Koch, K., Dommisse, A., Barthlott, W., and Gorb, S. (2007), "The Use of Plant Waxes as Templates for Micro- and Nanopatterning of Surfaces," *Acta Biomat.* **3**, 905–909.

Koch, K., Schulte, A. J., Fischer, A., Gorb, S. N., and Barthlott, W. (2008), "A Fast and Low-cost Replication Technique for Nano- and High-Aspect-Ratio Structures of Biological and Artificial Materials," *Bioinsp. Biomim.* **3**, 046002.

Koch, K., Bhushan, B., Jung, Y. C., and Barthlott, W. (2009), "Fabrication of Artificial Lotus Leaves and Significance of Hierarchical Structure for Superhydrophobicity and Low Adhesion," *Soft Matter* **5**, 1386–1393.

Lafuma, A. and Quéré, D. (2003), "Superhydrophobic States," *Nat. Mater.* **2**, 457–460.

Lamb, H. (1932), *Hydrodynamics*, Cambridge University Press, Cambridge, England.

Lee, S. -W. and Laibinis, P. E. (2000), "Directed Movement of Liquids on Patterned Surfaces Using Noncovalent Molecular Adsorption," *J. Am. Chem. Soc.* **122**, 5395–5396.

Lodge, R. A. and Bhushan, B. (2006), "Surface Characterization of Human Hair using Tapping Mode Atomic Force Microscopy and Measurement of Conditioner Thickness Distribution," *J. Vac. Sci. Technol. A* **24**, 1258–1269.

McHale, G., Aqil, S., Shirtcliffe, N. J., Newton, M. I., and Erbil, H. Y. (2005), "Analysis of Droplet Evaporation on a Superhydrophobic Surface," *Langmuir* **21**, 11053–11060.

Neinhuis, C., and Barthlott, W. (1997), "Characterization and Distribution of Water-Repellent, Self-Cleaning Plant Surfaces," *Annals of Botany* **79**, 667–677.

Niemietz, A., Wandelt, K., Barthlott, W., and Koch, K. (2009), "Thermal Evaporation of Multi-Component Waxes and Thermally Activated Formation of Nano-Tubules for Superhydrophobic Surfaces," *Prog. Org. Coat.* **66**, 221–227.

Noblin, X., Buguin, A., and Brochard-Wyart, F. (2004), "Vibrated Sessile Drops: Transition between Pinned and Mobile Contact Line Oscillations," *Eur. Phys. J. E* **14**. 395–404.

Nosonovsky, M. and Bhushan, B. (2005), "Roughness Optimization for Biomimetic Superhydrophobic Surfaces," *Microsyst. Technol.* **11**, 535–549.

Nosonovsky, M. and Bhushan, B. (2007a), "Multiscale Friction Mechanisms and Hierarchical Surfaces in Nano- and Bio-Tribology," *Mater. Sci. Eng.:R* **58**, 162–193.

Nosonovsky, M. and Bhushan, B. (2007b), "Hierarchical Roughness Makes Superhydrophobic Surfaces Stable," *Microelectronic Eng.* **84**, 382–386.

Nosonovsky, M. and Bhushan, B. (2007c), "Biomimetic Superhydrophobic Surfaces: Multiscale Approach," *Nano Lett.* **7**, 2633–2637.

Nosonovsky, M. and Bhushan, B. (2007d), "Hierarchical Roughness Optimization for Biomimetic Superhydrophobic Surfaces," *Ultramicroscopy* **107**, 969–979.

Nosonovsky, M. and Bhushan, B. (2008a), *Multiscale Dissipative Mechanisms and Hierarchical Surfaces: Friction, Superhydrophobicity, and Biomimetics*, Springer-Verlag, Heidelberg, Germany.

Nosonovsky, M. and Bhushan, B. (2008b), "Roughness-Induced Superhydrophobicity: A Way to Design Non-Adhesive Surfaces," *J. Phys.: Condens. Matter* **20**, 225009.

Nosonovsky, M. and Bhushan, B. (2008c), "Patterned Non-Adhesive Surfaces: Superhydrophobicity and Wetting Regime Transitions," *Langmuir* **24**, 1525–1533.

Nosonovsky, M. and Bhushan, B. (2008d), "Capillary Effects and Instabilities in Nanocontacts," *Ultramicroscopy* **108**, 1181–1185.

Nosonovsky, M. and Bhushan, B. (2008e), "Energy Transitions in Superhydrophobicity: Low Adhesion, Easy Flow and Bouncing," *J. Phys.: Condens. Matter* **20**, 395005.

Palacio, M. and Bhushan, B. (2010), "Normal and Lateral Force Calibration Techniques for AFM Cantilevers," *Crit. Rev. Solid State Mater. Sci.* **35**, 73–104.

Pompe, T. and Herminghaus, S. (2000), "Three-Phase Contact Line Energetics from Nanoscale Liquid Surface Topographies," *Phys. Rev. Lett.* **85**, 1930–1933.

Quere, D. (2004), "Surface Wetting Model Droplets," *Nat. Mater.* **3**, 79–80.

Reyssat, M., Pepin, A., Marty, F., Chen, Y., and Quere, D. (2006), "Bouncing Transitions on Microtextured Materials," *Europhys. Lett.* **74**, 306–312.

Richard, D., Clanet, C., and Quere, D. (2002), "Contact Time of a Bouncing Drop," *Nature* **417**, 811.

Rowan, S. M., Newton, M. I., and McHale, G. (1995), "Evaporation of Microdroplets and the Wetting of Solid Surfaces," *J. Phys. Chem.* **99**, 13268–13271.

Semal, S., Blake, T. D., Geskin, V., de Ruijter, M. L., Castelein, G., and De Coninck, J. (1999), "Influence of Surface Roughness on Wetting Dynamics," *Langmuir* **15**, 8765–8770.

Shibuichi, S., Onda, T., Satoh, N., and Tsujii, K. (1996), "Super-Water-Repellent Surfaces Resulting from Fractal Structure," *J. Phys. Chem.* **100**, 19512–19517.

Stelmashenko, N. A., Craven, J. P., Donald, A. M., Terentjev, E. M., and Thiel, B. L. (2001), "Topographic Contrast of Partially Wetting Water Droplets in Environmental Scanning Electron Microscopy," *J. Micros.* **204**, 172–183.

van Dijk, A. I. J. M., Bruijnzeel, L. A., and Rosewell, C. J. (2002), "Rainfall Intensity–Kinetic Energy Relationships: A Critical Literature Appraisal," *J. Hydrology* **261**, 1–23.

Yost, F. G., Michael, J. R., and Eisenmann, E. T. (1995), "Extensive Wetting Due to Roughness," *Acta Metall. Mater.* **45**, 299–305.

Zhang, X., Tan, S., Zhao, N., Guo, X., Zhang, X., Zhang, Y., and Xu, J. (2006), "Evaporation of Sessile Water Droplets on Superhydrophobic Natural Lotus and Biomimetic Polymer Surfaces," *Chem. Phys. Chem.* **7**, 2067–2070.

7 Fabrication and Characterization of Mechanically Durable Superhydrophobic Surfaces

7.1 Introduction

Superhydrophobic surfaces with low contact angle hysteresis can be used for water-repellency, self-cleaning/low adhesion, drag reduction in fluid flow, energy conservation, and energy conversion. The water repellant and self-cleaning surfaces are of interest in various applications, including self-cleaning windows, windshields, smart screens, camera lenses, exterior paint for buildings, utensils, roof tiles, textiles, solar panels, and exterior surfaces in transportation (Bhushan 2009; Bhushan and Jung 2011). These surfaces are also of interest for anti-icing and anti-fogging applications. Anti-icing or icephobic surfaces are important in various applications including airplane wings, wind turbine blades, boat hulls, automobile windshields and power lines. Anti-fogging surfaces are important for optical windows and eyeglasses. Self-cleaning/low adhesion and low drag surfaces also exhibit antifouling of interest in various applications including membranes for desalinization and water purification.

One of the emerging applications of self-cleaning is the camera lenses, camera and sensor enclosures and sensors used in autonomous vehicles. The rain, frost, fog, dirt, and dried salt will occlude the optical elements on the cameras and sensors and make them useless. It is not practical to clean them up frequently. The surfaces should be repellant to water, frost, fog and dirt. In the case of smart screens, finger touch resistance is also critical. A high level of visible light transmittance is required in the optical applications. In some sanitary application, such as toilets, coatings should be able to withstand high waterjet pressures in addition to mechanical scrub and liquid chemistry. For any industrial applications, mechanical durability is important for long life and the fabrication process should be scalable and economical.

By using smart materials and nanofabrication techniques, Jung and Bhushan (2009) produced mechanically durable carbon nanotube (CNT) composite hierarchical structures.

B. Bhushan, *Introduction to Biomimetics and Bioinspiration*,
https://doi.org/10.1007/978-3-031-62344-8_7

This was done by replication of a micropatterned silicon surface using an epoxy resin and by deposition of the CNT composite using a spray method. Ebert and Bhushan (2012a, b) produced mechanically durable hierarchical structures using micro/nanoparticles as well as nanoparticle composites with optical transparency. Ebert and Bhushan (2016) and Martin and Bhushan (2017) used ion etching and micropatterning techniques to fabricate mechanically durable superhydrophobic surfaces with optical transparency.

In this chapter, details of various approaches are presented to introduce mechanically durable surfaces using smart materials and fabrication techniques.

7.2 Characterization Techniques

For characterization of contact angle, contact angle hysteresis and surface roughness, experimental details can be found in Chap. 6. To measure mechanical durability, wear and friction studies were performed using an atomic force microscope (AFM) and a ball-on-flat tribometer. Waterfall/jet tests were also conducted to determine the loss of superhydrophobicity. Transmittance measurements were made to measure optical transparency. Details follow.

7.2.1 Mechanical Durability

To investigate the durability of fabricated surfaces, microscale wear tests were performed using a commercial AFM (Jung and Bhushan 2009; Bhushan 2017a, b). With the AFM in contact mode, the surfaces were worn using a 15 μm radius borosilicate ball mounted on a triangular Si_3N_4 cantilever with a nominal spring constant of 0.58 N m^{-1}. Wear scars with dimensions of 50×50 μm^2 were created and scanned for 1 cycle at two different loads of 100 and 10 μN. In order to analyze the changes in the morphology before and after wear tests, surface height maps were obtained in dimensions of 100×100 μm^2 using a square pyramidal Si(100) tip with a native oxide layer which has a nominal radius of 20 nm on a rectangular Si(100) cantilever with a spring constant of 3 N m^{-1} and at a natural frequency of 76 kHz in tapping mode.

In order to investigate durability at a high load, macroscale wear and friction tests on the fabricated surfaces were conducted using a ball-on-flat tribometer under reciprocating motion (Jung and Bhushan 2009; Bhushan 2013a, b). A sapphire ball with a diameter of 3 mm and surface finish of about 2 nm RMS was fixed on a stationary holder. A normal load of 10 mN was applied, and the frictional forces were measured with semiconductor strain gauges, which were then digitized and collected on a computer. Typical test conditions were: stroke length = 800 μm, average linear speed = 1 mm/s, temperature = 22 ± 1 °C, and relative humidity = $45 \pm 5\%$. Wear was characterized by imaging the resulting scar with an optical microscope with a CCD camera before and after wear tests.

The number of cycles to failure was determined by identifying the point where a sudden change in the friction force is observed.

7.2.2 Waterfall/Jet Tests

To investigate the durability of the fabricated surfaces in long-term exposure to water and different kinetic energies of water, a waterfall/jet test setup was used, as shown in Fig. 7.1 (Jung and Bhushan 2009). Specimens were fixed on the stage using a double-sided adhesive tape. They were placed 2 mm below the four holes in a pipe through which water flow occurred. In order to minimize flow interruption on the specimens, the runoff plate was tilted to 45°. Waterfall/jet experiments were carried out at two test conditions. In one test, water pressure was fixed at 10 kPa and specimens were exposed for 24 h. In a second test, in order to apply different kinetic energies of the water, the water pressure was controlled between 0 and 45 kPa. The exposure time applied to the specimens was 20 min for each experiment. During the tests, the change of static contact angle was measured using droplets of about 5 μL in volume (with radius of a spherical droplet about 1 mm) deposited on the substrate using a microsyringe. For contact angle hysteresis, the advancing and receding contact angles were measured at the front and back of the droplet moving along the tilted surface, respectively. The image of the droplet was obtained by a digital camcorder with a 10× optical and 120× digital zoom. Images obtained were analyzed using Imagetool® software (University of Texas Health Science Center) for the contact angle.

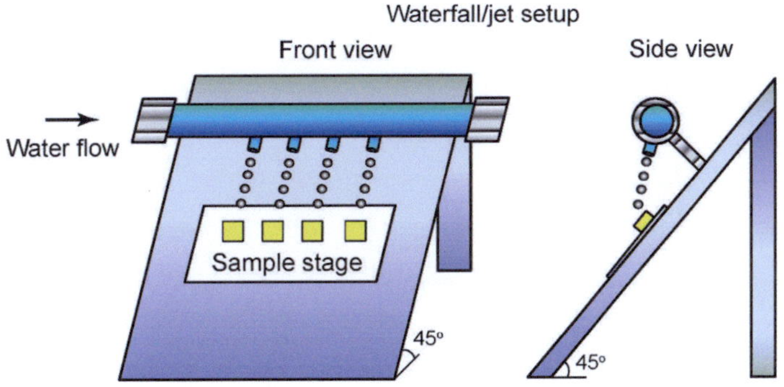

Fig. 7.1 Schematics of waterfall/jet setup shown in front and side views (Jung and Bhushan 2009)

7.2.3 Optical Transmittance Measurements

Optical transmittance measurements were performed using an incandescent light bulb as a point source, an Ocean Optics USB400 spectrometer with a 200 μm aperture width, and an intensified charge-coupled device (CCD) camera (Ebert and Bhushan 2012b). All transmittance data was obtained for a coating as a percentage of the transmittance of the uncoated substrate in the visible spectrum (400–700 nm wavelength).

7.3 Superhydrophobic Surfaces Using CNT Composites

Hierarchically structured superhydrophobic surfaces fabricated with CNT composites are presented. As a benchmark, structured surfaces with lotus wax were also prepared to compare the durability with that of CNT composite structures.

7.3.1 Fabrication Details

In order to fabricate nano- and hierarchical structures, Jung and Bhushan (2009) deposited CNT composite on flat epoxy resin and a microstructure using a spray technique, as shown in Fig. 7.2. CNTs were fabricated using catalyst-assisted chemical vapor deposition. An iron catalyst was used to initiate growth of nanotubes using natural gas as a carbon source and Ar/H_2 as a buffer gas at 750 °C. The contact angle of individual carbon nanotubes has been reported as 60° and higher. Microstructures were fabricated using a micropatterned Si surface with pillars of 14 μm diameter, 30 μm height, and 23 μm pitch by soft lithography, as described in the Chap. 6 (Bhushan et al. 2009; Jung

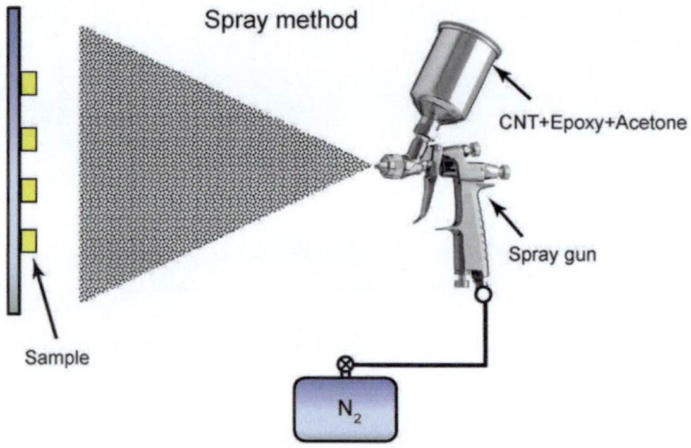

Fig. 7.2 Schematic of spray technique to deposit a mixture of CNT, epoxy, and acetone on surfaces (Jung and Bhushan 2009)

and Bhushan 2009). The first step of deposition of the CNT composite using the spray method was to disperse the CNT into a solvent in order to maintain a uniform distribution. Acetone was used as a solvent because it does not affect surface modification and is easily vaporized in ambient conditions. The dispersion process consisted of the sonication of 200 mgthe CNT in 100 mL of acetone for 4 min using a Branson Digital Sonifier with a frequency of 20 kHz at an amplitude of 80%. During this process, the mixture was exposed to ultrasonic pressure waves in a sonifier in order to disperse the CNT into smaller aggregates. To provide strong bonding between CNT and the substrate, 200 mg EPON epoxy resin 1002F was added to the mixture of CNT and acetone, and then the mixture was sonicated for 4 more minutes. Next, the sonified mixture was poured into a spray gun and sprayed onto the specimen surfaces. The conditions for uniform deposition of the CNT on the surfaces were optimized by adjusting the concentration of CNT in the solvent. After spraying the CNT on the surfaces, the CNT composite structures were then cured at 120 °C for 3 h (above the recommended curing temperature), which is above the melting point (80–88 °C) and below the burning point (180 °C) of EPON epoxy resin 1002F. At this temperature, the epoxy which initially covered the CNT was melted and moved to the interface between the CNT and substrates, resulting in exposed individual CNTs to provide the intended nanostructure that leads to high contact angle.

Figure 7.3 shows the scanning electron microscope (SEM) micrographs of nano- and hierarchical structures fabricated with CNTs (Jung and Bhushan 2009). SEM micrographs show an overview (left column), a detail in higher magnification (middle column), and a large magnification of the nanostructures with CNT (right column). All surfaces show a

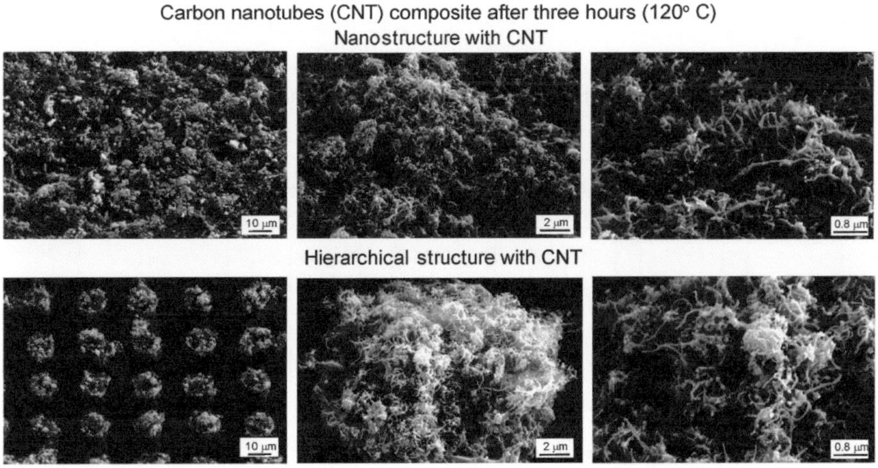

Fig. 7.3 SEM micrographs taken at 45° tilt angle, show three magnifications of nano- and hierarchical structures fabricated with CNT after three hours at 120 °C (Jung and Bhushan 2009)

homogeneous distribution of CNTs on the specimen. The CNTs were well dispersed and embedded on flat and microstructured surfaces for the desired nanostructure. The CNT diameter varied between 10 and 30 nm, and an aspect ratio varied between 160 and 200.

Lotus leaves have been the inspiration for the development of superhydrophobic and self-cleaning products. Therefore, as a benchmark for mechanical durability studies, Jung and Bhushan (2009) used nano-, micro-, and hierarchical structures created by the self-assembly of lotus wax, with the amounts of 0.8 μg/mm^2 deposited on flat epoxy resin and microstructure by thermal evaporation as shown in Fig. 6.39 in Chap. 6.

7.3.2 Contact Angle

To study the effect of CNT composite structures on superhydrophobicity, Jung and Bhushan (2009) measured the static contact angle and contact angle hysteresis on nano- and hierarchical structures with CNTs. For static contact angle and contact angle hysteresis, droplets of about 5 μL in volume (with the diameter of a spherical droplet about 2.1 mm) were deposited on the surface using a microsyringe. For contact angle hysteresis, the advancing and receding contact angles were measured at the front and back of the droplet moving along the tilted surface, respectively. Figure 7.4a shows that superhydrophobicity with a static contact angle of 166° and a contact angle hysteresis of 4° was found in the nanostructured surface with CNTs. After introducing the CNT nanostructure to the top of the micropatterned Si replica, the higher static contact angle of 170° and lower contact angle hysteresis of 2° were found for the hierarchical structures with CNTs. Both nano- and hierarchical structures created with CNTs showed superhydrophobic and self-cleaning surface, which has a static contact angle of more than 150° and contact angle hysteresis of less than 10° (Jung and Bhushan 2009).

For comparison, Fig. 7.4b presents data for nano-, micro-, and hierarchical structures using lotus wax. It shows that for the hierarchical structure with lotus wax, the highest

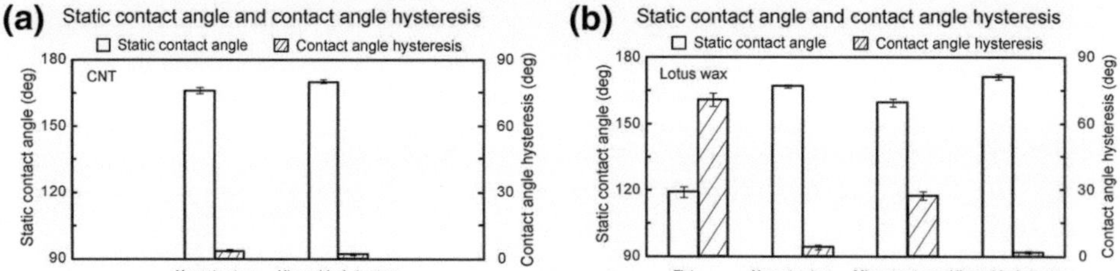

Fig. 7.4 Bar chart showing the measured static contact angle and contact angle hysteresis on **a** nano- and hierarchical structures fabricated with CNT after three hours at 120 °C, and **b** various structures fabricated with 0.8 μg/mm^2 of lotus wax after storage for seven days at 50 °C with ethanol vapor. The error bar represents ±1 standard deviation (Jung and Bhushan 2009)

static contact angles of 173° and lowest contact angle hysteresis of 1° were found. The recrystallized wax tubules were very similar to those of the lotus leaf, but are 0.5–1 μm longer. The static contact angle is higher and the contact angle hysteresis is lower than those reported for the lotus leaf (static contact angle of 164° and contact angle hysteresis of 3°) (Koch et al. 2009). Further, superhydrophobicity was found in the nanostructured surface with lotus wax with a static contact angle of 167° and a contact angle hysteresis of 5°. The microstructured surface with a lotus wax layer has a static contact angle of 160°, but shows a much higher contact angle hysteresis of 29° than found in hierarchical structures. Melting of the lotus wax led to a flat surface with a flat wax film and a much lower static contact angle of 119° and higher contact angle hysteresis of 71°. The data of a flat lotus wax film on a flat replica show that the lotus wax by itself is hydrophobic (Jung and Bhushan 2009).

7.3.3 Durability of Various Surfaces in Waterfall/Jet Tests

To investigate the durability of the fabricated surfaces in long-term exposure to water and how different kinetic energies of the water hitting the surface affect wetting properties, Jung and Bhushan (2009) conducted waterfall/jet tests on the surfaces fabricated with CNTs and lotus wax. The nano- and hierarchical structures with CNTs were exposed to water with pressure ranging from 0 to 45 kPa for 20 min. Figure 7.5a shows static contact angle and contact angle hysteresis as a function of water pressure. As water pressure hitting the surfaces increased, the static contact angle and contact angle hysteresis decreased and increased slightly, respectively, but a significant change was not found on both the nano- and hierarchical structures for superhydrophobicity. It can be interpreted that there was no deformation of CNT structures due to strong bonding with the substrate. As a result, superhydrophobic CNT composite structures showed good stability of wetting properties, not only from long-term exposure to water, but also high water pressure. Jung and Bhushan (2009) also conducted waterfall/jet tests on the flat, nano-, micro-, and hierarchical structures with lotus wax in order to compare durability of the created surfaces with CNTs and lotus wax. Figure 7.5b shows static contact angle and contact angle hysteresis as a function of water pressure exposure. As the water pressure increased up to 45 kPa, static contact angle and contact angle hysteresis remained almost constant for the flat surface and microstructure surface with lotus wax layer. However, as the water pressure increased up to 34 kPa, the static contact angle of nano- and hierarchical structures with lotus wax first decreased slightly, and then the contact angle started decreasing sharply. It was also observed that the corresponding large change in contact angle hysteresis was found above 34 kPa. As expected, it is observed that the nanostructure with lotus wax can be damaged by water with high pressure, resulting in a loss of superhydrophobicity.

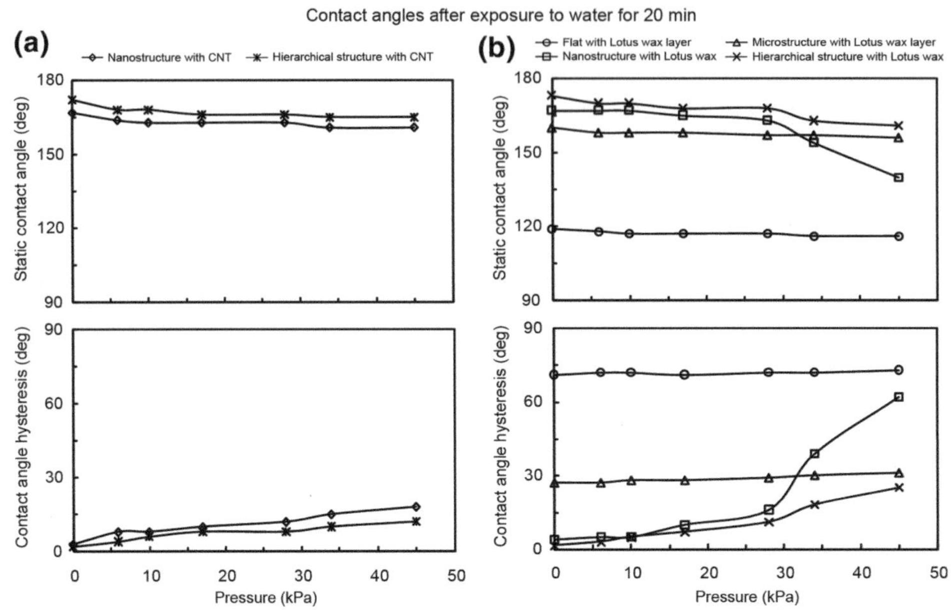

Fig. 7.5 Static contact angle and contact angle hysteresis as a function of water pressure of exposure for 20 min for the droplet with a 1 mm radius (5 μL volume) deposited on the surfaces that were made with **a** CNT and **b** lotus wax. The data are reproducible within ±5% (adapted from Jung and Bhushan 2009)

7.3.4 Durability of Various Surfaces in AFM and Ball-on-Flat Tribometer Tests

To investigate the mechanical durability of the nanostructure fabricated using CNTs, Jung and Bhushan (2009) conducted wear tests by creating 50×50 μm^2 wear scars with a 15 μm radius borosilicate ball for 1 cycle at two normal loads of 100 and 10 μN using an AFM. Figure 7.6 shows surface height maps before and after wear tests for nanostructures with CNTs. As the normal load of 100 nN was applied to the nanostructure with CNTs, the wear scar induced on the surface after the 100 nN normal load tests was not found or was very low, and it was also hard to quantify a wear depth on the nanostructure with CNTs scanned with a borosilicate ball. With increasing the normal load to 10 μN, it was found that the wear depth on the nanostructure with CNTs was not significantly changed, but the morphology of the CNTs differs slightly from that before the wear test. It can be interpreted that the individual CNTs might be expected to slide or bend with the borosilicate ball applied by high normal load of 10 μN during the test process.

For comparison, Jung and Bhushan (2009) also investigated the durability of the nanostructure fabricated using lotus wax by applying two normal loads of 100 nN and 10 μNusing AFM. Figure 7.6 shows surface height maps before and after wear tests

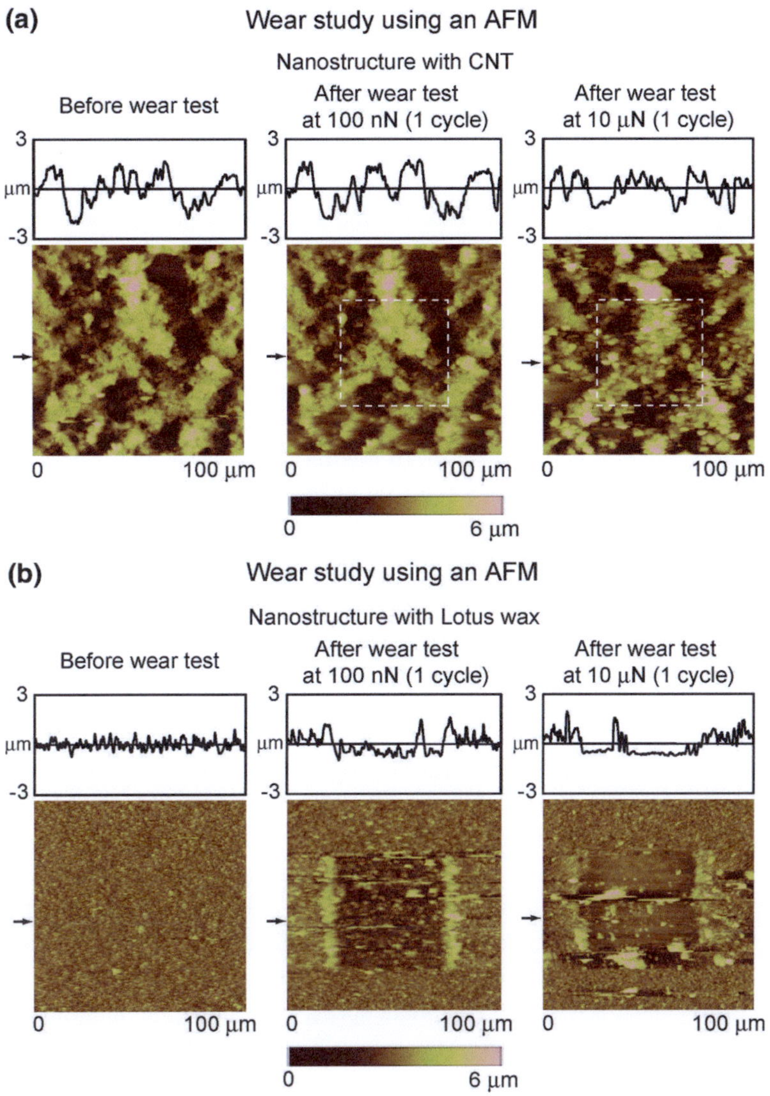

Fig. 7.6 Surface height maps before and after wear tests with a 15 μm radius borosilicate ball at 100 nN and 10 μN for nanostructures with **a** CNT and **b** lotus wax using an AFM (Jung and Bhushan 2009)

for nanostructures with lotus wax. As the normal load of 100 nN was applied to the nanostructure with lotus wax, the change in the morphology of the structured surface was observed, and a small amount of debris was generated compared to the surface before the wear test. This indicated that the wax nanostructure has weak mechanical strength, even with small loads. When the normal load was increased to 10 μN, the depth of the wear mark increased, and the nanostructure of the lotus wax was fully removed from the substrate. As expected, the nanostructure of the lotus wax exhibited greater wear compared to the nanostructure with the CNTs, as evidenced by debris build-up around the edge of the wear test region. The damage to the structured surface can cause the sticking of water droplets in the wear region, resulting in low static contact angle and high contact angle hysteresis.

In order to investigate the durability of structured surfaces at a high load, Jung and Bhushan (2009) conducted ball-on-flat tribometer experiments using a sapphire ball with a diameter of 3 mm at a normal load of 10 mN on the surfaces with CNTs. Figure 7.7a shows the coefficient of friction as a function of the number of cycles for the nano- and hierarchical structures with CNTs. The data are reproducible within ±5% based on three measurements. The coefficients of friction on both the nano- and hierarchical structures first increased slightly for 20 cycles. Such a trend can be due to the elastic bending or buckling of CNTs by contacting with a sapphire ball during the beginning of scan, resulting in an increase of the contact area. During the entire experiment, the coefficient of friction value of the nano- and hierarchical structures changed minimally, which indicates that the CNTs were not worn after 100 cycles. To investigate the change in the morphology of the surfaces after the wear test, optical microscope images were obtained before and after the wear test as shown in Fig. 7.8a. As expected, it was observed that there is no or low wear on the nano- and hierarchical structures after wear tests. No or low wear on the CNT composite structure possibly can be due to the significant increase in the mechanical strength and wear resistance from the uniform distribution and strong bonding of CNTs on flat epoxy resin and microstructure. The elastic bending or buckling exhibited by CNTs make them exceedingly tough materials that may be absorbing some of the force at contact, acting as a compliant spring moderating the impact of the ball on the surface (Dresselhaus et al. 2000; Meyyappan 2005; Chen et al. 2006).

Contact diameters and contact stresses of CNTs at three loads used in AFM and ball-on-flat tribometer tests were calculated (Jung and Bhushan 2009). Table 7.1a lists the physical properties of various specimens. It is assumed that contacts are single-asperity elastic contacts. Using Hertz analysis, the contact diameter is given by (Bhushan 2013a, b)

$$d = 2\left(\frac{3WR}{4E^*}\right)^{1/3} \tag{7.1}$$

where W is the total normal load, R is the asperity radius, and E^* is the effective elastic modulus. It should be noted that contact occurs on multiple asperities, and (7.1) gives an

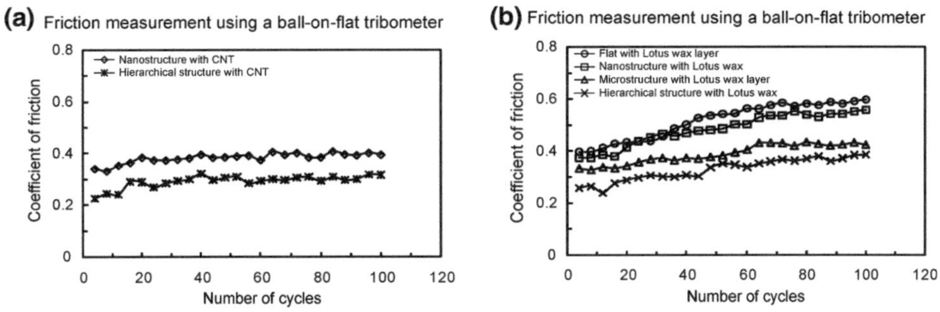

Fig. 7.7 Coefficient of friction as a function of number of cycles using a ball-on-flat tribometer using a sapphire ball with a diameter of 3 mm at 10 mN for the surfaces with **a** CNT and **b** lotus wax at room temperature (22 ± 1 °C) and ambient air (45 ± 5% RH). The data are reproducible within ±5% based on three measurements (Jung and Bhushan 2009)

approximate value. The calculated contact diameter and contact pressure are presented in Table 7.1b. Based on the contact pressures, deformation of CNTs appears to be elastic at the three loads applied by the borosilicate ball and sapphire ball.

In order to compare the durability of the created surfaces with CNT and with lotus wax at a high load, Jung and Bhushan (2009) conducted a wear study on the surfaces with lotus wax as well using a ball-on-flat tribometer. As shown in Fig. 7.7b, the coefficient of friction value of the surfaces with lotus wax exhibited a gradual increase when the sliding cycle increases up to about 70 cycles, and then remains constant. This indicates that the wax nanostructure and flat wax layer could be undergoing some wear due to weak bonding with the substrates. The change in the morphology of the surfaces with lotus wax was observed in optical microscope images as shown in Fig. 7.8b. As shown in the AFM study at low loads (Fig. 7.6b), it is observed that the flat wax layer and wax nanostructure on the flat and microstructure were fully removed from the surfaces. Therefore superhydrophobic CNT composite structures showed better mechanical durability than the structured surfaces made with lotus wax to withstand real world application requirements.

7.3.5 Summary

The hierarchical structure fabricated with CNTs shows a high static contact angle of 170° and a low contact angle hysteresis of 2°. As a benchmark, the structures created using lotus wax were used to compare the durability of CNT composite structures. It was found that superhydrophobic CNT composite structures showed stability of wetting properties in exposure to water. In contrast, it was observed that the nanostructure made with lotus wax can be damaged by water with high pressure, resulting in the loss of superhydrophobicity. From wear and friction studies, the CNT composite structure showed high wear resistance

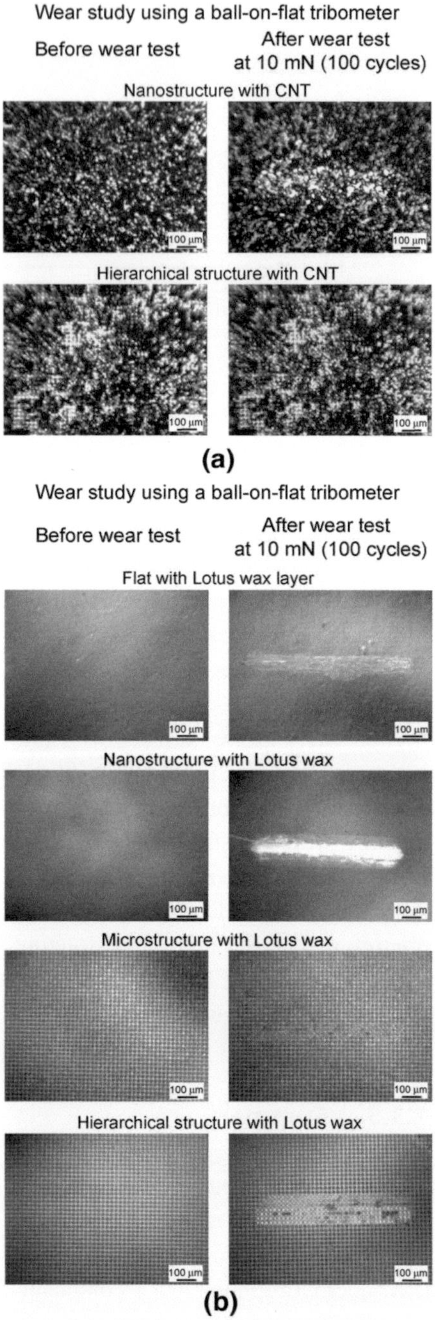

Fig. 7.8 Optical micrographs before and after wear test at 10 mN (100 cycles) using a ball-on-flat tribometer for the surfaces with **a** CNT and **b** lotus wax (Jung and Bhushan 2009)

Table 7.1 **a** Typical physical properties of various specimens and **b** calculated contact diameters and contact stresses at three loads used in AFM and ball-on-flat tribometer measurements. It is assumed that contacts are singe-asperity contact (Jung and Bhushan 2009)

(a)

	Borosilicate ball with a 15 μm radius	Sapphire ball with a 1.5 mm radius	Carbon nanotube
Elastic modulus (GPa)	70[a]	390[b]	1200[c]
Poisson's ratio	0.2[a]	0.23[b]	0.2[d]
Bending strength (GPa)			14.2[c]

(b)

	Contact diameter (μm)			Mean contact pressure (GPa)		
	Borosilicate ball at 100 nN	Borosilicate ball at 10 μN	Sapphire ball at 10 mN	Borosilicate ball at 100 nN	Borosilicate ball at 10 μN	Sapphire ball at 10 mN
	0.05	0.24	6.62	0.076	0.33	0.44

[a]Callister (2000)
[b]Bhushan and Gupta (1991)
[c]Wong et al. (1997)
[d]Zhang et al. (2008)

because of the uniform distribution and strong bonding of CNTs to the flat epoxy resin and microstructure.

However, it was found that the nanostructure with lotus wax can be damaged easily, even at a small load.

7.4 Superhydrophobic Surfaces Using Nanoparticle Composites with Hierarchical Structure

Superhydrophobic surfaces using silica particles have been created by drop casting, chemical deposition and sol-gel processes (Ming et al. 2005; Liu et al. 2006; Englert et al. 2006). Ebert and Bhushan (2012a) fabricated hierarchical structures resistant to mechanical wear with silica particles using a spray method. In their studies, hierarchically structured surfaces were produced by either spraying a nanoparticulate coating over a micropatterned surface or by spraying a coating with micro- and nanosized particles on a flat surface. The pitch between pillars for the micropatterns was known, and the average pitch between microparticles on a surface was determined through SEM imaging.

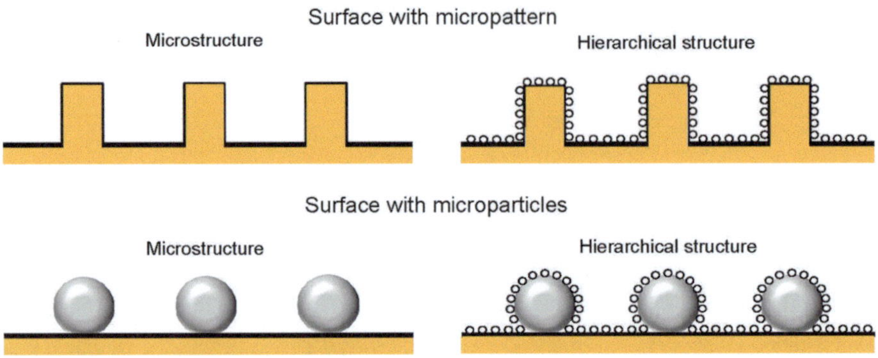

Fig. 7.9 Schematic examples of microstructured and hierarchically structured surfaces in which the microstructure is formed by both a uniform micropattern (top) and microparticles (bottom). In both cases, the hierarchical structure is achieved through addition of nanoparticles (adapted from Ebert and Bhushan 2012a)

Figure 7.9 illustrates the use of both particles and a patterned surface to form microstructures and hierarchical structures. For a known geometry of the particles, the transition to the Wenzel regime can be predicted to occur above a certain pitch value (Ebert and Bhushan 2012a).

7.4.1 Fabrication Details

Epoxy micropatterns were fabricated by replicating a Si micropatterned surface (D = 14 μm, H = 30 μm, P = 23 μm) by soft lithography as described in Chap. 6. To fabricate samples with hierarchical structure using a micropattern as the microscale roughness, nanoparticle-epoxy solution was sonicated and sprayed on the micropatterned surface. Silica nanoparticles of 10 nm (±1 nm) and 50 nm (±15 nm) were used. These were created through continuous flame hydrolysis of $SiCl_4$, and hydrophobized through silane treatment (Evonik-Degussa Corporation, Parsippany, New Jersey). In order to spray the particles onto the surfaces, they were first dispersed uniformly in solution. 200 mg of particles were sonicated in 100 mL of acetone for 4 min with a Branson Sonifier 450 A with a frequency of 20 kHz at 80% amplitude. For strong bonding of the nanoparticles to the substrates, 200 mg of EPON epoxy resin 1002F (Hexion Specialty Chemicals, Columbus, Ohio) was then added, and the solution was sonicated for another 4 min. The dispersion of particles in solution was then sprayed onto the sample surface using a spray gun. The samples were then annealed at 120 °C for 3 h. Annealing at this temperature (above the melting point and below the burning point of EPON 1002F) was done so that the resin would melt and move to the interface between the particles and the surface.

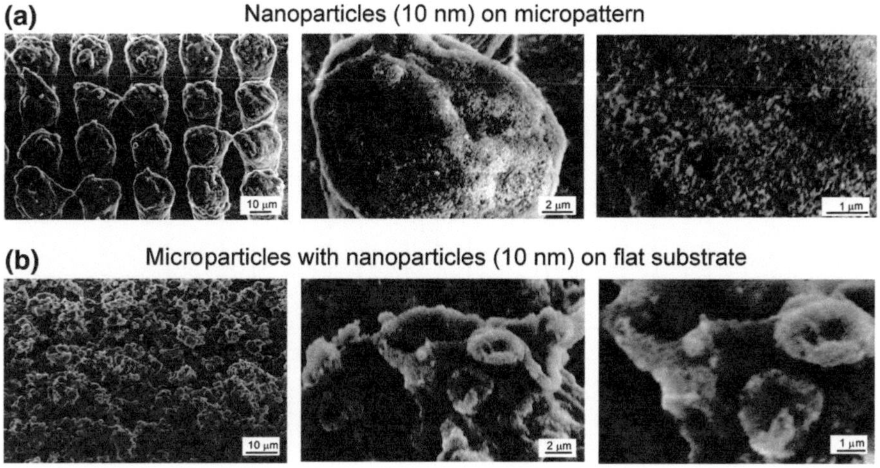

Fig. 7.10 SEM micrographs taken at 45° tilt angle showing **a** 10 nm nanoparticles on micropattern with a pitch value (P) of 23 μm at three magnifications (bottom), and **b** microparticles (10 μm) with nanoparticles (10 nm) with microparticle spray concentration of 1400 mg/L with resulting average pitch value (P_{avg}) of 21 μm (adapted from Ebert and Bhushan 2012a)

Figure 7.10a shows the SEM micrographs of hierarchical structures with silica nanoparticles and an epoxy micropattern (Ebert and Bhushan 2012a). The hierarchical structures are shown at three magnifications: the lowest shows the pitch between pillars, the middle shows an individual pillar, and the highest shows surface nanoscale roughness.

To fabricate samples using microparticles as the microscale roughness, the microparticles were sonicated in solution in the same manner and sprayed onto flat epoxy surfaces. The hydrophobic silica microparticles (diameter = 10 ± 5 μm) were obtained as a trimethylated silica gel powder from Dow Corning (Midland, Michigan). For hierarchically structured surfaces, an additional sonication and spray of nanoparticles was performed. Surfaces of varying pitch were created by changing the concentration of microparticles in the solution (400 mg/L or 1400 mg/L). The value of the average pitch between microparticles was found by determining the number of particles in a 1 mm^2 area in five different locations on the sample by examining SEM images. Figure 7.10b shows the SEM micrographs of surfaces using microparticles with concentration of 1400 mg/L resulting in an average pitch value (P_{avg}) of 21 μm (Ebert and Bhushan 2012a).

7.4.2 Contact Angle of Surfaces Using Micropattern

Contact angle and contact angle hysteresis data for the 10 and 50 nm particles on flat and micropatterned surfaces are shown in Fig. 7.11 (Ebert and Bhushan 2012a). Data for the micropattern without nanoparticles is also shown for reference. The contact angle and

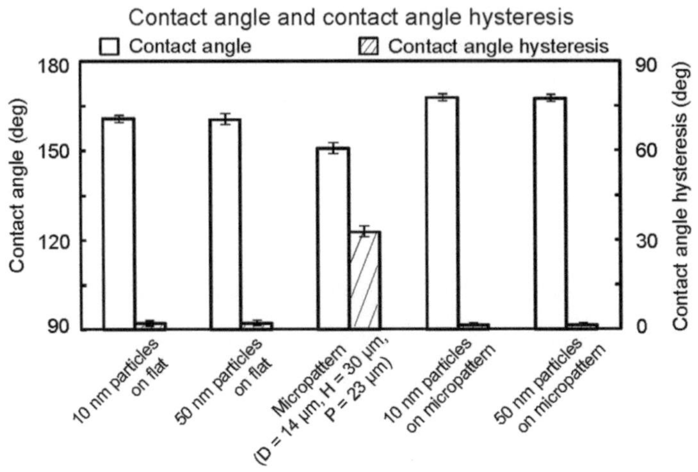

Fig. 7.11 Bar chart showing the measured static contact angle and contact angle hysteresis on nano-, micro- and hierarchically structured surfaces using both 10 and 50 nm particles. The error bars represent ±1 standard deviation (Ebert and Bhushan 2012a)

contact angle hysteresis of the epoxy micropattern alone were found to be 151° (±2°) and 33° (±2°), respectively. In the cases of both flat and micropatterned surfaces, the difference in the data between the 10 and 50 nm particles was insignificant. Nanoparticles on a flat surface (nanostructure) showed a contact angle of 161° (±2°) and contact angle hysteresis of 2° (±1°). For nanoparticles on the micropattern (hierarchical structure), the highest contact angle of 168° (±2°) and lowest contact angle hysteresis of 1° (±0.5°) were found.

To sum up, the data shows that one can achieve superhydrophobicity simply by micropatterning the surface, however, contact angle hysteresis is relatively high. Micropattern geometry can be optimized to achieve desirable contact angle and contact angle hysteresis values (shown in the next section). Nanostructured and hierarchically structured surfaces exhibit contact angle higher than that of flat surface as well as very low contact angle hysteresis (<2°).

7.4.3 Contact Angle of Surfaces Using Microparticles and Comparison to Micropatterns

The contact angle and contact angle hysteresis of surfaces with 10 μm particles alone on flat and 10 μm/10 nm particles combined on flat are shown in Fig. 7.12 (Ebert and Bhushan 2012a) across a range of average pitch values. In addition, the contact angle/contact angle hysteresis of epoxy micropatterns of varying pitch are shown in order to provide a comparison of the trends seen with microparticles versus micropatterns. In this comparison, micropatterns with pillars with $D = 5$ μm and $H = 10$ μm were used, as the size is more comparable to the microparticles. It is seen that for 10 μm particles alone on flat, the trends in contact angle and contact angle hysteresis with pitch are similar to those of the micropatterns. The micropatterns maintain a contact angle above 150° from the lowest pitch value until a sharp drop at a pitch of 60 μm. Similarly, the surfaces with microparticles show a steady contact angle above 150°, with a sharp drop at an average pitch of 40 μm. For contact angle hysteresis, the micropattern shows a steady decrease with increasing pitch, until a large increase occurs at 60 μm. For the 10 μm particles, an initial decrease is seen at low pitch, and very low contact angle hysteresis is seen until a large increase occurs, again at 40 μm.

The sudden changes in contact angle and contact angle hysteresis, occurring at a pitch of 60 μm for the micropatterns and 40 μm for the microparticles, represent the transition to the Wenzel regime. At these pitch values, the droop of the water droplet between microstructures is enough that the water fully penetrates between them, leaving no air pockets. For the micropatterns, the value of 60 μm is close to the value predicted by Eq. (6.4) in Chap. 6 (74 μm with $D = 5$ μm, $H = 10$ μm). The transition value for the surfaces with 10 μm particles is expected to be different since the pitch, diameter, and height are much less uniform as compared to the micropatterns. However, the similarity of the overall trends in contact angle and contact angle hysteresis with pitch indicates that the microparticles are reasonably mimicking a micropatterned surface. In addition, the data for the 10 μm/10 nm particles combined shows an overall increase in contact angle and decrease in contact angle hysteresis as compared to the 10 μm particles alone, indicating that the benefits of hierarchical structure can be realized using the 10 μm particles as the underlying microstructure. Contact angle as high as 166° (±2°) and contact angle hysteresis as low as 2° (±1°) were seen for these surfaces.

7.4.4 Durability of Various Surfaces in AFM and Ball-on-Flat Tribometer Tests

The results of the AFM wear experiment for 10 nm particles, as well as an epoxy resin, are shown in Fig. 7.13 (Ebert and Bhushan 2012a). Surface height maps before and after the wear experiment are displayed, as well as sample scans across the middle of the image (position indicated by an arrow) with RMS roughness within the wear area displayed. After 1 cycle at 10 μN with the borosilicate ball, the morphology of the surface with 10 nm particles was not significantly changed, and RMS of the sample scan within the wear area was nearly identical (270 nm after compared with 275 nm before). However, the after-image for the epoxy resin reveals some wear, as well as an overall swelling of the wear area. In addition, RMS of the sample scan increased from 25 to 177 nm within the wear area. This demonstrates that the durability of the nanoparticle-coated surfaces is superior to that of the surfaces with epoxy resin alone.

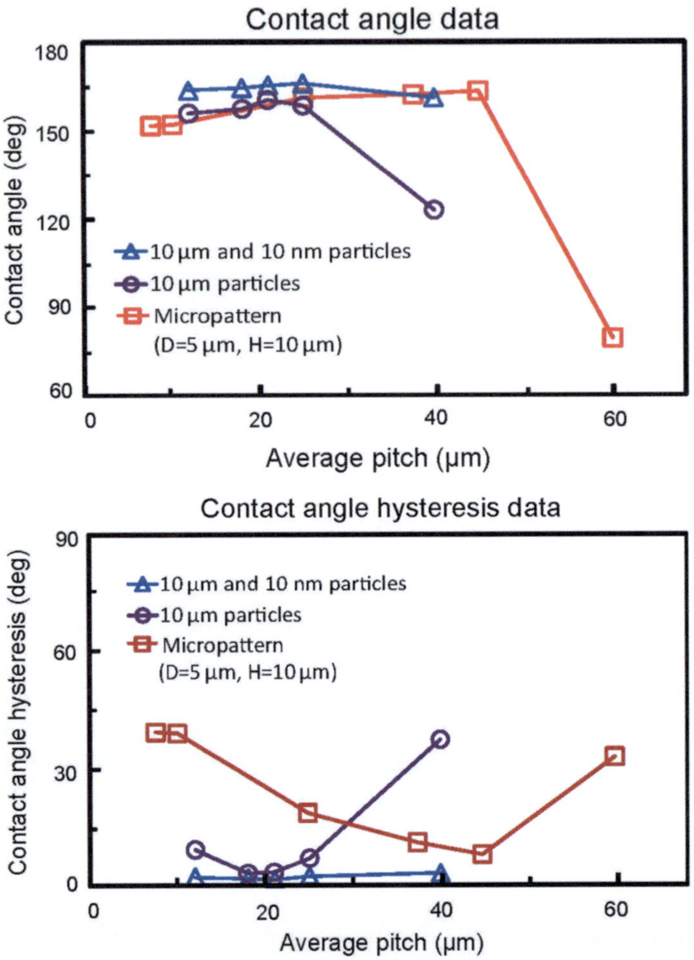

Fig. 7.12 Static contact angle and contact angle hysteresis as a function of average pitch for surfaces with 10 µm particles alone and with 10 µm/10 nm particles combined on flat. Epoxy micropattern data is also shown for comparison of microstructure behavior. Reproducibility is ±1° for CAH of 10 µm/10 nm particles combined, ±2° for CAH of 10 µm particles and epoxy micropatterns, and ±2° for all CA data (Ebert and Bhushan 2012a)

The ball-on-flat tribometer experimental results are shown in Fig. 7.14 for surfaces with 10 µm particles, 10 µm/10 nm particles combined, and an epoxy resin (Ebert and Bhushan 2012a). Before and after images as well as a scan across the wear scar are displayed for 10 µm particles, 10 µm with 10 nm particles, and epoxy resin on flat surfaces. After 100 cycles at 10 mN, the 10 µm particles alone showed minimal wear (mainly in the form of burnishing of the tops of particles), and a microscale roughness is preserved. The 10 µm and 10 nm particles combined show similar results of minimal

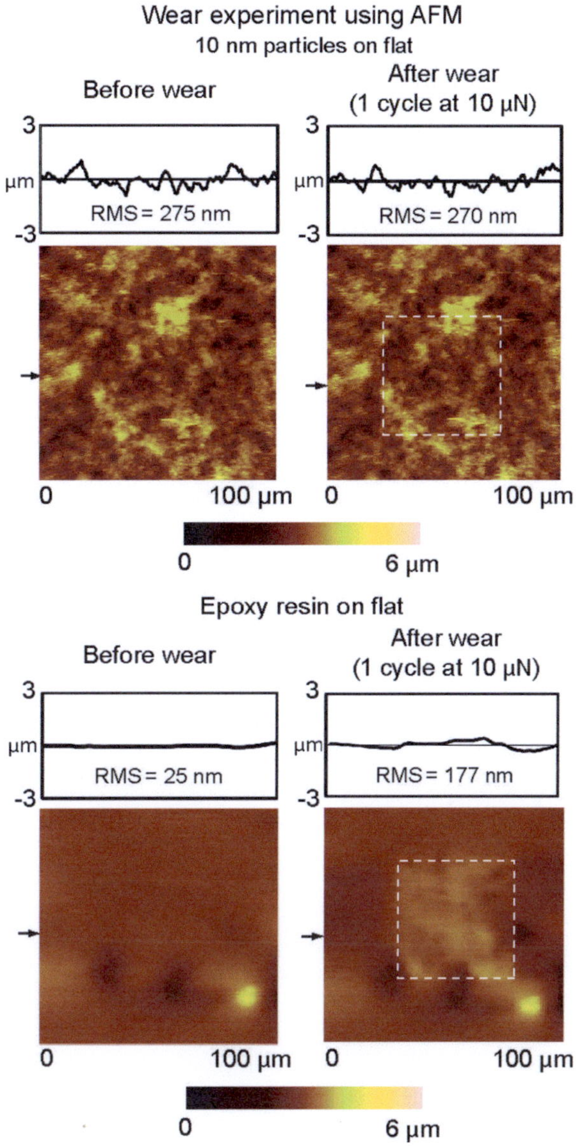

Fig. 7.13 Surface height maps and sample surface profiles (locations indicated by arrows) before and after AFM wear experiment with 15 μm radius borosilicate ball at load of 10 μN for 10 nm particles on flat surface (top) and epoxy resin on flat surface (bottom). RMS roughness values for surface profiles are displayed within surface profile boxes (Ebert and Bhushan 2012a)

wear. However, the after image of the epoxy resin reveals a well-defined groove. The scan shows buildup along the sides of the groove, and the groove depth can be seen as approximately 300 nm. Thus, the surfaces with microparticles also showed resistance to wear superior to that of the surfaces with epoxy resin alone.

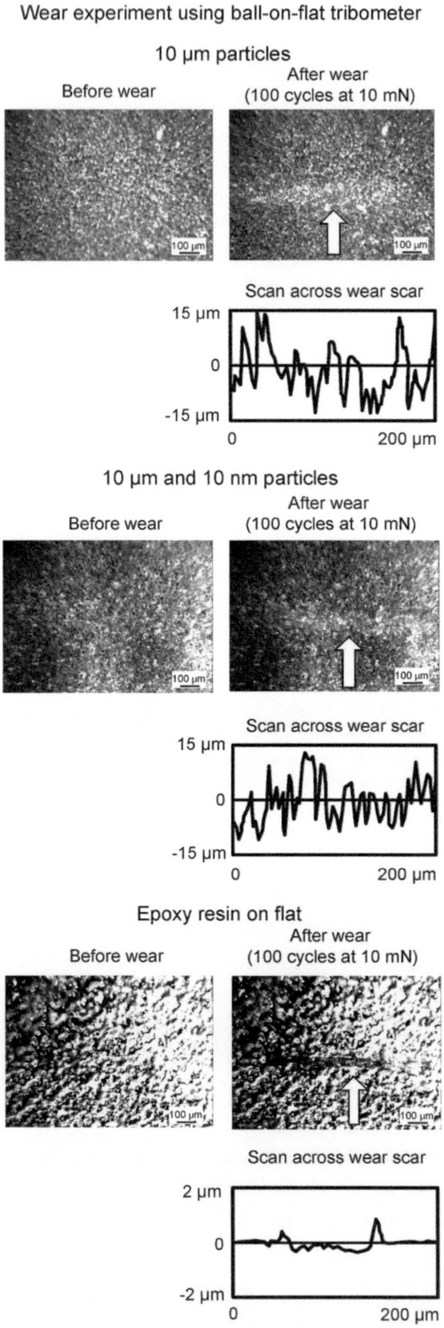

Fig. 7.14 Optical micrographs before and after wear experiment using ball-on-flat tribometer at 10 mN for 10 μm particles (top), 10 μm particles with 10 nm particles (middle), and epoxy resin on flat surface (bottom). Sample profiles across wear areas are shown below after-images (arrows indicate location of profile) (Ebert and Bhushan 2012a)

7.4.5 Summary

Ebert and Bhushan (2012a) produced mechanically durable, hierarchically structured surfaces exhibiting superhydrophobicity. They were produced either by spraying a nanoparticulate coating over a micropatterned surface or by spraying a coating with micro- and nanosized silica particles on a flat surface. A comparison was made between the use of microparticles and micropatterns in terms of wettability across different pitch values. Results show similar trends in CA/CAH with pitch value whether the microstructure was formed with particles or a micropattern. Hierarchical surfaces using micro- and nanoparticles exhibited a CA of 166° (±2°) and a CAH of 2° (±1°) at optimum pitch. For hierarchical surfaces using micropatterns with nanoparticles, a nearly identical CA of 168° (±2°) and CAH of 1° (±0.5°) were measured. In addition, the surfaces demonstrated wear resistance superior to the epoxy resin on multiple length scales in the AFM and tribometer experiments, showing ability to preserve both nanoscale and microscale roughness.

7.5 Superhydrophobic Surfaces Using Nanoparticle Composites for Optical Transparency

In applications such as self-cleaning windows, electronic touch screens (e.g., for mobile phones, tablets, and computers), camera lenses, optical devices, and solar panels, high optical transparency, as well as mechanical durability is required. For example, typical requirements for an automotive windshield are visible transmittance> 70%, haze< 1%, 10,000 cycles of wiper sliding, and 250 car wash cycles; the last two requirements representing 10 years of life.

The dual requirements of superhydrophobicity and transparency pose a challenge. The surface must be sufficiently rough to obtain high CA and low CAH, but the dimensions of the roughness features must be small enough to preserve high transmittance of light. The size of surface features should not exceed roughly one quarter of the wavelength of visible light (around 100 nm or less) (Nakajima et al. 1999; Manca et al. 2009; Karunakaran et al. 2011; Ebert and Bhushan 2012b). Micropatterned surfaces may compromise the substrate transparency. Coatings with a total thickness of less than one quarter of the wavelength of visible light are of interest.

Glass is the most common optical material for various applications, such as lenses, architectural windows, and electronic touch screens. Transparent polymers such as polycarbonate (PC) and poly(methyl methacrylate) (PMMA) also are used for wide-ranging applications, such as aircraft canopies, bullet-proof windows, solar cell panels, and many high-performance optical, electronic and medical devices.

SiO_2, ZnO, and ITO (Indium Tin Oxide) thin films are of interest for various applications. Select properties and potential benefits of SiO_2, ZnO, and ITO are shown in

Table 7.2 (Ebert and Bhushan 2012b). The three metal oxides have high transmittance for visible light due to low refractive indices (minimizing reflectance) and band gap wavelengths shorter than the visible range of 400–700 nm (minimizing visible-range absorption). SiO_2 in particular has extremely high visible transmittance. ZnO thin films can have a UV-protective effect, and have been shown to reduce photodegradation of PC (Moustaghfir et al. 2004). ZnO, when doped with other metals such as Al or Ga, can also be used for transparent conducting films (Noh et al. 2010; Park et al. 2010). ITO is the most commonly used material for transparent conducting films due to its combination of high visible transmittance and low electrical resistivity (Kim et al. 1999; Ederth et al. 2003; Cho et al. 2006). In addition, these particles have high hardness (Kucheyev et al. 2002; Vila et al. 2003; Zeng et al. 2003). Thus, SiO_2, ZnO, and ITO nanoparticles would seem to be suitable candidates for wear-resistant, transparent, superhydrophobic surfaces (Ebert and Bhushan 2012b).

Methylphenyl silicone resin was selected as a binder because it is known to be mechanically durable and offers strong adhesion between nanoparticles and substrate.

Ebert and Bhushan (2012b) used a dip coating technique to deposit nanoparticle composite coatings on glass, PC, and PMMA substrates using SiO_2, ZnO, and ITO

Table 7.2 Typical values for optical properties, electrical resistivity, and hardness for SiO_2, ZnO, and ITO particles (Ebert and Bhushan 2012b)

Particle	Band gap energy (band gap wavelength)	Refractive index	Hardness (GPa)	Electrical resistivity (Ω cm)	Potential benefits
SiO_2	8.9 eV (138 nm)[a]	1.5[b]	9[c]	1×10^{18}[d]	High visible transparency, high hardness
ZnO	3.3 eV (375 nm)[e]	2.0[f]	5 [g]	28 [h]	UV-protective effect
ITO	3.6 eV (345 nm)[i]	1.8[f]	7[j]	2×10^{-4}[k]	Transparent conducting oxide

[a]Schneider and Fowler (1976)
[b]Malitson (1965)
[c]Vila et al. (2003)
[d]Shackelford and Alexander (2001)
[e]Srikant and Clarke (1998)
[f]Yang et al. (2006)
[g]Kucheyev et al. (2002)
[h]Natsume and Sakata (2000)
[i]Hamberg and Granqvist (1984)
[j]Zeng et al. (2003)
[k]Kim et al. (1999)

nanoparticles in a methylphenyl silicone resin. Dip coating technique was selected because it is simple, versatile, and provides a uniform coating. In order to achieve high transparency, micropatterned substrates as in a previous study were not used. The samples were characterized in terms of wettability (contact angle/contact angle hysteresis) and optical transmittance in the visible spectrum. Wear resistance experiments in sliding and using a waterjet were performed.

7.5.1 Fabrication Details

Ebert and Bhushan (2012b) used soda-lime glass (2.2 mm thick), polycarbonate (Lexan, SABIC Innovative Plastics, 2.4 mm thick), and PMMA (Optix, Plaskolite Inc., 2 mm thick) to create 1 cm × 1 cm substrates. Silane-modified hydrophobic SiO_2 nanoparticles with average diameter of 55 nm (±15 nm) from Evonik Industries (AEROSIL RX 50), ZnO nanoparticles with average diameter of 70 nm (±30 nm) from Alfa Aesar (NanoTek Zinc Oxide), and ITO nanoparticles (In_2O_3:SnO_2 = 90:10) of average diameter 45 nm (±25 nm) from US Research Nanomaterials (US3855 Indium Tin Oxide Nanopowder) were used. Methylphenyl silicone resin was obtained from Momentive Performance Materials (SR355S Methylphenyl Silicone Resin), and was used as a binder.

The superhydrophobic SiO_2 particles obtained were already silane-modified. Conversely, the ZnO and ITO particles were not surface-modified as received. In order to hydrophobize them, they were treated in solution by octadecylphosphonic acid (ODP) (Aldrich). ODP can be used to functionalize metal oxides from hydrophilic to hydrophobic (Nishimoto et al. 2009). The process by which functionalization occurs is illustrated in Fig. 7.15 (Ebert and Bhushan 2012b). The exposed long-chain hydrocarbon tails of the ODP molecules result in a hydrophobic particle surface. Approximately 2 g of particles were added to a 100 mL ethanol solution with ODP concentration of 2 mM. The mixture was stirred vigorously for 10 min, covered, and left for 4 days at 20 °C. The solvent was then removed by evaporation, and the particles were heated at 100 °C for 1 h to improve ODP bonding and remove adsorbed water or remaining solvent.

Nanoparticles were dispersed in a 40%/60% tetrahydrofluorine/isopropyl alcohol (THF/IPA) (by volume) mixture to form the dip coating solution. Pure THF rapidly dissolves PC and PMMA, resulting in loss of transparency. However, if the THF concentration is low (about 5% by volume in IPA), the visible loss can be minimized. Pure IPA unfortunately does not evaporate quickly enough to leave a homogeneous coating on the substrate. Optimal concentrations of nanoparticles in the solvent were found to be approximately 10 mg/mL for SiO_2 particles, 35 mg/mL for ZnO particles, and 50 mg/mLfor ITO particles. Too low of a concentration resulted in loss of superhydrophobicity, while too high of a concentration resulted in visible agglomeration of particles on substrates, substantially reducing transparency. The nanoparticles were added to 30 mL of the THF/IPA solvent in a 100 mL glass beaker and sonicated for 4 min. Then, 150 mg of

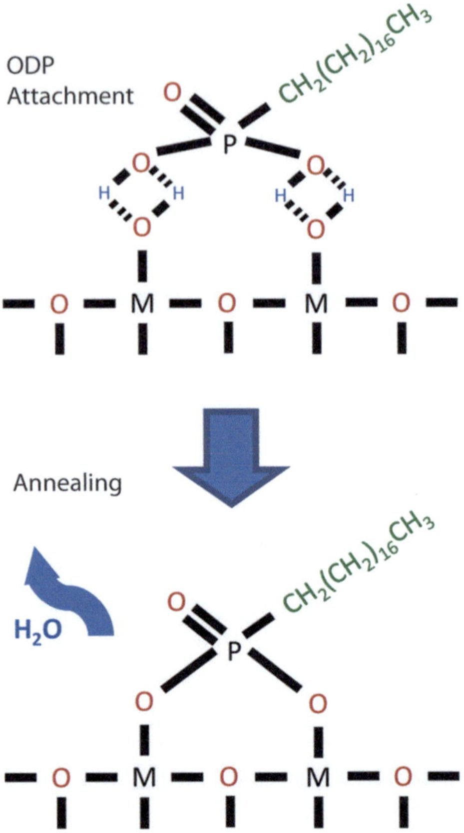

Fig. 7.15 Schematic of hydrophobization of a generic metal oxide surface with octadecylphosphonic acid (ODP). The ODP molecule first attaches to the hydrolyzed surface via hydrogen bonds. After annealing, the ODP is strongly bonded to the metal oxide surface (Ebert and Bhushan 2012b)

methylphenyl silicone resin were added and the mixture was sonicated for an additional 4 min. In the case of silicone resin, a lower concentration resulted in poor adherence of the particles to the substrate, and a higher concentration resulted in the loss of superhydrophobicity. After sonication, approximately 10 mL of fresh solvent was added at 40%/60% THF/IPA ratio. Substrates were dipped into the solution and immediately removed at a speed of 10 cm/min. Coated samples were then heated at 40 °C for 10 min to remove any remaining solvent. The samples required no chemical post-treatment or modification after dip coating (Ebert and Bhushan 2012b).

7.5.2 Surface Roughness and Morphology

Table 7.3 presents RMS roughness, P-V (peak-valley) distance, roughness factor (R_f), fractional area of liquid-area interface (f_{LA}), and coating thickness for samples with SiO_2, ZnO, and ITO. Surfaces had nanoscale roughness formed by nanoparticles bound

Table 7.3 Measured surface roughness values (RMS and P-V distance), coating thickness, and calculated R_f and liquid-air fractional area (f_{LA}) for samples with SiO_2, ZnO, and ITO particles ((Ebert and Bhushan 2012b)

Particles on sample	RMS (nm)	P-V distance (nm)	R_f	Coating thickness (nm)	f_{LA}
SiO_2	58 ± 3	137 ± 5	1.5	150 ± 10	0.94
ZnO	84	191	1.8	205	0.94
ITO	45	127	1.3	135	0.91

to the substrate with silicone resin (Ebert and Bhushan 2012b). The values of R_f and f_{LA} were calculated from AFM surface height maps (Burton and Bhushan 2006; Ebert and Bhushan 2012b). By using the Z-height of each data point in the AFM scan matrix, the real surface area can be approximated using simple geometry. Dividing this value by the two-dimensional scan area gives R_f. Using SPIP imaging software (Image Metrology), f_{LA} was estimated. R_f and f_{LA} can be used to identify Wenzel or Cassie-Baxter wetting regimes. Coating thicknesses were measured with a Tencor® stylus profiler on the step formed by partially coating a substrate, and found to be nearly equal to P-V distance.

Figure 7.16 shows SEM micrographs of sample surfaces using each of the three nanoparticles on glass substrates at two magnifications (Ebert and Bhushan 2012b). At lower magnification, the particles can be seen to form islands with width on the order of a few microns. ZnO tended to cover more of the substrate, but less evenly and with larger pockets of uncoated area. ITO tended to coat the most evenly and with smallest islands. At higher magnification, individual nanoparticles can be seen forming roughness on the nanoscale, which suggests the particles are well-dispersed in solution. This multiscale roughness is desirable for superhydrophobicity.

7.5.3 Contact Angle

Contact angle and contact angle hysteresis for samples using three different nanoparticles (SiO_2, ZnO, ITO) on three different substrates (glass, PC, PMMA) are shown in Fig. 7.17 (Ebert and Bhushan 2012b). All samples exhibited superhydrophobic and self-cleaning behavior, with contact angles nearly 170° and contact angle hysteresis as low as 1° in some cases. For all three particles, contact angle was slightly higher on PC and PMMA substrates than on glass. Contact angle hysteresis was slightly higher on glass substrates, except in the case of ZnO where it was unchanged at 1°. For all substrates, the samples with ITO particles had lower contact angle and higher contact angle hysteresis than those with SiO_2 and ZnO, although self-cleaning conditions were still met (contact angle $> 150°$ and contact angle hysteresis $< 10°$).

The f_{LA} values (shown in Table 7.3) were estimated to be 0.94, 0.94, and 0.91 for SiO_2, ZnO, and ITO samples, respectively. A lower value for ITO nanoparticles occurs

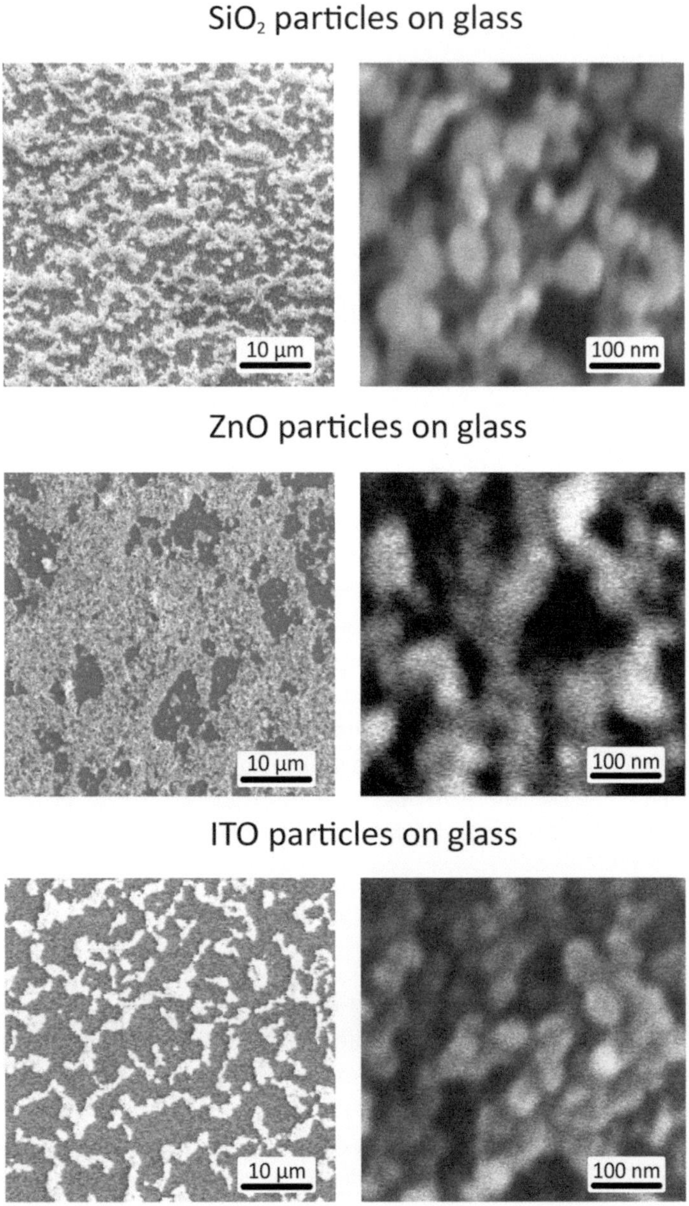

Fig. 7.16 SEM micrographs of SiO_2, ZnO, and ITO nanoparticle coatings on glass substrates at two magnifications each (Ebert and Bhushan 2012b)

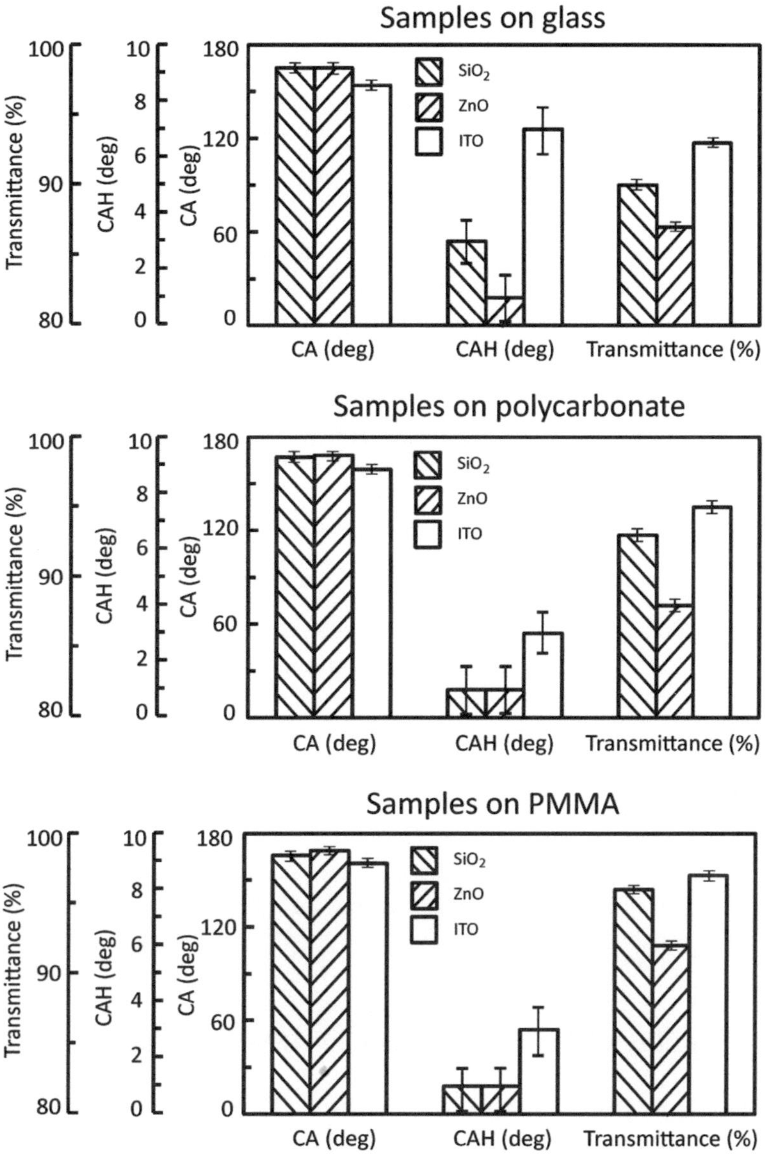

Fig. 7.17 Bar chart showing contact angle (CA), contact angle hysteresis (CAH), and visible transmittance for samples with SiO_2, ZnO, and ITO nanoparticles on glass, polycarbonate, and PMMA substrates. Error bars represent ±1 standard deviation (Ebert and Bhushan 2012b)

because of the formation of more evenly distributed microscale islands with lower height distribution compared to SiO_2 and ZnO, possibly due in part to smaller primary particle size. The data suggests that all interfaces were in the Cassie-Baxter regime.

7.5.4 Optical Transparency

The uncoated glass, PC, and PMMA substrates had visible-range transmittances of 92, 87, and 94%, respectively. Transmittances of the samples are reported as percentages of the transmittance of the uncoated substrate. Figure 7.17 shows the transmittance data in the visible spectrum and Fig. 7.18 shows transmittance spectra for SiO_2, ZnO, and ITO nanoparticles on all substrates (Ebert and Bhushan 2012b). For all three particles, samples on PMMA had higher transmittance than those on PC, and PC samples had higher transmittance than those on glass, even with transmittances normalized by substrate transmittance. For all substrates, the samples with ITO particles had higher transmittance than those with silica particles, and SiO_2 samples had higher transmittance than those with ZnO. The lower transmittance of the ZnO samples likely is due to their significantly higher roughness values as compared to ITO and SiO_2. Conversely, the higher transmittance of the ITO samples likely is due to their lower roughness, comparatively. Despite somewhat less favorable values for band gap and refractive index, the ITO samples had higher transmittance than SiO_2 samples, which suggests that in this case, roughness and coating thickness, which was on the order of 100 nm, played a larger role in transmittance than did inherent optical properties of particles (Ebert and Bhushan 2012b).

ITO-coated glass, PC, and PMMA samples with deposited water droplets can be seen in Fig. 7.19, showing superhydrophobicity and high transmittance of the coatings (Ebert and Bhushan 2012b). Blue dye was added to water for visual clarity of droplets. Goniometer image of a droplet on ITO-coated glass is shown for better view of a superhydrophobic contact angle.

7.5.5 Durability of Various Samples in AFM and Water Jet Tests

The data from the AFM wear experiment for SiO_2, ZnO, and ITO particles on glass, as well as silicone resin alone on glass, are shown in Fig. 7.20 (Ebert and Bhushan 2012b). Surface height maps before and after the wear experiment are displayed, as well as sample scans across the middle of the image (position indicated by arrow). Roughness values within the wear area (RMS and PV, before and after) are also displayed. Hardness of the silicone resin was measured with a microindenter (Micromet 3 Micro Hardness Tester) and found to be 1.3 GPa. The after-image of silicone resin alone reveals slight wear of the $50 \times 50\ \mu m^2$ area worn by the borosilicate ball. The wear mode appears to be adhesive, as there is a uniform removal of material (Bhushan 2013a, b). However, morphology was not

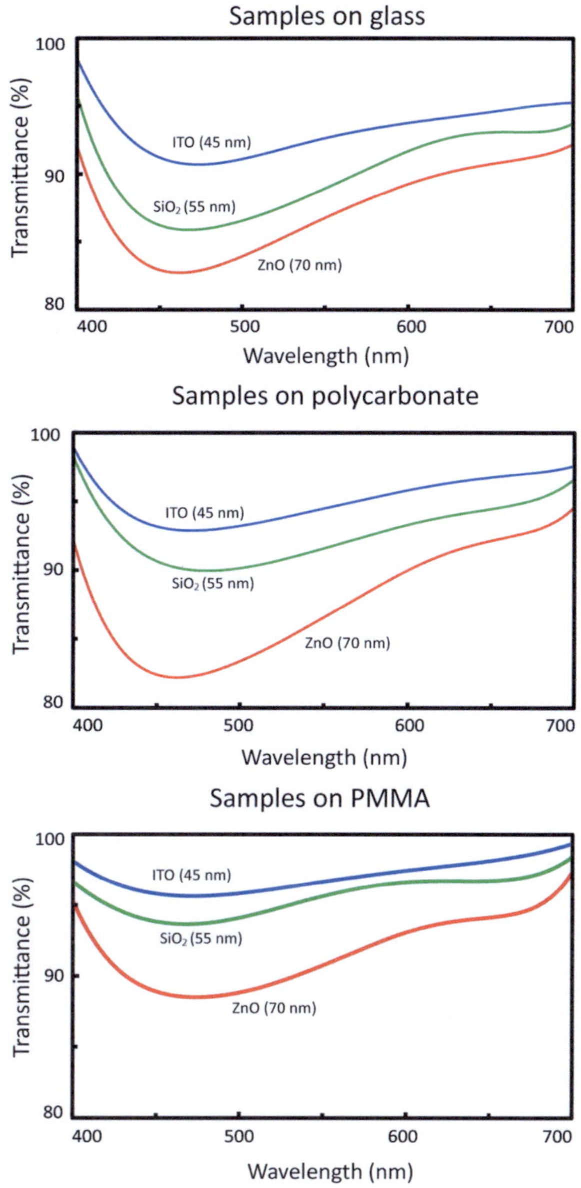

Fig. 7.18 Transmittance spectra in the visible range for samples with SiO_2, ZnO, and ITO nanoparticle coatings on glass, polycarbonate, and PMMA substrates. Data represent transmittance as a percentage of the transmittance of the uncoated substrate (Ebert and Bhushan 2012b)

significantly changed in the after-image for any of the three samples with nanoparticles, and RMS roughness and P-V distance values remained similar. The minimal wear of the silicone resin and preservation of nearly identical roughness and surface morphology for samples indicates mechanical strength of the silicone resin, sufficient hardness of

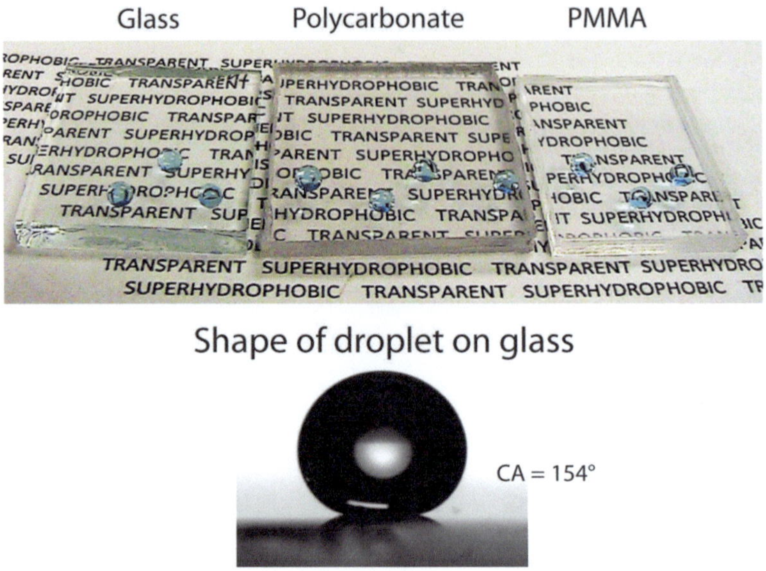

Fig. 7.19 Water droplets deposited on glass, polycarbonate, and PMMA with ITO nanoparticle coatings showing superhydrophobicity and high visible transmittance (top). Blue dye was added to water for visual clarity of droplets. Goniometer image of droplet is shown for glass substrate with ITO coating (bottom) (Ebert and Bhushan 2012b)

nanoparticles, and strong anchoring of particles in the silicone resin layer (Ebert and Bhushan 2012b).

The data from the water jet experiment is shown in Fig. 7.21 (Ebert and Bhushan 2012b). Samples were exposed to water jet for 20 min at each pressure ranging from 0 to 45 kPa. Contact angle and contact angle hysteresis data are displayed for silica nanoparticles as well as silicone resin alone on glass substrates. For the samples with SiO_2 nanoparticles, superhydrophobicity and self-cleaning properties were maintained, even at the highest pressure, with contact angle decreasing from 165° to 160°, and contact angle hysteresis increasing from 3° to 6°. At some intermediate pressures, contact angle hysteresis as low as 1° was measured. The wettability of the samples with silicone resin was likewise not significantly changed, with contact angle of 97° at 45 kPa compared to an initial value of 99°. Contact angle hysteresis for the silicone resin remained between 67° and 69° at all pressures. The results indicate wear resistance of the surfaces under impingement of water (Ebert and Bhushan 2012b).

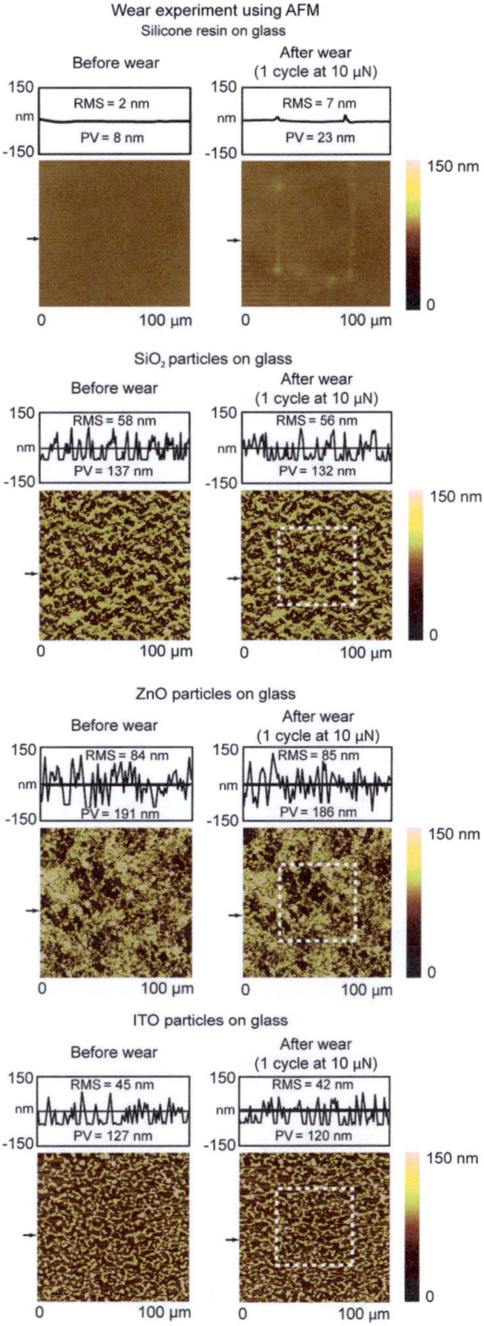

Fig. 7.20 Surface height maps and sample surface profiles (locations indicated by arrows) before and after AFM wear experiment using 15 μm radius borosilicate ball at load of 10 μN for glass samples with silicone resin alone, SiO_2 nanoparticles, ZnO nanoparticles, and ITO nanoparticles. RMS roughness and P-V distance values for surface profiles are displayed within surface profile boxes. Results shown are typical for all substrates (Ebert and Bhushan 2012b)

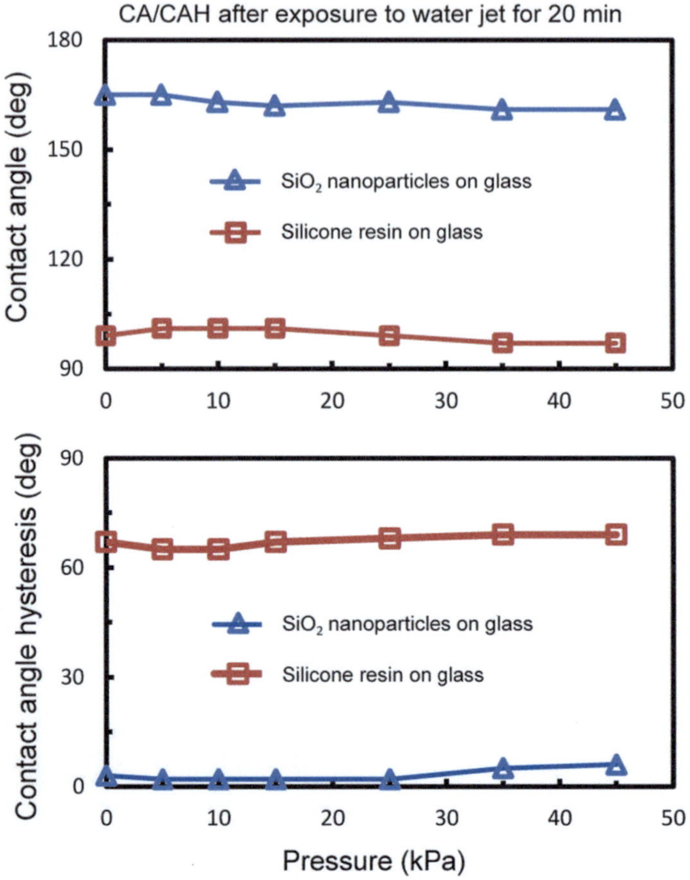

Fig. 7.21 Contact angle (CA) and contact angle hysteresis (CAH) as a function of water pressure after 20 min. exposure time for samples with SiO_2 nanoparticles on glass and silicone resin alone on glass. Reproducibility is ±1° for CAH of SiO_2 nanoparticles, ±2° for CAH of silicone resin alone on glass, and ±2° for all CA data. Results shown are typical for all samples. CA > 150° and CAH < 10° were maintained in all cases (Ebert and Bhushan 2012b)

7.5.6 Summary

To fabricate transparent, superhydrophobic surfaces on glass and plastic substrates with SiO_2, ZnO, and ITO nanoparticles, Ebert and Bhushan (2012b) used a dip coating technique. ITO samples had slightly lower contact angle and slightly higher contact angle hysteresis than SiO_2 or ZnO, which is likely the result of a comparatively lower liquid-air fractional area (f_{LA}). Roughness and coating thickness seemed to influence transmittance more than inherent optical properties of particles, which may be due to the proximity of roughness and thickness values to the 100 nm threshold for visible transparency. There appears to be little difference in the wettability, wear resistance, and optical transmission of the coatings fabricated using three different nanoparticles. Silica may be a preferred choice as it is known to be hard, provides wear resistance and is readily available in various particle sizes. Methylphenyl silicone resin used in this study is shown to be durable and offer strong adhesion between nanoparticles and substrate.

7.6 Superhydrophobic Surfaces Using Micropatterning, Nanoparticle Composite Coating and Ion Etching of PDMS for Optical Transparency

Polydimethylsiloxane (PDMS) is hydrophobic and optically transparent. It is an important material for biomedical devices, stretchable electronic circuits, and flexible microwave antennas (McClain et al. 2009; Gonzalez et al. 2009; Cheng and Wu 2010; Bhushan 2017a, b). Biomedical devices include microfluidic and nanofluidic devices due to their biocompatibility and chemical stability (Bhushan 2017a, b). Superhydrophobic PDMS has been used in catheters, passive valves as flow regulators, micropumps and microvalves where self-cleaning, antifouling and low drag properties are important. While PDMS is intrinsically hydrophobic, to produce a hydrophilic surface for applications such as analysis of biological samples and chemical synthesis, oxygen plasma treatment or other means are used (Chang 1994; Kim et al. 2000; Bodas and Khan-Malek 2007).

Superhydrophobic PDMS surfaces have been fabricated through plasma etching (Tserepi et al. 2006; Manca et al. 2008; Hwang et al. 2009; Tropmann et al. 2012; Ebert and Bhushan 2016), micropatterning, and by deposition of nanoparticle/binder composite coatings (Martin and Bhushan 2017). Deep reactive ion etching is an attractive technique to maintain high optical transparency of PDMS surfaces. An overview of superhydrophobic surfaces produced using micropatterning, nanoparticle/binder composite coating and ion etching follows.

7.6.1 Micropatterning and Nanoparticle/Binder Coating

7.6.1.1 Fabrication Details

A flat PDMS (Sylgard® 184, Dow Corning) sheet with thickness of 5 mm was prepared by mixing the curing agent and base at 1:10 weight ratio. The mixed, uncured PDMS was then degassed in a vacuum chamber for 15 min at room temperature to remove entrained air bubbles. Then, the mixture was fully cured in an oven operating at 70 °C for 2 h. The PDMS sheet was cut to dimensions of 10 by 10 mm to be used for substrates.

Micropatterned PDMS samples were fabricated using a soft-lithography technique described in Chap. 6 (Martin and Bhushan 2017). Bhushan and Jung (2011) studied the role of pitch for a given height of cylindrical pillars in Si micropatterned surfaces with a

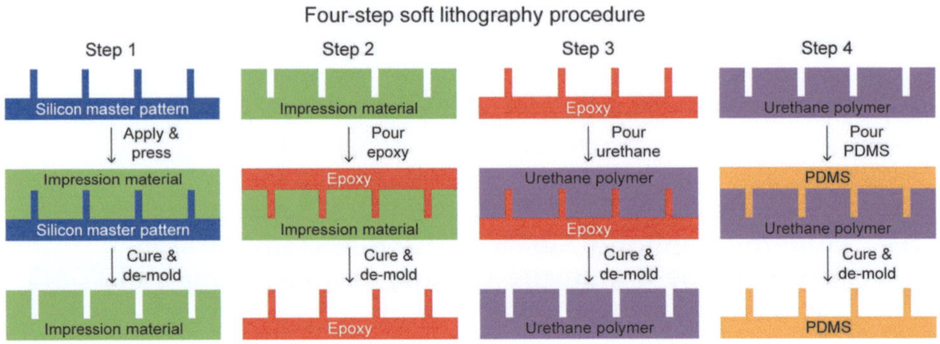

Fig. 7.22 Four-step soft lithography procedure to create micropatterned samples from silicone master patterns. First, impression material is applied to the silicon master patterns to create negative molds (step 1). Next, epoxy is poured into the impression mold to create a positive mold (step 2). Afterwards, urethane is poured to create a negative mold from the positive epoxy mold (step 3). Lastly, PDMS is poured to create final samples from the negative urethane molds (step 4). This procedure is necessary to ensure proper de-molding and sample replication (Martin and Bhushan 2017)

fluorosilane coating and given droplet radius on CA and CAH. They found an optimum pitch for highest CA and lowest CAH. Based on their study, a Si master pattern with pillars of 14 μm diameter, 30 μm height, and 126 μm pitch was chosen as the micropattern for superhydrophobicity and low CAH. A four-step soft lithography technique was needed because PDMS could not be de-molded from the impression material after curing and therefore intermediate replication steps using epoxy and urethane materials were required. This four-step soft lithography technique is shown in Fig. 7.22.

In step 1, hydrophilic vinyl polysiloxane impression material (Take 1®, Kerr), commonly used in dental applications, was applied to the micropattern using a cartridge dispensing gun with an attached mixing nozzle and pressed at room temperature for 30 min to create a negative replica. The impression material was chosen because of its ease of molding and demolding, and replication of pillar features. In step 2, a positive replica was made by pouring a liquid epoxy resin (Epoxydharz L, Nr. 236349, Conrad Electronics, Hirschau, Germany) with hardener (Härter S, Nr. 236365, Conrad Electronics, Hirschau, Germany) at a 60:40 weight ratio and allowing to cure for 24 h at room temperature. In step 3, liquid urethane polymer (Smooth-Cast® 300, Smooth-On) was poured into the epoxy replica at a 1:1 weight ratio and allowing to cure for 30 min at room temperature in order to form a negative replica. In step 4, PDMS was prepared by mixing the curing agent and base at 1:10 weight ratio and then poured into the urethane polymer mold. Afterwards, it was degassed in a vacuum chamber for 15 min at room temperature and cured in an oven operating at 70 °C for 2 h. The micropatterned PDMS was removed from the urethane polymer and used as substrates. The same urethane polymer negative mold was used to create additional micropatterned PDMS samples (Martin and Bhushan 2017).

A nanoparticle/binder composite coating consisting of silica nanoparticles and methylphenyl silicone resin was deposited on flat PDMS surface (Martin and Bhushan 2017). Before coating, PDMS substrates were cleaned in isopropyl alcohol (IPA, Fisher Scientific) with bath sonication (45 kHz frequency, Model FS5, Fisher Scientific) for 15 min. Then substrates were rinsed with DI water and allowed to dry. Next PDMS was chemically activated using ultraviolet-ozone (UVO) treatment for 90 min. PDMS is slightly hydrophobic, and activation was required for improved deposition and adhesion of the coating.

For the coating mixture, 600 mg of hydrophobic silica nanoparticles (NP, 10 nm diameter, Aerosil RX300) were dispersed in 30 mL of 40% tetrahydrofuran (THF, Fisher Scientific) and 60% IPA by volume. This mixture was sonicated using an ultrasonic homogenizer (20 kHz frequency at 35% amplitude, Branson Sonifer 450A) for 15 min. Then, 150 mg of methylphenyl silicone resin (SR355S, Momentive Performance Materials) was added. The mixture was then sonicated for an additional 15 min to form the final mixture.

1 mL of the coating mixture was deposited on the activated PDMS substrate via spray gun (Paasche®) from 10 cm away with compressed air at 210 kPa. An air spray technique was selected because it is simple, versatile and provides a rougher coating than by dip coating, needed for superhydrophibicity. The coated sample was placed in an oven at 70 °C for 5 min to remove the remaining solvent. The measured coating thickness was about 4 μm.

7.6.1.2 Contact Angle and Transmittance of Surfaces

Measurements on flat and micropatterned PDMS samples and a flat sample with nanoparticle/binder composite coating were made. Images of the droplets along with the measured contact angle (CA) and tilt angle (TA) values for water and hexadecane on flat and micropatterned PDMS as well as flat PDMS with coating are shown in Fig. 7.23. Untreated, flat PDMS is slightly hydrophobic with water CA of $113° \pm 2°$ and oleophilic with hexadecane CA of $52° \pm 2°$. By micropatterning the PDMS, roughness is introduced and enhances the hydrophobic and oleophilic properties, resulting in superhydrophobicity with water CA of $151° \pm 3°$ and superoleophilicity with wetting of hexadecane. In addition to simply micropatterning, superhydrophobicity was obtained with the nanoparticle/binder coating on a flat surface. Water CA of $158° \pm 2°$ was demonstrated because the flat PDMS had hexadecane CA under 90°, roughness enhanced the oleophilic property resulting in hexadecane wetting the surface.

Tilt angles decreased as the CA increased. Water droplets did not roll off the flat PDMS, but at $7° \pm 3°$ water droplets began to roll on the micropatterned surface. The

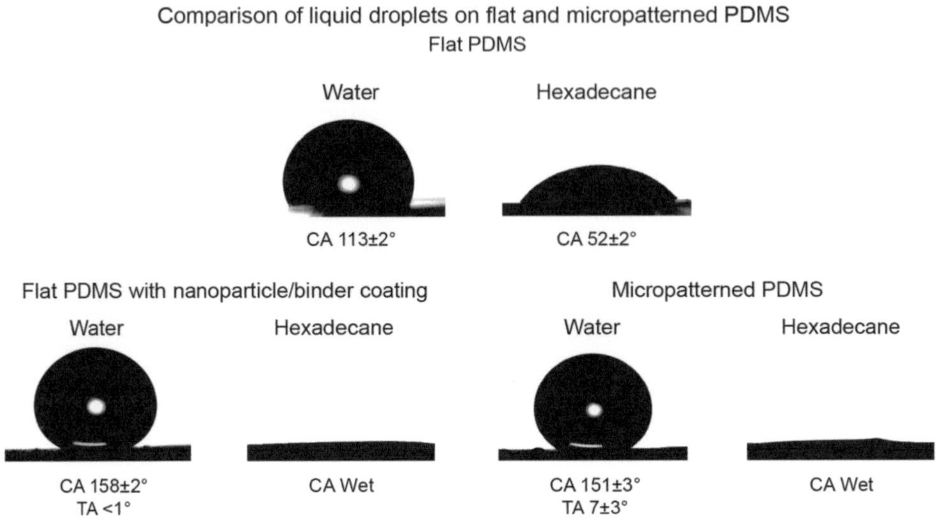

Fig. 7.23 Images for water and hexadecane droplets on flat and micropatterned PDMS and flat PDMS with nanoparticle/binder coating, CA and TA are listed under the images (adapted from Martin and Bhushan 2017)

nanoparticle/binder coating had the lower TA of less than 1°, which was lower than that of the micropatterned PDMS surface.

To measure transparency of samples, text was placed behind them and they were photographed on the front side. The text remained readable in all samples as shown in Fig. 7.24.

The edges of each sample are indicated with dashed line. The micropatterned samples have the micropattern in the center with flat PDMS surrounding it. PDMS is inherently transparent as shown with flat PDMS. For the superhydrophobic samples, a reduction in transparency is seen with the micropatterning and the nanoparticle/binder coating with micropatterning being superior. Although the thickness of the coating is less than the height of the micropatterning, the increased roughness of the coating leads to decreased transparency. The micropatterned PDMS is primarily flat except for the small areas occupied by the micropillars.

7.6.1.3 Summary

Transparent, wear-resistant, superhydrophobic surfaces of PDMS were fabricated. Superhydrophobicity with high CA (>150°) and low TA (<10°) was obtained either through micropatterning or a coating of hydrophobic SiO_2 nanoparticles and binder of methylphenyl silicone resin deposited using a spray technique.

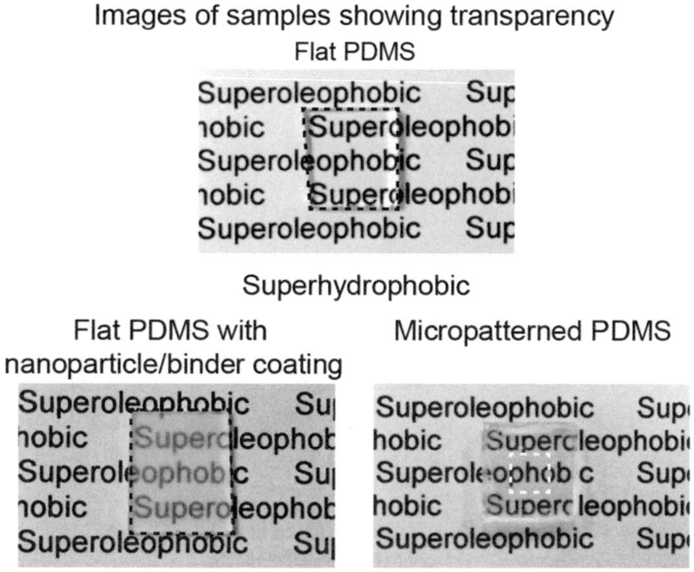

Fig. 7.24 Photographs showing transparency on flat PDMS, superhydrophbic samples from PDMS with nanoparticle/binder coating and micropatterned PDMS. The reduction in transparency in the coated samples is due to the SiO_2 nanoparticles and binder. Edges of each sample are shown in dashed lines (adapted from Martin and Bhushan 2017)

7.6.2 Ion Etching

Ebert and Bhushan (2016) fabricated, transparent, superhydrophobic surfaces via deep reactive ion etching (DRIE) using an O_2/CF_4 gas mixture with subsequent surface modification. DRIE is an extremely anisotropic process, making it useful for creating high aspect ratio surface roughness with appropriate process gases. In general, DRIE combines effects from both chemical and physical processes in order to etch. Free radicals react with the material chemically while accelerated ions bombard the surface causing physical removal of material (Bjørnsen et al. 2010). An O_2/CF_4 mixture is a particularly powerful etchant of polymers; when mixed, oxygen and tetrafluoromethane create the oxyfluoride ion (OF^-), which can cut the carbon-carbon bonds in the backbone of the polymer and remove the molecule.

To reduce surface energy on the etched surfaces, samples then underwent a plasma treatment with octafluorocyclobutane (C_4F_8), or were modified by perfluorooctyl-trichlorosilane (PFOTCS) through vapor deposition following surface activation by O_2 plasma. A relatively short C_4F_8 plasma treatment can result in deposition of a thin fluoropolymer layer. Fluorosilanes such as PFOTCS can form covalently bonded layers on PDMS after surface activation by oxygen plasma. In general, fluorinated materials are desired for ultra-repellent surfaces due to the very low polarizability of fluorine, which leads to weak adhesive forces and low surface energy.

7.6.2.1 Fabrication Details

The processes used to prepare samples with O_2/CF_4 plasma treatment followed by C_4F_8 plasma treatment or PFOTCS deposition are shown schematically in Fig. 7.25 (Ebert and Bhushan 2016). In the first step in both cases, the PDMS is etched for varying lengths of time using a mixture of O_2 (flow rate 25 sccm) and CF_4 (flow rate 75 sccm) as process gases. A flow ratio of about 1:3 O_2:CF_4 was found to yield the highest etch rate with this mixture (Garra et al. 2002). RF power level was 300 W and pressure was 30 mTorr. In the second step in (a) with C_4F_8 plasma, a thin fluorocarbon layer is formed on the etched PDMS using C_4F_8 gas (flow rate 100 sccm) for 30 s at 250 W, 250 mTorr. In the second step in (b) with PFOTCS vapor disposition, PFOTCS forms a covalently bonded fluorosilane layer on the surface. To achieve this, the O_2/CF_4 etched PDMS was first

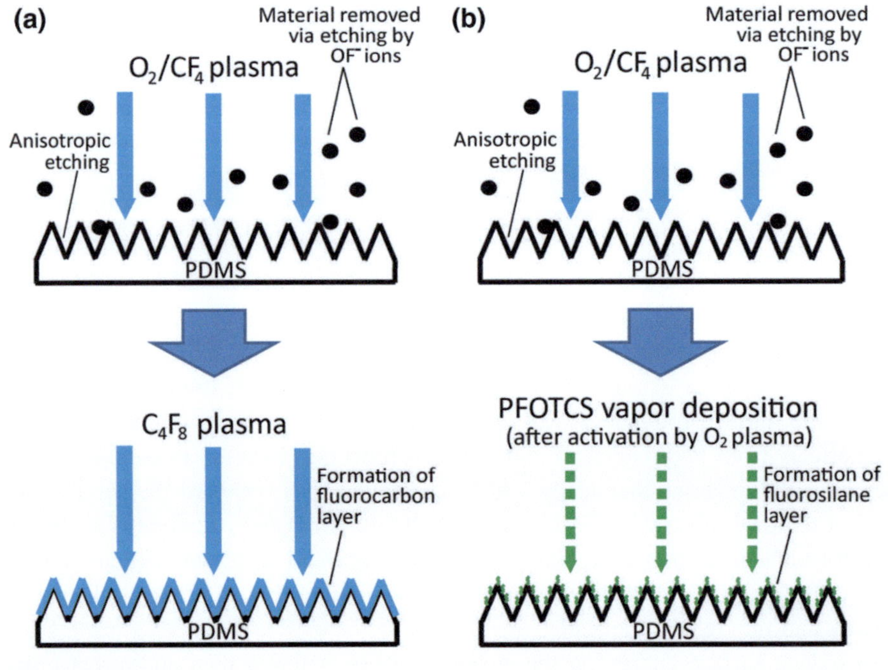

Fig. 7.25 Schematics of reactive ion etching of PDMS by O_2/CF_4 plasma and subsequent **a** C_4F_8 plasma treatment, and **b** vapor deposition of PFOTCS. In both cases, the surface was first etched anisotropically by the oxyfluoride (OF^-) ions formed in the O_2/CF_4 mixture, resulting in a rough surface. In the case of C_4F_8 treatment (**a**), a thin fluorocarbon layer is then formed on the surface. In the case of PFOTCS modification (**b**), after surface activation by O_2 plasma, a covalently bonded fluorosilane layer is then formed with exposed long-chain hydrophobic tails (Ebert and Bhushan 2016)

activated by O_2 plasma for 30 s at 100 W and 250 mTorr. This activation allows for Si atom of PFOTCS to bond to surface oxygen, leaving the hydrophobic tail of the molecule exposed. Immediately after O_2 plasma treatment, deposited droplets spread quickly on samples and had CA < 10°, indicating effective surface activation for PFOTCS attachment. Vapor deposition was then performed by placing samples in a small covered petri dish with a few drops of PFOTCS for 2 h at ambient conditions.

The anisotropy of the DRIE process can be maximized by increasing RF power and decreasing pressure. The lower pressure decreases the number of ions bombarding the surface, which decreases the uniformity of etching. The higher power causes ions to bombard the surface with greater energy, which helps to compensate for the decrease in ions in order to maintain sufficient etch rate. For this reason, high power level (300 W) and low pressure (30 mTorr) were used by Ebert and Bhushan (2016) for the etching of samples with O_2/CF_4 mixture.

7.6.2.2 Surface Roughness and Morphology

Table 7.4 presents RMS roughness and P-V (peak-valley) distance for samples with varying etch times under DRIE with O_2/CF_4 plasma only, samples etched by O_2/CF_4 plasma with subsequent C_4F_8 plasma treatment, and samples etched by O_2/CF_4 with subsequent PFOTCS vapor deposition. The etched surfaces had high aspect ratio nanoscale roughness because of the highly anisotropic nature of DRIE. The RMS roughness and P-V distance increased with etch time, reaching RMS = 302 nm and P-V = 567 nm at 25 min. The roughness of samples after the additional C_4F_8 plasma treatment, which forms a very thin fluorocarbon layer on the rough surface, decreased slightly for all samples, reaching RMS = 295 nm and P-V = 556 nm for samples etched 25 min by O_2/CF_4. The thickness of the fluorocarbon layer was measured to be approximately 25 nm using AFM with a step method. While the thickness was virtually uniform on a flat PDMS sample, the fluorocarbon formed on rough etched samples may be more prone to fill in gaps between asperities, decreasing measured roughness. In the case of the PFOTCS vapor deposition, which forms a covalently bonded fluorosilane layer, roughness was unchanged (Ebert and Bhushan 2016).

Figure 7.26 shows the SEM micrograph of an unetched PDMS sample at high magnification, as well as samples after 10 and 20 min etch time at three magnifications. Roughness features were noticeably larger after 20 min compared to 10 min, with increased space between peaks. This nanoscale roughness is important for superhydrophobicity and low CAH, but very high value will impede transmittance of light (Ebert and Bhushan 2016).

7.6.2.3 Contact Angle and Transmittance of Surfaces

Figure 7.27 and Table 7.4 show measured CA and CAH data for samples etched by O_2/CF_4 plasma only, by O_2/CF_4 with subsequent C_4F_8 plasma treatment, and by O_2/CF_4 with subsequent O_2 plasma surface activation and PFOTCS vapor deposition (Ebert and

Table 7.4 Measured surface roughness (RMS and P-V distance), contact angle (CA) and contact angle hysteresis (CAH) values for samples etched by O_2/CF_4 plasma only, samples etched by O_2/CF_4 with subsequent C_4F_8 plasma treatment, and samples etched by O_2/CF_4 plasma with subsequent PFOTCS vapor deposition (adapted from Ebert and Bhushan 2016)

O_2/CF_4 etch time (min)	RMS (nm)	P-V distance (nm)	CA (°) (±1°)	CAH (°) (±2°)
Samples etched by O_2/CF_4plasma				
0	~0	~0	108	20
5	29	78	116	46
10	75	186	151	18
15	126	292	153	15
20	207	431	156	14
25	302	567	155	14
Samples etched by O_2/CF_4plasmawith subsequent C_4F_8plasmatreatment				
0	~0	~0	111	15
5	27	75	123	42
10	72	181	163	9
15	120	282	166	7
20	199	419	167	7
25	295	556	166	8
Samples etched by O_2/CF_4plasmawith subsequent PFOTCSvapor deposition				
0	~0	~0	116	9
5	29	78	128	25
10	75	186	165	3
15	126	292	169	2
20	207	431	169	2
25	302	567	169	3

Bhushan 2016). The CA on flat, virgin PDMS was measured to be 108°. Etched samples became superhydrophobic for RMS roughness 75 nm and greater (corresponding to etch times 10, 15, 20, and 25 min), achieving apparent CA in the range 151°–156°. CAH for etched samples with RMS = 29 nm increased to 46° compared to 20° for unetched PDMS. For RMS values 75 nm and greater, CAH was reduced significantly to the range 14°–18°.

The CA on unetched PDMS after C_4F_8 treatment was measured to be 111°. Samples remained superhydrophobic for RMS roughness of 72 nm and greater, reaching higher apparent CA in the range 163°–167°. CAH was lower for the superhydrophobic samples after C_4F_8 treatment, achieving values in the range 7°–9°. The thin fluorocarbon layer from C_4F_8 plasma increased CA for superhydrophobic samples by 11°–13° and decreased

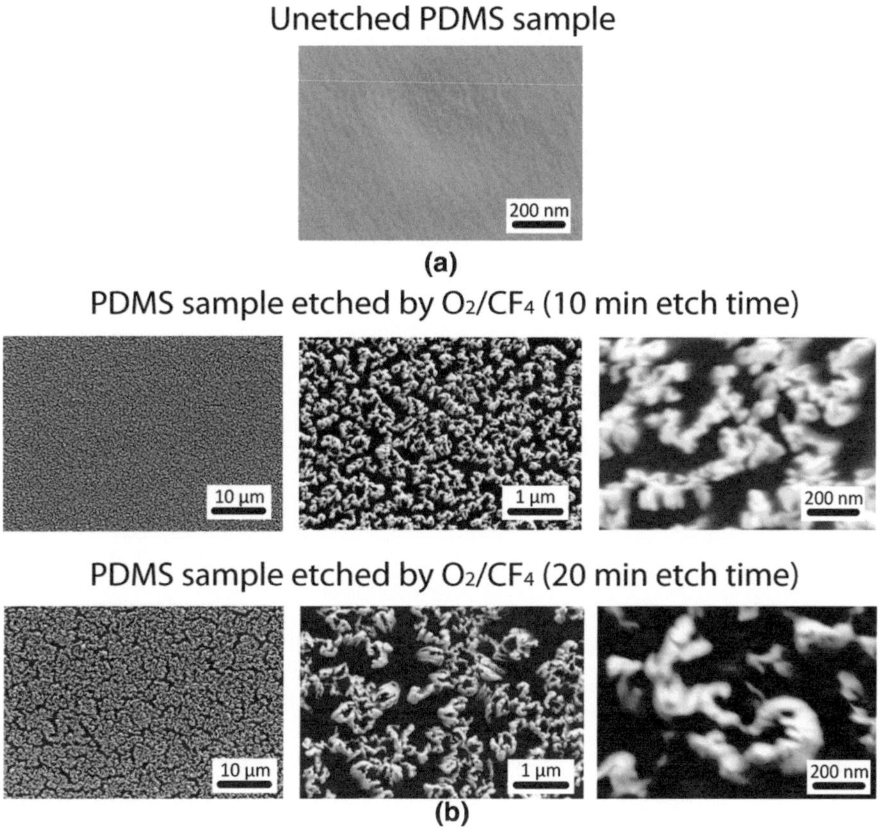

Fig. 7.26 SEM micrographs of **a** unetched, virgin PDMS sample at high magnification and **b** PDMS samples at three magnifications after 10 min and after 20 min etch time by O_2/CF_4 plasma (Ebert and Bhushan 2016)

CAH by 6°–9° after treatment. The reduction in CAH is perhaps a more significant benefit, as CAH less than 10° is often desired for many applications involving superhydrophobicity. The CA on unetched PDMS after PFOTCS modification by vapor deposition was measured to be 116°. Samples with roughness of 75 nm and greater achieved CA of 165°–169° and CAH as low as 2°. While the increase in CA with PFOTCS compared to C_4F_8 plasma is not significant, the further reduction in CAH to nearly zero is potentially beneficial (Ebert and Bhushan 2016).

Figure 7.28 displays the average transmittance across the visible range of samples plotted against RMS roughness (Ebert and Bhushan 2016). Transmittance is expressed as a percentage of the transmittance of a virgin PDMS sample. Etched samples remained highly transparent (>85%) up to RMS roughness value of 126 nm (achieved after 15 min etch time). For RMS roughness 207 and 302 nm, achieved after 20 and 25 min etch times,

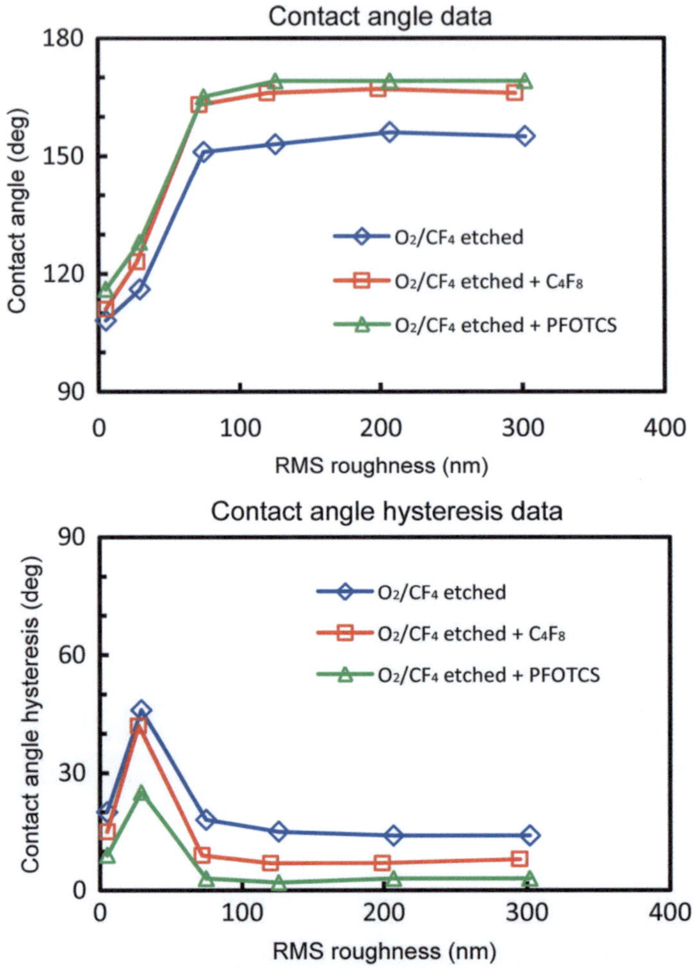

Fig. 7.27 Contact angle and contact angle hysteresis data on PDMS samples of varying RMS roughness values for samples etched by O_2/CF_4 plasma only, by O_2/CF_4 with subsequent C_4F_8 plasma treatment, and by O_2/CF_4 with subsequent PFOTCS vapor deposition. Reproducibility is $\pm 1°$ for CA data and $\pm 2°$ for CAH data (Ebert and Bhushan 2016)

respectively, samples became nearly or entirely opaque. This is likely due to the roughness values significantly exceeding one quarter the wavelength of visible light (around 100 nm for shortest wavelengths). Measured transmittance values were identical for a given etch time before and after 30-s treatment with C_4F_8 plasma or PFOTCS functionalization.

Samples were superhydrophobic and highly transparent after etch times of 10 and 15 min both before and after surface modification with either C_4F_8 plasma treatment or

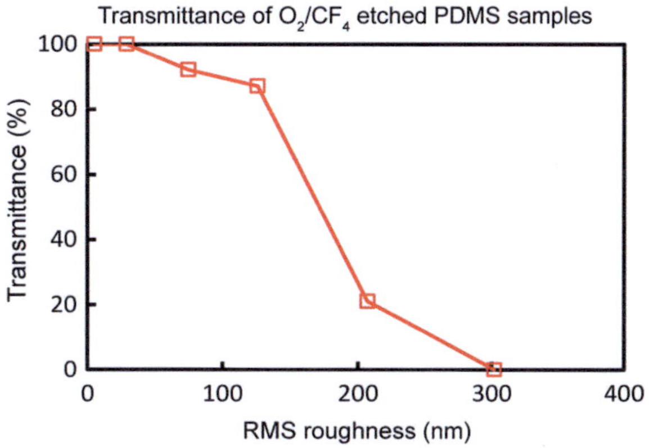

Fig. 7.28 Transmittance of O_2/CF_4 etched PDMS samples of varying RMS roughness values. Data represent the percentage of the baseline transmittance of a PDMS substrate, and is taken as the average across the visible range (λ = 400–700 nm) (Ebert and Bhushan 2016)

PFOTCS deposition. This agrees with other findings that superhydrophobicity and transparency can be simultaneously achieved with roughness approximately 100 nm (Ling et al. 2009; He et al. 2011; Ebert and Bhushan 2012b). An etched PDMS sample with RMS roughness 126 nm (etch time 15 min with C_4F_8 post-treatment) can be seen in Fig. 7.29 with deposited water droplet showing superhydrophobicity and high transmittance of light. Also displayed is a goniometer image of droplet on sample showing superhydrophobic contact angle (Ebert and Bhushan 2016).

7.6.2.4 Durability in AFM Tests

The results of the AFM sliding wear test are shown in Fig. 7.30. Surface height maps are displayed for flat, unetched PDMS as well as before and after wear for samples etched by O_2/CF_4 with C_4F_8 treatment. Roughness values within the wear area (RMS and P-V distance) are given for both before and after wear, as well as sample scans across the middle of the image (position indicated by arrow). Wear data for flat PDMS was difficult to obtain due to the compliance of the material. For etched PDMS, after cycle at load of 1 μN with a borosilicate ball, very little wear was evident in the surface height map and roughness did not change significantly. After cycle at a load of 10 μN, significant wear could be seen as material was pushed in the sliding direction of the borosilicate ball. Wear rate is generally inversely related to hardness and rigidity (Bhushan 2013a, b). For PDMS with 1:10 mixing ratio and cured at 85 °C, hardness is relatively low at about 46 (Shore A) or on the order of 400 MPa (Kim et al. 2013; Johnston et al. 2014). It is evident that while the high aspect ratio features formed on the PDMS through DRIE have some resistance to wear at lower load (on the order of 1 μN), they appear to be susceptible to wear at higher loads .

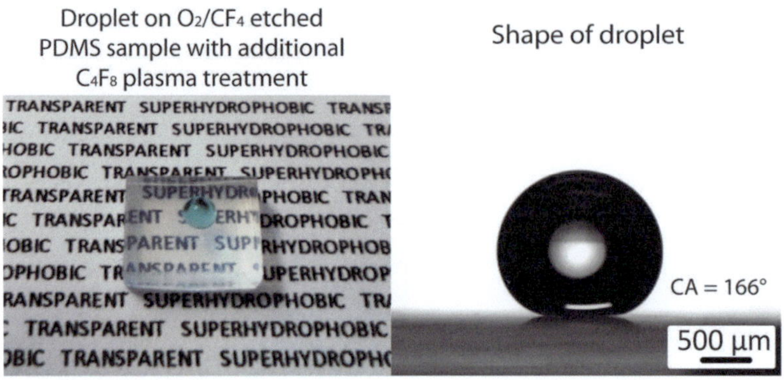

Fig. 7.29 Water droplet deposited on PDMS sample etched by O_2/CF_4 for 15 min with subsequent C_4F_8 plasma treatment showing superhydrophobicity and high visible transmittance (left). Blue dye was added to water for visual clarity. Goniometer image of droplet shown on right (Ebert and Bhushan 2016)

7.6.2.5 Summary

Superhydrophobic surfaces with high transparency were created on PDMS using a simple process using deep reactive ion etching and subsequent surface modification. The results demonstrated a range of surface roughness values that were high enough to result in superhydrophobicity but low enough to preserve high transmittance of visible light. A degree of mechanical wear resistance was seen at lower load, but wear was evident at higher load on the rough, etched surfaces of PDMS.

7.7 Superhydrophobic Paper Surfaces

Paper with superhydrophobic properties have been prepared by some investigators (Balu et al. 2008; Yang and Deng 2008; Obeso et al. 2013). A method proposed by Obeso et al. (2013) used biodegradable paper substrate. The precipitation of poly(hydroxybutyrate) (PHB) was used on the cellulose fibers of the paper using a phase separation process.

7.7.1 Fabrication Details

Obeso et al. (2013) immersed the paper in chloroform for 6 h, to extract any possible additives soluble in the solvent. Then, the samples were immersed for 6 h in 7.5% (w/v) solution of PHB in chloroform, followed by an immersion in a coagulation bath, formed by a mixture of 85/15 (v/v) of ethanol and water for 12 h. Finally, samples were dried at room temperature.

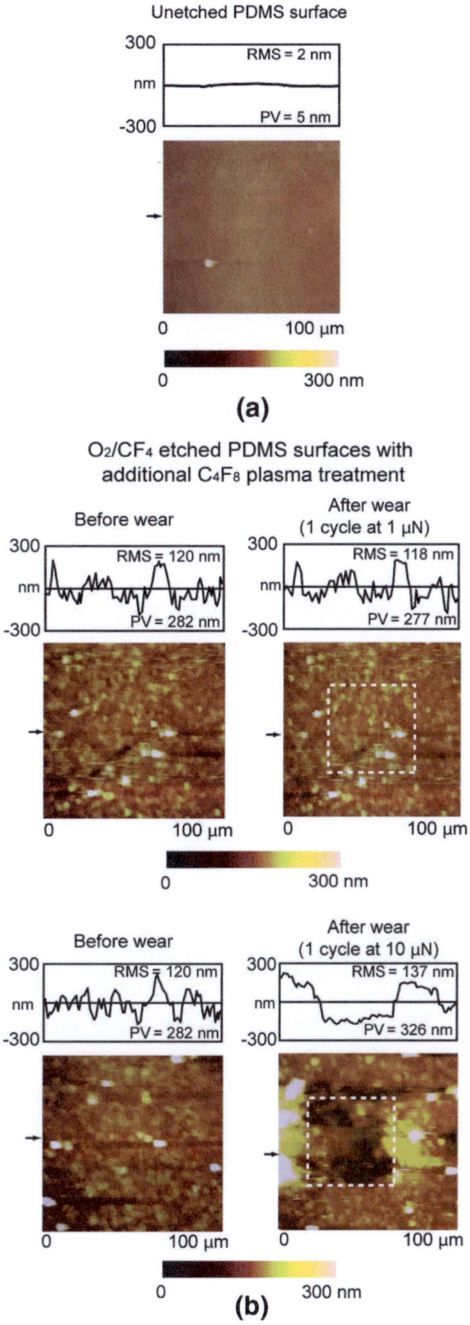

Fig. 7.30 Surface height maps and selected surface profiles (locations indicated by arrows) for **a** unetched, virgin PDMS sample and **b** before and after AFM wear experiment using 15-μm radius borosilicate ball at loads of 1 and 10 μN for PDMS samples etched by O_2/CF_4 for 15 min with subsequent C_4F_8 plasma treatment. RMS roughness and PV distance values for surface profiles are displayed within surface profile boxes (Ebert and Bhushan 2016)

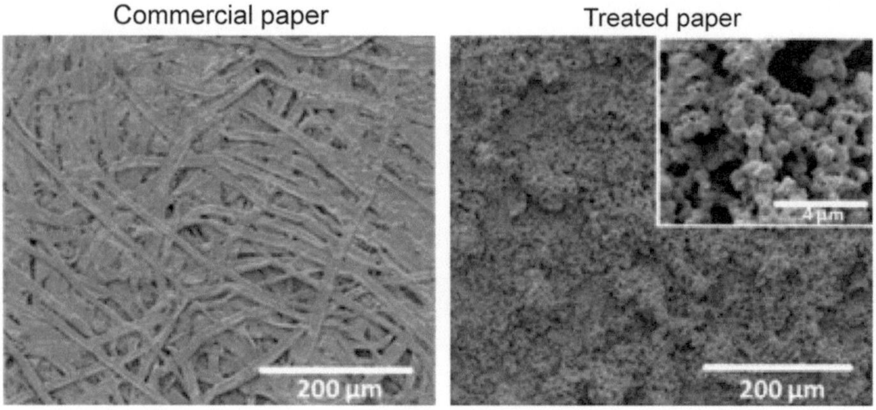

Fig. 7.31 SEM micrographs of a commercial paper and the paper treated with PHB (adapted from Obeso et al. 2013)

With the immersion of the paper sample in the coagulated bath of chloroform solution containing PHB, PHB started to penetrate into the paper fiber network. It induced the precipitation of the hydrophobic polymer, observed on the surface as a sphere-like aggregate structure, as shown in Fig. 7.31. It appeared that this hydrophobic agent attached to the entire substrate. The hierarchical structure of the treated paper is different from the original paper.

7.7.2 Contact Angle

Representative images of droplets on original and treated paper are shown in Fig. 7.32. The original paper had a CA of about 110°. After treatment with PHB, the CA increased to about 153°, being in the superhydrophobic regime.

The flow behavior of the water droplet was compared on original and treated paper. Figure 7.33 shows that on the smooth paper surface, the droplet size increases as it advances over the inclined surface, spreading in both lateral directions, with no spatial control, due to the progressive adsorption of water by the paper. In the superhydrophobic surface, the release of water droplets was hampered by the water repellence from the substrate (Obeso et al. 2013).

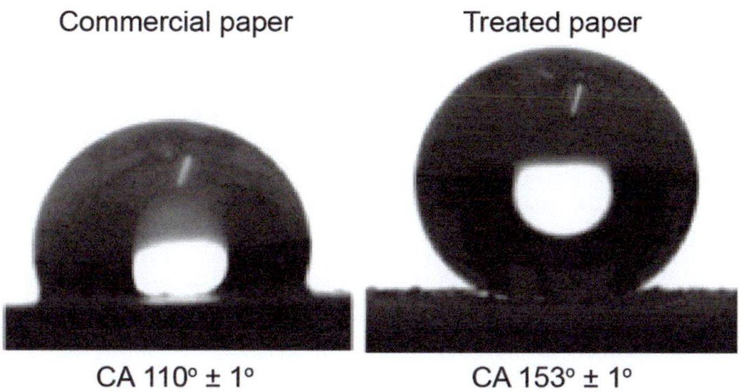

Fig. 7.32 Optical Images of a 5 μl water droplet sitting on a commercial paper and the paper treated with PHB (adapted from Obeso et al. 2013)

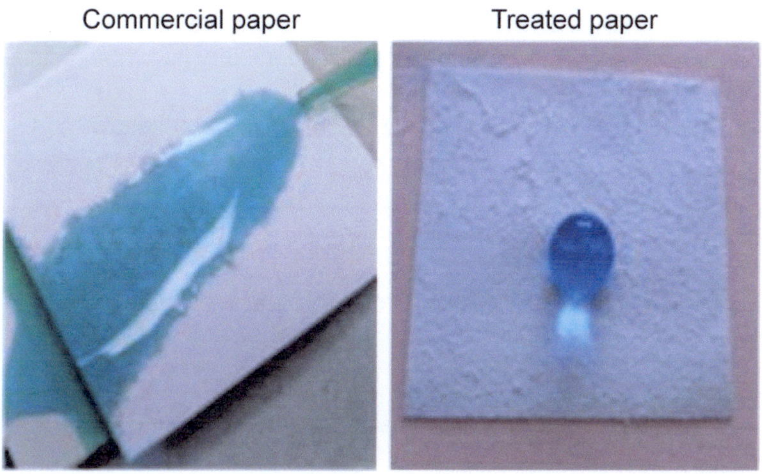

Fig. 7.33 Optical Images of 0.815 ml flow of blue colored water onto a commercial paper and the paper treated with PHB (adapted from Obeso et al. 2013)

7.7.3 Durability Test

To evaluate robustness of the treated paper, a mechanical test was performed by forcing down a finger on the paper. CA changed from about 153°–150° indicating that treated paper exhibited robustness (Obeso et al. 2013).

7.7.4 Summary

Superhydrophobic paper surface was fabricated using a simple phase separation approach using a natural origin polymer.

7.8 Closure

In this chapter, nanostructured and hierarchical structured, superhydrophobic surfaces were fabricated using soft lithography, photolithography, and spray technique. Mechanically durable surfaces were fabricated using nanoparticle composites with polymers mixed with carbon nanotubes, and silica, zinc oxide, or indium tin oxide nanoparticles to provide the surface roughness. Ion etching has also been used to provide the surface roughness. There appears to be little difference in the wettability and wear resistance of the coatings and silica appears to be the preferred choice as nanoparticles in various sizes are readily available. Methylphenyl silicone resin is shown to be mechanically durable. Superhydrophobicity, optical transparency, and mechanical durability have been demonstrated on various substrates.

Finally, a method to produce a superhydrophobic paper is presented.

References

Balu, B., Breedveld, V., and Hess, D. W. (2008), "Fabrication of Roll-off and Sticky Superhydrophobic Cellulose Surfaces via Plasma Processing," *Langmuir* **24**, 4785–4790.

Bhushan, B. (2009), "Biomimetics: Lessons from Nature – an Overview," *Phil. Trans. R. Soc. A* **367**, 1445–1486.

Bhushan, B. (2013a), *Principles and Applications of Tribology,* second ed., Wiley, New York.

Bhushan, B. (2013b), *Introduction to Tribology,* second ed., Wiley, New York.

Bhushan, B. (2017a), *Springer Handbook of Nanotechnology,* fourth ed., Springer International, Cham, Switzerland.

Bhushan B. (2017b), *Nanotribology and Nanomechanics – An Introduction,* fourth ed., Springer International, Cham, Switzerland.

Bhushan, B. and Gupta, B. K. (1991), *Handbook of Tribology: Materials, Coatings, and Surface Treatments*, McGraw-Hill, New York.

Bhushan, B. and Jung, Y. C. (2011), "Natural and Biomimetic Artificial Surfaces for Superhydrophobicity, Self-Cleaning, Low Adhesion, and Drag Reduction," *Prog. Mater. Sci.* **56**, 1–108.

Bhushan, B., Jung, Y. C., and Koch, K. (2009), "Micro-, Nano- and Hierarchical Structures for Superhydrophobicity, Self-Cleaning and Low Adhesion," *Phil. Trans. R. Soc. A* **367**, 1631–1672.

Bjørnsen, G., Henriksen, L., Ulvensøen, J. H., and Roots, J. (2010), "Plasma Etching of Different Polydimethylsiloxane Elastomers, Effects from Process Parameters and Elastomer Composition," *Microelectron. Eng.* **87**, 67–71.

Bodas, D. and Khan-Malek, C. (2007), "Hydrophilization and Hydrophobic Recovery of PDMS by Oxygen Plasma and Chemical Treatment – An SEM Investigation," *Sensors and Actuators B* **123**, 368–373.

Burton, Z. and Bhushan, B. (2006), "Surface Characterization and Adhesion and Friction Properties of Hydrophobic Leaf Surfaces," *Ultramicroscopy* **106**, 709–719.

Callister, W. D. (2000), *Materials Science and Engineering – An Introduction*, fifth ed., Wiley, New York.

Chang, C. M. (1994), *Surface Modification and Characterization of Polymers*, Hanser Publishers, New York, N.Y.

Chen, X. H., Chen, C. S., Xiao, H. N., Liu, H. B., Zhou, L. P., Li, S. L., and Zhang, G. (2006), "Dry Friction and Wear Characteristics of Nickel/Carbon Nanotube Electroless Composite Deposits," *Tribol. Int.* **39**, 22–28.

Cheng, S. and Wu, Z. (2010), "Microfluidic Stretchable RF Electronics," *Lab Chip* **10**, 3227–3234.

Cho, Y.-S., Yi, G.-R., Hong, J.-J., Jang, S. H., and Yang, S.-M. (2006), "Colloidal Indium Tin Oxide Nanoparticles for Transparent and Conductive Films," *Thin Solid Films* **515**, 1864–1871.

Dresselhaus, M. S., Dresselhaus, G., and Avouris, Ph., eds. (2000), *Carbon Nanotubes: Synthesis, Structure, Properties, and Applications*, Springer, Heidelberg, Germany.

Ebert, D. and Bhushan, B. (2012a), "Durable Lotus-Effect Surfaces with Hierarchical Structure Using Micro- and Nanosized Hydrophobic Silica Particles," *J. Colloid Interface Sci.* **368**, 584–591.

Ebert, D. and Bhushan, B. (2012b), "Transparent, Superhydrophobic, and Wear-Resistant Coatings on Glass and Polymer Substrates Using SiO_2, ZnO, and ITO Nanoparticles," *Langmuir* **28**, 11391–11399.

Ebert, D. and Bhushan, B. (2016), "Transparent, Superhydrophobic and Wear-resistant Surfaces Using Deep Reactive Ion Etching on PDMS Substrates," *J. Colloid Interface Sci.* **481**, 82–90.

Ederth, J., Heszler, P., Hultåker, A., Niklasson, G. A., and Granqvist, C. G., (2003), "Indium Tin Oxide Films Made from Nanoparticles: Models for the Optical and Electrical Properties," *Thin Solid Films* **445**, 199–206.

Englert, B. C., Xiu, Y., and Wong, C. P. (2006), "Deposition and Surface Treatment of Microparticles to Produce Lotus-Effect Surface," *11th International Symposium on Advanced Packaging Materials: Processes, Properties and Interfaces,* March 15–17, Atlanta, GA, 73–78.

Garra, J., Long, T., Currie, J., Schneider, T., White, R., and Paranjape, M. (2002), "Dry Etching of Polydimethylsiloxane for Microfluidics Systems," *J. Vac. Sci. Technol.* A **20**, 975–982.

Gonzalez, M., Axisa, F., Bossuyt, F., Hsu, Y., Vandevelde, B., and Vanfleteren, J. (2009), "Design and Performance of Metal Conductors for Stretchable Electronic Circuits," *Circuit World* **35**, 22–30.

Hamberg, I. and Granqvist, C. G. (1984), "Band-gap Widening in Heavily Sn-doped In_2O_3," *Phys. Rev. B* **30**, 3240–3249.

He, Z., Ma, M., Lan, X., Chen, F., Wang, K., Deng, H., Zhang, Q., and Fu, Q. (2011), "Fabrication of a Transparent Superamphiphobic Coating with Improved Stability," *Soft Matter* **7**, 6435–6443.

Hwang, S., Oh, D., Jung, P., Lee, S., Go, J., Kim, J., Hwang, K., and Ko, J. (2009), "Dry Etching of Polydimethylsiloxane using Microwave Plasma," *J. Micromech. Microeng.* **19**, 095010.

Johnston, I. D., McCluskey, D. K., Tan, C. K. L., and Tracey, M. C. (2014), "Mechanical Characterization of Bulk Sylgard 184 for Microfluidics and Microengineering," *J. Micromech. Microeng.* **24**, 035017.

Jung, Y. C. and Bhushan, B. (2009), "Mechanically Durable CNT-Composite Hierarchical Structures with Superhydrophobicity, Self-Cleaning, and Low-Drag," *ACS Nano* **3**, 4155–4163.

Karunakaran, R. G., Lu, C.-H., Zhang, Z., and Yang, S. (2011), "Highly Transparent Superhydrophobic Surfaces from the Coassembly of Nanoparticles ($\leq$100 nm)," *Langmuir* **27**, 4594–4602.

Kim, H., Gilmore, C. M., Pique, A., Horowitz, J. S., Mattoussi, H., Murata, H., Kafafi, Z. H., and Chrisey, D. B. (1999), "Electrical, Optical, and Structural Properties of Indium-Tin-Oxide Thin Films for Organic Light Emitting Devices," *J. Appl. Phys.* **86**, 6451–6461.

Kim, J., Chaudhury, M., and Owen, M. J. (2000), "Hydrophobic Recovery of Polydimethylsiloxane Elastomer Exposed to Partial Electrical Discharge," *J. Colloid Interface Sci.* **226**, 231–236.

Kim, M., Moon, B.-U., and Hidrovo, C. H. (2013), "Enhancement of the Thermo-mechanical Properties of PDMS Molds for the Hot Embossing of PMMA Microfluidic Devices," *J. Micromech. Microeng.* **23**, 095024.

Koch, K., Bhushan, B., Jung, Y. C., and Barthlott, W. (2009), "Fabrication of Artificial Lotus Leaves and Significance of Hierarchical Structure for Superhydrophobicity and Low Adhesion," *Soft Matter* **5**, 1386–1393.

Kucheyev, S. O., Bradby, J. E., Williams, J. S., and Jagadish, C. (2002), "Mechanical Deformation of Single-crystal ZnO," *J. Appl. Phys.* **80**, 956–958.

Ling, X. Y., Phang, I. Y., Vancso, G. J., Huskens, J., and Reinhoudt, D. N. (2009), "Stable and Transparent Superhydrophobic Nanoparticle Films," *Langmuir* **25**, 3260–3263.

Liu, Y., Chen, X., and Xin, J. H. (2006), "Super-Hydrophobic Surfaces from a Simple Coating Method: A Bionic Nanoengineering Approach," *Nanotechnology* **17**, 3259–3263.

Malitson, H. (1965), "Inter specimen Comparison of the Refractive Index of Fused Silica," *J. Opt. Soc. America* **55**, 1205–1209.

Manca, M., Cortese, B., Viola, I., Aricò, A., Cingolani, R., and Gigli, G. (2008), "Influence of Chemistry and Topology Effects on Superhydrophobic CF_4-Plasma-Treated Poly(dimethylsiloxane) (PDMS)," *Langmuir* **24**, 1833–1843.

Manca, M., Cannavale, A., De Marco, L., Aricò, A. S., Cingolani, R., and Gigli, G. (2009), "Durable Superhydrophobic and Antireflective Surfaces by Trimethylsilanized Silica Nanoparticles-Based Sol-Gel Processing," *Langmuir* **25**, 6357–6362.

Martin, S. and Bhushan, B. (2017), "Transparent, Wear-resistant, Superhydrophobic and Superoleophobic Poly(dimethylsiloxane) (PDMS) Surfaces," *J. Colloid Interface Sci.*, **488**, 118–126.

McClain, M., LaPlaca, M., and Allen, M. (2009), "Spun-cast Micromolding for Etchless Micropatterning of Electrically Functional PDMS Structures," *J. Micromech. Microeng.* **19**, 107002.

Meyyappan, M. (2005), *Carbon Nanotubes – Science and Applications*, CRC Press, Boca Raton, FL.

Ming, W., Wu, D., van Benthem, R., and de With, G. (2005), "Superhydrophobic Films from Raspberry-Like Particles," *Nano Lett.* **5**, 2298–2301.

Moustaghfir, A., Tomasella, E., Rivaton, A., Mailhot, B., Jacquet, M., Gardette, J. L., and Cellier, J. (2004), "Sputtered Zinc Oxide Coatings: Structural Study and Application to the Photoreception of the Polycarbonate," *Surf. Coatings Technol.* **180–181**, 642–645.

Nakajima, A., Fujishima, A., Hashimoto, K., and Watanabe, T. (1999), "Preparation of Transparent Superhydrophobic Boehmite and Silica Films by Sublimation of Aluminum Acetylacetonate," *Adv. Mater.* **11**, 1365–1368.

Natsume, Y. and Sakata, H. (2000), "Zinc Oxide Films Prepared by Sol-gel Spin-coating," *Thin Solid Films* **372**, 30–36.

Nishimoto, S., Kubo, A., Nohara, K., Zhang, X., Taneichi, N., Okui, T., Liu, Z., Nakata, K., Sakai, H., Murakami, T., Abe, M., Komine, T., and Fujishima, A. (2009), "TiO_2-based Superhydrophobic-Superhydrophilic Patterns: Fabrication via an Ink-jet Technique and Application in Offset Printing," *Appl. Surf. Sci.* **255**, 6221–6225.

Noh, J. H., Han, H. S., Lee, S., Kim, D. H., Park, J. H., Park, S., Kim, J. Y., Jung, H. S., and Hong, K. S. (2010), "A Newly Designed Nb-Doped TiO_2/Al-Doped ZnO Transparent Conducting Oxide Multilayer for Electrochemical Photoenergy Conversion Devices," *J. Phys. Chem. C* **114**, 13867–13871.

Obeso, C. G., Sousa, M. P., Song, W., Rodriguez-Perez, M. A., Bhushan, B., and Mano, J. F. (2013), "Modification of Paper Using Polyhydroxybutyrate to Obtain Biomimetic Superhydrophobic Substrates," *Colloids Surf. A* **416**, 51–55.

Park, T.-Y., Choi, Y.-S., Kang, J.-W., Jeong, J.-H., Park, S.-J., Jeon, D. M., Kim, J. W., and Kim, Y. C. (2010), "Enhanced Optical Power and Low Forward Voltage of GaN-based Light-Emitting Diodes with Ga-doped ZnO Transparent Conducting Layer," *Appl. Phys. Lett.* **96**, 051124.

Schneider, P. M. and Fowler, W. B. (1976), "Band Structure and Optical Properties of Silicon Dioxide," *Phys. Rev. Lett.* **36**, 425–428.

Shackelford, J. F. and Alexander, W., eds. (2001), *CRC Materials Science and Engineering Handbook*, third ed., CRC Press, Boca Raton, FL.

Srikant, V. and Clarke, D. (1998), "On the Optical Band Gap of Zinc Oxide," *J. Appl. Phys.* **83**, 5447–5451.

Tropmann, A., Tanguy, L., Koltay, P., Zengerle, R., and Riegger, L. (2012), "Completely Superhydrophobic PDMS Surfaces for Microfluidics," *Langmuir* **28**, 8292–8295.

Tserepi, A., Vlachopoulou, M., and Gogolides, E. (2006), "Nanotexturing of Poly(dimethylsiloxane) in Plasmas for Creating Robust Super-hydrophobic Surfaces," *Nanotechnology* **17**, 3977–3983.

Vila, M., Cáceres, D., and Prieto, C. (2003), "Mechanical Properties of Sputtered Silicon Nitride Thin Films," *J Appl. Phys.* **94**, 7868–7873.

Wong, E. W., Sheehan, P. E., and Lieber, C. M. (1997), "Nanobeam Mechanics: Elasticity, Strength, and Toughness of Nanorods and Nanotubes," *Science* **277**, 1971–1975.

Yang, H. and Deng, Y. (2008), "Preparation and Physical Properties of Superhydrophobic Papers," *J. Colloid Interface Sci.*, **325**, 588–593.

Yang, Y., Sun, X. W., Chen, B. J., Xu, C. X., Chen, T. P., Sun, C. Q., Tay, B. K., and Sun, Z. (2006), "Refractive Indices of Textured Indium Tin Oxide and Zinc Oxide Thin Films," *Thin Solid Films* **510**, 95–101.

Zeng, K., Zhu, F., Hu, J., Shen, L., Zhang, K., and Gong, H. (2003), "Investigation of Mechanical Properties of Transparent Conducting Oxide Thin Films," *Thin Solid Films* **443**, 60–65.

Zhang, Y. Y., Wang, C. M., and Tan, V. B. C. (2008), "Examining the Effects of Wall Numbers on Buckling Behavior and Mechanical Properties of Multiwalled Carbon Nanotubes via Molecular Dynamics Simulations," *J. Appl. Phys.* **103**, 053505.

Strategies for Superliquiphobic/Philic Surfaces

8

8.1 Introduction

Liquid repellent surfaces can be used for self-cleaning and antifouling from organic and biological contaminants both in air and underwater applications and can reduce fluid drag (Bhushan 2009). As a model surface in living nature for a liquid repellent surface in air, the upper side of the lotus leaf surface repels water (superhydrophobic) and is useful for self-cleaning and low adhesion applications (Barthlott and Neinhuis 1997; Bhushan and Jung 2011). As discussed in Chap. 4, the superhydrophobic properties of the leaf surfaces are achieved due to the presence of a hierarchical structure created by a microstructure formed by papillose epidermal cells covered with three dimensional (3-D) epicuticular hydrophobic wax nanotubules, shown in Fig. 8.1a. The wax layer makes the surface hydrophobic and the hierarchical structure makes the surface superhydrophobic. This structure causes water droplets to roll off the leaf surface and take contaminants with them to keep the leaf clean. The lower side of the lotus leaf does not contain 3-D wax crystals (Neinhuis and Barthlott 1997), and consists of rather flat, tabular, and slightly convex papillae (Koch et al. 2009). Therefore, the bottom surface is hydrophilic, but superoleophobic in water, with a contact angle of 155° with n-hexane oil, Fig. 8.1b (Cheng et al. 2011). The lotus leaf exhibits a so-called "Janus interface" (named for the two-faced Roman god), with superhydrophobicity on the upper side, and superoleophobicity under water on the lower side (Cheng et al. 2011).

Another model surface in living nature for superoleophobicity under water is provided by sea animals such as fish (including carp and trout), which are known to be well protected from contamination by oil pollution although they are wetted by water. Fish scales are covered with oriented scales that promote anisotropic flow from head to tail (Liu et al. 2009; Bixler and Bhushan 2012b). Shark skin, which is a model for a low drag surface, is covered by small individual tooth-like scales called dermal denticles

B. Bhushan, *Introduction to Biomimetics and Bioinspiration*,
https://doi.org/10.1007/978-3-031-62344-8_8

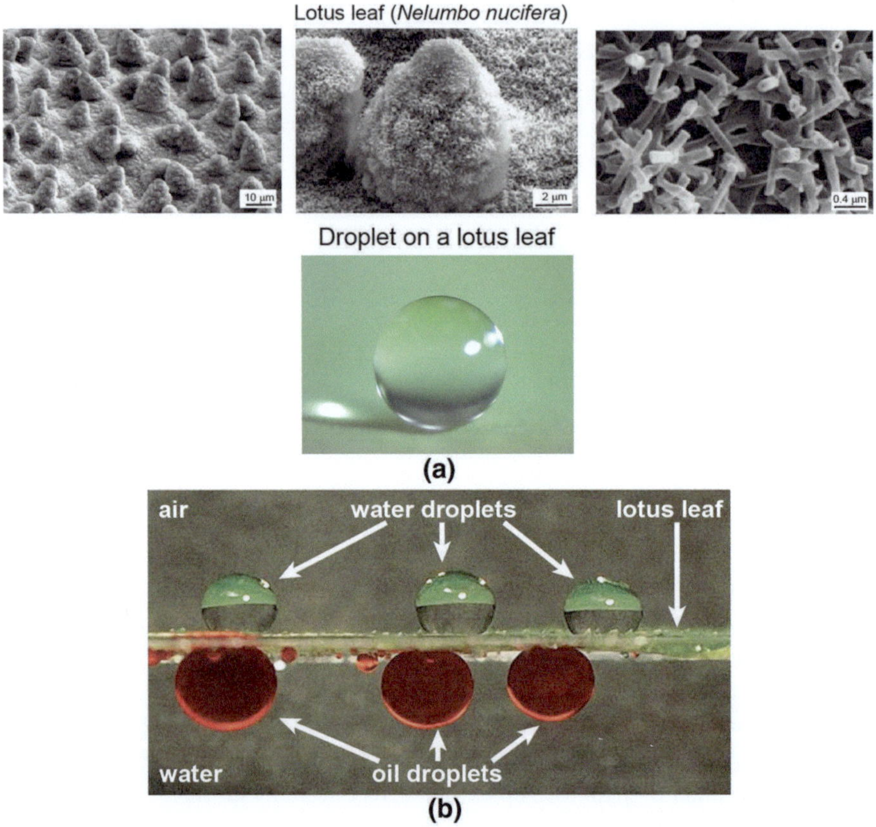

Fig. 8.1 **a** SEM micrographs (shown at three magnifications) of lotus (Nelumbo nucifera) leaf upper surface, consisting of microstructures formed by papillose epidermal cells covered with 3-D epicuticular hydrophobic wax nanotubules on the surface, which create hierarchical structure, and an image of water droplet sitting on lotus leaf (adapted from Bhushan et al. 2009); **b** lotus leaf floating on water surface. Spherical water droplets on its upper side and spherical oil (*n*-hexane dyed red) droplets on the lower side in water (adapted from Cheng et al. 2011)

(little skin teeth) shaped like small riblets, with longitudinal grooves (aligned parallel to the local flow direction of the water) (Bixler and Bhushan 2013a). Figure 8.2a shows shark skin replica and only three ribs on each scale. It is clearly visible that the V-shaped riblets' height varies between 200 and 500 μm, and their spacing varies between 100 and 300 μm (Jung and Bhushan 2009; Dean and Bhushan 2010). The contact angle of actual shark skin using water droplets with and without mucus was $<10°$ (superhydrophilic) and using paraffin oil droplets under water was 153° (superoleophobic), Fig. 8.2b (Bixler and Bhushan 2013b). Their superoleophobic characteristic in water and their microstructured surfaces lead to protection from marine fouling and play a role in the defense against

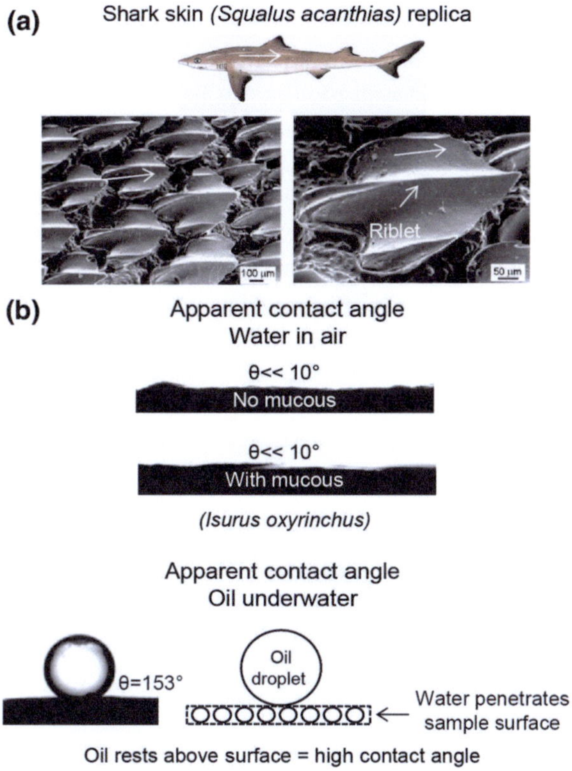

Fig. 8.2 **a** SEM micrograph of shark skin (*Squalus acanthias*) replica. The shark skin replica shows only three ribs on each scale. It is clearly visible that the V-shaped riblets' height varies between 200 and 500 μm and their spacing varies between 100 and 300 μm (adapted from Jung and Bhushan 2009); **b** water droplet images in air and paraffin oil droplet images under water for actual shark skin (*Isurus oxyrinchus*) (Bixler and Bhushan 2013b)

adhesion and growth of marine organisms, e.g., bacteria and algae (Genzer and Efimenko 2006; Bixler and Bhushan 2012a).

Hydrophilic surfaces exhibit superoleophobicity underwater. For these surfaces, the underwater surface is wetted by the water. In order for the underwater surface to become wet by oil, the water would first have to dewet the surface, which is difficult due to the very low receding angle of water on the hydrophilic surface. Thus, the hydrophilic surface exhibits superoleophobicity under water.

Model surfaces in living nature with superoleophobicity in air are rare (Brown and Bhushan 2016a). Certain species of leafhopper (Rakitov and Gorb 2013) and springtails (Hensel et al. 2013) are repellant to some low surface tension liquids. Integuments of leafhoppers (insects, *Hemiptera, Cicadellidae*) are coated with a thin layer of intricately structured particles known as brochosomes. They are hollow proteinaceous spheres of usually 200–700 nm diameter with honeycomb walls, Fig. 8.3a (adapted from Rakitov and Gorb 2013). The structure of brochosomal particles consists of re-entrant geometry (to be introduced later), which is repellant to water and some low surface tension liquids (diiodomethane); but is wettable with ethanol (surface tension ~22 mN/m). Nature provides this functionality to protect these sap-sucking insects from being trapped in the water and their own liquid exudates. Springtails (*Collembola*) are wingless arthropods that survive in rain-flooded habitats. Their skin consists of mushroom-shaped profiles with cavity sidewall profiles that have characteristic overhangs. The overhanging surface structures provide re-entrant geometry, Fig. 8.3b (Hensel et al. 2013). They form a plastron, protecting them against suffocation if immersed in water and low surface tension liquids, such as alkanes and olive oil. In both cases, re-entrant geometries are responsible for the oil repellency. However, a surface found in living nature to repel very low surface tension liquids (<30 mN/m) has yet to be found.

For liquid repellency, another model surface is the *Nepenthes* pitcher plant. These leaf surfaces are not superoleophobic. Pitcher plants exhibit their carnivorous nature and feature pitcher-shaped traps that capture and digest insects. Many pitcher plants capture prey by utilizing a waxy zone along the pitcher rim (peristome). This wax attaches to the adhesive pads of the insect, which can no longer adhere to the plant wall and falls

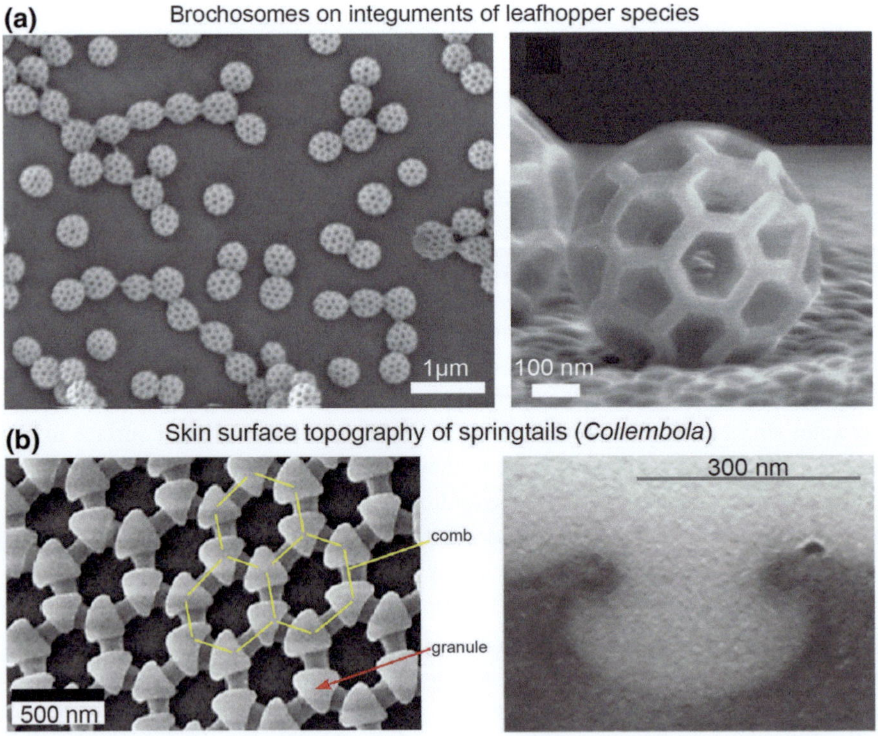

Fig. 8.3 **a** SEM micrograph of the structure of brochosomes on integuments of leafhopper species (adapted from Rakitov and Gorb 2013), and **b** SEM micrograph of the skin surface of springtail (*Orthonychiunis stachiamis*) showing granules at the intersections of the underlying comb structure (above) and sections through the skin, taken by TEM showing cavities with characteristic overhanging cross-sectional profiles (adapted from Hensel et al. 2013)

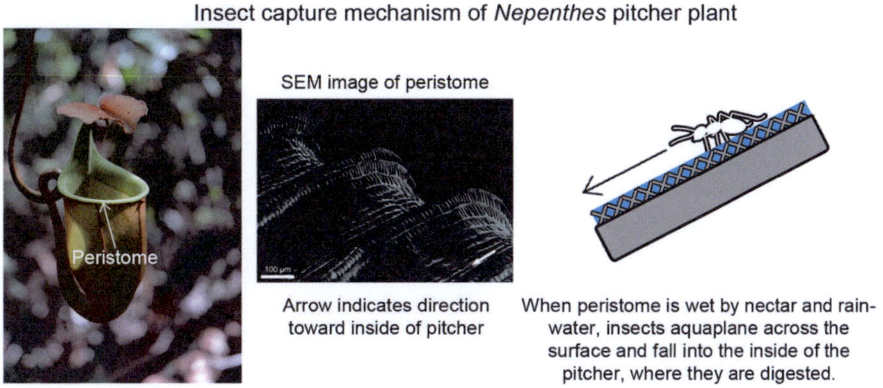

Fig. 8.4 Schematic to show mechanism for prey capture by *Nepenthes* pitcher plants. The rim of the pitcher (peristome) features surface features that become wet by nectar and rainwater. Insects landing on the wet peristome aquaplane across the surface and fall into the pitcher (adapted from Bohn and Federle 2004; Brown and Bhushan 2017a, b)

into the trap. Pitcher plants of the genus *Nepenthes* have been found to capture prey by a different mechanism, Fig. 8.4 (Bohn and Federle 2004; Brown and Bhushan 2017a, b). The peristome of these pitcher plants features a rectangular microstructure that is wetted by nectar and rainwater, resulting in a continuous liquid film over the surface of the rim. When wetted by a liquid that is immiscible with the liquid present on the rim, the peristome becomes extremely slippery and insects aquaplane across the surface and fall into the trap.

In this chapter, various liquids of interest are introduced first, followed by various strategies to obtain superoleophobicity in air, combinations of superoleophilicity/phobicity and liquid repellency. Then, a model advanced by Jung and Bhushan (2009) for predicting the oleophobic/philic nature of surfaces droplets is presented. To validate the model, it has been investigated how the water and oil droplets in water influence wetting behavior on flat surfaces, as well as micropatterened surfaces with varying pitch values.

8.2 Oils and Surfactant-Containing Liquids

Oil repellency is more difficult to achieve than water repellency, as the surface tension of oils is lower than that of water. The surface tension values of selected oils are presented in Table 8.1 and values as low as about 10 mN/m are reported (Brown and Bhushan 2016a).

Containers used today for soaps such as shampoos and detergents are not liquid repellent. One faces the frustration of struggling to get the last few drops of shampoo out of the bottle, or having a shampoo bottle lid become clogged with product. The primary

Table 8.1 Surface tension values of selected liquids (Brown and Bhushan 2016a)

Liquid	Surface tension (mN/m)
Water[a]	71.99
Diiodomethane[a]	50.80
Ethylene glycol[a]	47.70
Rapeseed oil	~35
Peanut oil	~32
Olive oil	~32
Soybean oil	~29
Hexadecane[a]	27.05
Tetradecane[a]	26.13
Dodecane[a]	25.35
Cyclohexane[a]	24.95
Decane[a]	23.37
Methanol[a]	22.70
Ethanol[a]	22.10
Octane[a]	21.14
Hexane[a]	18.43
Pentane[a]	15.48
Silicone oils[b]	16–20
Perfluorohexane (FC-72)[c]	10

[a]Haynes (2014),
[b]Anonymous (2015a),
[c]Anonymous (2015b)

goal is to create a bottle material that could allow all the shampoo to slide out of the container without leaving any residue. This ensures the customer is not wasting money by throwing away leftover shampoo and the discarded bottle itself is cleaner and more easily recyclable, which creates less waste. Soaps contain surfactants that consist of low surface tension components and chemically active groups. These molecules help soap clean, but they also make the soap stick to surfaces, including the inside of the soap bottle. To prevent this, one must make the bottle less sticky. One could add chemistry to the inside of the bottle to which soap does not stick.

Common ingredients in shampoos and laundry detergents are shown in Figs. 8.5 and 8.6, respectively (Martin et al. 2017). The shampoo primarily contains sodium lauryl sulfate and sodium laureth sulfate surfactants, while the laundry detergent primarily contains sodium alcoholethoxy sulfate and sodium alkylbenzene sulfonate surfactants. These surfactants act as detergents, wetting agents, and foaming agents. Surfactants are molecules that attach to a variety of surfaces and can lower the interfacial tension between phases.

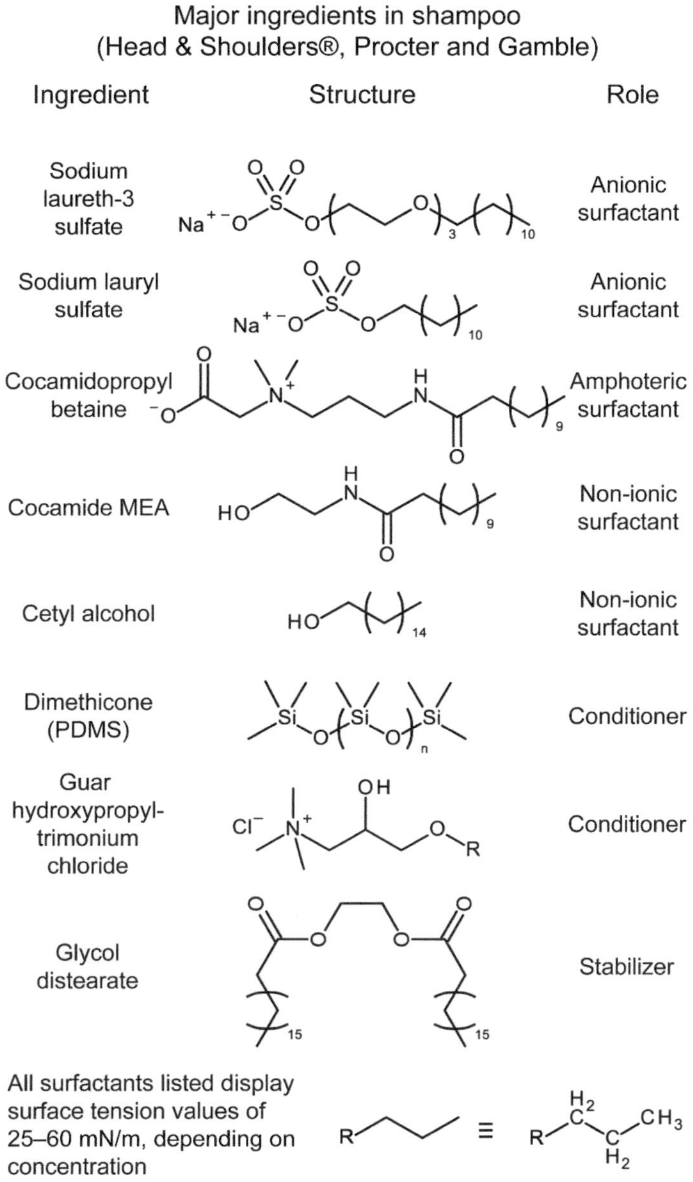

Fig. 8.5 Major ingredients in shampoo (Head and Shoulders®, Proctor and Gamble) with their structural formula and role in the product (Martin et al. 2017)

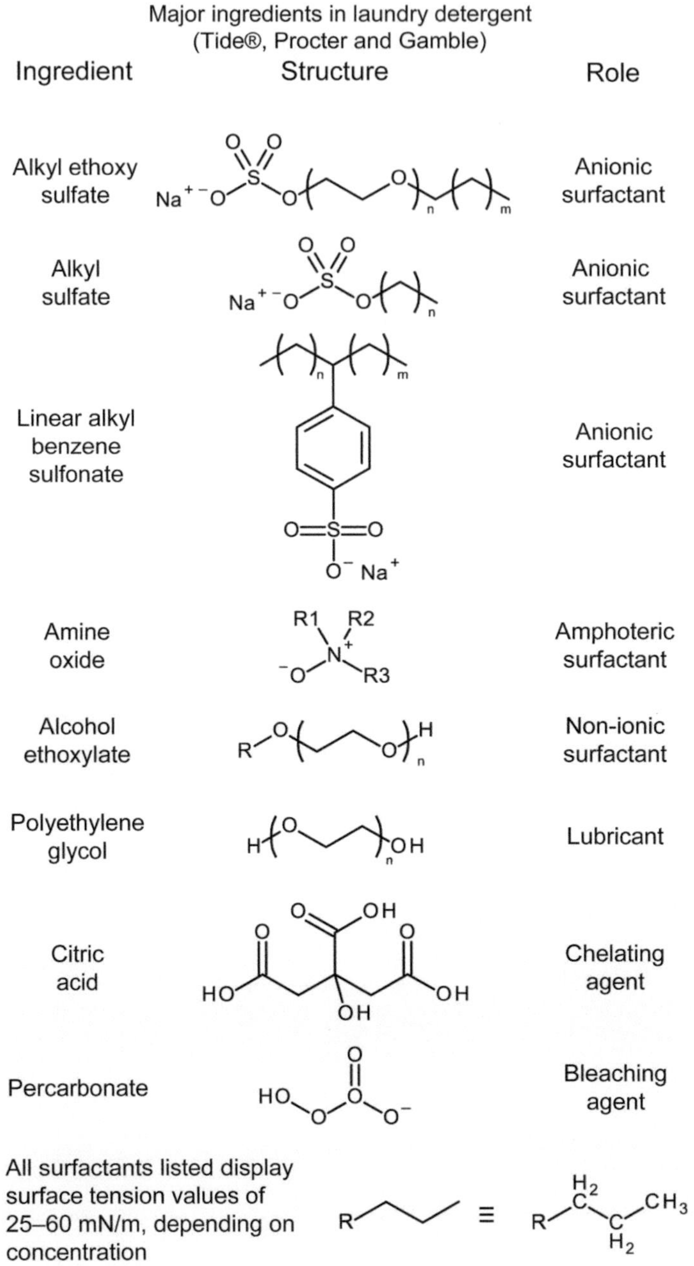

Fig. 8.6 Major ingredients in laundry detergent (Tide®, Proctor and Gamble) with their structural formula and role in the product (Martin et al. 2017)

They typically comprise a hydrophilic head group and a long hydrophobic tail. Common anionic surfactants feature a sulfate head group with sodium counter-ion. Both examples of shampoo and laundry detergent contain a mix of anionic, amphoteric, and non-ionic surfactants. This mixing of ionic and amphoteric surfactants is common in formulations for these types of products (Soontravanich et al. 2008). The anionic surfactants are typically the main surfactant for cleaning purposes, while the amphoteric surfactant serves as a foaming agent. All surfactants have the potential to lower the surface tension of the liquid, as well as adsorb at the interface of any solid with which the liquid comes into contact. These components typically have surface tension values of 25–60 mN/m, depending on concentration with higher surface tensions in the product if the surfactant is heavily diluted (Ross and Epstein 1958; Al-Sabagh et al. 2011). The surface tensions of surfactants are on the order of oils. However, a surfactant has polar head groups that are highly active and adhere to surfaces stronger than oils. These liquids readily coat the inside of plastic bottles, leading to wasted product and issues with recyclability. Figure 8.7 shows a fouled cap of a shampoo bottle.

Materials used for construction of plastic bottles typically are high density polyethylene (~70%), polyethylene terephthalate (~25%), and polypropylene (~5%). Bottle caps are typically made of polypropylene. Creating surfaces repellent to surfactant-containing liquids is important for the consumer packaging industry. To determine the applicability of superoleophobic coatings in real world applications such as consumer packaging of shampoos and laundry detergents, their repellency against shampoo and laundry detergent needs to be tested.

As temperature increases, the surface tension of all liquids decreases (Jasper 1972; Haynes 2014). The relationship between surface tension and temperature is given by the so-called Eötvös rule. The data for the selected liquids of interest is given in Fig. 8.8.

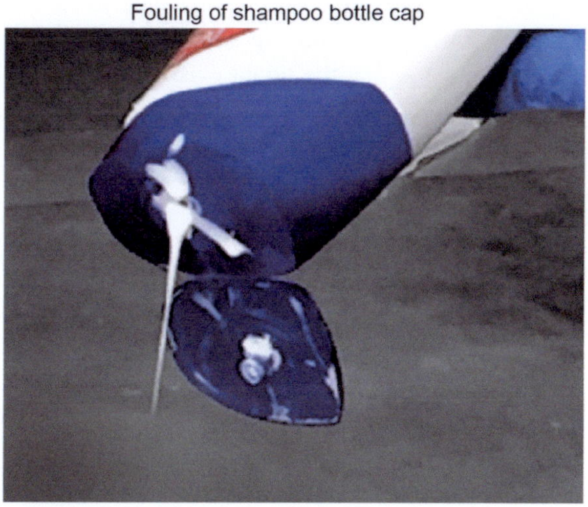

Fig. 8.7 Example of fouling on shampoo bottle cap. Shampoo contains surfactants which are highly active and readily adhere to surfaces

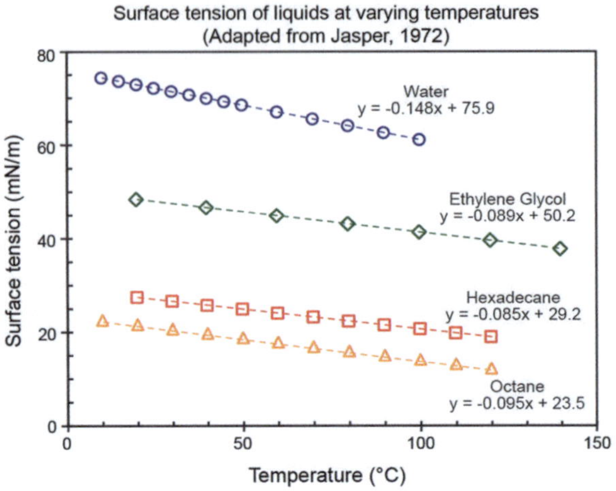

Fig. 8.8 Surface tension of water, ethylene glycol, hexadecane, and octane at varying temperatures. As temperature increases, surface tension decreases (adapted from Jasper 1972 and Haynes 2014)

Temperature is important because at higher temperatures, surface tension can decrease and surface can become fouled. The interior of a car on a warm day can reach on the order of 80 °C (Manning and Ewing 2009) and influence repellency of liquids from the interior surfaces.

8.3 Strategies to Achieve Superoleophobicity in Air and Liquid Repellency

The two approaches to achieve roughness-induced superoleophobicity in air and liquid impregnation to achieve liquid repellency are schematically shown in Fig. 8.9.

Roughness-induced superoleophobicity in air is commonly practiced. As discussed in Chap. 3, an issue of utilizing surface roughness and trapped air as seen in the Cassie-Baxter regime is that the liquid exists in a metastable state and liquid resting on such a surface can eventually penetrate into the roughness, transitioning from a Cassie-Baxter regime to a wetted state, or the Wenzel regime. In doing so, the surface becomes wet

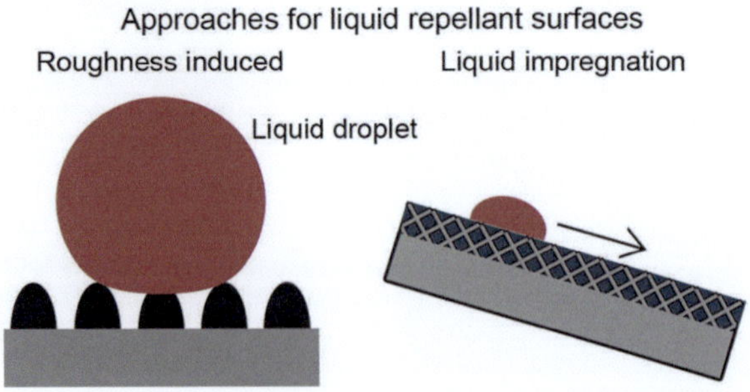

Fig. 8.9 Schematics of two approaches to achieve roughness-induced superoeleophibicity and liquid impregnation to achieve liquid repellency

by the liquid and is no longer repellent. Another method of creating liquid-repellent surfaces is to use liquid-impregnated surfaces inspired by the *Nepenthes* pitcher plant. By adding to the surface a non-volatile liquid layer with a surface tension lower than that of the liquid being repelled, repellent surfaces can be created. The repellency of the surface is dependent upon the properties of the liquid layer and its miscibility with the liquids being repelled. In liquid-impregnated surfaces, there is a danger of mixing and loss of impregnated liquid. The porous surfaces fabricated in the lab are generally soft and are expected to be less durable. On the other hand, roughness-induced superoliquiphobic surfaces can be fabricated using hard ceramic nanoparticles with a hard polymeric binder. Such surfaces are expected to be mechanically durable.

In roughness-induced superoleophobicity, in order for a nanostructured surface to exhibit superoleophobicity, its surface energy should be lower than the surface tension of the oil. Various roughness techniques are used. The nanoscale roughness is most commonly achieved through the use of nanoparticles. To repel liquids with the low surface tensions, a low surface energy material is also typically required. Fluorinated materials are commonly used as fluorine is very electronegative and has a low polarizability. This results in a low susceptibility to London dispersion forces, leading to weak intermolecular forces, weak cohesive and adhesive forces, and therefore low surface energies. A so called re-entrant geometry (to be described later) can be used as a way to increase the oil contact angle for a surface with surface energy too low to repel some very low surface tension liquids and particularly surfactant-containing liquids such as shampoos and laundry detergents (Nosonovsky and Bhushan 2008).

Various combinations of commonly used roughness and fluorination methods are shown in Fig. 8.10. Roughness methods used including etching, micropatterning, nanoparticles and polymer composites, and encapsulating nanoparticles in a substrate. Fluorination methods include using fluoropolymer, fluorosilane or fluorothiol, fluoroplasma, or fluorosurfactant. Fluorination methods generally require chemically active surfaces.

Various coating layers can be deposited by various deposition techniques.

8.3.1 Roughness Techniques

Selected techniques are summarized in Fig. 8.10 (left). The etching roughness method has been demonstrated on metals such as aluminum, copper, steel, titanium, and zinc using acid etching and/or electrochemical etching (Qian and Shen 2005; Qu et al. 2007; Song et al. 2013). Etching on various polymers including polystyrene and PMMA has been completed using plasma treatment (Flamm and Auciello 2012; Wang and Bhushan 2015). Etching is advantageous in that it typically treats an entire surface, regardless of geometries, and is an aggressive process. However, it can be difficult to control and is only suitable for small-scale experiments, making it unsuitable for scale-up. Micropatterning

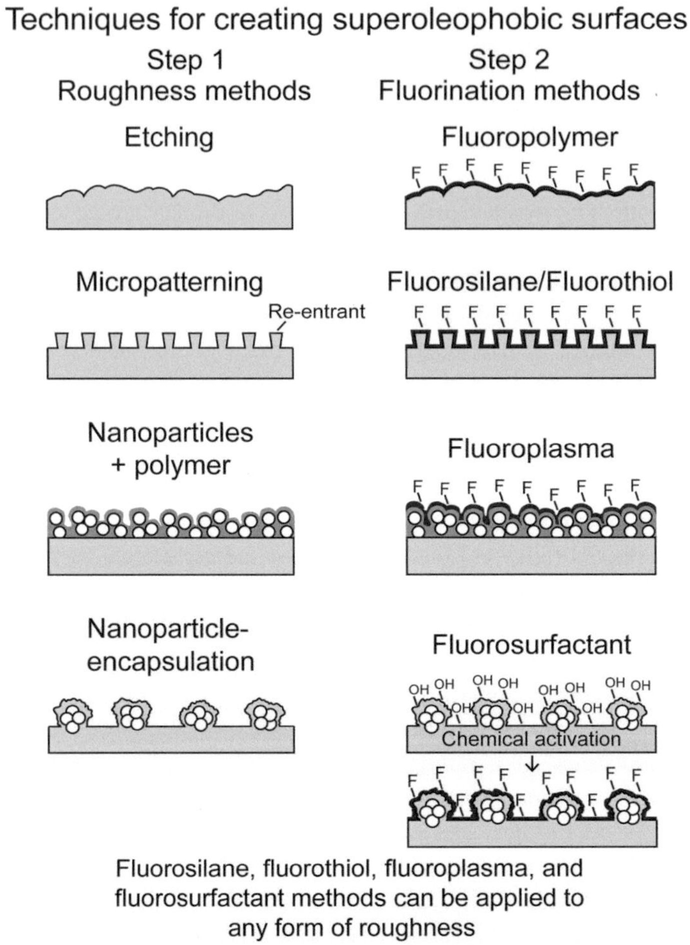

Fig. 8.10 Schematic of commonly used techniques for creation of roughness-induced, fluorinated, superoleophobic surfaces. Roughness methods (left) include etching, micropatterning, nanoparticles + polymer, and nanoparticle-encapsulation. These methods can introduce re-entrant geometries where surface features become narrower at the base. Various fluorination methods (right) are used. Fluoropolymer is straight-forward but mechanically durability is an issue, fluorosilane/fluorothiol layers are thin but require surface activation (e.g. via UV/plasma) for silanes or specific substrates (e.g. copper, silver) for thiols, fluoroplasma accomplishes surface texturing and fluorination in one step but is difficult to scale, and fluorosurfactant layers provide a surface that repels oils but attracts water for oil-water separation but requires specific surface chemistry for bonding (adapted from Brown and Bhushan 2016a; Martin et al. 2017)

allows for accurate replication of features with complex and possible re-entrant geometries. It has been used to create rectangular structures (Martin and Bhushan 2017), inverse trapezoidal pillars (Im et al. 2010), and mushroom-like pillars (Kang et al. 2012). Coatings consisting of the nanoparticle + polymer matrix are commonly used (Muthiah et al. 2013; Wang and Bhushan 2015; Martin and Bhushan 2017; Bhushan and Martin 2018). These are simple to fabricate and can be deposited on various substrates. Nanoparticle encapsulation roughness method has been developed by Brown and Bhushan (2016b, c). Since nanoparticles are incorporated into the substrate, they may provide durable surfaces. However, encapsulation techniques for a given substrate need to be developed.

8.3.2 Fluorination Techniques

In the fabrication of superoleophobic surfaces in air, fluorinated coatings are most commonly used over rough surfaces because of their low surface energy. Interaction between surfaces occurs due to polarization of molecules. Polarizability depends upon the mobility of electrons. A low polarizability signifies tightly bound electrons. Fluorine is the most electronegative element (nonpolar) on the periodic table with its small radius. Its atomic number is 9 with an electron configuration of $1s^1 2s^2 2p^5$. It consists of 5 electrons in 2p shell and only short of 1 electron. Fluorine is not easily polarized, because it wants to hold onto its 5 electrons, which are less mobile. The lack of polarizability means that fluorine is weakly susceptible to van der Waals forces and it is not easily attracted to or repelled by nearby objects. As a result, it is known to have weak intermolecular forces and therefore a lower energy.

Low adhesion of fluorinated materials can be a disadvantage as they do not adhere well to any nanoparticles or the substrates. Lack of adhesion may lead to poor mechanical durability, which presents challenges. There are several popular application techniques using various fluorinated materials summarized in Fig. 8.10 (right) (Brown and Bhushan 2016a). Details follow.

8.3.2.1 Fluoropolymers

Fluorinated polymers for low surface energy coatings are desirable as polymers. One drawback of fluoropolymers is the lack of binding to the nanoparticles which is needed to achieve superoleophobicity. To address adhesion and mechanical durability issues, hydrophilic nanoparticles may be used, which attach themselves to form a 3-D network. In another approach, the fluoropolymer can be deposited on a prefabricated substrate with the correct topography. However, if the coating is too thick, surface roughness features can be lost (Brown and Bhushan 2016a).

8.3.2.2 Fluorosilanes and Fluorothiols

Fluorosilanes are commonly used for creating superoleophobic surfaces as they can form thin, self-assembled layers that will not conceal small topographical features. Silanes typically require –OH groups on the surface (e.g., produced via ultraviolet/oxygen plasma) for a covalently bonded layer to form (Wang and Bhushan 2015; Brown and Bhushan 2016b, c, 2017a, b). The –OH groups on the coating surface react with Si–Cl groups of fluorosilane to form Si–O links.

Many silanes are sensitive to moisture that can lead to substrate corrosion and can form insoluble polymers in solution (Ulman 1996). When factored in with the requirement for surface activation, this can limit the use of silanes for industrial scale up.

Another organic compound capable of forming a self-assembled, fluorinated layer are thiols. However, these display long term instability toward oxidation (Lee et al. 1998) and only assemble on specific surfaces such as silver, gold, platinum, palladium, or copper. However, durability remains a challenge (Brown and Bhushan 2016a).

8.3.2.3 Fluoroplasma

Another method of fluorination is plasma deposition. Various plasma species can be used to etch and/or fluorinate a surface for oil-repellent applications. Plasma treatments are advantageous as they can accomplish surface texturing and fluorination in one step. They also can typically treat the entire surface, regardless of geometries. However, while these may be suitable for small-scale experiments, the use of specialty conditions and chemicals may be cost prohibitive for certain applications (Brown and Bhushan 2016a).

8.3.2.4 Fluorosurfactants (for Superhydrophilicity and Superoleophobicity)

Surfaces that repel oils also typically repel water, due to water having a higher surface tension (Table 8.1). However, it is possible to create a coating that repels oils but attracts water. This can be achieved through the use of a fluorosurfactant, which contains a high surface tension head group and a low surface tension tail group. When deposited, the fluorinated tails segregate at the air interface resulting in a low surface tension barrier that repels oils. However, when droplets of water are placed on the surface, they are able to penetrate down through the tail groups to reach the high surface tension groups below, and thus the coating appears hydrophilic. Such surfaces can be exploited for a variety of applications, including oil-water separation and anti-fouling (Brown and Bhushan 2015a, c).

It is also possible to use fluorosurfactants to create superomniphobic surfaces where all liquids, including water, are unable to penetrate (Brown and Bhushan 2016a).

8.3.3 Chemical Activation of Underlayer of a Coated Surface

Composite coatings generally need to be chemically activated to chemically bond the fluorinated layer to the underlayer. Techniques commonly used to create active hydroxyl groups (–OH) include oxygen plasma treatment (Wang and Bhushan 2015), ultraviolet (UV) (Brown and Bhushan 2016b, c, 2017a, b), and more aggressive ultraviolet-ozone or UV-ozone (UVO) (Martin and Bhushan 2017; Bhushan and Martin 2018). A comparison of UV-ozone and oxygen plasma activation techniques is presented in Table 8.2. UV or UV-ozone techniques do not require expensive plasma equipment. UV-ozone, being more aggressive than UV, can activate a coating in about an hour whereas UV treatment may take 12–24 h.

The UV technique uses irradiation at 254 nm, whereas the UVO method uses a combination of 185 and 254 nm wavelengths (Martin and Bhushan 2017). In the presence of 185 nm light, molecular oxygen can be dissociated into atomic oxygen, and atomic and molecular oxygen can be synthesized into ozone. In the presence of 254 nm light, ozone is decomposed into atomic and molecular oxygen (Vig 1985; Bhurke et al. 2007; Park et al. 2009). These chemical reactions are shown in Fig. 8.11a.

Schematics of the UV and UVO surface activation processes are shown in Fig. 8.11b (Martin and Bhushan 2017). In the UV process, 254 nm light breaks up organic bonds in

Table 8.2 Comparison of UV-ozone and oxygen plasma activation techniques

Method	Details	Pros	Cons
UV-ozone	Using combination of UV wavelengths (185 and 254 nm) to generate ozone to clean and react with surface, creating hydroxyl groups Reactive species O_3 (ozone), ·O (radical oxygen), ·OH (hydroxyl radical)	Can be done at ambient pressure and temperature Little to no etching	Takes longer than plasma to achieve same results Prolonged UV exposure can result in surface deformation (i.e. cracking) Not as effective for certain polymers/substrates
Oxygen plasma	Gas discharge (typically O_2 or air) at low pressure to clean, etch, and react with surface, creating hydroxyl groups Reactive species: O_3, ·O, O_2^- (superoxide), O_2^{2-}(peroxide), ·OH (air plasma only)	Flexible process where selection of gas, power, treatment time can determine surface cleaning, etching, and chemical modification Quick processing times for cleaning/chemical modification	Requires low pressure vacuum equipment Aggressive treatment conditions can lead to unwanted surface etching

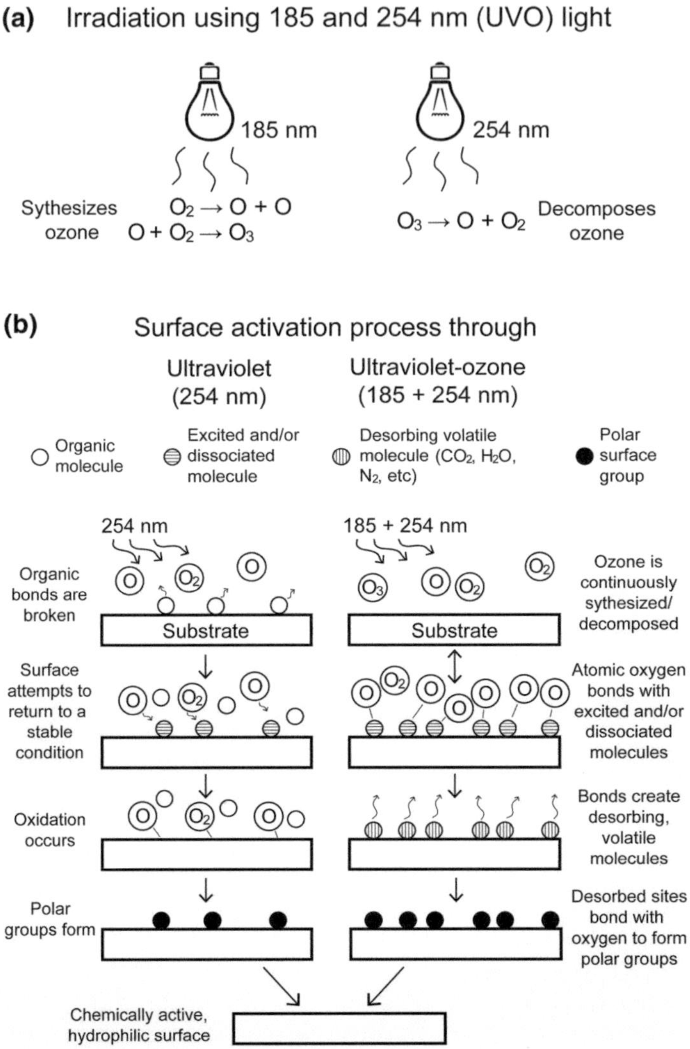

Fig. 8.11 Schematics of the irradiation using UV and UVO surface activation processes. **a** Irradiation using 185 nm and 254 nm (UVO) light. A wavelength of 185 nm decomposes molecular oxygen into atomic oxygen and also synthesizes ozone. In the presence of 254 nm ultraviolet light, ozone decomposes into atomic and molecular oxygen. **b** Comparison of ultraviolet and ultraviolet-ozone surface activation processes. In the UV process, 254 nm light breaks up organic molecules at the surface, and these excited and/or dissociated molecular sites attempt to return to a chemically stable condition by reacting with oxygen molecules. After oxidation, polar groups are formed, which leads to a chemically active, hydrophilic surface. In the UVO process, the combination of 185 and 254 nm light causes ozone to be continually synthesized and decomposed. Due to the irradiation, surface molecules are excited and/or dissociated, and therefore, react with atomic oxygen to form desorbing volatile molecules. These volatile molecules desorb from the surface. The desorbed sites bond with oxygen to form polar groups, which leads to a chemically active, hydrophilic surface (Martin and Bhushan 2017)

the surface layer, which causes the surface to attempt to return to a stable condition. When these open bond sites react with oxygen, oxidation occurs and highly active, polar surface groups are formed. These groups are commonly hydroxyl, carbonyl, and carboxylic acids and lead to a chemically active hydrophilic surface (Bhurke et al. 2007). In the UVO process, the combination of 185 and 254 nm light causes atomic oxygen to be continuously produced and ozone to be continuously synthesized and decomposed. The wavelengths of light excite and/or dissociate molecules at the surface and then react with atomic oxygen for desorbing, volatile molecules such as CO_2, H_2O, and N_2. The desorbing molecules create adsorption sites for oxygen to form highly active, polar surface groups, which lead to chemically active, hydrophilic surfaces. The UVO process is an order of magnitude faster that the UV process due to an extra wavelength that produces highly-reactive atomic oxygen and helps excite and/or dissociate molecules (Vig 1985; Efimenko et al. 2002; Martin and Bhushan 2017).

Figure 8.12 shows a schematic of the set up for UVO surface activation (Martin and Bhushan 2017). The UVO exposure was generated from a U-shaped, ozone-producing ultraviolet lamp (18.4 W, Model G18T5VH-U, Atlantic Ultraviolet Co.). The lamp outputs were expected to be a total of 5.8 W of 254 nm light, 0.4 W of 185 nm light, and 1.6 g/h of ozone in ambient conditions. In order to increase UV intensity, the samples can be placed in very close proximity (about 2 cm) at a location below the light source. The lamp was placed in an enclosure with width, height, and length dimensions of 17 by 15 by 45 cm and was connected to an electronic ballast (120 V, Model 10-0137, Atlantic Ultraviolet Co.) in order to provide the proper electrical conditions.

As an example, Fig. 8.13 shows various steps used in fabrication of a dual layer coating exhibiting superoleophobicity (to be discussed in detail in Chap. 11). Chemical reactions are shown before and after activation of a coating consisting of methylphenyl silicone resin with SiO_2 nanoparticles followed by perfluorotrichlorosilane deposition. Chemical activation creates –OH groups which react with Si–Cl groups to create strong Si–O links.

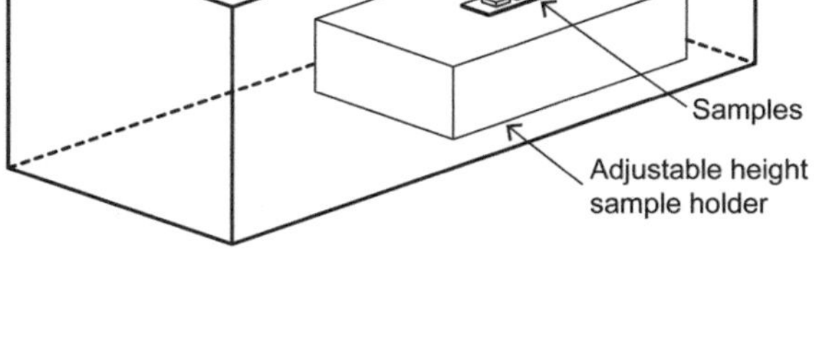

Fig. 8.12 Schematic of ultraviolet-ozone lamp set up using a U-shaped, ultraviolet lamp capable of ozone production (Martin and Bhushan 2017)

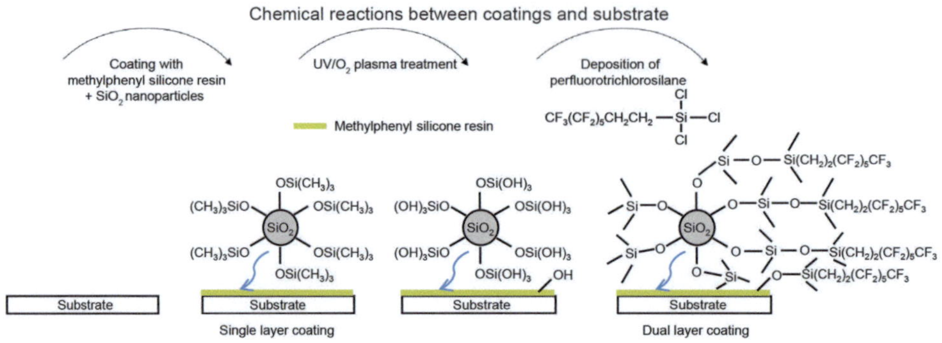

Fig. 8.13 Schematics of chemical reactions of dual layer coating during the deposition process. The chemical activation creates –OH groups which react with Si–Cl groups to create strong Si–O links during deposition of perfluorotrichlorosilane

8.3.4 Re-entrant Geometry

For a liquid with very low surface tension, it may not be possible to produce a surface with sufficiently low surface energy to realize a contact angle on a flat surface greater than 90°. By introducing structures with re-entrant curvatures, it is possible to achieve high contact angles, even if the contact angle on the flat surface (θ_{flat}) is less than 90° (Nosonovsky and Bhushan 2008; Brown and Bhushan 2016a). By pairing re-entrant geometries with low surface energy, fluorinated materials, superoleophobicity for liquids with very low surface tension can be achieved.

In a re-entrant geometry, each roughness feature is wider at some point than it is at another below, creating an overhang. To repel liquids with contact angles as low as 0°, a double re-entrant geometry may be necessary (Nosonovsky and Bhushan 2008; Liu and Kim 2014; Brown and Bhushan 2016a). Figure 8.14 shows three re-entrant geometries for demonstration.

These include inverse trapezoidal, cylindrical, and spherical. The top row shows one cylindrical surface and two trapezoidal surfaces with an inclination angle α (Brown and Bhushan 2016a, b). For two hypothetical liquids with $\theta_{flat} = 70°$ and 40°, imaginary solid-liquid-air interfaces are shown for the three cases. For the case of cylindrical features with $\theta_{flat} = 70°$, and the case of trapezoidal features with $\alpha = 25°$ and $\theta_{flat} = 40°$ ($\theta_{flat} + \alpha <$ 90°), the liquid-air interface is concave shaped. The Laplace pressure inside the droplet is negative, which results in a downward force, leading to a fully wetted interface. However, in the trapezoidal features with $\theta_{flat} = 70°$ and $\alpha = 25°$ ($\theta_{flat} + \alpha \geq 90°$), the liquid-air interface is convex shaped and the net Laplace force on the interface is upward, producing a composite interface (Cassie-Baxter regime) with a high contact angle. As long as $\theta_{flat} + \alpha \geq 90°$, a liquid with $\theta_{flat} < 90°$ will result in non-wetting ($\theta_{apparent} \geq 90°$). The middle row in Fig. 8.14 shows structure with re-entrant curvature (cylindrical, spherical, oval,

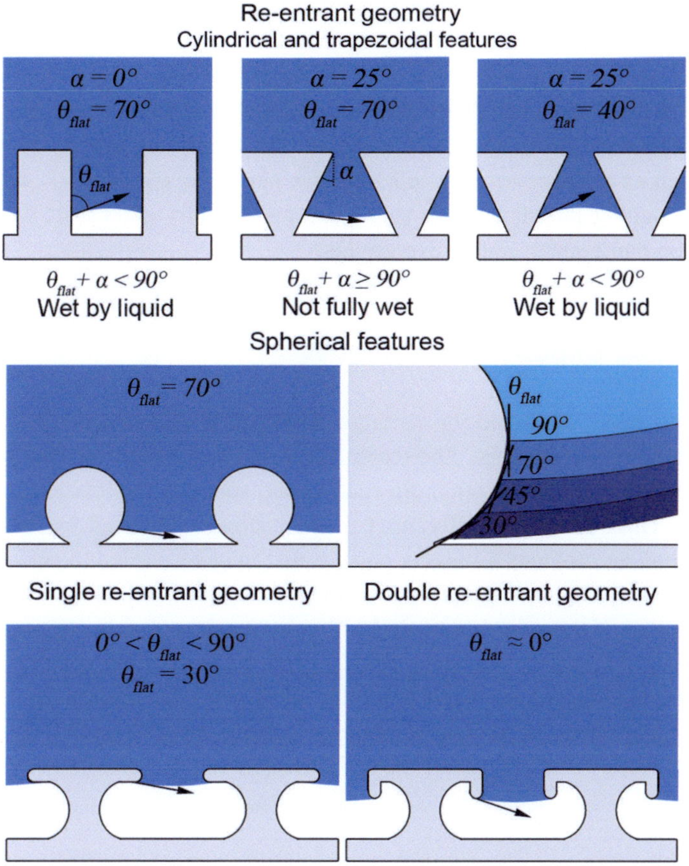

Fig. 8.14 Three types of hypothetical re-entrant geometries covered with liquids of various contact angles. The surfaces can become wet or form a composite interface, dependent upon the surface geometry and the liquid contact angle (adapted from Brown and Bhushan 2016a)

etc.), which can provide contact angles > 90° (non-wetting) for liquids with flat contact angles < 90°, since it is possible to draw tangents on the surface with liquid-air interface being convex shaped.

The bottom row in Fig. 8.14 shows single and double re-entrant geometries. The single re-entrant geometry (left) is able to support low flat angles, close to zero angle. They could theoretically support flat contact angles of 0°. However, the presence of positive pressure in the droplet means that a liquid-vapor interface shape with upward surface tension is required and it is not possible. This is possible with double re-entrant geometries (right). This geometry incorporates vertical overhangs normal to the surface in addition to the horizontal overhangs parallel to the surface found in the single re-entrant geometry (left). Tangents on the re-entrant curvature can be drawn beyond those, parallel to the flat surface and so a convex shaped liquid-air interface can be supported, even with a fully wetted liquid (Brown and Bhushan 2016a, b).

Various re-entrant geometries have been fabricated by researchers (Tuteja et al. 2008; Zhao et al. 2011; Lee et al. 2013; Liu and Kim 2014; Brown and Bhushan 2016b; Martin and Bhushan 2017; Martin et al. 2017; Bhushan and Martin 2018). These re-entrant geometries are necessary for repelling low surface tension liquids, and particularly surfactant-containing liquids such as shampoos and laundry detergents due to their low surface tension components and active groups.

8.3.5 Coating Deposition Techniques

Coatings have been deposited by a large number of techniques for fabrication of superliquiphobic/philic surfaces. The techniques commonly used for nanoparticle-binder composite coating include air spray, dip coating and spin coating because they are simple and versatile. There are some differences in the coatings produced by these techniques. Air spray technique provides a rougher surface that may be needed to achieve superliquiphobicity. Whereas dip coating and spin coating techniques provide a coating with more uniform thickness. If the substrate could be affected by the solvents present in the coating mixture, then spin coating or air spray may be preferable over dip coating.

The fluorinated layers have been deposited mostly by dip coating and vapor deposition. A vapor deposition technique provides a coating layer thinner than a dip coating technique.

8.3.6 Summary

To provide surface roughness, coatings consisting of nanoparticle/polymer matrix are commonly used. These are simple to fabricate and can be deposited on various substrates. For a low surface energy surface, fluorinated materials are popular due to their low adhesion. Fluoropolymers have the advantage of being able to form thick and uniform coatings. However, durability is typically poor. Fluorosilanes, fluorothiols, and fluorosurfactants are able to produce thinner films. However, they require surface activation or specific chemistry. Finally, fluorination via plasma treatments results in good surface coverage, though such treatments require specialty chemicals and conditions not typically suitable for scale-up.

In order to achieve high contact angle, even if the contact angle on the flat surface is low, even zero, re-entrant geometries can be used. A combination of fluorination and re-entrant geometry is desirable to achieve superliquiphobicity, particularly for low surface energy liquids (<20 mN/m) and surfactant-containing liquids.

Coatings are deposited by various deposition techniques including air spray, dip and spin coating.

8.4 Strategies to Achieve Combinations of Superliquiphilicity/Phobicity

Another area of interest is oil-water separation to clean up oil-spills as well as for water purification (Brown and Bhushan 2016d). For these applications, the coated surface should have affinity to one liquid and repel another. The coating can be applied on a porous membrane which repels one liquid and attracts another (Brown and Bhushan 2015a, b).

Superhydrophobic/superoleophilic porous surfaces repel water and let the oil go through. Alternatively, a superhydrophilic/superoleophobic porous surface will repel oil and let the water go through. Various approaches presented in Chap. 7 can be used to fabricate a superhydrophobic/superoleophilic surface. To fabricate a superhydrophilic/superoleophobic surface, a fluorosurfactant as a top functional layer can be used (Fig. 8.10). Fluorosurfactants contain a high surface tension head group and a low surface tension tail group. When deposited onto a surface, the fluorinated tails segregate at the air interface resulting in a low surface tension barrier that repels oils. However, when droplets of water are placed on such a surface, they are able to penetrate down through the tail groups to reach the high surface tension polar head groups below (Li et al. 2012), and thus the coating appears hydrophilic. Figure 8.15 schematically compares this so called "flip-flop" of surface properties to that of a typical "non-flip-flop" case of a fluorinated coating where penetration of neither water nor oil occurs (Brown and Bhushan 2015b). Such a "flip-flop" of surface properties can be exploited for a variety of applications, such as oil-water separation, anti-fouling, and water purification.

8.5 Model to Predict Oleophobic/Philic Nature of Surfaces

When an oil droplet is placed on a solid surface underwater, a three phase solid-water-oil interface exists. The nature of oleophobicity/philicity of an oil droplet in water can be determined from the values of surface energies of various interfaces and contact angles of water and oil in air. Jung and Bhushan (2009) performed experiments in a solid-water-oil interface. They found that some hydrophilic and oleophilic surfaces (solid-air-water interface and solid-air-oil interface) can switch into an oleophobic surface in water (solid-water-oil interface).

In this section, expressions are developed for CAs for various interfaces—solid-air-water, solid-air-oil, and solid-water-oil interfaces. If a water droplet is placed on a solid surface in air, the solid-air and water-air interfaces come together with a static CA, θ_W. The value of θ_W can be determined by minimizing the total energy of the system

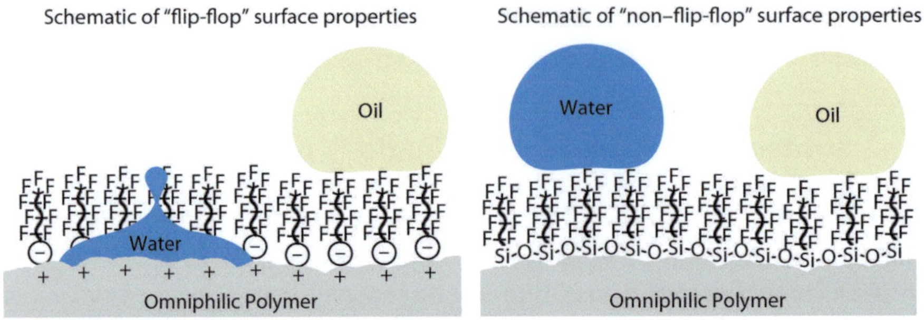

Fig. 8.15 Schematic of "flip-flop" versus "non-flip-flop" surface properties. For the "flip-flop" coating, water is able to penetrate down through the repellant surfactant tails of the functional layer (fluorosufactant) to the high surface tension portion of the coating while the bulky oil molecules are repelled. For the "non-flip-flop" coatings, water is unable to penetrate the functional layer (fluorosilane) (adapted from Brown and Bhushan 2015b)

(Adamson 1990; Israelachvili 1992), and is given by Young's equation for the CA, θ_W,

$$\cos\theta_W = \frac{\gamma_{SA} - \gamma_{SW}}{\gamma_{WA}} \tag{8.1}$$

where γ_{SW}, γ_{SA}, and γ_{WA} are surface tensions of the solid-water, solid-air, and water-air interfaces, respectively. If an oil droplet is placed on a solid surface in air, Young's equation for the CA, θ_O, can be expressed by,

$$\cos\theta_O = \frac{\gamma_{SA} - \gamma_{SO}}{\gamma_{OA}} \tag{8.2}$$

where γ_{SO}, γ_{SA}, and γ_{OA} are surface tensions of the solid-oil, solid-air, and oil-air interfaces, respectively. As predicted by (8.2), if γ_{SO} is higher than γ_{SA}, an oleophobic surface can be achieved.

To create an oleophobic surface in water, consider the solid-water-oil interface. If an oil droplet is placed on a solid surface in water, the CA of an oil droplet in water, θ_{OW}, is given by Young's equation (Israelachivili 1992),

$$\cos\theta_{OW} = \frac{\gamma_{SW} - \gamma_{SO}}{\gamma_{OW}} \tag{8.3}$$

where γ_{SO}, γ_{SW}, and γ_{OW} are the surface tensions of the solid-oil, solid-water, and oil-water interfaces, respectively. Combining (8.1), (8.2), and (8.3), the equation for the CA, θ_{OW}, of an oil droplet in water is given by Jung and Bhushan (2009),

$$\cos\theta_{OW} = \frac{\gamma_{OA}\cos\theta_O - \gamma_{WA}\cos\theta_W}{\gamma_{OW}} \tag{8.4}$$

As predicted by (8.4), an oleophobic surface can be created in the solid-water-oil interface on a hydrophilic surface ($\gamma_{SA} > \gamma_{SW}$) if $\gamma_{OA}\cos\theta_O$ is lower than $\gamma_{WA}\cos\theta_W$. Since the surface tension of oil and organic liquids is much lower than that of water, most hydrophilic surfaces can be made oleophobic in a solid-water-oil interface. For a surface that is hydrophobic ($\gamma_{SA} < \gamma_{SW}$) and oleophobic ($\gamma_{SA} < \gamma_{SO}$) in a solid-air-oil interface, an oleophobic surface in a solid-water-oil interface can be created if $\gamma_{OA}\cos\theta_O$ is higher than $\gamma_{WA}\cos\theta_W$ and vice versa. For a surface that is hydrophobic and oleophilic in a solid-air-oil interface, an oleophobic surface cannot be created in a solid-water-oil interface. Schematics of various cases for hydrophilic and hydrophobic surfaces and required surface properties are shown in Fig. 8.16, and the summary of surface properties required for philic/phobic nature in various interfaces is shown in Table 8.3.

8.6 Validation of Oleophobicity/Philicity Model for Oil Droplets in Air and Water

In this section, experiments are presented on how the oil droplets in air and in water in three phase interfaces influence wetting behavior of various surfaces.

8.6.1 Experimental Techniques

Jung and Bhushan (2009) used deionized (DI) water for the water droplet and hexadecane for the oil droplet for the measurement of the static CA in water. For analysis, the surface tensions of the water-air interface (γ_{WA}), oil-air interface (γ_{OA}), and oil-water interface (γ_{OW}) are 72 mN/m (Haynes 2014), 27.5 mN/m (Haynes 2014), and 51.4 mN/m (Tajima et al. 1980), respectively. The mass densities are 1000 and 773 kg/m^3 for water and hexadecane, respectively. Water and oil droplets of about 5 μL in volume (with radius of a spherical droplet about 1 mm) in an air environment were deposited on the specimen using a microsyringe. The process of wetting behavior of an oil droplet in water was obtained in a solid-water-oil interface system, as shown in Fig. 8.17 (Jung and Bhushan 2009). A specimen was first immersed in the water phase. Then, an oil droplet was deposited using a microsyringe from the bottom of the system, because the density of the oil (hexadecane) is lower than that of water. The image of the droplet was obtained by a digital camcorder with a 10× optical and 120× digital zoom. The images obtained were analyzed for the CA using Imagetool® software (University of Texas Health Science Center). The measurements were reproducible to within ±2°.

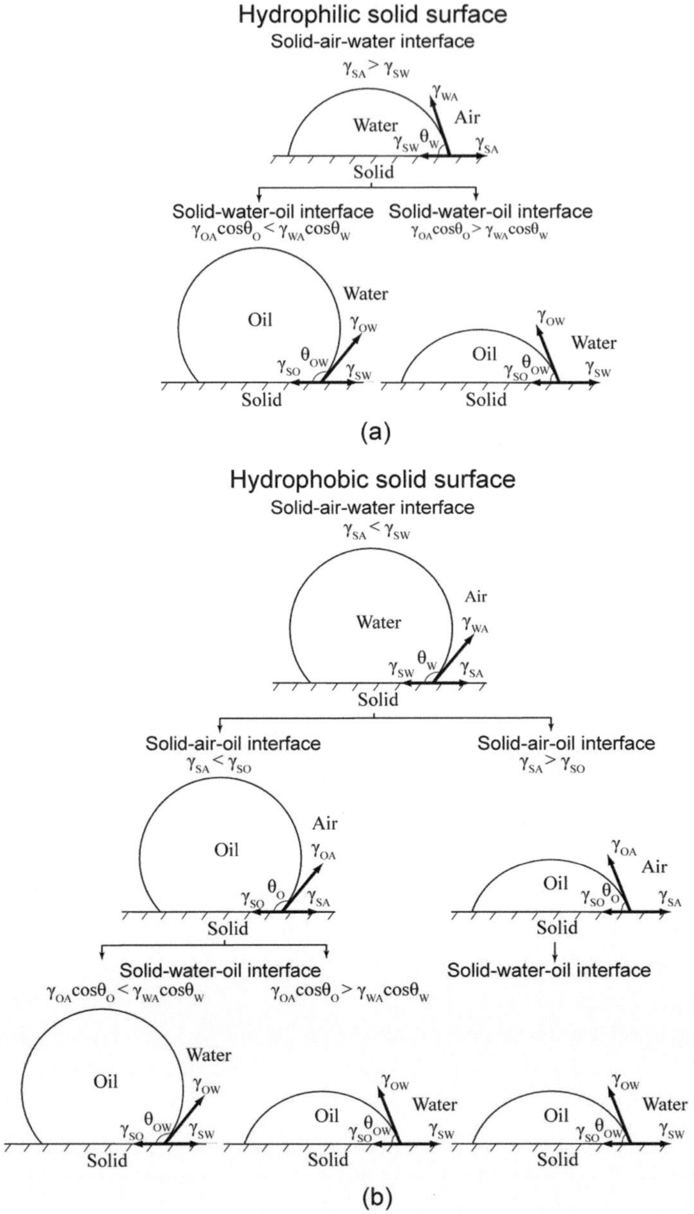

Fig. 8.16 Schematics of a droplet of liquid showing philic/phobic nature in three different phase interface on the surface—θ_W, θ_O, and θ_{OW} are static contact angles of water droplet, oil droplet, and oil droplet in water, respectively for a **a** hydrophilic solid surface, and **b** hydrophobic solid surface, and required surface properties (adapted from Jung and Bhushan 2009)

Table 8.3 Summary of surface properties required for philic/phobic nature in various interfaces (Jung and Bhushan 2009)

Solid-air-water interface		Solid-water-oil interface
Hydrophilic ($\gamma_{SA} > \gamma_{SW}$)	→	Oleophobic if $\gamma_{OA}\cos\theta_O < \gamma_{WA}\cos\theta_W$ Oleophilic if $\gamma_{OA}\cos\theta_O > \gamma_{WA}\cos\theta_W$
Solid-air-water interface	**Solid-air-oil interface**	**Solid-water-oil interface**
Hydrophobic ($\gamma_{SA} < \gamma_{SW}$) →	Oleophobic if $\gamma_{SA} < \gamma_{SO}$ →	Oleophobic if $\gamma_{OA}\cos\theta_O > \gamma_{WA}\cos\theta_W$ Oleophilic if $\gamma_{OA}\cos\theta_O < \gamma_{WA}\cos\theta_W$
	Oleophilic if $\gamma_{SA} > \gamma_{SO}$ →	Oleophilic

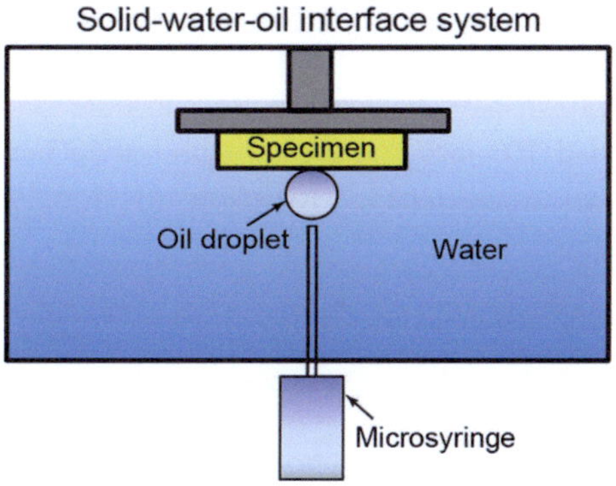

Fig. 8.17 Schematics of a solid-water-oil interface system. A specimen is first immersed in water phase, and then an oil droplet is gently deposited using a microsyringe, and the static contact angle in the system is measured (adapted from Jung and Bhushan 2009)

8.6.2 Fabrication of Oleophobic/Philic Surfaces

Micropatterned surfaces were fabricated using a two-step molding process. Microstructures were replicated using epoxy resin from master templates with varying pitch values. Master templates were created that used a flat Si surface with micropatterns fabricated by photolithography with pillars of 14 μm diameter and 30 μm height with different pitch values (21, 23, 26, 35, 70, 105, 126, 168 and 210 μm) (Jung and Bhushan 2009). To study surfaces with some oleophobicity, a surface coating that has a surface tension lower than that of oil is needed. For this purpose, Jung and Bhushan (2009) deposited *n*-perfluoroeicosane ($C_{20}F_{42}$) (268828, Sigma-Aldrich, USA) on the flat and micropatterned epoxy surfaces by thermal evaporation. The surface energy of *n*-perfluoroeicosane is 6.7 mJ/m^2 (6.7 mN/m) (Nishino et al. 1999). The specimens were mounted on a specimen holder with double-sided tape and placed in a vacuum chamber at 30 mTorr (4 kPa pressure), 2 cm above a heating plate loaded with 6000 μg *n*-perfluoroeicosane (Bhushan et al. 2009). The *n*-perfluoroeicosane was evaporated by heating it to 170 °C. In a vacuum chamber, the evaporation from the point source to the substrate occurs in a straight line; thus, the amount of sublimated material is equal in a hemispherical region over the point of the source (Bunshah 1994). In order to estimate the amount of sublimated mass, the surface area of the half sphere was calculated by using the formula $2\pi r^2$, whereby the radius (r) represents the distance between the specimen to be covered and the heating plate with the substance to be evaporated. The calculated amount of *n*-perfluoroeicosane deposited on the surfaces was 2.4 $\mu g/mm^2$ (amount of *n*-perfluoroeicosane loaded on a heating plate divided by surface area) (Jung and Bhushan 2009).

Lotus leaf inspired hierarchical structures were fabricated using fabrication of microstructured surfaces by a two-step molding process with the subsequent development of nanostructures on top by self-assembly of 0.2 $\mu g/mm^2$ of *n*-hexatriacontane ($C_{36}H_{74}$), a synthetic wax, deposited by thermal evaporation, as described in Chap. 6 (Bhushan et al. 2008, 2009). Figure 8.18a shows SEM micrographs taken at a 45° tilt angle, showing two magnifications of the micropatterned surface. Figure 8.18b shows the hierarchical structures fabricated with *n*-hexatriacontane platelets. The nanostructure is formed by three-dimensional platelets of *n*-hexatriacontane. Platelets are flat crystals, grown perpendicular to the substrate surface. The platelet thickness varied between 50 and 100 nm, and their length varied between 500 and 1000 nm. Surface energy of *n*-hexatriacontane is 31.4 mJ/m^2 (31.4 mN/m) and these surfaces are superhydrophobic and superoleophilic as will be reported later in Fig. 8.22.

8.6.3 Characterization of Oleophobic/Philic Surfaces

8.6.3.1 Wetting Behavior on Flat and Micropatterned Epoxy Surfaces

Jung and Bhushan (2009) performed experiments with water and oil droplets on hydrophilic, hydrophobic, and oleophilic surfaces in air and water to observe the wetting behavior of droplets for philic/phobic nature in air and water. Figure 8.19 shows optical micrographs of water and oil droplets in air and oil droplets in water on flat epoxy resin and micropatterned surfaces. In a solid-air-water interface with the application of a water droplet, the flat epoxy resin surface exhibited hydrophilicity and the micropatterned surface with 23 μm pitch exhibited superhydrophobicity. Jung and Bhushan (2009) reported evidence of air pocket formation between the pillars which results in a high static CA for a micropatterned surface. However, in a solid-air-oil interface with the application of an oil droplet, both surfaces exhibited oleophilicity. In the solid-water-oil interface system, in which the oil droplet sits on water trapped between the pillars, the flat epoxy resin and micropatterned surface with 23 μm pitch exhibit oleophobicity with CAs of 109° and 151°, respectively.

To study optimization of oleophobicity in the solid-air-water and solid-air-oil interfaces, the static CAs for water and oil droplets were measured on the micropatterned surfaces (Jung and Bhushan 2009). Figure 8.20 (top) shows the measured static CA as a function of pitch between the pillars for a water droplet (circle) and an oil droplet (cross) in air. The data are compared with predicted static CA values obtained using the Wenzel

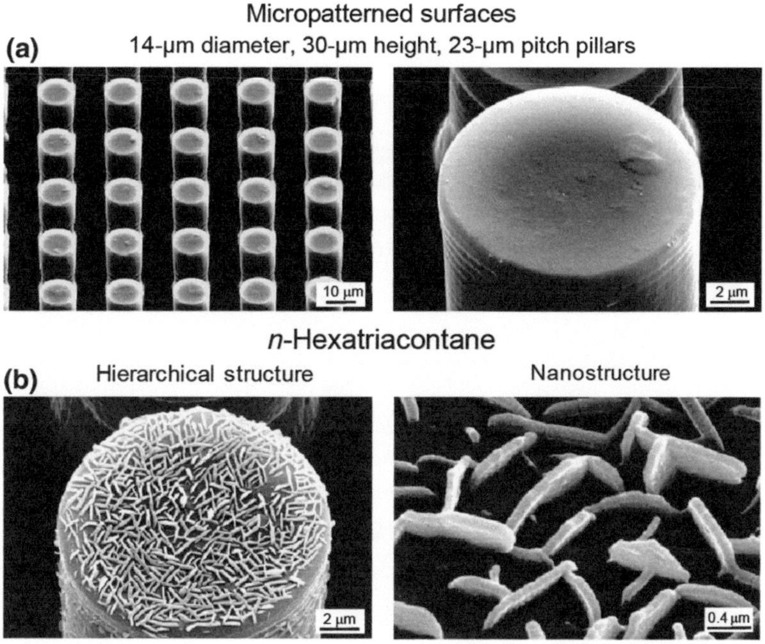

Fig. 8.18 SEM micrographs taken at a 45° tilt angle showing two magnifications of **a** the micropatterned surface, **b** hierarchical structure fabricated with three-dimensional platelets on the surface fabricated with 0.2 μg/mm^2 mass of *n*-hexatriacontane (adapted from Jung and Bhushan 2009)

and Cassie-Baxter equations (Chap. 6) (solid lines) with a measured value of θ_0 for the micropatterned surfaces. In a solid-air-water interface for a water droplet, the flat epoxy resin (pitch value = 0) showed a static CA of 76°. The static CA on micropatterned surfaces is higher than that of the flat surfaces. CA first increases with an increase in the pitch values, then starts to drop rapidly to a value slightly higher than that of the flat surfaces. In the first portion, it jumps to a high value of 150°, corresponding to a superhydrophobic surface, and continues to increase to 160° at a pitch of 26 μm because open air space between pillars increases with an increase in pitch. This open air space is responsible for the propensity of air pocket formation. The sudden drop at a pitch value of about 30 μm corresponds to the transition from the Cassie-Baxter to the Wenzel regime. The experimental observations for the transition are comparable to the value predicted from Wenzel and Cassie-Baxter equations (Jung and Bhushan 2009).

At a solid-air-oil interface for an oil droplet, the flat epoxy resin showed a static CA of 13°. As shown in Fig. 8.20 (top), with the application of an oil droplet, all micropatterned surfaces were found to be oleophilic, and the CA was lower than that of the flat surfaces. It increases with an increase in the pitch values, as predicted from Wenzel equation. As mentioned earlier, the surface tension of the oil-air interface is very low for hexadecane.

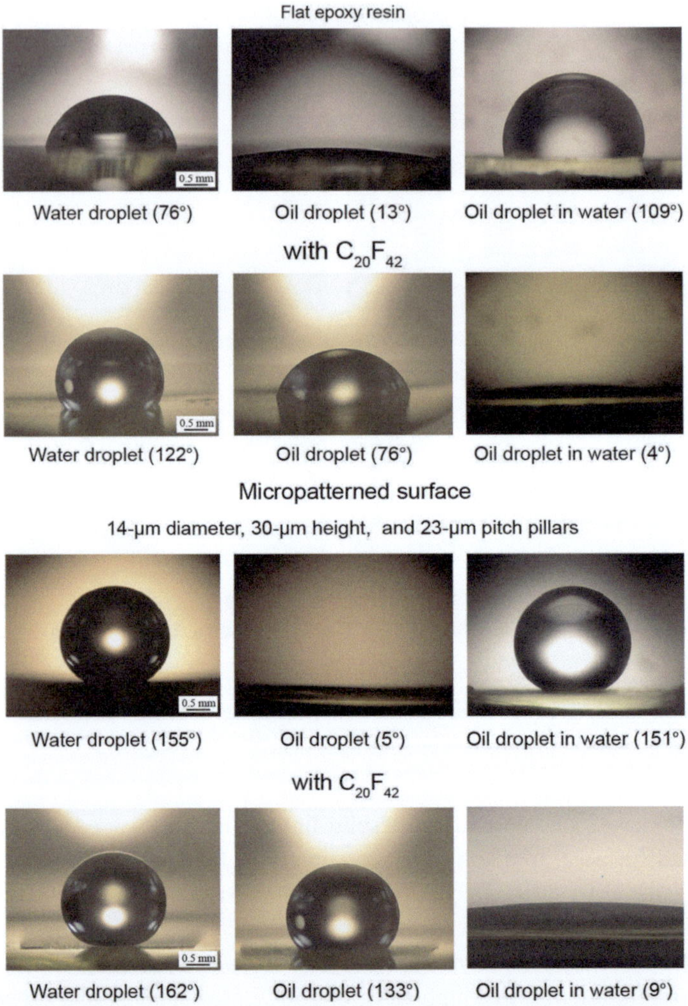

Fig. 8.19 Optical micrographs of droplets in air and under water on flat epoxy resin and micropatterned surface without and with $C_{20}F_{42}$. Left image: a water droplet is placed on a surface in air. Middle image: an oil droplet is placed on a surface in air. Right image: an oil droplet is placed on a solid surface in water (adapted from Jung and Bhushan 2009)

Therefore, it is observed that from (8.3) that the surface tension of solid-oil interface (γ_{SO}) is lower than that of solid-water interface (γ_{SW}), resulting in oleophilic state for all micropatterned surfaces (Jung and Bhushan 2009).

To study optimization of oleophobicity in a solid-water-oil interface, the static CAs for oil droplets in water were measured on the micropatterned surfaces (Jung and Bhushan, 2009). Figure 8.20 (bottom) shows the measured static CA as a function of pitch between

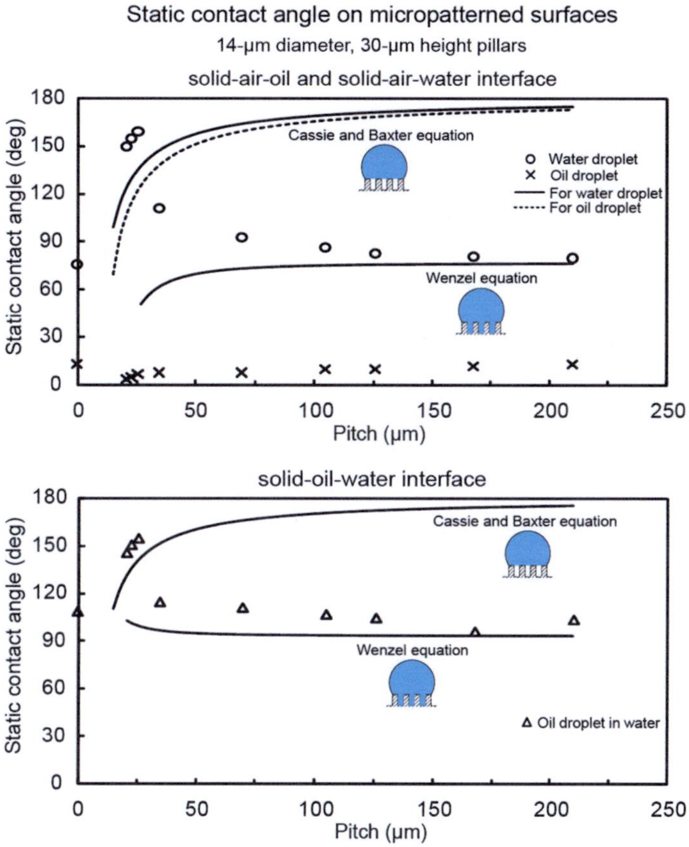

Fig. 8.20 Static contact angle as a function of pitch value for water droplet (circle) and oil droplet (cross) in air (top), and oil droplet in water (triangle) (bottom) compared with predicted static contact angle values obtained using Wenzel and Cassie-Baxter equations (solid lines) with a measured value of static contact angle for the micropatterned surfaces (adapted from Jung and Bhushan 2009)

the pillars for an oil droplet in water (triangle). The data are compared with the predicted static CA values obtained using the Wenzel and Cassie-Baxter equations (Chap. 6) (solid lines), with a measured value of θ_0 for the micropatterned surfaces. In a solid-water-oil interface, with the application of an oil droplet, the flat epoxy resin was found to be oleophobic and had a static CA of 109°. The static CA of micropatterned surfaces in the solid-water-oil interface showed a similar trend to that in the solid-air-water interface. As the pitch increases up to 26 μm, the static CA first increases gradually from 146° to 155°, because the oil droplet sits on water trapped between the pillars, and open space increases with an increase in pitch. Then, the CA starts decreasing rapidly due to the transition from the Cassie-Baxter to the Wenzel regime. The experimental observations for the transition are comparable to the values predicted from Wenzel and Cassie-Baxter equations. The micropatterned surfaces studied here were either hydrophilic or hydrophobic and both were oleophilic. In the solid-water-oil interface, they were oleophobic. However, hydrophilic surfaces became oleophobic in the solid-water-oil interface because $\gamma_{OA}\cos\theta_O$ is higher than $\gamma_{WA}\cos\theta_W$ (Jung and Bhushan 2009).

8.6.3.2 Wetting Behavior on Flat and Micropatterned Surfaces with $C_{20}F_{42}$

To study surfaces with some oleophobicity, *n*-perfluoroeicosane ($C_{20}F_{42}$), which has a lower surface tension than that of oil, was deposited on the surfaces, and experiments with droplets on hydrophobic and both oleophilic and oleophobic surfaces in air were performed (Jung and Bhushan 2009). Figure 8.19 shows the optical micrographs of droplets in three different phase interfaces on a flat epoxy resin and a micropatterned surface with $C_{20}F_{42}$. In a solid-air-water interface and a solid-air-oil interface, the water droplet and oil droplet showed CAs of 122° and 76° for the flat epoxy resin with $C_{20}F_{42}$ and CAs of 162° and 133° for the micropatterned surface with 23 μm pitch with $C_{20}F_{42}$, respectively. However, in a solid-water-oil interface with the application of an oil droplet in water, both surfaces exhibited oleophilicity and had contact angles of 4° and 9°, respectively. To explain why the oleophobic surfaces in air became oleophilic in water, the theoretical values for both surfaces were calculated using (8.4). For calculations, the surface tensions of the water-air interface (γ_{WA}), oil-air interface (γ_{OA}), and oil-water interface (γ_{OW}) were taken to be 72, 27.5, and 51.4 mN/m, and the contact angles for water and oil droplets in air were taken from the measured values. The theoretical values for the flat epoxy resin and the micropatterned surface with 23 μm pitch with $C_{20}F_{42}$ are 28° and 10°, respectively. These values are similar to those from the experiments. This indicates that the oleophobic surfaces become oleophilic in water (Jung and Bhushan 2009).

To study optimization of oleophobicity in two solid-air-water and solid-air-oil interfaces, the static CAs for water and oil droplets were measured on the micropatterned surfaces with different pitch values and with $C_{20}F_{42}$ (Jung and Bhushan 2009). Figure 8.21 shows the measured static CA as a function of pitch between the pillars for a water droplet (circle) and an oil droplet (cross) in air. The data are compared with the predicted static CA values obtained using the Wenzel and Cassie-Baxter equations (Chap. 6) (solid lines) with a measured value of θ_0 for the micropatterned surfaces with $C_{20}F_{42}$. In a solid-air-water interface for the water droplet, the flat epoxy resin with $C_{20}F_{42}$ showed a static CA of 122°. The static CA of micropatterned surfaces with $C_{20}F_{42}$ first increases from 158° to 169° with an increase in the pitch values, then drops rapidly at a pitch value of 110 μm. From comparison of the experimental data to the Wenzel and Cassie-Baxter equations, this corresponds to the transition from Cassie-Baxter to Wenzel regime. All surfaces with $C_{20}F_{42}$ had an increase in CA, and the transition took place at higher pitch value than that of the micropatterned surfaces (Fig. 8.20) (Jung and Bhushan 2009).

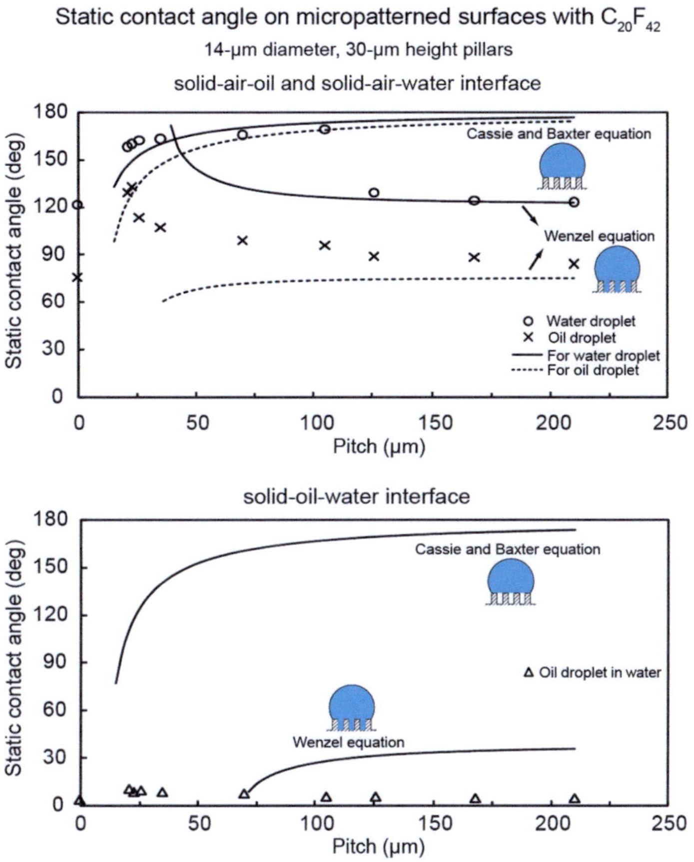

Fig. 8.21 Static contact angle as a function of pitch value for water droplet (circle) and oil droplet (cross) in air, and oil droplet in water (triangle) compared with predicted static contact angle values obtained using Wenzel and Cassie-Baxter equations (solid lines) with a measured value of static contact angle for the micropatterned surfaces with $C_{20}F_{42}$ (adapted from Jung and Bhushan 2009)

At a solid-air-oil interface for an oil droplet, the flat epoxy resin with $C_{20}F_{42}$ showed a static CA of 76°. As shown in Fig. 8.21, the highest CA of micropatterned surfaces with $C_{20}F_{42}$ was 133° at a pitch value of 23 μm. Then, it decreases with an increase in the pitch values, and these values are comparable with the values predicted by the Wenzel equations. The CAs of all micropatterned surfaces with $C_{20}F_{42}$ are higher than that of the flat surfaces (Jung and Bhushan 2009).

To study optimization of oleophobicity in a solid-water-oil interface, the static CAs for oil droplets in water were measured on the micropatterned surfaces with different pitch values and with $C_{20}F_{42}$ (Jung and Bhushan 2009). Figure 8.21 shows the measured static CA as a function of pitch between the pillars for an oil droplet in water (triangle).

The data are compared with the predicted static CA values obtained using the Wenzel and Cassie-Baxter equations (Chap. 6) (solid lines) with a measured value of θ_0 for the micropatterned surfaces with $C_{20}F_{42}$. In a solid-water-oil interface, the flat epoxy resin with $C_{20}F_{42}$ was oleophilic and had a static CA of 4°. All micropatterned surfaces with $C_{20}F_{42}$ were oleophilic and had CAs lower than 10°. The reason why hydrophobic and oleophobic surfaces in air became oleophilic in water can be explained from Fig. 8.16 and Table 8.3. The CA for a water droplet is higher than that for an oil droplet on all surfaces with $C_{20}F_{42}$, and the surface tension of the water-air interface (γ_{WA}) is higher than that of the oil-air interface (γ_{OA}). Therefore, it is observed that $\gamma_{WA}\cos\theta_W$ is higher than $\gamma_{OA}\cos\theta_O$, and then the surfaces become oleophilic in the solid-water-oil interface (Jung and Bhushan 2009).

8.6.3.3 Wetting Behavior on Nano- and Hierarchical Structures and Shark Skin Replica

To observe the wetting behavior of oil droplets for nano- and hierarchical structures found in lotus plant surfaces, experiments with the droplets on the surfaces in air and water were performed (Jung and Bhushan 2009). In addition, shark skin replicas without and with $C_{20}F_{42}$ coatings were tested. Shark skin is known to exhibit hydrophilicity and oleophobicity in water (Jung and Bhushan 2009).

Figure 8.22 shows the optical micrographs of water and oil droplets in air and oil droplets in water on a nanostructure and a hierarchical structure fabricated with 0.2 μg/mm^2 mass of *n*-hexatriacontane. Both nano- and hierarchical structures were superhydrophobic and had a static CA of 158° and 169° in the solid-air-water interface, respectively. However, they were oleophilic in the solid-air-oil interface because the surface energy of *n*-hexatriacontane is 31.4 mJ/m^2 (31.4 mN/m) (Wu 1979), and this value is higher than that of an oil droplet (hexadecane). In the solid-water-oil interface, nano- and hierarchical structures had a static CA of 10° and 5°, respectively. Based on Fig. 8.16 and Table 8.3, both surfaces are expected to be oleophilic in solid-water-oil interface (Jung and Bhushan 2009).

To study the surface structure on an aquatic animal, experiments with water and oil droplets on shark skin replica were performed in a three phase interface (Jung and Bhushan 2009). Figure 8.23 shows the optical micrographs of water and oil droplets in air and oil droplets in water on a shark skin replica without and with $C_{20}F_{42}$. First, the shark skin replica had CAs of 89° and ~0° for water and oil droplets, respectively. After the surface was coated with $C_{20}F_{42}$, the CAs of water and oil droplets became 142° and 115°, respectively. In the solid-water-oil interface with the application of an oil droplet in water, the shark skin replica was found to be oleophobic, and had a CA of 109°. Based on (8.4), the calculated value was 59° for the oil droplet in water on a shark skin replica. This difference may come from the open space under the scales of the shark skin replica (Fig. 8.2) responsible for the propensity of trapped water pocket formation. The shark skin replica with $C_{20}F_{42}$ was oleophilic and had a contact angle of ~0°. This state is the

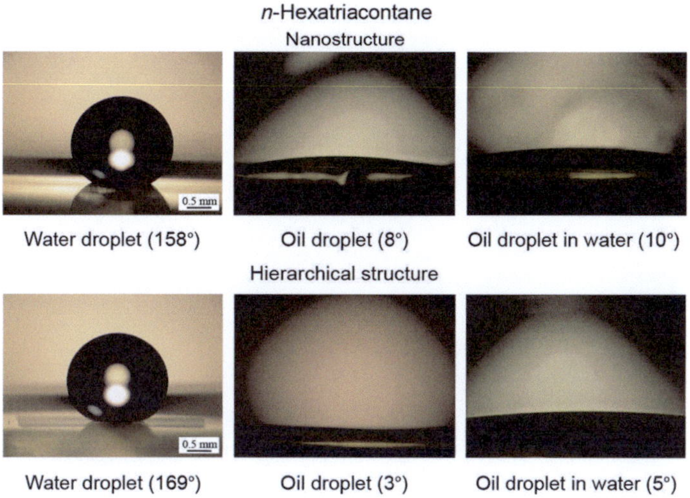

Fig. 8.22 Optical micrographs of droplets in three different phase interfaces on nanostructure and hierarchical structure fabricated with 0.2 μg/mm^2 mass of *n*-hexatriacontane. Left image: a water droplet is placed on a surface in air. Middle image: an oil droplet is placed on a surface in air. Right image: an oil droplet is placed on a solid surface in water (adapted from Jung and Bhushan 2009)

same as the micropatterned surfaces with $C_{20}F_{42}$ based on Fig. 8.16 and Table 8.3 (Jung and Bhushan 2009).

8.6.4 Summary

The wetting behavior of water and oil droplets for hydrophobic/philic and oleophobic/philic surfaces in air and water was studied. In underwater applications, oleophobicity/philicity of an oil droplet in water was studied on surfaces with different surface energies of various interfaces. Additionally, CAs of water and oil droplets in air were studied. Based on the model, it was found that for a hydrophilic surface, an oleophobic surface in the solid-water-oil interface can be created if $\gamma_{OA}\cos\theta_O$ is lower than $\gamma_{WA}\cos\theta_W$. For a hydrophobic surface and an oleophobic surface in solid-air-oil interface, an oleophobic surface in solid-water-oil interface can be created if $\gamma_{OA}\cos\theta_O$ is higher than $\gamma_{WA}\cos\theta_W$.

To validate the model for predicting the oleophobic/philic nature of the surfaces, flat and micropatterned surfaces with varying pitch values were produced by soft lithography. *n*-pefluoroeicosane ($C_{20}F_{42}$) with low surface energy (6.7 mN/m) was deposited by thermal evaporation to produce superhydrophobic and oleophobic flat and micropatterned surfaces. For water and oil droplets in water in three phase interfaces, the experimental observations showed that there was an agreement between the measured CA and the predicted values of models. It was also found that the transition from the Cassie-Baxter

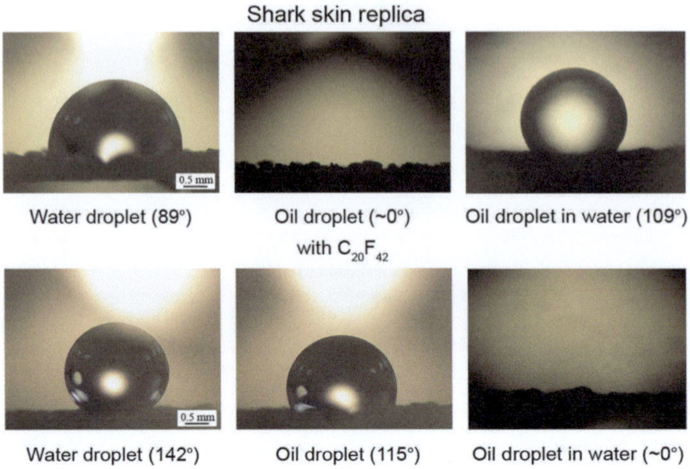

Fig. 8.23 Optical micrographs of droplets in three different phase interfaces on shark skin replica without and with $C_{20}F_{42}$. Left image: a water droplet is placed on a surface in air. Middle image: an oil droplet is placed on a surface in air. Right image: an oil droplet is placed on a solid surface in water (adapted from Jung and Bhushan 2009)

to the Wenzel regime can occur for hydrophobic and oleophobic micropatterned surfaces with larger distances between pillars.

The wetting behavior of the nano- and hierarchical structures found in lotus leaf surfaces and shark skin replicas was also investigated. It was found that the nano- and hierarchical structures with *n*-hexatriacontane were oleophilic due to the high surface energy of *n*-hexatriacontane. The structure of the shark skin replica showed a higher propensity for trapped water pocket formation, resulting in a higher CA than the theoretical value. In contrast, the shark skin replica with $C_{20}F_{42}$ had a CA of ~0° in the solid-water-oil interface, as predicted by the model. For self-cleaning and antifouling, an oleophobic surface can be created based on this study (Jung and Bhushan 2009).

8.7 Closure

Strategies to achieve superoleophobicity in air and liquid repellency are presented. Expressions for the surface properties for superoleophobic/philic surfaces have been developed for various interfaces—solid-air-water, solid-air-oil, and solid-water-oil interfaces. To validate the model, it has been investigated how the water and oil droplets in water influence the wetting behavior on various surfaces.

References

Adamson A. V. (1990), *Physical Chemistry of Surfaces*, Wiley, New York.

Al-Sabagh, A. M., Abd-El-Bary, H. M., El-Ghazawy, R. A., Mishrif, M. R. and Hussein, B. M. (2011), "Surface Active and Thermodynamic Properties of Some Surfactants Derived from Locally Linear and Heavy Alkyl Benzene in Relation to Corrosion Inhibition Efficiency," *Mater. Corros.* **62**, 1015–1030.

Anonymous (2015a), *Silicone Fluid Technical Data*, retrieved from http://www.shinetsusilicone-global.com/catalog/pdf/kf96_e.pdf, accessed July 26 2015.

Anonymous (2015b), *Fluorinert™ Electronic Liquid FC-72* retrieved from: http://multimedia.3m.com/mws/media/64892O/fluorinert-electronic-liquid-fc-72.pdf, accessed August 7 2015.

Barthlott, W. and Neinhuis, C. (1997), "Purity of the Sacred Lotus, or Escape from Contamination in Biological Surfaces," *Planta* **202**, 1–8.

Bhurke, A. S., Askeland, P. A., and Drzal, L. T. (2007), "Surface Modification of Polycarbonate by Ultraviolet Radiation and Ozone," *J. Adhes.* **83**, 43–60.

Bhushan, B. (2009), "Biomimetics: Lessons from Nature – An Overview," *Phil. Trans. R. Soc. A* **367**, 1445–1486.

Bhushan, B. and Jung, Y. C. (2011), "Natural and Biomimetic Artificial Surfaces for Superhydrophobicity, Self-Cleaning, Low Adhesion, and Drag Reduction," *Prog. Mater. Sci.* **56**, 1–108.

Bhushan, B. and Martin, S. (2018), "Substrate-independent Superliquiphobic Coatings for Water, Oil, and Surfactant Repellency: An Overview," *J. Colloid Interface Sci.* **526**, 90–105.

Bhushan, B., Koch, K., and Jung, Y. C. (2008), "Nanostructures for Superhydrophobicity and Low Adhesion," *Soft Matter* **4**, 1799–1804.

Bhushan, B., Jung, Y. C., and Koch, K. (2009), "Micro-, Nano- and Hierarchical Structures for Superhydrophobicity, Self-Cleaning and Low Adhesion," *Phil. Trans. R. Soc. A* **367**, 1631–1672

Bixler, G. D. and Bhushan, B. (2012a), "Biofouling: Lessons from Nature," *Phil. Trans. R. Soc. A* **370**, 2381–2417.

Bixler, G. D. and Bhushan, B. (2012b), "Bioinspired Rice Leaf and Butterfly Wing Surface Structures Combining Shark Skin and Lotus Effects," *Soft Matter* **8**, 11271–11284 (2012).

Bixler, G.D. and Bhushan, B. (2013a), "Fluid Drag Reduction with Shark-skin Riblet Inspired Microstructured Surfaces," (Invited), *Adv. Func. Mater.* **23**, 4507–4528.

Bixler, G. D. and Bhushan, B. (2013b), "Fluid Drag Reduction and Efficient Self-Cleaning with Rice Leaf and Butterfly Wing Bioinspired Surfaces," *Nanoscale* **5**, 7685–7710.

Bohn, H. F. and Federle, W. (2004), "Insect Aquaplaning: *Nepenthes* Pitcher Plants Capture Prey with the Peristome, a Fully Wettable Water-lubricated Anisotropic Surface," *Proc. Nat. Acad. Sci.* **101**, 14138–14143.

Brown, P. S. and Bhushan, B (2015a), "Mechanically Durable, Superoleophobic Coatings Prepared by Layer-by-Layer Technique for Anti-smudge and Oil–water Separation," *Sci. Rep. – Nature* **5**, 8701.

Brown, P. S. and Bhushan, B. (2015b), "Mechanically Durable, Superomniphobic Coatings Prepared by Layer-by-Layer Technique for Self-cleaning and Anti-smudge," *J. Colloid Interf. Sci.* **456**, 210–218.

Brown, P. S. and Bhushan, B. (2015c), "Bioinspired, Roughness-Induced, Water and Oil Super-philic and Super-phobic Coatings Prepared by Adaptable Layer-by-Layer Technique," *Sci. Rep. – Nature* **5**, 14030.

Brown, P. S. and Bhushan, B. (2016a), "Designing Bioinspired Superoleophobic Surfaces," *APL Mater.* **4**, 015703.

Brown, P. S. and Bhushan, B. (2016b), "Durable, Superoleophobic Polymer-nanoparticle Composite Surfaces with Re-entrant Geometry via Solvent-induced Phase Transformation," *Sci. Rep. – Nature* **6**, 21048.

Brown, P. S. and Bhushan, B. (2016c), "Durable Superoleophobic Polypropylene Surfaces," *Phil. Trans. R. Soc. A* **374**, 20160193.

Brown, P. S. and Bhushan, B. (2016d), "Bioinspired Materials for Water Supply and Management: Water Collection, Water Purification and Separation of Water from Oil," *Phil. Trans. R. Soc. A* **374**, 20160135.

Brown, P. S. and Bhushan, B. (2017a), "Liquid-impregnated Porous Polypropylene Surfaces for Liquid Repellency," *J. Colloid Interf. Sci.* **487**, 437–443.

Brown, P. S. and Bhushan, B. (2017b), "Mechanically Durable Liquid-impregnated Honeycomb Surfaces," *Sci. Rep. – Nature* **7**, 6083.

Bunshah, R. F. (1994), *Handbook of Deposition Technologies for Films and Coatings: Science, Technology and Applications*, Applied Science Publishers, Westwood, New Jersey.

Cheng, Q., Li, M., Zheng, Y., Su, B., Wang, S. and Jiang, L. (2011), "Janus Interface Materials: Superhydrophobic Air/solid Interface and Superoleophobic Water/solid Interface Inspired by a Lotus Leaf," *Soft Matter* **7**, 5948–5951.

Dean, B. and Bhushan, B. (2010), "Shark-Skin Surfaces for Fluid-Drag Reduction in Turbulent Flow: A Review," *Phil. Trans. Roy. Soc. A* **368**, 4775–4806.

Efimenko, K., Wallace, W. E. and Genzer, J. (2002), "Surface Modification of Sylgard-184 Poly(dimethylsiloxane) Networks by Ultraviolet and Ultraviolet/ozone Treatment," *J. Colloid Interface Sci.* **254**, 306–315.

Flamm, D. L. and Auciello, O. (2012), *Plasma Deposition Treatment, and Etching of Polymers: The Treatment and Etching of Polymers*, ed. R. d'Agostino, Academic Press, London.

Genzer, J. and Efimenko, K. (2006), "Recent Developments in Superhydrophobic Surfaces and Their Relevance to Marine Fouling: A Review," *Biofouling* **22**, 339–360.

Haynes, W. M. (2014), *CRC Handbook of Chemistry and Physics* 95th Ed., Taylor and Francis Group, Boca Raton, FL.

Hensel, R., Helbig, R., Aland, S., Braun, H.-G., Voig, A., Neinhuis, C. and Werner, C. (2013) "Wetting Resistance at Its Topographical Limit: The Benefit of Mushroom and Serif T Structures," *Langmuir* **29**, 1100–1112.

Im, M., Im, H., Lee, J.-H., Yoon, J.-B. and Choi, Y.-K. (2010), "A Robust Superhydrophobic and Superoleophobic Surface with Inverse-Trapezoidal Microstructures on a Large Transparent Flexible Substrate," *Soft Matter* **6**, 1401–1404.

Israelachvili, J. N. (1992), *Intermolecular and Surface Forces*, second edition, Academic Press, London, U. K.

Jasper, J. J., (1972), The Surface Tension of Pure Liquid Compounds," *J. Phys. Chem. Ref. Data* **1**, 841–1009.

Jung, Y. C. and Bhushan, B. (2009), "Wetting Behavior of Water and Oil Droplets in Three Phase Interfaces for Hydrophobicity/philicity and Oleophobicity/philicity," *Langmuir* **25**, 14165–14173.

Kang, S. M., Kim, S. M., Kim, H. N., Kwak, M. K., Tahk, D. H. and Suh, K. Y. (2012), "Robust Superomniphobic Surfaces with Mushroom-like Micropillar Arrays," *Soft Matter* **8**, 8563–8568.

Koch, K., Bhushan, B., and Barthlott, W. (2009), "Multifunctional Surface Structures of Plants: An Inspiration for Biomimetics," *Prog. Mater. Sci.* **54**, 137–178.

Lee, M.-T., Hsueh, C.-C., Freund, M. S. and Ferguson, G. S. (1998) "Air Oxidation of Self-Assembled Monolayers on Polycrystalline Gold: The Role of the Gold Substrate," *Langmuir* **14**, 6419–6423.

Lee, S. E., Kim, H. J., Lee, S. H., and Choi, D. G. (2013), "Superamphiphobic Surface by Nanotransfer Molding and Isotropic Etching," *Langmuir* **29**, 8070–8075.

Li, L., Wang, Y., Gallaschun, C., Risch, T. and Sun, J. (2012), "Why Can a Nanometer-Thick Polymer Coated Surface be More Wettable to Water than to Oil?" *J. Mater. Chem.* **22**, 16719–1672.

Liu, M. J., Want, S. T., Wei, Z. X., Song, Y. L, and Jiang, L. (2009), "Bioinspired Design of a Superoleophobic and Low Adhesive Water/Solid Interface," *Adv. Mater.* **21**, 665–669.

Liu, T. and Kim, C. J. (2014), "Turning a Surface Super-repellant Even to Completely Wetting Liquids," *Science* **346**, 1096–1100.

Manning, R. and Ewing, J. (2009), *Temperature in Cars Survey,* Royal Automobile Club of Queensland Limited (RACQ) Vehicle Testing Authority (Brisbane, Australia). https://www.racq.com.au/~/media/pdf/racq%20pdfs/cars%20and%20driving/driving/0814_temperature_in_cars_survey_2009.ashx.

Martin, S. and Bhushan, B. (2017), "Transparent, Wear-resistant, Superhydrophobic and Superoleophobic Poly(dimethylsiloxane) (PDMS) Surfaces," *J. Colloid Interface Sci.* **488**, 118–126 (2017).

Martin, S., Brown, P. S., and Bhushan, B. (2017), "Fabrication Techniques for Bioinspired, Mechanically-durable, Superliquiphobic Surfaces for Water, Oil, and Surfactant Repellency," *Adv. Colloid Interface Sci.* **241**, 1–23.

Muthiah, P., Bhushan, B., Yun, K. and Kondo, H. (2013), "Dual-Layered-Coated Mechanically-Durable Superomniphobic Surfaces With Anti-Smudge Properties," *J. Colloid Interface. Sci.* **409**, 227–236.

Neinhuis, C. and Barthlott, W. (1997), "Characterization and Distribution of Water-Repllent, Self-Cleaning Plant Surfaces,"*Annals of Botany* **79**, 667–677.

Nishino, T., Meguro, M., Nakamae, K., Matsushita, M., and Ueda, Y. (1999), "The Lowest Surface Free Energy Based on $-CF_3$ Alignment," *Langmuir* **15**, 4321–4323.

Nosonovsky, M. and Bhushan, B. (2008), *Multiscale Dissipative Mechanisms and Hierarchical Surfaces: Friction, Superhydrophobicity, and Biomimetics*, Springer, Heidelberg, Germany.

Park, E. J., Carroll, G. T., Turro, N. J. and Koberstein, J. T. (2009), " Shedding Light on Surfaces-Using Photons to Transform and Pattern Material Surfaces, *Soft Matter* **5**, 36–50.

Qian, B. and Shen, Z. (2005), "Fabrication of Superhydrophobic Surfaces by Dislocation-Selective Chemical Etching on Aluminum, Copper, and Zinc Substrates," *Langmuir* **21**, 9007–9009.

Qu, M., Zhang, B., Song, S., Chen, L., Zhang, J. and Cao, X. (2007), "Fabrication of Superhydrophobic Surfaces on Engineering Materials by a Solution-Immersion Process," *Adv. Funct. Mater.* **17**, 593–596.

Rakitov, R. and Gorb, S. N. (2013), "Brochosomal coats turn leafhopper (Insecta, Hemiptera, Cicadellidae) integument to superhydrophobic state," *Proc. R. Soc. B* **280**, 20122391, pp 1–9.

Ross, J. and Epstein, M. B. (1958), "Surface Tension and Surface Transition of Dilute Aqueous Solutions of Lauryl Alcohol and Sodium Lauryl Sulfate", *J. Phys. Chem* **62**, 533–535.

Song, J., Huang, S., Hu, K., Lu, Y., Liu, X. and Xu, W. (2013), "Fabrication of Superoleophobic Surfaces on Al Substrates," *J. Mater. Chem. A* **1**,14783–14789.

Soontravanich, S., Munoz, J. A. and Scamehorn, J. F. (2008), "Interaction Between an Anionic and an Amphoteric Surfactant. Part I: Monomer–micelle Equilibrium," *J. Surfact. Deterg.* **11**, 251–261.

Tajima, K., Tsutsui, T., and Murata, H. (1980), "Thermodynamic Relation of Interfacial Tensions in Three Fluid Phases," *Bull. Chem. Soc. Jpn.* **53**, 1165–1166.

Tuteja, A., Choi, W., Mabry, J. M., Mckinley, G. H., and Cohen, R. E. (2008), "Robust Omniphobic Surfaces," *Proc. Natl. Acad. Sci. U.S.A.* **105**, 18200–18205.

Ulman, A. (1996), "Formation and Structure of Self-Assembled Monolayers," *Chem. Rev.* **96**, 1533–1554.

Vig, J. R. (1985), "UV/ozone Cleaning Surfaces," *J. Vac. Sci. Technol. A* **3**, 1027–1034.
Wang, Y. and Bhushan, B. (2015), "Wear-Resistant and Antismudge Superoleophobic Coating on Polyethylene Terephthalate Substrate Using SiO_2 Nanoparticles," *ACS Appl. Mater. Interf.* **7**, 743–755.
Wu, S. (1979), "Surface-Tension of Solids-Equation of State Analysis," *J. Colloid Interface Sci.* **71**, 605–609.
Zhao, H., Law, K.-Y. and Sambhy, V. (2011), "Fabrication, Surface Properties, and Origin of Superoleophobicity for a Model Textured Surface," *Langmuir* **27**, 5927–5935.

9 Adaptable Fabrication Techniques for Mechanically Durable Superliquiphobic/Philic Surfaces

9.1 Introduction

Superliquiphobic and superliquiphilic surfaces have attracted interest in the scientific community and industry. These are being considered for applications which require self-cleaning/anti-smudge/fingerprint resistance, anti-icing, antifogging, optical transparency, low adhesion, drag reduction in fluid flow, and anti-fouling properties (Bhushan 2009, 2012, 2016, 2018, 2019a). Self-cleaning applications include self-cleaning windows, windshields, smart screens, camera lenses, exterior paint for buildings, utensils, roof tiles, textiles, solar panels, self-cleaning shampoo and detergent bottles, and exterior surfaces in transportation. Anti-smudge and fingerprint resistance properties are important in applications which require finger touch such as in smart devices and automotive and aerospace interior and exterior surfaces. Anti-icing properties are of interest in applications such as automobiles and aircraft surfaces which can be exposed in winter interest to temperatures as low as −60 °C. Antifogging properties are of interest in optical applications such as glass windows and eyeglasses. Optical transparency is of interest in electronic mobile devices and many other optical surfaces.

Low drag surfaces are of interest in both internal and external fluid flow applications. Major application is in transportation such as boat hulls, automobile skin and aircraft skin. Low adhesion/low drag surfaces also exhibit anti-fouling of interest in applications such as biomedical and food packaging as well as membranes for desalination and water purification. These surfaces can also be used for energy conservation because of low adhesion/drag, and energy conversion by using switchable superliquiphobic/philic surfaces (Bhushan 2018, 2019a; Li et al. 2019; Pan et al. 2019).

Surfaces with repellency of water or oil and affinity of the other can be used for oil–water separation for use in oil spill cleanup and water purification (Bhushan 2018, 2019b, 2020).

B. Bhushan, *Introduction to Biomimetics and Bioinspiration*,
https://doi.org/10.1007/978-3-031-62344-8_9

For industrial applications, surfaces should be durable. In addition, it is desired that the surfaces have self-healing properties for a long life. Species found in nature self-heal (Cremaldi and Bhushan 2018a). Self-healing may occur by intrinsic or extrinsic mechanisms. In intrinsic mechanism, self-healing materials consist of autonomous self-healing capabilities. In the extrinsic mechanism, self-healing occurs by chemical changes caused by an external stimulus. Hydrogels are commonly used materials, which heal by an external stimulus of humidity. Owing to high hydrophilicity, the hydrogels can absorb a large quantity of water and swell after exposure to humidity. Multanen and Bhushan (2019) demonstrated self-healing properties of nanoparticle-hydrogel coated surfaces.

Figure 9.1 shows the amphi-wettability landscape, first introduced by Bhushan (2016). On the horizontal axis, the water contact angle is plotted and on the vertical axis, oil contact angle is plotted. Four quadrants represent various combinations of water and oil repellency and affinity. The figure also shows various applications for surfaces lying in each quadrant.

Surfaces that repel water (hydrophobic) with low contact angle hysteresis are useful for self-cleaning applications (Bhushan and Jung 2011). Surfaces that attract water

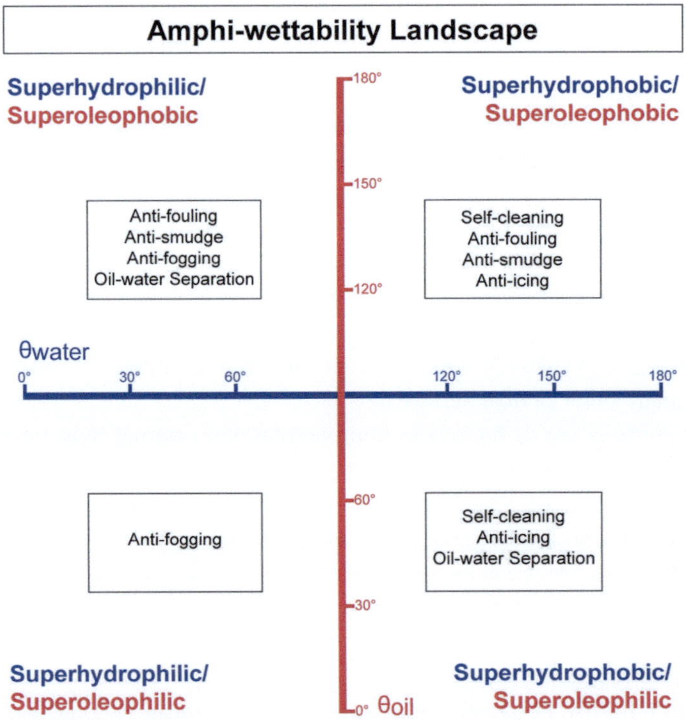

Fig. 9.1 Amphi-wettability landscape with four combinations of superliquiphilicity and superliquiphobicity (Adapted from Bhushan 2016)

(hydrophilic) have also been reported to be useful for self-cleaning because of spreading water into a thin film which cleans the surface (Nishimoto and Bhushan 2013). However, superhydrophobic surfaces with low contact angle hysteresis are known to be more efficient in self-cleaning.

Superhydrophobic surfaces with low contact angle hysteresis are also used for anti-icing as supercooled droplets of water can roll off the cold surface before freezing and any ice formed is weakly adhered due to an air cushion, as compared to hydrophilic surfaces (Cao et al. 2009; Wang et al. 2013; Brown and Bhushan 2015c; Cremaldi and Bhushan 2018a; Bhushan and Multanen 2019). After exposure to sub-zero temperatures for extended periods, superhydrophobic surfaces become wet. Condensation and frost formation may exhibit with high ice adhesion. For these applications, lower surface energy coatings (superoleophobic surfaces) with low contact angle hysteresis are desirable, which may prevent freezing during long exposures (Brown and Bhushan 2015c; Kreder et al. 2016; Cremaldi and Bhushan 2018b; Bhushan and Multanen 2019).

Hydrophilic surfaces are useful for anti-fogging applications as any water spreads out into a thin film, thereby maintaining high optical transmittance (Cebeci et al. 2006; Brown and Bhushan 2015c). Hydrophobic surfaces have also been reported for anti-fogging applications (Sun et al. 2014). However, hydrophilic surfaces are more favorable for anti-fogging, as hydrophobic surfaces require tilting the surface for the droplets to roll off, which maintains high optical transmittance.

Oleophobic surfaces with surface energy lower than that of surface tension of oil will repel both oil and water, and are useful for anti-fouling, for example, in medical and transport applications, where both the oil-repellency and nanostructure are of importance (Jung and Bhushan 2009; Bixler et al. 2014; Brown and Bhushan 2015b). Oil repellency also makes these surfaces ideal for anti-smudge applications where the oils from fingers are not deposited onto the surface and the surface remains clear (Muthiah et al. 2013; Wang and Bhushan 2015; Brown and Bhushan 2015a, b, c; Martin and Bhushan 2017; Bhushan and Martin 2018). Water repellency means these coatings can also be used in self-cleaning and anti-icing applications.

Coatings such as nonfluorinated silane with surface tensions lower than that of water (72 mN/m) but higher than that of oils (20–30 mN/m) will attract oils (oleophilicity) but repel water (hydrophobicity) and can be used to create oil–water separators (Brown and Bhushan 2015c; Bhushan 2018, 2019b, 2020). When applied to a porous substrate, the coating will allow the passage of oil but block the passage of water, resulting in their separation.

As mentioned earlier, surfaces with surface energy lower than that of surface tension of oils, will repel both oil and water. However, a hydrophilic/oleophobic coating that repels oils but attracts water can be created using a fluorosurfactant (Goddard 1986; Thünemann and Lochhaas 1999; Hutton et al. 2000; Brown and Bhushan 2015a; Saito et al. 2015). Brown and Bhushan (2015a) used a fluorosurfactant containing a high surface tension head group and a low surface tension tail group. When deposited onto a surface, the polar head group forms an electrostatic complex with the polyelectrolyte layer below and the fluorinated tails segregate at the air interface resulting in a low surface tension barrier that repels oils. When an oil–water mixture is poured over the surface, large bulky oil molecules are trapped at this fluorinated interface while smaller molecules can move easily to penetrate down through the tail groups to the hydrophilic region. The result is a "flip-flop" of surface properties and a coating that repels oils but is wet by water (Li et al. 2012; Brown and Bhushan 2015a, c). As discussed in Chap. 10, it is different from the "non–flip-flop" case of a fluorinated coating where penetration of neither water nor air occurs (Brown and Bhushan 2015c). Such a "flip-flop" of surface properties can be exploited for a variety of applications, such as oil–water separation and anti-fouling.

For oil–water separation, the use of hydrophilic/oleophobic coatings is preferable over alternative configurations where the surface is hydrophobic/oleophilic, because surface contamination by oil and other oil-based contaminants is common, and the porous material must then be cleaned or replaced, resulting in a drop in the separation efficiency. It is therefore preferable for the water phase to be allowed to pass through the mesh and the oil phase be repelled. Additionally, water is denser than oil and tends to sink to the bottom of a mixture meaning hydrophobic/oleophilic materials are not suitable for certain applications, such as gravity-driven separation (Bhushan 2018, 2019b, 2020). An overview of various common methods for oil spill cleanup and water purification is presented in Appendix 9.A (Bhushan 2018).

Superliquiphobic/philic properties have applications in many industries which include automotive, aerospace, electronics, packaging, and biomedical devices. When the liquid-repellent surfaces are also transparent, additional applications include smart screens for electronic display, camera lenses, window glass, and solar panels. A range of materials are used in these applications which require superoleophobic surfaces. Table 9.1 lists various substrates and some possible industrial applications. As an example, materials of interest for smart screens in electronic displays include soda-lime silica, polyethylene terephthalate (PET), and polycarbonate (PC). For plastic bottles, materials include polypropylene (PP), polyethylene terephthalate (PET), and high-density polyethylene (HDPE), and caps are generally made of PP. For automotive applications, materials include glass, PC, polydimethylsiloxane (PDMS), PET, nylon, HDPE, polyurethane, polymethyl methacrylate (PMMA), PP, steel, aluminum, and leather. For packaging, PP is commonly used. It is clear, liquid repellent coatings need to be deposited on various substrates for industrial applications (Bhushan 2013, 2017a, b). For all applications, surfaces should be mechanically durable and, in many cases, resistant to chemicals and/or light (Bhushan 2013, 2018, 2019a, b).

As discussed in Chap. 8, in living nature, an example for superhydrophobicity includes lotus leaf (*Nelumbo nucifera*) and for liquid repellency includes the pitcher plant (*Nepenthes*). This observation has led to two bioinspired approaches being pursued: roughness-induced and liquid impregnation. For roughness-induced superliquiphobicity, fluorinated materials are combined with surfaces having re-entrant geometries for

Table 9.1 Selected materials and some of their industrial applications (Bhushan 2018)

Material	Applications
Glass	Windshield, window glass, side/rear view mirrors, camera lenses, smart screens in electronic displays, solar panels
PP	Door liners, fabric, carpet, plastic caps, plastic bottles, plastic sheets
PDMS	Biomedical devices, catheters, drainage tubing, micropumps, microvalves, electronic circuits
PET	Plastic bottles, smart screens in electronic displays
Nylon	Automotive dash
PU	Automotive dash, automotive seats
PMMA	Smart screens for capacitive touch material in electronic displays, headlights
PE	Plastic bottles
PC	Catheters, smart screens in electronic displays, automotive dash
Leather	Seats and many others
Steel and Al	Many

improved repellency. Re-entrant geometries are shapes that have overhang structures where the surface features become narrower at the base. Re-entrant curvatures can be inverse trapezoidal, cylindrical, spherical, etc. and lead to higher CA than non-re-entrant geometries (Nosonovsky and Bhushan 2008; Brown and Bhushan 2016a). These geometries are necessary for repelling low surface tension liquids, and particularly surfactant-containing liquids such as shampoos and laundry detergents due to their low-tension components and active groups. To fabricate liquid repellent surfaces using liquid impregnation approach, the liquid used for impregnation of porous surface should be immiscible with the liquid being repelled (Brown and Bhushan 2017a, b; Bhushan 2018, 2019a).

The two bioinspired strategies—roughness-induced superliquiphobicity/philicity and liquid impregnation techniques—have been used to develop mechanically durable surfaces. In roughness-induced techniques, nanoparticles are used to provide roughness, and are divided into three categories: (1) nanoparticle-binder, (2) layer-by-layer technique and (3) nanoparticle-encapsulation technique (Brown and Bhushan 2015b, 2016b; Martin et al. 2017; Bhushan and Martin 2018).

In this chapter, various characterization techniques are presented first, followed by four fabrication techniques developed for mechanically durable superliquiphobic and superliquiphilic surfaces in air. Selected data for liquid repellency, self-cleaning, anti-smudge, fingerprint resistance, anti-icing, anti-fogging, optical transparency, drag reduction, and oil–water separation, primarily for glass and polypropylene substrates are presented. Finally, the four fabrication techniques are compared.

For completeness, Chap. 13 presents data on anti-biofouling properties of nanopatterned superhydrophobic surfaces with sharp edges. For anti-biofouling measurements, E. Coli bacteria was used. In Fig. 13.23, the percent reduced coverage of biofouling compared to that of flat uncoated surface was about 30%. A model has been proposed to explain this reduction (Bixler et al. 2014; Bhushan 2018, 2019a).

9.2 Characterization Techniques

Techniques used for characterization of the samples are described based on Wang and Bhushan (2015), Brown and Bhushan (2015a, c), Bhushan (2018, 2019a, b), Bhushan and Martin (2018), and Bhushan and Multanen (2019). These techniques include contact and tilt angle measurements, scanning electron microscope imaging, coating thickness, surfactant-containing liquid repellency, high temperature superliquiphobicity, wear resistance, self-cleaning, anti-smudge, fingerprint resistance, anti-fogging, anti-icing, optical transparency, and oil–water separation.

9.2.1 Contact Angle and Tilt Angle

Contact angle (CA) and tilt angle (TA) data were measured using a standard automated goniometer (Model 290, Ramé-Hart Inc.) using 5 μL DI water and hexadecane droplets and others deposited onto the samples using a microsyringe. Contact angle was measured by taking a static profile image of the liquid–air interface and was analyzed using DROPimage software. Tilt angle refers to the angle when the droplet just began to roll off the sample surface. All angle measurements were averaged over at least five measurements on different areas of the sample and reported as $\pm\sigma$ (Bhushan and Jung 2011).

9.2.2 Surface Morphology

Top down, scanning electron microscope images were taken to determine the topography of the glass and PP samples with the fluorinated nanoparticle-binder coating. To image the re-entrant geometry, SEM images were taken at a 70° angle. Samples were mounted with conductive tape and gold-coated prior to imaging using a Pelco sputter coater for the glass samples and a Leica ACE600 sputter coater for the PP samples. A 10 nm thick gold film was used for minimal charging effects (Brown and Bhushan 2016b, c).

The coating thickness was measured using a step technique. One half of the substrate was covered with a glass slide using double-sided sticky tape before coating and then removed after the coating procedure resulting in a step. A scanning area of 100 by 100 μm^2 including the step was imaged using a D3000 Atomic Force Microscopy (AFM) with a Nanoscope IV controller (Bruker Instruments) to obtain the coating thickness. A Si, n-type (Si_3N_4) tip (resonant frequency $f = 66$ kHz, spring constant $k = 3$ N/m, AppNano) operating in contact mode was used (Muthiah et al. 2013).

9.2.3 Surfactant-Containing Liquid Repellency

To test the repellency of surfaces towards surfactant-containing liquids, droplets of shampoo (Head and Shoulders®, Procter and Gamble Co.) and/or laundry detergent (Tide®, Procter and Gamble Co.) were placed onto surfaces tilted at ~ 25° (Brown and Bhushan 2016b, c). The resulting surface–liquid interaction was photographed. For repellency after prolonged contact with shampoo, coated glass samples were submerged into a container of shampoo for various periods of time. Before testing, the back side of the glass was covered with dark construction paper so that when the coating was photographed the shampoo on the back side would not be seen. The sample was submerged for one minute and before, during, and after photographs were taken. In another test, the sample was dipped into shampoo 10 times with 3 s contact with the shampoo each time. A photograph after each removal from the shampoo was taken.

9.2.4 High Temperature Superliquiphobicity

Liquiphobicity characteristics of the coating at high temperature were examined using a hot plate. A beaker containing water or hexadecane and a coated glass sample was placed on the hot plate. Temperatures were recorded using a digital thermometer placed in the beaker. The digital thermometer did not touch the sides of the beaker, and the liquid in the beaker was constantly stirred. Using a dropper, a droplet of liquid from the beaker was placed on the sample. An image of the droplet was taken, and the resulting CA was recorded. The droplet temperature was verified with the digital thermometer and was found to be the same as the temperature in the beaker.

9.2.5 Wear Resistance

Wear resistance of the surface has been examined both on macroscale using a ball-on-flat tribometer (Bhushan 2013) and scratching the surface with a tweezer, and on microscale using an atomic force microscope (AFM) in contact mode (Bhushan 2017b).

Macroscale Wear

For macroscale wear experiments, a 3 mm diameter sapphire ball was fixed in a stationary holder. A load of 10 mN was applied normal to the surface, and the tribometer was put into a reciprocating motion for 100 cycles (Bhushan 2013). The stroke length was at least 6 mm with an average linear speed of 1 mm/s. The surface was imaged before and after the experiment using an optical microscope with camera (Nikon Optihot-2) (Bhushan and Jung 2011).

To further investigate the effect of any wear damage on the wettability, hexadecane tilt angles were measured before and after the wear experiments (Brown and Bhushan 2015a; Wang and Bhushan 2015). The hexadecane droplets were dragged over the wear track to examine for any impediment of the motion of the droplet. In another measurement, the surface was tilted for the droplet to roll over the wear track and any increase in the tilt angle was measured. In a third measurement, the droplet was deposited directly onto the wear track and the surface was tilted to measure any increase in the tilt angle or pinning of the droplet.

In addition to ball-on-flat tribometer experiments, the coating was deliberately scratched with tweezers and wettability measurements were made on the scratched surface, as described above (Brown and Bhushan 2015a; Wang and Bhushan 2015).

Microscale Wear

For microscale wear experiments, a commercial AFM (Dimension 3000, Nanoscope IV controller, Bruker Instruments) was used (Bhushan 2017b). Surfaces were worn with a 30-μm diameter borosilicate glass ball mounted on a rectangular silicon cantilever with a nominal spring contact of 7.4 Nm^{-1}. Areas of 50×50 μm were worn for 1 cycle at a load of 10 μN. To analyze the change in morphology of the surface before and after the wear experiment, height scans of 100×100 μm in area were obtained using an Si, n-type (Si_3N_4) tip mounted on a rectangular cantilever with a nominal spring constant of 3 Nm^{-1} (AppNano) operating in tapping mode. Root mean square (RMS) roughness values before and after wear experiments were obtained.

Contact Pressures

Contact pressures for the tribometer and AFM wear experiments were calculated based on Hertz analysis (Bhushan 2013). Elastic moduli and Poisson's ratio of the coating and balls were needed.

The rule of mixtures was used for calculations for properties of the composite coating (Bhushan and Gupta 1991). An elastic modulus of about 3.5 MPa and a Poisson's ratio of 0.5 for methylphenyl silicone resin (Ellis and Smith 2008) were used and an elastic modulus of about 70 GPa and a Poisson's ratio of 0.2 for SiO_2 nanoparticles (Callister and Rethwisch 2013) were used. Calculated values for the composite coating were an elastic modulus of 32 GPa and Poisson's ratio of 0.36. An elastic modulus of 390 GPa and Poisson's ratio of 0.23 were used for the sapphire ball (Bhushan and Gupta 1991), and the mean contact pressure between the saphhire ball and the composite coating at a normal load of 10 mN was calculated as 65 MPa. An elastic modulus of 70 GPa (Callister and Rethwisch 2013) and Poisson's ratio of 0.2 (Bhushan and Gupta 1991) were used for

Table 9.2 Estimated contact stresses in wear/fingerprint resistance tests using the Hertz analysis

	Microscale wear with AFM	Macroscale wear with tribometer	Fingerprint resistance with rubber finger	Fingerprint resistance with actual thumb
Load	10 μN	10 mN	5 N	100 N
Wear test properties	Borosilicate ball $E = 70$ GPa[a] $v = 0.2$[a] $d = 30$ μm	Sapphire ball $E = 390$ GPa[b] $v = 0.23$[b] $d = 3$ mm	Rubber nub $E = 2$ MPa[c] $v = 0.5$[c] $d = 0.5$ mm	Skin ridge $E = 70$ GPa[d] $v = 0.5$[e] $d = 0.2$ mm[f]
Coating properties	Nanoparticle-binder composite coating $E = 32$ GPa $v = 0.36$[g]			
Mean contact pressure	115 MPa	65 MPa	1.0 MPa	70 MPa

[a]Callister and Rethwisch (2013)
[b]Bhushan and Gupta (1991)
[c]Mark (1999) Rubber used was styrene-butadiene (synthetic rubber). Estimated nubs across imprinted surface is 40 nubs based on imprint photographs
[d]Edwards and Marks (1995)
[e]Bhushan (2016)
[f]Stucker et al. (2001)
[g]Ellis and Smith (2008) Young's modulus and Poisson's ratio estimated using rule of mixture for composites where methylphenyl silicone has $E = 3.5$ MPa and $v = 0.5$ and SiO_2 NPs have $E = 70$ GPa and $v = 0.2$
(Adapted from Bhushan 2013; Bhushan and Martin 2018)

borosilicate glass ball, and the mean contact pressure between the borosilicate glass and the composite coating at normal load of 10 μN was calculated as 115 MPa. The data are summarized in Table 9.2.

9.2.6 Self-cleaning

The self-cleaning characteristics were examined by first contaminating the sample with silicon carbide particles and then comparing the removal of particles by water droplets before and after the experiment. Silicon carbide particles of size 10–15 μm were dispersed in a glass chamber (0.3 m diameter and 0.6 m high) by blowing 1 g of SiC powder for 10 s at 300 kPa. After dispersion, the particles were allowed to settle on the sample mounted on a 45° tilted stage for 30 min. The contaminated sample was then secured to a 10° stage and water droplets (total volume of 5 mL) were dropped onto the surface from 1 cm in height. The removal of particles by the water droplets was compared before and after tests. The ability for the water droplets to remove particles was quantified using image analysis software (SPIP 5.1.11, Image Metrology A/S, Horshølm, Denmark) (Bhushan and Jung 2011).

9.2.7 Finger Touch Tests

To characterize the coating repellency to fouling from finger touch, two tests are used. The first test was an anti-smudge test where the coating was contaminated and wiped with a hexadecane-impregnated micro-fiber wiping cloth. The second test was a fingerprint resistance test in which a rubber fingertip and an actual thumb were pressed onto the coating.

Anti-smudge

The anti-smudge characteristics were examined by contaminating the sample with silicon carbide, as described in the self-cleaning characterization techniques. The contaminated samples were then secured on a stage and a hexadecane-impregnated micro-fiber wiping cloth was attached to a horizontal glass rod (diameter of 5 mm) fixed on a cantilever above the sample. As the cloth was brought in contact with the sample, the microfiber cloth was set to rub the contaminated sample under a load of 5 g for 1.5 cm at a speed of about 0.2 mm/s. Images were taken using an optical microscope with a CCD camera (Nikon, Optihot-2). The removal and transfer of particles by the cloth were compared before and after the test (Bhushan and Muthiah 2013; Muthiah et al. 2013).

Fingerprint Resistance

The fingerprint resistance measurements were made using a rubber fingertip and an actual thumb (Bhushan and Martin 2018). Various artificial fingers are commercially available such as Abrex from Germany. For this study, a selected rubber fingertip (098,173, Cosco) was a nibbed rubber finger cover typically used to protect fingers from paper cuts and improved grip. The rubber fingertip was cut to be flat, taped to a glass slide, and placed on top of the coating with a 5 N load. A 5 N load was chosen based on experiments for the force that a single finger exhibits in a tapping mode on a standard telegraph key for college-aged subjects (Todor and Smiley-Oyen 1987). Other single finger tapping experiments using college-aged subjects (Inui et al. 1998) and pianists (Aoki et al. 2005) showed around 0.5 N tapping force, and so the greater tapping force of 5 N was used for the test.

In the fingerprint test with an actual thumb, a thumb was pressed into the coating (Bhushan and Martin 2018). The imprint force was estimated to be over 100 N based on thumb pushing experiments where adults pushed downwards with their thumb on a force plate and averaged over 100 N of force (Peebles and Norris 2003). The force was also checked by pushing with a thumb onto an analog scale showing forces that ranged

45–180 N with a typical force of 110 N. The fingerprint stresses were estimated to be 1.0 MPa using the rubber finger of diameter of 0.5 mm with an elastic modulus of 2 MPa and Poisson's ratio of 0.5 for synthetic rubber (Mark 1999) and 70 MPa using the actual thumb with diameter of 0.2 mm (Stucker et al. 2001) with an elastic modulus of 70 MPa (Edwards and Marks 1995) and Poisson's ratio of 0.5 (Bhushan 2016). Estimated contact pressure stresses are summarized in Table 9.2 (Bhushan and Martin 2018).

9.2.8 Anti-icing

Anti-icing behavior refers to the ability of a surface to prevent water from freezing on the surface. To study roll-off adhesion behavior of supercooled water droplets on frozen surfaces, samples were placed on an inclined block with a 10° slope in a freezer at either −18 or −60 °C and the fog was introduced to accelerate formation of ice crystals. To achieve −60 °C, dry ice pellets were used. To realize supercooled water, DI water in a sealed container was placed in the freezer at −18 °C for 2 h. DI water did not freeze, meaning that it was supercooled. Supercooled water droplets with 0.5–2 mm diameter were dropped from a height of 50 mm onto superhydrophilic surfaces as well as onto a glass surface for comparison. Surfaces were then photographed (Brown and Bhushan 2015c; Bhushan and Multanen 2019).

9.2.9 Anti-fogging

The anti-fogging characteristics of the surfaces were examined by placing the coated samples over boiling water for 5 s. The steam condensed on the coatings and was then photographed to determine the resulting optical transparency (Brown and Bhushan 2015c).

9.2.10 Optical Transparency

For qualitative measurements, text was placed behind the coated sample with a transparent substrate (such as glass) and the sample was photographed. The relative readability of the text was the qualitative measure of transparency (Ebert and Bhushan 2012).

For quantitative measurements, a line-of-sight light apparatus was assembled using a diffractive spectrometer (Acton, Princeton Instruments), an intensified CCD camera and an incandescent light bulb as a point source, which emitted a black-body type of spectrum across the 400–700 nm bandwidth of interest. The samples were placed within 1–2 mm of the incandescent light source. A pair of 50-mm diameter, 100-mm focal length plano-convex lenses was used to collect emission from the light source and focus it onto the

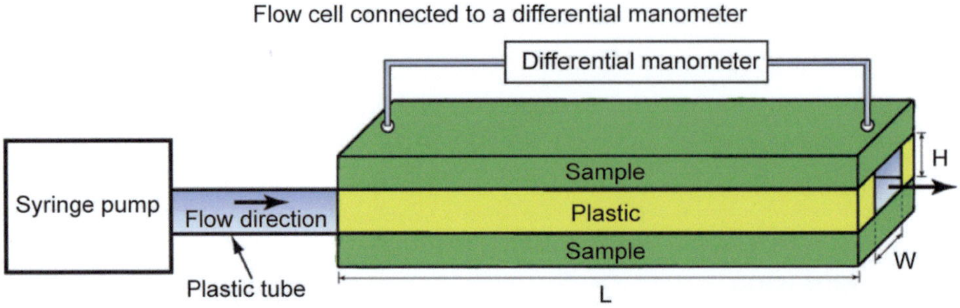

Fig. 9.2 Schematic of the experimental flow cell connected to a differential manometer (Adapted from Jung and Bhushan 2010)

entrance slits of the spectrometer. For a single camera exposure, the spectrometer bandwidth was approximately 80 nm, so the grating was stepped at ~ 60 nm intervals to sample the entire bandwidth with an overlap of about 40 nm between each grating position. A single camera exposure was acquired at each grating position. The spectra were then background subtracted and divided by the spectrum acquired from an uncoated glass slide and the data plotted as a function of wavelength (400–700 nm) (Brown and Bhushan 2015c).

9.2.11 Fluid Drag

For measurement of fluid drag, a closed rectangular flow channel was selected, Fig. 9.2 (Jung and Bhushan 2010; Bhushan 2018). The flow cell consisted of two samples sandwiched with a plastic spacer in between. Water was pumped into the cell using a syringe pump at flow rates ranging from 50 to 400 μl s^{-1} to achieve laminar to turbulent flow regimes. By measuring the amount of water and time for the water to flow from a starting to a final liquid level, the Reynolds number was calculated to determine flow regimes. Pressure drop was measured using a differential manometer. A lower pressure drop corresponds to a lower drag.

9.2.12 Oil–water Separation

Oil–water mixture was poured over a coated stainless-steel mesh to examine oil–water separation capability. Stainless steel meshes (#400) were first cleaned with acetone and IPA until they were found to be hydrophilic, then the coatings were deposited onto the meshes. The coated meshes were placed on top of beakers, and agitated mixtures of water and hexadecane were poured onto the coated meshes. In separate experiments, the meshes were inclined at an angle and the oil–water mixtures were poured over them. To improve contrast, Oil Red O and Blue 1 were used as oil and water dispersible dyes respectively. The use of dyes was not found to have any effect on the performance of the coating (Brown and Bhushan 2015a, c).

9.3 Nanoparticle-Binder Composite Coatings

Superoleophobic nanoparticle-binder coatings have been fabricated by various investigators. Table 9.3 summarizes coatings fabricated using various resin binders, nanoparticles, solvents, deposition methods, substrate and coating thicknesses with their CA, CAH, TA values and mechanical durability data (Adapted from Muthiah et al. 2013; Wang and Bhushan 2015; Martin and Bhushan 2017; Martin et al. 2017; Bhushan and Martin 2018).

Muthiah et al. (2013) were the first to develop a mechanically durable coating with finger touch resistance. They used a dual layer approach. Wang and Bhushan (2015) used the first layer to provide hierarchical roughness and mechanical durability while maintaining optical transparency, and the second layer to provide a low surface energy surface for superoleophobicity.

Wang and Bhushan (2015) used O_2 plasma treatment to activate the first layer. This required access to O_2 plasma and made the surface relatively rough which is not desirable for high transparency. Martin and Bhushan (2017) and Bhushan and Martin (2018) used UV-ozone (UVO) treatment which is economical and results in a very active surface and is adaptable to various substrates. To improve the transparency Martin and Bhushan (2017) and Bhushan and Martin (2018) used 10 nm diameter SiO_2 nanoparticles compared to 55 nm diameter used by Wang and Bhushan (2015).

Coatings fabricated using O_2 plasma and UVO treatments are introduced in this section. The wettability, surface morphology, repellency of surfactant-containing liquids, high temperature superliquiphobicity, wear resistance, self-cleaning, fingerprint resistance, transparency, and oil–water separation data are reported.

9.3.1 Experimental Details

Bhushan and Martin (2018) fabricated coatings consisting of hydrophobic SiO_2 nanoparticles and a binder of methylphenyl silicone resin with or without functional layers to obtain combinations of superhydrophobic/philic and superoleophobic/philic properties. The schematics of the coating methods are shown in Fig. 9.3. The superhydrophobic and superoleophilic method used nanoparticles and resin as shown in Fig. 9.3a. In addition to the nanoparticles and resin, the superhydrophilic and superoleophobic method required ultraviolet-ozone (UVO) activation and deposition of a fluorosurfactant as shown in Fig. 9.3b. Similarly, the superhydrophobic and superoleophobic method required UVO

Table 9.3 Literature review on binder and nanoparticle systems, deposition methods and substrates used in fabrication of superoleophobic surfaces with their contact angle (CA), contact angle hysteresis (CAH), and tilt angle (TA) values and mechanical durability (Adapted from Muthiah et al. 2013; Wang and Bhushan 2015)

Resin binder	Nanoparticles, solvent, deposition method, substrate, coating thickness	CA, CAH, TA (°)	Summary and comments
Fluorinated monoalkylphosphate	Anodized aluminum surface, ethanol, immersion coating (Tsujii et al. 1997)	CA (rapeseed oil) 150	A rough anodized aluminum substrate used
Fluorinated acrylic copolymer in water (Zony 8740 or Capstone ST-100—Dupont Co.)	TiO_2 nanoparticles (20–50 nm), water, spray coating on Si wafer (Hsieh et al. 2005)	CA (ethylene glycol) 144	Larger p-b ratio gives higher CA
	ZnO nanoparticles (50 nm), water/acetone, spray coating on glass (Steele et al. 2009)	CA (hexadecane) 154 CAH (hexadecane) 6	p-b ratio at 3.3 most superoleophobic and 1.6 most superhydrophobic surfaces
	SiO_2 nanoparticles (30–50 nm) by sol–gel method, water, spin coating on Si wafer, 6 μm thick (Hsieh et al. 2009)	CA (ethylene glycol) 165 CAH (ethylene glycol) 3	Based on p-b ratios, F/Si atomic ratio of 2.1 found to be optimum
	Carbon nanofibers (CNF) (100 nm dia., length ~ 130 μm), water/acetone + formic acid, spray coating on glass (Das et al. 2012)	CA (mineral oil) 164 TA (mineral oil) 9	CNF with heat transfer characteristics maybe used in icephobic surfaces
	Spray coating of fluoroacrylic + SiO_2 nanoparticles (55 nm), dip coating of PTFE amorphous, glass (Muthiah et al. 2013)	CA (hexadecane) 147 CAH (hexadecane) 12	p-b ratio at 0.6 most superomniphobic. Mechanically durable and anti-smudge

(continued)

Table 9.3 (continued)

Resin binder	Nanoparticles, solvent, deposition method, substrate, coating thickness	CA, CAH, TA (°)	Summary and comments
1H,1H,2H,2H-heptadecafluorodecyl modified polyhedral oligomeric silsesquioxane) (FluoroPOSS)	FluoroPOSS, poly(methyl methacrylate) (PMMA) and Asahiklin AK-225 (consists of 3,3-dichloro-1,1,1,2,2-pentafluoropropane and 1,3-dichloro-1,1,1,2,3-pentafluoropropane), electrospinning to produce fibers (bead size 10–20 μm) (Tuteja et al. 2008)	CA_{adv} (hexadecane) 156 TA (hexadecane) 5	A blend of hydrophilic PMMA and low energy fluoroPOSS is used to generate re-entrant surfaces
	Electrospun coating of cross-linked PDMS/ fluoroPOSS on stainless steel wire meshes, Asahiklin 225/methylformamide (Pan et al. 2013)	CA (hexadecane) > 150 TA (hexadecane) 1.5	Hierarchical scales of re-entrant texture. Chemical resistant to various liquids
(Heptadecafluoro-1,1,2,2-tetrahydrodecyl) triethoxysilane (Gelest Inc.) or 1H,1H,2H,2H-perfluorodecyltriethoxysilane (Degussa Chemical Company)	Fluorinated SiO_2 nanoparticles, isopropanol, spin coating, glass (Sheen et al. 2008)	CA (xylene) 141	Sol–gel method was used for fabrication of fluorinated SiO_2 nanoparticles
(Heptadecafluoro-1,1,2,2-tetrahydrodecyl) trimethoxysilane (Gelest Inc.) or 1H,1H,2H,2H-perfluorodecyltrimethoxysilane (Sigma-Aldrich)	SiO_2 nanoparticles (~150 nm), acetone, saturating cotton fabrics in solution and irradiating with microwave-assisted synthetic techniques (Lovingood et al. 2013)	CA (hexadecane) 137	Repel liquid and adsorb organic vapors from liquid droplets. High affinity for 3-hepten-2-one vapor

(continued)

Table 9.3 (continued)

Resin binder	Nanoparticles, solvent, deposition method, substrate, coating thickness	CA, CAH, TA (°)	Summary and comments
PTFE amorphous fluoropolymer (Teflon AF 2400—Dupont Co.)	Coaxial-electrospinning of polycaprolactam (PCL)/2,2,2-trifluoroethanol core and Teflon AF 2400/FC-75 (perfluorocarbon) sheath fibers (~2 μm in diameter) on aluminum-foil (Han and Steckl 2009)	CA (hexadecane) 138	Teflon AF 2400 is made electrospinnable by coaxial-electrospinning process using PCL
	PTFE amorphous fluoropolymer and FC-40 (perfluorocarbon) spin coating on polydimethylsiloxane (PDMS) surface with inverse-trapezoid (height = 12 μm, width = 26 μm, pitch = 40 μm) (Im et al. 2010)	CA (methanol) 135	Transparent, flexible, re-entrant surface was explored
n-perfluoroeicosane (268,828, Sigma-Aldrich)	Thermal evaporation on to micropatterned Si wafer with cylindrical pillars (diameter = 14 μm, height = 30 μm and pitch = 23 μm) (Jung and Bhushan 2009)	CA (hexadecane) 133	A model for predicting CA of water and oil droplets evaluated
1H,1H,2H,2H-Perfluorodecyltrichlorosilane (ABCR Germany)	Si nanofilaments grown on trichloromethylsilane treated glass, dry toluene, immersion coating (Zhang and Seeger 2011)	CA (hexadecane) 174	Optically transparent. Stable against outdoor conditions, ozone, UV light, basic and acid
Perfluorooctanoic acid	Spray coating of copper perfluorooctanoate from perfluorooctanoic acid and copper acetate on glass, ethanol, 150 μm thick (Yang et al. 2011)	CA (hexadecane) 155 CAH (hexadecane) 22	Stable to high speed rotation (4000 rps). Fragile to scratch but easy to repair by re-spray

(continued)

Table 9.3 (continued)

Resin binder	Nanoparticles, solvent, deposition method, substrate, coating thickness	CA, CAH, TA (°)	Summary and comments
Trichloro(1H, 1H, 2H, 2H-perfluorooctyl)silane, or tridecafluoro-1,1,2,2-tetrahydrooctyltrichlorosilane, or 1H,1H,2H,2H-perfluorooctyl-trichlorosilane (Sigma-Aldrich, Gelest Inc., ABCR, or Alfa Aesar Inc.)	Molecular vapor deposition on micropatterned Si wafer with cylindrical pillars (diameter = 2.7 μm, height = 7 μm and pitch = 6 μm) (Zhao et al. 2011)	CA (hexadecane) 158 TA (hexadecane) 10	Various fluorosilane coatings explored on patterned surface
	CVD of tetraethoxysilane on glass with carbon particles (40 nm) and calcination to form a silica shell first and then CVD of resin binder, ~8 μm thick (Deng et al. 2012)	CA (hexadecane) 156 TA (hexadecane) 5	Optically transparent. Limited resistant to sand abrasion (5 min). Thermally stable to 400 °C
	CVD of resin binder on SiO_2 aerogel (nanoparticle network), 1–2 mm thick (Jin et al. 2013)	CA_{adv} (mineral oil) 168 CAH (mineral oil) 38	Mechanically durable to sand abrasion. After sandpaper abrasion, CAH decreased to 14
	SiO_2 nanoparticles (55 nm), THF/IPA, dual-layered coating (dip coating of methylphenyl silicone, O_2 plasma, and vapor deposited perfluorotrichlorosilane (Wang and Bhushan 2015)	CA (hexadecane) 153 CAH (hexadecane) 4	p-b ratio of 3 most superoleophobic coating with low TA. Mechanically durable, anti-smudge, optically transparent

(continued)

Table 9.3 (continued)

Resin binder	Nanoparticles, solvent, deposition method, substrate, coating thickness	CA, CAH, TA (°)	Summary and comments
	SiO_2 nanoparticles (10 nm), THF/IPA, dual-layered coating (dip coating of methyphenyl silicone, O_2 plasma, and vapor deposited perfluorotrichlorosilane. (Martin and Bhushan 2017; Bhushan and Martin 2018)	CA (hexadecane) 156 TA (hexadecane) 2	p-b ratio of 3 most superoleophobic coating with low TA. Mechanically durable, anti-smudge, optically transparent
Fluorinated 3,4-ethylenedioxypyrrole (EDOP) monomer	Electropolymerization onto Si arrays of cylindrical micropillars (13 mm dia., 25 μm height, 40 μm distance between cylinders, photolithography and coated by gold first) (Darmanin et al. 2011)	CA_{adv} (hexadecane) 155 TA (hexadecane) 26	Super oil-repellency
	Electrochemical polymerization of monomer with micro/nano roughness on gold plate (Bellanger et al. 2012)	CA (hexadecane) 150 TA (hexadecane) 11	EDOP with different chain length found to affect surface morphology of polymer film
Fluorinated polyurethane	Fluorinated multi-walled carbon nanotubes (MWCNTs, 20–40 nm dia. and 30–50 mm length), acetone/toluene, spray coating on glass (Wang et al. 2012)	CA (hexadecane) 152	A coralline-like structure formed
Fluoropolymer (copolymer of ethylene, tetrafluoroethylene and perfluoromethylvinyl ether, and a cure site monomer) (Viton ETP-600S)	SiO_2 aggregates (92 m^2/g) (fluoroalkyl-functionalized), Asahiklin AK-225G (1,3-dichloro-1,2,2,3,3-pentafluoropropane), spray coating on Si wafer (Campos et al. 2012)	CA_{adv} (hexadecane) 155 TA (hexadecane) 25	Based on various particle-to-binder mass fractions, 80/20 (w/w) yielded maximum hexadecane CA values

p-B ratio—particle-to-binder ratio, CVD—chemical vapor deposition.
Surface tension of DI water = 72.0 mN/m, ethylene glycol = 48.0 mN/m, hexadecane = 27.0 mN/m, methanol = 22.1 mN/m (Rumble 2023–24), mineral oils, rapeseed oil = 30–35 mN/m (Muthiah et al. 2013), xylene = 29.5 mN/m (Pan et al. 2013).

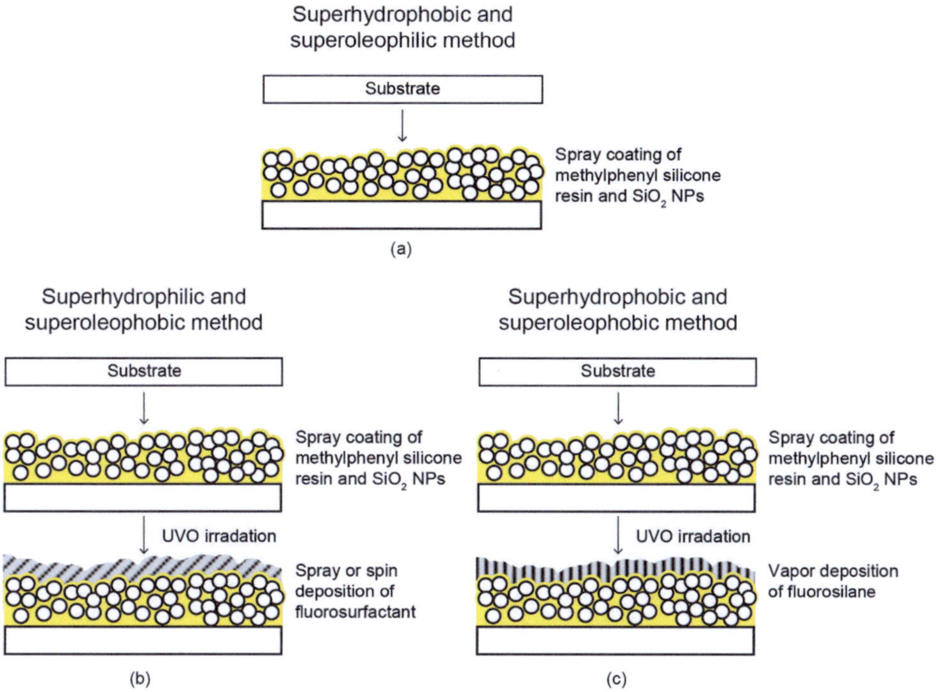

Fig. 9.3 Schematic of fabrication method for **a** superhydrophobic and superoleophilic surfaces by depositing nanoparticles and methylphenyl silicone resin binder to provide nanoroughness, followed by adding ultraviolet-ozone treatment to activate the surface and then **b** deposit fluorosurfactant to create superhydrophilic and superoleophobic surfaces, and **c** fluorosilane to create superhydophobic and superoleophobic surfaces (Adapted from Bhushan and Martin 2018)

activation and deposition of fluorosilane as shown in Fig. 9.3c. An ultraviolet (UV) method has also been used by Brown and Bhushan (2015a, b, c). However, since UVO activation creates a very active surface and is adaptable to various substrates and is preferred (Martin and Bhushan 2017; Bhushan and Martin 2018). Methylphenyl silicone resin was selected because it is known to be durable and offers strong adhesion between the nanoparticles and substrate. Hydrophobic, 10 nm SiO_2 nanoparticles were selected because they have high hardness for wear-resistance and high visible transmittance for transparency. Other commercial options for nanoparticles such as ZnO (zinc oxide) and ITO (indium tin oxide) of a similar diameter and surface treatment could be used in place of the SiO_2 nanoparticles with similar results (Ebert and Bhushan 2012).

For the coating mixture, 150–750 mg of hydrophobic silica nanoparticles (10 nm diameter, Aerosil RX300, Evonik Industries) were dispersed in 30 mL of 40% tetrahydrofuran (THF, Fisher Scientific) and 60% IPA by volume. This mixture was sonicated using an ultrasonic homogenizer (20 kHz frequency at 35% amplitude, Branson Sonifer 450A) for

15 min. Then, 150 mg of methylphenyl silicone resin (SR355S, Momentive Performance Materials) was added for a particle-to-binder ratio (p-b ratio) of 1.0–5.0. (It will be shown that a p-b ratio of 2.5 was optimum for wettability.) The mixture was then sonicated for an additional 15 min to form the final mixture (Martin and Bhushan 2017). In addition to methylphenyl silicone resin, an epoxy resin (EPON 1002F) was tested to see if similar properties could be obtained. Similar repellency was obtained; however, durability was an issue likely due to the higher hardness and therefore more brittle surface structures. Heating of the resins at 100 °C for one hour was also attempted, but little change was seen from the heated and unheated samples. The data on the epoxy and heating of the resins is not included here.

Various substrates were coated including glass, polypropylene (PP), poly(dimethylsiloxane) (PDMS), polyethylene terephthalate (PET), nylon 6/6, polyurethane (PU), poly(methyl methacrylate) (PMMA), polyethylene (PE), polycarbonate (PC), and stainless steel mesh. The typical substrate size for the glass and polymer samples was 25 by 10 mm. After choosing the substrate, 1 mL of the coating mixture was deposited via spray gun (Paasche®) from 10 cm away with compressed air at 210 kPa. The sample was transferred to an oven operating at 70°C for 5 min to remove the remaining solvent. These steps create the coating shown in Fig. 9.3a and are the initial steps in the coatings shown in Fig. 9.3b and c.

For the samples shown schematically in Fig. 9.3b and c, the samples were irradiated using UVO treatment with the samples placed 2 cm underneath the lamp source for 60 min. The UVO exposure was generated from a U-shaped, ozone producing, ultraviolent lamp (18.4 W, Model G18T5VH-U, Atlantic Ultraviolet Co.). It is expected that this lamp puts out a total of 5.8 W of 254 nm light, 0.4 W of 185 nm light, and 1.6 g/h of ozone in ambient conditions. The lamp was placed in an enclosure with width, height, and length dimensions of 17 by 15 by 45 cm and was connected to an electronic ballast (120 v, Model 10–0137, Atlantic Ultraviolet Co.) in order to provide the proper electrical conditions (Martin and Bhushan 2017).

For the superhydrophilic and superoleophobic method (Fig. 9.3b), 1 mL of a fluorosurfactant solution (Capstone FS-50, DuPont) diluted with ethanol to an overall fluorosurfactant concentration of 45 mg/mL was spin coated or spray coated onto the sample (Brown and Bhushan 2015a, b, c). For the superhydrophobic and superoleophobic method (Fig. 9.3c), one drop of trichloro(1H,1H,2H,2H-pefluorooctyl)silane (448,931, Sigma-Aldrich), referred to as fluorosilane, was vapor deposited on the sample using a closed container. The sample was attached to the top of the container via double-sided sticky tape with the surface facing down, and the drop was placed on the bottom of the container. This setup allowed the fluorosilane gas to uniformly coat the sample, and a vapor deposition time of 30 min was used (Wang and Bhushan 2015).

The chemical composition for each of these steps is shown in Fig. 9.4. The chemical composition of the methylphenyl silicone resin and hydrophobic SiO_2 nanoparticles is

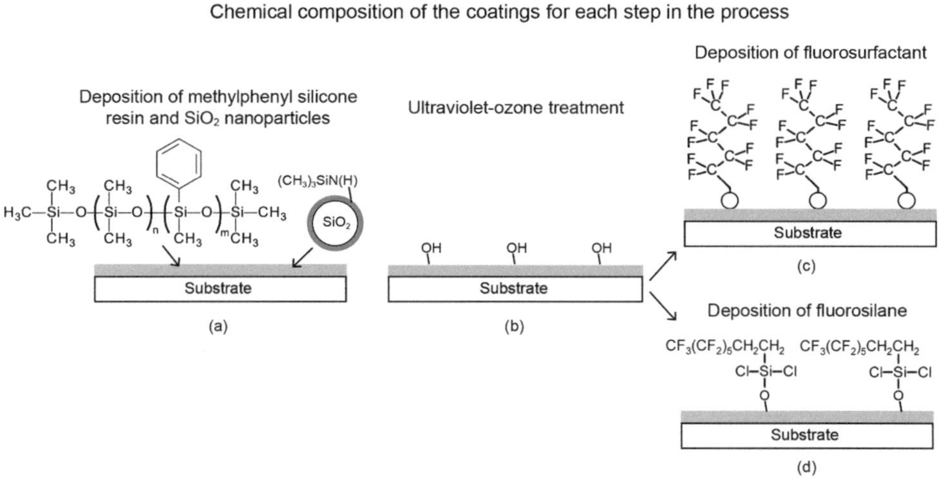

Fig. 9.4 Chemical composition for each step in the coating process: **a** deposition of the nanoparticles and binder for superhydrophobic and superoleophilic properties, **b** surface activation using ultraviolet-ozone treatment **c** deposition of fluorosurfactant where the hydrophilic head group of the fluorosurfactant is favorably attracted to the chemically activated surface resulting in the hydrophobic tail pointing away from the surface and superhydrophilic and superoleophobic properties, and **d** deposition of fluorosilane which bonds to the -OH groups for superhydrophobic and superoleophobic properties (Bhushan and Martin 2018)

shown in Fig. 9.4a. The surface activation step using UVO treatment is shown in Fig. 9.4b. The deposition of a fluorosurfactant or fluorosilane is shown in Fig. 9.4c or d, respectively.

Prior to Martin and Bhushan (2017) and Bhushan and Martin (2018), Wang and Bhushan (2015) used oxygen plasma treatment to activate nanoparticle-binder layer. They fabricated superoleophobic PET by depositing methylphenyl silicone resin with 55 nm SiO_2 nanoparticles with a particle-to-binder ratio of 3 on a PET substrate. PET substrate was selected as it is transparent and is of interest in smart screens in electronic displays. The sample was activated using O_2 plasma (Plasmalab System 100, Oxford Instruments) at an O_2 flow rate of 10 sccm and a power of 100 W for 10 min (chamber pressure 15 mTorr). The treated PET was then coated with perfluorotrichlorosilane (448,931, Sigma-Aldrich), hereafter referred to as fluorosilane, by vapor deposition as well as dip coating. For vapor deposition, fluorosilane with a volume of 0.1 mL was deposited in a vacuum chamber at 30 mTorr for 16 h duration. For dip coating, a solution with 2 wt% in hexane was deposited.

9.3.2 Characterization of Coatings Prepared Using Oxygen Plasma Treatment

To achieve superhydrophobicity and superoleophobicity, the hydrophobic SiO_2 nanoparticles with a methylphenyl silicon binder layer was activated using oxygen plasma treatment and fluorinated through vapor deposition of fluorosilane for a low surface energy coating.

The measured CA, CAH, and TA values of water, ethylene glycol and hexadecane on coated and uncoated PET samples with their relative durability based on AFM experiments are shown in Table 9.4 (Wang and Bhushan 2015). The uncoated PET is hydrophilic in water and oleophilic in ethylene glycol. The uncoated PET substrate became completely wet as a hexadecane droplet was deposited with a CA close to 0°.

In the initial experiments, for comparison, both hydrophobic and hydrophilic SiO_2 nanoparticles were used. The same procedure was followed with both types of particles. Lower CA with hydrophilic nanoparticles was obtained probably because of agglomeration of nanoparticles in the dip coating solution due to hydrogen bonding between hydroxyl groups (-OH) resulting in a non-uniform coating. It is known that, compared with hydrophilic SiO_2, hydrophobic SiO_2 have a better dispersion and homogeneity for lack of hydrogen bonds, and better wear resistance (Jalili and Moradian 2009; Tahmasebpoor et al. 2013). Thus, hydrophobic SiO_2 nanoparticles were selected.

For deposition of the first layer, PET was dip coated with methylphenyl silicone resin and hydrophobic SiO_2 to provide high durability. It is known that p-b ratio affects the wettability (Muthiah et al. 2013; Wang and Bhushan 2015), and it was optimized for high CA and low CAH. Dip coated PET with methylphenyl silicone resin alone showed wettability properties with CA $74° \pm 2°$ and $36° \pm 3°$, CAH $21° \pm 2°$ and $30° \pm 3°$ for water and ethylene glycol, respectively. The data provided a baseline. CA with water increased and CAH decreased as p-b ratio increases from 0 to 2.0, and then remains higher than 165° for p-b ratios of 2.0 to 4.0. This occurs because, with increasing concentration of nanoparticles, there is an increasing number of air pockets beneath the water droplets in the Cassie-Baxter wetting regime, leading to a high fraction of water–air contact area for p-b ratios of 0 to 2.0. When the fractional area between water and air has increased to a critical point, about 0.94 at a p-b ratio of 2.0, an increase of nanoparticle concentrations does not increase the fractional water–air area and CA and CAH do not change (Wang and Bhushan 2015). At the selected p-b ratio of 3.0, the coated surface exhibited a CA $165° \pm 2°$ and a CAH $2° \pm 1°$ for water, indicating superhydrophobicity. It exhibited a CA $140° \pm 3°$ and a CAH $9° \pm 2°$ for ethylene glycol, indicating oleophobicity. It exhibited a CA close to 0° for hexadecane, indicating superoleophilicity (Table 9.4). The lack of superoleophobicity was improved next by deposition of fluorosilane.

To verify the hydrophobicity and oleophobicity of fluorosilane, a single layer coating was deposited by dip coating and vapor deposition on PET substrate with a thickness of approximately 200 nm. The CA data in Table 9.4 show that the coated surface was

Table 9.4 Measured contact angle (CA), contact angle hysteresis (CAH), and tilt angle (TA) values and relative durability for coated PET samples at a particle-to-binder ratio (p–b) ratio of 3 (Adapted from Wang and Bhushan 2015)

Samples	DI water (°)			Ethylene glycol (°)			Hexadecane (°)			Relative durability based on AFM experiment	Comments
	CA	CAH	TA	CA	CAH	TA	CA	CAH	TA		
PET substrate	78 ± 2	20 ± 2	> 90	55 ± 2	22 ± 2	> 90	~0	N/A	> 90	Baseline	Hydrophilic
Single layer coating—Methylphenyl silicone resin + SiO₂											
Dip deposition of methylphenyl silicone resin + SiO_2 (~ 300 nm thick)	165 ± 2	2 ± 1	2 ± 1	140 ± 3	9 ± 2	> 90	~0	N/A	> 90	Excellent	Superhydrophobic, oleophobic/oleophilic
Single layer coating—fluorosilane											
Dip deposition of fluorosilane (~ 100 nm thick)	110 ± 2	13 ± 2	> 90	95 ± 3	15 ± 2	> 90	79 ± 2	16 ± 2	> 90	Good	Hydrophobic, oleophobic/oleophilic
Vapor deposition of fluorosilane (~ 200 nm thick)	106 ± 3	14 ± 2	> 90	92 ± 4	17 ± 2	> 90	76 ± 3	19 ± 2	> 90	Good	Hydrophobic, oleophobic/oleophilic
Dual layer coating—Methylphenyl silicone resin + SiO₂ + plasma treatment + fluorosilane											
Dip deposition of fluorosilane (~ 1 μm thick)	166 ± 2	2 ± 1	2 ± 1	151 ± 4	5 ± 3	4 ± 2	146 ± 3	8 ± 2	> 90	Medium	Superhydrophobic, superoleophobic
Vapor deposition of fluorosilane (~ 300 nm thick)	170 ± 2	2 ± 1	1 ± 1	165 ± 2	3 ± 2	2 ± 2	153 ± 2	4 ± 2	4 ± 2	Excellent	Superhydrophobic, superoleophobic

N/A—not applicable

hydrophobic and oleophobic in ethylene glycol, but slightly oleophilic in hexadecane. The durability was good.

Next, to achieve superoleophobicity and high durability, O_2 plasma treatment of the single layer was carried out followed by dip coating or vapor deposition of fluorosilane. After plasma treatment, the CA of water was less than 5° on the single layer coating, which suggested the introduction of (-OH) groups on the surface. Contact angle data are presented in Table 9.4. The dual layer coatings were superhydrophobic and superoleophobic in both ethylene glycol and hexadecane. However, only vapor deposited fluorosilane gave a low TA of 4 ± 2 for hexadecane. The relative durability of dip coating was moderate because of weaker bonding between condensed fluorosilane and the layer underneath. Whereas the relative durability of vapor deposited coating was excellent.

Wang and Bhushan (2015) reported RMS roughness of the coating as 24 ± 2 nm and P–V distance of 205 ± 10 nm. With a step technique using an AFM, the coating thickness of the first layer was measured to be approximately 300 nm, and the coating thickness of the second layer (vapor deposition) was also approximately 300 nm, with a total thickness of approximately 600 nm.

In summary, a superoleophobic coating was produced on PET substrates by dip coating with hydrophobic SiO_2 nanoparticles and methylphenyl silicone resin, followed by O_2 plasma treatment and vapor deposition of fluorosilane. The dual layer coating was found to be superhydrophobic and superoleophobic to ethylene glycol and hexadecane. Wang and Bhushan (2015) reported wear resistance on both the macroscale and microscale, as well as anti-smudge properties. They also reported the coating to be transparent, but some loss occurred due to the plasma treatment and a thicker coating. This can be improved by optimization of the plasma treatment parameters and coating thickness.

9.3.3 Characterization of Coatings Applied Using UV-O Treatment

To achieve superhydrophobicity and superoleophobicity, the hydrophobic SiO_2 nanoparticles with a methylphenyl silicone resin binder layer was activated using UV-O treatment and fluorinated through vapor deposition of fluorosilane for a low surface energy coating. In order to achieve superhydrophilicity and superoleophobicity, the nanoparticle-binder layer was fluorinated through spray or spin coating of fluorosurfactant (Bhushan and Martin 2018).

Data for the coatings is primarily presented on glass and PP. However, the same coating properties were achieved on different substrates by following a similar coating procedure. These coatings have also been successfully applied to poly(dimethylsiloxane) (PDMS), polyethylene terephthalate (PET), nylon 6/6, polyurethane (PU), poly(methyl methacrylate) (PMMA), polyethylene (PE), and polycarbonate (PC).

Wettability

As indicated earlier, particle-to-binder (p-b) ratio affects wettability and durability of the coating (Muthiah et al. 2013; Wang and Bhushan 2015; Bhushan and Martin 2018). An optimization of p-b ratio was performed using CA and TA of water and hexadecane with the nanoparticle-binder and fluorosilane coatings on glass substrate as shown in Fig. 9.5 (Bhushan and Martin 2018). A p-b range of 0.0–5.0 was tested to view a range of repellency properties. At a p-b ratio less than 2.0, the surface is not repellent to both liquids and exhibits high TA. By further increasing the concentration of nanoparticles, more air pockets are formed leading to more of the Cassie-Baxter wetting regime than the Wenzel wetting regime. At a p-b ratio greater than 2.0, the surface exhibits superhydrophobic and superoleophobic properties. An optimal p-b ratio of about 2.5 was chosen based on high CA, low TA, and high expected wear resistance.

The measured CA and TA values for water and hexadecane on glass and PP with various coatings are shown in Fig. 9.6a and Table 9.5 (Bhushan and Martin 2018). Untreated glass has a water CA of $22 \pm 2°$ and a hexadecane CA of $12 \pm 2°$, and untreated PP has a water CA of $96 \pm 1°$ and a hexadecane CA of $35 \pm 2°$ as shown in Table 9.5 (Bhushan and Martin 2018). By applying the coatings to these substrates, superhydrophobic/philic and superoleophobic/philic properties are obtained as shown in Fig. 9.6a (Bhushan and Martin 2018). The nanoparticle-binder coating results in water CA more than $164 \pm 2°$,

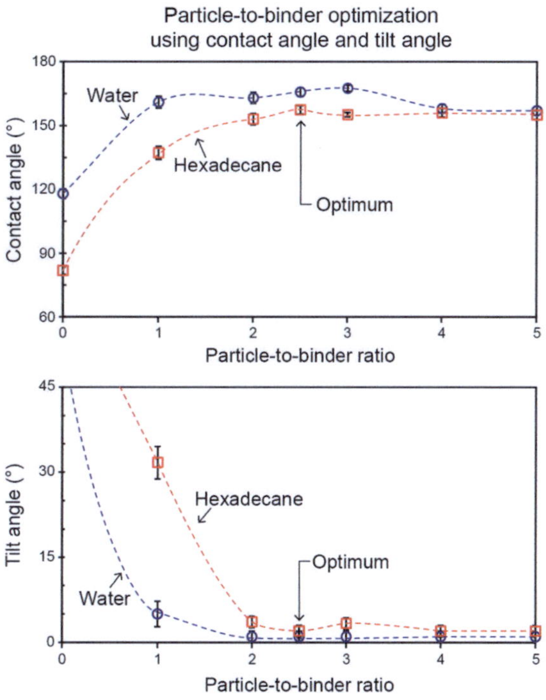

Fig. 9.5 CA and TA measured using water and hexadecane droplets on the superhydrophobic and superoleophobic coating on glass substrate as a function of particle-to-binder (p-b) ratio. A p-b ratio of about 2.5 was found to be optimal for repellency and durability (Adapted from Bhushan and Martin 2018)

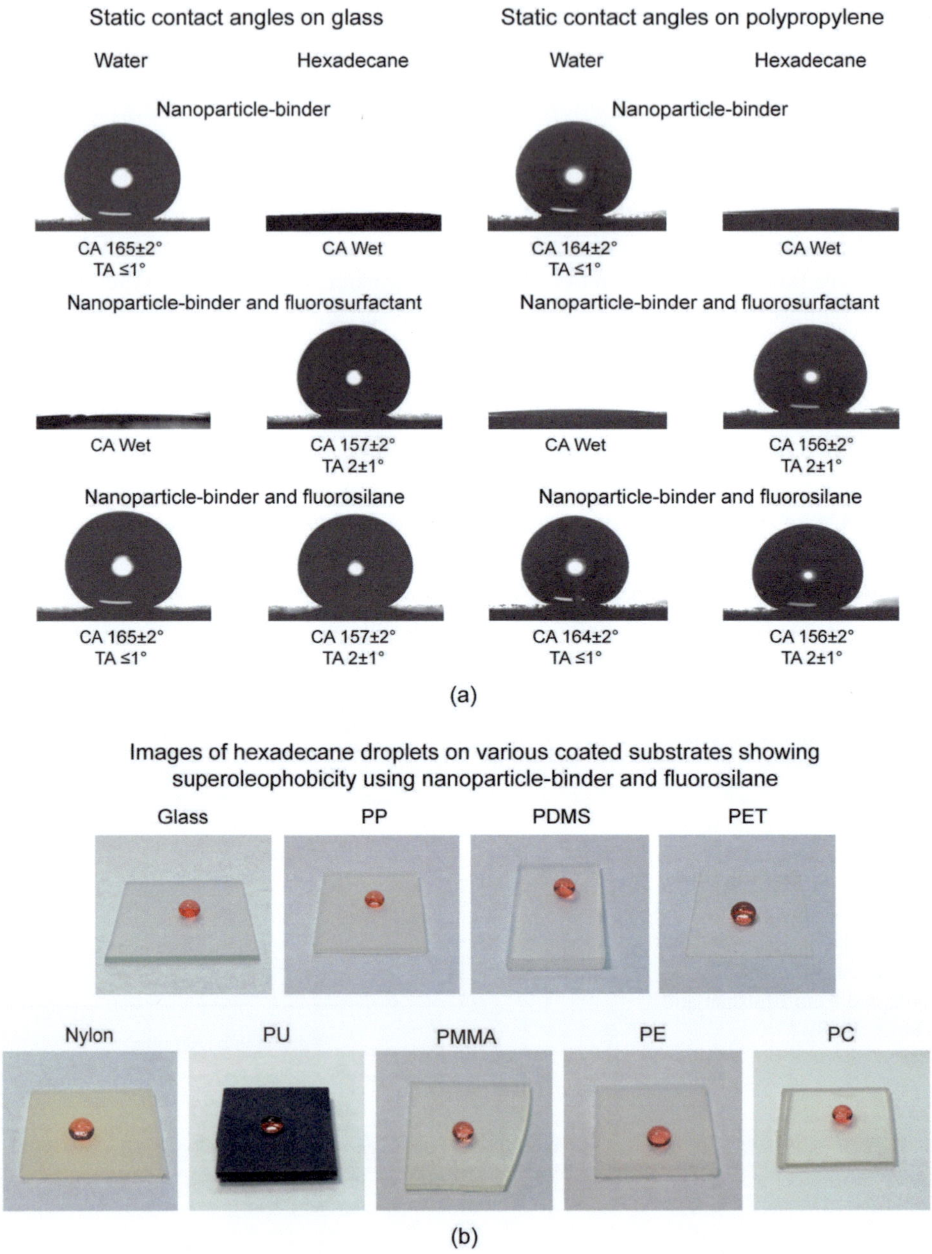

Fig. 9.6 **a** Optical images of water and hexadecane droplets deposited on glass and polypropylene substrates and coated samples to show repellency and wetting with the two liquids. Different coatings provide different repellency and wetting states which shows adaptability and flexibility in achieving them, and **b** optical images of hexadecane droplets on various coated substrates to show the substrate-independent ability. CA and TA measurements on these samples were within the standard deviation of ± 2 for CA and ± 1 for TA (Adapted from Bhushan and Martin 2018)

Table 9.5 Comparison of static contact angles and tilt angles for water and hexadecane droplets (Adapted from Bhushan and Martin 2018)

Technique	Contact angle (°)/Tilt angle (°)	
	DI water	Hexadecane
Glass substrate		
Untreated	22 ± 2/n/a	12 ± 2/n/a
Nanoparticle-binder (superhydrophobic/superoleophilic)	$165 \pm 2 / \leq 1$	Wet/n/a
Nanoparticle-binder and fluorosurfactant (superhydrophilic/superoleophobic)	Wet/n/a	$157 \pm 2/2 \pm 1$
Nanoparticle-binder and fluorosilane (superhydrophobic/superoleophobic)	$165 \pm 2 / \leq 1$	$157 \pm 2/2 \pm 1$
Polypropylene substrate		
Untreated	96 ± 1/n/a	35 ± 2/n/a
Nanoparticle-binder (superhydrophobic/superoleophilic)	$164 \pm 2 / \leq 1$	Wet/n/a
Nanoparticle-binder and fluorosurfactant (superhydrophilic/superoleophobic)	Wet/n/a	$156 \pm 2/2 \pm 1$
Nanoparticle-binder and fluorosilane (superhydrophobic/superoleophobic)	$164 \pm 2 / \leq 1$	$156 \pm 2/2 \pm 1$

Similar CA/TA data for coatings deposited on the additional substrates of PDMS, PET, Nylon, PU, PMMA, PE, and PC

water TA $\leq 1°$, and wetting with hexadecane. The nanoparticle-binder and fluorosurfactant coating results in wetting with water, hexadecane CA more than $156 \pm 2°$, and hexadecane TA of $2 \pm 1°$. The nanoparticle-binder and fluorosilane coating results in water CA more than $164 \pm 2°$, water TA $\leq 1°$, hexadecane CA in excess of $156 \pm 2°$, and hexadecane TA of $2 \pm 1°$.

Superoleophobicity has been obtained on a variety of substrates including PDMS, PET, nylon, PU, PMMA, PE, and PC, as shown in Fig. 9.6b (Bhushan and Martin 2018). These images show superoleophobicity, due to the resulting spherical geometry of a hexadecane droplet deposited on these surfaces. Water droplets were unable to be easily photographed because with the TA $\leq 1°$, the water droplets would quickly roll off the surface.

Surface Morphology

The AFM and SEM images of a coating with an optimum p-b ratio of 2.5 on glass and polypropylene substrates, respectively are shown in Fig. 9.7 (Bhushan and Martin 2018). RMS roughness of the coating was about 1 μm with a P–V distance of 5.5 μm. The measured coating thickness at p-b ratio of 2.5 was about 3 μm.

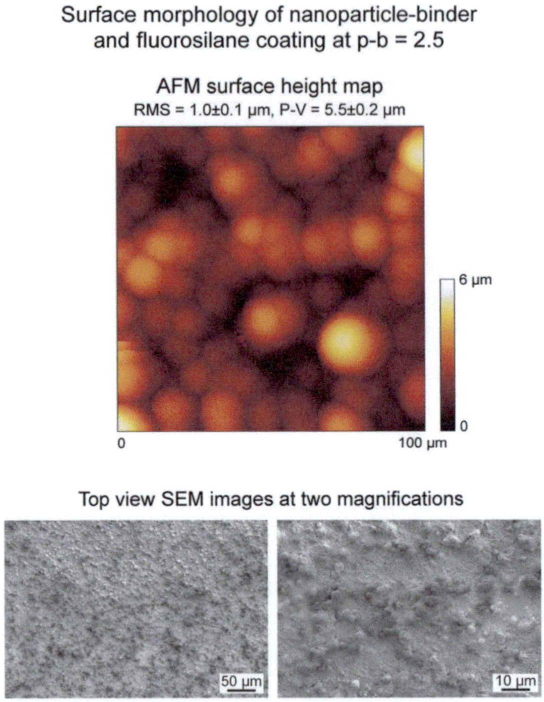

Fig. 9.7 AFM and SEM images of nanoparticle-binder and fluorosilane coating with an optimum p-b ratio of 2.5 on glass and polypropylene substrates, respectively (Adapted from Bhushan and Martin 2018)

To investigate roughness structure, SEM images of the glass and PP substrates with the coating with a top view and tilt view were taken as shown in Fig. 9.8 (Bhushan and Martin 2018). The coating was found to contain hierarchical structures showing re-entrant geometry (Martin et al. 2017). These re-entrant geometries helped support oil droplets for superoleophobicity. The micron-sized agglomerates of nanoparticles and binder are shown in a top-down view. This first level hierarchy exhibited structures of various sizes and shapes. The glass sample is shown with a p-b ratio of 4.0, and the PP sample is shown with a p-b ratio of 2.5. This difference in p-b ratio shows how altering the concentration of particles and binder influences the coating. In this case, the glass sample contains much rougher structures than the PP sample due to the greater number of particles (Bhushan and Martin 2018).

To view the re-entrant geometry of the structures, SEM images at a 70° from normal tilt angle were taken. These images show that the nanoparticles and binder generally combined into quasi-spherical shapes, which is one form of the re-entrant geometry. In addition, the nanoparticle and binder formed a rough outer surface of the micron-sized agglomerates, which is the second level of hierarchy. A rough surface would increase

Fig. 9.8 SEM images of the nanoparticle-binder coating on glass and PP substrates. The top-down view (shown at two magnifications) shows agglomerates of nanoparticles and binder that form micron-sized structures. The tilt view shows these structures have quasi-spherical re-entrant geometries that were found to be repellent to various liquids (Adapted from Bhushan and Martin 2018)

the number of air pockets and therefore increase the air–liquid contact area for increased liquid repellency (Bhushan and Martin 2018).

It is believed that the re-entrant structures are formed due to nanoparticle agglomeration. The small diameter of a nanoparticle leads to a very high surface-to-volume ratio. Since surface energy is inversely proportional to particle diameter, surface energy is a significant contributor to nanoparticle properties. The large surface energy would promote nanoparticle agglomeration and the resin would help bind the structure together. To minimize surface energy, these agglomerates would form tightly packed and quasi-spherical shapes (Baletto and Ferrando 2005). When these agglomerates are spray deposited, the resin binding and van der Waals attractive forces are great enough to resist the agglomerates breaking down due to the pressure of the spray gun and the collision energy from impacting the surface. In addition, smaller particles form agglomerates with a rougher surface that are more likely to give the re-entrant structure. A larger particle is less likely to form agglomerates and more likely to become partially embedded in the binder leading to hemispheres or spherical caps (Bhushan and Martin 2018).

Repellency of Surfactant-Containing Liquids

Droplets of shampoo and laundry detergent were found to roll or slide from coated glass and PP surfaces with no noticeable contamination as shown in Fig. 9.9 (Bhushan and Martin 2018). This is in contrast to the untreated glass and PP substrates where both shampoo and laundry detergent were found to spread and foul the surface.

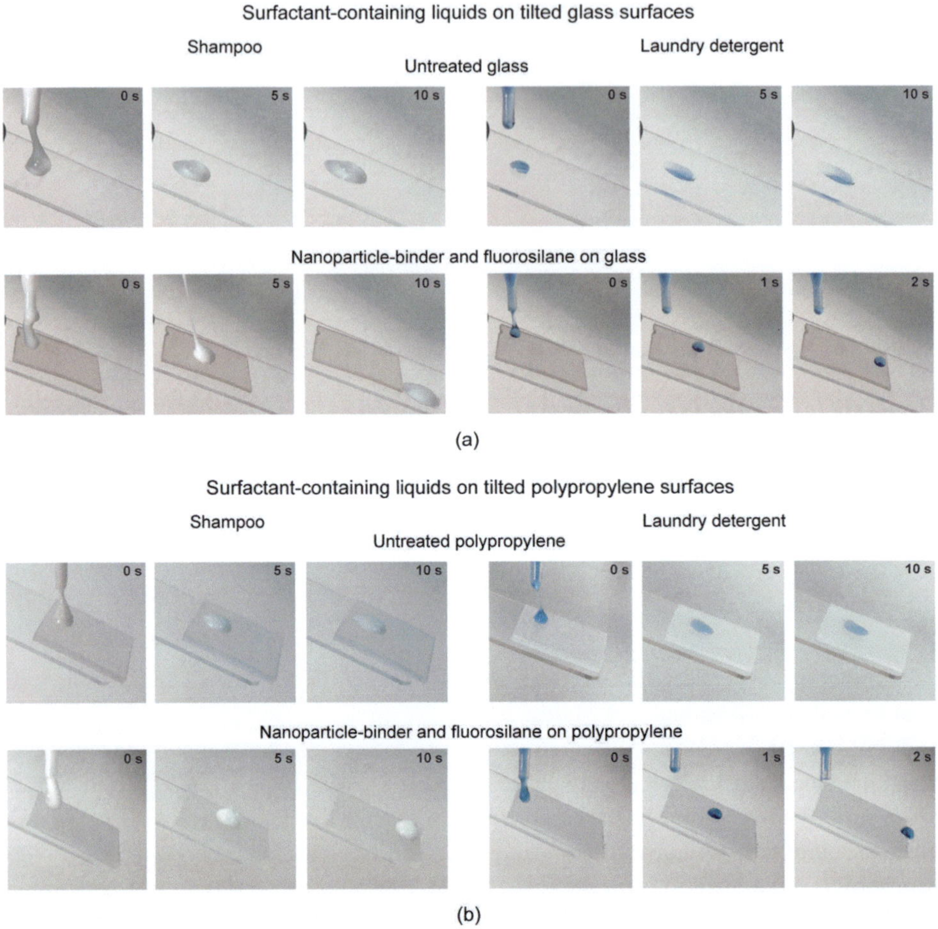

Fig. 9.9 Optical images demonstrating repellency of surfactant-containing liquids by comparing shampoo and laundry detergent droplets deposited on untreated and nanoparticle-binder and fluorosilane surfaces using **a** glass substrate and **b** polypropylene substrate. On the untreated substrates, the droplet spreads out and adheres to the surface. On the coated substrates, the droplet rolls and slides off the surface without any residue remaining (Adapted from Bhushan and Martin 2018)

For surfaces with roughness-induced repellency, surface topographic features can trap air between the substrate and liquid, reducing the solid–liquid interface and therefore increasing droplet CA and reducing TA. However, with a droplet of surfactant-containing liquid such as shampoo or laundry detergent is allowed to rest on these surfaces for a significant period, surfactant molecules may attach to the surface at the liquid–solid interface. These surfactant molecules may alter the surface chemistry of the solid surface, leading to a decrease in the CA and a collapsing of the composite interface between the

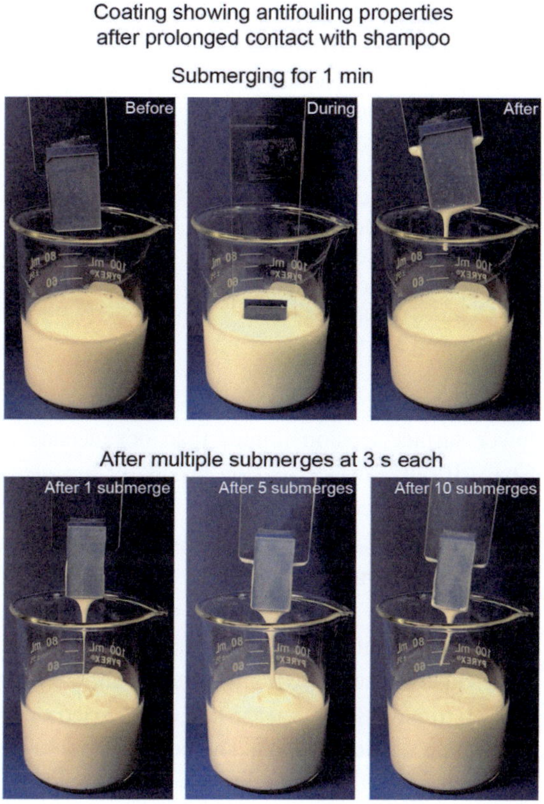

Fig. 9.10 Nanoparticle-binder and fluorosilane coating on glass substrate showing repellency after prolonged contact with shampoo. The coating was dipped into shampoo for one minute and withdrawn with minimal shampoo residue on the surface. The coating was repeatedly dipped into shampoo with minimal shampoo residue on the surface after 10 submerges (Adapted from Bhushan and Martin 2018)

droplet and the surface. The surfactant can also remain on the surface after the liquid is removed, which may lead to a permanent loss of repellency. Figure 9.10 shows that the coating can be dipped into shampoo for one minute and withdrawn with minimal shampoo residue on the surface (Bhushan and Martin 2018). The coating can be repeatedly dipped into shampoo with minimal shampoo residue on the surface after 10 submerges.

High Temperature Superliquiphobicity

The nanoparticle-binder and fluorosilane coating on glass substrate was tested for hot environments by placing the sample on a hot plate. Water and hexadecane droplets of increasing temperature were placed onto the heated sample and the CA were recorded as shown in Fig. 9.11 (Bhushan and Martin 2018). Droplet temperatures up to 80 °C were tested; however, it is expected that the substrate was subjected to even higher temperatures since it was in direct contact with the hot plate.

The coating exhibited superhydrophobicity and superoleophobicity at the maximum tested temperature. At 80 °C, the water CA was $159 \pm 3°$ and the hexadecane CA was $153 \pm 2°$. By testing hexadecane at a higher temperature, it shows that the coating repels

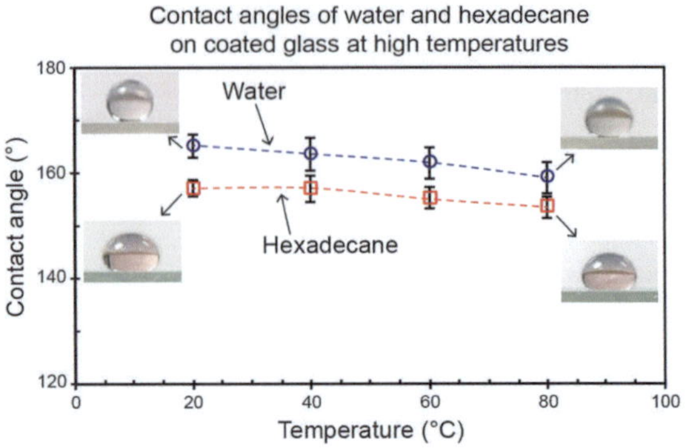

Fig. 9.11 Nanoparticle-binder and fluorosilane coating showing superhydrophobicity and superoleophobicity at droplet temperatures up to 80 °C. The coating repelled water and hexadecane at lower surface tensions due to the higher temperatures as well as staying resistant to the higher temperatures (Adapted from Bhushan and Martin 2018)

lower surface tension liquids than hexadecane at room temperature. Using the data in Chap. 10, the surface tension liquids of hexadecane at 20 °C is 27.5 mN/m, whereas at 80 °C the surface tension is 22.4 mN/m (Bhushan and Martin 2018).

Wear Resistance

Wear resistance of the coatings on the macroscale was investigated using a tribometer wear experiment and by scratching with tweezers. The optical images, after tribometer wear experiments showing a portion of the wear track, are shown in Fig. 9.12 (Bhushan and Martin 2018). The wear test was carried out for 100 cycles at 10 mN and resulted in burnishing of the coatings. The coating was still able to repel hexadecane over the defect suggesting that the coating was not destroyed.

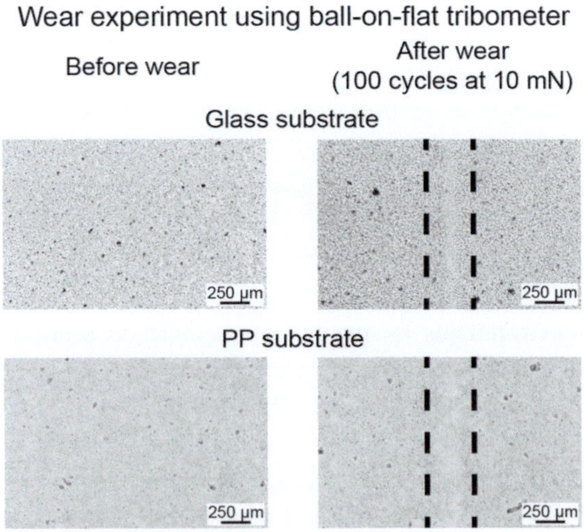

Fig. 9.12 Optical micrographs after wear experiment using ball-on-flat tribometer using a 3 mm diameter sapphire ball with 10 mN load on coated glass and PP substrates. Adapted from (Bhushan and Martin 2018)

Hexadecane TA on samples before and after the wear experiment and scratching with tweezers are shown in Fig. 9.13 (Bhushan and Martin 2018). Before the wear test, hexadecane droplets that were dragged across the surface were not obstructed in any way and rolled off the surface at a TA of $2 \pm 1°$. However, after the wear test, hexadecane droplets became pinned at the defect when dragged over the surface and rolled over at a higher TA of $5 \pm 1°$. The droplet placed on the defect was able to roll away at a TA of $17 \pm 2°$. The data suggests that damage to the surface increased pinning of droplets placed at the defect. However, when the droplets are moving, they are still repelled.

In contrast, on a deliberately destroyed coating (tweezers), a hexadecane droplet became highly pinned and required a TA of $53 \pm 4°$ for the droplet to roll off regardless of droplet starting position. Because the sample after the wear test still repelled hexadecane at a smaller TA than the destroyed coating (tweezers) sample, it showed that the coating has wear-resistant properties (Bhushan and Martin 2018).

To study wear resistance on microscale, the coated glass sample was scanned with an AFM using a 15 μm radius borosilicate glass ball mounted on a rectangular lever at a normal load of 10 μN over 50×50 μm region. Figure 9.14 (Bhushan and Martin 2018) shows AFM images over 100×100 μm scan area before and after the AFM wear experiments. After the AFM test, there is no burnishing of the coating (Bhushan and Martin 2018).

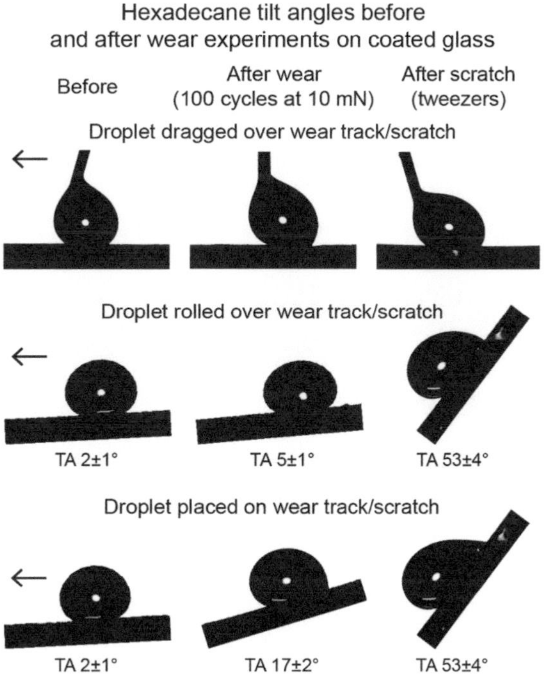

Fig. 9.13 Optical images of hexadecane droplets before and after wear/scratching. Droplets were dragged or tilted across the defect in the direction of arrows. Before the wear test, droplets rolled off the surface at $2 \pm 1°$ tilt angle. For the worn samples, droplets placed to the right of the wear track rolled over the defect at $5 \pm 1°$ tilt angle, and droplets placed directly on the wear track rolled over the defect at $17 \pm 2°$ tilt angle. Droplets on the scratched sample were pinned at the defect until $53 \pm 4°$ tilt angle regardless of droplet starting position (Adapted from Bhushan and Martin 2018)

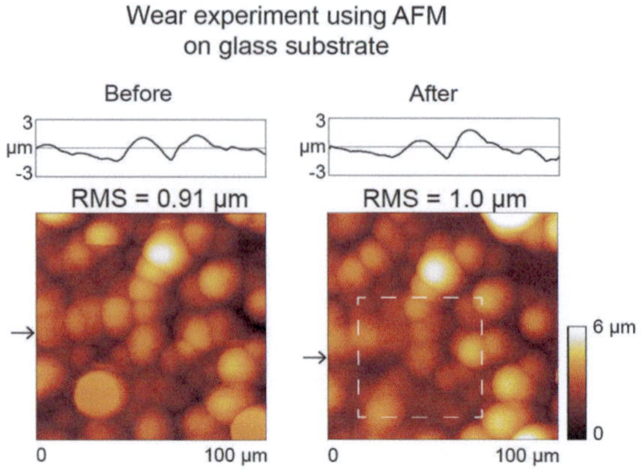

Fig. 9.14 AFM surface height maps and surface profiles (locations indicated by arrows) before and after AFM wear experiments with 15 µm radius borosilicate glass ball at a load of 10 µN on coated glass substrate (Adapted from Bhushan and Martin 2018)

Self-cleaning

The self-cleaning properties of the coating were assessed by first contaminating the surfaces with silicon carbide and seeing if contaminants could get removed by water droplets. Figure 9.15 shows optical micrographs of untreated and coated surfaces before and after cleaning (Bhushan and Martin 2018). The bar charts show measured percent particles removed by water droplets. On the untreated glass and polypropylene samples, most contaminants remain after the self-cleaning test. Whereas, on the coated samples, over 90% of the particles are removed due to the rolling action of water droplets collecting particles and removing them from the surface.

To further understand the mechanism of self-cleaning of superliquiphobic surfaces, water droplets were placed on a glass substrate and a superliquiphobic surface contaminated with red chili flakes and mounted on an inclined block with a 10º slope. Figure 9.16 shows video stills of droplets on the two surfaces (Bhushan and Multanen 2019). Water droplets slid over the glass substrate slowly leaving contaminants behind. Whereas the droplets rolled over with little impediment on a superliquiphobic surface, taking contaminants with them. These coatings are self-cleaning due to their high values of water CA and low CAH.

Finger Touch Tests

(a) Anti-smudge

For anti-smudge, untreated and nanoparticle-binder surfaces were contaminated with silicon carbide contaminants and a hexadecane-soaked cloth was used to wipe the surfaces. The anti-smudge data is shown in Fig. 9.17 (Bhushan and Martin 2018). For the oil-repellent surfaces, the particles were transferred to the cloth with no observable particles remaining on the surfaces. However, on the untreated samples, many particles remained on the sample and oil was transferred from the cloth to the sample. The anti-smudge

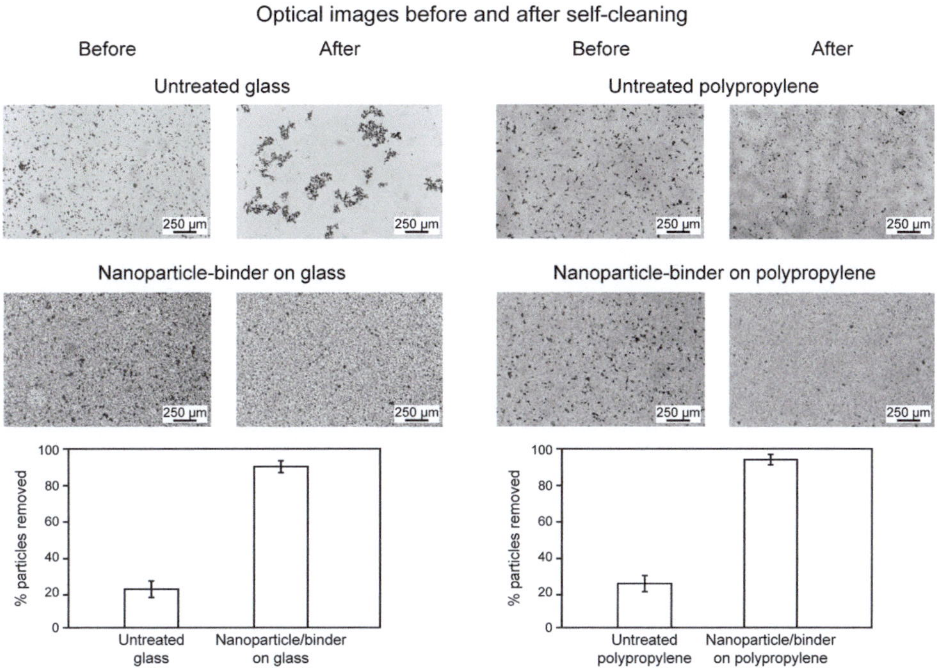

Fig. 9.15 Optical micrographs of contaminated coatings before and after self-cleaning test on untreated and nanoparticle-binder samples using glass and polypropylene substrates. Image analysis suggests > 90% removal of particles on the techniques (Adapted from Bhushan and Martin 2018)

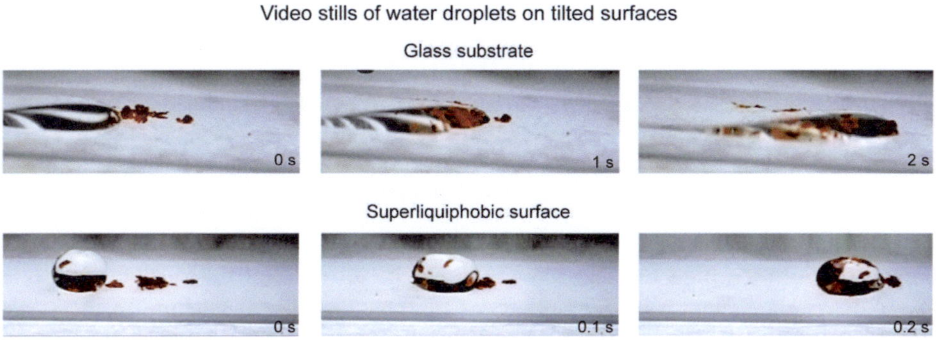

Fig. 9.16 Video stills of water droplets on inclined glass (hydrophilic) and superliquiphobic surfaces contaminated with red chili flakes. Water droplets roll over a superliquiphobic surface taking contaminants with them (Adapted from Bhushan and Multanen 2019)

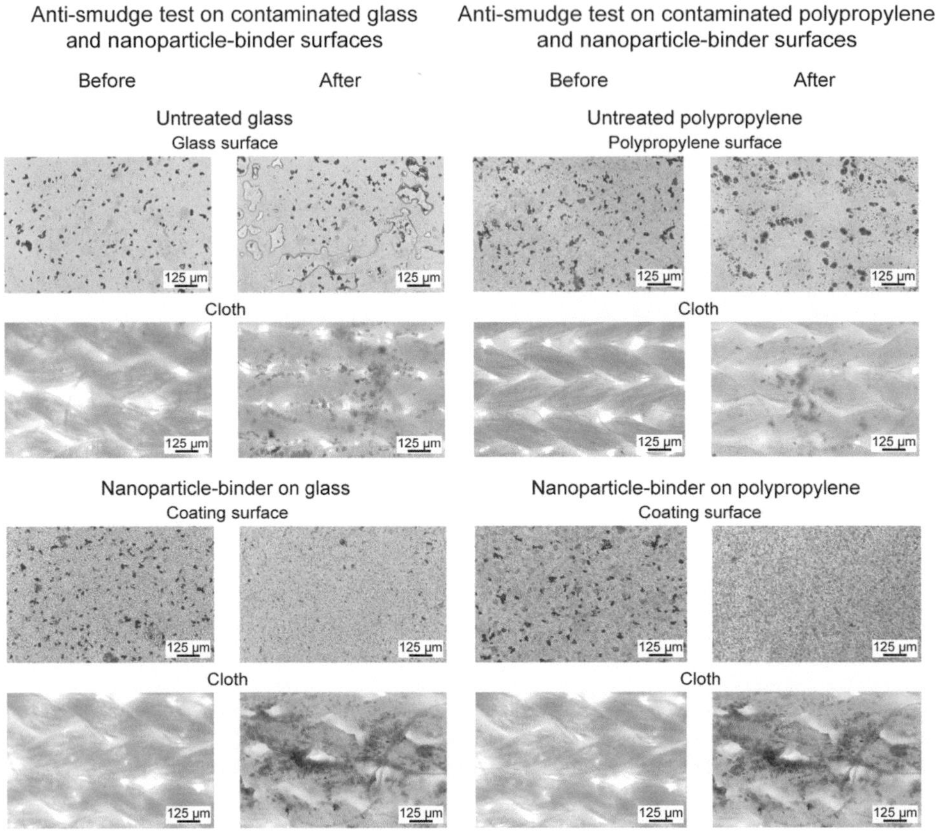

Fig. 9.17 Optical micrographs of contaminated surface and oil-impregnated microfiber cloth before and after anti-smudge test on untreated and nanoparticle-binder surfaces using glass and polypropylene substrates. Dark spots on coatings and cloth indicate silicon carbide particle contaminants. Untreated samples show oil transferred to the substrate and a few contaminants removed via cloth. Nanoparticle-binder samples show no oil transferred to the substrate and many contaminants removed via cloth (Adapted from Bhushan and Martin 2018)

property relies on a high CA and low CAH for the oil. The oil in the cloth can collect oleophilic contaminants from the surface of the coating without sticking to the surface (Bhushan and Martin 2018).

(b) Fingerprint Resistance

Optical images of the fingerprint test data using a rubber fingertip and thumb impressions are shown in Fig. 9.18 (Bhushan and Martin 2018). The p-b = 4.0 coating shows rubber nub impressions from the rubber fingertip test and a change in the light reflectivity of the

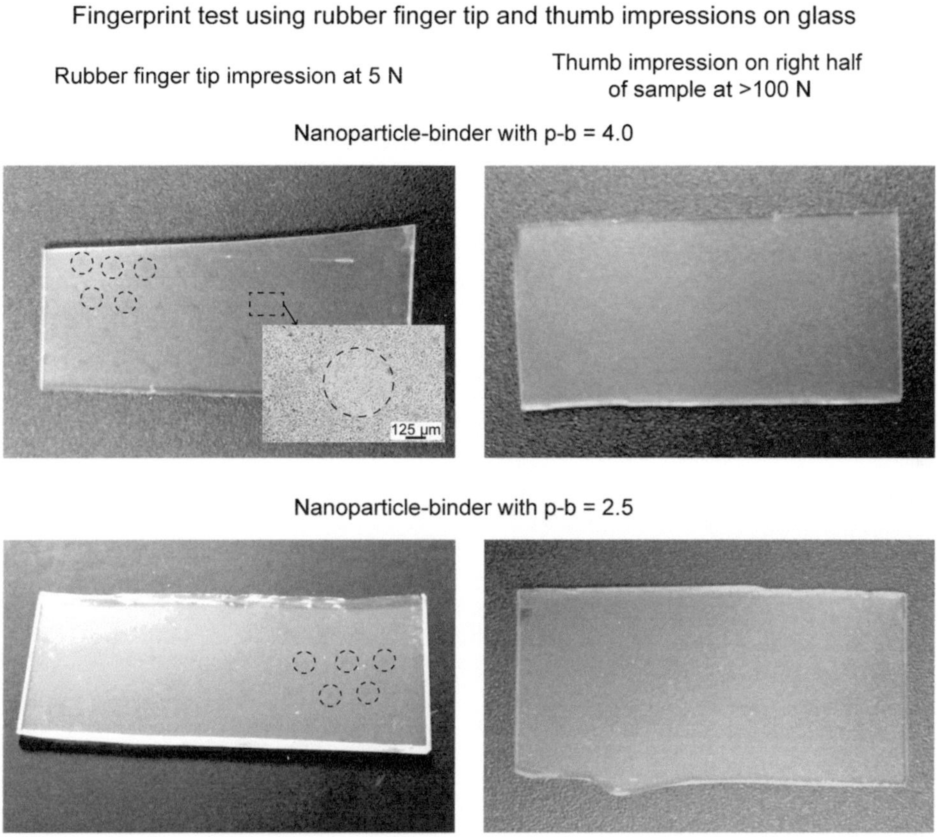

Fig. 9.18 Optical images showing durability to rubber fingertip and thumb impressions. The p-b = 4.0 coating shows rubber nub impressions from the rubber fingertip test and a change in the light reflectivity of the coating in the thumb impression test. The reduced p-b ratio coatings show minimal visible damage to the fingerprint resistance tests (Adapted from Bhushan and Martin 2018)

coating in the thumb impression test. The coating with decreased p-b ratio of 2.5 showed minimal visual damage to the fingerprint resistance tests. The CA and TA before and after fingerprint resistance tests were found to be virtually the same (Bhushan and Martin 2018).

Anti-icing

To measure anti-icing properties, coated surfaces were placed in tilted position in a freezer set at −60 °C for 2 h. Droplets of supercooled water were deposited on them (Bhushan and Multanen 2019). Figure 9.19a shows the supercooled droplets released onto surfaces at −60 °C. Frost was formed on both glass substrate and superhydrophobic surface, observed in higher magnification images. The supercooled water droplet froze and got stuck on the hydrophilic glass surface immediately after deposition on the sample. Supercooled water droplets rolled off the superhydrophobic surface before freezing on the base. The same results were found on superoleophobic surfaces. Figure 9.19b shows the side view of the supercooled water droplet on the superhydrophobic surface. Frost crystals can be clearly seen. Figure 9.19c shows the pathway of the droplet after it rolled off. Ice crystals were removed from the surface by the droplet. Ice crystals removal may have been facilitated due to the reduced ice adhesion strength caused by a reduced ice-solid contact area (Bhushan and Multanen 2019). This is in agreement with previous studies that show that ice adhesion strength on rough surfaces has good correlation with contact angle hysteresis or tilt angle, and therefore the ice-solid contact area (Kulinich and Farzaneh 2009).

Figure 9.20 shows video stills of supercooled water droplets falling from the height of 1 cm on surfaces at −60 °C (Bhushan and Multanen 2019). On the glass substrate,

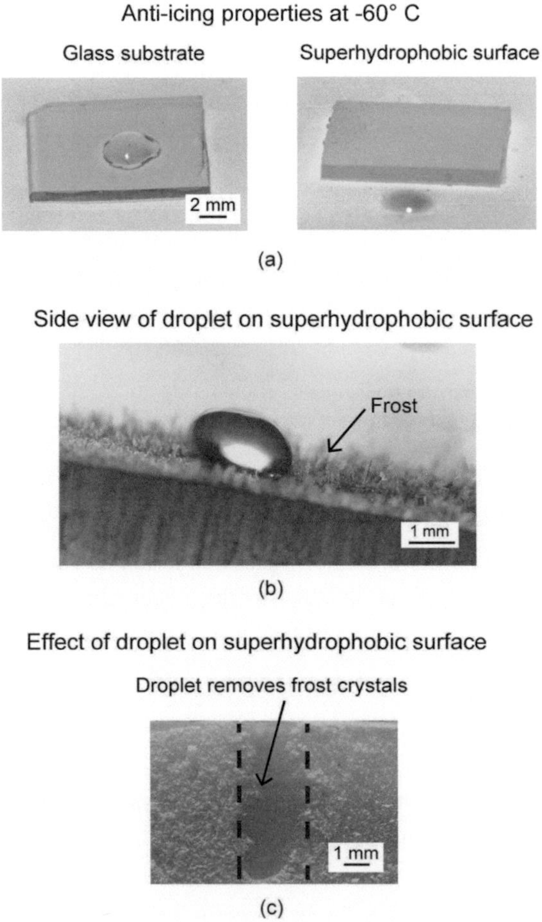

Fig. 9.19 Photographs showing the supercooled droplets on glass substrate and superhydrophobic surface at −60 °C. Frost was formed on both glass substrate and superhydrophobic surface **a** Water droplets froze on the glass substrate surface immediately after deposition. Water droplets rolled off the superhydrophobic surfaces, **b** Side view of the superhydrophobic surface shows frost crystals, **c** After rolling of the droplet, a track is observed which suggests that rolling of the supercooled water droplet removed frost crystals from the superhydrophobic surface (Adapted from Bhushan and Multanen 2019)

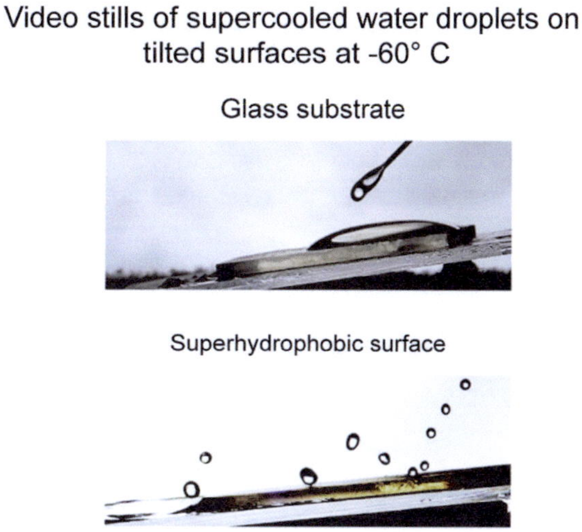

Fig. 9.20 Video stills of supercooled droplets on glass substrate and superhydrophobic surface frozen at −60 °C. Water droplet got stuck on the glass surfaces immediately after deposition, whereas they continued to bounce off superhydrophobic surfaces (Adapted from Bhushan and Multanen 2019)

droplets got stuck immediately. On superhydrophobic surfaces, water droplets bounced and rolled off the surfaces. As stated earlier, ice crystals have lower adhesion strength to superhydrophobic surfaces, and can be shed by taking advantage of external forces, such as gravity or aerodynamic forces, to overcome the ice-surface adhesion forces. Size of the droplets was 0.5–2 mm with an approach velocity of 0.1–0.2 m/s. After hitting the surface, droplets moved over an inclined superhydrophobic surface at 0.08–0.15 m/s.

Bhushan and Multanen (2019) also studied the effect of freeze–thaw cycles on anti-icing behavior. The contact angles and tilt angles did not change after 100 freeze–thaw cycles. They also studied the effect of freezing time on icephobicity. After freezing the surface for 100 h, the rolling behavior of supercooled water droplets did not change.

Figure 9.21 shows a schematic illustration of how icephobicity is maintained after frost (Bhushan and Multanen 2019). On the left side, a water droplet is sitting on the superhydrophobic surface at ambient. Spherical shape is maintained by air pockets formed on the surface. On the right side, a supercooled water droplet is placed on the frozen surface partially covered by snowflakes; air pockets are retained on the surface, providing the spherical shape of the water droplet. Furthermore, as stated earlier, even if some frost forms on the surface, the ice adhesion strength is lower on a superhydrophobic rough surface with low TA. The underlying surface is still hydrophobic and the action of the water droplet rolling across the surface is sufficient to remove the weakly adhered ice on a rough surface.

Optical Transparency

Many applications of liquid-repellent surfaces require optical transparency of the coating. When text was placed behind the coatings on glass and PET, the text remained readable suggesting that the coatings display characteristics of transparency as shown in Fig. 9.22

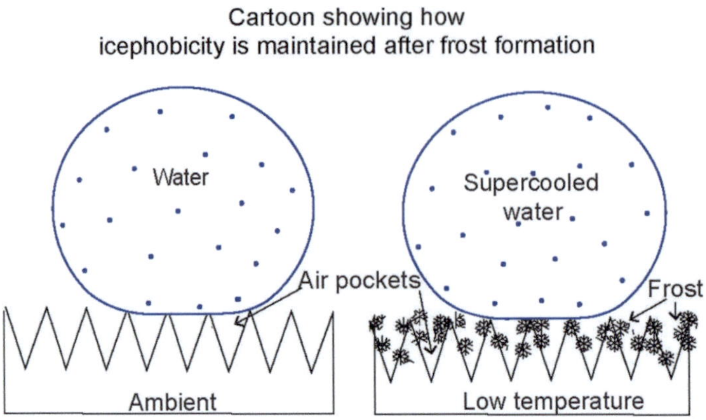

Fig. 9.21 Schematic showing a water droplet on the superhydrophobic surface, and a supercooled water droplet on the frosted surface. Air pockets remain even after the surface became partially covered with snowflakes (Adapted from Bhushan and Multanen 2019)

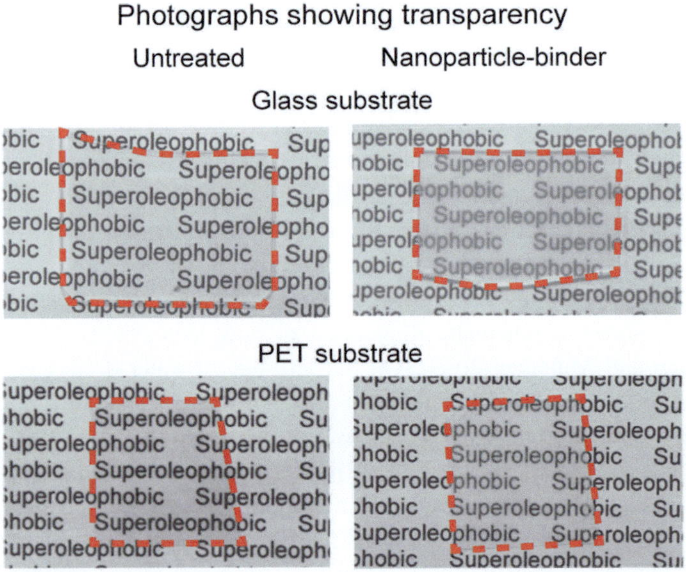

Fig. 9.22 Optical images showing optical transparency for uncoated and nanoparticle-binder on glass and PET substrates. The reduction in transparency in the coated samples is due to the SiO_2 nanoparticles and binder. The edges of each sample are shown in dashed lines (Adapted from Bhushan and Martin 2018)

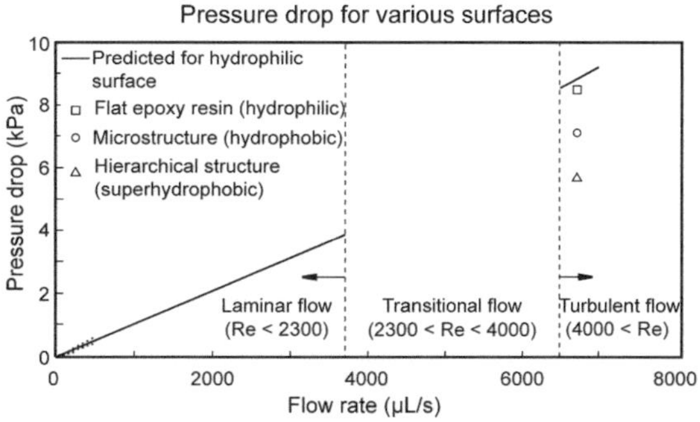

Fig. 9.23 Pressure drop as a function of water flow rate in the channel lined with various surfaces. Superhydrophobic surfaces provided lowest pressure drop in turbulent flow corresponding to the lowest drag (Jung and Bhushan 2010)

(Bhushan and Martin 2018). The edges of each sample are indicated with dash line. PP is an opaque polymer and so PET was chosen for transparency because PET is a transparent polymer. Further improvement in this design may be possible, potentially by decreasing the thickness of the nanoparticle layer, (Bhushan and Martin 2018).

Fluid Drag

Pressure drop measurements were made during water flow in a flow cell lined with samples having different degrees of hydrophobicity/philicity (Jung and Bhushan 2010; Bhushan 2018). Measurements were made on a flat epoxy surface (hydrophilic), a flat surface with a thin wax layer having microstructure (hydrophobic), and a micropatterned surface with a thin wax layer having hierarchical structure (superhydrophobic) in laminar to turbulent flow regimes. Figure 9.23 shows pressure drop data as a function of flow rate for various samples. Data for the flat surface was similar to the values predicted for a hydrophilic surface. The superhydrophobic surface provided the lowest pressure drop of about 33% in turbulent flow. A lower pressure drop corresponds to a lower drag. It is believed that air pockets underneath the fluid are responsible for lower pressure drag.

Oil–water Separation

With water and oil, coatings that attract one liquid (-philic) and repel the other liquid (-phobic) are suitable for oil–water separation (Brown and Bhushan 2015a, c). The nanoparticle-binder coating exhibits superhydrophobic and superoleophilic properties, whereas the nanoparticle-binder with fluorosurfactant coating exhibits superhydrophilic and superoleophobic properties. Due to their different responses to water and oil, these coatings can function as oil–water separators. Agitated oil–water mixtures were poured onto coated stainless steel meshes suspended horizontally over beakers. In both cases, the -philic component quickly passed through the mesh, while the -phobic component remained on top of the mesh. When the mesh was tilted, the -phobic component rolled across the top of the mesh and was collected in another beaker (Bhushan and Martin 2018).

Summary

A nanoparticle-binder coating method is of interest due to its ease of deposition, flexibility in substrate application, and desirable properties. Nanoparticle-binder coating exhibits superhydrophobicity/philicity, superoleophobicity/philicity, resistance to hot environmental properties, wear resistance, self-cleaning, finger touch resistance, anti-icing, optical transparency, drag reduction, and oil–water separation capabilities.

9.4 Layer-By-Layer Technique

Amphi-wettability landscape has been introduced earlier with four combinations of superliquiphilicity and superliquiphobicity, and their potential applications. As discussed earlier, to develop superoleophobic surfaces, low surface energy fluorinated coatings are needed. These coatings do not attach easily to other components in a composite or with the underlayer. Therefore, mixing fluorinated compounds with other components in a nanocomposite in a so-called "single pot" technique does not provide a durable coating. Mechanical durability of nanocomposite coatings consisting of fluorinated compounds has been a major issue for industrial applications. Furthermore, some functionality in the fluorinated layer and/or underlayer is desirable and individual layers allow fabrication of various combinations of superliquiphilicity and superliquiphobicity.

Brown and Bhushan (2015a, b, c) used a so called "layer-by-layer" technique for water and oil superliquiphilicity and superliquiphobicity, in which components were kept separate and deposited individually. Layers of oppositely charged species were deposited one after another to create a multi-layered coating that was bound by electrostatic interaction. Many differently charged species can be used to create a multi-layered coating. Thus, the technique is flexible and adaptable and various combinations can be used to realize a combination of superliquiphilicity and superliquiphobicity for a range of liquids, including water and oil.

Brown and Bhushan (2015a, b, c) used two polyelectrolyte layers with a silica nanoparticle layer in between. The nanoparticle layer was used to increase the roughness to realize air pocket formation and increase the hardness of the coating to improve durability. Intermediate polyelectrolyte layers were used to help bind the particles to the surface. The final (top) layer in the coating, referred to as the functional layer, contains the desired surface chemistry and can be easily swapped to produce a different functionality. Dependent upon the deposition of a functional top layer, they were able to achieve various combinations of superliquiphilicity and superliquiphobicity. Deposition of a separate functional layer provides correct functionality at the air interface without compromising the durability of the bulk coatings.

Brown and Bhushan (2015c) produced coatings with four combinations of water and oil repellency or affinity through layer-by-layer deposition, as shown in Fig. 9.24. Experiments were performed for wettability, durability, and functionality of all coatings for

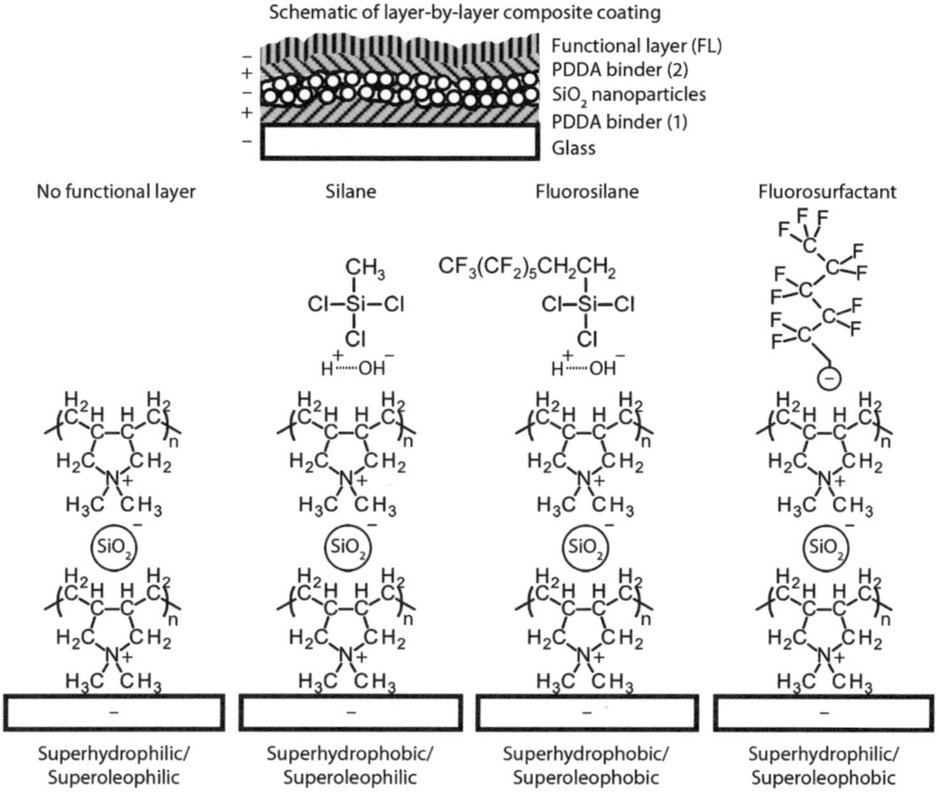

Fig. 9.24 Schematic of the four layer-by-layer composite coatings. Each layer is deposited separately. Also shown are the chemical composition and charge of each layer. The functional layer (FL) is deposited last and provides the desired surface chemistry (Adapted from Brown and Bhushan 2015c)

a variety of applications, including transparency, anti-fogging, anti-icing, self-cleaning, anti-smudge, and oil–water separation. Details follow.

9.4.1 Experimental Details

Brown and Bhushan (2015c) deposited various coatings on a glass substrate, a substrate of interest for optical and electronic display screens applications. The four coatings described in this study are comprised of various layers deposited separately, as shown in Fig. 9.24, each of which aids the creation of a mechanically durable, functional coating. In the case of the superhydrophilic/superoleophilic coating, no additional functional layer is added to leave the high surface tension polymer layer exposed. For superhydrophobic/superoleophilic coatings, a non-fluorinated silane is used to repel water but not oils, which have lower surface tensions (surface tension 20–30 mN/m). For superhydrophobic/superoleophobic coatings, a fluorinated silane is used to repel both water and oils. Finally, for superhydrophilic/superoleophobic coatings, a fluorosurfactant is used to yield the desired "flip-flop" of surface properties required. In all cases, the inclusion of a nanoparticle layer enhances the surface properties of the functional layer to result in superliquiphilic/phobic surfaces.

Polydiallyldimethylammonium chloride (PDDA) was chosen as the polymer base layer as it has a high cationic charge density and has been shown to bind strongly to both glass substrates (Du et al. 2010; Lee and Ahn 2013) and SiO_2 nanoparticles. The specific molecular weight range (100,000–200,000) was chosen to balance mechanical properties and ease of deposition (viscosity). Untreated, hydrophilic SiO_2 nanoparticles were used to enhance the roughness of the coating. The negatively charged surface silanol groups ensure good adhesion to the positively charged polymer layers. Additionally, SiO_2 nanoparticles are known to have the high hardness (Ebert and Bhushan 2012) and wear resistance necessary for the creation of a mechanically durable coating (Bhushan 2013). Particles 7 nm in diameter were selected with the goal of creating a transparent coating. The material selected for the final, functional layer varied depending upon the desired surface properties. Two different silanes (non-fluorinated and fluorinated silanes) were selected to provide the desired repellency and for their ability to form self-assembled layers via vapor deposition. Silanes have been shown to condense on hydrophilic polymer layers in the past due to the presence of absorbed water (Xie et al. 2010). For the superhydrophilic/superoleophobic coating ("flip-flop" coating), a fluorosurfactant was selected for its oil repellency (low surface tension tail) and its ability to complex to a positively charged polyelectrolyte layer (high surface tension head group).

Glass slides (Fisher Scientific) cut to dimensions of 25 mm by 10 mm were used as substrates. PDDA (MW 100,000–200,000, Sigma Aldrich) was dissolved in distilled water (DS Waters of America Inc.) at various concentrations. Hydrophilic silica nanoparticles (NP, 7 nm diameter, Aerosil 380, Evonik Industries) were dispersed in acetone (Fisher Scientific Inc.) using an ultrasonic homogenizer (Branson Sonifier 450A, 20 kHz frequency at 35% amplitude) at a concentration of 15 mg mL^{-1}. The fluorosurfactant solution (FL, Capstone FS-50, DuPont) was diluted with ethanol (Decon Labs Inc) so that the overall fluorosurfactant concentration was 45 mg mL^{-1}. Coatings were deposited via spray gun (Paasche) operated with compressed air at 210 kPa. The gun was held 10 cm from the glass slide. First, PDDA solution (52 mg mL^{-1}, 2 mL) was spray coated and any excess was removed from the surface via bursts of compressed air from the spray gun. Second, the SiO_2 NP solution (15 mg mL^{-1}, 3 mL) was spray coated. Third, a second PDDA layer was deposited (8 mg mL^{-1}, 1 mL). After this, the samples were transferred to an oven operating at 140 °C for 1 h. Finally, the functional layer was deposited either via spray coating or vapor deposition under atmospheric conditions. For spray coating, the fluorosurfactant solution (1 mL) was spray coated and the samples were allowed to dry in air. For vapor deposition, one drop of either methyltrichlorosilane (methylsilane, Sigma Aldrich) for superhydrophobic/superoleophilic coatings or trichloro(1H,1H,2H,2H-perfluorooctyl) silane (fluorosilane, Sigma Aldrich) for superhydrophobic/superoleophobic coatings was deposited next to the samples, which were covered and left for 6 h (Brown and Bhushan 2015c).

The first layer of PDDA had a thickness of about 200 nm and acted as an anchor layer to the glass substrate. The second layer contained SiO_2 nanoparticles with a thickness of about 350 nm and acted as the roughness layer, enhancing the overall liquid–solid interactions. The third layer was a second polymer layer (PDDA (2), with a thickness of about 50 nm, which helps to bind the nanoparticle layer, improving adhesion and mechanical durability. Then a final functional layer was deposited to provide the desired surface functionality. For the superhydrophilic/superoleophilic coating, there was no separate functional layer. For superhydrophobic/superoleophilic and superhydrophobic/superoleophobic coatings, the final layer was a silane layer with a thickness of about 25 nm, which condensed onto the hydrophilic PDDA (2) layer and provided either water-repellency (methylsilane) or water- and oil repellency (fluorosilane). For the superhydrophilic/superoleophobic coating, the final layer is a fluorosurfactant layer with a thickness of about 30 nm, which complexes with the positively charged PDDA (2) layer and provides the oil-repellency. Deposition of a separate functional layer ensures the correct functionality at the air interface without compromising the durability of the bulk coating. The total coating thickness was about 630 nm (Brown and Bhushan 2015c).

9.4.2 Results and Discussion

Wettability

Figure 9.25 and Table 9.6 show water and hexadecane contact angles and tilt angles for all four coatings (Adapted from Brown and Bhushan 2015c). The superhydrophilic/superoleophilic coating was instantly wetted by both water and oil. The superhydrophobic/superoleophilic coating was wetted by oil while repelling water. The superhydrophobic/superoleophobic coating repelled both liquids with a low tilt angle. The superhydrophilic/superoleophobic coating repelled oil but was wetted by water.

Oil repellency of both superoleophobic coatings was further tested with oils with various surface tensions by Brown and Bhushan (2015a, b). Figure 9.26 shows that coatings remained superoleophobic for hexadecane, tetradecane, dodecane, decane, and octane, with surface tensions of 27.05, 26.13, 25.35, 23.37, and 21.14, respectively. There were only slight increases in tilt angles for the lower chain length oils, due to their lower surface tensions Brown and Bhushan (2015a, b).

The oil repellency of the superhydrophilic/superoleophobic coating, in addition to wetting by water, is due to the fluorosurfactant containing a low surface tension fluorinated

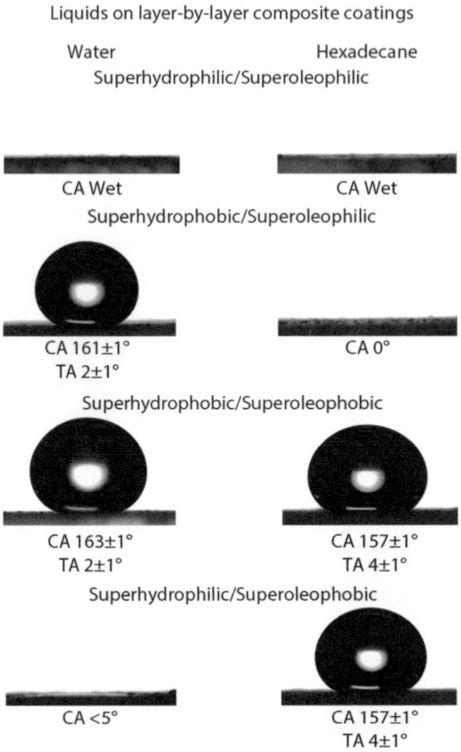

Fig. 9.25 Optical images of water and hexadecane droplets deposited on the four layer-by-layer composite coatings. CA and TA values are shown for all images (Adapted from Brown and Bhushan 2015c)

Table 9.6 Comparison of contact and tilt angles for water and hexadecane droplets deposited on the four layer-by-layer composite coatings (Adapted from Brown and Bhushan 2015c)

Functional layer	Surface properties	Water		Hexadecane	
		Contact angle (°)	Tilt angle (°)	Contact angle (°)	Tilt angle (°)
None	Superhydrophilic/ Superoleophilic	~0	N/A	~0	N/A
Silane	Superhydrophobic/ Superoleophilic	161 ± 1	2 ± 1	~0	N/A
Fluorosilane	Superhydrophobic/ Superoleophobic	163 ± 1	2 ± 1	157 ± 1	4 ± 1
Fluorosurfactant	Superhydrophilic/ Superoleophobic	< 5	N/A	157 ± 1	4 ± 1

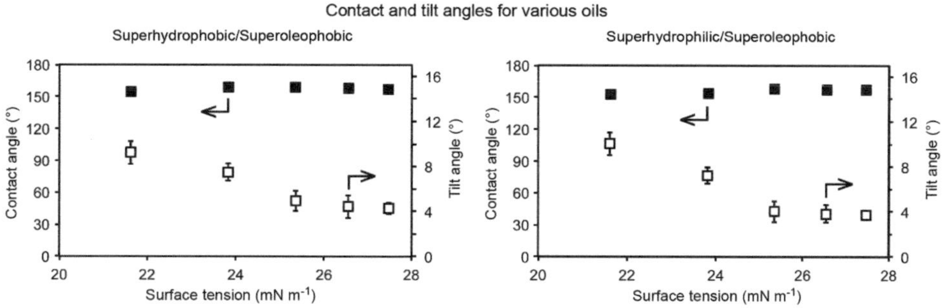

Fig. 9.26 Static contact angle and tilt angles for various liquids on the superhydrophobic/superoleophobic and superhydrophilic/superoleophobic surfaces as a function of liquid surface tension. Oils used: octane (21.14 mN/m), decane (23.37 mN/m), dodecane (25.35 mN/m), tetradecane (26.13 mN/m), and hexadecane (27.05 mN/m) (Adapted from Brown and Bhushan 2015a, b)

tail and a high surface tension head group, complexed with a hydrophilic polyelectrolyte. During spray coating, the polar head group forms an electrostatic complex with the polyelectrolyte layer below and the fluorinated tails orient themselves at the air interface. Large, bulky oil molecules are trapped at this fluorinated interface while smaller water molecules can more easily penetrate down through the thin layer (ca. 30 nm) to the hydrophilic region where the surfactant head group complexes with the polyelectrolyte layer. Water droplets (5 μL) were found to immediately (less than 2 s) wet the surface, which is desirable for anti-icing, anti-fogging, and oil–water separation (Brown and Bhushan 2015c).

Surface Morphology

AFM images of the coated surfaces were taken and will be presented later. RMS roughness was on the order of 430 nm. The total coating thickness was about 630 nm (Brown and Bhushan 2015c).

SEM images of the coating on a glass substrate are shown in Fig. 9.27. The coating was found to contain hierarchical structures showing re-entrant geometry (Martin et al. 2017). These re-entrant geometries helped support oil droplets for superoleophobicity. The micron-sized agglomerates of nanoparticles and binder are shown in a top-down view. This first level hierarchy exhibited structures of various sizes and shape. To view the re-entrant geometry of the structures, SEM images at a 70 ° from tilt angle were taken. These images show that the nanoparticles and binder generally combined to quasi-spherical shapes, which is one form of the re-entrant geometry. In addition, the nanoparticle and binder formed a rough outer surface of the micro-sized agglomerates, which is the second level of hierarchy. A rough surface would increase the number of air pockets and therefore increase the air–liquid contact area for increased liquid repellency.

Fig. 9.27 SEM micrographs of the layer-by-layer coating on glass substrate with superhydrophobic/superoleophobic properties. The top-down view (shown at two magnifications) shows agglomerates of nanoparticles and binder that form micron-sized structures. The tilt view shows these structures have quasi-spherical re-entrant geometries that were found to be repelling various liquids (Adapted from Martin et al. 2017)

Repellency of Surfactant Containing Liquids

Droplets of shampoo and laundry detergent were found to roll or slide from coated glass surfaces with no noticeable contamination as shown in Fig. 9.28. This is in contrast to the untreated glass substrate where both shampoo and laundry detergent were found to spread and foul the surface.

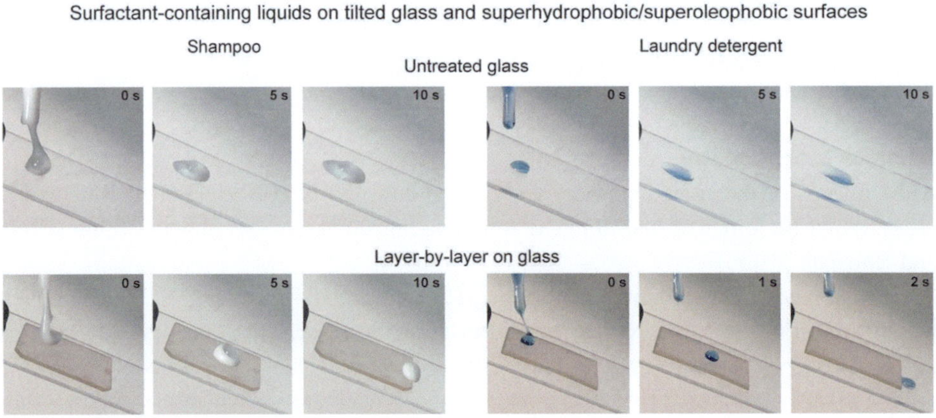

Fig. 9.28 Photographs demonstrating repellency of surfactant-containing liquids by comparing shampoo and laundry detergent droplets deposited on untreated and layer-by-layer coated surfaces using glass substrate. On the untreated substrates, the droplet spreads out and adheres to the surface. On the coated substrates, the droplet rolls and slides off the surface without any residue remaining (Adapted from Martin et al. 2017)

Wear Resistance

AFM and tribometer wear experiments were used to measure wear resistance of the coatings. As an illustration, wear data for flat PDDA (with no SiO_2 nanoparticles) with a functional layer coating, a superhydrophobic/superoleophobic coating, and a superhydrophilic/superoleophobic coating are shown in Fig. 9.29 (Brown and Bhushan 2015c). AFM images show a $100 \times 100\ \mu m^2$ scan area with the wear location ($50 \times 50\ \mu m^2$) in the center of each image. The optical images show a portion of the wear track from the tribometer experiments. For the soft PDDA/FL coating (ca. 225 nm thick), there is significant wear with both AFM and tribometer experiments, causing observable damage to the surface. In contrast, both layer-by-layer composite coatings survived the AFM wear experiment with no observable defects.

For the tribometer experiments, there is some noticeable burnishing to the coating. However, it is minimal when compared to the PDDA/FL coating. Higher magnification images confirmed that the layer-by-layer composite coatings morphology is similar before and after the wear test, and there is no removal of the coating from the substrate. This is in contrast to the PDDA/FL coating that was completely destroyed by the wear test, revealing the substrate underneath. Similar results were found for the other two coatings investigated in their study, not shown here (Brown and Bhushan (2015a, b, c). This suggests that the hard SiO_2 nanoparticle layer (underneath ca. 75 nm thick PDDA/FL layers) helps improve the durability of the coating. For both superoleophobic coatings, it was demonstrated that they are able to maintain their oil-repellency after wear testing, with hexadecane droplets rolling over and from the wear scar with little to no impediment (Brown and Bhushan (2015a, b).

To further demonstrate the benefits of the layered structure on the mechanical durability of the coating, a fluorosurfactant-containing, superhydrophilic/superoleophobic coating was fabricated using a "one-pot" technique, where all the materials were mixed (at the same concentrations used in the layer-by-layer technique) and deposited together. This coating, which was found to be similar in terms of thickness and roughness as the layer-by-layer composite coating, was then subjected to the same ball-on-flat tribometer experiment as described above. The coating was found to have significantly poorer adhesion to the glass substrate than the layer-by-layer composite coating, most likely due to the presence of the low surface tension material throughout the coating instead of as only the top layer in the layer-by-layer coating (Brown and Bhushan 2015c).

As far as thermal stabilities of various functional layers, methylsilane and fluorosilane exhibit thermal stabilities of about 66° and 83°, respectively. The thermal stability of the fluorosurfactant layer is higher, about 175 °C.

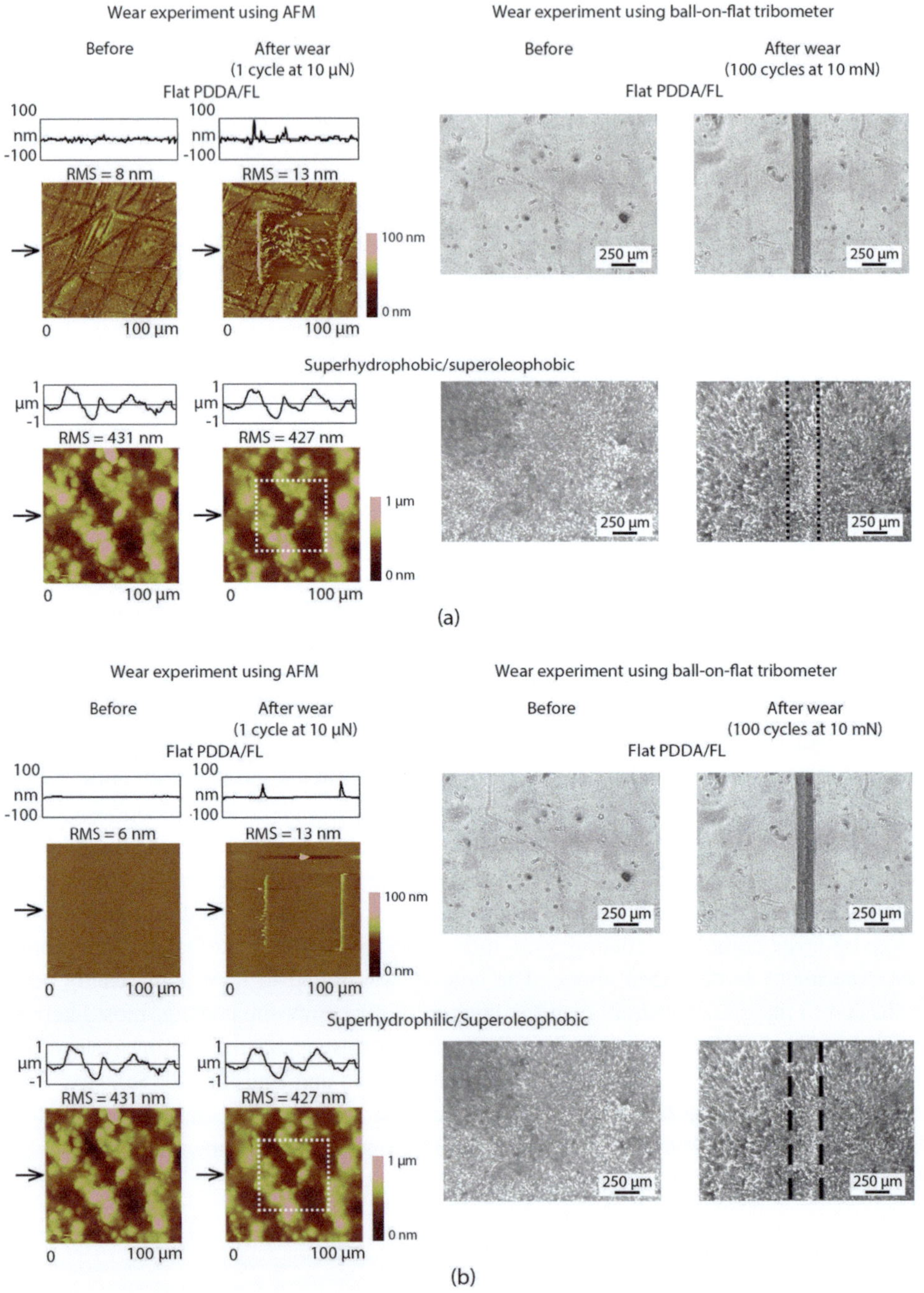
Wear experiment using AFM
Wear experiment using ball-on-flat tribometer
Before
After wear
(1 cycle at 10 μN)
Before
After wear
(100 cycles at 10 mN)
Flat PDDA/FL
Flat PDDA/FL
100
nm
-100
RMS = 8 nm
RMS = 13 nm
100 nm
0 nm
0
100 μm
250 μm
Superhydrophobic/superoleophobic
1
μm
-1
RMS = 431 nm
RMS = 427 nm
1 μm
0 nm
(a)
Wear experiment using AFM
Wear experiment using ball-on-flat tribometer
Before
After wear
(1 cycle at 10 μN)
Before
After wear
(100 cycles at 10 mN)
Flat PDDA/FL
Flat PDDA/FL
RMS = 6 nm
RMS = 13 nm
Superhydrophilic/Superoleophobic
RMS = 431 nm
RMS = 427 nm
(b)

◀**Fig. 9.29** AFM surface height maps and surface profiles (locations indicated by arrows) before and after AFM wear experiment with 15 μm radius borosilicate ball at a load of 10 μN (left) and optical micrographs before and after wear experiments using ball-on-flat tribometer at 10 mN (right) **a** for flat PDDA/FL and superhydrophobic/superoleophobic coatings and **b** flat PDDA/FL and superhydrophilic/superoleophobic layer-by-layer composite coatings. RMS roughness values obtained using an AFM are displayed. Similar results were obtained for two remaining layer-by-layer composite coatings (Adapted from Brown and Bhushan 2015c)

Self-cleaning

To measure the self-cleaning properties of two superhydrophobic coatings, the coatings were contaminated with silicon carbide particles, as shown in Fig. 9.30 (Brown and Bhushan 2015c). A stream of water droplets was then used to clean the surface. For comparison, a flat PDDA/FL coating (without SiO_2 nanoparticles) was also tested. For the superhydrophobic/superoleophilic and superhydrophobic/superoleophobic coatings, the vast majority of the particles were removed by the action of water droplets rolling across the repellent surfaces, collecting particles in the process. In contrast, the flat PDDA/FL surface remained contaminated after cleaning with water. Superhydrophobic coatings are self-cleaning due to their high values of water contact angle and low hysteresis. Water droplets deposited onto these samples can roll over the coating with little impediment, collecting fewer hydrophobic contaminants as they go (Brown and Bhushan 2015c).

Anti-smudge

To measure the anti-smudge properties of the two superoleophobic coatings, a hexadecane-soaked cloth was used to wipe the surfaces contaminated with silicon carbide, as shown in Fig. 9.31 (Brown and Bhushan 2015c). For comparison, a flat PDDA/FL coating (without SiO_2 nanoparticles) was also tested. For the superhydrophobic/superoleophobic and superhydrophilic/superoleophobic coatings, the particles were transferred to the cloth with no observable particles remaining on the surfaces. In contrast, the flat PDDA/FL surface remained contaminated after wiping with the cloth. Similar to the self-cleaning experiments with water, the anti-smudge property relies on a high contact angle and low hysteresis for the oil. The oil in the cloth is able to collect oleophilic contaminants from the surface of the coating without sticking to the surface (Brown and Bhushan 2015c).

Anti-icing

To measure anti-icing properties, all four coatings were placed in a freezer set at –18 °C for 2 h. The samples were tilted, and droplets of supercooled water were deposited onto them, as shown in Fig. 9.32 (Brown and Bhushan 2015c). On the superhydrophilic coatings, the droplets spread out and froze on the sample surface. On the superhydrophobic coatings, droplets rolled off the surface to freeze on the bottom of the freezer. This occurs because the water droplets are in the Cassie-Baxter state. The formation of a composite

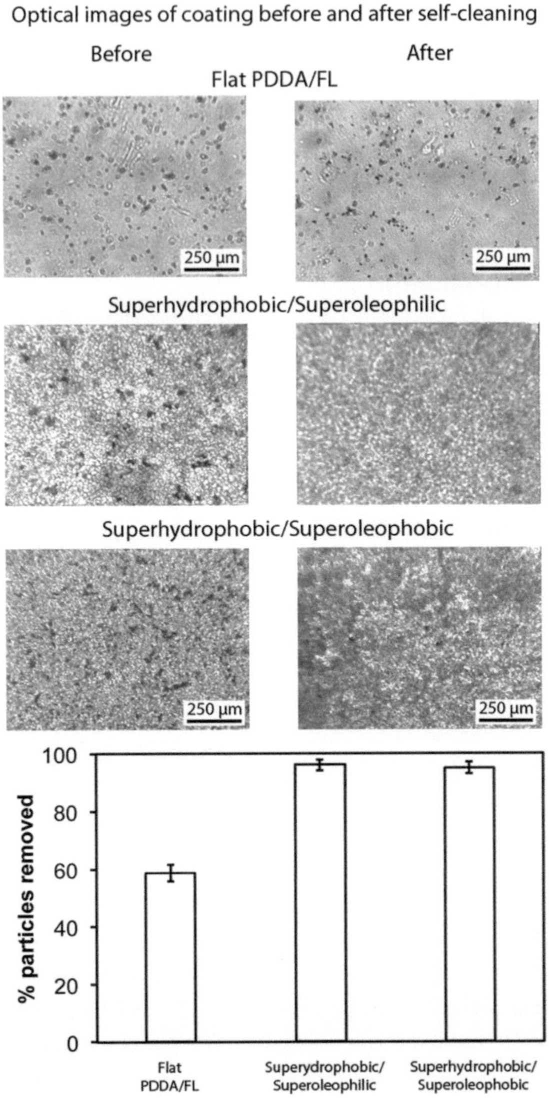

Fig. 9.30 Optical micrographs of contaminated coatings before and after self-cleaning test on flat and two superhydrophobic layer-by-layer composite coatings. Dark spots on coatings and cloth indicate silicon carbide particle contaminants. Image analysis suggests a > 90% removal of particles on the two composite coatings (Adapted from Brown and Bhushan 2015c)

interface minimizes the contact with the cooled substrate and ensures a low hysteresis, so droplets can roll from the tilted surface (Brown and Bhushan 2015c).

As indicated in Chap. 3, the work of adhesion is related to $(1 + \cos \theta_0)$, where θ_0 is the static contact angle. Therefore, for low work of adhesion, the surface should have high static contact angle, preferably exhibiting superhydrophobicity. It was also reported that contact angle hysteresis or tilt angle is a measure of energy dissipation during the flow of a droplet. Therefore, for low ice adhesion (typically 150 kPa or lower) to facilitate the

Fig. 9.31 Optical micrographs of contaminated coatings and oil-impregnated microfiber cloth before and after smudge test on flat and the two superoleophobic layer-by-layer composite coatings. Dark spots on coatings and cloth indicate silicon carbide particle contaminants (Adapted from Brown and Bhushan 2015c)

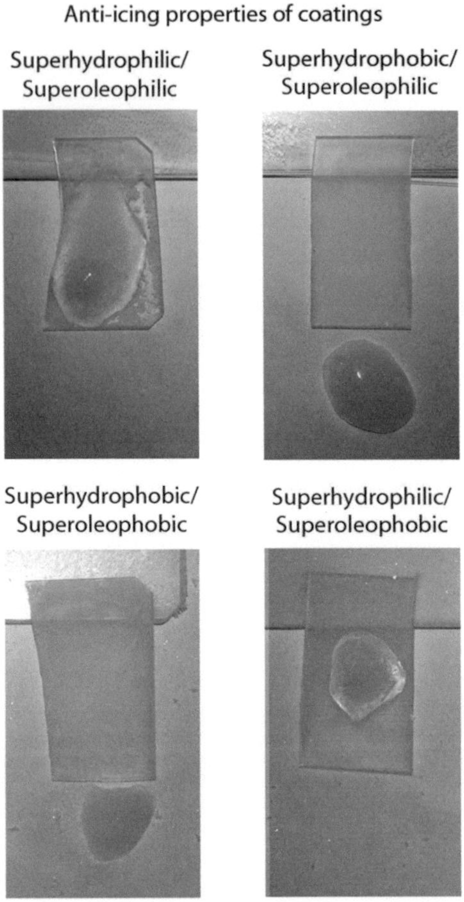

Fig. 9.32 Photographs of the four layer-by-layer composite coatings after freezing and deposition of supercooled water. The water immediately froze upon contact with the hydrophilic coatings while the droplets were able to roll off the hydrophobic coatings before freezing (Adapted from Brown and Bhushan 2015c)

flow of a droplet, desirable for anti-icing, the surface should be superhydrophobic with a low tilt angle. As indicated earlier, some oleophobicity may also be desirable.

Anti-fogging

To measure the anti-fogging properties, all four coatings were placed directly above a source of boiling water for 5 s. The samples were then photographed to assess their transparency, as shown in Fig. 9.33 (Brown and Bhushan 2015c). Both superhydrophilic coatings were found to retain their transparency, with text remaining visible through the condensed water layer. Condensed droplets on superhydrophilic surface spread out into a thin film and evaporate rapidly to maintain transparency. In contrast, on the superhydrophobic coatings, the formation of discrete droplets of water results in samples that are opaque.

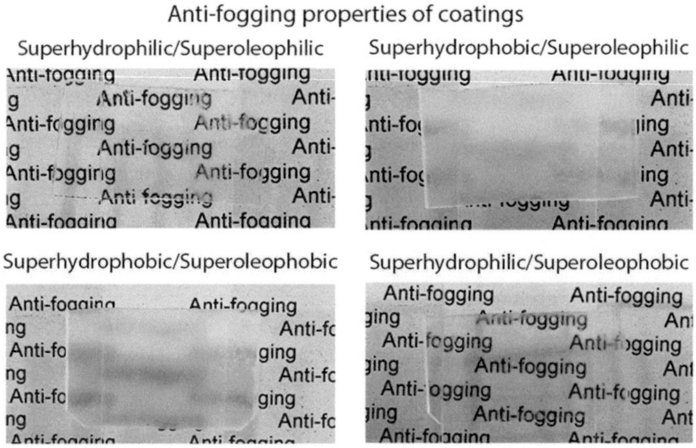

Fig. 9.33 Photographs of the four layer-by-layer composite coatings after exposure to water vapor. The hydrophilic coatings maintain optical transparency due to the formation of a thin water film on the surface. The hydrophobic coatings become opaque due to the formation of discrete water droplets on the surface (Adapted from Brown and Bhushan 2015c)

Optical Transparency

To test for transparency, text was placed directly behind the layer-by-layer composite coating samples, data shown for superhydrophilic/superoleophobic and superhydrophilic/superoleophilic coatings, shown in Fig. 9.34a (Brown and Bhushan 2015c). The text remained legible, suggesting that the coating displays characteristics of transparency. To measure degree of transparency, the transmission of visible light through the coatings was measured using a diffractive spectrometer. It was found to vary between 58 and 93% of that of uncoated glass, depending upon the wavelength and the specific coating, as shown in Fig. 9.34b. The superhydrophilic/superoleophilic coating was the most transparent, with transmittance of 70–93% over the visible spectrum. This coating had no functional layer whereas the other three coating types did have functional layers. A level of 70% visible light transmittance is acceptable for certain automotive applications (Thomsen et al. 2005). Further improvement in transparency can be made by decreasing the thickness of the nanoparticles and functional layers or reducing particle agglomeration (Brown and Bhushan 2015c).

Oil–Water Separation

The superhydrophilic/superoleophobic and superhydrophobic/superoleophilic coatings exhibit different responses to water and oil, and therefore are suitable for use as oil–water separators. Agitated oil–water mixtures were poured onto coated meshes suspended over beakers, as shown in Fig. 9.35 (Brown and Bhushan 2015c). For the superhydrophilic/superoleophobic-coated mesh, the water component of the mixture passed through, while

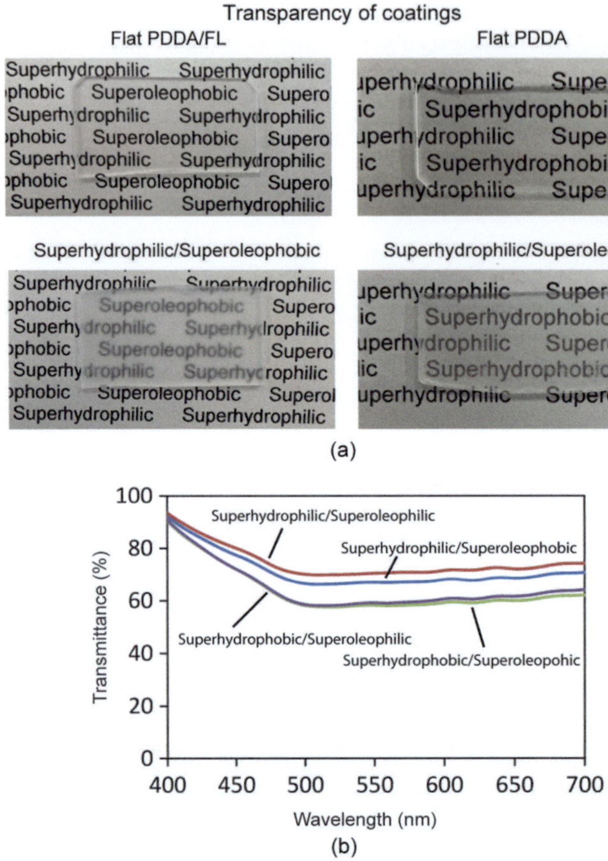

Fig. 9.34 **a** Photographs of flat, superhydrophilic/superoleophobic and superhydrophilic/superoleophilic layer-by-layer composite coatings. The flat coating appears optically transparent. Any reduction in transparency for the composite coating compared to the flat coating is due to the nanoparticles and functional layers. **b** Shows optical transmittance as a function of wavelength of light for all four composite coatings (Adapted from Brown and Bhushan 2015c)

the oil remained on top. However, for the superhydrophobic/superoleophilic-coated mesh, the opposite occurred, with the oil component passing through the mesh and the water remaining on top. Placing both meshes on an inclined plane resulted in the simultaneous collection of oil and water in two separate beakers (Brown and Bhushan 2015c).[1] The oil–water separation technique can be used for eco-friendly oil spill cleanup and water purification.

[1] For this research, authors received 2015 Institution of Chemical Engineers (UK) Global Award. https://youtu.be/3gqkirE-RG8

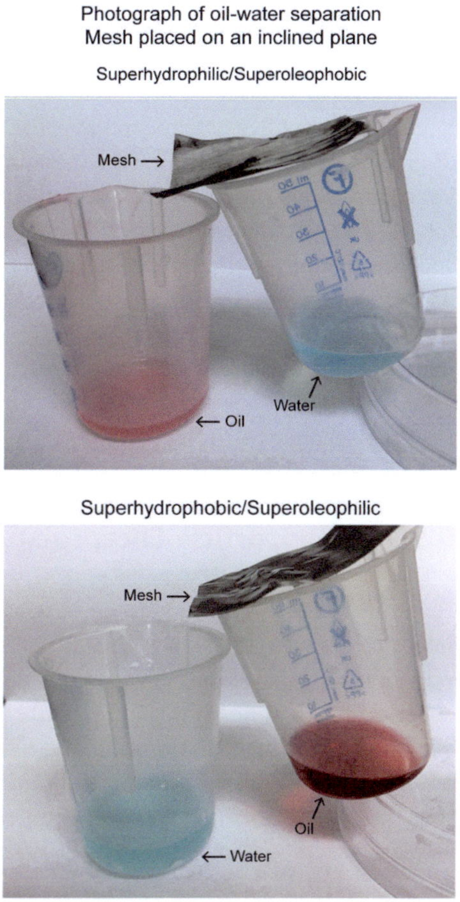

Fig. 9.35 Photographs of the superhydrophilic/superoleophobic and superhydrophobic/superoleophilic layer-by-layer composite coated stainless steel meshes acting as oil–water separators. On the superhydrophilic/superoleophobic coated mesh, oil collects on top of the mesh while water passes through. In contrast, on the superhydrophobic/superoleophilic coated mesh, oil passes through the mesh while the water remains on the top surface. If the meshes are placed at an angle and oil and water collected simultaneously in separate beakers. Oil and water dyes used to enhance contrast (Adapted from Brown and Bhushan 2015c)

9.4.3 Summary

Coatings with four possible combinations of water and oil repellency and affinity were produced using a layer-by-layer technique. In this technique, the charged layer-by-layer method was used for durability, plus the addition of a functional layer on top for the desired surface properties. The superoleophobic coatings display oil contact angles of $> 150°$ and tilt angles $< 5°$ and the superhydrophobic coatings display water contact angles of $> 160°$ and tilt angles $< 2°$. One coating combines both superoleophobic and superhydrophobic properties, while others can be used to mix and match oil and water repellency and affinity (Brown and Bhushan 2015a, b, c).

The coatings were found to be wear resistant, with micro- and macrowear experiments not causing any considerable damage due to the hard SiO_2 nanoparticles and the electrostatic interaction between the base layers. Additionally, these surfaces were found to display characteristics of transparency, with an averaged transmission of 75% and text remaining visible through the coating. This level of transparency is acceptable for certain automotive applications.

Superhydrophilic/superoleophilic coatings could find use in anti-fogging. Superhydrophobic/superoleophilic coatings could be used for self-cleaning, anti-fouling, anti-icing, and oil–water separation. The superhydrophobic/superoleophobic coating is suitable for self-cleaning, anti-fouling, anti-smudge, and anti-icing. Finally, the superhydrophilic/superoleophobic coating could be used for anti-fouling, anti-smudge, anti-fogging, and oil–water separation. This particular coating could be useful in anti-biofouling, where superoleophobicity, superhydrophilicity, and nanostructuring all contribute to reducing microorganism attachment. Additionally, when superhydrophobic/superoleophilic and superhydrophilic/superoleophobic coatings were applied to a porous substrate, they were found to separate oil from water.

9.5 Nanoparticle-Encapsulation Technique

Polymers are used in various products. Water and oil-repellent plastic surfaces are of interest for a wide range of industrial and consumer applications including medical, transportation, aerospace, energy, and construction. Liquid repellency, mechanical durability, self-cleaning, and finger touch resistance are among the properties of interest. To achieve superoleophobicity, nanotexturing of polymer surfaces and a fluorinated layer are needed. Mechanical durability is important if these surfaces are to be feasible for applications. Various methods for creating nanotextured polymer surfaces typically rely on techniques such as plasma etching (Gogolides et al. 2010) or photolithography (Nie and Kumacheva 2008), top-down approaches that are expensive and time-consuming (Mijatovic et al. 2005). The nanoparticle-binder coatings have been deposited as described earlier which results in an interface which may be susceptible to interfacial failure. Alternatively, the roughness of a polymer surface can be altered through the encapsulation of nanoparticles (Brown and Bhushan 2016b, c).

For a polymer surface, a suitable solvent is selected using solvent compatibility charts. Solvents with dissolved nanoparticles are used to attack the surface and encapsulate nanoparticles. Details on superoleophobic polycarbonate and polypropylene surfaces follow, based on Brown and Bhushan (2016b, c).

9.5.1 Polycarbonate Surfaces

Polycarbonate is used in various applications including catheters, smart screens in electronic displays and automotive dash. Polycarbonate dissolves readily in acetone. It is known to undergo solvent-induced phase transformation when exposed to acetone and

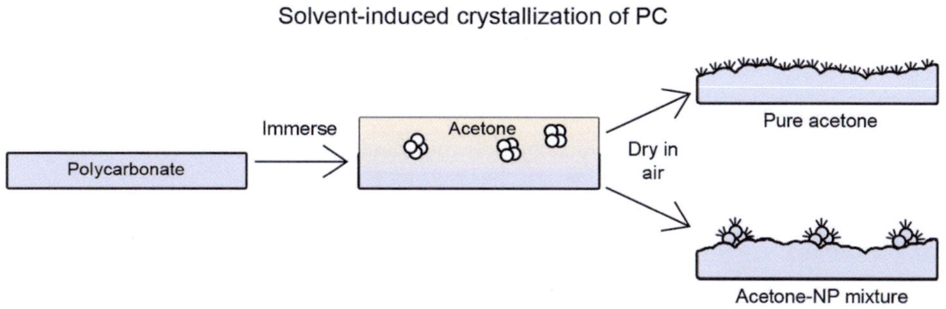

Fig. 9.36 Schematic of solvent-induced crystallization of polycarbonate using either pure acetone or an acetone-NP mixture (Adapted from Brown and Bhushan 2016b)

the resulting crystallization leads to a rough, superhydrophobic surface (Zhao et al. 2006; Cui et al. 2012.

To create a superoleophobic surface, polycarbonate was treated with an acetone–nanoparticle (acetone–NP) mixture (Brown and Bhushan 2016b). It was found that the nanoparticles become incorporated into the polymer during swelling and the subsequent polymer crystallization occurs, Fig. 9.36 (Brown and Bhushan, 2016b). By creating polymer–NP composite surfaces in this way, re-entrant geometries were achieved as the nanoparticle agglomerates were impregnated into the surface and near-surface region of the polymer. Since the nanoparticles are only incorporated near the interface, this technique could be advantageous compared to other polymer nanocomposite fabrication techniques where nanoparticles are distributed throughout the polymer, affecting the properties of the bulk material. Roughened polymer surfaces were then fluorinated via chemical vapor deposition of fluorosilane. Surfaces were characterized for wettability and wear resistance. Details based on Brown and Bhushan (2016b) follow.

Experimental Details

Polycarbonate sheets (PC, Lexan 9030, SABIC Innovative Plastics IP BV) cut to dimensions of 25 by 10 mm were used throughout. Hydrophilic silica nanoparticles (NP, 7 nm diameter, Aerosil 380, Evonik Industries) were dispersed in acetone (Fisher Scientific Inc.) using an ultrasonic homogenizer (Branson Sonifier 450A, 20 kHz frequency at 35% amplitude) at various concentrations and optimized for wettability. The polycarbonate samples were immersed in the acetone-NP mixture for 5 min before being removed and allowed to dry in air. To activate the surface for silane attachment, samples were UV irradiated for 40 min (15 W, $\lambda_{max} = 254$ nm) or treated with O_2 plasma (Plasmalab System 100, Oxford Instruments) at an O_2 flow rate of 20 sccm and a power of 40 W for 2 min. Both techniques were found to result in similar levels of activation and only UV irradiated samples are reported. Samples were fluorinated via chemical vapor deposition of a silane. One drop of trichloro(1H,1H,2H,2H-perfluorooctyl) silane (fluorosilane,

Sigma Aldrich) was deposited next to the samples which were covered and left for 6 h (Brown and Bhushan 2016b).

Results and Discussion

The wettability, wear resistance, and optical transparency of nanostructured surfaces are presented in this section.

(a) Wettability

Untreated polycarbonate (flat PC) is slightly hydrophilic with water contact angles of $76 \pm 1°$, Table 9.7 (Brown and Bhushan 2016b). After immersion in acetone for 5 min followed by drying in air, the surface became opaque; suggesting crystallization of the polymer had taken place. The solvent-induced phase transformation led to a hierarchical structure of crystalline nanostructures atop micron-sized mounds, Fig. 9.37a. This roughness was sufficient to create a superhydrophobic surface with water droplets displaying contact angles of $158 \pm 1°$ with low tilt angles due to the ability of the hierarchical surface to trap air and thus create a composite interface. However, due to the low contact angles for hexadecane on flat polycarbonate (ca. 12°), the acetone-treated surface is superoleophilic with hexadecane fully wetting the surface. Activation of the surface by UV irradiation followed by fluorination via silane attachment led to hexadecane angles of $76 \pm 2°$ for the flat surface and $100 \pm 2°$ for the acetone-treated surface, Table 9.7. This low angle is due to the lower surface tension of hexadecane being unable to support a composite interface on the hierarchical structure of the acetone-treated polycarbonate (Brown and Bhushan 2016b).

To create superoleophilic surfaces, nanoparticles were added to the acetone solvent used during the solvent-induced phase transformation of polycarbonate. The resulting

Table 9.7 Comparison of static contact angles and tilt angles for water and hexadecane droplets deposited on nanoparticle-encapsulated polycarbonate surfaces (Adapted from Brown and Bhushan 2016b)

Surface	Water		Hexadecane	
	Contact angle (°)	Tilt angle (°)	Contact angle (°)	Tilt angle (°)
Flat PC	76 ± 1	N/A	12 ± 2	N/A
Flat PC + fluorosilane	110 ± 2	N/A	76 ± 2	N/A
Acetone treated PC	158 ± 1	3 ± 1	~0	N/A
Acetone treated PC + fluorosilane	163 ± 2	2 ± 1	100 ± 2	N/A
PC–NP composite surface	30 ± 2	N/A	~0	N/A
PC–NP composite surface + fluorosilane	165 ± 2	2 ± 1	154 ± 2	5 ± 2

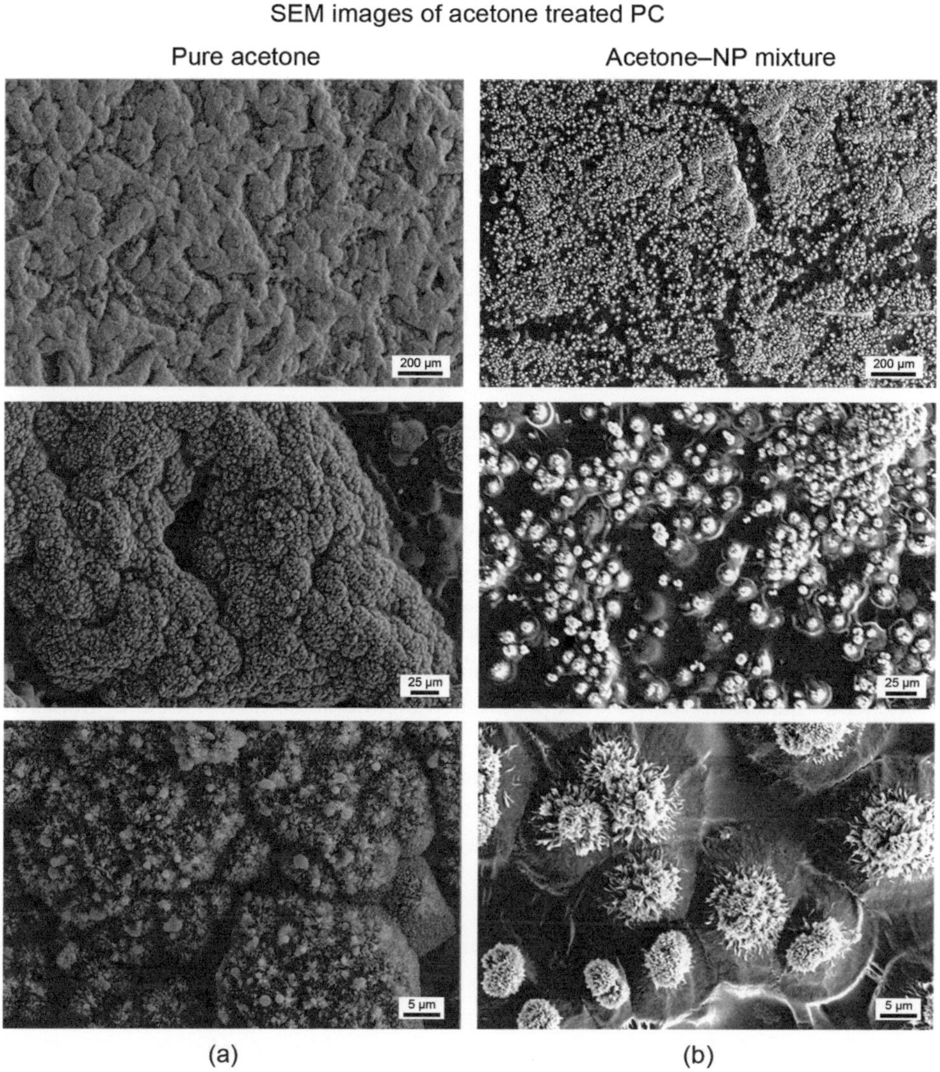

Fig. 9.37 SEM images of solvent-induced crystallization of polycarbonate using either pure acetone or an optimized acetone-NP mixture. **a** For pure acetone, the resulting surface contains good coverage of nanostructures atop micron-sized mounds. **b** For the actone-NP mixture, the nanostructures are limited to discrete micron-sized spherulites due to the nanoparticle aggregates acting as nucleation sites for the polycarbonate crystallization (Adapted from Brown and Bhushan 2016b)

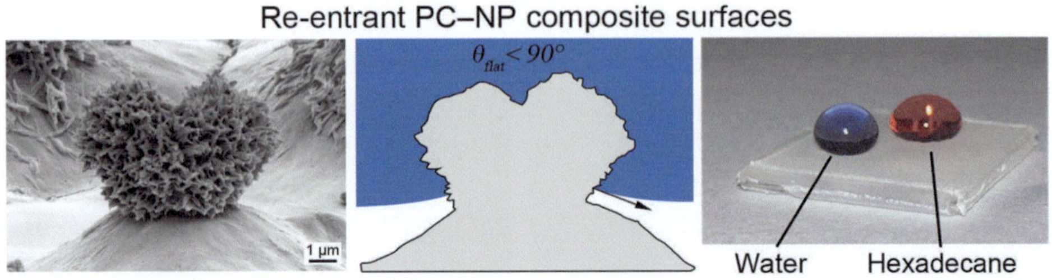

Fig. 9.38 SEM images of the PC–NP composite surface displaying re-entrant geometry. This re-entrant surface, once fluorinated, was found to be repellent towards both hexadecane and water (Adapted from Brown and Bhushan 2016b)

surface was found to be dramatically different from that of the acetone-treated polycarbonate. Instead of a consistent coverage of nanostructures atop micron mounds, the nanotexturing was instead limited to discrete micron-sized spherulites, Fig. 9.37b. This is believed to be due to the presence of nanoparticle agglomerates, which are incorporated into the surface and near-surface during polymer swelling and act as nucleation sites for polycarbonate crystallization during evaporation of the solvent. The result is re-entrant, hierarchical structures as shown in Fig. 9.38 (Brown and Bhushan 2016b).

The water contact angle of the polycarbonate-nanoparticle (PC–NP) composite surface was $30 \pm 2°$ indicating the presence of the hydrophilic silica nanoparticle agglomerates on the surface. After UV activation and fluorosilane treatment, the surface became superhydrophobic and superoleophobic with contact angles of $165 \pm 2°$ and $154 \pm 2°$ for water and hexadecane respectively, Table 9.7 (Brown and Bhushan 2016b). The presence of the re-entrant, hierarchical structures enables the formation of a favorable liquid–vapor interface for lower surface tension liquids, such as hexadecane, that exhibit contact angles < 90° on the corresponding flat surface.

To optimize the tilt angle of the surface, the concentration of nanoparticles in the acetone solvent was varied. An optimum hexadecane tilt angle of 5° was found on the fluorinated PC–NP composite surface resulting from an acetone–NP mixture concentration of around 8 mg mL^{-1}, Table 9.7. At this concentration, discrete spherulites were formed and their re-entrant geometry resulted in good oil repellency and a low hexadecane tilt angle (Brown and Bhushan 2016b).

(b) Recrystallization

Brown and Bhushan (2016b) carried out combined atomic force microscopy/infrared spectroscopy (AFM-IR) measurements to further analyze the polycarbonate-nanoparticle (PC-NP) composite surface. They confirmed that the nanoparticle agglomerates become imbedded into the polymer surface, Fig. 9.39. Infrared bands characteristic of polycarbonate were present on both the untextured surface (region A) and the re-entrant hierarchical structures (regions B-D). The major infrared peak classifications were found to be as

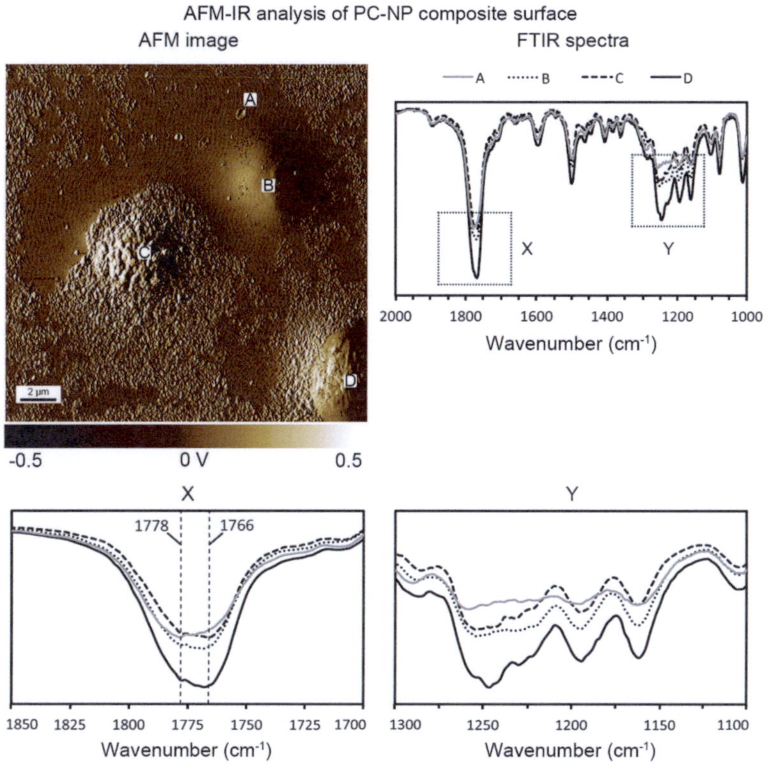

Fig. 9.39 AFM-IR analysis of PC-NP composite surface. The AFM map displays the untextured region (**A**) and the re-entrant hierarchical structures (**B–D**) selected for IR analysis. The FTIR spectra confirm that polycarbona te is present in all regions and that the polymer crystallinity increases on the re-entrant hierarchical structures, suggesting solvent-induced polymer recrystallization occurs primarily on the nanoparticle agglomerates (Adapted from Brown and Bhushan 2016b)

follows: C = O stretch (1775 cm^{-1}), C = C stretch (1496 cm^{-1}), and C–O–C bands (1300–1100 cm^{-1}). Further analysis of the carbonyl stretches revealed that it consists of two peaks at 1778 and 1766 cm^{-1}, the relative intensities of which are known to reflect the amount of non-crystalline and crystalline polycarbonate, respectively (Brown and Bhushan 2016b). By comparing the infrared spectra from the untextured region with that from the re-entrant hierarchical structures, it was found that the relative intensity of the 1766 cm^{-1} band increased, suggesting an increase in crystallinity. Sharper, more defined peaks from the re-entrant hierarchical structures compared to the untextured polymer also suggest an increase in crystallinity. Brown and Bhushan (2016b) concluded that these observations support the theory that solvent-induced recrystallization of the polycarbonate occurs predominately at the nanoparticle agglomerates, likely due to the agglomerates acting as nucleation points.

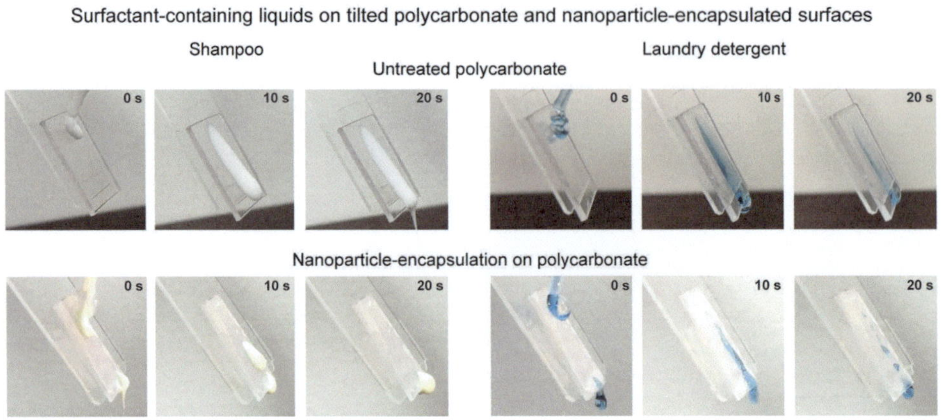

Fig. 9.40 Photographs demonstrating repellency of surfactant-containing liquids by comparing shampoo and laundry detergent droplets deposited on untreated and nanoparticle-encapsulated polycarbonate surfaces. On the untreated substrates, the droplet spreads out and adheres to the surface. On the coated substrates, the droplet rolls and slides off the surface without any residue remaining (Adapted from Bhushan 2018)

(c) Repellency of Surfactant-Containing Liquids

Droplets of shampoo and laundry detergent were found to roll or slide from coated glass and PP surfaces with no noticeable contamination as shown in Fig. 9.40 (Brown 2016). This is in contrast to the untreated glass and PP substrates where both shampoo and laundry detergent were found to spread and foul the surface.

(d) Wear Resistance

Wear resistance of the nanoparticle-encapsulated polycarbonate surface was investigated through the use of ball-on-flat tribometer wear experiments. Optical images showing a portion of the wear track are displayed in Fig. 9.41 (Brown and Bhushan 2016b). The initial wear experiments were carried out with a load of 20 mN, however few defects were found on the surface after this experiment and the coating remained oil repellent. To further test the durability of the surface, the load was increased to 45 mN. This increased load resulted in an observable wear scar, upon which hexadecane droplets were found to pin when rolled over the wear scar location. However higher magnification images confirmed that the surface features were not completely destroyed and, after re-application of the fluorosilane, the surface was found to regain its oil repellency suggesting that only the low surface tension material was removed at the higher load and that any damage to the underlying polymer–nanoparticle composite structures was minimal (Brown and Bhushan 2016b).

These surfaces were found to maintain their repellency towards water and oil after 6 months of storage with no noticeable degradation in their liquid repellent properties.

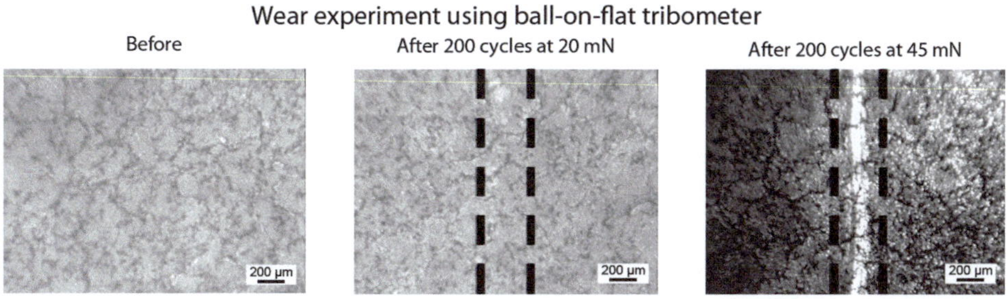

Fig. 9.41 Optical micrographs before and after wear experiments using ball-on-flat tribometer at 20 mN and 45 mN loadings for fluorinated PC–NP composite surfaces (Adapted from Brown and Bhushan 2016b)

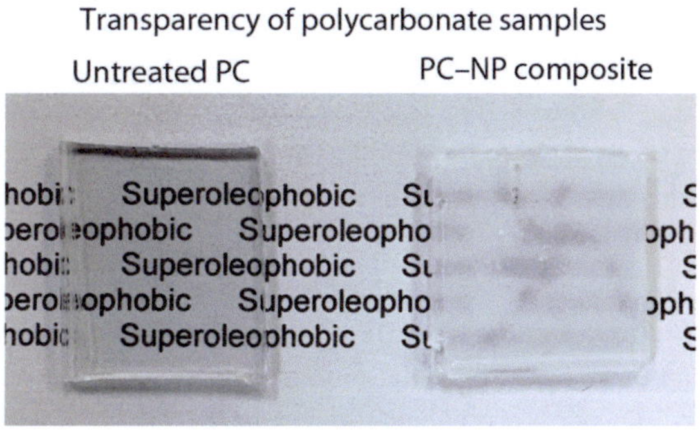

Fig. 9.42 Photographs showing optical transparency for untreated and nanoparticle-encapsulated-polycarbonate surfaces. The text is blurred for the treated sample (Adapted from Bhushan 2018)

(e) Optical Transparency

For transparency, text was placed behind the nanoparticle-encapsulated polycarbonate as well as untreated polycarbonate substrate and the samples were photographed, Fig. 9.42 (Brown 2016). The text was blurred for the treated sample whereas the treated substrate was transparent. Agglomerate encapsulation compromised the sample transparency.

Summary

Mechanically durable, superoleophobic polycarbonate surfaces were created involving the impregnation of polycarbonate with silica nanoparticles during solvent-induced phase transformation. Following treatment, re-entrant structures were present on the surface due to the nanoparticle agglomerates acting as nucleation sites for polymer crystallization. After fluorosilane treatment, the surfaces displayed both high contact angles and low tilt angles for water and hexadecane and repellency to shampoo and detergent. These surfaces were found to be durable (Brown and Bhushan 2016b).

9.5.2 Polypropylene Surfaces

Polypropylene is used in a wide variety of applications including consumer packaging, automotive parts, and textiles. Many of these applications require water and oil-repellency, such as in packaging, where it is desirable for the product to be repelled from and not foul the container, reducing waste and improving the customer experience. Polypropylene is commonly found in the bottles and caps used for shampoos, laundry detergents, and soaps. Such products contain surfactants to aid in cleaning, however they attach to the walls of the packaging. The polypropylene surface could be treated to make repellent to water, oils, and low surface tension liquids as well as surfactants (Brown and Bhushan 2016c).

Water repellent polypropylene surfaces have been fabricated by solvent casting a polypropylene film from a mixture of good (xylene) and bad (methyl ethyl ketone) solvents (which dissolve and precipitate the polymer respectively) (Erbil et al. 2003) or by dipping the substrate in a xylene–NP solution (Contreras et al. 2014). Polypropylene surfaces that are repellent to water and oils were fabricated by incorporation of nanoparticles into the polymer surface by Brown and Bhushan (2016c). Unlike polycarbonate, polypropylene does not undergo solvent-induced phase transformation when exposed to a solvent. When exposed to hot xylene solvent, it melts which causes softening and dissolves nanoparticles present in the solvent. Nanoparticles can be spin coated with a xylene-nanoparticle-polypropylene mixture and then heated. Polypropylene in the solution helps bind the particles to the polypropylene substrate.

The polypropylene sheet was spin coated with a xylene–nanoparticle–polypropylene (xylene–NP–PP) mixture heated to 135 °C, as shown in Fig. 9.43 (Brown and Bhushan 2016c). By creating polymer–NP composite surfaces in this way, re-entrant geometries suitable for oil repellency was achieved. Nanoparticles incorporated deeper into the polymer would have a less pronounced effect on the surface topography. Small, hydrophilic SiO_2 nanoparticles (Aerosil 380, Evonik Industries) were used as these have been shown to be susceptible to irreversible aggregation, resulting in non-spherical, micron-sized clusters (Pham et al. 2007). To achieve the oil repellency, the polymer-nanoparticle composite surface was then activated using UV irradiation and treated with fluorosilane.

It is expected that this technique can be used on various substrates in addition to polypropylene. Details based on Brown and Bhushan (2016c) follow.

Experimental Details

Samples with dimensions of 25 by 10 mm were cut from a polypropylene sheet (PP, ASTM D4101-0112, SPI) and were used. Silica nanoparticles (NP, 7 nm diameter, Aerosil 380, Evonik Industries) were dispersed in xylene (Mallinckrodt Baker, Inc.) using an ultrasonic homogenizer (Branson Sonifier 450A, 20 kHz frequency at 35% amplitude) at various concentrations. Then, 0.6 g of polypropylene was added to the mixture. The xylene–NP–PP mixture was heated to a temperature of 135 °C in a glass flask fitted

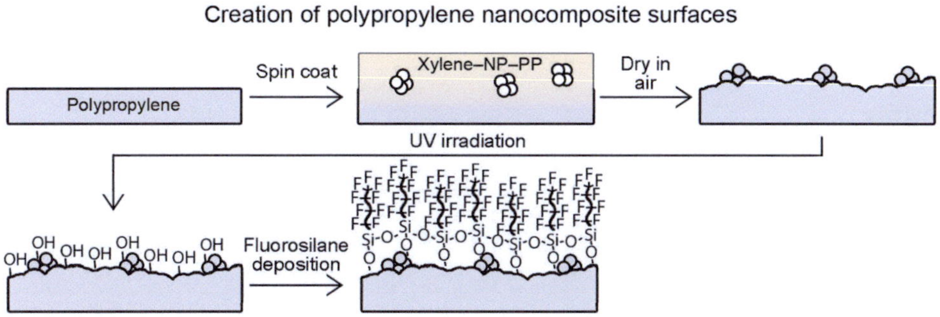

Fig. 9.43 Schematic showing fabrication of superoleophobic polypropylene (PP) using a xylene–NP–PP mixture (Adapted from Brown and Bhushan 2016c)

with a water-cooled condenser to prevent solvent evaporation and 1 mL of the mixture was added to the polypropylene surface via spin coating and the surface was allowed to dry in air. To activate the polymer surface for silane attachment, samples were UV irradiated for at least 12 h (15 W, $\lambda_{max} = 254$ nm). Samples were fluorinated via chemical vapor deposition of a silane, which was required to achieve superoleophobicity. One drop of trichloro(1H,1H,2H,2H-perfluorooctyl) silane (fluorosilane, Sigma-Aldrich) was deposited next to the samples which were covered and left for 2 h (Brown and Bhushan 2016c).

Results and Discussion

(a) Wettability

Untreated polypropylene (flat PP) is found to be slightly hydrophobic with water contact angles of $96 \pm 1°$, Table 9.8 (Brown and Bhushan 2016c). After treatment with the xylene–NP–PP mixture followed by drying in air, the surface became superhydrophobic, with water contact angle of $161 \pm 2°$ and tilt angles of $2 \pm 1°$. However, due to the low contact angles for hexadecane on flat polypropylene ($35 \pm 2°$), the xylene-treated surface is superoleophilic with hexadecane fully wetting the surface.

For superoleophobicity, a low surface energy material is required. After UV activation (to facilitate silane attachment) and fluorosilane treatment, the surface became superhydrophobic and superoleophobic with contact angles of $172 \pm 1°$ and $159 \pm 1°$ for water and hexadecane respectively, Table 9.8. To optimize the wettability of the surface, the concentration of nanoparticles in the xylene solvent was varied. The hexadecane tilt angle was used as a measure of optimizing the surface treatment, Fig. 9.44 (Brown and Bhushan 2016c). An optimum hexadecane tilt angle of $4 \pm 1°$ was found on the fluorinated PP–NP composite surface resulting from a xylene–NP mixture concentration of around 8–9 mg mL^{-1}. Above this concentration, hexadecane tilt angle began to rise due to contact line pinning on the surface features.

Table 9.8 Comparison of static contact angles and tilt angles for water and hexadecane droplets deposited on nanoparticle-encapsulated polypropylene surfaces (Adapted from Brown and Bhushan 2016c)

Surface	Water		Hexadecane	
	Contact angle (°)	Tilt angle (°)	Contact angle (°)	Tilt angle (°)
Flat PP	96 ± 1	N/A	35 ± 2	N/A
Flat PP + fluorosilane	110 ± 2	N/A	76 ± 2	N/A
PP–NP composite surface	161 ± 2	2 ± 1	~0	N/A
PP–NP composite surface + fluorosilane	172 ± 1	1 ± 1	159 ± 1	4 ± 1

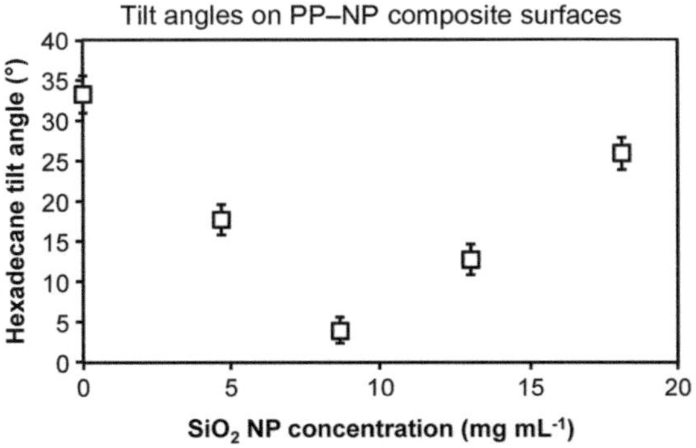

Fig. 9.44 Hexadecane tilt angles on fluorinated PP–NP composite surfaces as a function of SiO_2 NP concentration used during xylene–NP–PP treatment (Adapted from Brown and Bhushan 2016c)

(b) Surface Morphology

Inclusion of NPs in the xylene mixture leads to the formation of a rough surface. The PP dissolved in the xylene helps to bind the NPs to the substrate after spin coating. The resulting rough surface contains a hierarchical surface morphology with re-entrant, overhanging geometry, Fig. 9.45 (Brown and Bhushan 2016c). This morphology is sufficient to create a superhydrophobic state in water due to the ability of the hierarchical surface to trap air and thus create a composite interface. The combination of fluorosilane with the re-entrant geometry and hierarchical structures obtained from NP incorporation enables the surface to become superoleophobic.

(c) Shampoo Repellency

To test the surface repellency to shampoo containing surfactants, shampoo was poured over the superoleophobic polypropylene. During testing it was found that the superoleophobic polypropylene surface was able to repel the shampoo, with the viscous liquid sliding down the surface with no visible product remaining on the substrate, as seen in Fig. 9.46 (Brown and Bhushan 2016c). However, shampoo deposited onto untreated

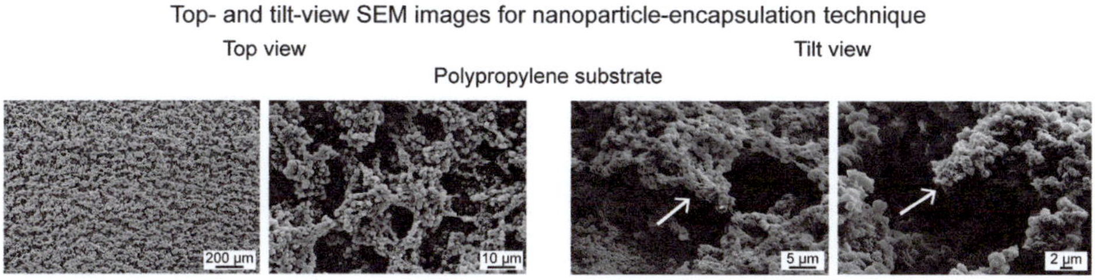

Fig. 9.45 SEM images of superoleophobic polypropylene after spin coating of an optimized xylene–NP–PP mixture. **a** Top view and **b** side view showing re-entrant, overhang geometry (sample tilted 70°) (Adapted from Brown and Bhushan 2016c)

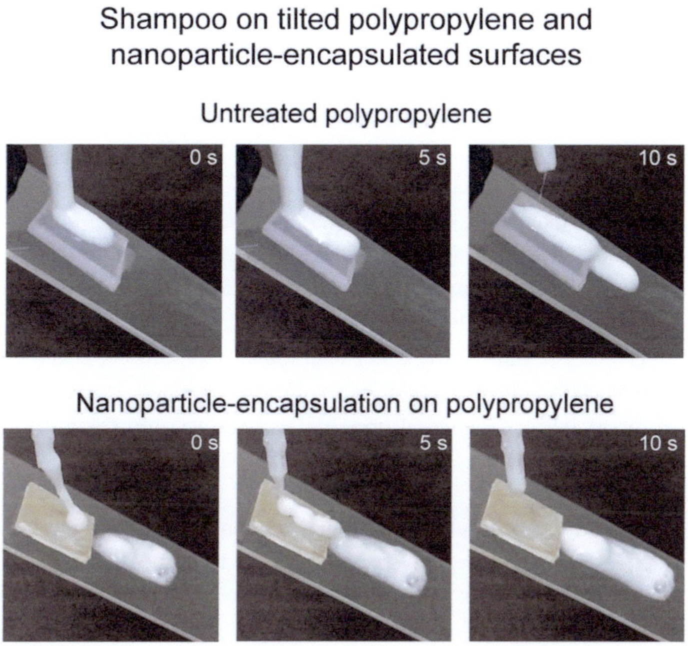

Fig. 9.46 Photographs demonstrating shampoo repellency of nanoparticle-encapsulated polypropylene compared to an untreated sample (Adapted from Brown and Bhushan 2016c)

polypropylene remains for a longer period of time and fouls the surface. Superoleophobic polypropylene surface was found to be also repellent to laundry detergent.

(d) Wear Resistance

Wear resistance of the nanoparticle-encapsulated-polypropylene surface was investigated using tribometer wear experiments and the resulting optical images, showing a portion

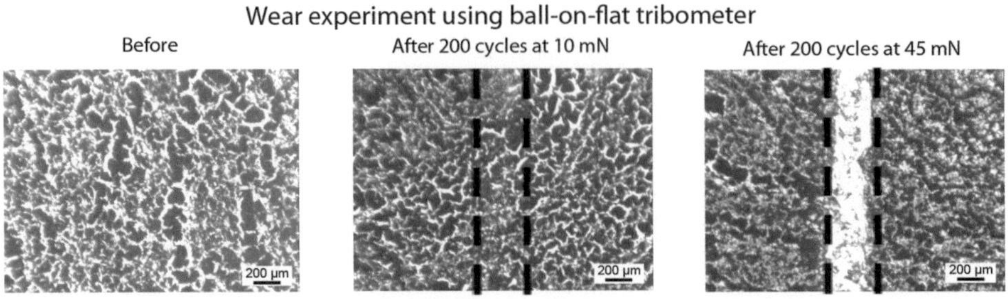

Fig. 9.47 Optical micrographs before and after wear experiments using ball-on-flat tribometer using a 3-mm diameter sapphire ball at 10 mN and 45 mN loadings for fluorinated PP–NP composite surfaces (Adapted from Brown and Bhushan 2016c)

of the wear track, are displayed in Fig. 9.47 (Brown and Bhushan 2016c). The initial wear experiments were carried out with a load of 10 mN, however no observable defect was found on the surface after this experiment and the surface remained oil repellent, suggesting strong fluorosilane attachment to the polymer. Therefore, to further test the durability of the surface, the load was increased to 45 mN. This increased load resulted in an observable wear scar, upon which hexadecane droplets were found to pin (tilt angle increases) when rolled over the wear scar location. However, the images confirmed that the surface features were not completely destroyed, and the surface could regain its oleophobicity after re-application of the fluorosilane layer. It is believed that these surfaces are significantly more durable than many other examples of superoleophobic surfaces, which typically rely on weakly adhered coatings or delicate surface structures. Many studies typically either report poor mechanical properties or fail to report any durability data.

(e) Self-cleaning

When the nanoparticle-encapsulated-polypropylene surface was contaminated with silicon carbide particles, most of the contaminants could be removed in a self-cleaning test where water droplets roll across the repellent surfaces, collecting particles in the process, Fig. 9.48 (Adapted from Martin et al. 2017). After the self-cleaning test, > 90% of the particles are removed. On the untreated polypropylene samples, most of the contaminants remain after the self-cleaning test. These coatings are self-cleaning due to their high water CA and low CAH. Water droplets deposited onto these samples are able to roll over the coating with little impediment, collecting less hydrophobic contaminants as they go (Brown and Bhushan 2015b, c, 2016c).

(f) Anti-smudge

For anti-smudge, the surfaces were contaminated as in the self-cleaning test, and a hexadecane-soaked cloth was used to wipe the surfaces (Bhushan and Muthiah 2013). The anti-smudge data for the nanoparticle-encapsulated-polypropylene sample are shown

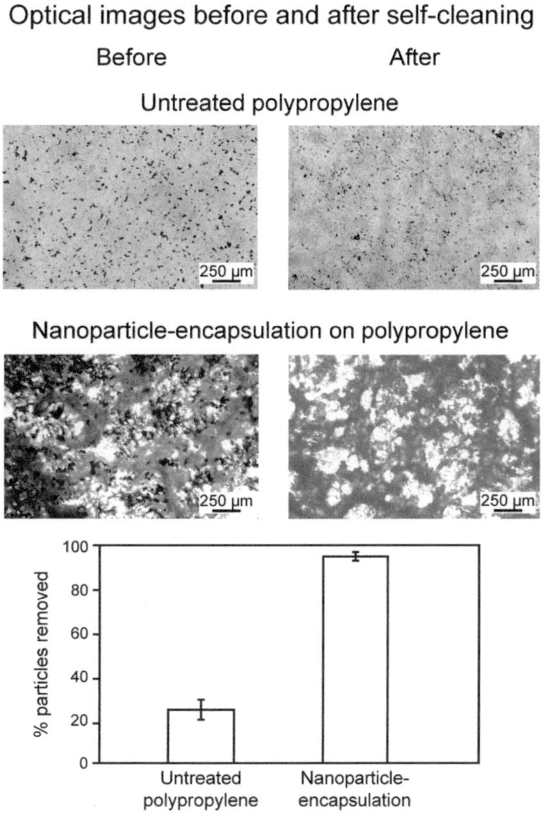

Fig. 9.48 Optical micrographs of contaminated surfaces before and after self-cleaning test on untreated and nanoparticle-encapsulated surfaces. Dark spots on coatings indicate silicon carbide particle contaminants. Image analysis suggests > 90% removal of particles on the treated samples (Adapted from Martin et al. 2017)

in Fig. 9.49. For the oil repellent surfaces, the particles were transferred to the cloth with no observable particles remaining on the surfaces. However, on the untreated polypropylene samples, many particles remained on the sample and oil was transferred from the cloth to the sample. Like the self-cleaning experiments with water, the anti-smudge property relies on a high CA and low CAH for the oil. The oil in the cloth is able to collect oleophilic contaminants from the surface of the coating without sticking to the surface (Brown and Bhushan 2016c).

(g) Optical Transparency

For transparency, text was placed behind the nanoparticle-encapsulated polypropylene as well as untreated polypropylene substrate and the samples were photographed, Fig. 9.50 (Brown 2016). The text was blurred for the treated sample whereas the treated substrate was transparent.

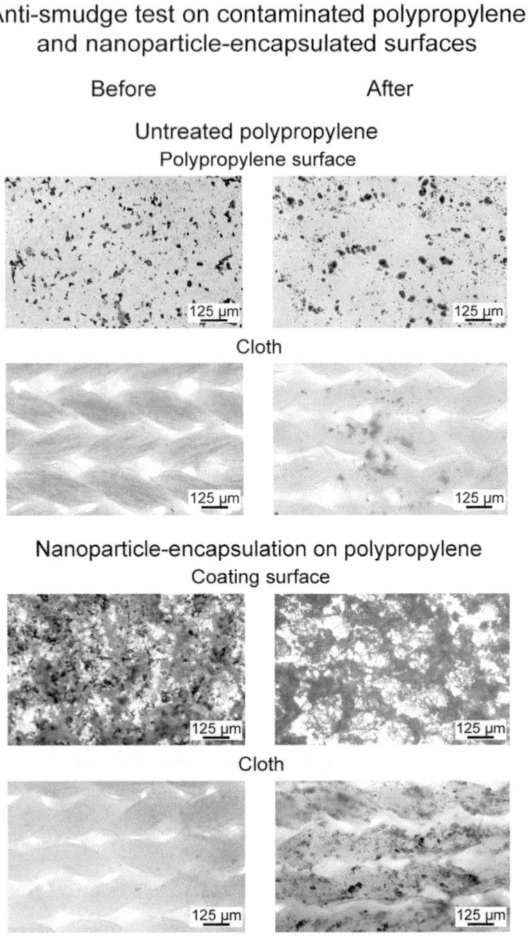

Fig. 9.49 Optical micrographs of contaminated surface and oil-impregnated microfiber cloth before and after anti-smudge test on untreated polypropylene and the nanoparticle-encapsulated surfaces. Dark spots on coatings and cloth indicate silicon carbide particle contaminants (Adapted from Martin et al. 2017)

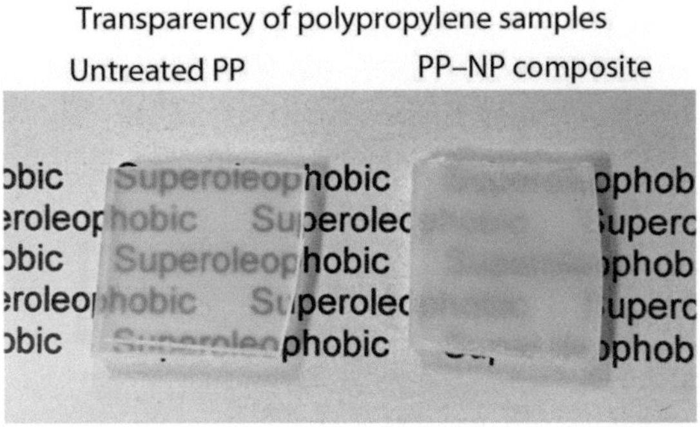

Fig. 9.50 Photographs showing optical transparency for untreated and nanoparticle-encapsulated-polypropylene surfaces. The text is blurred for the treated sample (Adapted from Bhushan 2018)

Summary

Mechanically durable, superoleophobic polypropylene surfaces were created with encapsulated nanoparticles by using the spin coating of a xylene–nanoparticle–polypropylene mixture. Following UV activation and fluorosilane coupling, the low surface energy material, combined with the re-entrant surface structures, results in a surface that displays both high contact angles and low tilt angles for water and hexadecane and is repellent to shampoo. These surfaces were found to be wear resistant with self-cleaning and anti-smudge properties (Brown and Bhushan 2016c).

9.6 Liquid Impregnation Technique

Roughness-induced liquiphobicity is commonly used. One drawback of utilizing surface roughness and trapped air as seen in the Cassie-Baxter regime is that the liquid exists in a metastable state and liquid resting on such a surface can eventually penetrate the roughness, transitioning from a Cassie-Baxter regime to a wetted state, referred to as the Wenzel regime. In doing so, the surface becomes wet by the liquid and is no longer repellent.

Another method of creating liquid-repellent surfaces is to take inspiration from the *Nepenthes* pitcher plants, as described earlier. Pitcher plants are well known for their carnivorous nature and feature pitcher-shaped traps that have evolved to capture and digest insects. The peristome of these pitcher plants features a regular microstructure that is wetted by nectar and rainwater, resulting in a continuous liquid film over the surface of the rim. When wet, the peristome becomes extremely slippery and insects aquaplane across the surface and fall into the trap (Brown and Bhushan 2017a, b).

To create a liquid-impregnated slippery surface, a porous surface is impregnated with a liquid that is immiscible with the liquid to be repelled. A liquid slide readily over another liquid surface resulting in a slippery surface. For fabrication of such surfaces, a porous structure needs to be created into which the impregnating liquid can wick. Second, the impregnating liquid (lubricating liquid) must preferentially wet the solid surface. Untreated, the porous surface likely has a much higher surface energy than the fluorinated lubricating liquid (surface energy < 20 mN m^{-1}). This means that, although the lubricating liquid coats the surface, there is little interaction between the two and a higher energy liquid that is added to the surface can displace the lubricating liquid. This can be mitigated by altering the chemistry of the porous structure through addition of a fluorosilane to match the chemistry of the lubricating liquid (Wong et al. 2011). This ensures that the solid–liquid interaction of the lubricating liquid will not be replaced by a more favorable solid–liquid interaction of the liquid being repelled. Finally, the lubricating liquid and the liquid to be repelled must be immiscible. If the two liquids are immiscible, the result is atomically flat liquid–liquid contact, which causes the repelled droplet to slide across the lubricating film with very little tilting of the surface (Brown and Bhushan 2017a, b).

In an early example, Wong et al. (2011) utilized random Teflon nanofibers or epoxy-molded nanoposts as the porous solid surface; however, their composition, fragility, and cost makes them unsuitable for certain real-world applications. Other examples requiring specific substrates included electrodeposited polypyrrole nanostructures (Kim et al. 2012) and nano-textured alumina (Ma et al. 2013). A third example involved spray coating wax to form the porous layer (Smith et al. 2013). The wax layer is sufficiently porous to be impregnated by the liquid layer, which is sprayed separately. The focus on spray coating means that this technique could be used in manufacturing processes; however, the durability of the porous wax coating remains unclear. Porous structures are likely to be more fragile than their planar equivalents and the long-term durability of the porous layer is crucial to ensure the lubricating liquid remains immobilized on the surface and is able to repel the liquid of interest.

Brown and Bhushan (2017a, b) presented two approaches to fabricate mechanically durable liquid-repellent polypropylene and polystyrene surfaces through the creation of a liquid-impregnated porous polymer surface. It is expected that polystyrene porous structures can be produced on a variety of substrates including polypropylene which makes this approach adaptable. As indicated earlier, polypropylene is a common polymer used in a wide variety of applications including consumer packaging, automotive parts, and textiles.

9.6.1 Porous Polypropylene Surface Created Using Solvent-Nonsolvent Mixture

Porous polypropylene surfaces have been created using a solvent–nonsolvent polymer solution by Erbil et al. (2003), Franco et al. (2008), and Brown and Bhushan (2017a). When the solvent–nonsolvent solution is added to polypropylene, it leads to separation into polymer-rich and polymer-poor phases, resulting in a porous structure after solvent evaporation. Brown and Bhushan (2017a) created porous polypropylene surfaces (which are hydrophobic) and treated with UV irradiation and fluorosilane coupling to ensure the lubricating liquid will remain impregnated within the polymer structure and no preferential dewetting will occur when the liquid to be repelled is added to the surface. The chemically modified porous polypropylene surfaces were then dipped into the lubricating liquid.

In order to create the porous polymer layer, polypropylene was dissolved in a good solvent, xylene, at 135 °C. Cyclohexanone (nonsolvent) was then added and a drop of the solution was cast onto a polypropylene substrate and dried in an oven at 70 °C. The presence of the nonsolvent leads to separation into two phases, one polymer-rich and the other polymer-poor (Erbil et al. 2003). This phase separation results in the crystallization of the polymer into the porous structure observed after solvent evaporation and drying. The porous surface was then activated using UV irradiation and treated with fluorosilane

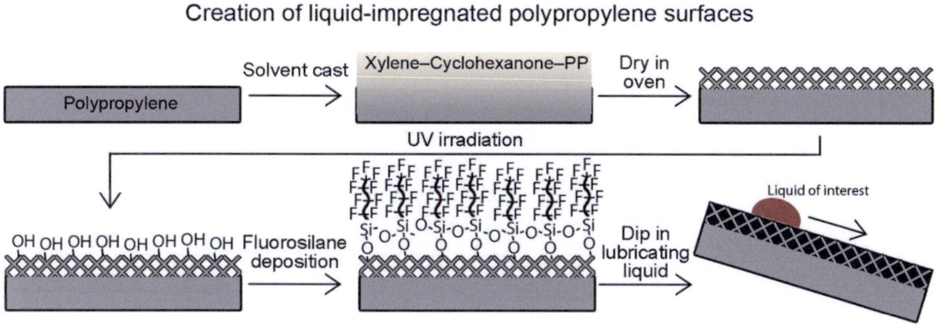

Fig. 9.51 Schematic showing the creation of liquid-impregnated porous polypropylene surfaces using a solvent–nonsolvent technique (Adapted from Brown and Bhushan 2017a)

to ensure that the impregnating liquid would preferably wet the surface. The sample was then dipped in the impregnating liquid to result in a liquid repellent surface, Fig. 9.51 (Brown and Bhushan 2017a).

Details based on Brown and Bhushan (2017a) follow.

Experimental Details

Samples cut to dimensions of 15 by 15 mm were cut for a polypropylene sheet (PP, ASTM D4101-0112, SPI) and used. 0.2 g of polypropylene was dissolved in 10 mL xylene (Mallinckrodt Baker, Inc.) heated to 135 °C in a glass flask fitted with a water-cooled condenser to prevent solvent evaporation. Once the polymer was fully dissolved, 4 mL cyclohexanone (Sigma-Aldrich) was added, and the mixture was stirred for 1 h. Then, a droplet of the solution was added to the polypropylene surface and the surface was dried in an oven at 70 °C for 1 h. To activate the polymer surface for silane attachment, samples were UV irradiated for at least 12 h (15 W, $\lambda_{max} = 254$ nm). Samples were fluorinated via chemical vapor deposition of a silane (to lower surface energy). One drop of trichloro(1H,1H,2H,2H-perfluorooctyl) silane (fluorosilane, Sigma-Aldrich) was deposited next to the samples which were covered and left for 2 h. The sample was then dipped into the impregnating liquid, in this instance a perfluoropolyether (Krytox GPL 102, Dupont) with a chemical structure of $F\text{-}(CF(CF_3)\text{-}CF_2\text{-}O)_n\text{-}CF_2CF_3$ where n = 10–60, a surface tension of 16–20 mN m^{-1}, and a viscosity of 38 cSt (Brown and Bhushan 2017a).

Results and Discussion

(a) Wettability

Untreated polypropylene (flat PP) was found to be slightly hydrophobic with water contact angles of $96 \pm 1°$. Once dried, the porous polymer layer was found to have a water

Table 9.9 Comparison of static contact angles and tilt angles for water and hexadecane droplets deposited on liquid-impregnated polypropylene surfaces (Adapted from Brown and Bhushan 2017a)

Surface	Water		Hexadecane	
	Contact angle (°)	Tilt angle (°)	Contact angle (°)	Tilt angle (°)
Flat PP	96 ± 1	N/A	35 ± 2	N/A
Porous PP	122 ± 2	N/A	17 ± 2	N/A
Porous PP + fluorosilane	136 ± 2	N/A	97 ± 2	N/A
Porous PP + fluorosilane + lubricating liquid	115 ± 2	2 ± 1	70 ± 1	4 ± 2

contact angle of $122 \pm 2°$ due to the increase in surface roughness, Table 9.9 (Brown and Bhushan 2017a). As indicated earlier, in order to create the liquid-impregnated surfaces, the chemistry of the porous polymer layer must first be altered to ensure favorable wetting with the lubricating liquid and no preferential dewetting when another liquid is added to the surface. To achieve this, the porous polypropylene was first activated via UV irradiation to facilitate silane attachment. Following fluorosilane treatment, the porous polypropylene surface displayed water contact angles of $136 \pm 2°$ and hexadecane contact angles of $97 \pm 2°$, Table 9.9 (Brown and Bhushan 2017a). This lowering of the surface energy is required to ensure that the lubricating liquid added in the next step will remain impregnated in the porous structure and will not be preferentially dewetted by the liquid to be repelled.

In the final step to create the liquid-impregnated surfaces, the porous polypropylene was dipped into the lubricating liquid. The surface exhibited water contact angles of $115 \pm 2°$ and hexadecane contact angles of $70 \pm 2°$, Table 9.9 (Brown and Bhushan 2017a). However, due to the presence of the lubricating liquid, the surface displayed very low tilt angles of $2 \pm 1°$ and $4 \pm 2°$ for water and hexadecane respectively, Table 9.9. Because the low tilt angles are a product of the homogeneity of the liquid–liquid interface, the surface tension of the liquid being repelled has little effect on the repellency of the surface. This is in contrast to lotus-inspired superoleophobic surfaces, where decreasing the surface tension of the oil being repelled results in an increase in the tilt angle (Brown and Bhushan 2015a).

If the porous polypropylene surface is not chemically modified by fluorosilane prior to the liquid-impregnation step, the resulting surface is not repellent to certain liquids, such as hexadecane. This is because, in order for the liquid-impregnated surfaces to be liquid-repellent, the lubricating film must have a strong affinity for the porous surface and must not be displaced by the liquid being repelled (Wong et al. 2011). In the absence of the fluorosilane treatment, droplets of hexadecane can displace the lubricating liquid and wet the porous structure, causing a loss of repellency. The fluorosilane ensures that the impregnating liquid is stable, and droplets of water and hexadecane remain on top of the lubricating film and can slide off with low tilt angles.

The low tilt angle means that liquid droplets placed on the surface are able to slide around the surface with very little tilting. In Fig. 9.52, droplets of hexadecane were added to untreated polypropylene and liquid-impregnated porous polypropylene surfaces (Brown and Bhushan 2017a). On the untreated surface, the hexadecane droplet spread readily and then spread further when the surface was tilted. In contrast, on the liquid-impregnated surface, the hexadecane droplet did not spread and retained a more spherical shape. In addition, as the surface was tilted, the hexadecane droplet slid across the surface with very little resistance. The red dye present in the hexadecane droplet helped to confirm that the vacated area of the surface was not contaminated by the hexadecane. If it was tilted for a sufficiently long time, the droplet was completely removed from the slippery surface.

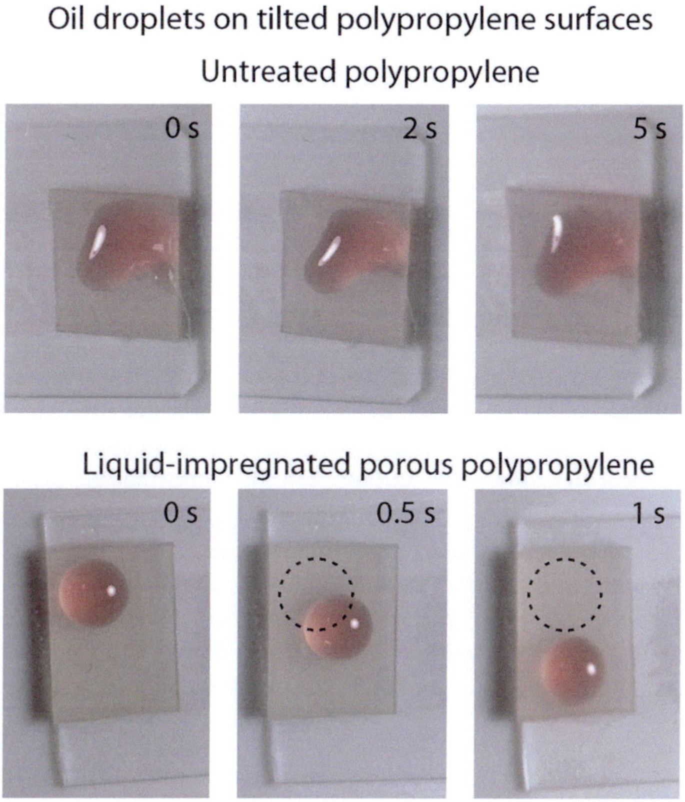

Fig. 9.52 Photographs demonstrating hexadecane repellency of liquid-impregnated porous polypropylene compared to an untreated sample (Adapted from Brown and Bhushan 2017a)

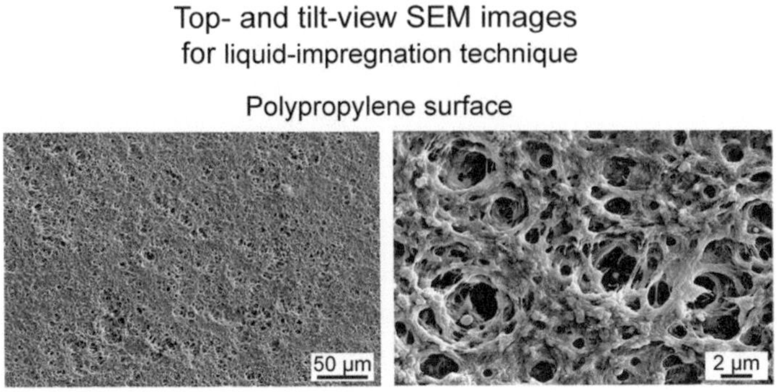

Fig. 9.53 SEM images of porous polypropylene after solvent casting from a solvent–nonsolvent polymer solution (Adapted from Martin et al. 2017)

(b) Surface Morphology

The SEM images of a porous polypropylene surface after solvent casting from a solvent-nonslovent polymer solution is shown in Fig. 9.53. When cast, the presence of the nonsolvent leads to separation into two phases, one phase being polymer-rich and the other phase being polymer-poor. This phase separation results in the crystallization of the polymer into the porous structure observed after solvent evaporation and drying.

(c) Repellency of Surfactant-Containing Liquids

To test the surface repellency against shampoo and laundry detergent, tests were performed using these liquids, as shown in Fig. 9.54 (Brown and Bhushan 2017a; Martin et al. 2017). When droplets of the two liquids were added to samples of untreated plastic, gravity helped them spread over the surface with the liquids pinned at the location they were added. This is in comparison to the liquid-impregnated samples, upon which both liquids were able to slide over with no contamination. In the case of shampoo, the highly viscous liquid slid from the surface in the shape of a droplet with a pronounced trailing edge. For the laundry detergent, the liquid spread more due to its lower viscosity. Though the bulk of the liquid quickly left the surface, a thin film of detergent briefly remained before slowly dewetting the surface due to gravity and leaving the sample clean (Brown and Bhushan 2017a).

(d) Wear Resistance

Wear resistance of the porous polymer layer was investigated through the use of ball-on-flat tribometer wear experiments. The resulting optical images, showing a portion of the wear track, are displayed in Fig. 9.55 (Brown and Bhushan 2017a). The images confirm that the surface features are not completely destroyed with minimal burnishing of the surface in the wear location. It is believed that these surfaces are significantly more durable than many other examples of liquid-impregnated surfaces, which typically rely on

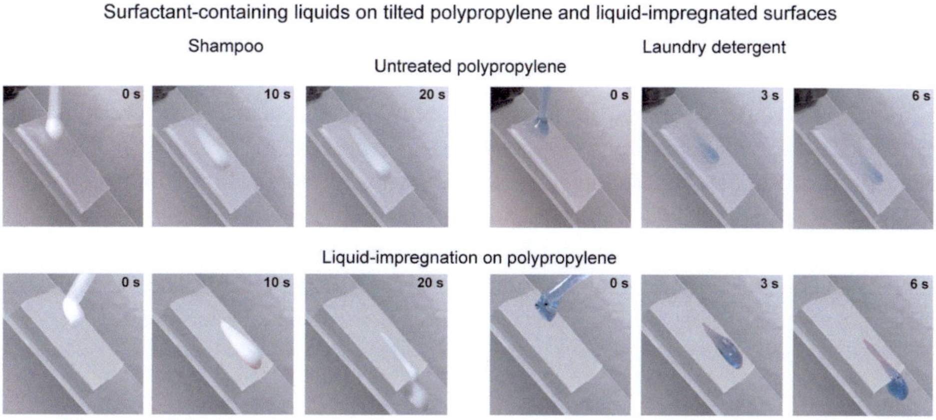

Fig. 9.54 Photographs demonstrating repellency of shampoo and laundry detergent by liquid-impregnated porous polypropylene compared to an untreated sample (Adapted from Brown and Bhushan 2017a; Martin et al. 2017)

poorly adhered wax coatings (Smith et al. 2013) or delicate surface structures (Wong et al. 2011). The durability of the porous layer is crucial since it ensures the lubricating liquid remains in place, impregnated within the pores. The porous structures required for this liquid repellency technique will be susceptible to higher contact stress than typical flat surfaces and so it is crucial that mechanical durability is reported (Brown and Bhushan 2017a).

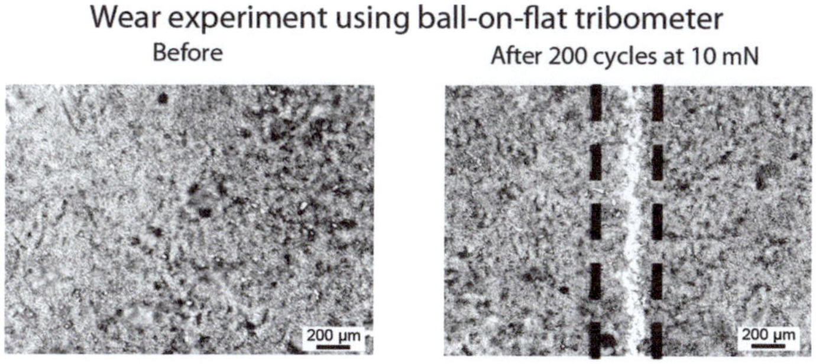

Fig. 9.55 Optical micrographs before and after wear experiments using ball-on-flat tribometer using a 3-mm diameter sapphire ball at 10 mN loading on a porous polypropylene surface (Adapted from Brown and Bhushan 2017a)

Summary

Liquid-impregnated slippery surfaces were created via the deposition of a porous polymer layer by utilizing a solvent–nonsolvent solution. It is believed that this method is more industrially viable and produces durable porous structures than those reported by other researchers. Following UV activation and fluorosilane coupling to reduce the surface energy of the pores, the porous substrate was dipped in a lubricating liquid, which became impregnated within the porous structure. This lubricating liquid provides the liquid-repellency through immiscible liquid–liquid contact, resulting in very low tilt angles and therefore good repellency with droplets of water, hexadecane, shampoo, and laundry detergent sliding from the surface with ease (Brown and Bhushan 2017a).

9.6.2 Porous Polystyrene Surfaces Created Using Breath Figures

One approach to create porous structure is to utilize breath figures (Rayleigh 1911; Widawski et al. 1994). Breath figures are two-dimensional hexagonally packed arrays of water droplets condensed onto a cooled surface. Such a breath figure can form on a drying polymer film surface, so long as the solvent used to cast the film is immiscible with water. Evaporation of the solvent leads to evaporative cooling of the film surface, resulting in water condensation and the formation of a breath figure array. Droplet coalescence is limited either by the increasing viscosity of the drying polymer film or by precipitation of the polymer at the water–solvent interface (Stenzel et al. 2006). This array of water droplets acts as a template for the drying polymer solution leading to the formation of the porous surface structure once the solvent and water have fully evaporated. Based on their appearance, such breath figure-templated surfaces are commonly called honeycomb surfaces (Brown and Bhushan 2017b).

Polystyrene is a common polymer used in the creation of honeycomb surfaces (Srinivasarao et al. 2001; Peng et al. 2004; Ferrari et al. 2011). A polystyrene coating was cast in a volatile, water-immiscible solvent on a glass substrate and allowed to dry in a humid environment (Brown and Bhushan 2017b). The resulting honeycomb surface was then activated using UV irradiation and treated with fluorosilane to ensure the lubricating liquid would remain impregnated within the polymer structure and no preferential dewetting would occur when the liquid to be repelled was added to the surface. The chemically modified honeycomb surfaces were then dipped into the lubricating liquid to realize a liquid-impregnated surface with superliquiphobicity, Fig. 9.56 (Brown and Bhushan 2017b). It is expected that these polystyrene honeycomb structures can be produced on a variety of substrates including polypropylene.

Details based on Brown and Bhushan (2017b) follow.

Experimental Details

Glass slides (Fisher Scientific) and a polypropylene sheet (PP, ASTM D4101-0112, SPI) cut to dimensions of 15 by 15 mm were used as substrates. 0.2 g of polystyrene

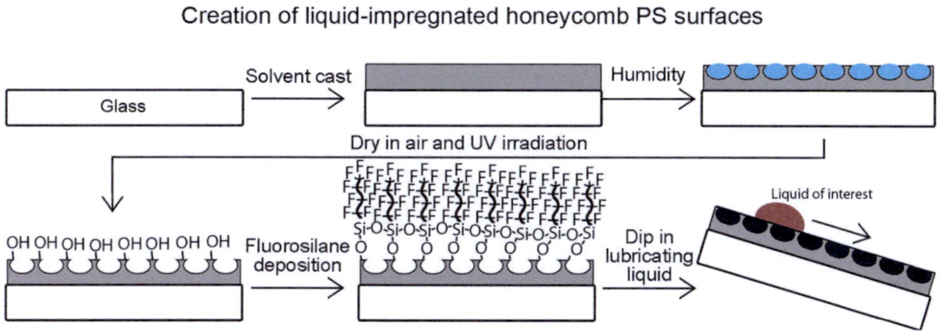

Fig. 9.56 Schematic showing various steps used in fabrication of liquid-impregnated honeycomb surfaces. Polymer dissolved in a water-immiscible, volatile solvent is first cast onto a substrate in a humid environment. Evaporative cooling leads to condensation of water and the formation of a breath figure array of droplets. The droplet array acts as a sacrificial template for the drying polymer film. The surface is then activated using UV irradiation followed by fluorosilane treatment and impregnated with a liquid, immiscible to liquid being repelled (Adapted from Brown and Bhushan 2017b)

(Mw ~ 350,000, Sigma-Aldrich) was dissolved in a water immiscible volatile solvent, 10 mL chloroform (Mallinckrodt Baker, Inc.) at room temperature. Once the polymer was fully dissolved, a droplet of the solution was added to the glass surface at room temperature and ambient humidity (~ 50% RH) and the surface was dried in a humid environment at room temperature. Evaporative cooling of the drying polymer film results in the condensation of water droplets and the formation of a breath figure. This array of water droplets acts as a sacrificial template for the drying polymer film. Once evaporation of the solvent and water is complete, it results in a porous, honeycomb surface structure, Fig. 9.57 (Brown and Bhushan 2017b).

Fig. 9.57 Optical images of a honeycomb surface on glass after solvent casting polystyrene from a water-immiscible, volatile solvent in a humid environment (Adapted from Brown and Bhushan 2017b)

Table 9.10 Comparison of static contact angles and tilt angles for water and hexadecane droplets deposited on liquid-impregnated polystyrene surfaces (Adapted from Brown and Bhushan 2017b)

Surface	Water		Hexadecane	
	Contact angle (°)	Tilt angle (°)	Contact angle (°)	Tilt angle (°)
Flat PS	94 ± 1	N/A	31 ± 2	N/A
Honeycomb PS	107 ± 2	N/A	11 ± 2	N/A
Honeycomb PS + fluorosilane	138 ± 2	N/A	107 ± 2	N/A
Honeycomb PS + fluorosilane + lubricating liquid	109 ± 2	2 ± 1	70 ± 1	4 ± 2

For the lubricating liquid to fully penetrate the porous surface, the chemistry of the honeycombs was altered, ensuring favorable wetting and no preferential dewetting when another liquid is added on top of the lubricating liquid layer. The polystyrene honeycomb coating was activated via UV irradiation for 30 min (15 W, $\lambda_{max} = 254$ nm) to activate the surface for silane attachment. Next, samples were fluorinated via chemical vapor deposition of a silane (to lower surface energy), which was required in order to ensure preferential wetting by the lubricating liquid. One drop of trichloro(1H,1H,2H,2H-perfluorooctyl) silane (fluorosilane, Sigma-Aldrich) was deposited next to the samples which were covered and left for 2 h. Following fluorosilane treatment, the polystyrene honeycomb surface displayed water contact angles of $138 \pm 2°$ and hexadecane contact angles of $107 \pm 2°$. This altering of the surface energy is necessary to ensure that the lubricating liquid, in this case a lower surface tension fluorinated oil, will remain impregnated in the honeycomb structure and will not be preferentially replaced by the liquid to be repelled (Brown and Bhushan 2017a, b).

As a last step, the honeycomb surface was dipped into the impregnated lubricating liquid, in this instance a perfluoropolyether (Krytox GPL 102, Dupont) with a chemical structure of $F\text{-}(CF(CF_3)\text{-}CF_2\text{-}O)_n\text{-}CF_2CF_3$ where n = 10–60, a surface tension of 16–20 mN m^{-1}, and a viscosity of 38 cSt. Following this, the liquid-impregnated surface exhibited water contact angles of $109 \pm 2°$ and hexadecane contact angles of $70 \pm 2°$. Due to the presence of the lubricating liquid, the surface displays very low tilt angles of $2 \pm 1°$ and $4 \pm 2°$ for water and hexadecane respectively, Table 9.10 (Brown and Bhushan 2017b). Because the low tilt angles are a product of the homogeneity of the liquid–liquid interface, the surface tension of the liquid being repelled should have little effect on the repellency of the surface.

Results and Discussion

(a) Wettability

Flat polystyrene (PS) was found to be slightly hydrophobic with water contact angles of $94 \pm 1°$. Fully dried, the polystyrene honeycomb surface was found to have a water contact angle of $107 \pm 2°$ due to the increase in surface roughness, Table 9.10 (Brown and Bhushan 2017b). Contact angles of liquid-impregnated surface was $109 \pm 2°$ and $70 \pm 2°$ for water and hexadecane, respectively. However, tilt angles were low, $2 \pm 1°$ and $4 \pm 2°$ for water and hexadecane, respectively.

The low tilt angle means that liquid droplets placed on the surface are able to slide over the surface easily. In Fig. 9.58, droplets of hexadecane were added to fluorosilane-treated honeycomb and liquid-impregnated honeycomb surfaces (Brown and Bhushan 2017b). Hexadecane droplets on the fluorosilane-treated surface were not easily removed when the surface was titled due to high hysteresis and droplet pinning. In contrast, as the liquid-impregnated surface was tilted, the hexadecane droplet slid across the surface with very little resistance. The red dye present in the hexadecane droplet helped to confirm that the vacated area of the surface was not contaminated by the hexadecane.

(b) Wear Resistance

Ball-on-flat tribometer wear experiments were performed at assess the mechanical durability of the polystyrene honeycomb surface. The resulting optical images, showing a portion of the wear track, are displayed in Fig. 9.59 (Brown and Bhushan 2017b). The images confirm that the polymer coating was not removed from the glass substrate. The

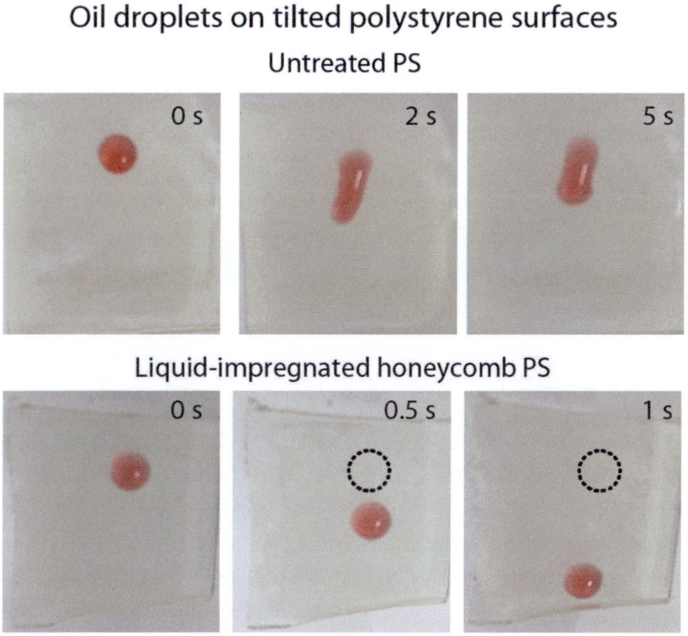

Fig. 9.58 Photographs demonstrating hexadecane repellency of liquid-impregnated honeycomb surface compared to a fluorinated honeycomb sample (Adapted from Brown and Bhushan 2017b)

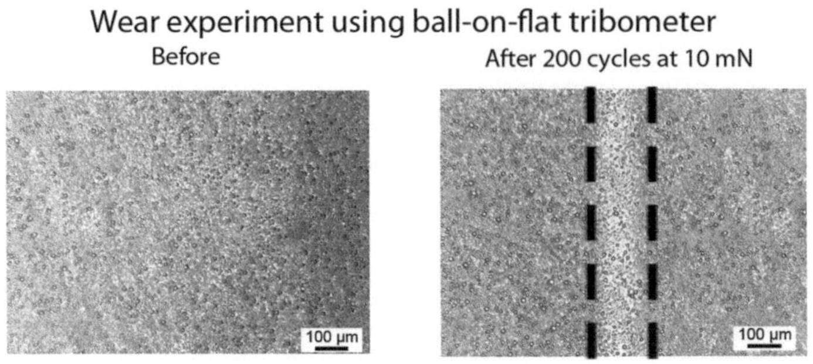

Fig. 9.59 Optical micrographs before and after wear experiments using ball-on-flat tribometer using a 3-mm diameter sapphire ball at 10 mN loading on a honeycomb surface (Adapted from Brown and Bhushan 2017b)

density of the honeycomb structure appeared to decrease in the wear location due to plastic deformation of the polymer. However, the porous structure is not completely destroyed, allowing for the impregnating liquid to remain in the wear region.

After wear experiments carried out on honeycomb surfaces containing the lubricating liquid, no change was found in the repellent properties of the surface. Droplets of hexadecane slid over the wear location with no noticeable degradation in the repellency (Brown and Bhushan 2017b).

Summary

Liquid-repellent, slippery surfaces have been created on glass substrates via the formation of a honeycomb structure using breath figures. Following UV activation and fluorosilane coupling to reduce the surface energy of the honeycombs, the substrate was dipped in a lubricating liquid, which became impregnated within the pores. This lubricating liquid layer repelled other liquids placed on the surface through immiscible liquid–liquid contact. The honeycomb structure was found to be mechanically durable (Brown and Bhushan 2017b).

9.7 Comparison of Various Roughness-Induced and Liquid Impregnation Techniques for Superoleophobicity

Three roughness-induced and one liquid impregnation techniques for creating superoleophobic surfaces are described in Sects. 9.3 to 9.6. Roughness-induced techniques include: nanoparticle-binder, layer-by-layer, and nanoparticle-encapsulation. Schematics of the four techniques are summarized in Fig. 9.60. In this section, data for these four techniques are compared. Based on the comparative analysis, pros and cons of various techniques are presented. This overview is based on Martin et al. (2017).

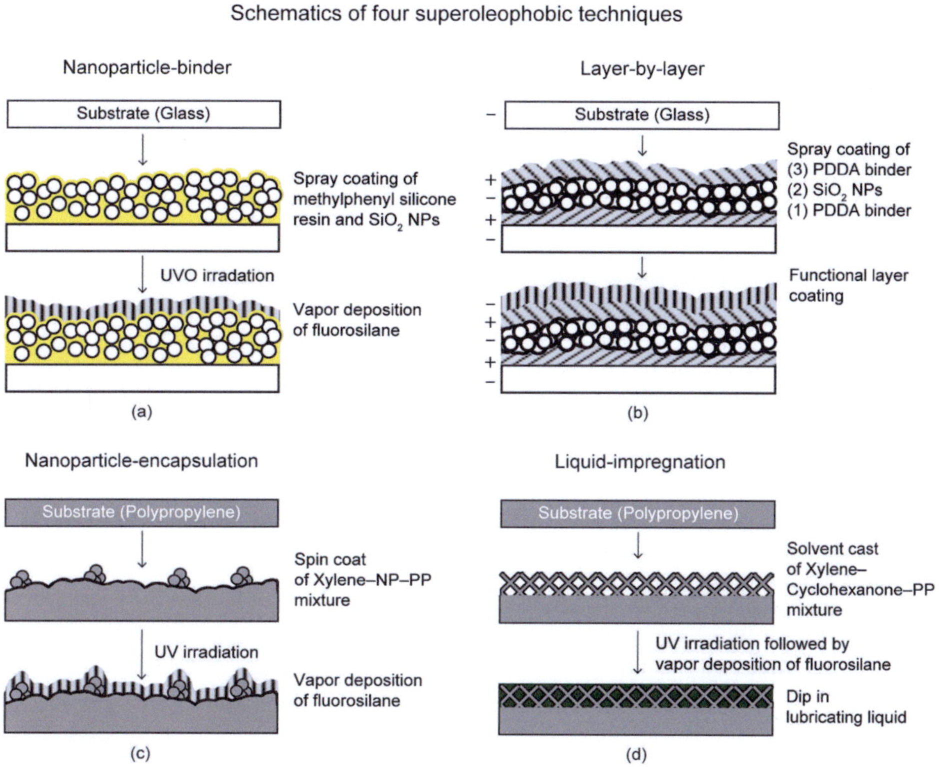

Fig. 9.60 Schematic of four superoleophobic techniques: **a** Nanoparticle-binder (Adapted from Bhushan and Martin 2018), **b** layer-by-layer (Adapted from Brown and Bhushan 2015c), **c** Nanoparticle-encapsulation (Adapted from Brown and Bhushan 2016c), and **d** Liquid-impregnation (Adapted from Brown and Bhushan 2017a)

9.7.1 Comparison of Data

The wettability, surface morphology, repellency of surfactant-containing liquids, wear resistance, self-cleaning, anti-smudge, anti-fouling, anti-icing, anti-fogging, and optical transparency for each of the four techniques are compared. Additionally, surfaces which can be used for oil–water separation are discussed.

Wettability

The CA and TA data using water and hexadecane droplets for the surfaces produced by three roughness-induced techniques is presented in Table 9.11. Hexadecane data are also presented in Fig. 9.61 (Martin et al. 2017). The table presents data on untreated glass and untreated PP as well as the data for coatings deposited on these substrates.

Table 9.11 Comparison of static contact angles and tilt angles for water and hexadecane droplets on the four superoleophobic techniques (Adapted from Martin et al. 2017)

Technique	Contact angle (°)/Tilt angle (°)	
	DI water	Hexadecane
Glass substrate		
Untreated	22 ± 2/n/a	12 ± 2/n/a
Nanoparticle-binder	165 ± 2 / ≤ 1	156 ± 2/2 ± 1
Layer-by-layer (Brown and Bhushan 2015c)	163 ± 1/2 ± 1	157 ± 1/4 ± 1
Polypropylene substrate		
Untreated (Brown and Bhushan 2016c)	96 ± 1/n/a	35 ± 2/n/a
Nanoparticle-encapsulation (Brown and Bhushan 2016c)	172 ± 1/1 ± 1	159 ± 1/4 ± 1
Liquid-impregnation (Brown and Bhushan 2017a)	115 ± 2/2 ± 1	70 ± 2/4 ± 2

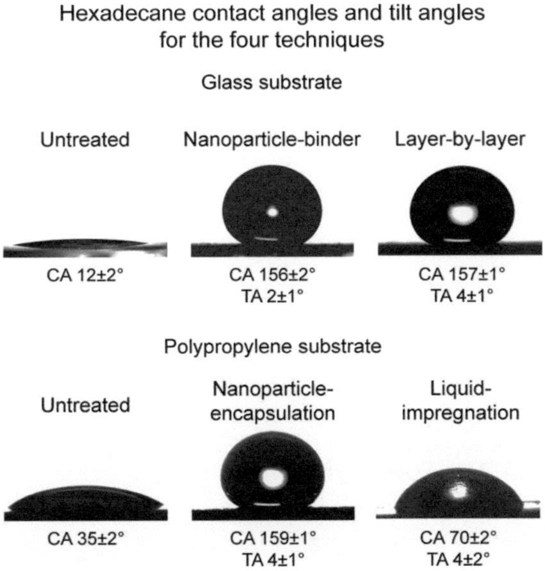

Fig. 9.61 Hexadecane contact angles and tilt angles for the four superoleophobic surfaces compared to untreated glass and polypropylene substrates (Adapted from Brown and Bhushan 2015c, 2016c, 2017a; Bhushan and Martin 2018)

Untreated glass substrates show CA for water and hexadecane of 22 ± 2° and 12 ± 2° respectively, Table 9.11. For the nanoparticle-binder and layer-by-layer coatings on glass substrates, the hexadecane CA were found to be > 150°, Fig. 9.61 (Brown and Bhushan 2015c; Bhushan and Martin 2018). In addition, the hexadecane TA were < 5° on both samples. Therefore, both could be superoleophobic, in addition to superhydrophobic (high CA and low TA for water) (Brown and Bhushan 2015b).

Untreated PP substrates were slightly hydrophobic with water CA of $96 \pm 1°$ but oleophilic with hexadecane CA of $35 \pm 2°$, Table 9.11. After treatment with the xylene–NP–PP mixture followed by drying in air, UV activation, and fluorosilane treatment, the surface becomes superhydrophobic and superoleophobic with CA of $172 \pm 1°$ and $159 \pm 1°$ for water and hexadecane respectively, Fig. 9.61 (Brown and Bhushan 2016c). Low hexadecane TA of $4 \pm 1°$ were also found on the fluorinated PP–NP composite surface. These high CA and low TA are a product of a composite interface with the droplet resting partially on pockets of trapped air (Brown and Bhushan 2015c, 2016c; Martin et al. 2017; Bhushan and Martin 2018).

Liquid-impregnated surfaces comprising porous PP dipped into a lubricating liquid were found to exhibit water CA of $115 \pm 2°$ and hexadecane CA of $70 \pm 2°$, Table 9.11 and Fig. 9.61 (Brown and Bhushan 2017a). These numbers are lower than found on the roughness-induced surfaces. However, due to the presence of the lubricating liquid, the surface displays very low TA of $2 \pm 1°$ and $4 \pm 2°$ for water and hexadecane respectively. Because the low TA are a product of the homogeneity of the liquid–liquid interface, the surface tension of the liquid being repelled has little effect on the repellency of the surface. This is in contrast to roughness-induced superoleophobic surfaces, where decreasing the surface tension of the oil being repelled results in an increase in the TA (Brown and Bhushan 2015a, b, c; Martin and Bhushan 2017).

Surface Morphology

To achieve superoleophobicity, specific surface topography is required. To repel very low surface tension liquids and surfactants, re-entrant geometries to support local CA of $< 90°$, need to be incorporated into the coating or surface treatment. In all roughness-induced cases, hierarchical, re-entrant geometries were found, as shown earlier.

Repellency of Surfactant-Containing Liquids

For the nanoparticle-binder and layer-by-layer coatings, droplets of shampoo and laundry detergent were found to roll or slide from the surface with no noticeable contamination, as reported earlier. This is in contrast to the untreated glass substrates where both shampoo and laundry detergent were found to spread and foul the surface (Martin et al. 2017).

For nanoparticle-encapsulated surfaces, shampoo was found to slide across the surface. However, it was found that laundry detergent was less repellent. This difference is probably due to the different viscosities of the two surfactant-containing liquids. The laundry detergent is less viscous and so is potentially more readily able to wet the solid surface and collapse the liquid–air interface of the Cassie-Baxter state (Brown and Bhushan 2017a; Martin et al. 2017).

For the liquid-impregnated samples, both surfactant-containing liquids were able to slide over with no contamination. In the case of shampoo, the highly viscous liquid slid from the surface in the shape of a droplet with a pronounced trailing edge. For the laundry detergent, the liquid was found to spread more due to its lower viscosity. It was found that these samples required more tilting than the roughness-induced superoleophobic surfaces due to the lower contact angles of the droplets and so a slower sliding action on the former compared to the rolling action on the latter (Brown and Bhushan 2017a; Martin et al. 2017). However, impregnated liquid will be lost after some time, so the useful lifetime is expected to be low (Martin et al. 2017).

Wear Resistance

Wear resistance of the samples produced by the four techniques was investigated using ball-on-flat tribometer and AFM wear experiments and the representative images are compared in Fig. 9.62 (Brown and Bhushan 2015c, 2016c, 2017a; Martin et al. 2017; Bhushan and Martin 2018). These experiments show the surface resistance of the coatings to wear and abrasion. Surfaces produced using nanoparticle-binder and nanoparticle-encapsulation techniques were too rough to be imaged by AFM. However, similar microscale durability is expected due to similar nanoparticle hardness.

In the case of the nanoparticle-binder and layer-by-layer coatings on glass, the ball-on-flat tribometer wear experiments were carried out using a 3 mm diameter sapphire ball at a load of 10 mN for 100 cycles, Fig. 9.62a. After the wear experiment, there is some noticeable burnishing to the coatings; however, damage to the coatings is minimal in both cases. In the case of nanoparticle-encapsulated surfaces, there was no noticeable wear scar, even after 200 cycles. It is believed that these surfaces are more durable than others. Other surfaces typically rely on poorly adhered coatings, delicate surface structures. Finally, for the liquid-impregnated samples, the images confirm that the surface features are not completely destroyed with minimal burnishing of the surface in the wear location. While not as durable as the nanoparticle-encapsulated samples as well as the other two techniques, it is believed that these surfaces are more durable than many other examples of liquid-impregnated surfaces, which typically rely on poorly adhered wax coatings (Smith et al. 2013) or delicate surface structures (Wong et al. 2011). The durability of the porous layer is required since it ensures the lubricating liquid remains in place, impregnated within the pores. The porous structures required for this liquid repellency technique may be susceptible to higher contact stress than more typical oil-repellent surfaces and so it is crucial that mechanical durability data is reported.

For the AFM wear experiment on the layer-by-layer technique, the layer-by-layer coating survived the AFM wear experiment with no observable defects, Fig. 9.62b (Brown and Bhushan 2015c). Similar microscale durability is expected for the other nanoparticle techniques due to their incorporation of hard nanoparticles. However, AFM wear experiments were not performed due to the other surfaces being too rough to image.

To sum up, because of incorporation of nanoparticles, samples developed using roughness-induced techniques exhibit higher durability than liquid impregnated samples. Among roughness-induced, nanoparticle-encapsulation technique exhibits the best durability.

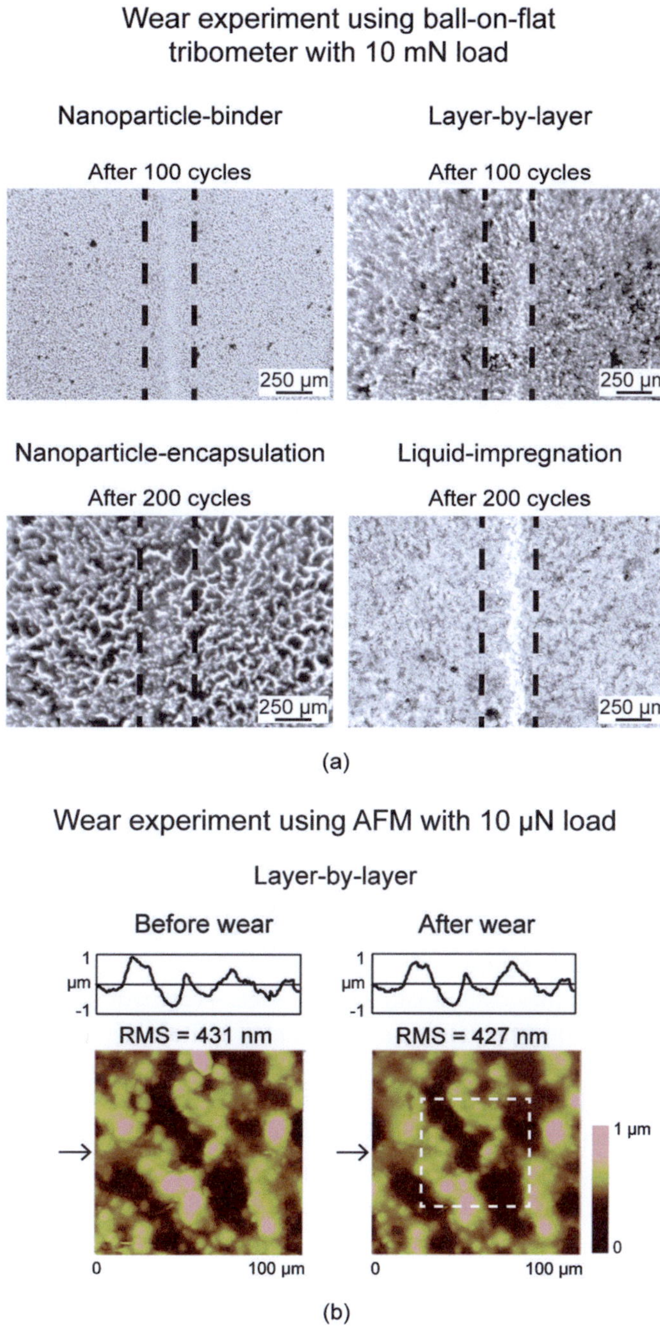
Wear experiment using ball-on-flat
tribometer with 10 mN load
Nanoparticle-binder
After 100 cycles
250 μm
Layer-by-layer
After 100 cycles
250 μm
Nanoparticle-encapsulation
After 200 cycles
250 μm
Liquid-impregnation
After 200 cycles
250 μm
(a)
Wear experiment using AFM with 10 μN load
Layer-by-layer
Before wear
After wear
1
μm
-1
RMS = 431 nm
RMS = 427 nm
1 μm
0
0
100 μm
(b)

◀**Fig. 9.62** Wear experiments using ball-on-flat tribometer and AFM. **a** Optical images after macrowear experiment with a 3 mm diameter sapphire ball at 10 mN load for all samples prepared by four fabrication techniques **b** Surface height maps and sample surface profiles (locations indicated by arrows) before and after AFM microwear experiment on layer-by-layer sample using 30 μm diameter borosilicate ball at 10 μN load. Other nanoparticle techniques were too rough to image using AFM and therefore are not reported (Adapted from Brown and Bhushan 2015c, 2016c, 2017a; Bhushan and Martin 2018)

Self-cleaning, Anti-smudge, and Antifouling

The water- and oil-repellent properties of surfaces make roughness-induced techniques suitable for self-cleaning, anti-smudge, and antifouling applications. When contaminated with silicon carbide, the vast majority of the contaminants can be removed in a self-cleaning test where water droplets roll across the repellent surfaces, collecting particles in the process, Fig. 9.63 (Brown and Bhushan 2015c; Martin et al. 2017; Bhushan and Martin 2018). After the self-cleaning test, > 90% of the particles are removed on the superhydrophobic coatings. On the untreated glass and polypropylene samples, the majority of the contaminants remain after the self-cleaning test. These coatings are self-cleaning due to their high water CA and low CAH (Martin et al. 2017). The liquid-impregnation technique does not exhibit the self-cleaning action because it does not have high water CA and so water droplets do not roll over the surface collecting contaminants.

Anti-smudge data has been presented on the roughness induced samples and they all performed well. Because the liquid impregnation technique does not exhibit high hexadecane CA, the hexadecane-soaked cloth is likely to transfer oil to the surface making it less suitable for anti-smudge applications.

Finally, the roughness-induced coatings could be used for anti-biofouling, where superoleophobicity and nanostructuring can contribute to reducing microorganism attachment (Bixler et al. 2014).

Anti-icing and Anti-fogging

Superhydrophobic coatings exhibit anti-icing whereas superhydrophilic coatings exhibit anti-fogging. For superhydrophobic coatings, droplets roll off the surface to freeze on the bottom of the freezer and provide anti-icing. The droplets rolling off occurs because the water droplets are in the Cassie-Baxter state. The formation of a composite interface minimizes the contact with the cooled substrate and ensures a low hysteresis so droplets can roll from the tilted surface (Brown and Bhushan 2015c; Bhushan and Multanen 2019). For superhydrophilic coatings, condensed droplets spread out into a thin film and evaporate rapidly to provide anti-fogging properties. Formation of a continuous water layer maintains transparency (Brown and Bhushan 2015c).

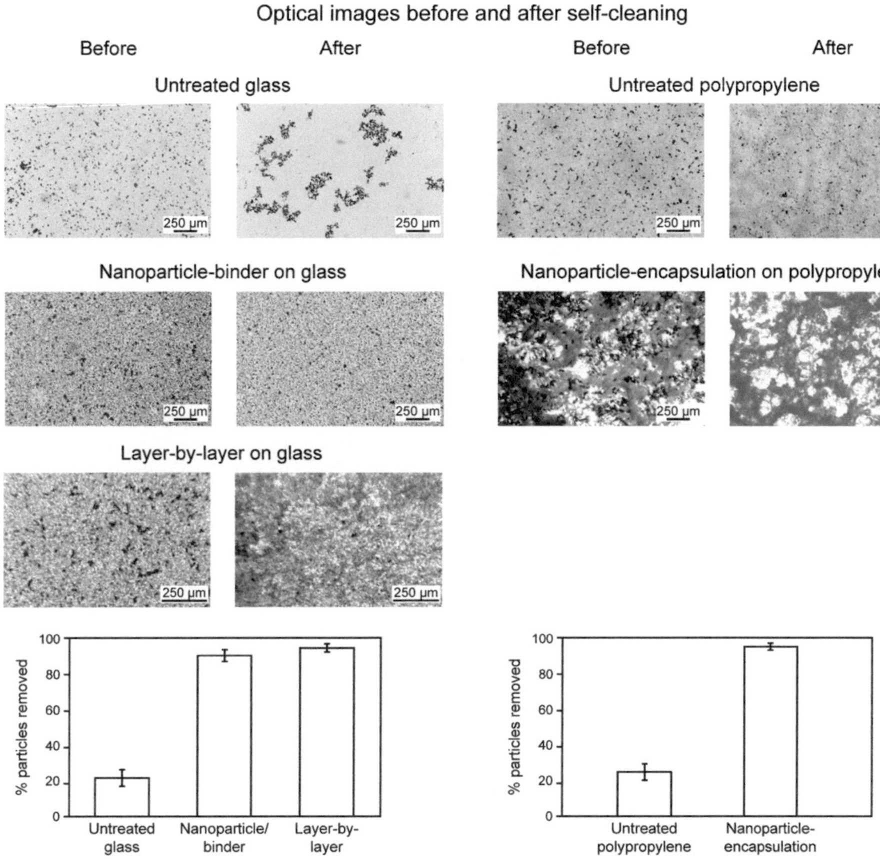

Fig. 9.63 Optical micrographs of contaminated coatings before and after self-cleaning test. Glass samples include untreated, nanoparticle-binder, and layer-by-layer. Polypropylene samples include untreated and nanoparticle-encapsulation. Dark spots on coatings indicate silicon carbide particle contaminants. Image analysis suggests > 90% removal of particles from the contaminated surfaces (Adapted from Brown and Bhushan 2015c; Martin et al. 2017; Bhushan and Martin 2018)

Optical Transparency

Many applications of oil-repellent surfaces rely on the transparency of the coating. Text remained legible when it is placed behind the nanoparticle-binder and layer-by-layer coatings, suggesting that the coatings display characteristics of transparency, Fig. 9.64 (Brown and Bhushan 2015a; Martin et al. 2017; Bhushan and Martin 2018).

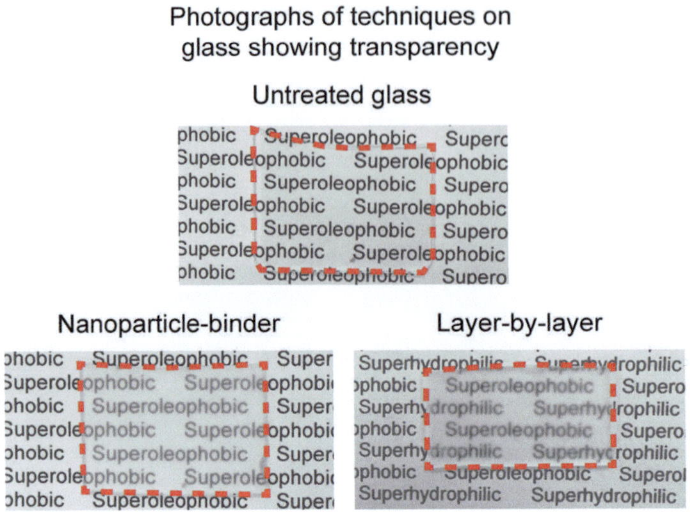

Fig. 9.64 Photographs of untreated glass compared to nanoparticle-binder and layer-by-layer techniques to show optical transparency (Adapted from Brown and Bhushan 2015a; Martin et al. 2017; Bhushan and Martin 2018)

Oil–water Separation

Superhydrophobic/superoleophilic and superhydrophilic/superoleophobic coatings exhibit a different response to water and oil and therefore are suitable for use in oil–water separation.

9.7.2 Summary and Outlook

Four fabrication techniques have been developed for superliquiphobic surfaces repellent to water, oil, and surfactant-containing liquids. All four techniques result in surfaces that are repellent to water and hexadecane. The high contact angles and low tilt angles for the nanoparticle-binder, layer-by-layer, and nanoparticle-encapsulation techniques result in self-cleaning and anti-smudge properties. Because liquid impregnation is only repellent to these liquids at lower contact angles than the other three techniques, it does not exhibit the self-cleaning and anti-smudge properties.

All techniques demonstrate some repellency toward surfactant-containing liquids. In the case of three roughness-induced techniques, re-entrant geometry allows repulsion of liquids. However, if a surfactant-containing droplet is allowed to rest on the surface, eventually surfactant molecules can alter the surface chemistry of the solid surface, leading to a decrease in the local contact angle and a collapsing of the composite interface between the droplet and the surface. The liquid-impregnation method may perform well because shampoo and laundry detergent may not foul these surfaces even after extended contact, so long as the lubricating liquid remains within the porous structure. However, loss of lubricating liquid is a serious concern.

All roughness-induced techniques exhibit wear resistance because of incorporation of nanoparticles. The nanoparticle-encapsulation technique exhibited the best wear resistance. Nanoparticles were directly incorporated into the surface during solvent-induced softening of the polymer material, resulting in durable surface properties. In contrast, the nanoparticle-binder and layer-by-layer techniques are applied as a coating. The robustness and durability of the surface properties achieved from a coating will be dependent upon their adhesion to the substrate. Selecting suitable coating materials and activating the substrate for coating attachment can lead to high wear resistance.

For transparency required for optical applications, nanoparticle-binder and layer-by-layer techniques can provide surfaces with the highest transmittance of visible light.

All four techniques can be easily fabricated. However, the nanoparticle-binder and nanoparticle-encapsulation techniques require the fewest steps. The nanoparticle-binder and the nanoparticle-encapsulation technique method are two-step spray techniques. However, both the nanoparticle-binder and nanoparticle-encapsulation methods require surface activation before fluorosilane deposition. In contrast, the layer-by-layer technique does not require surface activation for fluorosilane deposition but requires three spray steps and possible surface activation on hydrophobic substrates for the first layer. The nanoparticle-encapsulation technique needs to be tailored for a particular substrate. The difficulty in fabricating liquid-impregnated surfaces arises from forming the porous structure. It is also expected that liquid immiscibility between impregnated liquid and repelling liquid may be lost with time thus compromising life (Martin et al. 2017).

Finally, nanoparticle-binder and layer-by-layer techniques can be used for variety of substrates. Between the two, nanoparticle-binder technique requires the fewest steps whereas layer-by-layer technique provides flexibility in tailoring to various properties and better control.

9.8 Closure

Various fabrication techniques and selected characterization data are presented for mechanically durable superoleophobic/philic coatings. These techniques include nanoparticle-binder, layer-by-layer, and nanoparticle encapsulation. Then two liquid impregnated techniques are described. These coated surfaces are found to be repellent to various liquids including surfactant-containing liquids.

In addition to liquid repellency, some surfaces are found to exhibit self-cleaning, anti-smudge, finger touch resistance, anti-fogging, anti-icing, high optical transmittance, low drag, and anti-biofouling properties.

Appendix 9.A: Oil–Water Separation for Oil Spill Cleanup and Water Purification (Bhushan 2018, 2020)

Appendix 9.A.1: Introduction

Contamination of ground water and ocean water is of concern. Various industrial accidents on the ground and offshore occur periodically which lead to water contamination. Table 9.12 presents a list of notable offshore oil spills (Brown and Bhushan 2016d; Bhushan 2018). The Deepwater Horizon in the Gulf of Mexico in 2010 was the largest offshore oil spill in history with some 206 million gallons of oil being released in the Gulf. In addition, the emergence of fracking in the US, where water-based fluids (containing sand and chemicals) are injected under high pressure to fracture rocks for the release of previously inaccessible oil and gas, has led to a large increase in domestic oil production, Fig. 9.65 (Brown and Bhushan 2016d; Bhushan 2018). However, the process has also led to an increase in the amount of oil-contaminated wastewater. It is estimated that, between 2005 and 2014, some 248 billion US gallons of water were used for shale gas and oil extraction in the United States (Kondash and Vengosh 2015). In addition to this large amount of wastewater that must be processed before release, there is the potential for accidental contamination of ground and surface water.

Oil contamination occurs as part of routine operation in various industrial machinery. Examples include industrial lubricating oil effluents in electric transformers, air brakes of commercial vehicle systems, industrial compressors, and various components in mining

Table 9.12 List of notable offshore oil spills (Adapted from Brown and Bhushan 2016d; Bhushan 2018)

Spill	Location	Year	Amount spilled (million US gallons)
Deepwater Horizon§	Gulf of Mexico	2010	206
Gulf War†	Persian Gulf	1991	240
ABT Summer*	Angola	1991	80
MT Haven*	Italy	1991	42
Valdez*	Alaska	1989	11
Odyssey*	Canada	1988	43
Nowruz†	Persian Gulf	1983	80
Castillo de Bellver*	South Africa	1983	79
Ixtoc I†	Gulf of Mexico	1979	140
Atlantic Empress*	West Indies	1979	88
Amoco Cadiz*	France	1978	69

†Etkin and Welch, 1997; § from Ramseur, 2015; * ITOPF, 2015

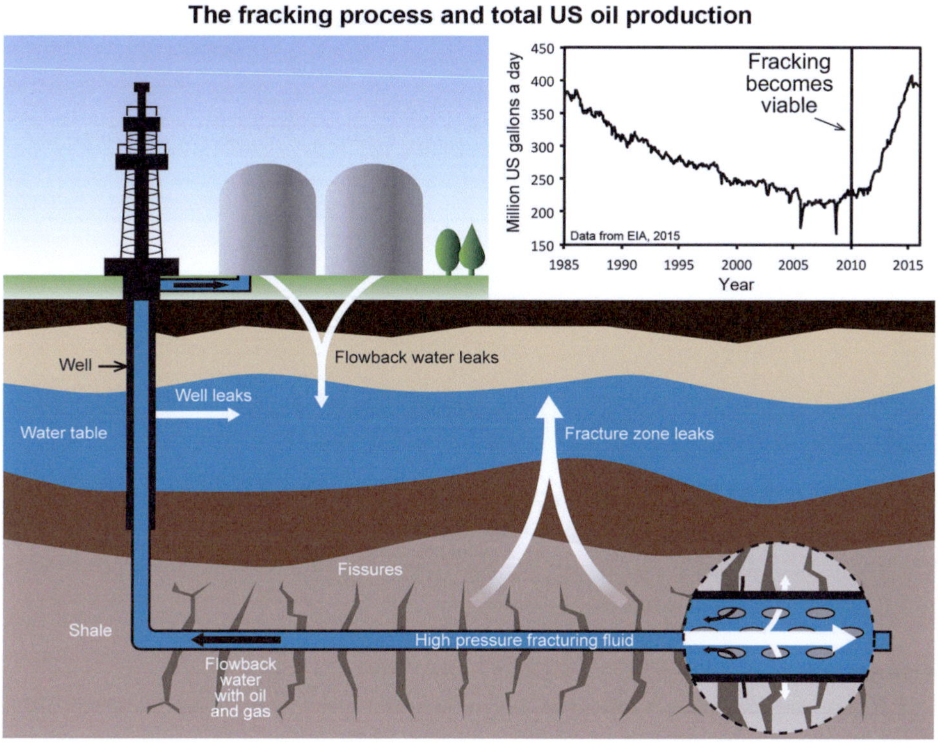

Fig. 9.65 Schematic of the fracking process. High pressure fluid (typically water, sand, and chemicals) is pumped underground, creating fissuresand releasing previously inaccessible pockets of oil and gas, which are then removed along with the fluid. Leaks of contaminated fracking fluid into groundwater are possible from the fracture zone, the well, and flowback water storage containers. Inset: US crude oil production. Increase in production from 2010 is due to emergence of fracking, primarily occurring onshore in the lower 48 states (Adapted from Brown and Bhushan 2016d; Bhushan 2018)

and paper manufacturing industries. Oil contaminants need to be captured in some applications to maintain efficient operation of machinery and avoid environmental impact. Separating oil from industrial wastewater is of interest to allow water re-use.

Finally, removal of oil and other organic contaminants from water for human consumption is critical. This is particularly important in the parts of the world with high contamination levels and limited water resources.

Oil spills cleanup and separation of oil from water for industrial re-use and human consumption remains an important environmental challenge.

Fig. 9.66 Photograph of the sorbent sucking up oil spill (Adapted from Bhushan 2018)

Appendix 9.A.2: Common Methods for Oil Spill Cleanup

Various methods are used for oil spill cleanup (Dave and Ghaly 2011; Fingas 2011; Brown and Bhushan 2016d). A summary follows.

Dispersants

Dispersants are chemical compounds that reduce the interfacial tension between oil and water to promote the breakup of globs of oil into smaller droplets. Dispersants are toxic and harmful to marine life, as well as expensive, and they decrease biodegradability of the oil (Nwaizuzu et al. 2015). Dispersants are sometimes used to prevent oil from getting to the shore, however, oil that has been treated with dispersants that does get to shore will sink deep into the ground where it cannot easily be degraded or captured. Lastly, dispersants can cause oil to sink where it is both harder to capture, and slower to degrade due to the lower temperature.

Controlled Burning

Controlled burning is another method of disposing of oil. It is quick and effective; however, burning releases a large amount of greenhouse gases as well as toxic compounds. Burning must be done far away from population centers to prevent health effects of the toxic gases released. To burn effectively, there must be at least 3 mm of oil on the surface of water.

Sorbents

Sorbents or sponges are materials that will soak up oil on contact, Fig. 9.66 (Fingas 2011; Bhushan 2018). Absorbent materials include zeolites, organoclays, or natural fibers such as straw, cellulose, wool, or human hair. The absorbent materials often are non-selective, absorbing both oil and water, lowering their efficiency (Zhu et al. 2011; Bayat et al. 2015).

They also can sink when saturated with oil, making it difficult to recover. After sucking up oil, sorbents would either must be thrown away or cleaned, making them incompatible with continuous processes. This method is very slow and is not suitable for cleanup of large surface area.

Skimmers

Skimmers are floating devices that "skim" oil off the surface. Figure 9.67 shows three different forms of oleophilic surface skimmers and their oil collection wells (Fingas 2011; Bhushan 2018). Their efficiency is dependent upon favorable conditions including calm waters and slow oil speeds. They are prone to clogging with trash in the water and cannot be used in rough water. In the case of absorbent belt skimmers, the collected oil needs to be physically removed from the surface during oil removal. If the oil is not removed, surfaces could become fouled, making recycling of the surfaces impossible.

Booms

Booms are long floating objects that can be dragged across the water to corral oil and prevent it from spreading. Figure 9.68 shows various configurations used in boom deployment (Fingas 2011; Bhushan 2018).

To sum up, various methods used today use toxic chemicals which are environmentally unfriendly or slow and tedious requiring significant time and expense.

Appendix 9.A.3: Proposed Bioinspired Net

Various purification membranes used for conventional water purification may be capable of removing the chemical contaminants resulting from environmental contamination and fracking, where the amount of contaminating oil is relatively low. However, their low water flux and tendency to suffer from surface fouling can make them unsuitable for cleanup of larger oil spills.

One approach for oil–water separation uses dual pH- and ammonia-vapor-responsive electrospun nanofibrous membranes developed by Ma et al. (2017). Also see Bhushan (2018).

In another approach, Brown and Bhushan (2015a, c) developed a coated steel mesh which allows oil–water separation in continuous fluid flow. Surfaces that are superoleophobic in air are typically also superhydrophobic due to water having a higher surface tension. Such surfaces would therefore be unsuitable for oil–water separation since the interactions with both liquids is similar. However, it is possible to create a coating that repels oils but has an affinity for water. This was achieved using a fluorosurfactant, which contains a high surface energy head group and a low surface energy tail group, as introduced in Chap. 8. Fluorosurfactants are typically used as additives in paint or cleaners. However, the presence of an ionic head group in many examples allows for complexation,

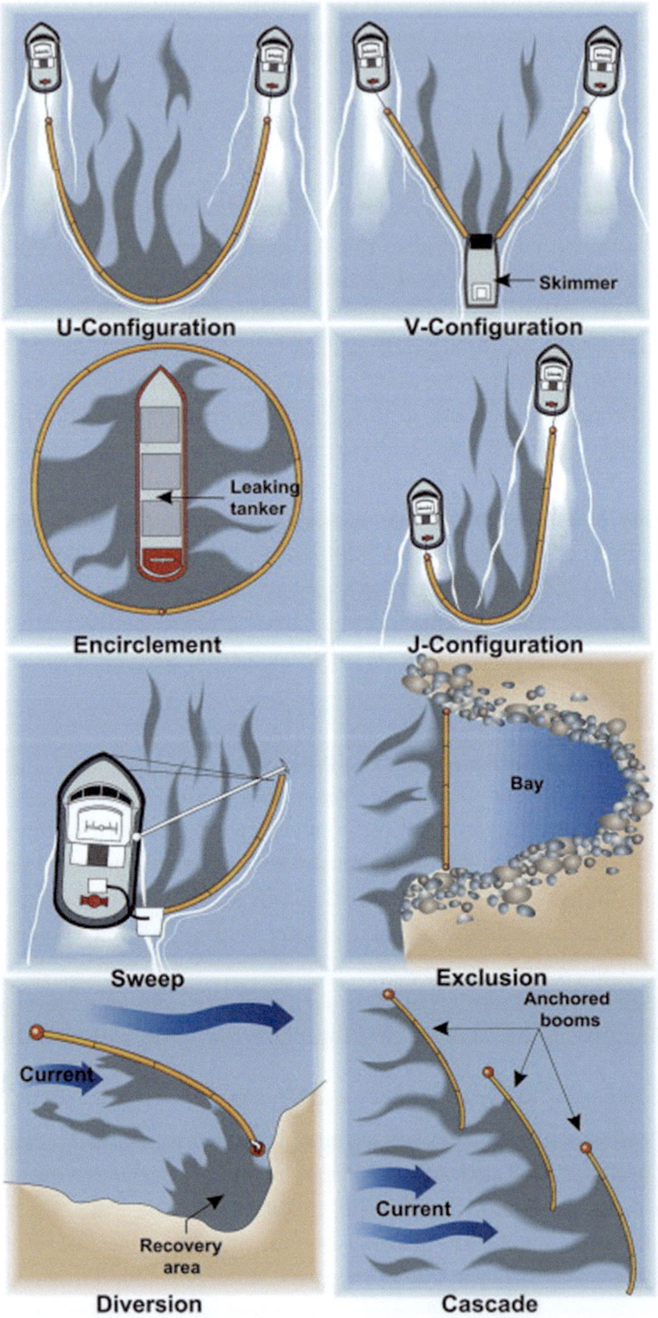

Fig. 9.67 Schematics of various configurations used in boom deployment (Adapted from Fingas 2011; Bhushan 2018)

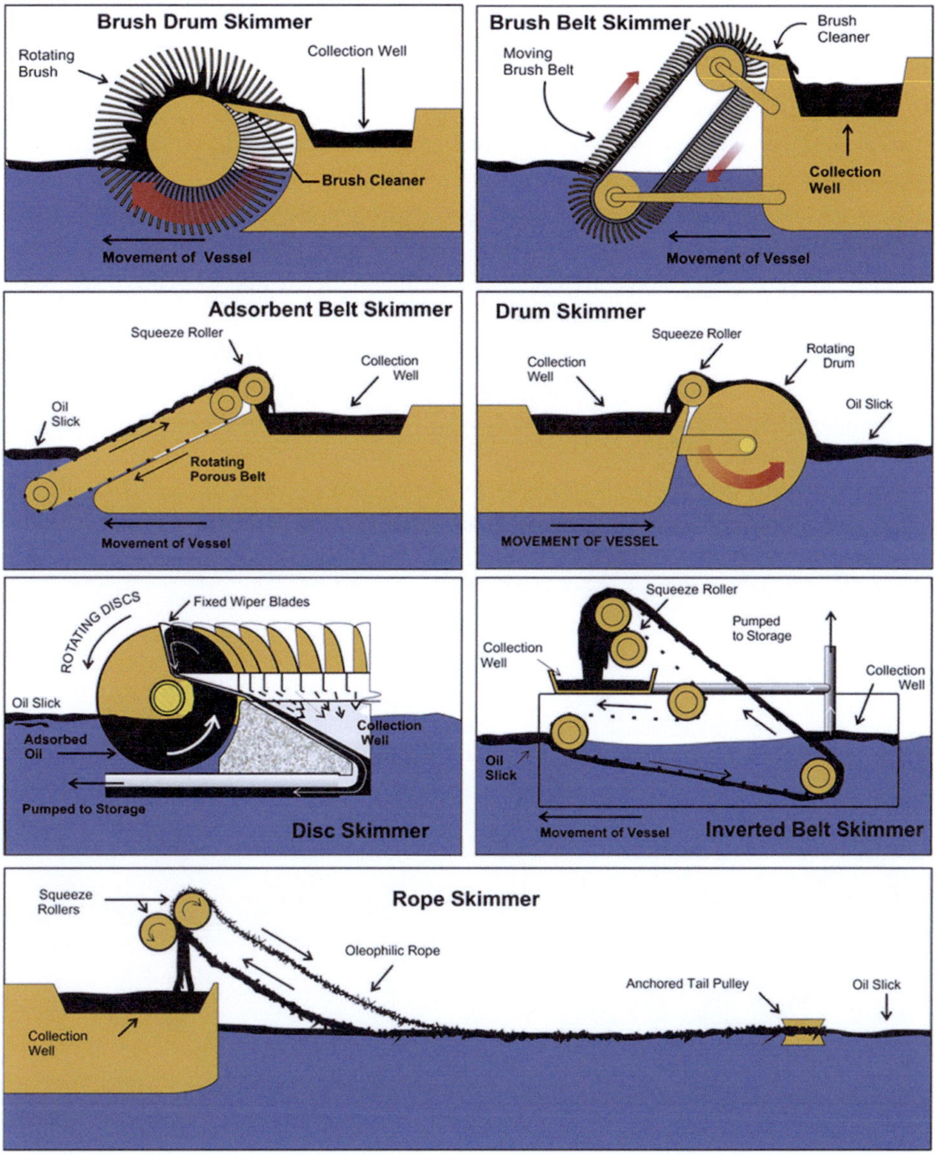

Fig. 9.68 Schematics of the principles of three different forms of oleophilic surface skimmers (Adapted from Fingas 2011; Bhushan 2018)

in either solution or on a coating, with polyelectrolytes. After deposition, the fluorinated tails segregate at the air interface, resulting in a low surface energy barrier that repels oils. However, when droplets of water are placed on the surface, they can penetrate down through the tail groups to reach the high surface energy polar head groups below. The coating therefore appears to be hydrophilic while also being oleophobic (Brown and Bhushan 2015a, 2015c).

Steel nets coated with the coating described above could aid oil spill cleanup by trapping the oil in the net while allowing the water to pass through. Figure 9.69 shows a schematic of how these bioinspired nets can be incorporated into currently used booms (Brown and Bhushan 2016d; Bhushan 2018, 2019b). It also provides a comparison between absorbent booms and bioinspired nets. The booms absorb the oil from the surface of the water but require favorable conditions and must be regularly cleaned to recover the oil. The bioinspired nets will capture the oil, where it can be recovered by pumping, while allowing water to pass through.

Such materials and methods are also of relevance in a wide range of applications including treatment of wastewater from metal, textile, or food processing industries, and for scenarios where contaminant water must be removed from oil, for instance in engine components (Brown and Bhushan 2016d).

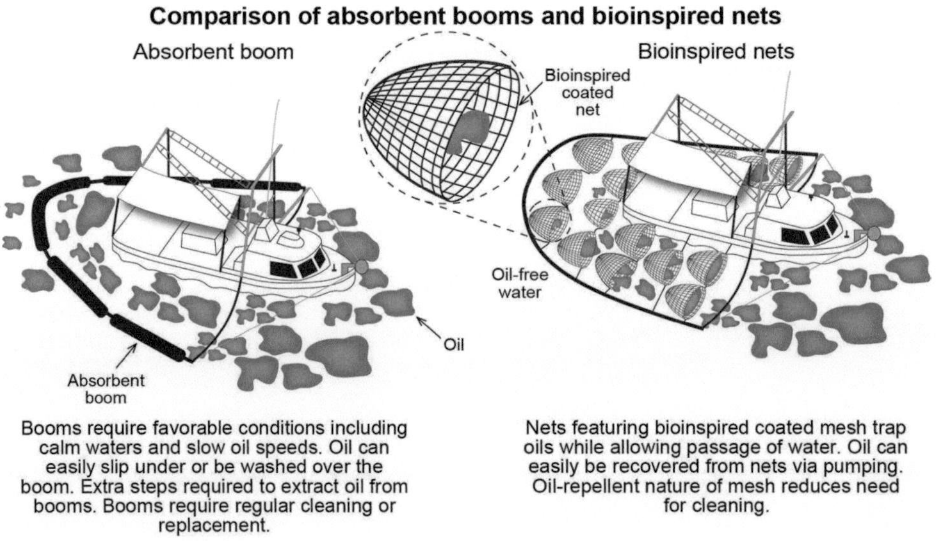

Fig. 9.69 Comparison of absorbent booms commonly used during oil spill remediation and bioinspired nets. The booms absorb the oil from the surface of the water but require favorable conditions and must be regularly cleaned to recover the oil. The bioinspired nets will capture the oil, where it can be recovered by pumping, while allowing water to pass through (Adapted from Brown and Bhushan 2016d; Bhushan 2018, 2019)

Appendix 9.A.4: Summary

Contamination of water by oil remains a problem. Oil is constantly threatening sources of water. Large-scale oil spills in disasters such as Deepwater Horizon show us that, so long as oil is sought, the danger will be there. Bioinspired surfaces hold promises for development of new techniques for cleaning up the oil spills and water purification.

References

Aoki, T., Furuya, S., and Kinoshita, H. (2005), "Finger-tapping Ability in Male and Female Pianists and Nonmusician Controls," *Motor Control* **9**, 23–39.

Baletto, F. and Ferrando, R. (2005), "Structural Properties of Nanoclusters: Energetic, Thermodynamic, and Kinetic Effects," *Rev. Mod. Phys.* **77**, 371–423.

Bayat, A., Aghamiri, S., Moheb, A., and Vakili-Nezhaad, G. R. (2015), "Oil Spill Cleanup from Sea Water by Sorbent Materials," *Chem. Eng. Technol.* **28**, 1525–1528.

Bellanger, H., Darmanin, T. and Guittard, F. (2012), "Surface Structuration (Micro and/or Nano) Governed by the Fluorinated Tail Lengths toward Superoleophobic Surfaces," *Langmuir* 28, 186–192.

Bhushan, B. (2009), "Biomimetics: Lessons from Nature—An Overview," *Phil. Trans. R. Soc. A* **367**, 1445–1486.

Bhushan, B. (2012), *Biomimetics: Bioinspired Hierarchical-structured Surfaces for Green Science and Technology*, Springer International, Cham, Switzerland.

Bhushan, B. (2013), *Introduction to Tribology*, second ed., Wiley, New York.

Bhushan, B. (2016), *Biomimetics: Bioinspired Hierarchical-Structured Surfaces for Green Science and Technology*, second ed., Springer International, Cham, Switzerland.

Bhushan, B. (2017a), *Springer Handbook of Nanotechnology*, fourth ed., Springer International, Cham, Switzerland.

Bhushan, B. (2017b), *Nanotribology and Nanomechanics: An Introduction*, fourth ed., Springer International, Cham, Switzerland.

Bhushan, B. (2018), *Biomimetics: Bioinspired Hierarchical-Structured Surfaces for Green Science and Technology*, third ed., Springer International, Cham, Switzerland.

Bhushan, B. (2019a), "Lessons from Nature for Green Science and Technology: An Overview and Bioinspired Superliquiphobic/philic Surfaces," *Phil. Trans. R. Soc. A* **377**, 20180274.

Bhushan, B.(2019b), "Bioinspired Oil-water Separation Approaches for Oil Spill Clean-up and Water Purification," *Phil. Trans. R. Soc. A* **377**, 20190120.

Bhushan, B. (2020), *Bioinspired Water Harvesting, Purification, and Oil-Water Sepa*ration, Springer International, Cham, Switzerland.

Bhushan, B. and Gupta, B. K. (1991), *Handbook of Tribology: Materials, Coatings and Surface Treatments*, McGraw-Hill, New York.

Bhushan, B. and Jung, Y. C. (2011), "Natural and Biomimetic Artificial Surfaces for Superhydrophobicity, Self-Cleaning, Low Adhesion, and Drag Reduction," *Prog. Mater. Sci.* **56**, 1–108.

Bhushan, B. and Martin, S. (2018), "Substrate-independent Superliquiphobic Coatings for Water, Oil, and Surfactant Repellency: An Overview," *J. Colloid Interface Sci.* **526**, 90–105.

Bhushan, B. and Multanen, V. (2019), "Designing Liquid Repellant, Icephobic and Self-Cleaning Surfaces with High Mechanical and Chemical Durability," *Phil. Trans. R. Soc. A* **377**, 20180270.

Bhushan, B. and Muthiah, P. (2013), "Anti-Smudge Screening Apparatus for Electronic Touch Screens," *Microsyst. Technol.* **19**, 1261–1263.

Bixler, G. D., Theiss, A., Bhushan, B. and Lee, S. C. (2014), "Anti-fouling Properties of Microstructured Surfaces Bio-inspired by Rice Leaves and Butterfly Wings," *J. Colloid Interface Sci.* **419**, 114–133.

Brown, P. S. (2016), Personal communication.

Brown, P. S. and Bhushan, B. (2015a), "Mechanically Durable, Superoleophobic Coatings Prepared by Layer-by-Layer Technique for Anti-smudge and Oil-water Separation," *Sci. Rep.—Nature* **5**, 8701.

Brown, P. S. and Bhushan, B. (2015b), "Mechanically Durable, Superomniphobic Coatings Prepared by Layer-by-Layer Technique for Self-cleaning and Anti-smudge," *J. Colloid Interface Sci.* **456**, 210–218.

Brown, P. S. and Bhushan, B. (2015c), "Bioinspired, Roughness-Induced, Water and Oil Super-philic and Super-phobic Coatings Prepared by Adaptable layer-by-Layer Technique," *Sci. Rep.—Nature* **5**, 14030.

Brown, P. S. and Bhushan, B. (2016a), "Designing Bioinspired Superoleophobic Surfaces," *APL Materials* **4**, 015703.

Brown, P. S. and Bhushan, B. (2016b), "Durable, Superoleophobic Polymer-nanoparticle Composite Surfaces with Re-entrant Geometry via Solvent-induced Phase Transformation," *Sci. Rep.—Nature* **6**, 21048.

Brown, P. S. and Bhushan, B. (2016c), "Durable, Superoleophobic Polypropylene Surfaces," *Phil. Trans. R. Soc. A* **374**, 20160193.

Brown, P. S. and Bhushan, B. (2016d), "Bioinspired Materials for Water Supply and Management: Water Collection, Water Purification and Separation of Water from Oil," *Phil. Trans. R. Soc. A* **374**, 20160135.

Brown, P. S. and Bhushan, B. (2017a), "Liquid-impregnated Porous Polypropylene Surfaces for Liquid Repellency," *J. Colloid Interface Sci.* **487**, 437-443.

Brown, P.S. and Bhushan, B (2017b), "Mechanically Durable Liquid-impregnated Honeycomb Surfaces," *Sci. Rep.—Nature* **7**, 6083.

Callister, W. D. and Rethwisch, D. G. (2013), *Materials Science and Engineering—An Introduction*, ninth ed., Wiley, New York.

Campos, R., Guenthner, A. J., Meuler, A. J., Tuteja, A., Cohen, R. E., McKinley, G. H., Haddad, T. S. and Mabry, J. M. (2012), "Superoleophobic Surfaces Through Control of Sprayed-on Stochastic Topography," *Langmuir* **28**, 9834–9841.

Cao, L., Jones, A. K., Sikka, V. K., Wu, J. and Gao, D. (2009), "Anti-Icing Superhydrophobic Coatings," *Langmuir* **25**, 12444–12448.

Cebeci, F. Ç., Zhizhong, W., Zhai, L., Cohen, R. E., and Rubner, M. F. (2006), "Nanoporosity-Driven Superhydrophilicity: A Means to Create Multifunctional Antifogging Coatings," *Langmuir* **22**, 2856–2862.

Contreras, C. B., Chagas, G., Strumia, M. C., and Weibel, D. E. (2014), "Permanent superhydrophobic polypropylene nanocomposite coatings by a simple one-step dipping process," *Appl. Surf. Sci.* **307**, 234–240.

Cremaldi, J. C. and Bhushan, B. (2018a), "Bioinspired Self-healing Materials: Lessons from Nature," *Beilstein J. Nanotechnol.* **9**, 907–935.

Cremaldi, J. and Bhushan, B. (2018b), "Fabrication of Bioinspired, Self-Cleaning Superliquiphilic/phobic Stainless Steel Using Different Pathways," *J. Colloid Interface Sci.* **518**, 284–297.

Cui, Y., Paxson, A. T., Smyth, K. M., and Varanasi, K. K. (2012), "Hierarchical Polymeric Textures via Solvent-induced Phase Transformation: A Single-step Production of Large-area Superhydrophobic Surfaces," *Colloids Surf. A* **394**, 8–13.

Darmanin, T., Guittard, F., Amigoni, S., Tafin de Givenchy, E., Noblin, X., Kofman, R. and Celestini, F. (2011), "Superoleophobic Behavior of Fluorinated Conductive Polymer Films Combining Electropolymerization and Lithography," *Soft Matter* **7**, 1053–1057.

Das, A., Schutzius, T. M., Bayer, I. S. and Megaridis, C. M. (2012), "Superoleophobic and Conductive Carbon Nanofiber/Fluoropolymer Composite Films," *Carbon* **50**, 1346–1354.

Dave, D. and Ghaly, A. (2011), "Remediation Techniques for Marine Oil Spills: A Critical Review and Comparative Analysis," *Am. J. Environ. Sci.* **7**, 423–440.

Deng, X., Mammen, L., Butt, H.-J. and Vollmer, D. (2012), "Candle Soot as a Template for a Transparent Robust Superamphiphobic Coating," *Science* **335**, 67–70.

Du, X., Li, X. and He, J. (2010), "Facile Fabrication of Hierarchically Structured Silica Coatings from Hierarchically Mesoporous Silica Nanoparticles and Their Excellent Superhydrophilicity and Superhydrophobicity" *ACS Appl. Mater. Interfaces* **2**, 2365–2372.

Ebert, D. and Bhushan, B. (2012), "Transparent, Superhydrophobic, and Wear-Resistant Coatings on Glass and Polymer Substrates Using SiO_2, ZnO, and ITO nanoparticles," *Langmuir* **28**, 11391–11399.

Edwards, O. and Marks, R. (1995), "Evaluations of Biomechanical Properties of Human Skin," *Clin. Dermatol.* **13**, 375–380.

Ellis, B. and Smith, R., Eds. (2008), *Polymers: A Property Database*, second ed., CRC Press.

Erbil, H. Y., Demirel, A. L., Avci, Y., and Mert, O. (2003), "Transformation of a Simple Plastic into a Superhydrophobic Surface," *Science* **299**, 1377–1380.

Etkin, D. S. and Welch, J. (1997) "Oil Spill Intelligence Report International Oil Spill Database: Trends in Oil Spill Volumes and Frequency," *Int. Spill Conf*. 949–952.

Ferrari, E., Fabbri, P., and Pilati, F. (2011), "Solvent and Substrate Contributions to the Formation of Breath Figure Patterns in Polystyrene Films," *Langmuir* **27**, 1874–1881.

Fingas, M. (ed.), (2011), *Oil Spill Science and Technology: Prevention, Response, and Cleanup*, Elsevier, Amsterdam.

Franco, J. A., Kentish, S. E., Perera, J. M., and Stevens G. W. (2008), "Fabrication of a superhydrophobic polypropylene membrane by deposition of a porous crystalline polypropylene coating," *J. Membr. Sci.* **318**, 107–113.

Goddard, E. D. (1986) "Polymer−Surfactant Interaction. Part II. Polymer and Surfactant of Opposite Charge," *Colloids Surf.* **19**, 301−329.

Gogolides, E., Vlachopoulou, M., Tsoungeni, K., Vourdas, N., and Tserepi, A. (2010), "Micro and Nano Structuring and Texturing of Polymers Using Plasma Processes: Potential Manufacturing Applications," *Int. J. Nanomanuf.* **6**, 152–163.

Han, D. and Steckl, A. J. (2009), "Superhydrophobic and Oleophobic Fibers by Coaxial Electrospinning," *Langmuir* **25**, 9454–9562.

Hsieh, C.-T., Chen, J.-M., Kuo, R.-R., Lin, T.-S. and Wu, C.-F. (2005), "Influence of Surface Roughness on Water- and Oil-Repellent Surfaces Coated With Nanoparticles," *Appl. Surf. Sci.* **240**, 318–326.

Hsieh, C.-T., Wu, F.-L. and Chen, W.-Y. (2009), "Super Water- and Oil-Repellencies from Silica-Based Nanocoatings," *Surf. Coat. Technol.* **203**, 3377–3384.

Hutton, S. J., Crowther, J. M. and Badyal, J. P. S. (2000), "Complexation of Fluorosurfactants to Functionalized Solid Surfaces: Smart Behavior," *Chem. Mater.* **12**, 2282–2286.

Im, M., Im, H., Lee, J.-H., Yoon, J.-B. and Choi, Y.-K. (2010), "A Robust Superhydrophobic and Superoleophobic Surface with Inverse-Trapezoidal Microstructures on a Large Transparent Flexible Substrate," *Soft Matter* **6**, 1401–1404.

Inui, N., Ichihara, T., Minami, T., and Matsui, A. (1998), "Interactions: Timing and Force Control of Finger-tapping Sequences," *Percept. Mot. Skills* **86**, 1395–1401.

ITOPF (2015) "Oil Tanker Spill Statistics," see http://www.itopf.com/knowledge-resources/data-statistics/statistics

Jalili, M. M. and Moradian, S. (2009), "Deterministic Performance Parameters for an Automotive Polyurethane Clearcoat Loaded With Hydrophilic or Hydrophobic Nano-Silica," *Prog. Org. Coat.* **66**, 359–366.

Jin, H., Tian, X., Ikkala, O. and Ras, R. H. A. (2013), "Preservation of Superhydrophobic and Superoleophobic Properties Upon Wear Damage," *ACS Appl. Mater. Interfaces* **5**, 485–488.

Jung, Y. C. and Bhushan, B. (2009), "Wetting Behavior of Water and Oil Droplets in Three-Phase Interfaces for Hydrophobicity/philicity and Oleophobicity/philicity," *Langmuir* **25**, 14165–14173.

Jung, Y. C. and Bhushan, B. (2010), "Biomimetic Structures for Fluid Drag Reduction in Laminar and Turbulent Flows," *J. Phys.: Condens. Matter* **22**, 035104.

Kim, P., Wong, T.-K., Alvarenga, J., Kreder, M. J., Adorno-Martinez, W. E., and Aizenberg, J. (2012), "Liquid-Infused Nanostructured Surfaces with Extreme Anti-Ice and Anti-Frost Performance," *ACS Nano* **6**, 6569–6577.

Kondash, A. and Vengosh, A. (2015) "Water Footprint of Hydraulic Fracturing," *Environ. Sci. Technol. Lett.* **2**, 276–280.

Kreder, M. J., Alvarenga, J., Kim, P. and Aizenberg, J. (2016), "Designing of anti-Icing Surfaces: Smooth Textured or Slippery?," *Nature Rev.* **1**, 1–15.

Kulinich, S. A. and Farzaneh, M. (2009), "How Wetting Hysteresis Influences Ice Adhesion Strength on Superhydrophobic Surfaces," *Langmuir* **25**, 8854–8856.

Lee, K. K. and Ahn, C. H. (2013), "Superhydrophilic Multilayer Silica Nanoparticle Networks on a Polymer Microchannel Using a Spray Layer-by-Layer Nanoassembly Method" *ACS Applied Mater. Interfaces* **5**, 8523–8530.

Li, L., Wang, Y., Gallaschun, C., Risch, T. and Sun, J. (2012), "Why Can a Nanometer-Thick Polymer Coated Surface be More Wettable to Water than to Oil?" *J. Mater. Chem.* **22**, 16719−1672.

Li, F., Kong, W., Bhushan, B., Zhao, X., and Pan, Y. (2019), "Ultraviolet-Driven Switchable Superliquiphobic/superliquiphilic Coating for Separation of Oil-water Mixtures and Emulsions and Water Purification," *J Colloid and Interface Sci.* **557**, 395–407.

Lovingood, D. D., Salter, W. B., Griffith, K. R., Simpson, K. M., Hearn, J. D., and Owens, J. R. (2013), "Fabrication of Liquid and Vapor Protective Cotton Fabrics," *Langmuir* **29**, 15043–15050.

Ma, W., Higaki, Y., Otsuka, H., and Takahara, A. (2013), "Perfluoropolyether-infused nano-texture: a versatile approach to omniphobic coatings with low hysteresis and high transparency," *Chem. Commun.* **49**, 597–599.

Ma, W., Samal, S. K., Liu, Z., Xiong, R., De Smedt, S. C., Bhushan, B., Zhang, Q., and Huang, C. (2017), "Dual pH- and Ammonia-vapor-responsive Electrospun Nanofibrous Membranes for Oil-water Separations", *J. Membrane Sci.* **537**, 128–139.

Mark, J. E. (1999), *Polymer Data Handbook*, second ed., Oxford University Press, Oxford, U.K.

Martin, S. and Bhushan, B. (2017), "Transparent, Wear-resistant, Superhydrophobic and Superoleophobic Poly(dimethylsiloxane) (PDMS) Surfaces," *J. Colloid Interface Sci.* **488**, 118–126.

Martin, S., Brown, P. S., and Bhushan, B. (2017), "Fabrication Techniques for Bioinspired, Mechanically-durable, Superliquiphobic Surfaces for Water, Oil, and Surfactant Repellency," *Adv. Colloid Interf. Sci.* **241**, 1–23.

Mijatovic, D., Eijkel, J. C. T., and van den Berg, A. (2005), "Technologies for Nanofluidic Systems: Top-down vs. Bottom-up—A Review," *Lab Chip* **5**, 492–500.

Multanen, V. and Bhushan, B. (2019), "Bioinspired Self-healing, Superliquiphobic and Self-cleaning Hydrogel- coated Surfaces with High Durability," *Phil. Trans. R. Soc. A* **377**, 20190117.

Muthiah, P., Bhushan, B., Yun, K. and Kondo, H. (2013), "Dual-Layered-Coated Mechanically-Durable Superomniphobic Surfaces With Anti-Smudge Properties," *J. Colloid Interface Sci.* **409**, 227–236.

Nie, Z. and Kumacheva, E. (2008), "Patterning Surfaces with Functional Polymers," *Nature Mater.* **7**, 277–290.

Nishimoto, S. and Bhushan, B. (2013), "Bioinspired Self-cleaning Surfaces with Superhydrophobicity, Superoleophobicity, and Superhydrophilicity," *RSC Advances* **3**, 671–690.

Nosonovsky, M. and Bhushan, B. (2008), *Multiscale Dissipative Mechanisms and Hierarchical Surfaces: Friction, Superhydrophobicity, and Biomimetics*, Springer, Heidelberg, Germany.

Nwaizuzu, C., Joel, O. F., and Sikoki, F. D. (2015) "Evaluation of Oil Spill Dispersants with a focus on their toxicity and Biodegradability," Society of Petroleum. Engineers (https://doi.org/10.2118/178321-MS).

Pan, S., Kota, A.K., Mabry, J.M. and Tuteja, A. (2013), "Superomniphobic Surfaces for Effective Chemical Shielding," *J. Am. Chem. Soc.* **135**, 578–581.

Pan, Y., Kong, W., Bhushan, B., and Zhao, X. (2019), "Rapid, Ultraviolet-induced, Reversibly Switchable Wettability of Superhydrophobic/Superhydrophilic Surfaces," *Beilstein J. Nanotechnology* **10**, 866–873.

Peebles, L. and Norris, B. (2003), "Filling 'Gaps' in Strength Data for Design," *Appl. Ergon.* **34**, 73–88.

Peng, J., Han, Y., Yang, Y., and Li, B. (2004), "The Influencing Factors on the Macroporous Formation in Polymer Films by Water Droplet Templating," *Polymer* **45**, 447–452.

Pham, K. N., Fullston, D., and Sagoe-Crentsil, K. (2007), "Surface Charge Modification of Nano-Sized Silica Colliod," *Aust. J. Chem.* **60**, 662–666.

Ramseur, J. L. (2015) "Deepwater Horizon Oil Spill: Recent Activities and Ongoing Developments," see https://www.fas.org/sgp/crs/misc/R42942.pdf

Rayleigh, L. (1911), "Breath Figures," *Nature* **86**, 416–417.

Rumble, J. C. (2023–2024), *CRC Handbook of Chemistry and Physics* 95th Ed., Taylor and Francis Group, Boca Raton, FL.

Saito, T., Tsushima, Y. and Sawada, H. (2015), "Facile Creation of Superoleophobic and Superhydrophilic Surface by Using Fluoroalkyl End-Capped Vinyltrimethoxysilane Oligomer/Calcium Silicide Nanocomposites: Development of these Nanocomposites to Environmental Cyclical Type-Fluorine Recycle through Formation of Calcium Fluoride," *Colloid Polym. Sci.* **293**, 65–73.

Sheen, Y., Huang, Y., Liao, C., Chou, H. and Chang, F. (2008), "New Approach to Fabricate an Extremely Super-Amphiphobic Surface Based on Fluorinated Silica Nanoparticles," *J. Polym. Sci. Part B: Polym. Phys.* **46**, 1984–1990.

Smith, J. D., Dhiman, R., Paxson, A. T., Love, C. J., Solomon, B. R., and Varanasi, K. K. (2013), "Self-lubricating Surfaces for Food Packaging and Food Processing Equipment," U.S. Patent no. 8,535,779B1, Sep. 13.

Srinivasarao, M., Collings, D., Philips, A., and Patel, S. (2001), "Three-Dimensionally Ordered Array of Air Bubbles in a Polymer Film," *Science* **292**, 79–83.

Steele, A., Bayer, I. and Loth, E. (2009), "Inherently Superoleophobic Nanocomposite Coatings by Spray Atomization," *Nano Lett.* **9**, 501–505.

Stenzel, M. H., Barner-Kowollik, C., and Davis, T. P. (2006), "Formation of Honeycomb-structured, Porous Films via Breath Figures with Different Polymer Architectures," *J. Polym. Sci. Part A: Polym. Chem.* **44**, 2363–2375.

Stucker, M., Geil, M., Keck, S., Hoffman, K., Röchling, A., Memmel, U. and Altmeyer, P. (2001) "Interpapillary Lines-The Variable Part of the Human Fingerprint," *J. Forensic Sci.* **46**, 857-861.

Sun, Z., Liao, T., Liu, K., Jiang, L., Kim, J. H., and Dou, S. X. (2014), "Fly-Eye Inspired Superhydrophobic Anti-fogging Inorganic Nanostructures," *Small* **10**, 3001–3006.

Tahmasebpoor, M., de Martín, L., Talebi, M., Mostoufi, N., and van Ommen, J. R. (2013), "The Role of the Hydrogen Bond in Dense Nanoparticle-Gas Suspensions," *Phys. Chem. Chem. Phys.* **15**, 5788–5793.

Thomsen, S. V., Hulme, R., Landa, L. & Landa, K. (2005), "High Visible Transmission and Infrared/Ultraviolet Radiation Absorption; for Automotive Windows (Windshields, Sidelites, Backlites and Sunroofs) and in Architectural Windows," US Patent 20050020430 A1, 27 Jan. 2005.

Thünemann, A. F. and Lochhaas, K. H. (1999), "Surface and Solid-State Properties of a Fluorinated Polyelectrolyte−Surfactant Complex," *Langmuir* **15**, 4867−4874.

Todor, J. I. and Smiley-Oyen, A. L. (1987), "Force Modulation as a Source of Hand Differences in Rapid Finger Tapping," *Acta Psychol.* **65**, 65–73.

Tsujii, K., Yamamoto, T., Onda, T. and Shibuichi, S. (1997), "Super Oil-Repellent Surfaces," *Angew. Chem. Int. Ed. Engl.* **36**, 1011–1012.

Tuteja, A., Choi, W., Mabry, J. M., Mckinley, G. H., and Cohen, R. E. (2008), "Robust Omniphobic Surfaces," *Proc. Natl. Acad. Sci. U.S.A.* **105**, 18200–18205.

Wang, Y. and Bhushan, B. (2015), "Wear-Resistant and Antismudge Superoleophobic Coating on Polyethylene Terephthalate Substrate Using SiO_2 Nanoparticles," *ACS Appl. Mater. Interf*. **7**, 743–755.

Wang, X., Hu, H., Ye, Q., Gao, T., Zhou, F. and Xue, Q. (2012), "Superamphiphobic Coatings with Coralline-Like Structure Enabled by One-Step Spray of Polyurethane/Carbon Nanotube Composites," *J. Mater. Chem.* **22**, 9624–9631.

Wang, Y., Xue, J., Wang, Q., Chen, Q. and Ding, J. (2013), "Verification of Icephobic/Anti-icing Properties of a Superhydrophobic Surface," *ACS Appl. Mater. Interf*. **5**, 3370–3381.

Widawski, G., Rawiso, M., and François, B. (1994), "Self-organized Honeycomb Morphology of Star-polymer Polystyrene Film," *Nature* **369**, 387–389.

Wong, T.-S., Kang, S. H., Tang, S. K. Y., Smythe, E. J., Hatton, B. D., Grinthal, A. and Aizenberg, J. (2011), "Bioinspired Self-repairing Slippery Surface with Pressure-stable Omniphobicity," *Nature* **477**, 443–447.

Xie, Y., Hill, C. A. S., Xiao, Z., Militz, H. and Mai, C. (2010) "Silane Coupling Agents used for Natural Fiber/Polymer Composites: A Review," *Composites: Part A* **41**, 806–819.

Yang, J., Zhang, Z., Men, X., Xu, X. and Zhu, X. (2011), "A Simple Approach to Fabricate Superoleophobic Coatings," *New J. Chem.* **35**, 576–580.

Zhang, J. and Seeger, S. (2011), "Superoleophobic Coatings with Ultralow Sliding Angles Based on Silicone Nanofilaments," *Angew. Chem. Int. Ed.* **50**, 6652–6656.

Zhao, H., Law, K.-Y. and Sambhy, V. (2011), "Fabrication, Surface Properties, and Origin of Superoleophobicity for a Model Textured Surface," *Langmuir* **27**, 5927–5935.

Zhao, N., Weng, L., Zhang, Xie, Q., Zhang, X., and Xu, J. (2006), "A Lotus-leaf-like Superhydrophobic Surface Prepared by Solvent-Induced Crystallization," *ChemPhysChem* **7**, 824–827.

Zhu, G., Pan, Q., and Liu, F. (2011) "Facile Removal and Collection of Oils from Water Surfaces through Superhydrophobic and Superoleophilic Sponges," *J. Phys. Chem. C* **115**, 17464–17470.

Fabrication and Characterization of Mechanically Durable Superliquiphobic Engineering Surfaces

10

10.1 Introduction

Superliquiphobic surfaces are of interest for liquid repellence, self-cleaning, anti-icing, anti-smudge, and antifouling applications (Nosonovsky and Bhushan 2008; Bhushan 2018, 2019). The surfaces should be mechanically durable for commercial applications.

In this chapter, details of fabrication and characterization of superoleophobic and self-cleaning, engineering surfaces are presented which include aluminium, steel, titanium, cotton fabric, and synthetic leather.

10.2 Superoleophobic Aluminum Surfaces

Aluminium and its alloys present various properties, including low density, excellent heat and electrical conductivities, high specific strength, natural abundance, and good castability. They find widespread applications in aerospace, automotive industries, architecture, and household goods (Davis 1993; Hatch 1984). It is desirable to create mechanically durable superhydrophobic and superoleophobic aluminum surfaces with self-cleaning, anti-smudge, antifouling, and corrosion resistance properties, which can find promising applications in aircraft components, marine components, automotive components, machinery, and cooking utensils (Kaufman and Rooy 2004; Peng and Bhushan 2016).

Various preparation methods have been developed to produce superhydrophobic/ superoleophobic aluminum surfaces. Chemical etching and anodization have been used to produce superhydrophobic aluminum surfaces. For the chemical etching method, chemical etchants include boiling water (Ren et al. 2004; He et al. 2012), NaOH (Liu et al. 2013a; Tudu et al. 2019a), HCl (Qian and Shen 2005), $ZnNO_3 \cdot 6H_2O$ (Zhang et al. 2008; Lu et al. 2010), HNO_3 combined with $Cu(NO_3)_2$ (Fu and He 2008; Liu et al. 2013b),

B. Bhushan, *Introduction to Biomimetics and Bioinspiration*,
https://doi.org/10.1007/978-3-031-62344-8_10

$AlCl_3$ combined with triethanolamine (Liu et al. 2011), $La(NO_3)_3$ (Li et al. 2014a), and $FeCl_3$ (Maitra et al. 2014). For the anodization method, electrolytes include oxalic acid (Kim et al. 2012) and mixed solutions of NaCl and fluorosilane (Song et al. 2012). In the above-mentioned papers, nanoplatelet, microstep, nanoreticula, microflower, and nanowire structures have been created to achieve superhydrophobicity.

To produce superoleophobic aluminum surfaces, either a one-step or a two-step technique has been used to create hierarchical structures. In one-step techniques, superoleophobic surfaces were achieved by anodization in H_2SO_4, followed by fluorinated monoalkylphosphate dip coating (Tsujii et al. 1997), chemical etching in perfluorocarboxylic acid (Meng et al. 2008), and fluorinated nanoparticles (Nanda et al. 2017). There are various two-step techniques applied to fabricate superoleophobic surfaces. These include chemical etching in HCl and then in boiling water, followed by fluorosilane coating (Peng et al. 2014); chemical etching in HCl and then in $HNO_3 + Cu(NO_3)_2$, followed by fluorosilane coating (Peng and Bhushan 2016); electrochemical etching in Na_2SO_4 combined with anodization in oxalic acid, and subsequent fluorosilane coating (Wu et al. 2009a); and chemical etching in HCl combined with anodization in H_2SO_4, followed by fluorosilane coating (Barthwal et al. 2013). These two-step techniques have been used to create the microstep structure first, followed by nanoplatelet, nanowire forest or nanoporous structure, or deposition of nanoparticles on the top of the microstep structure to form hierarchical structure, with subsequent fluoro-compound modification. Only Peng et al. (2014) and Peng and Bhushan (2016) have reported some mechanical durability data.

In this section, based on Peng and Bhushan (2016) and Nanda et al. (2017), we present data on mechanically durable superoleophobic aluminum surfaces. Peng and Bhushan (2016) developed a two-step technique which involves etching to produce nanoreticula structure followed by deposition of a fluorinated coating. Nanda et al. (2017) developed a single-step technique which involves the use of fluorinated nanoparticles. The surfaces were characterized for wettability, wear resistance, self-cleaning, anti-fogging, anti-smudge properties, corrosion resistance, and propensity of air pocket formation.

10.2.1 Two-Step Technique Using Etching and Fluorination

A two-step technique that involves etching followed by deposition of a fluorinated coating has been developed to produce a superoleophobic surface. The surface consists of a combined microstep and nanoreticula structure modified with fluorosilane to provide the ability to form air pockets and a low surface energy surface.

The fabrication technique for superoleophobic aluminum surfaces is shown in Fig. 10.1 (Peng and Bhushan 2016). The microstep structure was first obtained from HCl etching. Then, the microstructured surface was immersed in aqueous solutions of HNO_3 and $Cu(NO_3)_2$ to construct a nanoreticula structure, to achieve hierarchical structure. The

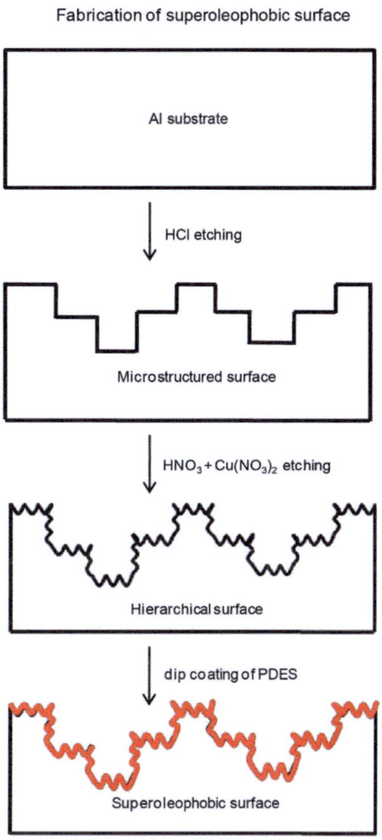

Fig. 10.1 Schematic of the fabrication process of the superoleophobic hierarchical surface. The aluminum sample was etched first in HCl to get the microstep structure, then immersed in solution of HNO_3 and $Cu(NO_3)_2$ to obtain the nanoreticula structure in order to form a hierarchical surface. The hierarchical surface when modified with fluorosilane, resulting in superoleophobicity to hexadecane (Adapted from Peng and Bhushan 2016)

hierarchical surface was modified with fluorosilane to realize superoleophobicity. Various properties of the surfaces were measured. An overview of the research based on Peng and Bhushan (2016) follows.

Experimental Details

An aluminum plate (composition: wt 98% Al, 0.3% Si, 1.7% O) was cleaned by ultrasonication in acetone, ethanol, and DI water, in sequence, before use. The cleaned plate was first immersed in 2.5 M HCl solution for 10 min at room temperature to achieve the microstep structure. Then, it was rinsed with DI water and dried. The microstructured aluminum surface was next etched in aqueous solutions of nitric acid (HNO_3, 65% mass fraction) and copper nitrate trihydrate ($Cu(NO_3)_2 \cdot 3H_2O$) with a concentration of 9.7 g/L at 80 °C for 6 min, determined by optimization for the highest CA and lowest CAH. After the two-step etching process, hierarchical structure was obtained (Peng and Bhushan 2016).

A separate nanostructured surface and microstructured surface were also fabricated, to compare data with the hierarchical surface. The nanostructured surface was fabricated through one-step etching in HNO_3 and $Cu(NO_3)_2$, while the microstructured surface was obtained from one-step HCl etching. Finally, the finished samples of nanostructured, microstructured, and hierarchical surfaces were dipped into 1.0 wt% ethanol solution of (1H, 1H, 2H, 2H-Perfluorodecyltriethoxysilane) PDES (Sigma-Aldrich #658,758) for 2 h by dip coating technique and then cured at 100 °C for 30 min (Peng and Bhushan 2016).

CA, CAH, TA, surface morphology, roughness, P–V distance, wear resistance on macroscale, self-cleaning, anti-smudge, and corrosion resistance properties were measured. Experimental details have been presented in the previous chapter.

In this study, evidence of air pocket formation was also obtained. To assess the ability of the surfaces to trap air, the samples were photographed while submerged under water to document the formation of any plastron air layers. If a plastron air layer is formed, the surface will appear highly reflective, due to the large difference in refractive index for water and air, and the smooth interface formed between the two as the air becomes trapped by the surface roughness (Zimmerman et al. 2008; Poetes et al. 2010; Peng and Bhushan 2016).

To study corrosion resistance, the electrochemical measurements were made using 3.5 wt % aqueous solution of NaCl at room temperature using an electrochemical workstation (IM6ex, Zahner, Germany) (Anonymous 1994, 2005; Peng and Bhushan 2016). These were conducted using a three-electrode configuration with saturated calomel as the reference electrode, platinum as the counter electrode, and the samples with an exposed area of 1 cm^2 as the working electrode. The potentiodynamic polarization test was carried out at a scanning rate of 5 mVs^{-1} and potential-current density curves were obtained.

Results and Discussion

The properties of the nanostructured and microstructured surfaces, as compared to the hierarchical surface, are presented in this section.

(a) Surface Morphology

Figure 10.2 shows the scanning electron microscope (SEM) micrographs for the nanostructured, microstructured, and hierarchical surfaces. When the aluminum plate was etched only by a mixture HNO_3 and $Cu(NO_3)_2$ solution, the aluminum surface became uneven. The high-magnification SEM image indicated that a large area of nanoreticula porous structure was formed with a nanoporous diameter about 30–40 nm (Peng and Bhushan 2016).

The reaction process for the formation of nanoreticula structure is believed to be as follows. In the chemical etchant solution of HNO_3 and $Cu(NO_3)_2$, copper nitrate precipitates into copper ions (Cu^{2+}). Copper ions react with electrons from the aluminum plate and reduce to copper particles, which deposit on the aluminum surface. This process results in the formation of micro-orifices, which in turn accelerates the erosion speed

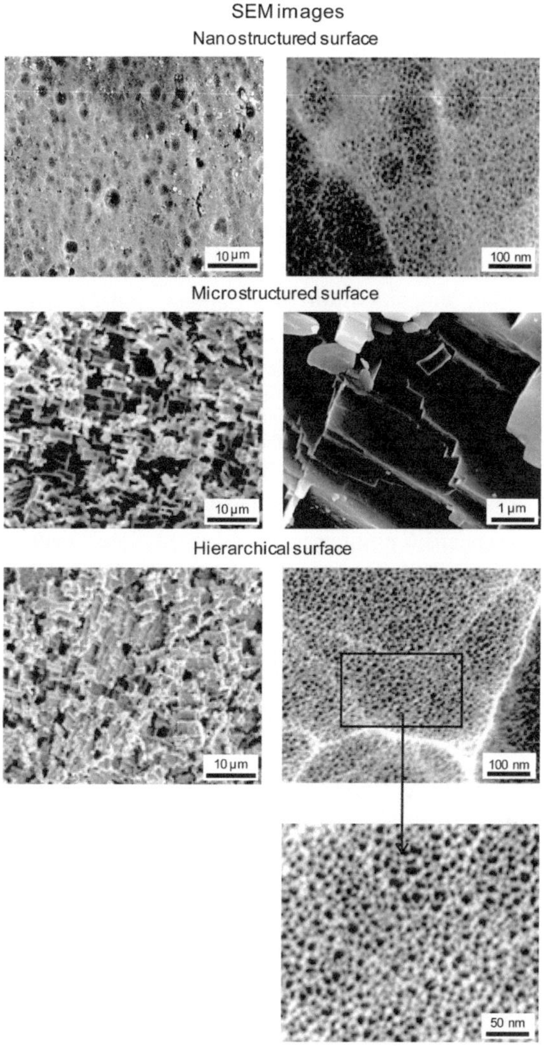

Fig. 10.2 Low and high-magnification SEM images of the nanostructured, microstructured, and hierarchical surfaces. The nanostructured surface shows a reticular porous structure with diameter of about 20–40 nm. The microstructured surface shows smooth microstep morphology and the length of each step is about 3–4 μm. A combination of microstep and nanoreticula structure formed on the hierarchical surface. The microsteps on the hierarchical surface are no longer smooth and have a large area of nanoreticula structure (Adapted from Peng and Bhushan 2016)

of the surrounding aluminum. Consequently, a large area of reticular porous structure was formed (Peng and Bhushan 2016).

When the aluminum plate was etched only by HCl, it resulted in a microstep structure on the surface. The low-magnification SEM image demonstrated that the surface exhibited microstep topography. As indicated by the high-magnification SEM image, the surface of each microstep was smooth. The length of the step was about $3 - 4$ μm. Many dislocation defects that possess relatively higher energy exist in common crystalline metals. These dislocation sites are prone to attack by chemical etchants and dissolve first. The selective corrosion effect led to a large number of micro-pits on the surface and resulted in the formation of a step morphology (Qian and Shen 2005; Vander Voort 1984; Peng and Bhushan 2016).

When the two-step chemical etching process was applied on the aluminum plate, a hierarchical structure was produced. As shown in the low-magnification SEM image, the microstep structure was first formed through HCl etching. However, when the microstructured surface was further etched in mixed HNO_3 and $Cu(NO_3)_2$ solution, the microstep surface was full of the nanoreticula structure, and was no longer smooth. This variation could be observed from the high-magnification SEM images of the hierarchical surface. Higher magnification SEM images indicated that the nanoreticula diameter was about 20 – 30 nm (Peng and Bhushan 2016).

It should be noted that when both a smooth aluminum plate and a microstructured aluminum surface with microstep morphology are immersed in HNO_3 and $Cu(NO_3)_2$ solution, the required reaction time for forming nanoreticula structures on each surface is very different. It takes a longer etching time (about 1 h) to form nanoreticula structure on a smooth aluminum surface, while the microstructured surface, as mentioned previously, only needs 6 min immersion time to obtain nanoreticula topography. The number of dislocation defects on the microstructured aluminum surface increased dramatically compared with the smooth aluminum plate. These dislocations were destroyed faster by chemical etchant attack, and thus, further accelerated the dissolution of surrounding aluminum, which reduced the necessary etching time (Peng and Bhushan 2016).

Surface roughness and P–V distance of the nanostructured, microstructured, and hierarchical surfaces were measured using a stylus profiler. RMS of 2.0 and 8.3 μm, and P–V distance of 11 and 42 μm were obtained for the nanostructured and microstructured surfaces, respectively. The hierarchical structure achieved a rougher surface, with a RMS and P–V data of 9.6 and 49 μm, respectively (Peng and Bhushan 2016).

(b) Wettability

The measured CA, CAH, and TA values for water, ethylene glycol, and hexadecane on the nanostructured, microstructured, and hierarchical surfaces before and after PDES coating are presented in Table 10.1 (Peng and Bhushan 2016). All uncoated surfaces were hydrophilic and oleophilic. Depositing a PDES coating made all surfaces superhydrophobic and two of the three nanostructured and hierarchical surfaces superoleophobic. The nanostructured surface exhibits superhydrophobicity and oleophobicity to water and ethylene glycol with CAs of $152° \pm 2°$ and $105° \pm 2°$, and CAHs of $4° \pm 2°$ and $12° \pm 2°$, respectively. Hexadecane droplets, which have a lower surface tension as compared to ethylene glycol and water, behave differently on the samples. When a hexadecane droplet wets the nanostructured surface, it shows oleophilicity, with a CA of $58° \pm 2°$ and CAH of $22° \pm 2°$. The microstructured surface exhibits superoleophobicity to water and ethylene glycol with CAs of $158° \pm 2°$ and $151° \pm 2°$, and CAHs of $2° \pm 2°$ and $12° \pm 2°$, respectively. For hexadecane, it shows only oleophobicity with a CA of $144° \pm 2°$ and CAH of $16° \pm 2°$.

Table 10.1 Contact angle (CA), contact angle hysteresis (CAH), and tilt angle (TA) of treated aluminum surfaces with and without PDES dip coating, using various liquid droplets (Peng and Bhushan 2016)

	DI water (deg)			Ethylene glycol (deg)			Hexadecane (deg)		
Samples	CA	CAH	TA	CA	CAH	TA	CA	CAH	TA
Nanostructured surface without coating	36 ± 2	20 ± 2	N/A	~4	N/A	N/A	~0	N/A	N/A
Nanostructured surface (with coating)	152 ± 2	4 ± 2	4 ± 1	105 ± 2	12 ± 2	N/A	58 ± 2	22 ± 2	N/A
Microstructured surface without coating	~0	N/A	N/A	~0	N/A	N/A	~0	N/A	N/A
Microstructured surface (with coating)	158 ± 2	2 ± 2	2 ± 1	151 ± 2	12 ± 2	20 ± 1	142 ± 2	16 ± 2	N/A
Hierarchical surface without coating	~0	N/A	N/A	~0	N/A	N/A	~0	N/A	N/A
Hierarchical surface (with coating)	160 ± 2	2 ± 1	1 ± 1	157 ± 2	5 ± 2	5 ± 2	152 ± 2	12 ± 2	14 ± 2

N/A—not applicable

In comparison, the hierarchical surface is superhydrophobic and superoleophobic to water and ethylene glycol, as well as hexadecane, as seen by CAs of 160° ± 2°, 157° ± 2°, and 152° ± 2°, and CAHs of 2° ± 1°, 5° ± 2°, and 14° ± 2°, respectively. These results indicate that only the hierarchical structure achieves superoelophobicity to oils with low surface tension, such as hexadecane (Peng and Bhushan 2016).

(c) Wear Resistance

Optical images of the nanostructured, microstructured, and hierarchical surfaces before and after wear experiments using ball-on-flat tribometer are presented in Fig. 10.3 (Peng and Bhushan 2016). The wear scar is identified by two lines. The optical images show that both the nanostructured and microstructured surfaces show limited burnishing after 100 cycles of wear at 10 mN. The hierarchical surface with dual structure characteristics also shows little burnishing trace. It appears that the wear tests do not cause noticeable damage on the hierarchical surface.

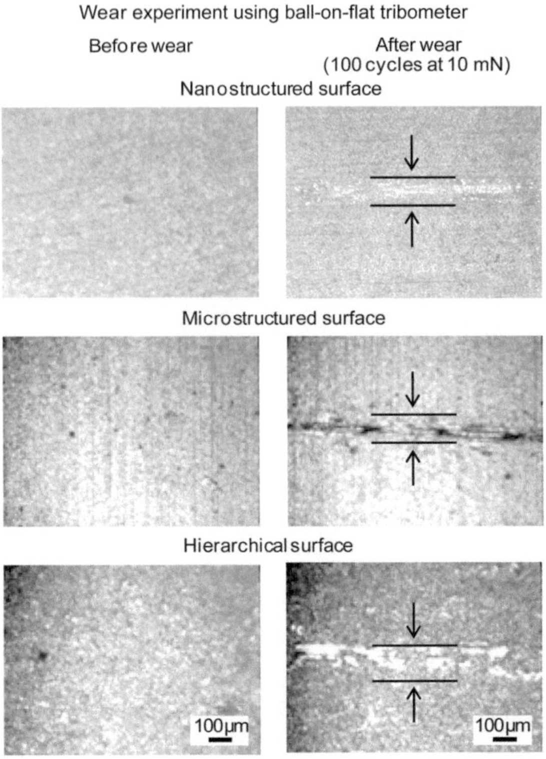

Fig. 10.3 Optical images of the nanostructured, microstructured, and hierarchical surfaces before and after wear experiments using ball-on-flat tribometer at 10 mN. A sapphire ball with a radius of 1.5 mm was used and the tribometer was set into reciprocating for 100 cycles with an average linear speed of 1 mm/s. The images show very little burnishing induced on the nanostructured, microstructured, and hierarchical surfaces (Adapted from Peng and Bhushan 2016)

(d) Self-Cleaning

To measure self-cleaning efficiency, the nanostructured, microstructured, and hierarchical surfaces were exposed to wash experimentation. The optical images for the nanostructured, microstructured, and hierarchical surfaces before and after wash experiments are presented in Fig. 10.4. These optical images are analyzed with SPIP image software to quantify the percentage of SiC particles removed. After the wash experiment, there were still some particles remaining on the nanostructured surface, while the microstructured and hierarchical surfaces became totally clean. The histogram indicates that the self-cleaning efficiency of nanostructured, microstructured, and hierarchical surfaces are 60%, 96%, and 99%, respectively. These results demonstrate that the microstructured and hierarchical surfaces present a more effective contaminant removal ability as compared to the nanostructued surface. Specifically, the combination of micro-nanostructure amplifies the self-cleaning ability of the hierarchical surface, which makes the surface achieve the highest efficient particle removal ability (Peng and Bhushan 2016).

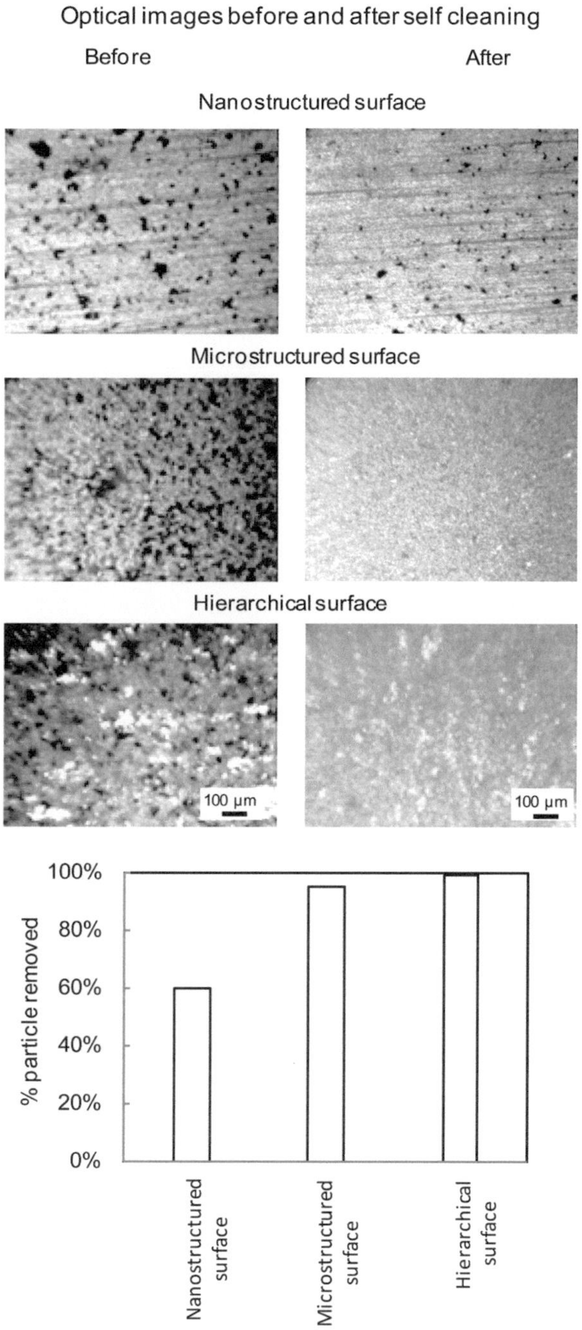

Fig. 10.4 Optical images of the nanostructured, microstructured, and hierarchical surfaces before and after self-cleaning experiments. Silicon carbide (SiC) nanoparticles were removed partially from the nanostructured surface, while the microstructured and hierarchical surfaces became totally clean after the experiments. Particle analysis software was used to quantify the amount of particle removal (Adapted from Peng and Bhushan 2016)

(e) Anti-Smudge

Optical images of the contaminated samples and hexadecane-impregnated microfiber cloth before and after smudge tests on the microstructured and hierarchical surfaces are presented in Fig. 10.5 (Peng and Bhushan 2016). The nanostructured surface was not tested because of its known poor self-cleaning properties. Dark spots on the sample surface and cloth indicate silicon carbide particles. It can be seen that the cloth can only wipe a few particles on the microstructured surface, while particles on the hierarchical surface can be totally removed by the cloth. The microstructured surface cannot realize anti-smudge properties since it is unable to repel hexadecane. This demonstrates the anti-smudge properties of the hierarchical surface.

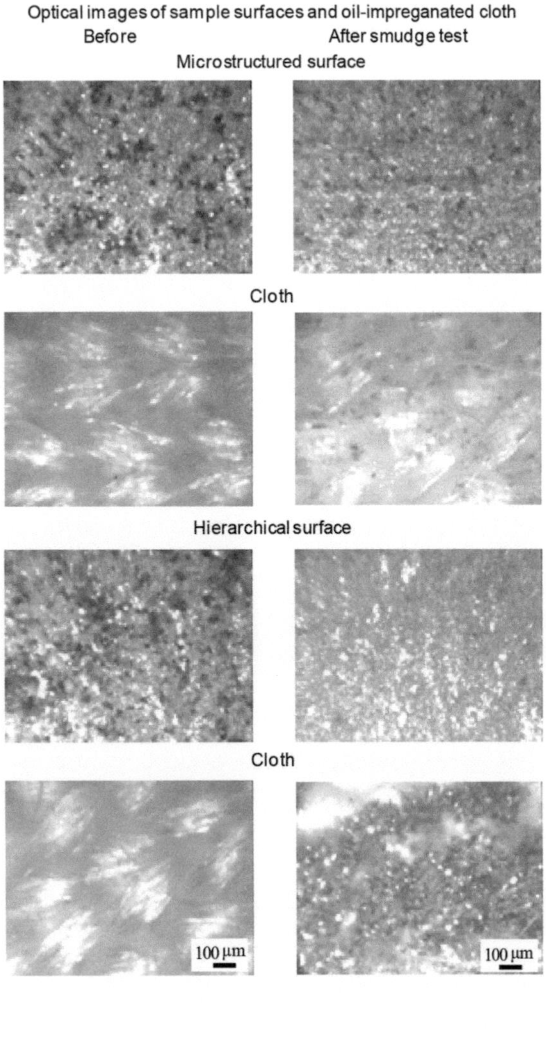

Fig. 10.5 Optical images of the contaminated samples and hexadecane-impregnated microfiber cloth before and after smudge tests on the microstructured and hierarchical surfaces. Dark spots on the sample surfaces are particle contaminants (400 mesh particle size by Aldrich, with sizes ranging from 10–15 μm). The results demonstrate that the cloth can only wipe a few particles on the microstructured surface, while particles on the hierarchical surface can be totally removed by the cloth (Adapted from Peng and Bhushan 2016)

(f) Evidence of Air Pockets Formation

To verify the formation of air pockets, the samples were submerged in water and photographed for any reflectivity. The optical images are shown in Fig. 10.6 (Peng and Bhushan 2016). The treated samples were found to be highly reflective under water, which indicated the formation of a plastron air layer. In contrast, the untreated surface did not become reflective under water, indicating that no air layer was formed. Air layers were present on the nanostructured, microstructured, and hierarchical surfaces. However, the air layer appeared more uniform and stable on the hierarchical surface, especially compared to the nanostructured surface. This suggests that a hierarchical structure is better at trapping air. The data suggests that reticular nanostructure is efficient in air pocket formation. As an added benefit of air pocket formation, due to the presence of these plastron layer, these surfaces could maintain a clean, reflective surface under water (Peng and Bhushan 2016).

(g) Corrosion Resistance

The corrosion resistance of superoleophobic aluminum surface was assessed by electrochemical experiments. The potentiodynamic polarization curves of the flat aluminum and superoleophobic aluminum surface in 3.5 wt % NaCl solutions are shown in Fig. 10.7 (Peng and Bhushan 2016). The vertical axis represents electrical potential, and the horizontal axis represents the logarithm of current density. For each sample, as seen in

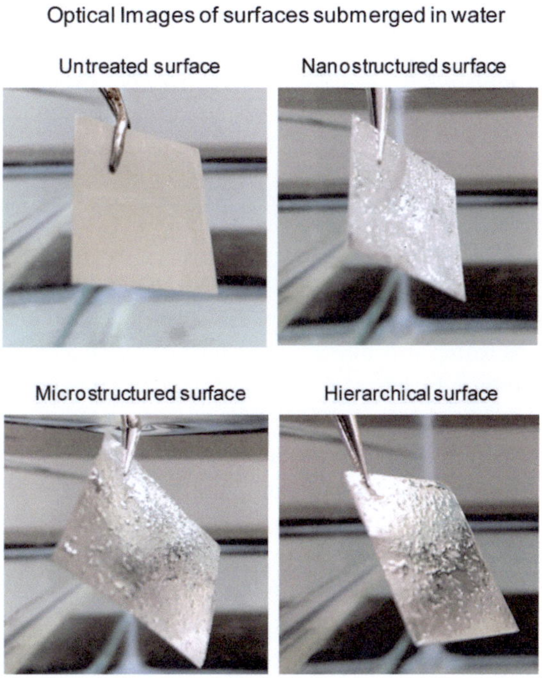

Fig. 10.6 Optical images of an untreated aluminum surface, and the nanostructured, microstructured, and hierarchical surfaces when submerged into water. The nanostructured, microstructured, and hierarchical surfaces became reflective compared with the untreated sample, indicating the formation of an air layer on the surfaces. However, the air layer appeared more uniform and stable on the hierarchical surface, especially compared to the nanostructured surface, which suggests that a hierarchical structure is better at trapping air (Adapted from Peng and Bhushan 2016)

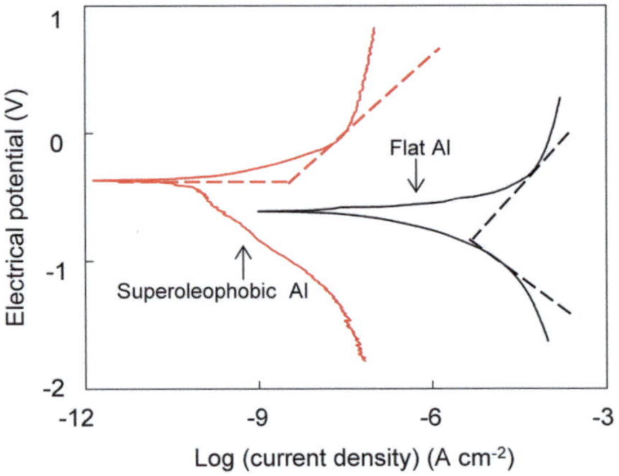

Fig. 10.7 Potentiodynamic polarization curves of the flat aluminum and superoleophobic aluminum surfaces in 3.5 wt % NaCl aqueous solution (Peng and Bhushan 2016)

Fig. 10.7 the current density decreases during cathodic reaction phase and increases during the anodic reaction phase as the electrical potential is increased. The values of corrosion potential and corrosion current density can be calculated by the Tafel analysis, which is performed by extrapolating the linear portions of a logarithmic current density versus potential back to their intersection. The intersection point corresponds to corrosion potential and corrosion current density. Typically, a lower corrosion current density and a higher corrosion potential corresponds to a lower corrosion rate and an improved corrosion resistance (Anonymous 2005; Zhang et al. 2008). In the case of a superoleophobic aluminum surface, the bottom part of the curve is not smooth and only the top curve is used. By comparing the two curves, it is observed that the corrosion potential of a superoleophobic aluminum surface is higher, and the corrosion density is lower than that of flat Al. This indicates that the superoleophobic aluminum surface is more corrosion resistant than flat Al.

Summary

The microstep structure and the nanoreticula structure believed to be beneficial to trap air-pockets, were combined to form a hierarchical structure on aluminum surfaces. The microstep structure was first formed by HCl etching. Then, the nanoreticula structure was produced on top of the microstep structure through HNO_3 and $Cu(NO_3)_2$ etching. The hierarchical structure combined with subsequent fluorosilane dip coating made the surface superoleophobic to hexadecane with a CA of $152° \pm 2°$, CAH $14° \pm 2°$, and TA $14° \pm 2°$. The superoleophobic aluminum surfaces were mechanically durable and exhibited self-cleaning and anti-smudge properties (Peng and Bhushan 2016).

10.2.2 Single Step Technique Using Fluorinated Nanoparticles

A single step technique has been developed which involves SiO_2 nanoparticles modified by 1H, 1H, 2H,2H-Perfluorooctyltrichlorosilane (PFOTS) using a dip coating technique. PFOTS was used because of its ability to form self-assembled monolayers (SAM) rapidly during functionalization of nanoparticles (Barthwal et al. 2013). A detailed schematic of the fabrication steps is shown in Fig. 10.8 (Nanda et al. 2017). An overview of the research, based on Nanda et al. (2017), follows.

Experimental Details

Aluminum sheets (Al 100, > 99% purity) cut to dimension of 50 mm by 20 mm were used as the substrates. Prior to the coating, the aluminum substrate was cleaned with DI water in an ultra-sonicator bath for 10 min. After cleaning, substrates were washed with isopropanol and dried at 40ºC for 1 h. The sol–gel solution of trichloro (1H,1H, 2H,2H perfluorooctyl) silane (PFOTS) ($\geq$ 97% purity, Sigma-Aldrich) was prepared by adding 0.2 ml of PFOTS of 1% (v/v) in 20 ml toluene (Merck). After stirring for 10 min,

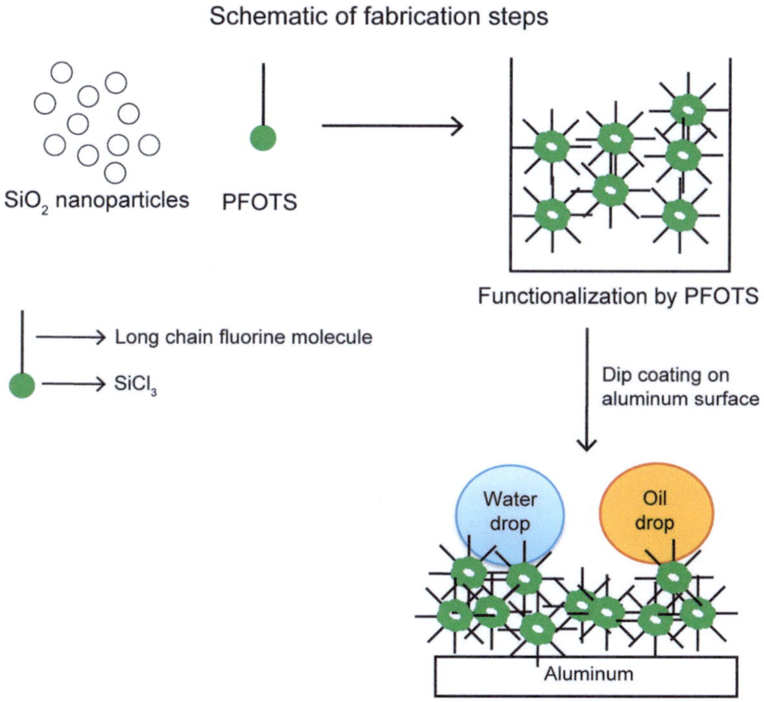

Fig. 10.8 Schematic showing fabrication steps of the coating on an aluminum surface with PFOTS-modified SiO_2 nanoparticles resulting in a nanostructured surface (Adapted from Nanda et al. 2017)

40 mg of hydrophilic silica nanoparticles (10–20 nm particle size, ~97% purity, CAS #7631-86-9, Sigma-Aldrich) were dispersed in the sol–gel solution and stirred further using a hot plate magnetic stirrer for 15 min. For proper dispersion of the nanoparticles and functionalization of nanoparticles, the mixture was further ultra-sonicated using an ultrasonic bath for 10 min. Next, the aluminum substrate was dip coated in the prepared solution for 10 s using a dip coater. The coated samples were air dried for 10 min and then were dried in an oven at 100°C for 1 h (Nanda et al. 2017).

The following chemical reaction takes place between PFOTS molecules (R-$SiCl_3$) and–OH ions present in the SiO_2 nanoparticles (He et al. 2011).

$$R - SiCl_3 + 3(-OH) \rightarrow R - Si - O_{3-} + 3\,HCl,\ \text{where}\ R = CF_3(CF_2)_5CH_2CH_2 \tag{10.1}$$

The hydrophilic head ($SiCl_3-$) of a PFOTS molecule undergoes the above reaction with the –OH ions of the SiO_2 nanoparticles, keeping the oleophobic tail ($CF_3(CF_2)_5CH_2CH_2$) upward to functionalize the nanoparticles. This reaction continues until all the –OH groups have reacted. When dip coating the aluminum substrate, these modified nanoparticles then were deposited on the surface forming a physisorption bond with the surface, changing the wettability from hydrophilic to superliquiphobic (Nanda et al. 2017).

Functionalization of SiO_2 nanoparticles by low surface energy PFOTS was verified by Fourier transform infrared spectroscopy (FTIR) spectra, as shown in Fig. 10.9 (Nanda et al. 2017). The peak at 900 cm^{-1} represents the Si–OH stretching vibrations, and the peak at ~ 1100 cm^{-1} corresponds to the Si-O-Si stretching vibrations of the SiO_2

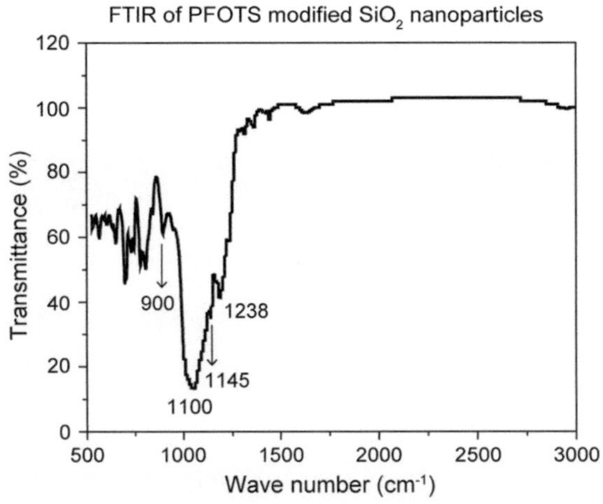

Fig. 10.9 FTIR spectrum of PFOTS-modified SiO_2 nanoparticles. Peaks at 1145 cm^{-1} and 1238 cm^{-1} reveal the presence of –CF2- functional groups (Adapted from Nanda et al. 2017)

nanoparticles (Ke et al. 2011). Peaks at 1145 cm^{-1} and 1238 cm^{-1} correspond to the $-CF_2-$ functional group (Chen et al. 2016). These groups are found in modified SiO_2 nanoparticles due to the attachment of the long chain $CF_3(CF_2)_5CH_2CH_2$ group of PFOTS.

Results and Discussions

Contact angle and tilt angle of water and oil (ethylene glycol, glycerol, and hexadecane) droplets were measured. Surface morphology of the coated surfaces was obtained using SEM imaging. The impact behavior of liquid droplets on the coated surface at different velocities was studied. Thermal stability, chemical stability, and mechanical durability tests were carried out. Finally, self-cleaning and anti-fogging properties of the coating were measured (Nanda et al. 2017).

(a) Surface Morphology

SEM images of uncoated and coated samples are presented in Fig. 10.10 (Nanda et al. 2017). The data shows that uncoated aluminum was smooth and island-like microstructures were formed after coating with modified nanoparticles. On these islands, flower-like structures can be seen at high magnification, suggesting hierarchical morphology with

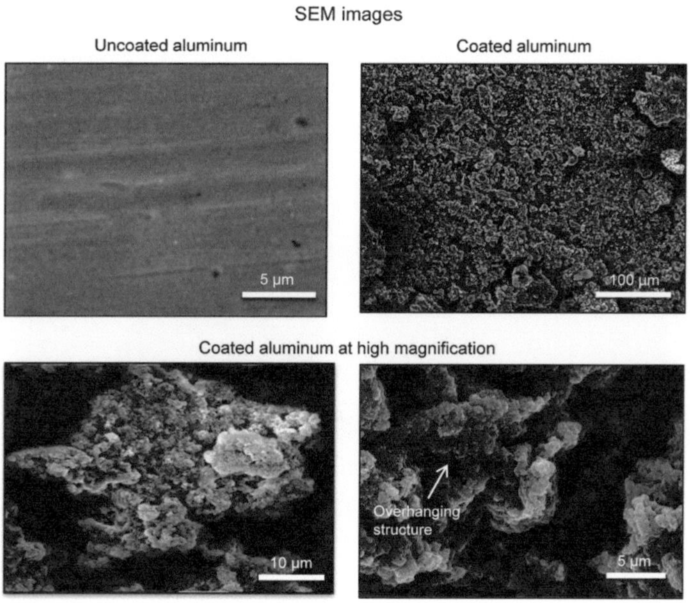

Fig. 10.10 SEM images of uncoated and coated aluminum surfaces at two magnifications. Modified silica nanoparticles increase the roughness on the surface and make a flower-like structure. At higher magnification, overhanging-like structure is seen, which helps in making the surface superliquiphobic (Adapted from Nanda et al. 2017)

overhanging geometry (referred to as re-entrant geometry) (Nosonovsky and Bhushan 2008; Bhushan 2018, 2019). Surface roughness measured using a stylus profiler of the coated surface was 3.80 ± 0.40 μm R_a, and 0.40 ± 0.06 μm R_a for uncoated aluminum.

(b) Wettability

The uncoated aluminum substrate was hydrophilic and oleophilic with water CA of $79° \pm 5°$, and oil CA of $70° \pm 5°$ for ethylene glycol, $73° \pm 6°$ for glycerol and hexadecane $14° \pm 3°$, as shown in Fig. 10.11 (Nanda et al. 2017). For the coated sample, the water CA was $168° \pm 3°$ with a tilt angle < $3° \pm 0.5°$. The oil CA of ethylene glycol, glycerol, and hexadecane on the coated sample were $165° \pm 4°$ with a TA < $4° \pm 0.5°$, $166° \pm 4°$ with a TA < $4° \pm 0.5°$ and $155° \pm 4°$ with a TA of $6° \pm 0.5°$, respectively (Nanda et al. 2017).

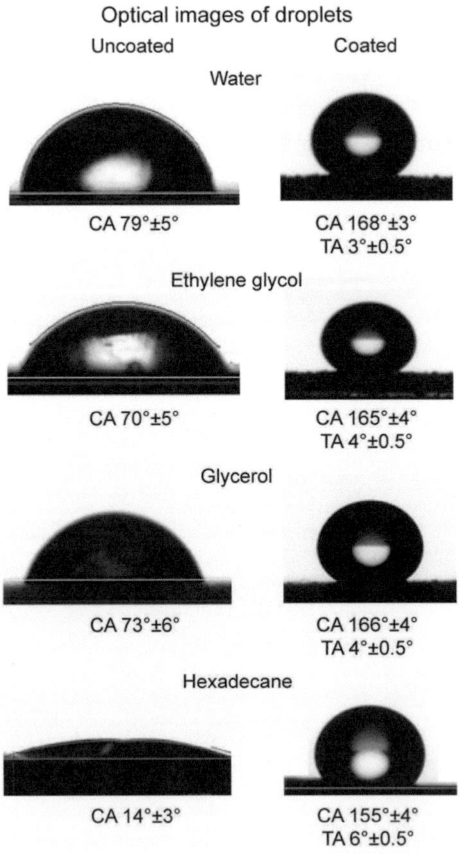

Fig. 10.11 Optical images of droplets of water, ethylene glycol, glycerol and hexadecane and contact angles (CA) and tilt angles (TA) on uncoated and coated aluminum surfaces. It is observed that the coated surface is superliquiphobic (Adapted from Nanda et al. 2017)

(c) Impact Study of Water and Oil Droplets

It is desirable to maintain superoleophobicity during impact of water and oil droplets. Nanda et al. (2017) performed experiments using water, ethylene glycol, glycerol, and hexadecane with different impact velocities. For water at low impact velocity, different types of impact dynamics are observed. At velocities of 0.4 and 0.6 m/s, partial bouncing and complete bouncing are observed, respectively, as shown in Fig. 10.12a. With an increase in velocity to 1.0 m/s and above, splashing, followed by bouncing, and breaking of the water droplet was observed. Due to the presence of air pockets, the surface repelled water without wetting. It was concluded that the Cassie-Baxter state of the surface remained unchanged on impact of water droplets with a wide range of velocities.

For ethylene glycol droplets, partial bouncing occurred at low speeds but sticking occurred at high speeds, Fig. 10.12b. Glycerol and hexadecane droplets exhibited sticking at all speeds tested (Nanda et al. 2017).

(d) Thermal Stability

To measure thermal stability of the modified SiO_2 nanoparticles. thermogravimetric analysis (TGA) was performed. The TGA curve shown in Fig. 10.13 shows that thermal decomposition of the coating began at 322.7 °C and ended at 535.2 °C (Nanda et al. 2017). This is due to the decomposition of PFOTS molecules above 330 °C (Devaprakasam et al. 2004). The thermal stability of the coating was further examined by annealing the sample at different temperatures (from 40 °C to 450 °C) for 1 h. Water CA and hexadecane CA remained unchanged up to 400 °C. With an increase in temperature, hexadecane CA started to decrease and reached below 150° at 420 °C, whereas the water CA decreased below 150° at 450 °C.

(e) Chemical Stability

The chemical stability of the coating was evaluated. Coated samples were immersed in solutions of pH values of 2, 5, 8, and 11 for 24 h. The water CAs and hexadecane CAs were measured at intervals of 1 h. The pH 2 solution is very corrosive and easily damaged the surface. The coating was destroyed within 20 min of immersion. For other solutions, the coating was found to be stable for 24 h of immersion. After 24 h of immersion, the water CA and hexadecane CA at pH 5 reduced to $145° \pm 4°$ and $141° \pm 4°$, respectively. For pH 8 water CA and hexadecane CA decreased to $148° \pm 3°$ and $145° \pm 3°$, respectively. For pH 11 water CA and hexadecane CA decreased to $143° \pm 3°$ and $139° \pm 4°$, respectively. The tilt angle increased to more than 45° in all pH values tested. This may be due to the islands and overhanging structures on the surface, which are unable to protect it from etching after a long immersion (Nanda et al. 2017).

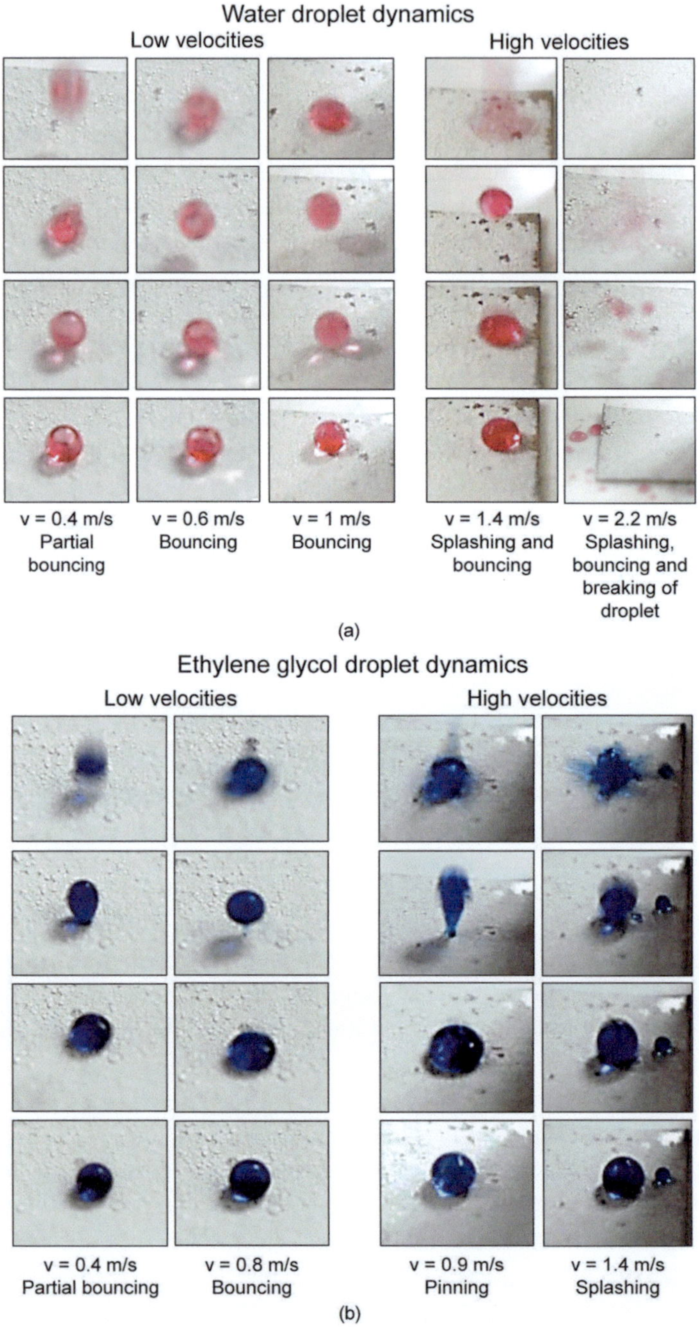

Fig. 10.12 Optical images of the dynamic impact study of **a** water and **b** ethylene glycol droplets at low and high impact velocities. It is observed that the liquid droplets bounce from the surface at a wide range of velocities (Adapted from Nanda et al. 2017)

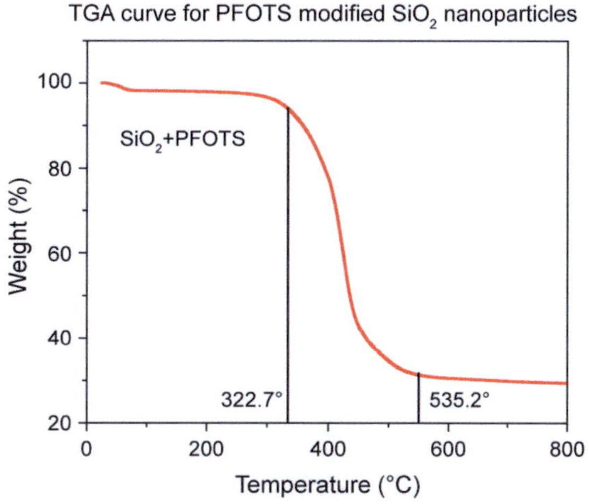

Fig. 10.13 The TGA plot of the PFOTS-modified SiO_2 nanoparticles reveals that the coating is thermally stable up to about 322 °C, and decomposition of the coating begins at 322.7 °C and ends at 535.2 °C (Adapted from Nanda et al. 2017)

(f) Mechanical Durability

Mechanical durability is an important property of coating. The mechanical durability was examined by abrasion, surface bending/folding, and water-jet impact tests. For the abrasion test, the sample was rubbed on sandpaper with 0.5 N load for 6 cm distance. The test procedure and photographs of liquid droplets on the surfaces before and after 10 cycles of abrasion tests are presented in Fig. 10.14a (Nanda et al. 2017). The coating maintained its superliquiphobicity up to 10 cycles without having any scratches. After a larger number of cycles, contact angle started to decrease. After 30 cycles, the water contact angle decreased to $105° \pm 11°$ and hexadecane contact angle decreased to $70° \pm 12°$ with tilt angles greater than 30° for both the water and hexadecane. Thus, the superoleophobic property was lost after 30 cycles.

Bending and folding tests were also carried out by bending the samples into different shapes, and contact angles were measured. In these studies, the coated samples were bent at 90° into a V shape and kink shape and folded and unfolded six times. No change was observed in water CA and TA and hexadecane CA and TA as the water and hexadecane easily rolled off the surface, as shown in Fig. 10.14b (Nanda et al. 2017).

To examine the effect of continuous impact of water or oil on the surface, a water jet test was performed (Nanda et al. 2017). Water was sprayed at an angle of about 35° on the surface with an impact velocity of 2.6 m/s for 2 min using a water syringe, as shown in Fig. 10.14c (Nanda et al. 2017). The water jet bounced easily back in the opposite direction when it came into contact with the surface. This is due to the presence of low-surface-energy PFOTS that tends to repel water from its surface. The water jet test confirms the good mechanical strength.

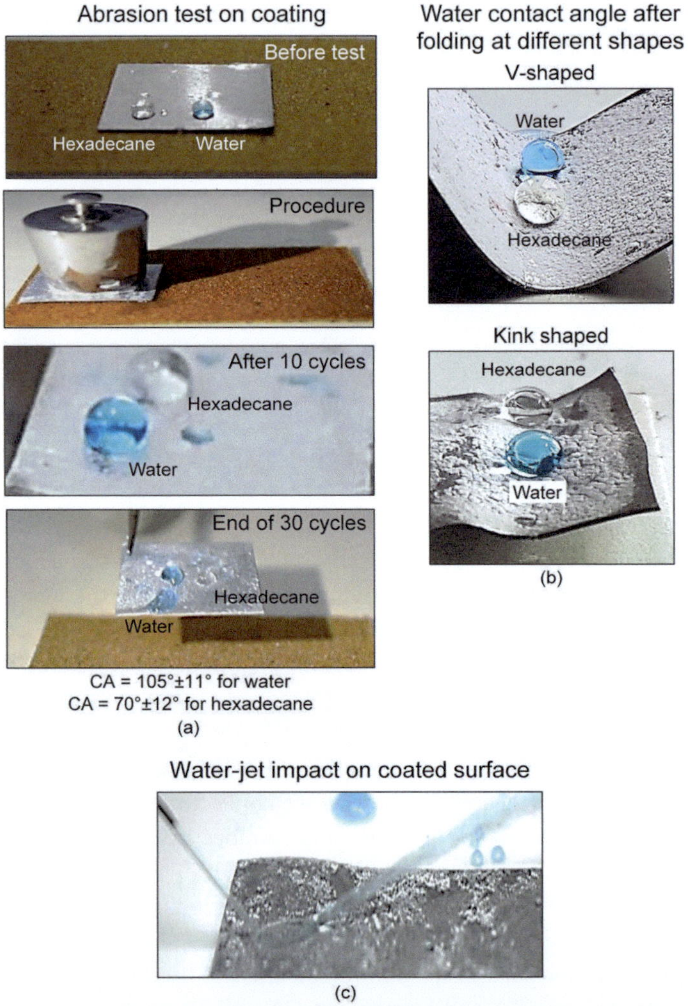

Fig. 10.14 **a** Optical images of water and hexadecane droplets on surfaces after abrasion. The coating withstood up to 30 abrasion cycles by sandpaper with 0.5 N load **b** Optical images of the water and hexadecane droplets at V shaped and kink-shaped bends. Water and hexadecane droplets formed a spherical shape at the bending and folding area, indicating its mechanical stability, and **c** Optical images of the water jet impact test showing that water bounces back in the opposite direction from the surface due to its superoleophobic nature (Adapted from Nanda et al. 2017)

(g) Self-Cleaning

For the self-cleaning tests, charcoal dirt was spread on the surface of the uncoated and coated aluminum surfaces. Excess charcoal was removed. When water and hexadecane were poured on the uncoated surface, neither water nor hexadecane would clean the surface, as they immediately spread out over the surface, as shown in Fig. 10.15 (Nanda et al. 2017). In the case of the coated sample, water and hexadecane poured on the surface rolled down and taking the dirt away, exhibiting the self-cleaning nature of the coating.

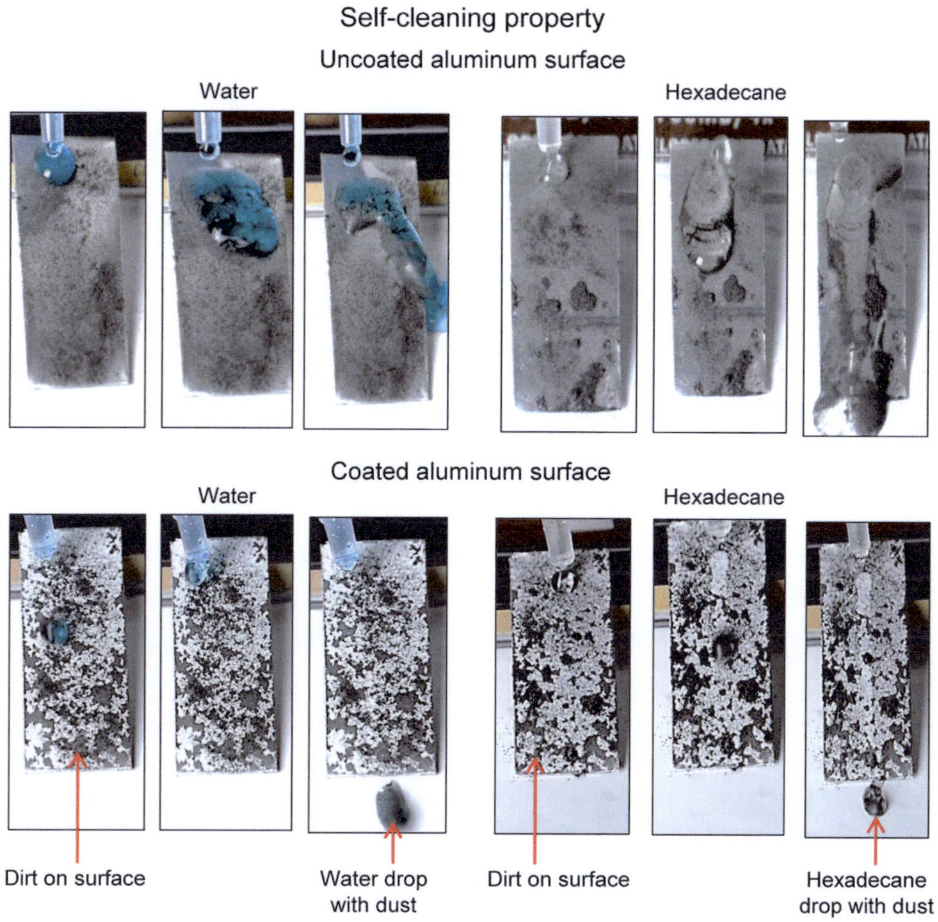

Fig. 10.15 Optical images of uncoated and coated aluminum surfaces, where water and hexadecane droplets rolled down the surface of the coated sample, taking dirt away with them suggesting a self-cleaning property. The uncoated sample remained dirty (Adapted from Nanda et al. 2017)

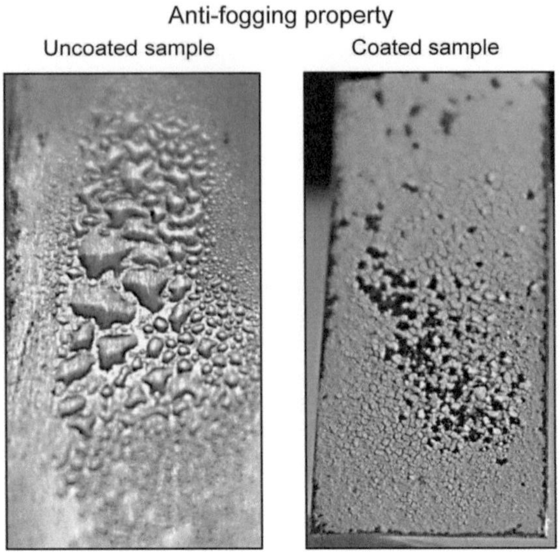

Fig. 10.16 Optical images of uncoated and coated aluminum surfaces after condensation of water vapor. It is observed that water droplets were stuck to the uncoated aluminum surface, but in the case of the coated aluminum surface, only tiny droplets were formed suggesting an anti-fogging property (Adapted from Nanda et al. 2017)

(h) Anti-Fogging

In anti-fogging tests, water vapor was condensed on both uncoated and coated samples as shown in Fig. 10.16 (Nanda et al. 2017). It was observed that water starts to accumulate immediately on the surface of uncoated aluminum, whereas, for the coated sample, a few small water drops were seen on the surface, indicating the anti-fogging property of the coating. The superoleophobic property was regained after complete drying. Repeated fogging tests were carried out by drying the sample at 40 °C and recondensing the water vapor. It was observed that the coating retained its superoleophobic property up to 4 cycles of the fogging test (Nanda et al. 2017).

Summary

In this study, a superoleophobic coating was synthesized by a single-step dip coating method on an aluminum surface using PFOTS-modified SiO_2 nanoparticles. Surface morphology analysis shows the presence of hierarchical morphology with overhanging geometry on the treated surface, and high contact angles (water: $168 \pm 3°$, ethylene glycol: $165 \pm 4°$, glycerol: $166 \pm 4°$ and hexadecane: $155 \pm 4°$) and low tilt angles (water: $3 \pm 0.5°$, ethylene glycol: $4 \pm 0.5°$, glycerol: $4 \pm 0.5°$ and hexadecane: $6 \pm 0.5°$). TGA analysis revealed that the decomposition of PFOTS modified SiO_2 nanoparticles took place at about 322 °C. The coating also was found to be thermally stable up to 400 °C without losing its superoleophobic property. Various thermal, chemical and mechanical durability tests were conducted, and the coating was found to be durable (Nanda et al. 2017).

10.3 Superoleophobic Stainless Steel Surfaces

Iron based alloys are abundant and are commonly used metals in structural and many other applications (Cobb 2010). They provide a range of physical and chemical properties of interest, at low cost. Stainless steel has chromium content greater than about 10.5 wt. % (Bhushan and Gupta 1991). The chromium imparts corrosion resistance on stainless steel through its ability to form highly adherent, impervious chromium oxide layers on the surface. Stainless steel can withstand the harsh environmental demands of chemical transport, and its use is prevalent in the food industry due to its ability to be more easily cleaned and disinfected. Some unprotected steels can rust due to atmospheric corrosion from moisture. In addition, many applications of steel, such as in oil processing and transport, involve contact with low surface tension liquids that can easily foul the surface of steel. It is desirable to treat the steel surface to be repellent to water and oils (superliquiphobic) and to reduce corrosion and fouling. Therefore, liquid repellent and antifouling properties are of interest for various applications. It is expected that liquid repellency will improve corrosion resistance of steel surfaces. Some of the applications include self-cleaning tabletops, tar repellent railroad car transports, anti-icing airplane bodies and windmills, and drag reduction in pipe flows.

To produce superliquiphobic steel surfaces, a combination of hierarchical morphology and surface chemistry is required. Researchers have utilized a variety of chemical and physical methods to achieve them. Preparation methods include chemical etching and coating, mechanical abrasion and coating, laser machining and coating, plasma etching and coating, and coatings only (Cremaldi and Bhushan 2018).

Superhydrophobicity has been achieved through various pathways. A variety of chemicals such as nitric acid, hydrofluoric acid, iron trichloride, cupric chloride, and sulfuric acid, have been used to create roughness which was then combined with a hydrophobic coating such as methylchlorosilane or a fluoro-compound (Qu et al. 2007; Li et al. 2012; Boinovich et al. 2013; Li et al. 2014b; Lu et al. 2014; Latthe et al., 2015; Cremaldi and Bhushan 2018; Nanda et al., 2019; Meena et al. 2020). Mechanical abrasion in the form of sandblasting is another technique commonly used to create roughness followed by a hydrophobic coating (Beckford and Zou 2011; Frank et al. 2014; Cremaldi and Bhushan 2018). Femtosecond laser machining combined with silanization has been shown as well (Wu et al. 2009b). Her et al. (2012) made use of plasma etching to create roughness followed by siloxane deposition. Another route to creating roughness has been the coating deposition. Coatings include self-assembled fluorosilane (Xu et al. 2008), electrodeposited Zn (Brassard et al. 2015), and dip coated nanoparticle-binder composite (Cremaldi and Bhushan 2018).

Superoleophobicity has also been achieved through a variety of pathways. Motlagh et al. (2013) used mechanical abrasion with nanocomposite spray coating to achieve superliquiphobicity. However, the resulting surfaces were only superliquiphobic with high surface tension liquids such as ethylene glycol. Huang et al. (2017) achieved superliquiphobicity by depositing Ni electroless coating followed by a fluorinated coating. They did not provide any data on the durability of the coating. Cremaldi and Bhushan (2018) achieved superliquiphobicity by depositing a nanoparticle-binder composite coating with an overlay of fluorosilane. Coated surfaces exhibited self-cleaning and anti-icing properties as well as mechanical durability and corrosion resistance.

In this section, an overview of the research based on Cremaldi and Bhushan (2018) is presented. Superhydrophilic, superhydrophobic, and superliquiphobic stainless steel 304 L surfaces were produced using various pathways including sandblasting, chemical etching, and nanocomposite coatings to create roughness. Various pathways provide competing methods of imparting superliquiphilicity/phobicity. Procedures used to create various wetting regimes on stainless steel 304 L were then verified on stainless steel 430. Finally, to accentuate the improvement, stainless steel 430 (less corrosion resistant than 304 L) was used to test corrosion resistance imparted by the superliquiphobic surface modification (Cremaldi and Bhushan 2018).

10.3.1 Experimental Details

To produce superliquiphilic/phobic stainless-steel surfaces, roughness and the desired surface chemistry are necessary. To provide roughness onto the surface, the three pathways utilized on the stainless steel substrates are shown in Fig. 10.17 (Cremaldi and Bhushan 2018). The first two, shown on the left, are sandblasting and chemical etching. These two pathways aim to create roughness out of the substrate itself. Etching utilizes acids to eat away at the substrate's surface at different rates, leaving behind roughness. Sandblasting does not remove the substrate but deforms it using a highly pressurized particulate spray. The third pathway involves deposition of a nanoparticle-binder coating.

Once the roughness has been created, the surface chemistry must be controlled to achieve superliquiphilicity/phobicity. Steel is naturally hydrophilic, so simply imparting roughness moves it toward being superhydrophilic. When using the nanoparticle-binder coating system, choosing a hydrophilic nanoparticle in a methylphenyl silicone binder result in superhydrophilicity (Cremaldi and Bhushan 2018).

To create a superhydrophobic surface, in one of the approaches, coating with hydrophobic nanoparticles in a binder can be used. Although methylphenyl silicone resin is hydrophilic with a water contact angle of 74° (Wang and Bhushan 2015), the exposed hydrophobic nanoparticles interact with the water and create the desired superhydrophobicity. In another approach, a coating without nanoparticles can be added on roughened and as received surfaces. However, in the cases of sandblasting, chemical etching, or through use of the existing roughness in as-received samples, multiple requirements must be met to create superhydrophobicity. First, the coating must be hydrophobic. Second, the coating must retain the roughness in the substrate. In the choice of coating materials, use of a polymer precursor versus polymer resin dissolved in solution has a drastic effect on

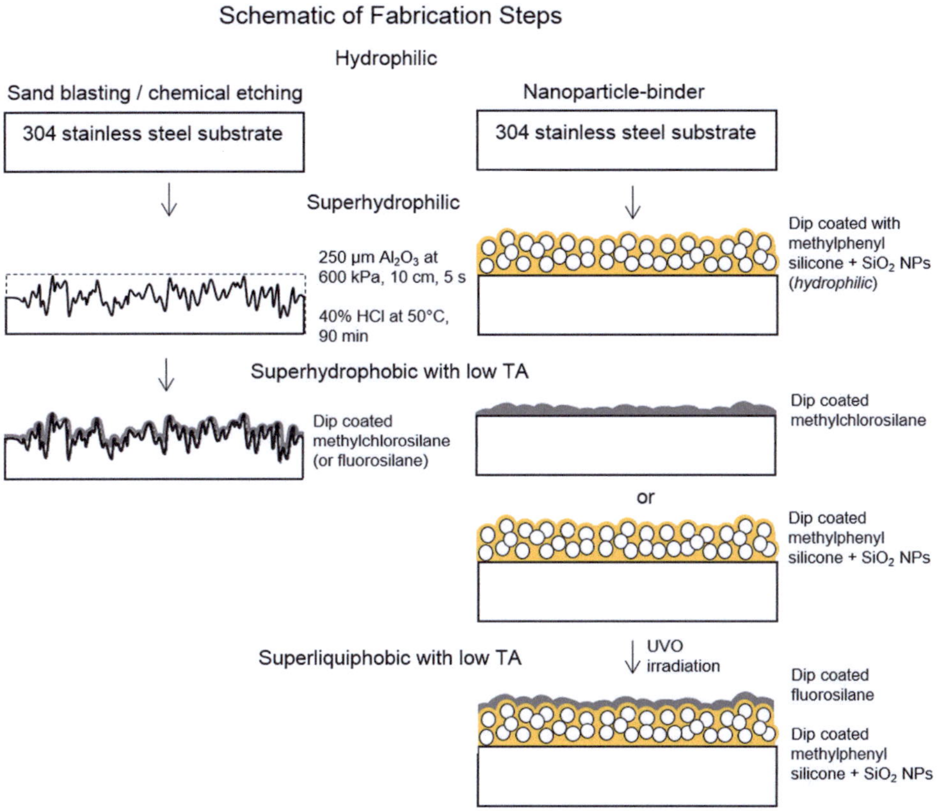

Fig. 10.17 Schematic of the fabrication steps in stainless steel surfaces with varying wettability through chemical etching/sandblasting or through the use of a nanoparticle-binder coating and fluorosilane coating. The resultant surfaces can be hydrophilic, superhydrophilic, superhydrophobic, or superliquiphobic (Adapted from Cremaldi and Bhushan 2018)

the retention of roughness during the coating process, as shown in Fig. 10.18 (Cremaldi and Bhushan 2018).

Chlorosilanes are a polymer precursor and undergo hydrolysis and then condensation to form polymers (Materne et al. 2012; Arkles et al. 2014; Cremaldi and Bhushan 2018). When a hydroxylated substrate is present during this hydrolysis and condensation, the polymerization includes the substrate and a polysiloxane coating forms, as shown in Fig. 10.19a (Cremaldi and Bhushan 2018). Because this reaction occurs uniformly on top of the substrate, the coating conforms to the rough surface and no roughness is lost. Furthermore, the substrate often provides the water molecules necessary to initiate the reaction. As seen in Fig. 10.19b, reacting chlorosilanes in the absence of a substrate can form resins such as polydimethylsiloxane (PDMS) and methylphenyl silicone (Cremaldi

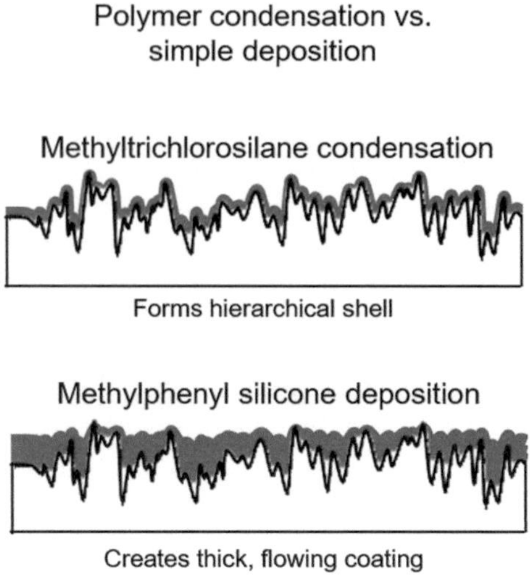

Fig. 10.18 Schematics show the difference in dip coating versus coating through a condensation reaction on a rough surface. Dip coating allows material to accumulate inside of the roughness, reducing it. Polymer condensation forms the coating uniformly over the surface, retaining the roughness morphology and is desirable (Adapted from Cremaldi and Bhushan 2018)

and Bhushan 2018). These resins may then be dissolved and used for dip coating. However, this coating method may result in the loss of roughness, as shown schematically in Fig. 10.18. The reason is that the substrate is dipped into the (resin) solution and then the solvent evaporates off to form the polymer coating. In the time it takes the solvent to evaporate the fluid gathers in the recesses of the roughness and smoothens the sample. Therefore, Cremaldi and Bhushan (2018) chose to use chlorosilane condensation (methylchlorosilane and fluorosilane) because it met both requirements. The resulting coatings were hydrophobic due to the low energy hydrocarbon chains (methyl- and perfluorooctyl- for methylchlorosilane and fluorosilane, respectively) oriented towards the surface, and the roughness from the substrate was retained through the condensation reaction.

To achieve superliquiphobicity, modification of the surface chemistry was necessary. Creating roughness on the substrate from sandblasting and chemical etching provided sufficient roughness to achieve superhydrophilicity and superhydrophobicity but did not result in the re-entrant geometries necessary to achieve superliquiphobicity (Nosonovsky and Bhushan 2008; Brown and Bhushan 2016). Use of the nanoparticle-binder coating, however, did result in re-entrant geometries (Bhushan and Martin 2018; Cremaldi and Bhushan 2018; Bhushan. 2018, 2019). Fluorination of the superhydrophobic nanoparticle-binder surface reduced the surface energy low enough to achieve superliquiphobicity.

Substrate Materials

In the research reported here, stainless steel 304 L (SS 304 L, rolled finish, standard composition) and stainless steel 430 (SS 430, 1292T15, McMaster Carr) were used as substrate materials (Cremaldi and Bhushan 2018). SS 304 is the most widely used 300-series (austenitic) stainless steel and the "L" represents low carbon content (< 0.3%) and

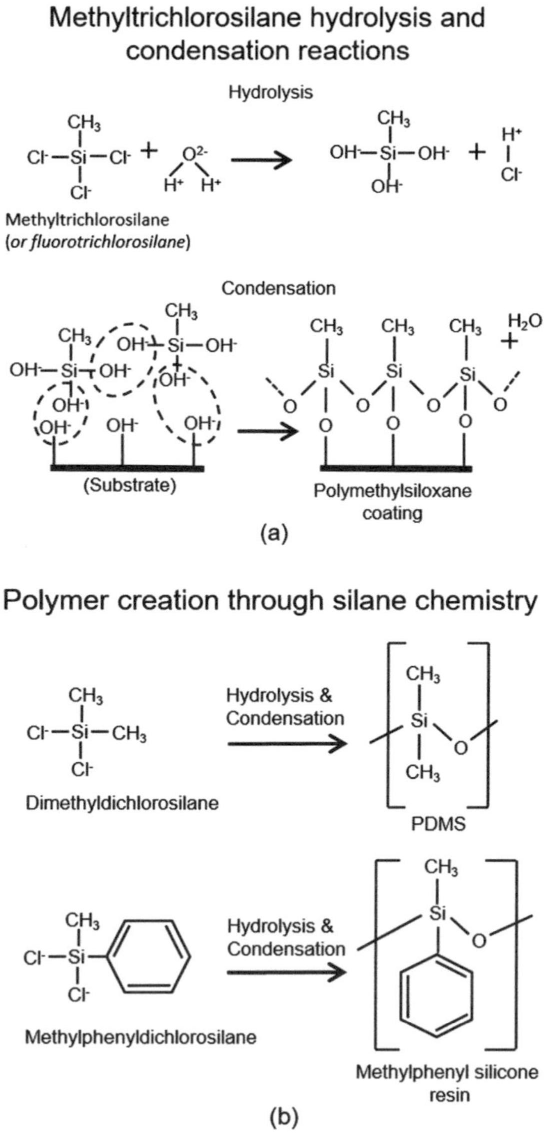

Fig. 10.19 **a** Schematic showing the mechanism of a condensation reaction, in which the chlorosilanes first undergo hydrolysis and then condenses to form a polymer coating. This mechanism causes the coating to grow at a uniform rate and maintain the morphology of the substrate, and **b** Schematic showing the mechanism of resin creation from silane precursors. PDMS can be created through hydrolysis and condensation of dimethyldichlorosilane. Methylphenyl silicone resin can be created through hydrolysis and condensation of methylphenyldichlorosilane (Adapted from Cremaldi and Bhushan 2018)

is related to weldability. SS 430 is a less corrosion resistant, ferritic alloy. SS 430 was used to show transferability of the techniques as well as a display of corrosion resistance. The composition of the two alloys is presented in Table 10.2 (Krauss 2005). SS 304 L was cleaned by sonicating in acetone and ethanol and then air dried prior to use. SS 430 was wet polished to 1500 grit (3 M Imperial Wetordry Sandpaper), cleaned with water, sonicated in acetone and ethanol, and then left to air dry (Cremaldi and Bhushan 2018).

Table 10.2 Chemical compositions of stainless steel 304 L and 430 (Krauss, 2005)

Type	% Cr	% Ni	% C	% Mn	% Fe
304 L	18.0–20.0	8.0–12	0.03 max	2.0	Balance
430	16.0–18.0	–	0.08 max	1.0	Balance

Sandblasting

Sheets of the SS 304 L were sandblasted to introduce roughness. By using 250 μm alumina (Al_2O_3) particles at 600 kPa from approximately 10 cm. Sandblasting for longer than 5 s (test to 30 s) did not noticeably increase the roughness of the substrate. After sandblasting, samples were sonicated in acetone, then ethanol, and left to air dry (Cremaldi and Bhushan 2018).

Chemical Etching

SS 304 L was etched using a 40% hydrochloric acid (HCL, A144, Fisher Scientific) solution at 50 °C for 90 min (Walker and Tarn 1991; Cremaldi and Bhushan 2018). SS 430 was etched using a 50% sulfuric acid (H_2SO_4, ACS Grade, Ricca) at room temperature for 90 min (Cremaldi and Bhushan 2018). Both samples were re-passivated using 50% nitric acid, (HNO_3, ACS Grade, Mallinckrodt) at 50 °C for approximately 10 s (ASTM standard A967/A967M–13). The reduced time (from ASTM recommendations) was due to the re-passivation process eating away the desired roughness at the recommended times. After etching, samples were sonicated in DI water (multiple cycles) and left to air-dry.

The effects of time and temperature on etching are shown through optical and atomic force microscope (AFM) images in Fig. 10.20 (Cremaldi and Bhushan 2018). The as received SS 304 L has an RMS of 144 nm and maximum peak-to-valley (P–V) distance of 1.7 μm. SS 304 is known for its corrosion resistance. Chemical etching for 120 min at room temperature (RT) had almost no effect on the surface other than revealing grain boundaries. The acidic solution was then heated to 50 ºC to increase the etch rates, and one can see the noticeable difference in using a heated acid for a shorter time in the third row of Fig. 10.20 where the RMS increases to 207 nm. The limited change to P–V in etching for 120 min at RT or 30 min at 50 ºC may be attributed to acid attacking the more vulnerable protrusions first. Combining longer etch times with heated HCl, 90 min at 50 ºC, increases the RMS to 539 nm and P–V to 3.3 μm. The increased etching time at a raised temperature creates roughness from more corrosion-resistant metals in the alloy, resulting in protrusions. This combination of time and temperature resulted in the desired hierarchical roughness (Cremaldi and Bhushan 2018).

Condensation Procedure

Methylchlorosilane (methylchlorosilane, 99%, Aldrich) was condensed on sandblasted, chemically etched, and as received stainless steel surfaces to impart superhydrophobicity (Cremaldi and Bhushan 2018). A smooth glass slide was also used as a reference surface.

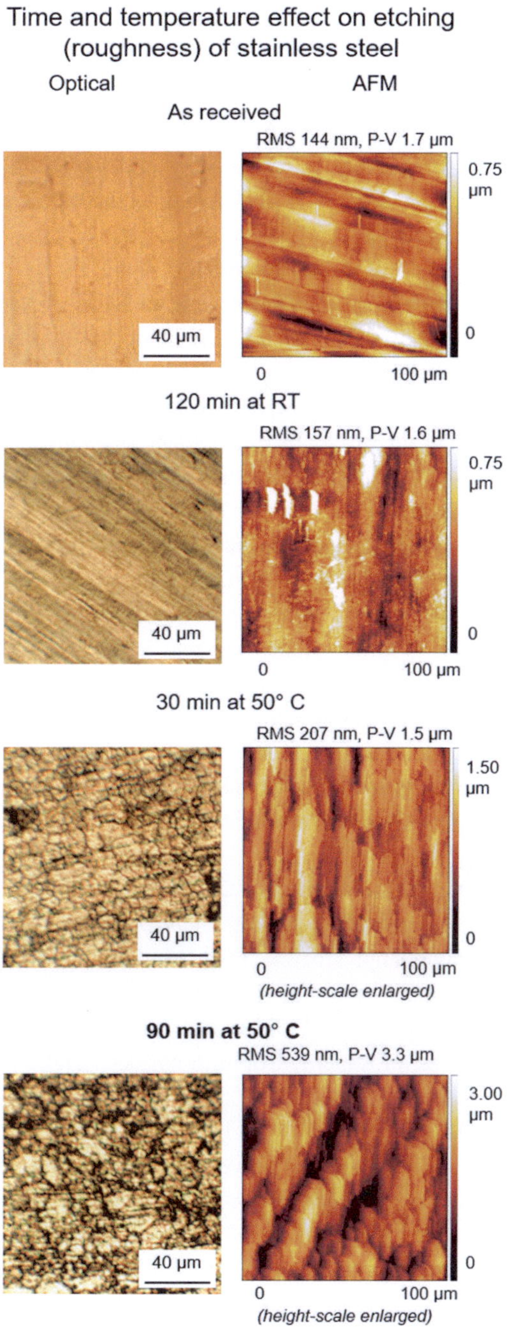

Fig. 10.20 Optical micrographs and AFM images showing the effects of time and temperature on the etching of stainless steel 304 L, with no noticeable change in the roughness observed after 90 min at 50 °C (Adapted from Cremaldi and Bhushan 2018)

Surfaces were immersed in 2% methylchlorosilane by volume in hexane (Fisher Chemical) for 2 h. The samples were then dehydrated in an oven at 100 °C for approximately 10 min. Methylchlorosilane on glass achieved a water contact angle of 140°, showing that some level of nanostructure was created during the polymerization. In another approach to impart superhydrophobicity and superliquiphobicity, fluorosilane was condensed on sandblasted, chemically etched, or as received and nanoparticle-binder coated surfaces, respectively.

For chlorosilane deposition (both methylchlorosilane and fluorosilane are chlorosilanes) on a steel surface, a discussion of vapor deposition or a dip coating technique is needed (Cremaldi and Bhushan 2018). The condensation reaction through which the methylchlorosilane layer is created the same reaction mechanism through which fluorosilane molecules functionalize the nanoparticles, where the time component regulates the coating thickness. In Fig. 10.19a, one can see that the reaction byproducts are water and HCl (Cremaldi and Bhushan 2018). Stainless steel is corrosion resistant, but it is not corrosion proof. Exposure to the HCl in the vapor phase deposition processes largely corrodes the substrate. Furthermore, attempts to protect the surface with a thick polymer underlayer did not prevent rust formation. Rust formation competes with the condensation reaction and weakens the adhesion of coatings. It was found that rust is the prime issue with application of chlorosilanes through vapor deposition. When using dip coating, however, no apparent rust was visible. Therefore, Cremaldi and Bhushan (2018) used dip coating for both methylchlorosilane and fluorosilane condensation. In the case of methylchlorosilane, the two-hour coating process allowed for a thick coating on the order of a micron to form. In the case of fluorosilane, the much shorter, five-minute deposition and low reactivity of the fluoro-groups produced a much thinner coating on the order of tens of nanometers.

Nanoparticle-Binder Coating

The nanoparticle-binder nanocomposite consisted of silica nanoparticles and a methylphenyl silicone resin binder polymer binder, which was spray coated onto the substrate (Wang and Bhushan 2015; Cremaldi and Bhushan 2018). Hydrophilic silica nanoparticles (NP, 7 nm diameter, Aerosil 380, Evonik Industries) were used in the creation of superhydrophilic surfaces and hydrophobic silica nanoparticles (NP, 10 nm diameter, Aerosil RX300, Evonik Industries) were used to create superhydrophobic surfaces. Nanoparticles to binder ratio was optimized to about 4.5 for the highest CA and lowest TA. The solvent system of 40% tetrahydrofuran (THF, HPLC Grade, Fisher Chemical) and 60% isopropyl alcohol (IPA, Fisher Scientific) in the selected nanoparticle-to-binder ratio was found to be optimal in allowing NP to agglomerate and dry to form re-entrant geometries (Ebert and Bhushan 2012; Cremaldi and Bhushan 2018).

To create the nanoparticle-binder solution, 675 mg of NP were dispersed in 30 mL of solvent (40% THF, 60 IPA) and ultrasonicated for 15 min. 150 mg of methylphenyl silicone was then added and the solution was once again sonicated for 15 min. Solution was spray coated onto substrates at 10 cm and 210 kPa, using ≈ 1 mL solution per 1 cm^2 substrate (Cremaldi and Bhushan 2018).

Fluorosilane Coating

To achieve superhydrophobicity, trichloro (1H, 1H,2H,2H-perfluorooctyl) silane (fluorosilane, 448,931, Sigma-Aldrich) condensation was used by dip coating to fluorinate sandblasted or chemical etched surfaces. To achieve superliquiphobicity, the coating was applied on hydrophobic nanoparticle-binder coated samples. Coated sample surfaces were first activated (hydroxylated) using UVO (18.4 W, Model G18T5VH-U, Atlantic Ultraviolet Co.) for 60 min. Samples were then immersed in a 2% fluorosilane solution in hexanes for 5 min and dehydrated in an oven at 100 °C for approximately 5 min (Cremaldi and Bhushan 2018).

Characterization of Samples

The treated samples were characterized for wettability (CA and TA), self-cleaning, anti-icing, and wear and corrosion resistance (Cremaldi and Bhushan 2018).

10.3.2 Results and Discussion

Based on Cremaldi and Bhushan (2018), the wetting properties of SS 304 L and SS 430 with various surface modifications are presented. The wetting properties of SS 304 L created through sandblasting, chemical etching, and nanoparticle-binder coatings with and without fluorosilane coating are discussed first. Next self-cleaning and anti-icing properties are discussed. Afterwards, the mechanical durability of the nanoparticle-binder coatings (as the lone path to superliquiphobicity) was tested using a reciprocating wear test and a bending test. Finally, corrosion resistance measurements were made to test if corrosion resistance was improved by the liquid repellency. SS 430 was used in corrosion test as it is more susceptible to corrosion than the SS 304 L.

Wettability

Optical images of droplets and contact angles and tilt angles are shown in Fig. 10.21 on SS 304 L with various wettability fabricated using surface treatments: sandblasting, chemical etching, and a nanoparticle-binder coating (Cremaldi and Bhushan 2018). As received SS 304 L had a water CA = 69° with no movement of the droplet. The SS surface was completely wetting to hexadecane oil. Once sandblasted, the hydrophilicity of the surface increases, showing a CA = 50°. However, not enough roughness is imparted to make the surface superhydrophilic.

When the sandblasted surface was coated using a methylchlorosilane condensation, the surface becomes superhydrophobic with the CA increasing to 156° and the tilt angle dropping to 5°. Chemical etching of the surface imparted hierarchical roughness as

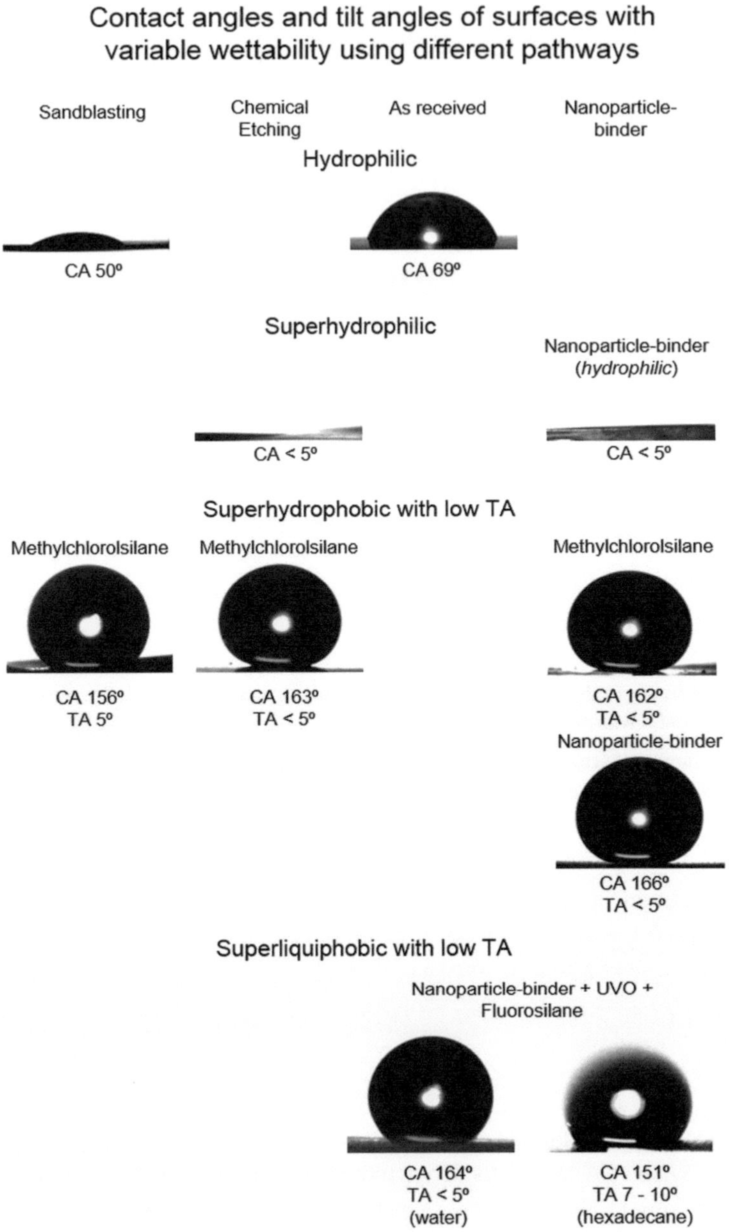

Fig. 10.21 Droplet images and contact angles (CA) and tilt angles (TA) for as received, hydrophilic stainless steel 304 and the effect of surface morphologies and chemistry created through sandblasting, chemical etching, nanoparticle-binder coatings, and fluorosilane overcoat (Adapted from Cremaldi and Bhushan 2018)

seen in Fig. 10.20 (Cremaldi and Bhushan 2018). This roughness changed the surface from hydrophilic to superhydrophilic, decreasing the water CA to less than 5°. Once a hydrophobic coating such as methylchlorosilane was added to the hierarchical roughness, superhydrophobicity was achieved with a CA = 156° and TA < 5°. The SS 304 L as received has enough microscale roughness, when combined with nanoscale roughness of the self-assembled methylchlorosilane (CA = 140° on glass), creates superhydrophobicity as well. Methylchlorosilane on SS 304 L gave a CA = 162° and TA < 5°. Use of a nanoparticle-binder coating allowed one to create both superhydrophilic and superhydrophobic surfaces depending on the type of nanoparticle used. Use of a hydrophilic nanoparticle gives a CA of < 5°. Use of a hydrophobic nanoparticle gives a CA = 166° and TA < 5° (Cremaldi and Bhushan 2018).

The superhydrophobic nanoparticle-binder surface can be further modified to be superliquiphobic as well by activation with UVO and treatment with fluorosilane. The superliquiphobic nanoparticle-binder surface had a water CA = 164° and TA < 5° and a hexadecane oil CA = 151° and TA = 7–10°. Still video images in Fig. 10.22 show the behavior of water (blue) and hexadecane (red) droplets released on the surfaces at a 10° incline (Cremaldi and Bhushan 2018). Water droplets quickly rolled off the superhydrophobic and superliquiphobic surfaces while wetting the hydrophilic and superhydrophilic surfaces. Hexadecane droplets wetted all the surfaces except for the superoleophobic surface (Cremaldi and Bhushan 2018).

To sum up, Figs. 10.21 and 10.22 show wettability results from treating SS 304 L surfaces through sandblasting, chemical etching, and nanoparticle-binder coating with and without fluorosilane (Cremaldi and Bhushan 2018). Superhydrophilicity can be achieved either through chemical etching/sandblasting, or hydrophilic nanoparticle-binder coating. Superhydrophobicity can be achieved either through chemical etching/sandblasting or no roughening + methylchlorosilane or through hydrophobic nanoparticle-binder coating on the as-received samples. Superliquiphobicity can be achieved through modification of the superhydrophobic nanoparticle-binder surface with UVO and fluorosilane treatment. The contact angles and tilt angles for various SS 304 L surfaces are summarized in Table 10.3 (Cremaldi and Bhushan 2018).

Self-cleaning

Hydrophilic, superhydrophobic, and superliquiphobic surfaces were contaminated with SiC microparticles, cleaned with water, and then analyzed before and after the water cleaning for particle removal. Optical images of the contaminated surfaces before and after cleaning, and a quantitative measure of cleaning by measuring the percent of particles removed, are presented in Fig. 10.23 (Cremaldi and Bhushan 2018). The sliding behavior of the water on the hydrophilic surface clumped together the debris but did not remove it. In optical images, one can observe the clumped-together piles of particulate, which represents about a 23% removal rate. The rolling behavior of the water cleared debris from the superhydrophobic and superliquiphobic surfaces, representing a 100% and 95% particle removal rate respectively (Cremaldi and Bhushan 2018).

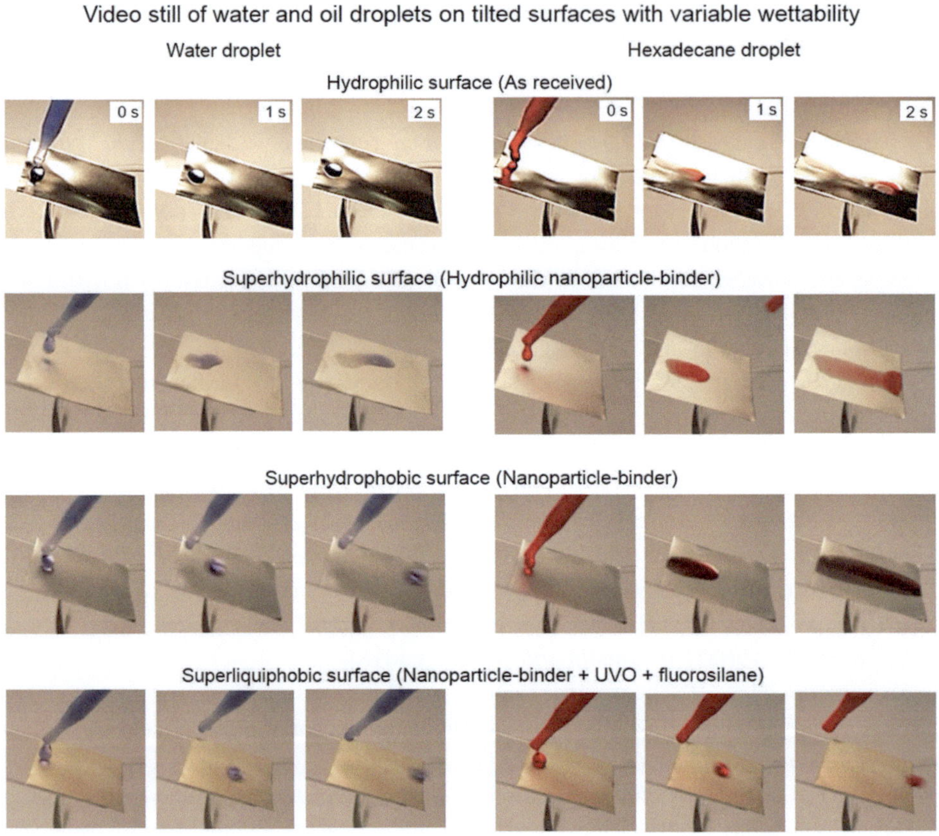

Fig. 10.22 Video still images showing water and hexadecane oil droplet behavior on tilted, as received and nanoparticle-binder coated surfaces varying in wettability: hydrophilic, superhydrophilic, superhydrophobic, and superliquiphobic (Adapted from Cremaldi and Bhushan 2018)

Table 10.3 Contact angles and tilt angles of treated steel surfaces with DI water and hexadecane droplets (Adapted from Cremaldi and Bhushan, 2018)

Sample wetting	Contact Angle (°) / Tilt angle (°)				
	DI water			Hexadecane	
	Sandblasting	Chemical etching	Nanoparticle-binder	Chemical etching	Nanoparticle-binder
Hydrophilic	50°/ > 90°	69°/ > 90°		< 10°/ > 90°	
Superhydrophilic	N/A	< 5°/N/A	< 5°/N/A	< 5°/N/A	< 5°/N/A
Superhydrophobic	156°/ < 5°	163°/ < 5°	166°/ < 5°	N/A	< 5°/N/A
Superliquiphobic	N/A	N/A	> 164°/ < 5°	N/A	155°/7–10°

N/A Not Applicable

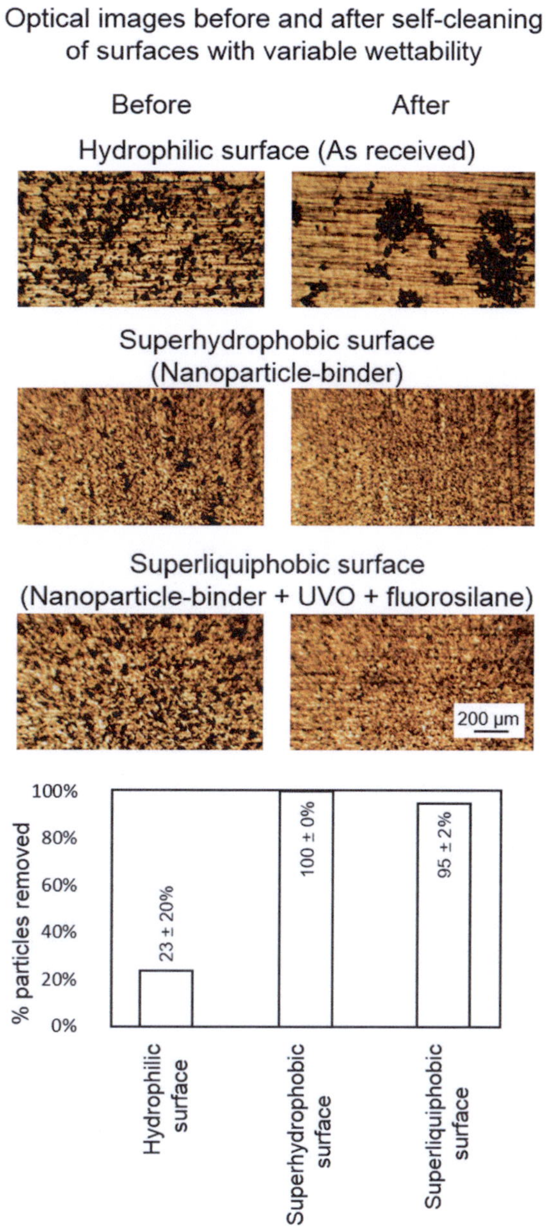

Fig. 10.23 Optical images and % particles removed showing the self-cleaning behavior of surfaces varying in wettability. Water introduced onto a contaminated hydrophilic surface caused the particulate to agglomerate but remained on the surface. The rolling behavior of water introduced onto superhydrophobic and superliquiphobic surfaces caused the water to pick up and almost completely remove all particulate from those surfaces (Adapted from Cremaldi and Bhushan 2018)

Anti-icing

Anti-icing property refers to the ability of a surface to prevent water from freezing on the surface. To test for anti-icing, the samples were mounted on a glass sample holder. Supercooled water droplets were released onto hydrophilic, superhydrophilic,

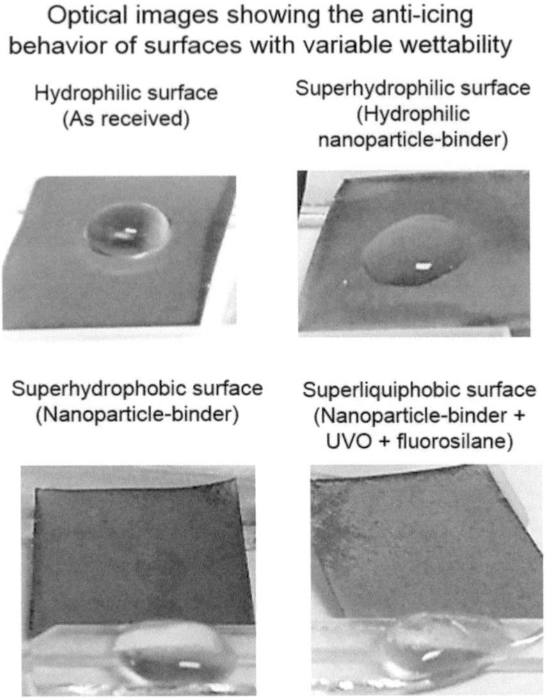

Fig. 10.24 Optical images of droplets showing the anti-icing behavior on surfaces varying in wettability. On tilted hydrophilic and superhydrophilic surfaces, the supercooled water froze immediately after contact. On superhydrophobic and superliquiphobic surfaces, the supercooled water droplets rolled off the surface before they can freeze (Adapted from Cremaldi and Bhushan 2018)

superhydrophobic, and superliquiphobic surfaces at a 10° incline. Optical images of the supercooled water droplets on various surfaces are shown in Fig. 10.24 (Cremaldi and Bhushan 2018). The supercooled water froze on the hydrophilic and superhydrophilic surfaces immediately after deposition. Water rolled off the superhydrophobic and superoleophobic surfaces before freezing. The droplets can be seen frozen on the glass sample holder (Cremaldi and Bhushan 2018).

Mechanical Durability

Mechanical durability of the treated samples was measured by performing sliding wear tests using a ball-on-flat tribometer and a bending test.

(a) Sliding Wear Using ball-on-flat Tribometer

The wear resistance of the coating was first tested using a ball-on-flat tribometer. Figure 10.25a shows optical micrographs of the various surfaces before and after 100 cycles of wear at 10 mN load (Cremaldi and Bhushan 2018). The methylphenyl silicon (binder) by itself, superhydrophobic nanoparticle-binder, and superoleophobic nanoparticle-binder + fluorosilane were tested. The methylphenyl silicone by itself wore completely through to the underlying SS substrate with the 10 mN load. The added strength from the nanoparticles kept the coating from wearing completely through in the case of superhydrophobic and superoleophobic surfaces. The asperities were knocked off in both cases, however the burnished coating remained and protected the surface (Cremaldi and Bhushan 2018).

Wettability of the superoleophobic surface after the wear test was investigated by dragging and rolling hexadecane droplets across the wear track, data shown in Fig. 10.25b (Cremaldi and Bhushan 2018). A hexadecane droplet dragged over the wear track stuck to the wear track initially and caused the deformation in the droplet shape (Cremaldi and Bhushan 2018). However, the droplet eventually was released and did not leave any liquid remaining. In the case of rolling, the hexadecane droplets rolled across the wear track at angles of 8–10°, only a slight increase of the static TA of the original surface (7–10°) (Cremaldi and Bhushan 2018).

(b) Bending

In the bending test, superoleophobic samples were bent to angles of 105° and 60°. Water (dyed blue) and hexadecane (dyed red) droplets were then placed ono the bend to show the effects of bending on the wettability of the coating. Figure 10.26 shows both water and hexadecane drop resting in the bend in perfectly spherical shapes (Cremaldi and Bhushan 2018). The data shows the retention of the coating integrity, with no apparent loss of wettability after being bent.

Corrosion Resistance of SS 430

It is expected that liquid repellency will improve corrosion resistance of the steel surfaces. Corrosion resistance was measured on the less corrosion resistant SS 430. After polishing to 1500 grit, the RMS of SS 430 was 44 nm and P–V was 1.1 μm. SS 430 underwent similar treatments to SS 304 L in chemical etching by H_2SO_4 and use of a nanoparticle-binder coating. After etching, the RMS increased to 866 nm and the P–V increased to 5.7 μm.

Figure 10.27 shows the wettability regimes of SS 430 (Cremaldi and Bhushan 2018). The SS 430 as received had a water CA of 70° with no movement of the droplet. The SS surface was completely wetted by hexadecane oil. Chemical etching of the surface changed the surface from hydrophilic to superhydrophilic, decreasing the water CA to 8°. Coating the etched surface with methylchlorosilane achieved superhydrophobicity with a CA of 163° and TA $< 5°$. The (unpolished) SS 430 as received had enough microscale roughness, when combined with the nanoscale roughness of the self-assembled methylchlorosilane (CA 140° on glass), created superhydrophobicity as well, with CA of 159° and TA $< 5°$. Use of a nanoparticle-binder coating allows one to create both superhydrophilic and superhydrophobic surfaces depending on the type of nanoparticle used. Use of a hydrophilic nanoparticle gave a CA of $< 5°$. Use of a hydrophobic nanoparticle gave a CA of 166° and TA $< 5°$ (Cremaldi and Bhushan 2018).

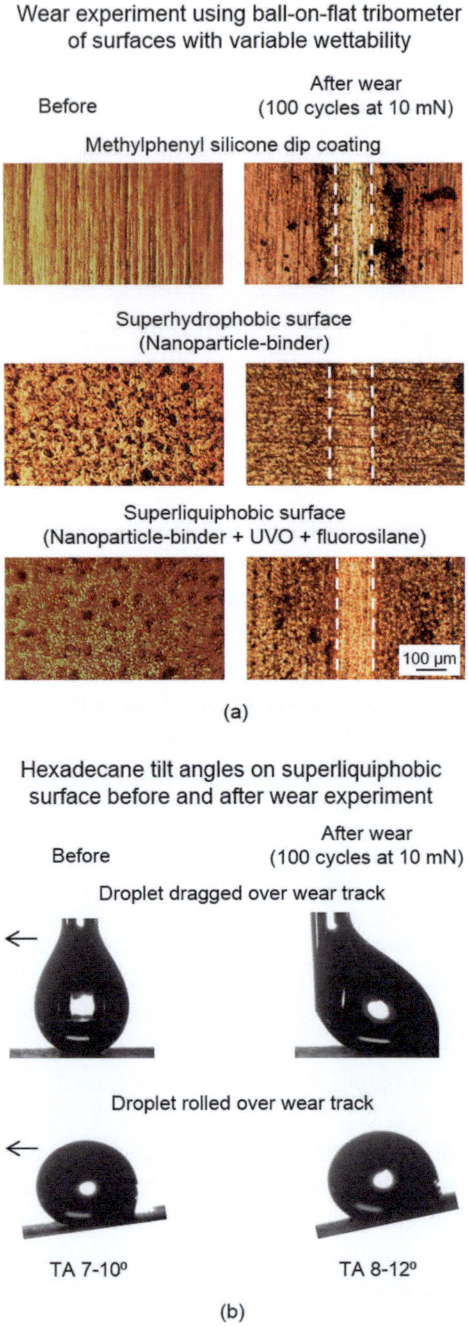

Fig. 10.25 **a** Optical micrographs showing the effects of wear on a nanoparticle-binder coating (superhydrophobic surface) and with fluorosilane overcoat (superoleophobic surface) using a ball-on-flat tribometer. The control surface with binder wore completely through while the nanoparticle-binder coatings showed only burnishing. **b** Hexadecane droplet images showing the effect of wear on the superliquiphobic surfaces. Droplets dragged over the wear track were caught but eventually released and droplets rolled over the wear at an angle comparable to the tilt angle of the non-worn sample (Adapted from Cremaldi and Bhushan 2018)

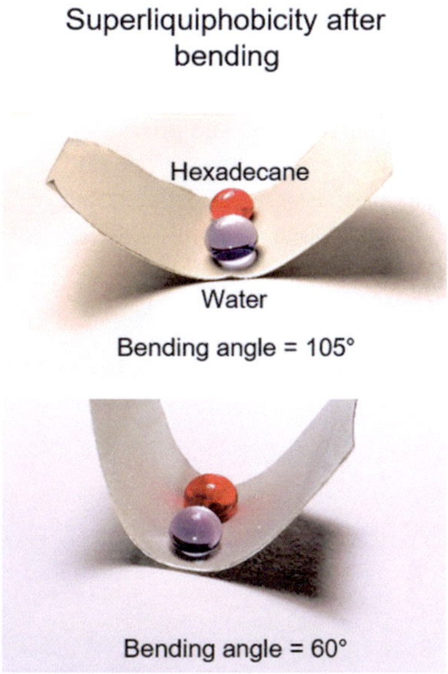

Fig. 10.26 Droplet images showing the effect of bending on superliquiphobic surfaces on the nanoparticle-binder coating, which retained their functionality as shown by the spherical oil and water droplets resting on the 105° and 60° bends (Adapted from Cremaldi and Bhushan 2018)

The superhydrophobic nanoparticle-binder coated surface was subjected to activation with UVO and treatment followed with fluorosilane to make it superliquiphobic. It had a water CA of 166° and TA < 5° and a hexadecane CA of 166° and TA of 10° (Fig. 10.27).

To sum up, for SS 430 similar to what is reported for SS 304 L, superhydrophilicity can be achieved through chemical etching or nanoparticle-binder coating with hydrophilic particles on as received samples. Superliquiphobicity can be achieved through etching + methylchlorosilane or nanoparticle-binder coating. Finally, superliquiphobicity can be achieved through modification of the hydrophobic nanoparticle-binder surface with UVO and fluorosilane treatment (Cremaldi and Bhushan 2018).

To test for the corrosion resistance imparted by the superoleophobicity, samples were immersed in an HCl acid bath which is known to corrode many stainless steels. Uncoated (as received) and coated samples were immersed in a 1% HCl solution for 3 days. Figure 10.28 shows images of the samples once they were removed from the acidic solution (Cremaldi and Bhushan 2018). The uncoated SS 430 corroded, with a black oxide film forming over the entirety of the surface. The superliquiphobic nanoparticle-binder coating minimized contact with the acid and prevented corrosion of the underlying steel substrate. The coating itself remained in good condition, and the underlying SS 430 substrate (photographed after coating removal) looked pristine after the immersion in acid (Cremaldi and Bhushan 2018).

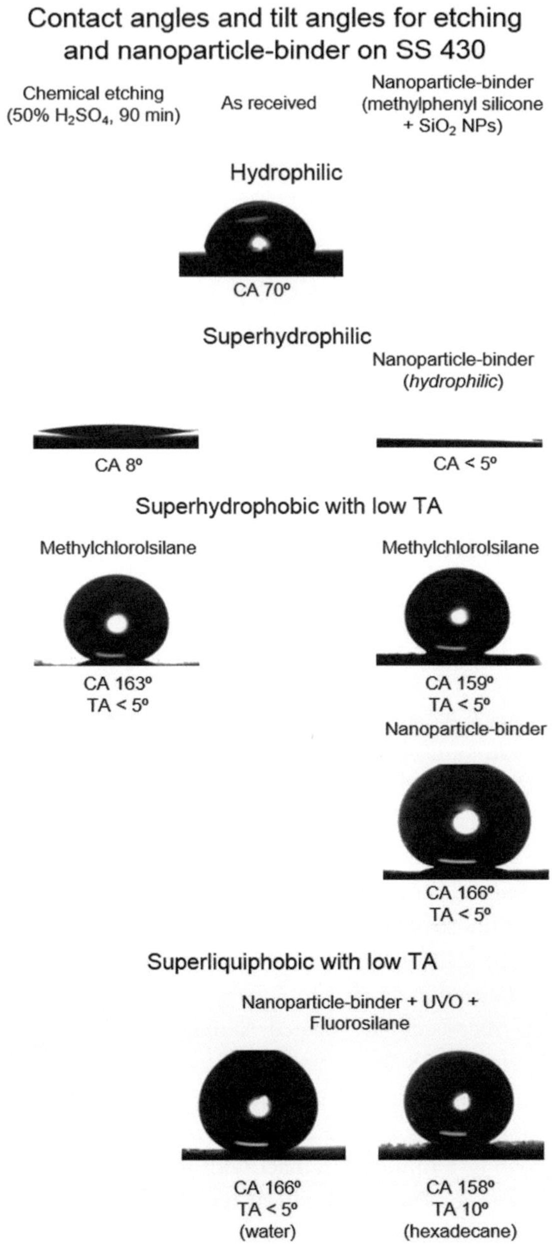

Fig. 10.27 Droplet images and contact angles (CA) and tilt angles (TA) for as received, stainless steel 430 and after chemical etching, nanoparticle-binder coatings, and fluorosilane coating (Adapted from Cremaldi and Bhushan 2018)

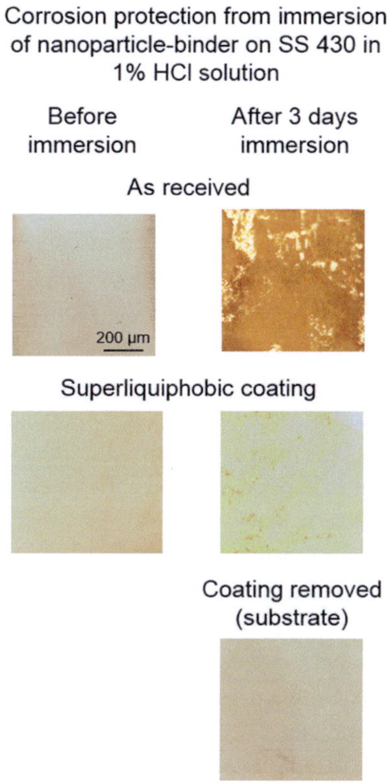

Fig. 10.28 Optical images showing the corrosion protection afforded a stainless steel 430 surface by a superliquiphobic nanoparticle-binder coating immersed in 1% hydrochloric acid for 3 days. The uncoated sample was completely corroded, shown by a black oxide layer. The coated samples showed almost no change in the coating and the underlying substrate (after removal of the coating) (Adapted from Cremaldi and Bhushan 2018)

10.3.3 Summary

In this study, various pathways are presented to change the hydrophilic stainless-steel surfaces to be superhydrophilic, superhydrophobic, or superliquiphobic. Creation of these functional surfaces requires hierarchical roughness and chemistry. Roughness was created using various pathways including sandblasting, chemical etching, and nanocomposite coatings. Surface chemistry was controlled using methylchlorosilane, nanoparticles in methylphenyl silicone, and fluorosilane treatment. The broad approach allowed for direct comparisons of these pathways. Resulting treatments can create stainless steel surfaces with a hexadecane contact angle of 155° and tilt angle of 7–10°. Discussion of rust-avoidance and coating through condensation reactions is provided. Self-cleaning behavior, anti-icing behavior, wear resistance, and bending resistance are demonstrated on stainless steel 304 L. Stainless steel 430, which is more corrosion prone than stainless steel 304 L, was then used to demonstrate transferability of the treatments and corrosion resistance imparted through superliquiphobicity (Cremaldi and Bhushan 2018).

10.4 Superoleophobic Titanium Surfaces

Titanium is a lustrous, silver colored, transition metal. It is the fourth most abundant structural metal in earth's crust after aluminum, iron, and magnesium (Lütjering and Williams 2007). Titanium is commonly used as an alloy and one of the most commonly used alloys is Ti-6Al-4 V. The titanium alloy has a comparable tensile strength to that of stainless steel 304 and is half as dense, and also it is about twice as stiff as aluminum, leading to its high tensile strength to density ratio, Table 10.4 (Gurera and Bhushan 2019). Other notable properties of the titanium alloy include high corrosion resistance, high fatigue resistance, ability to withstand moderately high temperatures up to 500 °C, and biocompatibility. Consequently, it has found applications in many areas including aerospace, marine, automotive, and biomedical industries.

Titanium is a liquiphilic material. There have been several attempts in making titanium surface superoleophobic for various applications. The preparation methods used to make a titanium surface superhydrophobic include laser texturing (Li et al. 2008), thermal oxidation (Guo et al. 2014), sandblasting (Fleming and Zou 2013), chemical etching (Wang et al. 2007; Gao et al. 2015), and anodizing (Huang et al., 2014) along with fluorination on top. The preparation methods used to make titanium surface superoleophobic include anodizing (Lu et al. 2013; Sun et al. 2015), two-step anodizing (Wang et al. 2010), anodizing and nanoparticle-binder coating (Li et al. 2016), chemical etching and anodizing (Lee and Hwang 2017), and nanoparticle-binder coating (Gurera and Bhushan 2019), along with a fluorinated coating on top. Two of the commonly used to introduce roughness are chemical etching and anodizing. In these studies, important characteristics of interest include self-cleaning, anti-icing, and durability.

In this section, various methods based on nanoparticle-binder technique are presented to prepare surfaces with superhydrophilicity, superhydrophobicity, and superoleophobicity, based on Gurera and Bhushan (2019). The various wettability were achieved using different pathways which include chemical etching, anodizing, and nanoparticle-binder coating. A fluorinated compound was added as the last step of each pathway to obtain the

Table 10.4 Physical properties of some commonly used materials (Davis, 1998; Callister, 2007; Gurera and Bhushan 2019)

Metal	Density (kg/m^3)	Elastic modulus (GPa)	Tensile strength (MPa)	Strength-to-density ratio (kPa.m^3/kg)
Titanium (Ti-6Al-4V)	4430	120	900	200
Stainless steel 304	7800	200	600–900	75–115
Aluminum	2700	70	40–50	15

oil repellency. Data for contact and tilt angles for DI water and hexadecane, self-cleaning, anti-icing, and mechanical durability are presented.

10.4.1 Experimental Details

Ti-6Al-4 V was used in the study reported here. Surface roughness and surface chemistry are the determining factors for a surface's wettability. In this comparative study, three different pathways were chosen to alter the surface roughness—chemical etching, anodizing and nanoparticle-binder coatings, Fig. 10.29 (Gurera and Bhushan 2019). The surface chemistry was altered by coating the top with a low surface energy material. Two examples of such materials include fluorosilane and methylchlorosilane to achieve superoleophobicity and superhydrophobicity, respectively.

Contact angles (CA) and tilt angles (TA), self-cleaning, anti-icing and wear resistance measurements were made (Gurera and Bhushan 2019). Experimental details have been presented in the previous chapter.

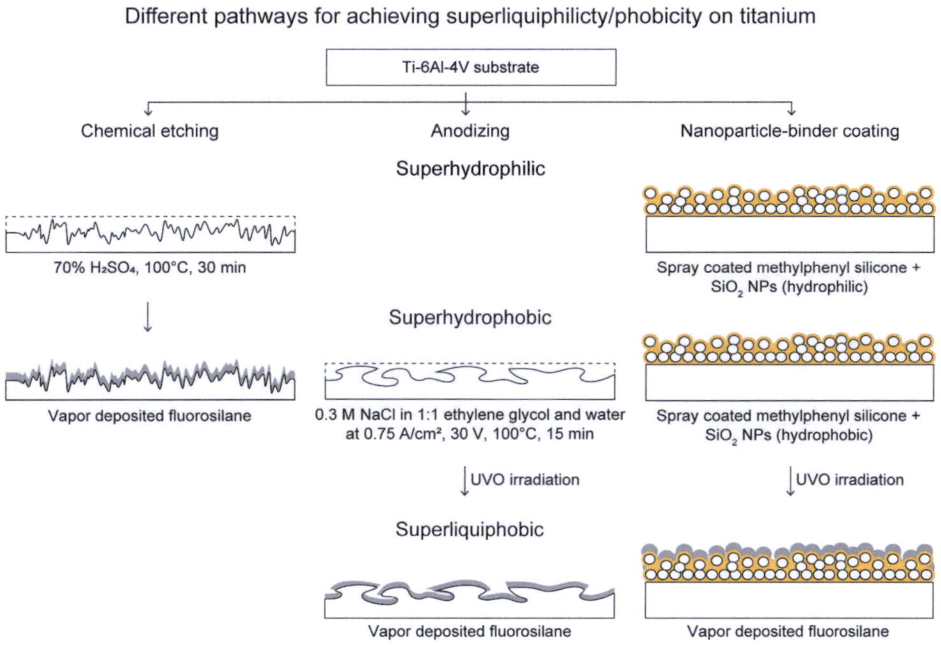

Fig. 10.29 Schematic of the fabrication steps followed for creating Ti-6Al-4 V with various wettability using chemical etching, anodizing, and nanoparticle-binder coating pathways. The resultant surfaces can be superhydrophilic, superhydrophobic, or superoleophobic (Adapted from Gurera and Bhushan 2019)

10.4.2 Results and Discussion

Titanium is a hydrophilic material with a water CA of $65° \pm 2°$ and hexadecane CA of $21° \pm 1°$. All surfaces were characterized for contact and tilt angles for DI water and hexadecane, self-cleaning, anti-icing, and mechanical durability (Gurera and Bhushan 2019).

Wettability

Figure 10.30 presents a summary of CA and TA of various wettability of titanium using different pathways, along with their goniometer images (Gurera and Bhushan 2019). As mentioned before, titanium is naturally hydrophilic. Only by altering its roughness it can be made, superhydrophilic via chemical etching and nanoparticle-binder coating pathways. All three pathways can result in superhydrophobicity with low water TA. In the case of chemical etching, both surface roughness and surface energy were required to be changed for superhydrophobicity. However, only roughness was required to be changed in case of anodizing and nanoparticle-binder pathways. On further adding the low surface energy to them, both the pathways resulted in superoleophobicity with low hexadecane TA. Surface morphology.

Self-cleaning

Figure 10.31 shows data on self-cleaning properties of surfaces with different wettability (Gurera and Bhushan 2019). The top figure presents optical images of the surfaces before and after the self-cleaning experiment. The bottom figure presents a quantitative comparison of the self-cleaning abilities of the surfaces. As expected, superhydrophobic and superoleophobic surfaces with low TA are found to be self-cleaning. The superhydrophobic and superoleophobic surfaces were able to clean 1.5 times more dirt than was cleaned by the hydrophilic surface. Among the self-cleaning surfaces, the nanoparticle-binder coated surface was able to clean more dirt out than the surface created via chemical etching and anodizing pathways. This is because the size of dirt particles used was less than or equal to the distance between the two roughness asperities. Hence the dirt particles may remain stuck between them.

Anti-icing

Figure 10.32 shows optical images of surfaces after the anti-icing test (Gurera and Bhushan 2019). Anti-icing is another characteristic property of surfaces with low TA. Liquids at lower temperatures become more viscous and it takes them longer to flow. Water gets twice as viscous at 0 °C as it was at 20 °C (Pritchard and Mitchell 2015). It may give water time to start freezing on a cooled surface. However, it was not the case with the surfaces. Supercooled water droplets roll over the cooled superhydrophobic and superoleophobic surfaces. However, the droplet remains stuck and starts to freeze on the hydrophilic surface.

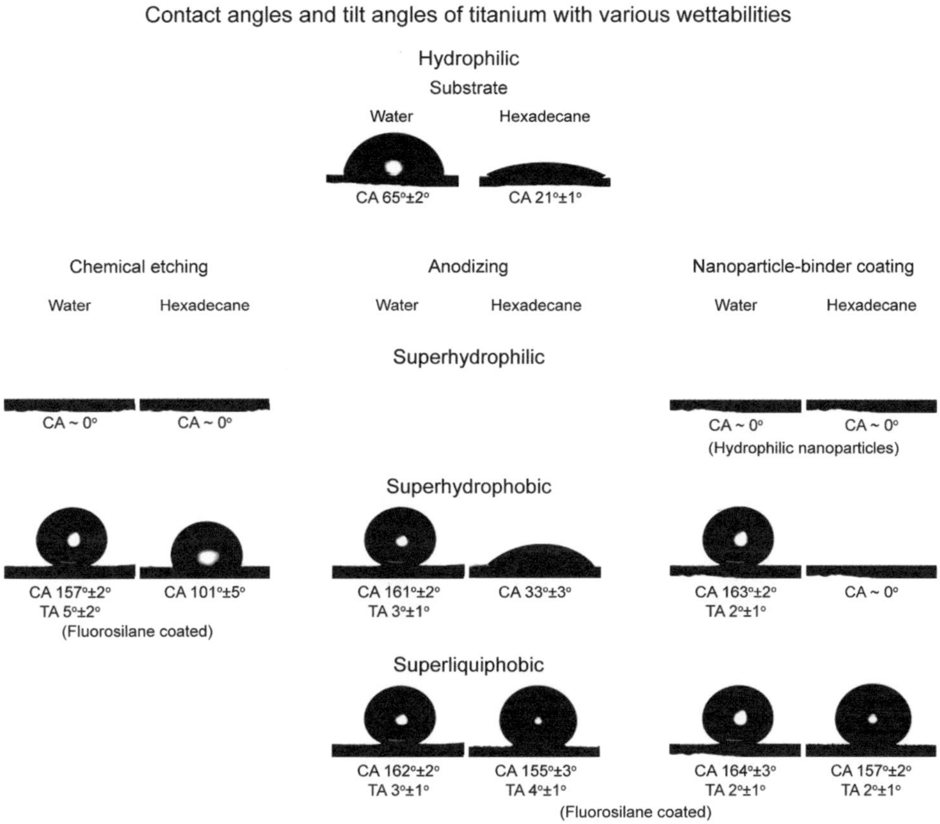

Fig. 10.30 Contact angles (CA) and tilt angles (TA) of water and hexadecane of titanium with various wettability (Adapted from Gurera and Bhushan 2019)

Wear Resistance

Figure 10.33 shows data on wear tests on the surfaces with different wettability (Gurera and Bhushan 2019). The top figure shows the optical images of the wear tracks, and the bottom figure shows data on loss of TA after the wear tests. The superhydrophobic and superoleophobic surface did not show much degradation after the wear test as shown in Fig. 10.33a. In the context of performance, there was some pinning observed when sliding over the wear track as shown in top of Fig. 10.33b. However, the drops still rolled over the wear track at well below 10°. When the drops were placed on the track and made to roll over, there was an increase in the TA up to 30°.

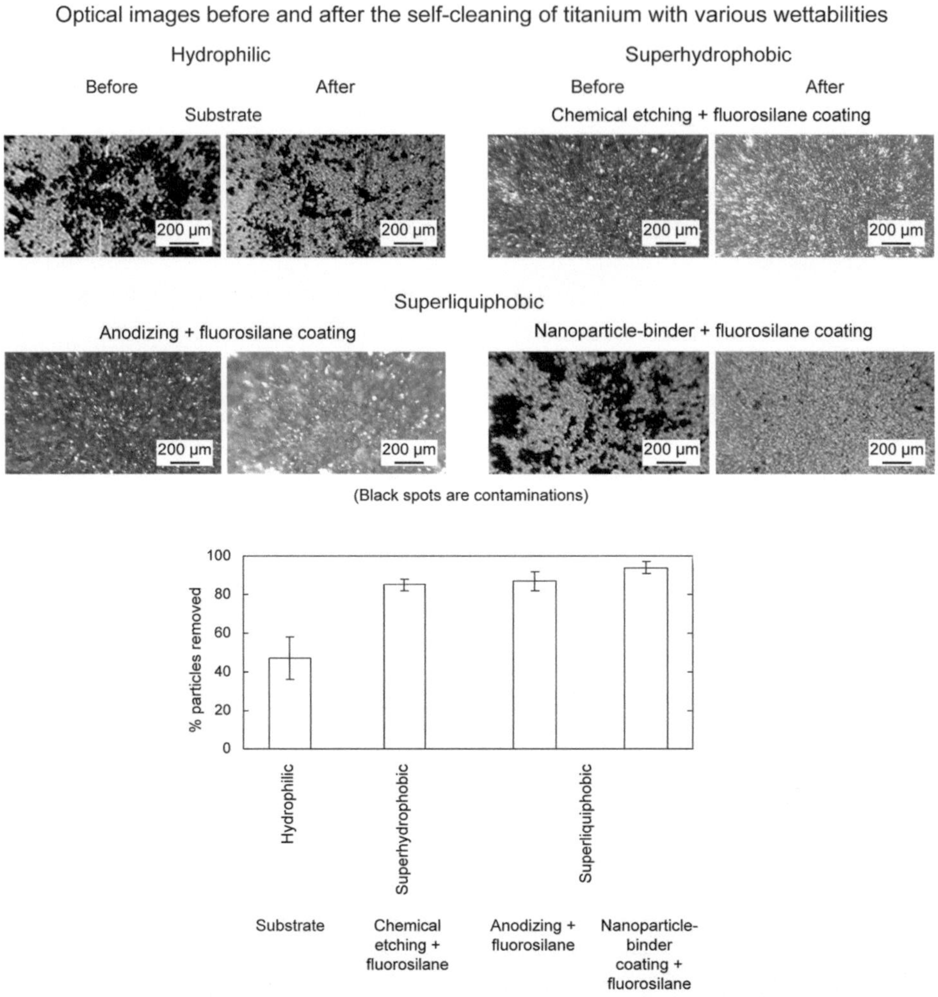

Fig. 10.31 Optical micrographs of contaminated coatings before and after self-cleaning test on titanium with various wettability using different pathways. Optical image analysis suggests > 80% removal of particles for samples with different pathways (Adapted from Gurera and Bhushan 2019)

10.4.3 Summary

A comparative study is presented for providing various wettability to titanium using different pathways. The various wettability includes superhydrophilicity, superhydrophobicity, and superoleophobicity. The various pathways include chemical etching, anodizing, and nanoparticle-binder coating. By altering roughness alone, superhydrophilicity was

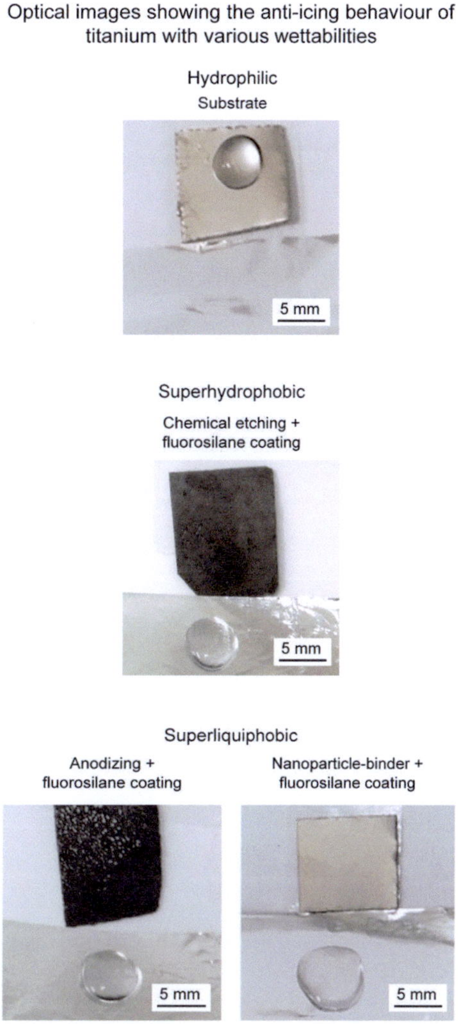

Fig. 10.32 Optical images showing the anti-icing behavior on surfaces with various wettability. The supercooled water droplet immediately freezes onto the hydrophilic surface after coming in contact with the surface. Whereas the supercooled water droplets roll off the superhydrophobic and superoleophobic surfaces even before they can freeze (Adapted from Gurera and Bhushan 2019)

achievable using chemical etching and nanoparticle-binder coated pathways. Superhydrophobicity can also be achieved just by altering roughness in the case of anodizing and nanoparticle-binder coating pathways. However, an additional low energy coating was required in case of chemical etching to achieve the superhydrophobicity. The additional step of creating a low energy coating takes the anodizing and nanoparticle-binder coating pathway even further, to superoleophobicity.

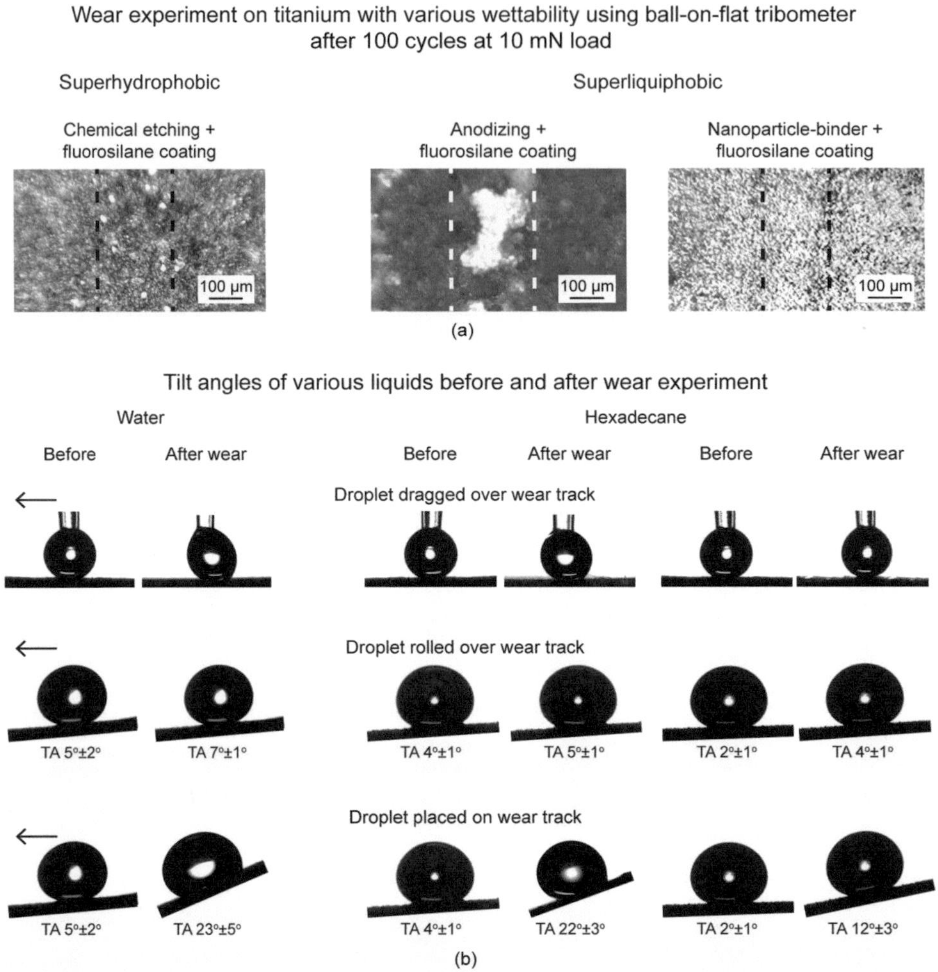

Fig. 10.33 **a** Optical microscope images showing the effects of wear using a ball-on-flat tribometer on titanium with various wettability fabricated using different pathways. **b** Optical images of liquid droplets before and after the wear test. Water droplets were used for the superhydrophobic samples and hexadecane droplets were used for superoleophobic samples. Droplets were dragged or tilted across the wear track in the direction of the arrows in two ways. Once by placing it to the right of the track and once by placing it directly on the track. Before the wear test, droplets rolled off the surface at 1°–7° tilt angles. For the worn samples, droplets placed to the right of the wear track rolled over the defect at 3°–8° tilt angles, and droplets placed directly on the wear track rolled over the defect at 9°–28° tilt angles (Adapted from Gurera and Bhushan 2019)

10.5 Superoleophobic Cotton Surfaces

Textiles are natural or synthetic fibrous materials used primarily in clothing to provide protection and comfort to the human body. Natural fibers used for textiles include plant fibers, such as cotton and flax, and animal fibers, such as wool and silk (Merritt 1924). Synthetic fibers include regenerated plant cellulose fibers, such as rayon, and petroleum-based fibers, such as polyester, polyacrylic and polyolefin. Apart from clothing, textiles are also used in household, medical and industrial applications, such as hygiene, agricultural, automotive, aerospace, and marine applications (Schindler and Hauser 2004; Perera et al. 2013).

Cotton fabric materials are commonly used in various household, medical and industrial applications due to their unique properties such as good feel, biodegradability, renewability, and being environment friendly. It is difficult to keep the fabric clean due to the attraction of dust particles. Hydroxyl groups (−OH) present in the cotton fabric makes it hydrophilic. Due to this hydrophilic nature, cotton fabric can quickly get wetted by water and various liquids including oil. Due to its porous nature, it can absorb water, moisture, and oil. Staining is a major problem for textiles. Although many liquids could remove the stain from the cotton fabrics, complete cleaning is extremely difficult. Wettability increases the probability of stains and bacterial growth, and subsequently limits its usage. For practical applications, cotton fabric should work well in ambient conditions as well as in harsh environments such as during washing, high temperature, acidic/alkaline/organic contact, and UV irradiation. Treatment of cotton fabric is of interest to provide desired properties including superoleophobicity, self-cleaning and stain resistance (Bhushan 2018).

Treatment of cotton fabric surfaces is challenging due to the inherent heterogeneous roughness and low thermal stability. Various researchers have used solution-immersion methods to provide s superoleophobicity, self-cleaning, oil–water separation and anti-bacterial properties (Perera et al. 2013; Lin et al. 2014; Zeng et al. 2015; Aslanidou et al. 2016; Przybylak et al. 2016; Liu et al. 2017; Panda et al. 2018; Ren et al. 2018; Chauhan et al. 2019; Li et al. 2019; Tudu et al. 2019b). Perera et al. (2013) applied silver nanoparticles on the cotton fabric by immersing in $AgNO_3$ solution followed by irradiation using an UV treatment. Chauhan et al. (2019) fabricated a superhydrophobic coating on cotton fabric by immersion in hexadecyltrimethoxysilane (HDTMS) solution. Li et al. (2019) fabricated superhydrophobic/superoleophobic coating on cotton fabric for oil–water separation by immersion in a fluorosurfactant and hydrophobic Al_2O_3 nanoparticles solution. Tudu et al. (2019b) fabricated a superoleophobic coating on cotton fabric by immersion in TiO_2 nanoparticles, perfluorodecyl-triethoxysilane and tetraethylorthosilicate solution. Many treated fabrics provided superoleophobicity, self-cleaning, stain resistance, mechanical durability, chemical and thermal stability, UV-protection, anti-bacterial properties, and oil–water separation properties.

In this section, a facile method to fabricate superoleophobic cotton fabric by immersion in TiO_2 nanoparticles, perfluorodecyltriethoxysilane (PFDTS) and tetraethylorthosilicate (TEOS) solution is described, based on Tudu et al. (2019b). Data on various tests are reported which include assessing self-cleaning and stain resistance, mechanical, chemical, and thermal stability, and anti-bacterial properties.

10.5.1 Experimental Details

Figure 10.34 shows a schematic of the fabrication process of superoleophobic cotton fabric (Tudu et al. 2019b). As received cotton fabric was cleaned ultrasonically to remove wax and other impurities with distilled water with acetone for 30 min and subsequently dried at 60° C. 0.2 ml of PFDTS and 0.1 ml of TEOS were mixed in 20 ml acetone with 10 ml of ammonia. The prepared solution was mixed in a magnetic stirrer for 15 min. 300 mg of TiO_2 nanoparticles were added in the anhydrous ethanol. The prepared TiO_2 and ethanol solution was mixed for 15 min and sonicated for 30 min. After that, the prepared PFDTS and TEOS solution was poured into the TiO_2 solution and mixed for six hours. The fabric was immersed in the solution for one hour and then dried at 120 ºC for 1.5 h.

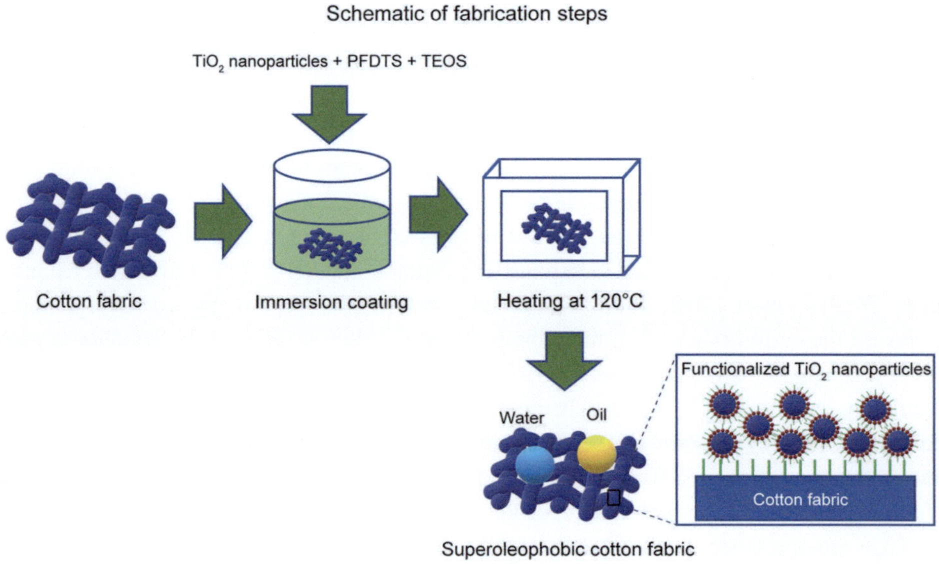

Fig. 10.34 Schematics showing various steps during fabrication of superoleophobic cotton fabric (Adapted from Tudu et al. 2019b)

Surface morphology and chemical characterization were performed using Fourier transform infrared spectroscopy (FTIR) analysis. To characterize for superoleophobicity. contact angles were measured using various liquids with a range of surface tensions (Table 10.6). Liquids included water, tea, turmeric water, honey, glycerol, milk, ethylene glycol, aniline, toluene, chloroform, hexadecane, and n-hexane. Self-cleaning properties were studied by contaminating the surface with dust particles and then cleaning it with water droplets. The stain resistance was studied by putting colored water droplets on the surface.

For mechanical stability, the cotton fabric was subjected to a water jet to examine coating durability. In this study, a water jet was injected on the surface at an angle of 25° for one min. Also, the mechanical stability was studied by regular washing with different solutions (hot water, chloroform, toluene, detergent water) using sonication method for seven days. After washing, the prepared coated samples were dried, and the contact angles were measured for any changes in wettability. For chemical stability tests, coated samples were dipped in different pH solutions ranging from pH 2 to pH 13. Samples were taken out at regular intervals to measure their contact angles. For thermal stability tests, coated samples were subjected to different annealing temperatures ranging from 50 to 300 °C for one hour. After one hour of annealing, the coated samples were cooled down at room temperature. Then any physical changes, color changes and deposition of residue on the surface were examined. Also, the water contact angles were measured.

For anti-bacterial properties, Escherichia coli (*E. coli*) bacteria was used. At first, roughly 20 ml sterilized nutrient agar was poured into the petri dish and left it for 30 min to allow it to solidify before inoculation. 500 μl of the bacterial suspension with a concentration of 1 × 105 colony forming unit (CFU)/ml was applied all over the surface of each agar prepared petri dish. Then holes were made and both coated and uncoated samples were placed inside these holes. Petri dishes for bacterial growth were left in the incubator at 37 °C for 24 h and 72 h. After 24 h and 72 h, the growth of bacteria was checked in each petri dish.

10.5.2 Results and Discussion

Surface morphology and chemical characterization of coated samples was carried out. Various tests were performed to assess wettability, self-cleaning and stain resistance, mechanical, chemical, and thermal stability, and anti-bacterial properties as well as oil–water separation capability (Tudu et al., 2019b).

Surface Morphology and Chemical Characterization

Figure 10.35 shows the optical images of uncoated and coated fabrics along with their water and oil static contact angles and tilt angles (Tudu et al. 2019b). High magnification

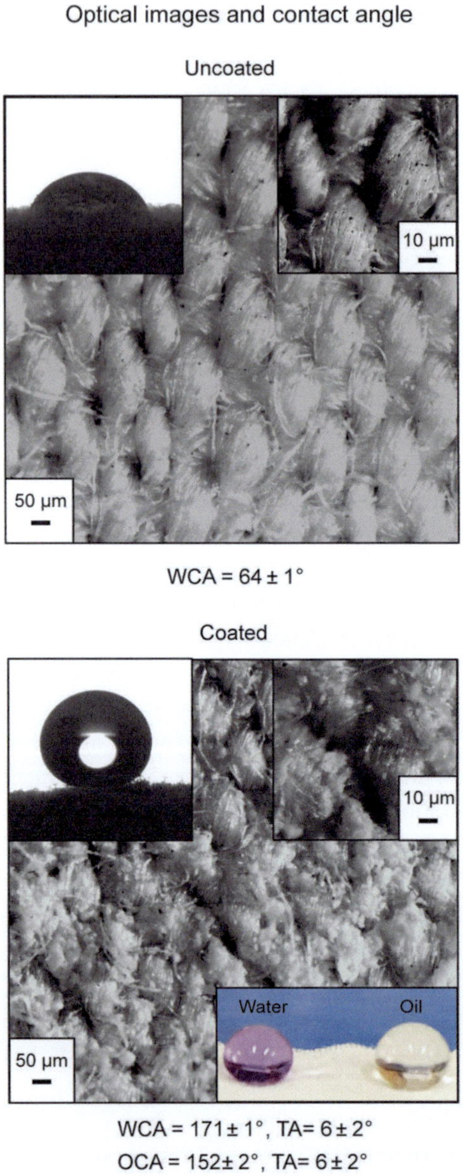

Fig. 10.35 Optical images at two magnifications of coated and uncoated cotton fabric with oil and water contact angles and tilt angles (Adapted from Tudu et al. 2019b)

images (inserts) show that fabric threads consist of surface microstructure and the multi-scale roughness which creates hierarchical structure.

Figure 10.36 shows FTIR spectra of a coated sample, which shows functional groups present on the surface (Tudu et al. 2019b). Several peaks were found. Peaks at a wavenumber of 670 cm^{-1} and 3094 cm^{-1} represent the presence of –CH– bond. Peaks 1033 cm^{-1}

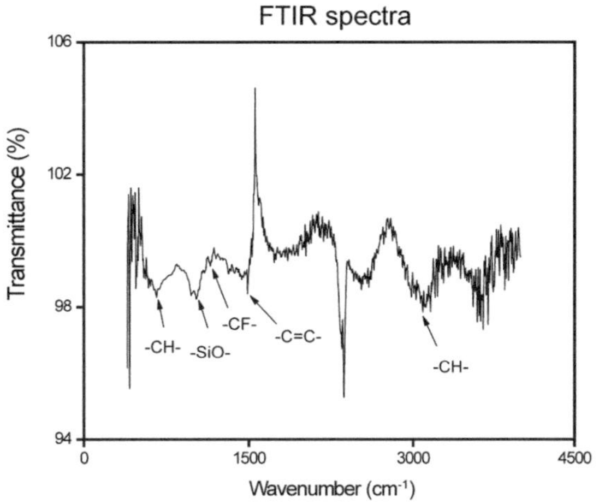

Fig. 10.36 FTIR spectra of superoleophobic cotton fabric showing presence of functional groups −CH−, −SiO−, −CF−, and -C = C- (Adapted from Tudu et al. 2019b)

and 1164 cm^{-1} represent the presence of −SiO− and −CF− bonds on the surface, respectively. The peak at 1492 cm^{-1} represents the presence of −C = C- bond. From FTIR spectra, it is confirmed that the hydrophobic tail of PFDTS is attached to the surface.

Wettability, Self-cleaning and Stain Resistance

Figure 10.37 shows the optical images of selected liquid droplets on the coated fabric along with their contact angles (Tudu et al. 2019b). Table 10.5 shows the wettability of the coated sample with different liquids with surface tension ranging from 72.8 to 18.43 mN m^{-1}) (Tudu et al. 2019b). The water static contact angle of the uncoated sample was found to be $64 \pm 1°$ and that of the coated sample was found to be $171 \pm 2°$ and tilt angle $6 \pm 2°$. Oil (hexadecane) contact angle of the coated sample was found to be $152 \pm 2°$and tilt angle of $6 \pm 2°$. Also, wettability was checked with different liquids. Prepared superoleophobic cotton fabric repels all liquids with surface tension greater than 27 mN m^{-1}, with contact angles more than 150°.

Figure 10.38 shows optical images of uncoated and coated fabrics after contamination and cleaning with water droplets (Tudu et al. 2019b). Both uncoated and coated samples were covered with sand dust particles. In the case of the uncoated sample, when water droplets fall on the surface, they do not remove contaminant particles but start accumulating. This is because of the hydrophilic nature of uncoated cotton fabric which holds the water droplet and doesn't allow particles to move ahead. However, in the case of the coated sample, the water droplets roll down on the surface. While rolling downward they collect the dust particles and move downward leaving a trail of clean surface, showing the self-cleaning nature of coated cotton fabric.

Figure 10.39 shows the stain resistance property of uncoated and coated cotton fabric (Tudu et al. 2019b). It was observed that water droplets were immediately absorbed by uncoated fabric. In the case of a coated fabric, the droplet remained in a spherical shape.

Liquid droplets on coated fabric

Fig. 10.37 Optical images of selected liquid droplets on the coated fabric along with their contact angles (Adapted from Tudu et al. 2019b)

Table 10.5 Contact angles and reported surface tensions of various liquids used (Adapted from Tudu et al. 2019b)

Liquid	Contact angle (°)	Surface tension (mN m^{-1})
Water	171 ± 2	72.8
Tea	162 ± 3	72.8
Turmeric water	168 ± 2	72.8
Honey	153 ± 2	68.0
Glycerol	168 ± 1	63.4
Milk	160 ± 3	51.8
Ethylene glycol	161 ± 2	47.7
Aniline	159 ± 2	43.4
Toluene	155 ± 2	28.4
Chloroform	152 ± 1	27.5
Hexadecane	152 ± 2	27.05
n-Hexane	110 ± 3	18.4

After drying, stain was formed around the water droplet position on the uncoated surface. In the case of the coated sample, the water droplet gets evaporated and leaves behind only small patches of colored particles. This color patch can also be removed by simply blowing air or spraying water. Eventually a clean surface is obtained. This demonstrates the stain resistance ability of coated cotton fabric.

Table 10.6 Contact angle of coated cotton fabric after 7 days washing in different solvents used (Adapted from Tudu et al. 2019b)

Solvent	Water contact angle after 7 days (°)
Hot water	168 ± 3
Chloroform	151 ± 1
Toluene	152 ± 2
Detergent water	168 ± 2

Fig. 10.38 Optical images of coated and uncoated cotton fabric after contamination after cleaning with water droplets, demonstrating the self-cleaning property of coated fabric (Adapted from Tudu et al. 2019b)

Mechanical, Chemical and Thermal Stability

Figure 10.40 shows optical images of surfaces subjected to a water jet on both uncoated and coated samples to evaluate the mechanical stability of the coating (Tudu et al. 2019b). When the water jet impacts the uncoated surface, water spreads on the surface due to its hydrophilic nature. When the water jet was injected at an angle on the coated surface, it repelled the water jet in the opposite direction due to its superhydrophobic nature. A similar study was done on a coated sample using an oil jet. After impact, the oil jet also bounced off in the opposite direction away from the coated sample.

Additionally, to test mechanical stability, coated samples were washed with different solvents and hot water to examine durability of the coating. Table 10.6 shows the water contact angles of coated samples after washing in various solvents (Tudu et al. 2019b). The water contact angles were found to be more than 150° even after seven days of washing. This confirms that washing has a very low impact on the superoleophobicity of the coating.

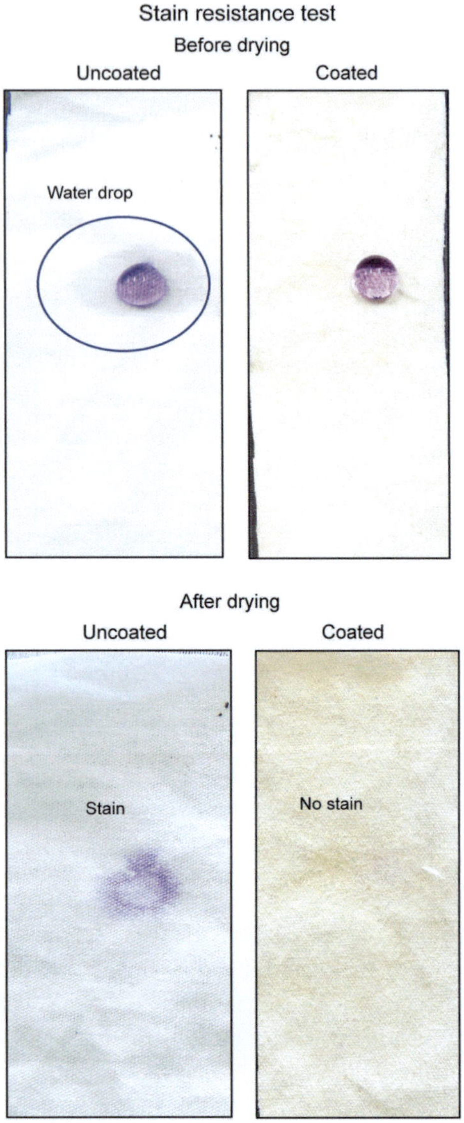

Fig. 10.39 Optical images of coated and uncoated cotton fabric after putting a colored water drop as well as after drying, demonstrating the stain resistance property of coated fabric (Adapted from Tudu et al. 2019b)

Chemical stability of the coated samples was examined by immersing the coated samples in different pH solutions. The water contact angles of the coated samples, after 10 and 20 h immersing in pH 2 and pH 13 solutions, were found to be less than 150°. This is due to the high acidic and alkaline nature of the solutions; therefore, the coating could not survive after immersion and the contact angle decreased. The wetting characteristics of coated samples remained constant even after immersion for 40 days in pH 5, pH 8 and pH 11 solutions, showing high chemical stability of coating.

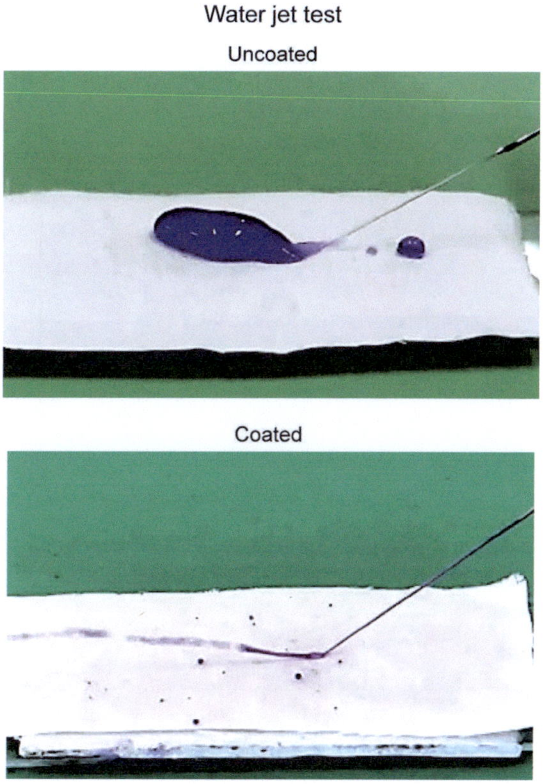

Fig. 10.40 Optical images of coated and uncoated cotton fabric after subjecting to a water jet, demonstrating the water repelling property of coated fabric (Adapted from Tudu et al. 2019b)

Thermal stability of the coating was investigated by annealing the coated samples at different annealing temperatures. Figure 10.41 shows water contact angles of coated samples after 1 h annealing in temperature range of 50°-300 °C (Tudu et al. 2019b). The water contact angle was found to be more than 150° after annealing in temperature range of 50–250 °C. However, after annealing at 300 °C, water contact angle decreased to about 71 °C. This could be due to boiling PFDTS at this high temperature.

Anti-bacterial Properties

A qualitative analysis for growth of bacteria around the samples was carried out to assess anti-bacterial activity of the coating against the *E. coli* bacteria. Figure 10.42 shows the optical images of bacterial growth around the uncoated and coated samples in the petri dish (Tudu et al. 2019b). In the air, bacteria start to grow around the sample and spread all over the petri dish. After 24 h of incubation, bacteria growth around the start sample was visible whereas no bacteria growth was noticed around the coated sample. Furthermore, the petri dishes were left in the incubator for 72 h. After that, high growth of bacteria was found around the uncoated sample, but very low growth of bacteria was observed around

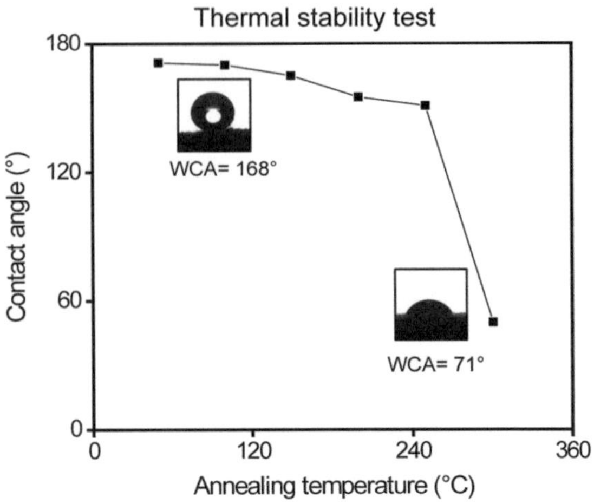

Fig. 10.41 Water contact angle versus annealing temperature showing thermal stability of coated cotton fabric in the temperature range of 50–250 °C (Adapted from Tudu et al. 2019b)

the coated sample. Thus, overall growth of bacteria on coated fabric was slower than on the uncoated fabric and this demonstrated the anti-bacterial activity of the coated fabric.

10.5.3 Summary

The superoleophobic cotton fabric was prepared by simple immersion technique with a binary silane mixture of PFDTS and TEOS, and TiO_2 nanoparticles. Coated fabric shows the static water contact angle of $171 \pm 2°$ and tilt angle $6 \pm 2°$. It exhibits the oil (hexadecane) contact angle of $152 \pm 2°$ and tilt angle of $6 \pm 2°$. Various tests demonstrate that coated samples exhibit self-cleaning and stain resistance, mechanical, chemical, and thermal stability, and anti-bacterial properties.

10.6 Superoleophobic Synthetic Leather Surfaces

Leather has been an important material since the beginning of the human race (Kite and Thomson 2006; Harris and Veldmeijer 2014). Due to its desirable traits, including high tensile strength, flexibility, resistance to tearing, puncturing and abrasion, and permeability by air and water, there are many applications including footwear, furnishings, automotive industry, and clothing.

Genuine or natural leather is expensive. To reduce cost and tailor flexibility, there has been a growing demand for synthetic leather. Two types of synthetic leather commonly used in commercial applications are polyurethane (PU) based and polyvinyl chloride (PVC) based. Various layers present in the synthetic leather are shown in Fig. 10.43. Table 10.7 summarizes a list of plasticizers, stabilizers and fillers used in making the

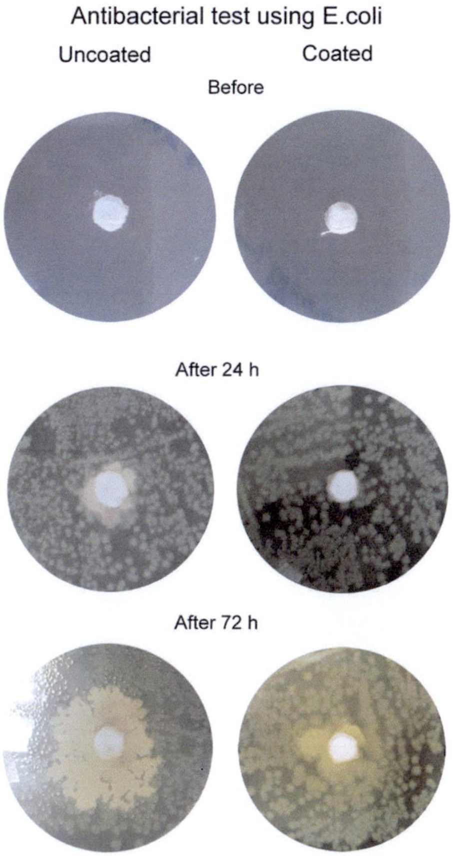

Fig. 10.42 Optical images of anti-bacterial activity of uncoated and coated cotton fabric after 24 h and 72 h incubation using *E. coli*. Growth of bacteria around the uncoated fabric is observed after 24 h incubation, whereas bacteria start growing around the coated fabric after 72 h incubation, showing antibacterial properties of coated cotton fabric (Adapted from Tudu et al. 2019b)

polymer solution for fabrication of synthetic leather (Gurera and Bhushan 2018). The list includes surfactants present in the plasticizer that make the surface chemically active (Kim and Hsieh 2001).

For some applications, it is required that leather surfaces exhibit liquid repellency, self-cleaning, and low adhesion to reduce stickiness particularly at elevated temperatures. An attempt was made to make natural leather superhydrophobic by using a spray coating of silica nanoparticles and an epoxy binder (Ma et al. 2015). However, the coating lost its superhydrophobicity in just one hour. In another study, synthetic leather surfaces were successfully made superliquiphobic by using a multilayered nanoparticle composite structure (Gurera and Bhushan 2018). An overview of the research based on Gurera and Bhushan (2018) follows.

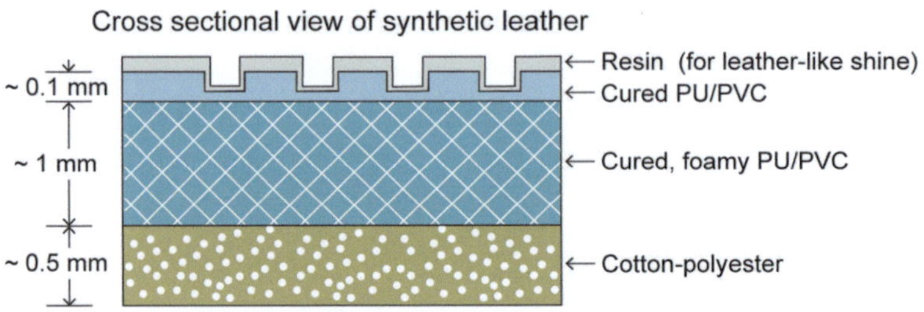

Fig. 10.43 Cross sectional view of synthetic leather showing various layers on a cotton-polyester substrate (Adapted from Gurera and Bhushan 2018)

Table 10.7 Typical composition of the polymer solution used for fabrication of synthetic leather (Gurera and Bhushan 2018)

Components (% composition by volume)		Examples
Polymer (~55%)		Polyurethane (PU), polyvinyl chloride (PVC)
Plasticizer (~40%)		Non-ionic surfactants, butyl benzyl phthalate, dialkyl phthalate, triethyl phosphate
Stabilizer (~1%)		
Filler (~5%)	Riser	Dialkyl phthalate
	Releasing agent	Polysiloxane
	Surface modifier	Polysiloxane
	Coloring dye dispenser	Anionic surfactant, amine organic compound, phosphoric ester

10.6.1 Experimental Details

A multilayered nanoparticle composite structure was used to achieve superliquiphobicity in synthetic leather, shown in Fig. 10.44 (Wang and Bhushan 2015; Bhushan and Martin 2018; Gurera and Bhushan 2018). A spray coating of a nanocomposite layer of hydrophobic silica nanoparticles and methylphenyl silicone resin was used to achieve hierarchical structure which exhibited superhydrophobicity. An overcoat of fluorosilane was applied to the nanocomposite layer to make the surface superoleophobic. To attach the fluorosilane coating, the underlayer was chemically activated using ultraviolet-ozone (UVO) treatment.

Synthetic leather consists of a grooved surface to simulate the feel of natural leather. To fill the grooves present on synthetic leather, a thick undercoat layer was applied before deposition of the nanoparticle composite layer. A thick undercoat helps in filling up the grooves. It was found that if a spray coating method is used, the sides of the grooves in

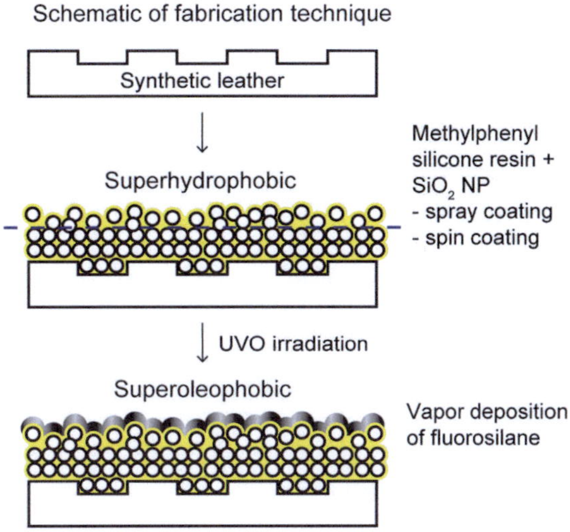

Fig. 10.44 Schematic of various steps for fabrication of superhydrophobic and superoleophobic synthetic leather (Gurera and Bhushan 2018)

the synthetic leather remain uncoated, and the surface remains liquiphilic. When a droplet of liquid is placed on such a surface, the uncoated sides of the synthetic leather grooves pull the droplet across the surface. To get a uniform undercoat, spin coating was used. Dip coating can also be used to provide a thick uniform surface coating. However, long immersion in the dip coating process damaged the synthetic leather (Gurera and Bhushan 2018).

Synthetic PU and PVC leather (FurFabric, China, and SyFabrics, Los Angeles, CA, respectively) were used in the study. To prepare the nanocomposite coating, methylphenyl silicone resin (SR355S, Momentive Performance Materials) and hydrophobic silica nanoparticles (NP, 10 nm diameter, Aerosil RX300) were mixed in a solvent and spray coated. The solvent used in this study was isopropyl alcohol (IPA, Fisher Scientific). In some studies, solvents consisting of 40% tetrahydrofuran (THF, Fisher Scientific) and 60% IPA have been used (Ebert and Bhushan 2012; Wang and Bhushan 2015). However, THF reacts with the leather substrate and is not desirable. The coating solution consisted of 375 mg of nanoparticles and 150 mg of resin (with a particle-to-binder ratio of 2.5) dispersed in 30 mL of solvent. The fluorosilane used in this study was trichloro (1H, 1H,2H,2H-perfluorooctyl) silane (448,931, Sigma-Aldrich) and was vapor deposited (Gurera and Bhushan 2018).

First, an undercoat was applied by spin coating on a 10 mm × 10 mm sample using 1 mL of the coating mixture. It was dried in an oven at 50 °C for 5 min. Second, 1 mL of the coating mixture was spray coated and dried in an oven at 50 °C for 5 min. Next, the sample was chemically activated by using the UVO treatment for 60 min. The samples were placed 20 mm underneath the light source. Lastly, one droplet of the fluorosilane was vapor deposited on the sample using a closed container. The sample was attached to the top of the container via double-sided sticky tape with the surface facing down. The droplet was placed on the bottom of the container. The sample was placed 20 mm above the droplet and vapor deposition occurred for 30 min (Gurera and Bhushan 2018).

The coatings were characterized for liquid repellency, self-cleaning, high temperature exposure and wear resistance. To test whether the sample retained superhydrophobicity at high temperatures, coated leather samples were put in an oven for 5 min at various temperatures. Immediately after they were taken out of the oven, CA and TA were measured (Gurera and Bhushan 2018). A maximum temperature of 80 °C was selected as it is the typical temperature inside an automobile on a hot day (Manning and Ewing 2009).

10.6.2 Results and Discussion

Wettability

Water and hexadecane droplet images and CA and TA values for PU and PVC leather, untreated and with various coatings are presented in Fig. 10.45a (Gurera and Bhushan 2018). Uncoated synthetic leathers are hydrophilic. Untreated PU leather has water CA of $76° \pm 5°$ and hexadecane completely wets it. PVC leather has a water CA of $62° \pm 6°$ and hexadecane CA of $19° \pm 2°$. Water droplets on untreated substrates were pinned. After deposition of the nanoparticle-binder coating, both leathers became superhydrophobic with TA of about 2°. As expected, the coatings were superoleophilic. After the application of the fluorosilane coating, surfaces became superoleophobic with hexadecane CA of about 157° and TA of about 3°. However, the superoleophobicity was found to be non-uniform (Gurera and Bhushan, 2018).

Figure 10.45b presents video still images of the liquid droplets after different periods on PU leather, untreated and with various coatings. Water was dyed blue and hexadecane was dyed red. The sample was tilted at 10° angle. For untreated leather, water droplets were stuck on the surface, and hexadecane droplets completely wetted the surface. For nanoparticle-binder coating, water droplets rolled off the surface. Water droplets (not shown) and hexadecane droplets also rolled off the surface in the case of nanoparticle-binder/fluorosilane coating (Gurera and Bhushan 2018).

Next, the effect of storage time on water CA for PU leather was studied. The initial water CA of about 163° remained stable up to three days and then it came down to about 150° in 5 days and about 140° in seven days. As stated earlier, superoleophobicity was non-uniform. We next discuss mechanisms responsible for these observations. It has been reported that plasticizers present in the synthetic leather (Table 10.7) reduce the water contact angle (Brown et al. 2017). They can leach out through the coating via capillary action shown in Fig. 10.46. and make the leather surface chemically active with its affinity to liquids. This is also corroborated by the fact that a similar coating did not lose its superliquiphobicity on a pure polyurethane sample (Bhushan and Martin 2018). Therefore, leaching out of the plasticizers is believed to be responsible for decrease in water CA with time and nonuniform superoleophobicity.

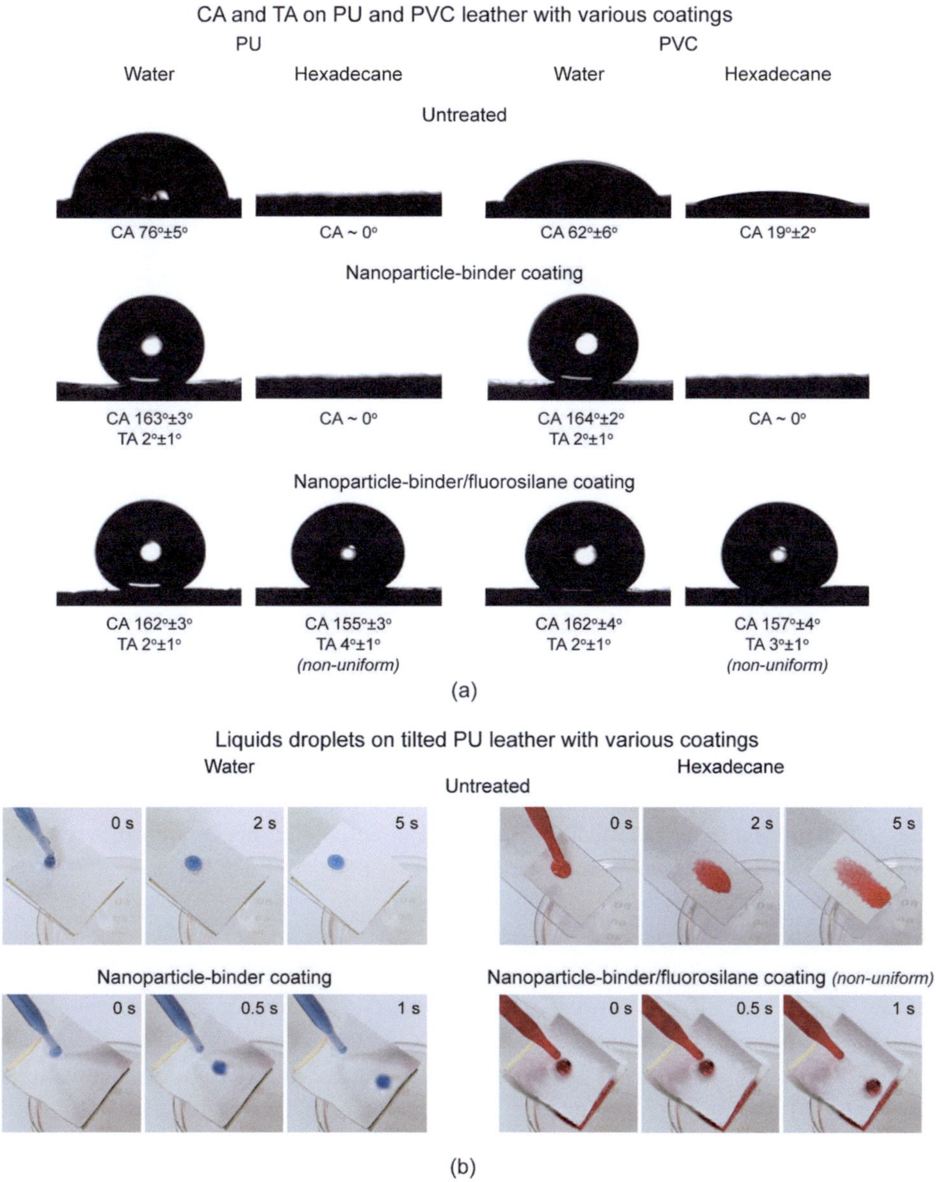

Fig. 10.45 **a** Water and hexadecane droplet images on PU and PVC leather, untreated and with nanoparticle-binder coating and nanoparticle-binder/fluorosilane coating, along with their contact angle (CA) and tilt angle (TA) values. **b** video still images of liquid droplets after different periods on PU leather demonstrate repellency of water and hexadecane on nanoparticle-binder and nanoparticle-binder/fluorosilane coated PU leather (Gurera and Bhushan 2018)

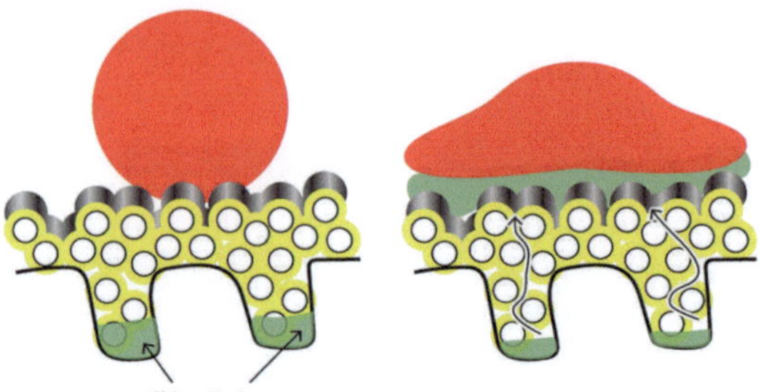

Fig. 10.46 Schematic of potential reason for unstable superliquiphobicity of PU leather (Adapted from Gurera and Bhushan 2018)

Self-cleaning

For self-cleaning tests, the nanoparticle-binder coated PU leather samples were first contaminated and then photographed before and after cleaning of the contaminated surface. Figure 10.47 (top) shows black spots on the contaminated nanoparticle-binder coated surface before and after cleaning. It is clear that most of the particles were removed by cleaning (Gurera and Bhushan 2018). Figure 10.47 (bottom) shows a quantitative comparison of self-cleaning abilities of the untreated and nanoparticle-binder coated surfaces. In the case of untreated synthetic leather, about 10% of the particles were removed. Whereas in the nanoparticle-binder coating, about 90% of the particles were removed. The untreated sample is less efficient because it is liquiphilic, hence a droplet will slide on it instead of rolling (Gurera and Bhushan 2018).

High Temperature Exposure

To test whether coated leather loses its non-wettability properties at high temperature, contact angles and tilt angles were measured on the PU leather at room temperature and after exposure to high temperatures of 70 °C and 80 °C. Figure 10.48 shows water CA and TA on the nanoparticle-binder coated PU leather at 22 °C (room temperature), 70 °C and 80 °C (Gurera and Bhushan 2018). The CA of about 160° and TA about 2° at room temperature did not change up to 70 °C. CA and TA were also measured after the droplet set for 5 min and no change was observed. At 80 °C, the leather started to lose its superhydrophobicity. The polymer in the leather reaches its glass transition temperature, which is about 80 °C for PVC and may be responsible for the drop in contact angle (Drobny 2007). That may be a reason why one feels leather being sticky inside the automobile on a hot day with a temperature of about 80 °C (as reported earlier).

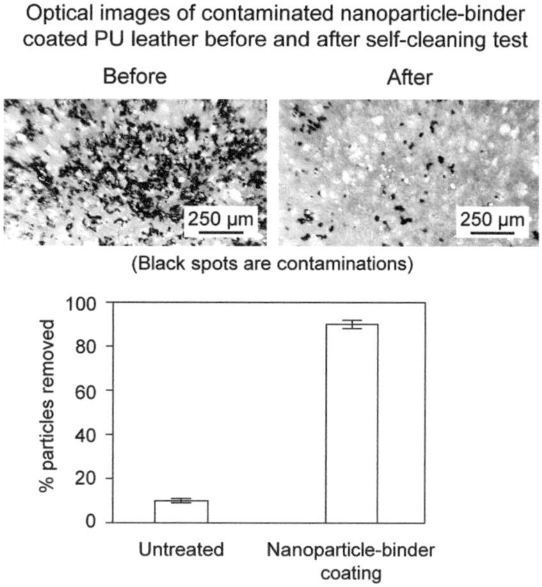

Fig. 10.47 Optical images of contaminated nanoparticle-binder coated PU leather before and after self-cleaning test. Also shown is the quantitative comparison of self-cleaning abilities of the untreated and nanoparticle-binder coating; error bars represent a standard deviation from the mean values of three measurements (Adapted from Gurera and Bhushan 2018)

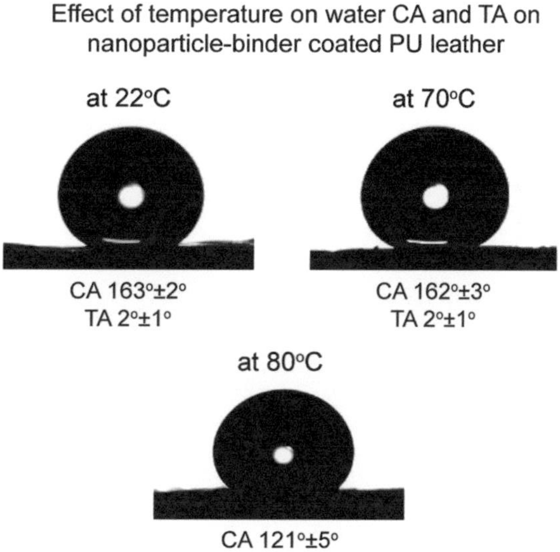

Fig. 10.48 Effect of temperature on water repellency on the nanoparticle-binder coated PU leather. Water droplet images and contact angle (CA) and tilt angle (TA) at three temperatures are shown (Adapted from Gurera and Bhushan 2018)

Wear Resistance

Wear resistance of the coating was tested using a ball-on-flat tribometer. Figure 10.49a shows optical micrographs of the untreated and nanoparticle-binder coated PU leather after the wear experiment using a ball-on-flat tribometer (Gurera and Bhushan 2018). The dotted lines show the boundary for the wear track. For the untreated leather, there were

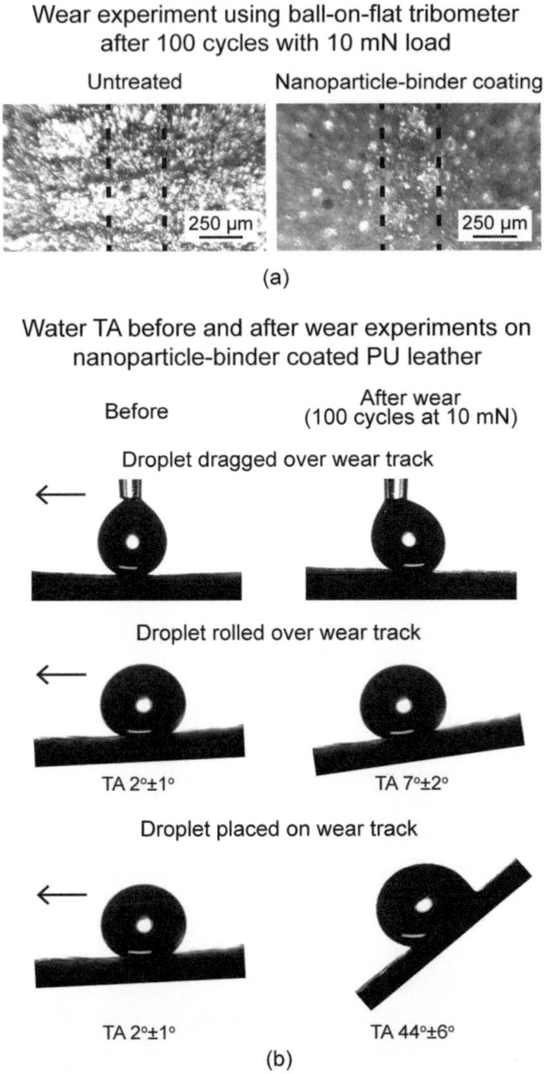

Fig. 10.49 **a** Optical micrographs of the untreated and the nanoparticle-binder coated PU leather after wear experiment using ball-on-flat tribometer with a 3 mm diameter sapphire ball at 10 mN load after 100 cycles. **b** Optical images of water droplets on the PU leather before and after wear, and their tilt angles (TA). Droplets were dragged, rolled over, or placed on a wear track orthogonal to the track as shown by arrows (Adapted from Gurera and Bhushan 2018)

no visible track marks, because the polymer is compliant. For the nanoparticle-binder coating, some burnishing was observed.

To check for any loss of liquid repellency of coated sample after wear, water droplets were dragged over, rolled over, or placed on wear tracks, and TA measurements were made, and droplet images were taken. The data are presented in Fig. 10.40b (Gurera and Bhushan 2018). When the droplet was dragged orthogonally to the wear track, the droplet was pinned somewhat over the track. When a water droplet was allowed to roll across the wear track, TA increased from 2° to 7°. A TA of 7° still maintains self-cleaning properties.

When a water droplet was placed in a wear tack and allowed to roll along the track, TA increased to 44°, suggesting some loss of the coating performance (Gurera and Bhushan 2018).

10.6.3 Summary

Superhydrophobicity and self-cleaning properties in synthetic leather were achieved by applying the nanoparticle-binder coating, with a CA of about 160° and TA of about 2°. Superoleophobicity was achieved by applying a fluorosilane coating on top of nanoparticle-binder coating, hexadecane CA of about 157° and TA of about 3° was achieved. However, the superoleophobicity was non-uniform. A possible explanation is that plasticizer, used in the manufacturing of synthetic leather, leaches out and makes the surface chemically active. About 90% of the contaminants were removed on the coated leather in self-cleaning tests. The coated leather maintained liquid repellency up to about 70 °C. The coated leather was found to be wear resistant using a ball-on-flat tribometer. Superhydrophobic synthetic leather is expected to be of interest in a wide variety of applications (Gurera and Bhushan 2018).

10.7 Closure

Fabrication techniques and characterization data for superliquiphobic and self-cleaning engineering surfaces are presented which include aluminum, steel, titanium, cotton fabric, and synthetic leather. These surfaces were found to exhibit mechanical durability and chemical and thermal stability.

References

Anonymous (1994), ASTM standard 3–89, (1994), Metals Test Methods and Analytical Procedures, *Annual Book of ASTM standards* (v.03.02, Wear and Erosion; Metal Corrosion), ASTM. Philadelphia.

Anonymous (2005), ISO 17475:2005, Corrosion of Metals and Alloys-Electrochemical Test Methods-Guidelines for Conducting Potentiostatic and Potentiodynamic Polarization Measurements. Americal National Standards Institute.

Arkles, B., Maddox, A., Singh, M., Zazyczny, J., Matisons, J. (2014), *Silane Coupling Agents: Connecting Across Boundaries,* third ed., Gelest Inc., Morrisville, PA.

Aslanidou, D., Karapanagiotis, I., and Panayiotou, C. (2016), "Superhydrophobic, Superoleophobic Coatings for the Protection of Silk Textiles," *Prog Org Coat.* **97**, 44–52.

Barthwal, S., Kim, Y. S. and Lim, S.-H. (2013), "Mechanically Robust Superamphiphobic Aluminum Surface with Nanopore-Embedded Microtexture," *Langmuir* **29**, 11966–11974.

Beckford, S. and Zou, M. (2011), "Micro/nano Engineering on Stainless Steel Substrates to Produce Superhydrophobic Surfaces" *Thin Solid Films* **520**, 1520–1524.

Bhushan, B. (2018), *Biomimetics: Bioinspired Hierarchical-Structured Surfaces for Green Science and Technology*, third ed., Springer International, Cham, Switzerland.

Bhushan, B. (2019), "Lessons from Nature for Green Science and Technology: An Overview and Bioinspired Superliquiphobic/philic Surfaces," Phil. Trans. R. Soc. A **377**, 20180274.

Bhushan, B. and Gupta, B. K. (1991), *Handbook of Tribology: Materials, Coatings, and Surface Treatments*, McGraw-Hill Book Company, New York.

Bhushan, B. and Martin, S. (2018), "Substrate-independent Superliquiphobic coatings for Water, Oil, and Surfactant Repellency: An Overview," *J. Colloid Interface Sci.* **526**, 90–105.

Boinovich, L. B., Emelyanenko, A. M., Ivanov, V. K., and Pashinin, A. S. (2013), "Durable Icephobic Coating for Stainless Steel," *ACS Appl. Mater. Inter.* **5**, 2549–2554.

Brassard, J., Sarkar, D., Perron, J., Audibert-Hayet, A., and Melot, D. (2015), "Nano-Micro Structured Superhydrophobic Zinc Coating on Steel for Prevention of Corrosion and Ice Adhesion," *J. Colloid Interface Sci.* **447**, 240–247.

Brown, P. S. and Bhushan, B. (2016), "Designing Bioinspired Superoleophobic Surfaces," *APL Materials* **4**, 015703.

Brown, R., Russell, S., May, S., Regan, F., and Chapman, J. (2017), "Reproducible Superhydrophobic PVC Coatings; Investigating the Use of Plasticizers for Early Stage Biofouling Control," *Adv. Eng. Mater.* **19**, 1700053.

Callister, W. D. (2007), *Materials Science and Engineering: An Introduction*, 7th ed., Wiley, Hoboken, New Jersey.

Chauhan, P., Kumar, A., and Bhushan, B. (2019), "Self-cleaning, Stain-resistant and Anti-bacterial Superhydrophobic Cotton Fabric Prepared by Simple Immersion Technique," *J. Colloid Interface Sci.* **535**, 66–74.

Cobb, H. M. (2010), *The History of Stainless Steel*, ASM International, Materials Park, OH.

Cremaldi, J. and Bhushan, B. (2018), "Fabrication of Bioinspired, Self-Cleaning Superliquiphilic/phobic Stainless Steel Using Different Pathways," *J. Colloid Interface Sci.* **518**, 284–297.

Davis, J. R. (Ed.) (1993), *ASM Special Handbook, Aluminum and Aluminum Alloys*, ASM International, Materials Park, Ohio.

Davis, J. R. (Ed.) (1998), *Metals Handbook*, second ed., ASM International, Materials Park, Ohio, U.S.A.

Devaprakasam, D., Sampath, S., and Biswas, S. K. (2004), "Thermal Stability of Perfluoroalkyl Silane Self-Assembled on a Polycrystalline Aluminum Surface," *Langmuir* **20**, 1329–1334.

Drobny, J. G. (2007), *Handbook of Thermoplastic Elastomers*, William Andrew Publishing, Norwich, NY.

Ebert, D. and Bhushan, B. (2012), "Transparent, Superhydrophobic, and Wear-Resistant Coatings on Glass and Polymer Substrates Using SiO_2, ZnO, and ITO Nanoparticles," *Langmuir* **28**, 11391–11399.

Fleming, R. A., and Zou, M. (2013), "Silica Nanoparticle-Based Films on Titanium Substrates with Long-Term Superhydrophilic and Superhydrophobic Stability," *Appl. Surf. Sci.* **280**, 820–827.

Frank, M. A., Boccaccini, A. R., and Virtanena, S. (2014), "A Facile and Scalable Method to Produce Superhydrophic Stainless Steel Surface," *Appl. Surf. Sci.* **311**, 753–757.

Fu, X. and He, X. (2008), "Fabrication of Super-Hydrophobic Surfaces on Aluminum Alloy Substrates," *Appl. Surf. Sci.* **255**, 1776–1781.

Gao, X., Tong, W., Ouyang, X., and Wang, X. (2015), "Facile Fabrication of a Superhydrophobic Titanium Surface with Mechanical Durability by Chemical Etching," *RSC Adv.* **5**, 84666–84672.

Guo, M., Kang, Z., Li, W., and Zhang, J. (2014), "A Facile Approach to Fabricate a Stable Superhydrophobic Film with Switchable Water Adhesion on Titanium Surface," *Surf. Coat. Technol.* **239**, 227–232.

Gurera, D. and Bhushan, B. (2018), "Fabrication of Bioinspired Superliquiphobic Synthetic Leather with Self-cleaning and Low Adhesion," *Colloids Surf. A* **545**, 130–137.

Gurera, D. and Bhushan, B. (2019), "Fabrication of Bioinspired, Self-cleaning, Anti-icing, Superliquiphilic/phobic Titanium using Different Pathways," *Phil. Trans. R. Soc. A* **377**, 20180273.

Harris, S., and Veldmeijer, A. J. (Eds.) (2014), *Why Leather?: The Material and Cultural Dimensions of Leather*, Sidestone Press, Leiden, Netherlands.

Hatch, J. E. (1984), *Aluminum-Properties and Physical Metallurgy*, Am. Soc. Metals, Metals Park, Ohio.

He, Z., Ma, M., Lan, X., Chen, F., Wang, K., Deng, H., Zhang, Q. and Fu, Q. (2011), "Fabrication of a Transparent Superamphiphobic Coating with Improved Stability," *Soft Matter* **7**, 6435–6443.

He, M., Zhou, X., Zeng, X., Cui, D., Zhang, Q., Chen, J., Li, H., Wang, J., Cao, Z. and Song, Y. (2012), "Hierarchically Structured Porous Aluminum Surfaces for High-Efficient Removal of Condensed Water," *Soft Matter* **8**, 6680–6683.

Her, E. K., Ko, T. J., Lee, K. R., Oh, K. H., and Moon, M.-W. (2012), "Bioinspired Steel Surfaces with Extreme Wettability Contrast," *Nanoscale* **4**, 2900–2905.

Huang, J., Lai, Y., Wang, L., Li, S., Ge, M., Zhang, K., Fuchs, H., and Chi, L. (2014), "Controllable Wettability and Adhesion on Bioinspired Multifunctional TiO_2 Nanostructure Surfaces for Liquid Manipulation," *J. Mater. Chem. A* **2**, 18531–18538.

Huang, L., Song, J., Lu, Y., Chen, F., Liu, X., Jin, Z., Zhao, D., Carmalt, C. J., and Parkin, I. P. (2017), "Superoleophobic Surfaces on Stainless Steel Substrates Obtained by Chemical Bath Deposition," *Micro. Nano. Lett.* **12**, 76–81.

Kaufman, J. G. and Rooy, E. L (Eds.). (2004), *Aluminum Alloy Castings: Properties, Processes, and Applications*. ASM International, Materials Park, Ohio.

Ke, Q., Fu, W., Jin, H., Zhang, L., Tang, T., and Zhang, J. (2011), "Fabrication of Mechanically Robust Superhydrophobic Surfaces Based on Silica Micro-nanoparticles and Polydimethylsiloxane," *Surf. Coat. Technol.* **205**, 4910–4914.

Kim, C., and Hsieh, Y.-L. (2001), "Wetting and Absorbency of Nonionic Surfactant Solutions on Cotton Fabrics," *Colloids Surf. A* **187–188**, 385–397.

Kim, Y., Lee, S., Cho, H., Park, B., Kim, D. and Hwang, W. (2012), "Robust Superhydrophilic/Hydrophobic Surface Based on Self-Aggregated Al_2O_3 Nanowires by Single-Step Anodization and Self-Assembly Method," *ACS Appl. Mater. Interfaces* **4**, 5074–5078.

Kite, M., and Thomson, R. (2006), *Conservation of Leather and Related Materials*, Butterworth-Heinemann, Oxford, UK.

Krauss, G. (2005), *Steels: Processing, Structure, and Performance*, ASM International, Materials Park, Ohio.

Latthe, S. S., Sudhagar, P., Devadoss, A., Kumar, A. M., Liu, S., Terashima, C., Nakata, K. and Fujishima, A. (2015), "A Mechanically Bendable Superhydrophobic Steel Surface with Self-Cleaning and Corrosion-Resistant Properties," *J. Mater. Chem. A* **3**, 14263–14271.

Lee, J.-W., and Hwang, W. (2017), "Simple Fabrication of Superoleophobic Titanium Surfaces via Hierarchical Microhorn/Nanoporous Structure Growth by Chemical Acid Etching and Anodization," *J. Alloys Compd.* **728**, 966–970.

Li, B., Zhou, M., Yuan, R., and Cai, L. (2008), "Fabrication of Titanium-Based Microstructured Surfaces and Study on Their Superhydrophobic Stability," *J. Mater. Res.* **23**, 2491–2499.

Li, L., Breedveld, V., and Hess, D. W. (2012), "Creation of Superhydrophobic Stainless Steel Surfaces by Acid Treatments and Hydrophobic Film Deposition," *ACS Appl. Mater. Inter.* **4**, 4549–4556.

Li, L., Huang, T., Lei, J., He, J., Qu, L., Huang, P., Zhou, W., Li, N. and Pan, F. (2014a), "Robust Biomimetic-Structural Superhydrophobic Surface on Aluminum Alloy," *ACS Appl. Mater. Interfaces* **7**, 1449–1457.

Li, Y., Wan Y., Dong Z., and Zhang J. (2014b), "Excellent Friction-reducing Performance of Superhydrophobic Steel Surface in Dry Sliding," RSC *Adv.* **4**, 20548–20553

Li, Q., Tang, F., Wang, C., and Wang, X. (2016), "Facile Fabrication of Wear-Resistant Multifunctional Surfaces on Titanium Alloy Substrate by One-Step Anodization and Modification with Silicon Dioxide Nanoparticles," *J. Sol-Gel Sci. Technol.* **80**, 318–325.

Li, F., Bhushan, B., Pan, Y., and Zhao, X. (2019), "Superoleophobic/superhydrophilic Functionalized Cotton for Efficient Separation of Immiscible Oil-Water Mixtures and Oil-Water Emulsions," *J. Colloid Interface Sci.* **548,** 123–130.

Lin, J., Zheng, C., Ye, W. J., Wang, H. Q., Feng, D. Y., and Huan, B. W. (2014), "A Facile Dip-coating Approach to Prepare SiO_2/fluoropolymer Coating for Superhydrophobic and Superoleophobic Fabrics with Self-cleaning Property," *J. Appl. Polym. Sci.* **125**, 112–117

Liu, L., Zhao, J., Zhang, Y., Zhao, F. and Zhang, Y. (2011), "Fabrication of Superhydrophobic Surface by Hierarchical Growth of Lotus-Leaf-Like Boehmite on Aluminum Foil," *J. Colloid Interface Sci.* **358**, 277–283.

Liu, L., Feng, X. and Guo, M. (2013a), "Eco-Friendly Fabrication of Superhydrophobic Bayerite Array on Al Foil Via an Etching and Growth Process," *J. Phys. Chem. C* **117**, 25519–25525.

Liu, Y., Liu, J., Li, S., Liu, J., Han, Z. and Ren, L. (2013b), "Biomimetic Superhydrophobic Surface of High Adhesion Fabricated with Micronano Binary Structure on Aluminum Alloy," *ACS Appl. Mater. Interfaces* **5**, 8907–8914.

Liu, B., Tian, T., Yao, J., Huang, C., Tang, W., Xiang, Z., Xu, X., and Min, J. (2017), "Superhydrophobic Organosilicon-based Coating System by a Novel Ultraviolet-curable Method," *Nanomaterial and Nanotechnology* **7**, 1–10.

Lu, S., Chen, Y., Xu, W. and Liu, W. (2010), "Controlled Growth of Superhydrophobic Films by Sol-Gel Method on Aluminum Substrate," *Appl. Surf. Sci.* **256**, 6072–6075.

Lu, Y., Song, J., Liu, X., Xu, W., Xing, Y., and Wei, Z. (2013), "Preparation of Superoleophobic and Superhydrophobic Titanium Surfaces via an Environmentally Friendly Electrochemical Etching Method," *ACS Sustain. Chem. Eng.* **1**, 102–109.

Lu, Y., Sathasivam, S., Song, J., Chen, F., Xu, W., Carmalt, C. J., and Parkin, I. P. (2014), "Creating Superhydrophobic Mild Steel Surfaces for Water Proofing and Oil–Water Separation," *J. Mater. Chem. A* **2**, 11628–11634.

Lütjering, G., and Williams, J. C. (2007), *Titanium*, 2nd ed., Springer-Verlag, Berlin, Germany.

Ma, J., Zhang, X., Bao, Y., and Liu, J. (2015), "A Facile Spraying Method for Fabricating Superhydrophobic Leather Coating," *Colloids Surf. A* **472**, 2 1–25.

Maitra, T., Antonini, C., auf der Mauer, M., Stamatopoulos, C., Tiwari, M. K. and Poulikakos, D. (2014), "Hierarchically Nanotextured Surfaces Maintaining Superhydrophobicity under Severely Adverse Conditions," *Nanoscale* **6**, 8710–8719

Manning, R., and Ewing, J. (2009), *Temperature in Cars Survey*, RACQ Vehicle Technologies Department, Brisbane, Australia.

Materne, T., de Buyl, F., and Witucki, G. L. (2012), *Organosilane Technology in Coating Applications: Review and Perspective*, Dow Corning Corporation, Midland, MI.

Meena, M. K., Tudu, B. K., Kumar, A., and Bhushan, B. (2020), "Development of Polyurethane Based Superhydrophobic Coatings on Steel Surfaces," *Phil Trans R. Soc. A* **378**, 20190446.

Meng, H., Wang, S., Xi, J., Tang, Z. and Jiang, L. (2008), "Facile Means of Preparing Superamphiphobic Surfaces on Common Engineering Metals," *J. Phys. Chem.C* **112**, 11454–11458.

Merritt M.J. (1924), *The Textile Fibers; Their Physical, Microscopical, and Chemical Properties*, Wiley, New York.

Motlagh, N. V., Birjandi, F. C., Sargolzaei, J., and Shahtahmassebi, N. (2013), "Durable, Superhydrophobic, Superoleophobic and Corrosion Resistant Coating on the Stainless Steel Surface Using a Scalable Method," *Appl. Surf. Sci.* 283, 636–647.

Nanda, D., Varshney, P., Satapathy, M., Mohapatra, S. S., Bhushan, B., and Kumar, A. (2017), "Single Step Method to Fabricate Durable Superliquiphobic Coating on Aluminum Surface with Self-cleaning and Anti-fogging Properties" *J. Colloid Interface Sci.* **507**, 397–409.

Nanda, D., Sahoo, A., Kumar, A., Bhushan, B. (2019), "Facile Approach to Develop Durable and Reusable Superhydrophobic/superoleophilic Coatings for Steel Mesh Surfaces," *J. Colloid Interface Sci.* **535**, 50–57.

Nosonovsky, M. and Bhushan, B. (2008), *Multiscale Dissipative Mechanisms and Hierarchical Surfaces: Friction, Superhydrophobicity, and Biomimetics*, Springer, Heidelberg, Germany.

Panda, A., Varshney, P., Mohapatra, S. S., Kumar, A. (2018), "Development of Liquid Repellent Coating on Cotton Fabric by Simple Binary Silanization with Excellent Self-cleaning and Oil-water Separation Properties," *Carbohydr. Polym.* 181, 1052–1060.

Peng, S. and Bhushan, B. (2016), "Mechanically Durable Superoleophobic Aluminum Surfaces with Microstep and Nanoreticula Hierarchical Structure for Self-Cleaning and Anti-smudge Properties," *J. Colloid Interface Sci.* **461**, 273–284.

Peng, S., Yang, X., Tian, D. and Deng, W. (2014), "Chemically Stable and Mechanically Durable Superamphiphobic Aluminum Surface with a Micro/Nanoscale Binary Structure," *ACS Appl. Mater. Interfaces* **6**, 15188–15197.

Perera, S., Bhushan, B., Bandara, R., Rajapakse, G., Rajapakse, S., and Bandara, C. (2013), "Morphological, Antimicrobial, Durability, and Physical Properties of Untreated and Treated Textiles using Silver-nanoparticles," *Colloids Surf. A: Physicochem. Eng. Aspects* **436**, 975–989.

Poetes, R., Holtzmann, K., Franze, K. and Steiner, U. (2010) "Metastable Underwater Superhydrophobicity," *Phys. Rev. Lett.* **105** 166104, 1–4.

Pritchard, P. J., and Mitchell, J. W. (2015), *Introduction to Fluid Mechanics*, 9th ed., Wiley, Hoboken, New Jersey.

Przybylak, M., Maciejewski, H. Dutkiewicz, A., Dabek, I., and Nowicki, M. (2016), "Fabrication of Superhydrophobic Cotton Fabrics by a Simple Chemical Modification," Cellulose **23**, 2185–2197.

Qian, B. and Shen, Z. (2005), "Fabrication of Superhydrophobic Surfaces by Dislocation-Selective Chemical Etching on Aluminum, Copper, and Zinc Substrates," *Langmuir* **21**, 9007–9009.

Qu, M., Zhang, B., Song, S., Chen, L., Zhang, J. and Cao, X. (2007), "Fabrication of Superhydrophobic Surfaces on Engineering Materials by a Solution-Immersion Process," *Adv. Funct. Mater.* **17**, 593–596.

Ren, S.,Yang, S. and Zhao, Y. (2004), "Nano-Tribological Study on a Super-Hydrophobic Film Formed on Rough Aluminium Substrates," *Acta Mech. Sin.* **20**, 159–164.

Ren, G., Song, Y., Li, X., Wang, B., Zhou, Y., Wang, Y., Ge, B., and Zhu, X. (2018), "A Simple Way to an Ultra-robust Superhydrophobic Fabric with Mechanical Stability, UV Stability, and UV Shielding Property," *J. Colloid Interface Sci.* **522**, 57–62.

Schindler W.D. and Hauser P.J., eds. (2004), *Chemical Finishing of Textiles*, Woodhead Publishing, Cambridge, UK.

Song, J., Xu, W. and Lu, Y. (2012), "One-Step Electrochemical Machining of Superhydrophobic Surfaces on Aluminum Substrates," *J. Mater. Sci.* **47**, 162–168.

Sun, Y., Wang, L., Gao, Y., and Guo, D. (2015), "Preparation of Stable Superamphiphobic Surfaces on Ti-6Al-4V Substrates by One-Step Anodization," *Appl. Surf. Sci.* **324**, 825–830.

Tsujii, K., Yamamoto, T., Onda, T. and Shibuichi, S. (1997), "Super Oil-Repellent Surfaces," *Angew. Chem. Int. Ed. Engl.* **36**, 1011–1012.

Tudu, B. K., Kumar, A., and Bhushan, B. (2019a), "Facile Approach to Develop Anti-corrosive Superhydrophobic Aluminum with High Mechanical, Chemical and Thermal Durability," *Phil. Trans. R. Soc. A* **377**, 20180272.

Tudu, B. K., Kumar, A., and Bhushan, B. (2019b), "Fabrication of Superoleophobic Cotton Fabric for Multi-purpose Applications," *Phil. Trans. R. Soc. A* **377**, 20190129.

Vander Voort, G. F. (1984), *Metallography: Principles and Practice*, McGraw-Hill, New York.

Walker, P. and Tarn, W. H. (Eds.) (1991), *CRC Handbook of Metal Etchants*, CRC Press, Boca Raton, FL.

Wang, Y., and Bhushan, B. (2015), "Wear-Resistant and Antismudge Superoleophobic Coating on Polyethylene Terephthalate Substrate Using SiO_2 Nanoparticles," *ACS Appl. Mater. Interfaces* **7**, 743–755.

Wang, D., Guo, Z., Chen, Y., Hao, J., and Liu, W. (2007), "In Situ Hydrothermal Synthesis of Nanolamellate CaTiO3 with Controllable Structures and Wettability," *Inorg. Chem.* **46**, 7707–7709.

Wang, D., Wang, X., Liu, X., and Zhou, F. (2010), "Engineering a Titanium Surface with Controllable Oleophobicity and Switchable Oil Adhesion," *J. Phys. Chem. C* **114**, 9938–9944.

Wu, W., Wang, X., Wang, D., Chen, M., Zhou, F., Liu, W. and Xue, Q. (2009a), "Alumina Nanowire Forests Via Unconventional Anodization and Super-Repellency Plus Low Adhesion to Diverse Liquids," *Chem. Commun.* **9**, 1043–1045.

Wu, B., Zhoua, M., Lia, J., Yea, X., Lia, G., and Cai, L. (2009b), "Superhydrophobic Surfaces Fabricated by Microstructuring of Stainless Steel using a Femtosecond Laser," *Appl. Surf. Sci.* **256**, 61–66.

Xu, W., Liu, H., Lu, S., Xi, J., and Wang, Y. (2008), "Fabrication of Superhydrophobic Surfaces with Hierarchical Structure through a Solution-Immersion Process on Copper and Galvanized Iron Substrates," *Langmuir* **24**, 10895–10900.

Zeng, Z., Wang, H., Zhou, H., and Lin, T. (2015), "Self-cleaning, Superhydrophobic Cotton Fabrics with Excellent Washing Durability, Solvent Resistance and Chemical Stability Prepared from an SU-8 Derived Surface Coating," *RSC Adv.* **5**, 61044–61050.

Zhang, F., Zhao, L., Chen, H., Xu, S., Evans, D. G. and Duan, X. (2008), "Corrosion Resistance of Superhydrophobic Layered Double Hydroxide Films on Aluminum," *Angew. Chem. Int. Ed.* **47**, 2466–2469.

Zimmerman, J., Reifler, F. A., Fortunato, G., Gerhardt, L.-C. and Seeger, S. (2008) "A Simple, One-Step Approach to Durable and Robust Superhydrophobic Textiles," *Adv. Funct. Mater.* **18**, 3662–3669.

Shark Skin Surfaces for Fluid-Drag Reduction in Turbulent Flows 11

11.1 Introduction

Nature has created ways of reducing drag in fluid flow, evident in the efficient movement of fish, dolphins, and sharks. The mucus secreted by fish reduces drag as they move through water, protects the fish from abrasion by making the fish slide across objects rather than scrape, and prevents disease by making the surface of the fish difficult for microscopic organisms to adhere to, minimizing biofouling (Shephard 1994). Applications of drag reducing polymers has been long known. For example, by adding as little as a few hundred parts per million guar, a naturally occurring polymer, friction in pipe flow can be reduced by up to two-thirds. Other synthetic polymers are known to provide an even larger benefit (Hoyt 1975; Frings 1988). The compliant skin of the dolphin is also known to reduce drag. By responding to the pressure fluctuations across the surface, a compliant material on the surface of an object in a fluid flow has been shown to be beneficial. A drag reduction of about 7% has been reported (Choi et al. 1997).

Fast swimming sharks also experience low drag. Lower drag is necessary for shark survival, since it allows sharks to swim faster in order to catch prey (Reif 1985; Bechert et al. 1997a; Dean and Bhushan 2010; Bixler and Bhushan 2013a). The low drag also protects against fouling as these sharks swim through water. The tiny scales covering the skin of fast swimming sharks, known as dermal denticles (skin teeth), are shaped like small riblets and aligned in the direction of fluid flow are believed to be responsible for low drag. Slower sharks are covered in dermal denticles as well, but they are not shaped like riblets and do not provide any drag reduction benefit. The shape of riblets on fast swimming sharks varies greatly, even at different locations on the same species of shark (Reif 1985; Dean and Bhushan 2010).

B. Bhushan, *Introduction to Biomimetics and Bioinspiration*,
https://doi.org/10.1007/978-3-031-62344-8_11

It is believed that sharks stay clean due to these microstructured riblets, flexion of the dermal denticles, and a mucous layer (Bechert et al. 1997a, 2000b; Collins and Brebbia 2004; Dean and Bhushan 2010; Bixler and Bhushan 2013a). Because low drag allows for faster movement through water, the increased fluid flow velocity on the skin reduces microorganism settlement time and promotes antifouling (Kesel and Liedert 2007; Ralston and Swain 2009; Dean and Bhushan 2010). In addition, microorganisms larger than the spacing between riblets are unable to effectively adhere to and ultimately colonize the skin, which further promotes antifouling (Schumacher et al. 2007; Scardino 2009; Brennan et al. 2010; Bixler and Bhushan 2012a).

Low drag surfaces are of interest in both internal and external flow applications. Examples of internal flow include pipelines, catheters, and micro/nanochannels such as micro/nanofluidic based sensors. A major example of external flow is transportation—ships, airplanes, and automobiles. Low drag surfaces have been the subject of much experimentation using shark skin riblet-inspired microtextured surfaces (Dean and Bhushan 2010; Bixler and Bhushan 2013a). Experimental results indicate that shark skin inspired surfaces effectively reduce drag in open and closed channel flow. In open channel flow, drag has been measured using water (Reidy and Anderson 1988; Gillcrist and Reidy 1989; Walsh 1990a; Neumann and Dinkelacker 1991; Rohr et al. 1992; Wen et al. 2014), oil (Bechert et al. 1997b, 2000a; Buttner and Schulz 2011; Gruneberger and Hage 2011), and air (Walsh 1982; Walsh and Lindemann 1984; Bechert et al. 1985, 2000a; Wilkinson and Lazos 1988; Coustols 1989; Walsh and Anders 1989; Subaschandar et al. 1999; Viswanath 1999, 2002; Han et al. 2003; Sareen et al. 2011). Similarly, in closed channel flow, drag also has been measured using water (Reidy and Anderson 1988; Liu et al. 1990; Rohr et al. 1992; Jung and Bhushan 2010; Bixler and Bhushan 2013b), oil (Bixler and Bhushan 2013c), and air (Nitschke 1984; Enyutin et al. 1995; Bixler and Bhushan 2013b). Both drag (for open channel) and pressure drop (for closed channel) measurements characterize the efficiency of riblet drag reduction.

A variety of riblet geometries, configurations, material properties including wettability, fluid viscosity (water, oil, and air), and flow conditions (laminar and turbulent flow) have been used in experiments. Geometries include blade, sawtooth, scalloped, and bullnose geometries with continuous and segmented (aligned and staggered) configurations in water, oil, and air. Both open and closed pipes/channels with various dimensions have been used. Open channel oil experiments with metal riblets show drag reduction of nearly 10% (Bechert et al. 1997b), while closed channel water experiments with polymer riblets show pressure drop reduction of up to 34% (Bixler and Bhushan 2013b). Shark skin replicas have also been evaluated, and have reduced pressure drop by up to 30% in closed channel water flow (Jung and Bhushan 2010; Bixler and Bhushan 2012b).

Larger scale drag reducing efforts have included the experimental 3 M Corp. (Minneapolis, MN) vinyl riblets, as well as the commercially available Speedo brand FastSkin® fabric racing swimsuit. Such riblet technology has captured the attention of the National Aeronautics and Space Administration (NASA), U.S. Department of Energy, U.S. Navy,

Airbus, Boeing, and Olympic competitors. In the 1984 Los Angeles Olympics and 1987 America's Cup, 3 M riblets were applied to U.S. boats, which presumably helped to secure victories. The 2008 Beijing Olympics witnessed the benefit of the Speedo Fast-Skin® swimsuit, when American Michael Phelps broke eight Olympic records, seven world records, and won eight gold medals, the most of any athlete in any single Olympic showing. The swimsuits reduced drag up to 4% for men and 3% for women (Krieger 2004).

In this chapter, based on Dean and Bhushan (2010), Bixler and Bhushan (2013a, b), and Martin and Bhushan (2014, 2016a, b), we present mechanisms of fluid drag and an overview of experimental and fluid flow modeling studies to understand the mechanisms responsible for drag reduction in shark skin and optimization of riblet geometries for low drag, followed by applications of riblets in commercial applications.

11.2 Fluid Drag Reduction

In this section, we first present mechanisms of fluid drag, which include pressure and skin friction drag, followed by the mechanisms of low drag shark skin.

11.2.1 Mechanisms of Fluid Drag

Fluid drag comes in several forms, the most basic of which are pressure drag and skin friction drag (Dean and Bhushan 2010; Bixler and Bhushan 2013a). Pressure drag is the drag associated with the energy required to move fluid out from in front of an object in the flow and then back in place behind the object. Much of the drag associated with walking through water is pressure drag, as the water directly in front of a body must be moved out and around the body before it can move forward. The magnitude of pressure drag can be reduced by creating streamlined shapes. Skin friction or viscous drag is caused by the interactions between the fluid and a surface parallel to the flow, as well as the attraction between molecules of the fluid. Skin friction drag is similar to the motion of a deck of cards sliding across a table. The frictional interactions between the table and the bottom card as well as between each successive card mimic the viscous interactions between molecules of fluid. Moving away from the surface of an object in fluid flow, each fluid layer has a higher velocity until a layer is reached where the fluid has velocity equal to the mean flow. Fluids of higher viscosity—the attraction between molecules—have higher apparent friction between fluid layers, which increases the thickness of the fluid layer distorted by an object in a fluid flow. For this reason, more viscous fluids have relatively higher drag than less viscous fluids. A similar increase in drag occurs as fluid velocity increases (Batchelor 1970; Blevins 1984; Davies 2002; Munson et al. 2012; Pritchard and Mitchell 2015).

The above discussion of friction drag assumes all neighboring fluid molecules move in the same relative direction. Figure 11.1 shows an image of the transition between laminar flow and turbulent flow, in which molecules move in swirling and cross-stream motions such that an average velocity is maintained in the direction of flow. Laminar and turbulent boundary layers help describe fluid flow, where turbulent boundary layers lead to higher skin friction drag. The boundary layer is the fluid layer adjacent to the surface, with the innermost layer called the viscous sublayer. In this layer, laminar flow appears smooth and orderly, while turbulent flow appears random and chaotic. Additionally, laminar flow is controlled by viscous forces between the fluid molecules, whereas turbulent flow is controlled by inertial forces. Turbulent vortices in the viscous sublayer naturally translate in cross-flow and streamwise directions, which leads to vortices intermingling and ejecting from the viscous sublayer. This movement increases momentum transfer and shear stresses, which results in higher drag (Kline et al. 1967; Munson et al. 2012). Therefore, laminar flow is preferred for low drag. However, in many real-world fluid flow applications, the boundary layers naturally transition from laminar to turbulent. This transition can be described with the dimensionless Reynolds number (*Re*), which is the ratio of inertial forces to viscous forces, or (Munson et al. 2012; Pritchard and Mitchell 2015):

$$Re = \frac{VD}{\nu} \tag{11.1}$$

where V is the fluid mean flow velocity, D is the characteristic length which is the hydraulic diameter (for closed channel) or a distance downstream from the leading edge (for open channel), and ν the fluid kinematic viscosity. For a rectangular closed channel, $D = 4A/WP$, where A is the cross sectional area and WP is the wetted perimeter. Transition from laminar to turbulent flow depends on parameters in (11.1) as well as surface roughness and freestream disturbances, where such features may “trip” the boundary layer to become turbulent. Such transition occurs around $Re = 2300$ and fully turbulent at

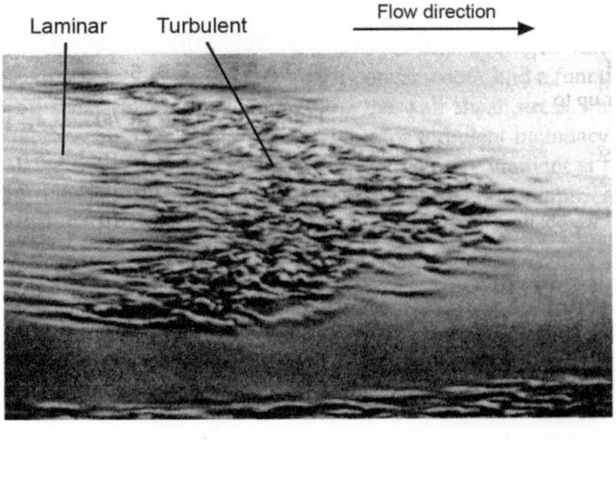

Fig. 11.1 Transition between laminar and turbulent flow in fluid over a flat plate (adapted from Munson et al. 2012)

about Re = 4000 for closed channel flow. The transition occurs at Re = 500,000 for open channel flow over a flat plate (Munson et al. 2012; Pritchard and Mitchell 2015).

Turbulent flow has highly chaotic flow characteristics. The turbulent action leads to generation of vortices. The velocity distribution, vorticity (the curl of the flow velocity), and the concept of fluid circulation (line integral of velocity field along a fluid element) are used to characterize vortices. Vortices are regions of fluid where the flow is rotating along an axis line in a roughly circular or spiral pattern and can take the form of streamwise vortices. Figure 11.2 shows two streamwise vortices with a low speed streak between vortices (Martin and Bhushan 2016a). These pairs of counter-rotating vortices repeat themselves. The counter-rotating streamwise pair does not occur as often as a simple streamwise vortex, also shown in Fig. 11.2.

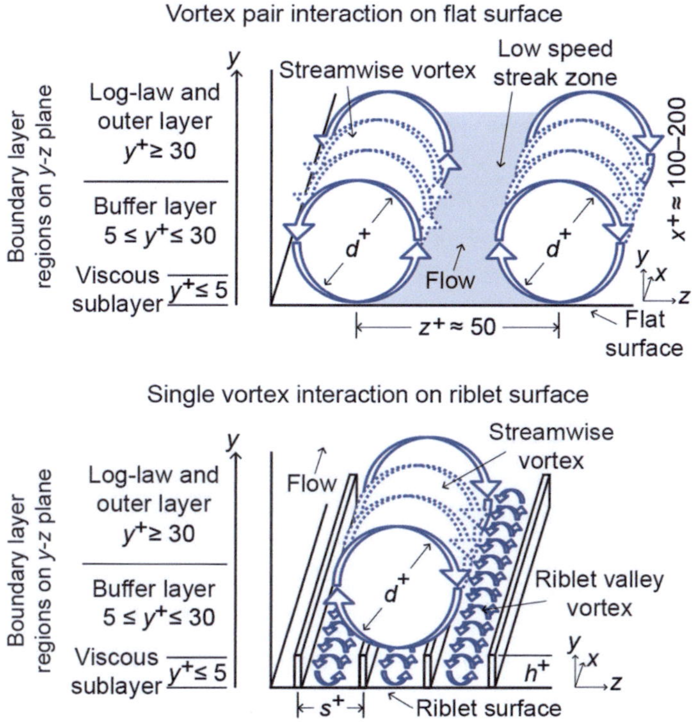

Fig. 11.2 Model of vortex pair interaction on flat surface and single vortex interaction on riblet surface. A streamwise vortex has a diameter of d^+ that ranges from 5 to 115 nondimensional units with an average diameter of 30 nondimensional units for the streamwise vortices with the center of the vortex in the buffer layer $5 \leq y^+ \leq 30$. Due to the sreamwise vortices lifted above the riblets, vortices within the riblet valleys are formed. The longitudinal length of the streamwise vortex generally ranges from 100 to 200 nondimensional units, x^+, y^+, z^+, and d^+ are nondimensional parameters (Martin and Bhushan 2016a)

Since vortices are not generated in laminar flow, this method of drag reduction is not possible. Instead, a drag increase can result due to the greater surface area on the riblet surface (Goldstein et al. 1995). In the transitional regime, a drag reduction is possible due to the possibility of vortex structures being created; however, a drag reduction through this mechanism is best obtained through fully turbulent flow.

11.2.2 Shark Skin and Riblets Present

A small amount of mucus is present on the skin of the sharks (Bechert et al. 1986). However, it is not enough to provide drag benefits, as in the case of fish. It is the microstructured skin that provides reduction in skin friction drag during turbulent flow. The skin of fast swimming sharks (such as Mako, *Isurus oxyrinchus*) is covered with scales called dermal denticles (little skin teeth), which contain specially sized and spaced riblets (microscopic grooves) oriented parallel to the swimming direction. Dermal denticles from both the Mako and spiny dogfish shark species are shown in Fig. 11.3, with the flow direction indicated by arrows. The Mako shark has scales with some gaps and no offsets (referred to as aligned-segmented riblets), whereas the spiny dogfish has both gaps and offsets (referred to as staggered-segmented riblets). The shape and size of the scales, as well as the number of riblets, varies between shark species. Riblets can also vary between different locations on the same shark (Reif 1985; Jung and Bhushan 2010). Typically, the scales are generally diamond shaped 50–100 μm in height and length and width on the order of 200–500 μm. The riblets are scalloped, with the grooves spaced 30–100 μm apart.

When a shark swims fast during turbulent flow, vortices form on the surface that cause high shear stresses across the entire surface, and can lead to high drag. The vortices increase the shear stress at the wall (τ_0) by increasing the local velocity gradient, $\frac{\partial u}{\partial y}$, given by fluid shear stress equation

$$\tau_0 = \rho \nu \left.\frac{\partial u}{\partial y}\right|_{y=0} \tag{11.2}$$

where ρ is the mass density. The microstructured riblets with longitudinal grooves encourage anisotropic flow as well as control vortices on the skin naturally present in turbulent flow. The riblets lift and presumably pin any vortices generated in the viscous sublayer. The low velocity fluid in the riblet valleys causes minimal shear stress across most of the surface, which leads to lower drag. With appropriately sized riblets, for a given velocity, vortices are lifted above the surface and presumably pinned at the riblet tips. Lifting reduces the total shear stress since vortices contact just the small riblet tips (as opposed to the total surface area). Pinning is believed to reduce the cross-stream motion of fluid, which provides anisotropic flow, and ejection of vortices from the viscous sublayer, which reduces momentum transfer (Bechert et al. 1985, 1997a, b, 2000a, b; Lowrey and Harasha

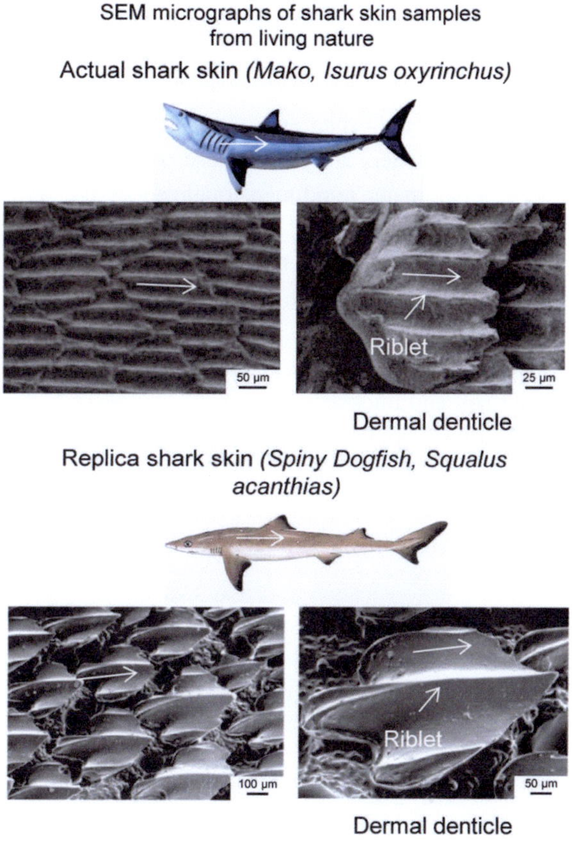

Fig. 11.3 SEM micrographs of shark skin samples from living nature. Actual Mako (top) (adapted from Bixler and Bhushan 2012b) and replica of spiny dogfish (bottom) (adapted from Jung and Bhushan 2010) shark skin micrographs are shown at two magnifications. These highlight the diamond-shaped scales (dermal denticles) and scalloped riblets which are responsible for low drag and antifouling properties. Arrows indicate fluid flow direction. Approximate scale dimensions based on spiny dogfish shark are: scale length = 450 μm, scale width = 375 μm, gap length = 175 μm, gap width = 100 μm, riblet spacing = 125 μm, and riblet length = 50 μm

1991; Luchini et al. 1991; Lee and Lee 2001; Dean and Bhushan 2010; Deyuan et al. 2011; Oeffner and Lauder 2012; Bixler and Bhushan 2013a). Turbulent vortices are also present in fluids besides water; therefore, the shark skin inspired riblets are expected to provide a similar drag-reducing benefits with any fluid, such as air and oil.

11.3 Experimental Studies of Riblet-Inspired Surfaces

Experiments have been performed on shark skin replicas and shark skin inspired artificial riblets fabricated in the lab (Jung and Bhushan 2010; Dean and Bhushan 2012; Bixler and Bhushan 2012b, 2013a, b, c, d). Such riblets represent those found on actual shark skin. The purpose of riblets is to lift and pin vortices while minimizing drag (skin friction) due to the riblets themselves. Since riblets protrude into the flow channel, the increased surface area equates to increased drag. The riblets should lift and pin the vortices, as well as encourage anisotropic flow along the surface. Such efficient flow is found by considering each valley between riblets and minimizing the wetted perimeter, since increased wetted perimeter leads to increased drag. In general, skin friction increases with higher Reynolds numbers.

To present drag data at various geometrical parameters and flow conditions, dimensionless parameters are used. These parameters are measured in wall units, where a wall unit is a parameter nondimensionalized by the wall shear stress velocity. These parameters are denoted by the ($^+$) symbol, which allows for better comparison of experiments with various riblet geometries and flow conditions. As shown in (11.2), a characteristic length multiplied by mean flow velocity and divided by kinematic viscosity provides the Reynolds number, which is dimensionless. Similarly, dimensionless numbers based on riblet parameters can also be calculated. Important riblet parameters include spacing (s), height (h), and thickness (t).

The dimensionless numbers based on spacing and height of the riblets are defined as (Walsh 1982; Bechert et al. 1997b; Bixler and Bhushan 2013a):

$$s^+ = \frac{sV_\tau}{\nu} \tag{11.3}$$

$$h^+ = \frac{hV_\tau}{\nu} \tag{11.4}$$

The dimensionless number based on thickness is defined here as (Bixler and Bhushan 2013a)

$$t^+ = \frac{tV_\tau}{\nu} \tag{11.5}$$

where V_τ is the wall shear stress velocity; s^+, h^+ and t^+ are referred to as dimensionless riblet spacing, height, and thickness parameters, respectively. Wall shear stress velocity is simply a quantity that has dimensions of velocity and is in the form that allows for different wall shear stresses to be compared (Munson et al. 2012).

Considering kinetic energy, the wall shear stress expression $\tau_o = \rho V_\tau^2$ provides the wall shear stress velocity as,

$$V_\tau = \left(\frac{\tau_o}{\rho}\right)^{1/2} \tag{11.6}$$

A simple force balance shows that fluid shear stress τ_o imposed on the wall of a uniform surface by the friction of the fluid in a circular pipe is equal to the friction factor multiplied with the pipe dynamic pressure (Blevins 1984; White 2006),

$$\tau_o = \frac{\rho V^2}{2} C_f \tag{11.7}$$

where C_f is the Fanning friction factor or skin friction coefficient, and is given as

$$C_f = \frac{f}{4} \tag{11.8}$$

A so-called Blasius formula is one of the first curve fits of the friction factor as a function of *Re* of air flowing in a smooth pipe in turbulent flow (White 2006),

$$C_f \sim \frac{0.0791}{Re^{1/4}} \tag{11.9}$$

Combining (11.1), (11.7), and (11.9), the following expression is obtained for turbulent flow in a circular pipe,

$$\tau_o = 0.03955\nu^{-1/4}\rho V^{7/4}D^{-1/4} \tag{11.10}$$

Determining optimal riblet dimensions for various flow conditions is possible by considering dimensionless parameters. For instance, the riblet spacing for a given fluid can be determined by rewriting (11.3) where $s = s^+ \nu/V_\tau$. This equation indicates that for a constant s^+ value, the required riblet spacing increases as kinematic viscosity ν increases or velocity decreases (Bechert et al. 1992, 1997b, 2000a).

We first provide some flow visualization studies to study drag mechanisms, followed by an overview of the various riblet geometries and various fabrication techniques used to create riblets, as well as their advantages and disadvantages. Finally, we describe the experimental procedures for open and closed channel experimentation using water, oil, and air, followed by results and discussion. Most of the data presented is based on Bixler and Bhushan (2013a, b).

11.3.1 Flow Visualization Studies

Visualizing the interaction between riblets and vortices provides insight into drag mechanisms. Figure 11.4 shows flow visualization experimental results using smoke from atomized oil burned in air, both with and without riblets, and at two different velocities (V) (Lee and Lee 2001). Flow visualization studies also have been carried out by Kline et al. (1967). The figure shows the vortices generated at two velocities V of 3 and 5 m/s for $h/s = 0.5$. These values produce s^+ values of 25.2 and 40.6, respectively, and h^+ values of 12.6 and 20.3, respectively. Vortex sizes were approximately 35–50 wall units in diameter for both cases. It was reported that $V = 3$ m/s case showed drag decreasing, whereas the $V = 5$ m/s case showed drag increasing. In the $V = 3$ m/s case, the vortex diameter was larger than s^+, which lead to lifting the vortex, and thus lowering drag. However, for the $V = 5$ m/s case, the vortex diameter was about the same as s^+, therefore providing less vortex lifting and leading to higher drag.

An s^+ near 15 with $h/s = 0.5$ corresponds to approximately three riblets per vortex, where vortices are believed to remain lifted and presumably pinned at or near the riblet

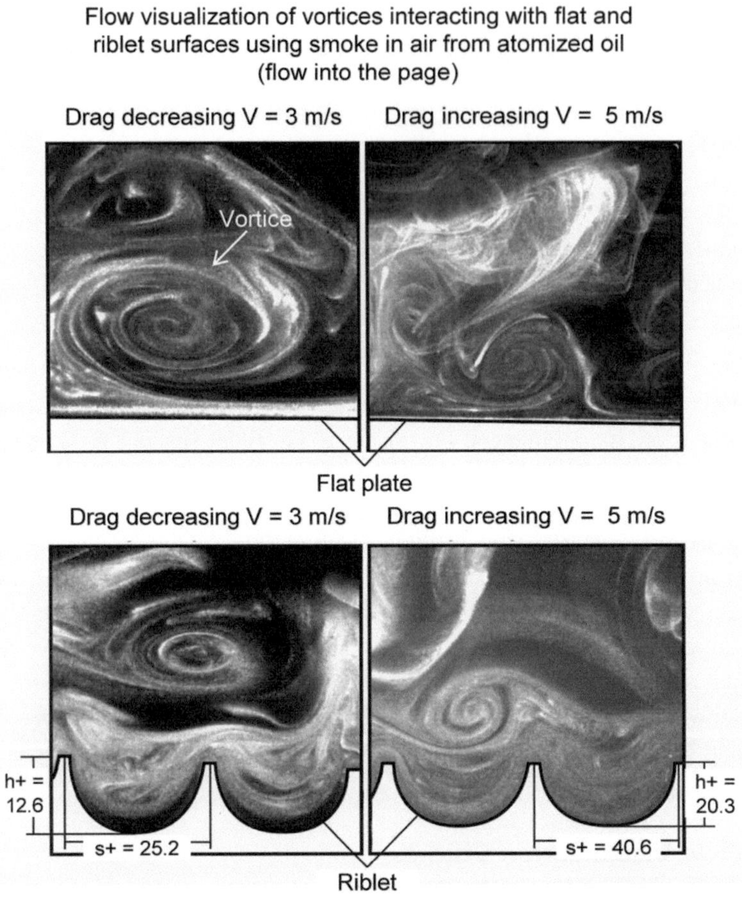

Fig. 11.4 Flow visualization images using smoke from atomized oil burned in air to study the turbulent vortices behavior with and without riblets at two velocities (adapted from Lee and Lee 2001). Top images show the vortices on a flat plate with a relatively large amount of surface contact area, which leads to higher drag. Bottom images show the vortices are lifted above the riblet tips, with a relatively small amount of surface contact area, which leads to lower drag. The dimensionless parameters for height (h^+) and spacing (s^+) aid in the understanding of riblet design for any flow condition. Such parameters are based on riblet spacing or height, fluid kinematic viscosity, and wall shear stress velocity

tips. Both s^+ and h^+ information is useful in order to determine relationships between drag, fluid viscosity, and riblet dimensions. Such information allows for scaling riblets for drag reduction and comparing experimental results.

11.3.2 Riblet Geometries and Configurations

Shark skin dermal denticles and riblets exhibit rather complex three dimensional geometries and configurations, therefore fabrication of such microstructured features is challenging. Additionally, in order to study the dimensional effects of riblets in fluid flow, experiments with varying sizes of relatively durable fluid-tolerant riblets are necessary. However, due to the micro-sized nature of the riblets, this poses difficulty in producing cost effective riblet samples, especially for larger sample sets. The most important characteristics of the riblets have been identified, and the samples have been fabricated for experimentation.

Riblet sizes range from species to species and from location to location on sharks. For reference, the spiny dogfish shark *Squalus acanthias* has triangular shaped riblets with a base width of 100–300 μm, summit radius of about 15 μm, height of 200–500 μm, and spacing of 100–300 μm (Jung and Bhushan 2010). Using actual riblets for inspiration, simplified artificial riblet samples include various combinations of blade, sawtooth, scalloped, and bullnose geometries with continuous and segmented (aligned and staggered) configurations. It is believed that actual shark skin most closely resembles scalloped riblets with a staggered-segmented configuration. Schematics of simplified riblets fabricated for experimentation are illustrated in Fig. 11.5 (Bixler and Bhushan 2013a), with typical geometries and configurations. Dimensions of interest include riblet height (h), spacing (s), thickness (t), and valley width (vw); and their importance will be discussed in the results section. Drag reduction data in open and closed channel experiments, and geometries and configurations are summarized in Tables 11.1, 11.2 and 11.3 (Bixler and Bhushan 2013a).

11.3.3 Riblet Fabrication

Riblets have been fabricated with a variety of polymers and metals using various fabrication and assembly processes. Feasible riblet fabrication techniques include using metal shims (Bechert et al. 1997b), machined acrylic (Bechert et al. 1997b; Jung and Bhushan 2010; Dean and Bhushan 2012), machined aluminum (Walsh and Lindemann 1984), extruded/coextruded polymer (Marentic and Morris 1992), embossed polymer (Marentic and Morris 1992), soft lithography (Jung and Bhushan 2010; Bixler and Bhushan 2012b), photolithography (Brennan et al. 2010), wet and dry etching (Brennan et al. 2010), grinding (Denkena et al. 2010), rolling (Hirt and Thome 2007), laser etching (Dickinson and Proudley 1991; Bixler and Bhushan 2013b, c) and additive manufacturing (Bhushan and Caspers 2017). In general, blade and scalloped riblets are less durable than sawtooth riblets, and continuous riblets are easier to fabricate than segmented. Embossed polymer sheets fabricated from masters appear to be a cost effective way to scale up.

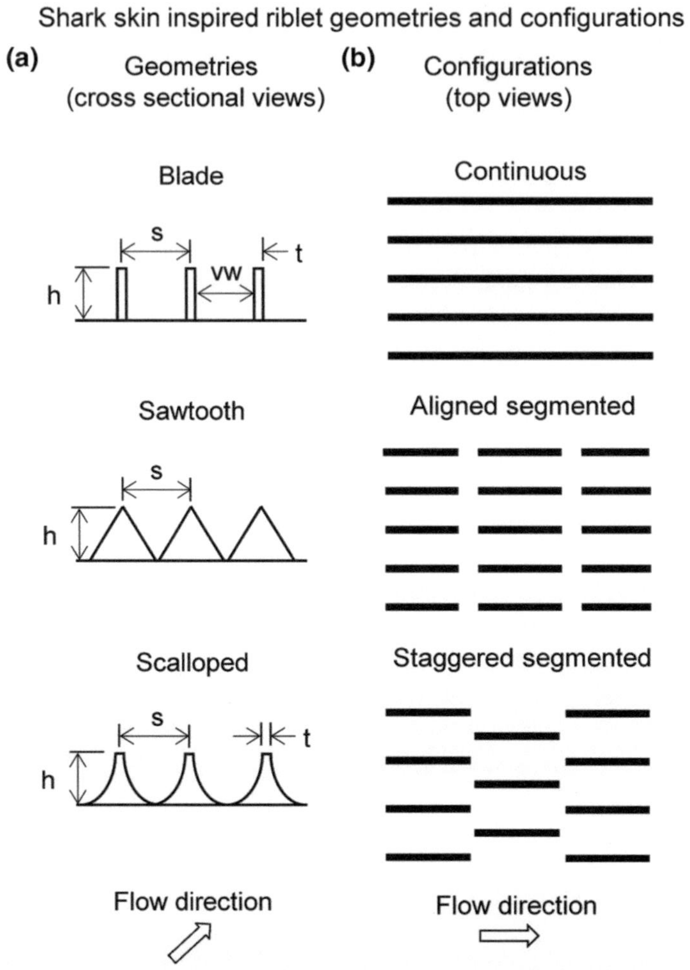

Fig. 11.5 Typical shark skin inspired riblet geometries and configurations that have been experimentally evaluated. Shown are **a** blade, sawtooth, and scalloped geometries as well as **b** continuous, aligned-segmented, and staggered-segmented configurations. Actual shark skin most closely resembles the scalloped geometry and staggered segmented configuration (Bixler and Bhushan 2013a)

The 3 M Corp (Minneapolis, MN) produces an experimental vinyl adhesive backed riblet sheet, which has been used in some fluid drag studies. The riblet cross sectional shape consists of equilateral triangles (sawtooth geometry) (Fig. 11.5). The height (h) equals the spacing (s), which range from 44 to 150 μm. These sawtooth riblets have many advantages such as ease of installation onto large flat or curved surfaces and high durability. The sheets are reported to be either coextruded through a die capable of a

Table 11.1 Drag reduction data in open channel riblet experimentation (Bixler and Bhushan 2013a)

Fluid	Riblet design	Riblet configuration	Riblet material	Max turbulent drag reduction (%)	Reference
Water	Sawtooth	Continuous	Polymer	8	Reidy and Anderson (1988)
	Sawtooth	Continuous	Vinyl	9	Rohr et al. (1992)
	Sawtooth	Continuous	Vinyl	6	Walsh (1990a)
	Sawtooth	Continuous	Vinyl	13	Neumann and Dinkelacker (1991)
	Sawtooth	Continuous	Polymer	7	Gillcrist and Reidy (1989)
Oil	Blade, sawtooth and scalloped	Continuous	Brass and Plexiglas	9.9	Bechert et al. (1997b)
	Blade	Staggered-segmented	Brass	7	Bechert et al. (2000a)
	Blade	Continuous	Titanium and nickel	4.9	Buttner and Schulz (2011)
	Sawtooth	Continuous	Polyurethane	7.6	Gruneberger and Hage (2011)
Air	Blade	Continuous	Metal and polymer	8.5	Wilkinson and Lazos (1988)
	Blade	Staggered-segmented	Epoxy	7	Bechert et al. (2000a)
	Sawtooth, scalloped and bullnose	Continuous	Aluminum and vinyl	8	Walsh (1982)
	Sawtooth, scalloped and bullnose	Continuous	Aluminum and vinyl	8	Walsh and Lindermann (1984)

Table 11.2 Drag reduction data in closed channel riblet experimentation (Bixler and Bhushan 2013a)

Fluid	Riblet design	Riblet configuration	Riblet material	Max turbulent pressure drop/drag reduction	Reference
Water	Blade	Aligned-segmented	Acrylic	23[a]	Jung and Bhushan (2010)
	Blade and sawtooth	Aligned-segmented and continuous	Vinyl and acrylic	22[a]	Bixler and Bhushan (2013b)
	Sawtooth	Continuous	Vinyl	9	Rohr et al. (1992)
	Sawtooth	Continuous	Polymer	28	Reidy and Anderson (1988)
	Sawtooth	Continuous	Vinyl	7	Lui et al. (1990)
Oil	Blade	Aligned-segmented and continuous	Acrylic	7[a]	Bixler and Bhushan (2013a)
Air	Blade	Continuous	Polymer	3	Nitschke (1984)
	Blade and sawtooth	Aligned-segmented and continuous	Vinyl and acrylic	11[a]	Bixler and Bhushan (2013b)
	Sawtooth	Continuous	Epoxy	7	Enyutin et al. (1995)

[a]Micro-sized channel

patterned surface or, more likely, embossed with a negative master mold using heat and/or pressure (Marentic and Morris 1992).

In addition to riblet geometries and configurations, researchers have examined the effect of applying coatings to alter wettability properties. For instance, studies suggest that superhydrophobic surfaces exhibit lower water drag (Ou et al. 2004; Jung and Bhushan 2010; Daniello et al. 2009; Martell et al. 2009, 2010; Bhushan and Jung 2011; Bhushan 2012; Bixler and Bhushan 2012b, 2013b) and self-cleaning (Bhushan and Jung 2011; Bhushan 2012; Bixler and Bhushan 2012b) properties. These are believed to promote antifouling. One study suggests that superoleophilic as well as superoleophobic surfaces can reduce oil drag (Bixler and Bhushan 2013c). Furthermore, one study suggests that

Table 11.3 Drag reduction data in sawtooth riblet airfoil experimentation (Bixler and Bhushan 2013a)

Reynolds number	Airfoil cross section description	Airfoil type	Sawtoot hriblet size with h = s (μm)	Riblets applied to longitudinal location/chord length (%)	Strip applied to longitudinal location/chord length (%)	Angle of attack (deg)	Max drag reduction (%)	Reference
17,000	Symmetrical	NACA 0012 	180	0–100	n/a	0	4.3	Han et al. (2003)
250,000	Symmetrical	NACA 0012 	23, 76, 152	10–100	n/a	0	13.3	Caram and Ahmed (1991)
530,000–790,000	Thin	LC 100D 	76, 152	20–95	2.5	0	2.7	Coustols (1989)
750,000	Thin	GAW-2 	114	12–96	10	0–6	6	Sundaram et al. (1999)
1,000,000	Symmetrical	NACA 0012 	76, 152	12–96	10	0–6	13	Sundaram et al. (1996)
1,000,000	Thin	GAW-2 	76	12–96	10	0–12	10	Subaschandar et al. (1999)
1,000,000	Thin	NACA 0012 	76	12–96	10	0–12	14	Subaschandar et al. (1999)
1,000,000	Thick	NREL S807 	114	5–100	5	0	5	Wetzel and Farokhi (1996)
1,000,000–1,850,000	Thick	DU 96-W-180 	44, 62, 100, 152	40–100	n/a	0	5	Sareen et al. (2011)

(continued)

Table 11.3 (continued)

Reynolds number	Airfoil cross section description	Airfoil type	Sawtoot hriblet size with h = s (μm)	Riblets applied to longitudinal location/chord length (%)	Strip applied to longitudinal location/ chord length (%)	Angle of attack (deg)	Max drag reduction (%)	Reference
3,000,000	Thick	ADA-S1 	18	15–100	6	−0.5 to 1	10	Viswanath and Mukund (1995)
3,300,000	Thin	CAST 7 	17, 23, 33, 51	15–100	n/a	0	3.3	Coustols and Schmitt (1990)
2,000,000–6,000,000	Symmetrical	NACA 0012 	44, 100, 152	0–100	5	0	7	Present work
4,900,000–22,300,000	n/a	Conical nose with cylindrical body 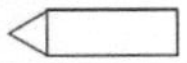	33, 51, 76	87 coverage	n/a	0	4	Coustols and Cousteix (1994)

a nanostructured coating can improve the smoothness of milled or laser etched riblets, which may be desirable for air flow (Bixler and Bhushan 2013b).

11.3.4 Drag Measurement Techniques

In this section, we present the various methods to evaluate riblets in water, oil, and air in both open and closed channel experiments. For each of these experimental set ups, descriptions of the various drag measurement techniques are provided.

11.3.4.1 Open Channel

Measurement of drag in an open channel is conducted with water and oil channels, as well as in wind tunnels. Various methods include using force balances and wake traverses, depending on the fluid and application. For instance, force balances measure water, oil, and air drag, whereas wake traverses measure wind tunnel airfoil drag (Caram and Ahmed 1992; Viswanath 1999). Riblets have also been evaluated on airplanes during flight tests, with riblets covering a portion of the wing and fuselage sections. Total fuel consumption is measured during the riblet flight tests in order to evaluate the total amount of drag reduction (Viswanath 1999).

A typical force balance method for open channel experimentation is illustrated in Fig. 11.6a, where flat sample plates mounted on a force balance are lowered into the moving liquid (typically water or oil) (Bixler and Bhushan 2013a). The flow rate is maintained with a flow channel set-up, as illustrated in Fig. 11.6b, with much research conducted using the so-called Berlin oil channel (Bechert et al. 1992). Propeller driven flow provides a constant current in order to measure the drag resistance force on the sample plates. The percentage drag reduction is calculated using the dimensionless expression $\Delta\tau/\tau_o$ (%), where $\Delta\tau$ is the difference between measured shear stresses on the riblet sample (τ) and the smooth (τ_o) surfaces.

Typical wake traverse and force balance methods for wind tunnel experimentation are also illustrated in Fig. 11.6a. Many experiments have been conducted in subsonic tunnels, whereas fewer in the transonic or supersonic tunnels. Accounting for compressibility effects, speed is defined as the Mach number (Ma), with subsonic (Ma < 0.75), transonic (0.75 < Ma < 1.2), and supersonic (1.2 < Ma < 5) designations. A transonic wind tunnel design is illustrated in Fig. 11.6b, which highlights the transverse wake probe and infrared camera locations. High pressure air in holding tanks is suddenly released into the settling chamber, which then flows across the airfoil sample at a known constant speed. Drag is measured using a traversing wake probe, which measures pressure by traversing downwind from the airfoil during experimentation. The probe measures the pressure in the airfoil wake, and when compared with a smooth airfoil is used to calculate the percentage drag reduction. The drag coefficient (C_d) is calculated with the wake integral expression using velocity calculated from the wake pressure profile (Lee et al. 1978; Barlow et al.

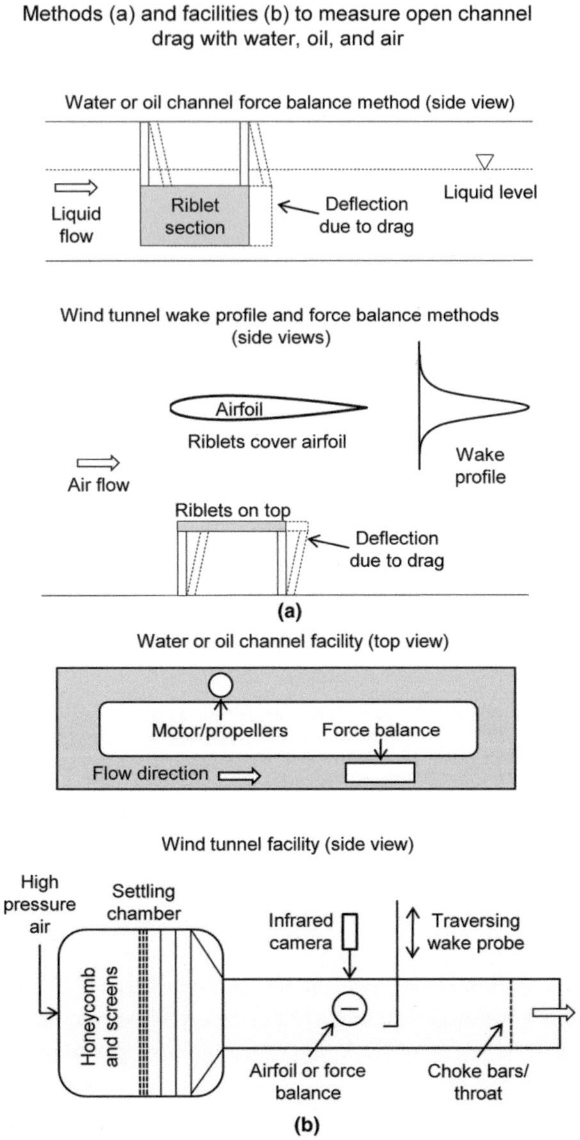

Fig. 11.6 Open channel methods and facilities to measure drag using water, oil, and air. **a** In water or oil channel flow, riblets are applied to a force balance plate that is lowered into the moving fluid (top). The deflection of the force balance translates into the total amount of drag. In airfoil experiments (bottom), the wake profile determines the total amount of drag. Drag may also be measured with riblets applied to a flat plate or airfoil using a force balance in the wind tunnel. **b** Open flow channels filled with water or oil are used with the force balance (top). Shown is the location of the motor and propellers that produce a constant flow. Wind tunnel schematic (bottom) illustrates set up for the wake traverse method. As air flows through the wind tunnel, the wake traverse lowers at a specified rate to measure the wake profile. Also shown is the infrared camera position used to visualize turbulence on the airfoil (Bixler and Bhushan 2013a)

1999; Anderson 2010). The percentage of drag reduction is calculated using the dimensionless expression ΔC_d (%), which is the difference between the drag coefficient on the riblet sample and the smooth surfaces. C_d is expressed as (White 2006),

$$C_d = \frac{F_d}{0.5\rho V^2 A_w}. \tag{11.11}$$

where F_d is the drag force and A_w is the total surface area in contact with the fluid (or wetted area).

In wind tunnel experiments, riblet covered airfoils have been evaluated over a wide range of speeds and angles of attack. Military aircraft commonly experience transonic and supersonic flight. However, commercial aircraft usually cruise between Mach 0.78 and 0.9 (Barlow et al. 1999; Anonymous 2007; Anderson 2010). In the lab, achieving transonic or supersonic flow requires specialized wind tunnels that are relatively expensive to operate. The airfoil angle of attack is the angle between level horizontal and the chord lines. This angle varies during flight in order to maintain lift. At lower speed a higher angle of attack is required, and at higher speeds, a lower angle of attack is required. Experiments have been conducted between 0° and 12° angle of attack in order to cover the majority of cruise conditions (Anonymous 2007; Anderson 2010).

A summary of conditions for open channel flat plate experiments is shown in Table 11.1 outlining the various riblet geometries, configurations, and materials that have been evaluated. Also, Table 11.3 shows wind tunnel experiments that have been conducted with various airfoils, conditions, riblet coverage, and trip strip locations. Knowing that riblets may increase drag in laminar flow, riblets were only applied in the turbulent region of an airfoil, thereby leading to improved drag reduction. Trip strips were installed in many experiments in order to force transition at a particular location (Barlow et al. 1999; Anderson 2010).

11.3.4.2 Closed Channel

Measuring drag in closed channel flow is conducted with water, oil, and air experiments using pipes and channels with both circular and rectangular cross sections. Various closed channel experiments have been conducted using riblets applied to their interiors. The fluid flows through the sample test area and the pressure difference between two points was measured with a pressure manometer. In such cases, a lower pressure drop indicates drag reduction. Apparatuses for both split round pipe and rectangular channel designs are illustrated in Fig. 11.7a (Bixler and Bhushan 2013a). Drag reduction is presented as either negative $\Delta\tau/\tau_o$ (%), where $\Delta\tau$ is the difference between shear stresses on the riblet sample surface (τ) and the smooth surfaces (τ_o), or with negative $\Delta P/P_o$ (%) values, where ΔP is the difference in pressure drop between the riblet sample (P) and the smooth (P_o) surfaces.

Generally, the greatest challenge for technology applications is affixing riblets inside of the closed channel. For larger pipes (or macro-sized closed channels), flexible sheets such as the 3 M vinyl adhesive backed riblets can be applied to the interiors. Smaller pipes (or micro-sized closed channels) must be split in order to apply the flexible 3 M riblet sheets. However, when riblets are not in the form of flexible sheets, they need to be fabricated via processes such as milling, soft lithography, or laser etching. Such fabrication generally requires flat and not curved surfaces. For applications using these fabrication techniques,

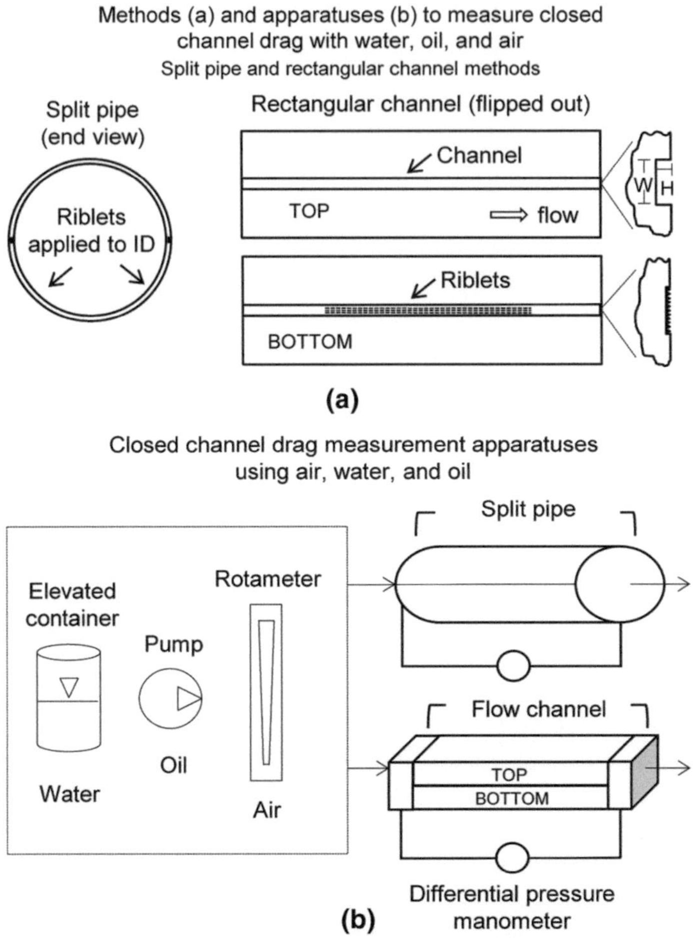

Fig. 11.7 Closed channel methods and apparatuses to measure drag using water, oil, and air. **a** Split designs are necessary in order to apply riblets inside micro-sized closed channels. Left schematic shows a round split pipe, while the right shows a rectangular channel sandwich design. The rectangular channel allows for riblets to be fabricated on a flat surface, which may be desirable (adapted from Bixler and Bhushan 2013b). **b** Drag is determined via pressure drop reduction using closed channels lined with riblet samples. Flow is regulated with the elevated container (water), pump (water or oil), and laboratory air connected to a rotameter (air). The pressure drop is measured with a manometer connected to both ends of the flow channel. Interface views are shown highlighting the top and bottom halves that are sandwiched together (Bixler and Bhushan 2013a)

the split rectangular channel design is more feasible, as it allows riblets to be applied inside the channel prior to final assembly.

During experiments, a fluid is introduced through a variety of different methods depending on its viscosity and desired flow rates, as shown in Fig. 11.7b (Bixler and Bhushan 2013a). For instance, experiments with water have been conducted with both elevated bottle apparatuses (Jung and Bhushan 2010; Bixler and Bhushan 2013b), as well as syringe and gear pumps (Jung and Bhushan 2010; Dean and Bhushan 2012; Bixler and Bhushan 2013b, c). Experiments have been conducted with air flow by connecting a rotometer to a laboratory air supply (Jung and Bhushan 2010; Bixler and Bhushan 2013b) in order to achieve particular flow velocities. In the case of water and oil experiments, average flow velocity can be calculated from the volumetric flow rate and channel cross-sectional area. A summary of experimental closed channel conditions is shown in Table 11.2 outlining the various riblet geometries, configurations, and materials that have been evaluated. Drag is determined via pressure drop reduction, which is measured with a manometer connected to both ends of the flow channel.

Pressure drop predictions for a flat milled channel surface were made for comparison to experimental data for riblet surfaces. These predictions were made using the methodology presented in Chap. 14 on the Rice Leaf and Butterfly Wing Effect.

11.3.5 Riblet Results and Discussion

In this section, we present open and closed channel drag measurement results using water, oil, and air flow (Bixler and Bhushan 2013a). Results are discussed and compared between the various measurement techniques, riblet geometries, riblet configurations, and flow regimes. This comparison helps to provide an understanding of the data on riblet drag reduction.

Open and closed channel results can be presented similarly, including drag compared to riblet geometry and dimensions. Comparing and studying such parameters allows for the development of lessons learned in order to optimize riblets. Micro-sized closed channel results are presented with pressure drop reduction at various Reynolds numbers based on the channel dimensions. This differs from open channel results in order to account for hydraulic diameter, due to the interaction of vortices from neighboring walls (Bixler and Bhushan 2013a). Therefore both open and closed channel results are presented in forms that best represent their findings.

11.3.5.1 Open Channel

A summary of experimental results are presented in Table 11.1 for flat plate with open channel flow using water, oil, and air, and Table 11.3 for wind tunnel airfoil flow. Several studies have included varying geometries and configurations. Drag reduction is presented as negative $\Delta\tau/\tau_o$ (%) and ΔC_d (%) values at various Reynolds numbers. The data indicate that properly designed riblets can reduce drag.

Flat Plate Experiments

Water channel experiments conducted at a range of s^+ values indicate maximum drag reduction of nearly 6% using continuous sawtooth riblets on a flat plate, as shown in Fig. 11.8 (Walsh 1990a; Bixler and Bhushan 2013a). Results also show that the greatest drag reduction occurs at an s^+ of 10–15 (Walsh 1990b), which correlates well with oil channel findings presented next (Bechert et al. 1997b). Extensive research has been conducted with oil channel flow, since oil allows for larger scale riblets, which drastically eases the challenge of riblet fabrication. These experiments included blade, sawtooth, and scalloped riblet geometries with continuous and staggered-segmented configuration, all using white paraffin oil ($\nu = 1.2 \times 10^5$ m^2/s). This research is highlighted in Fig. 11.9a, showing a wide range of *h/s* values that are important for determining the optimal drag reducing geometry (Bixler and Bhushan 2013a). A summary of oil channel experiments is provided in Fig. 11.9b (Bechert et al. 1997b). Figure 11.9c shows the effect of continuous and staggered-segmented configurations of blade riblets (Bechert et al. 1997b, 2000a).

Results indicate maximum drag reduction of nearly 9% is possible with blade riblets where $h/s = 0.5$, $t/s = 0.04$, and $s^+ \approx 15$. This drag reduction further improves with a lower *t/s* value of 0.02, where maximum drag reduction is nearly 10% (Bechert et al. 1997b). An optimum s^+ for various values of *h/s* suggests that a narrow range of h^+ is desirable. Furthermore, sawtooth riblets with $h/s = 0.98$ and $\alpha = 54°$ provide the greatest drag reduction of nearly 6%, indicating that the smaller α is beneficial. Also, scalloped riblets with $h/s = 0.7$ provide the greatest drag reduction of nearly 7%. Results also

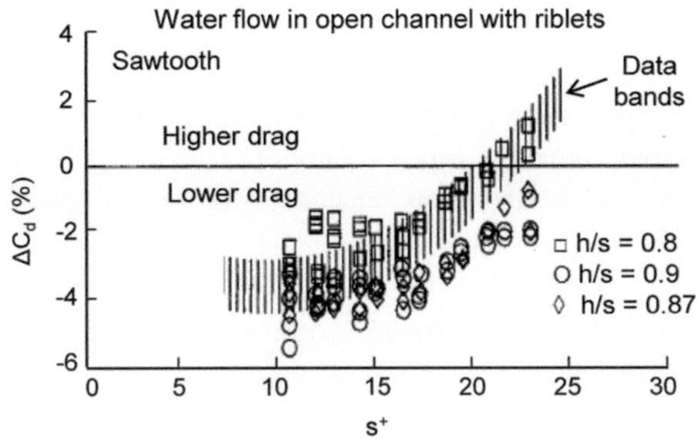

Fig. 11.8 Open channel experimental results with riblets in water flow (adapted from Walsh 1990a). Drag reduction is presented as negative ΔC_d (%) values, which is the percentage drag coefficient reduction between the smooth and riblet samples. Riblet samples with various *h/s* values are compared, with the maximum drag reduction near 6%. Riblets have sawtooth geometry and continuous configuration (Bixler and Bhushan 2013a)

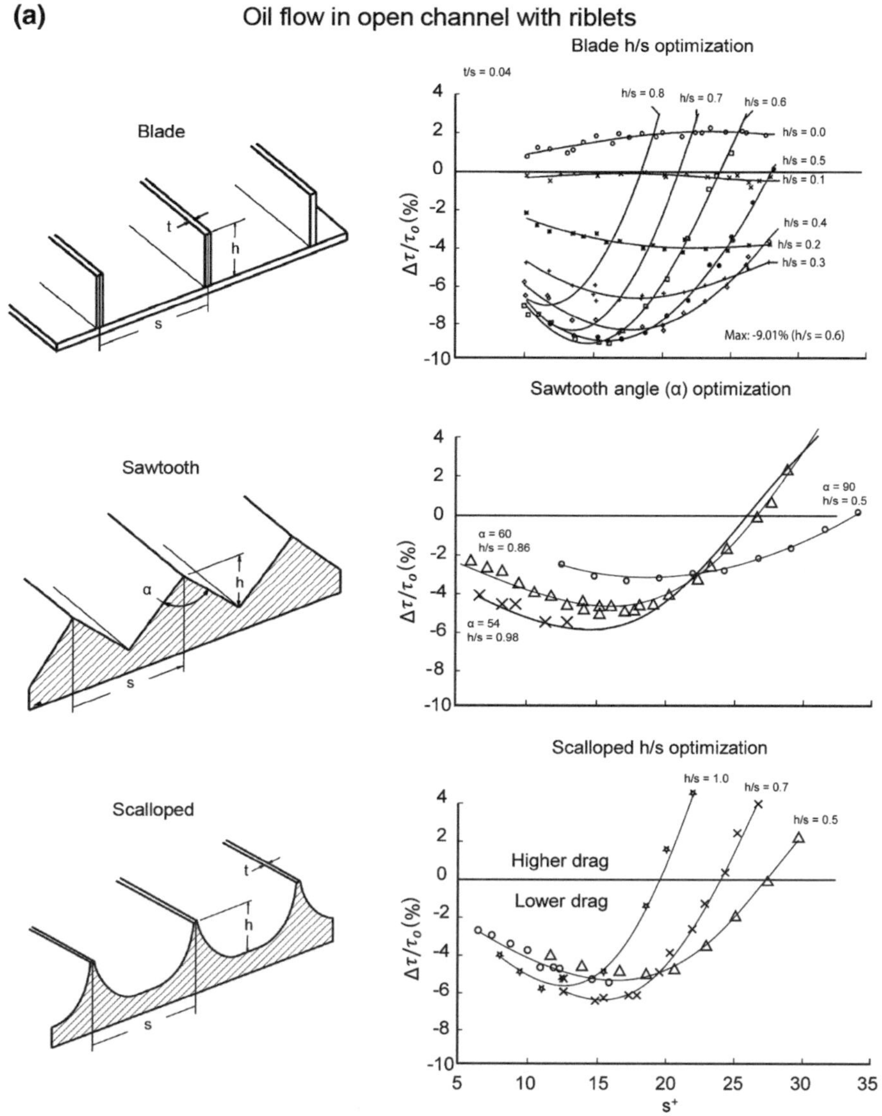

Fig. 11.9 Open channel experimental results with riblets in oil flow (adapted from Bechert et al. 1997b, 2000a). Drag reduction is presented as negative $\Delta\tau/\tau_o$ (%) values, where $\Delta\tau$ is the difference between shear stresses on the riblet sample (τ) and the smooth (τ_o) surfaces. **a** Various experiments with blade, sawtooth, and scalloped riblets were conducted in order to discover the optimal geometry for maximum drag reduction. Varied parameters include the riblet geometry, *h/s*, *t/s*, and s^+, which allows for a better comparison between experiments. Oil channel experiments were conducted using white paraffin oil. **b** Maximum drag reduction results from the various blade, sawtooth, and scalloped experiments are compiled. Results indicate that the blade geometry with $h/s \approx 0.5$ and $t/s \approx 0.02$ produces the maximum drag reduction of 10% near $s^+ \approx 15$. **c** Experimental results comparing continuous and staggered segmented blade riblets, indicating that latter configuration shows less drag reduction

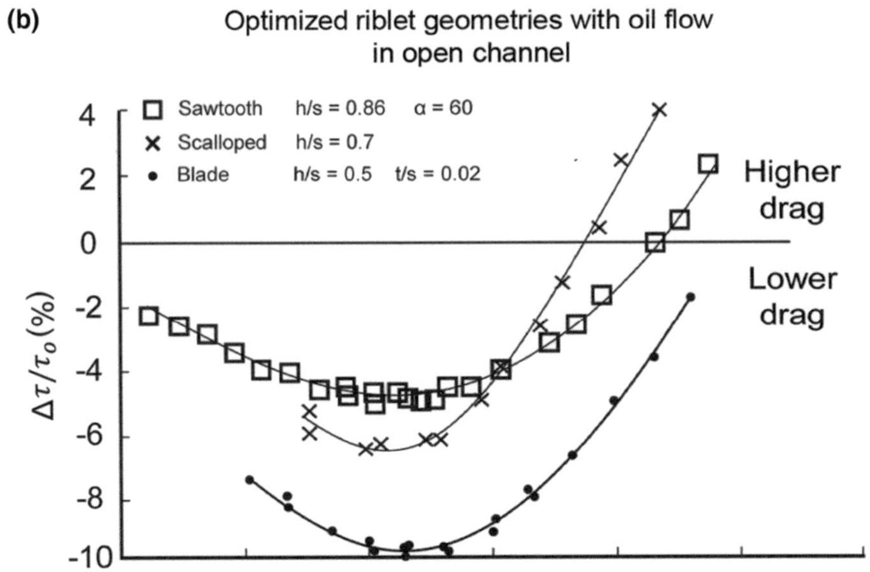

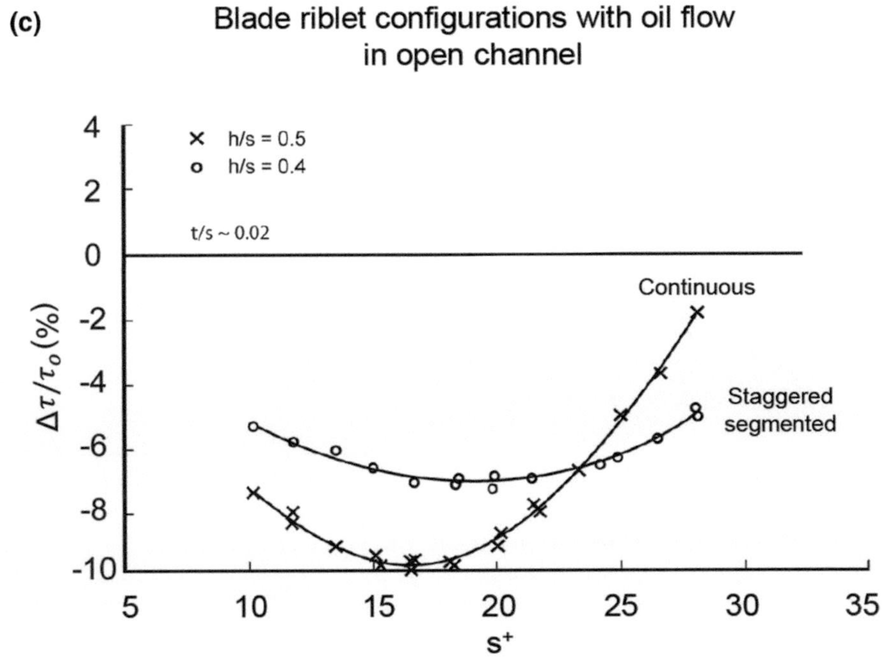

Fig. 11.9 (continued)

indicate that staggered-segmented trapezoidal shaped riblets provided less drag reduction compared to continuous blade riblets (about 6% vs. 10%) (Bechert et al. 1997b, 2000a). However, the staggered-segmented configuration provides drag reduction (albeit less magnitude) for a larger s^+ range, as compared to the continuous riblets.

The data presented above suggests that riblets with lower *h/s* values show less drag reduction but over a wider range of s^+ values (e.g. blade riblets where $h/s = 0.2–0.3$ versus $h/s = 0.5$). This is presumably due to the vortices interacting with larger area on the riblets and with each other. Previously presented flow visualization indicates that for higher flow velocity (where s^+ and h^+ increase and $h/s = 0.5$), the vortices appear not to be lifted, and shear stresses at the riblet tips are believed to increase. With constant *h* and increasing *s* values (thus lowering *h/s*), it is believed that the vortices fit in the riblet valleys (for instance at $h/s = 0.2–0.3$ versus $h/s = 0.5$). Furthermore, increasing *h/s* increases wetted surface area between the riblets, which leads to increased friction and thus drag (Bixler and Bhushan 2013a).

Choosing an optimal *h/s* based on the aforementioned experimental results depends on particular applications; specifically, whether or not the flow velocities are known or constant. For instance, riblets may be optimized for drag reduction over a wider range of s^+ values where flow velocity conditions are varying. Conversely, riblets may be optimized for maximum drag reduction over a smaller s^+ range, where flow velocity is relatively constant, such as aircraft at cruising speed. In addition to *h/s*, the *t/s* ratio affects drag, and a smaller ratio is desirable. Furthermore, research suggests that drag reduction is possible with sawtooth and scalloped riblets, but such geometries require differing *h/s* values. It should be noted that the blade riblets are the most fragile, followed by scalloped and sawtooth. Therefore, clear tradeoffs are necessary when designing riblets for commercial applications (Bixler and Bhushan 2013a).

Finally, the effect of yaw angle on drag reduction on sawtooth riblets in air flow was studied by Walsh and Lindeman (1984). Yaw angle is the angle between the average flow direction and the riblet orientation. It was reported that yaw angle has a deleterious effect on the drag reduction benefit. While a small drag reduction can be seen up to a yaw angle of 15°, drag increases above a yaw angle of 30°.

Airfoil Experiments

The 3 M vinyl riblet sheets have been fabricated in large sizes that consist of continuous sawtooth riblets with various aspect ratios (to be introduced later). Many experiments have been conducted with airfoils covered with these riblet sheets in wind tunnels. Experimental results are shown in Fig. 11.10a for riblets applied to the wind turbine DU 96-W-180 airfoil (Sareen et al. 2011), as well as the aircraft wing NACA 0012 airfoil (Bixler and Bhushan 2013a). Results indicate that smaller height riblets provide the greatest drag reduction, around 2% for the DU 96-W-180 airfoil (Sareen et al. 2011) and 7% for the NACA 0012 airfoil (Bixler and Bhushan 2013a). Larger sized riblets and higher Reynolds numbers provide less drag reduction, and, in some cases, the drag increases compared to

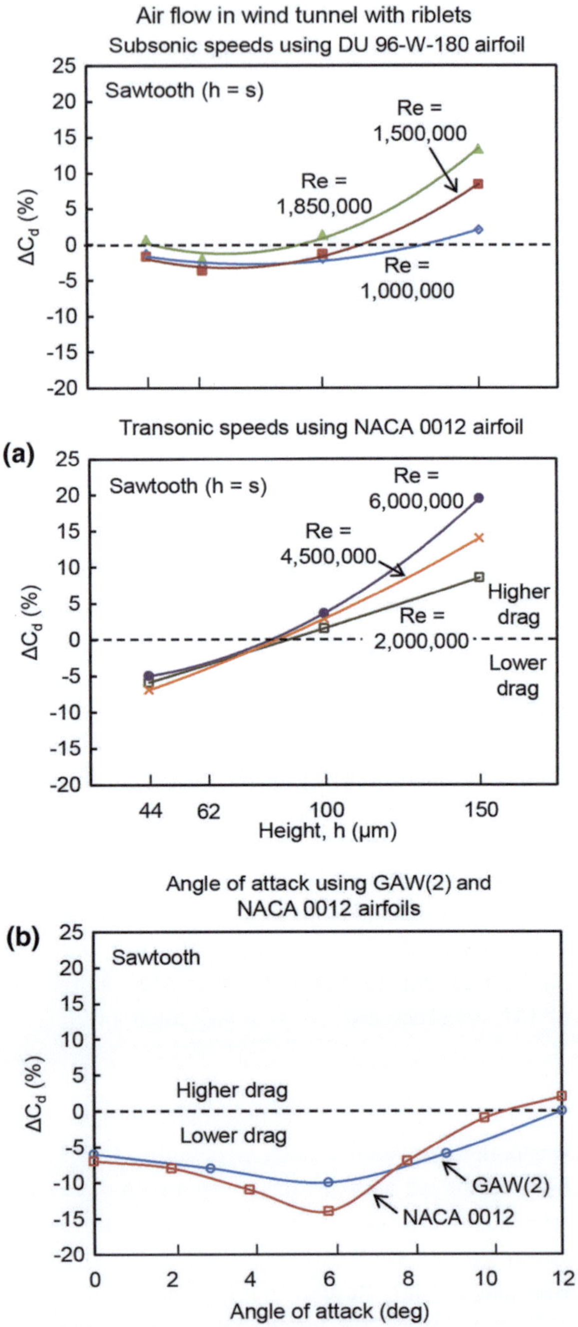

Fig. 11.10 Wind tunnel experimental results with riblets on airfoils. Drag reduction is presented as negative ΔC_d (%) values, which is the percentage drag coefficient reduction between the smooth and riblet samples. **a** Drag reduction is shown with wind turbine DU 96-W-180 (adapted from Sareen et al. 2011) and symmetrical NACA 0012 (Bixler and Bhushan 2013a) airfoils with various 3 M Corp. sawtooth riblets. The top and bottom figures show Reynolds numbers in the subsonic and transonic ranges, respectively. Maximum drag reduction is near 7% for the riblets with $h = s = 44$ μm in transonic flow. **b** Riblet drag reduction is related to angle of attack, as shown with two airfoils with angles of attack up to 12°. The greatest drag reduction of nearly 14% is shown at 6° angle of attack (adapted from Subaschandar et al. 1999)

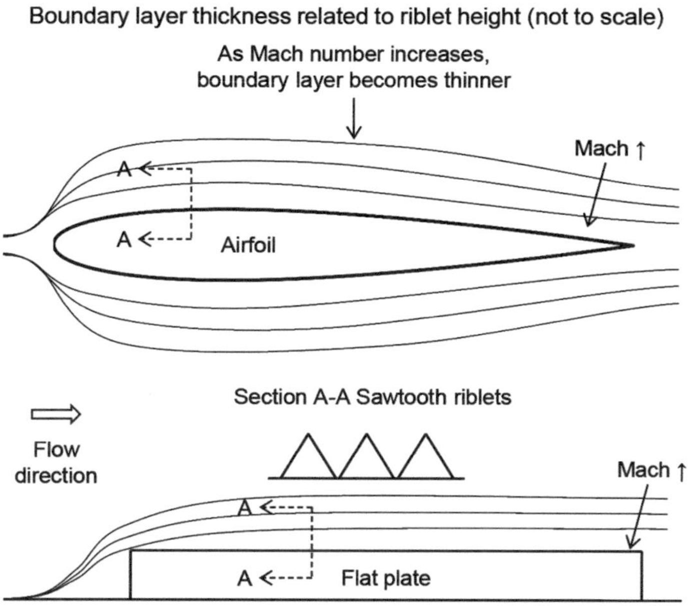

Fig. 11.11 Boundary layer thickness related to riblet height. As the Mach number increases, the boundary layer thickness decreases. It is believed for drag reduction that riblets should remain inside the viscous sublayer in order to effectively control the turbulent vortices. Shown is an illustration (not to scale) explaining the relationship (Bixler and Bhushan 2013a)

the smooth airfoil. The Reynolds numbers for experiments provide a wide spectrum of subsonic (Ma < 0.75 with Re = 1,000,000) to transonic (0.75 < Ma < 1.2 with Re = 6,000,000) velocities. This allows for comparison airfoils, Reynolds numbers, and riblet sizes. An additional variable of interest is the angle of attack, shown in Fig. 11.10b using sawtooth riblets. The greatest drag reduction of nearly 14% occurs at a 6° angle of attack for the NACA 0012 airfoil (Subaschandar et al. 1999).

Boundary layer thickness can be used to describe flow over an airfoil and flat plate, and its effect on the magnitude of drag reduction. As shown in Fig. 11.11, with increasing Mach number (and thus Reynolds number), the turbulent boundary layer thickness decreases. Riblets are effective by remaining inside the turbulent viscous sublayer in order to properly lift and presumably pin any vortices. However, if the riblet protrudes outside of the viscous sublayer, the increased surface area of the riblets will increase skin friction drag. Therefore, it is believed that riblets must remain within the viscous sublayer in order to provide benefit (Bixler and Bhushan 2013a).

11.3.5.2 Closed Channel

A summary of closed channel experimental results using water, oil, and air are presented in Table 11.2. Several studies have included varying geometries and configurations that have been conducted with laminar through turbulent flow. Drag reduction is presented as negative $\Delta\tau/\tau_o$ (%) or pressure drop values. Results indicate that properly designed riblets can reduce drag.

Comparison of Open and Macro-sized Closed Channels

Figure 11.12 shows a comparison of open and macro-sized closed channel results. The data show that drag reduction is similar. These results represent cases when neighboring walls are sufficiently far apart, as shown with pipe diameters approximately 25 mm or greater. However, when channel dimensions are relatively small (as shown in Fig. 11.7a with micro-sized closed channel where $H = 0.7$ mm and $W = 3.2$ mm), neighboring wall effects are believed to play an important role in riblet drag reduction. A set of experiments were conducted with such micro-sized closed channels, and results are presented in the following section.

Molded Shark Skin Replica and Machined Micro-sized Channel

Representative pressure drop data for the flat epoxy resin replica of scales from the skin of the spiny dogfish (*Squalus acanthias, L. Squelidae*), along with predicted data for a hydrophilic surface, are shown in Fig. 11.13a (Jung and Bhushan 2010). A decrease in pressure drop of about 30% in turbulent flow versus a smooth surface is observed.

Role of Micro-sized Channel Dimensions

Closed channel experiments with various riblet geometries in water, oil, and air have been conducted (Jung and Bhushan 2010; Dean and Bhushan 2012; Bixler and Bhushan 2012b; 2013a, b, c, d). In addition to data presented here, further data are presented on shark skin replicas in Chap. 14 on the Rice Leaf and Butterfly Wing Effect. Jung and Bhushan (2010) and Dean and Bhushan (2012) fabricated segmented blade-type riblets on acrylic, that were machined using an end mill with a 130 μm diameter bit. Representative pressure drop data for a flat acrylic sheet and machined riblets are shown in Fig. 11.13b, along with predicted data for a hydrophilic surface. A decrease of about 23% is in turbulent flow versus a smooth surface is observed. The static contact angle and pressure drop data for shark skin replica and flat acrylic sheet and machined riblets are summarized in Table 11.5 (to be presented later).

Laser etching has been used to fabricate continuous and segmented blade-type riblets with micro-sized dimensions on cast acrylic sheets (Bixler and Bhushan 2013b). The baseline size for riblets was derived from the spiny dogfish shark *Squalus acantias*, whose riblet dimensions were provided earlier. These values served as the basis for laser-etched blade riblet samples with an expanded range of dimensions. The top image of Fig. 11.14 shows the segmented blade configuration. Sample riblet dimensions with various thicknesses (t), valley widths (vw), spacing (s), gaps (g), lengths (L), and heights (h) were selected. Blade riblet studies suggest that h/s of about 0.5, small t/s, and dimensionless riblet spacing near 15 optimizes drag reduction (Bechert et al. 1997a; Dean and Bhushan

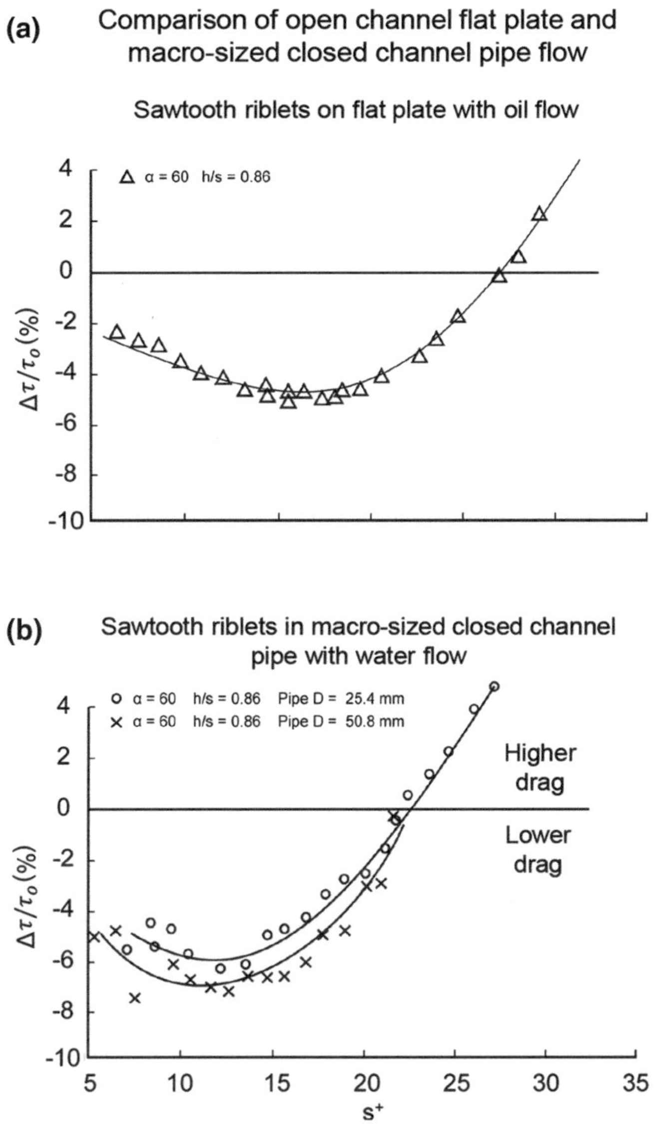

Fig. 11.12 Comparison of open channel flat plate and macro-sized closed channel pipe flow (where neighboring wall effects are assumed negligible). Drag reduction is presented as negative $\Delta\tau/\tau_o$ (%) values, where $\Delta\tau$ is the difference between shear stresses on the riblet sample (τ) and the smooth (τ_o) surfaces. Samples consist of sawtooth geometry where the angle $\alpha = 60°$, as defined in Fig. 11.9a. **a** Experimental results are shown with sawtooth riblets on a flat plate (Bechert et al. 1997b) and **b** in round pipe with diameter D (Liu et al. 1990). Maximum drag reduction is similar between the presented flat plate and macro-sized pipe flow

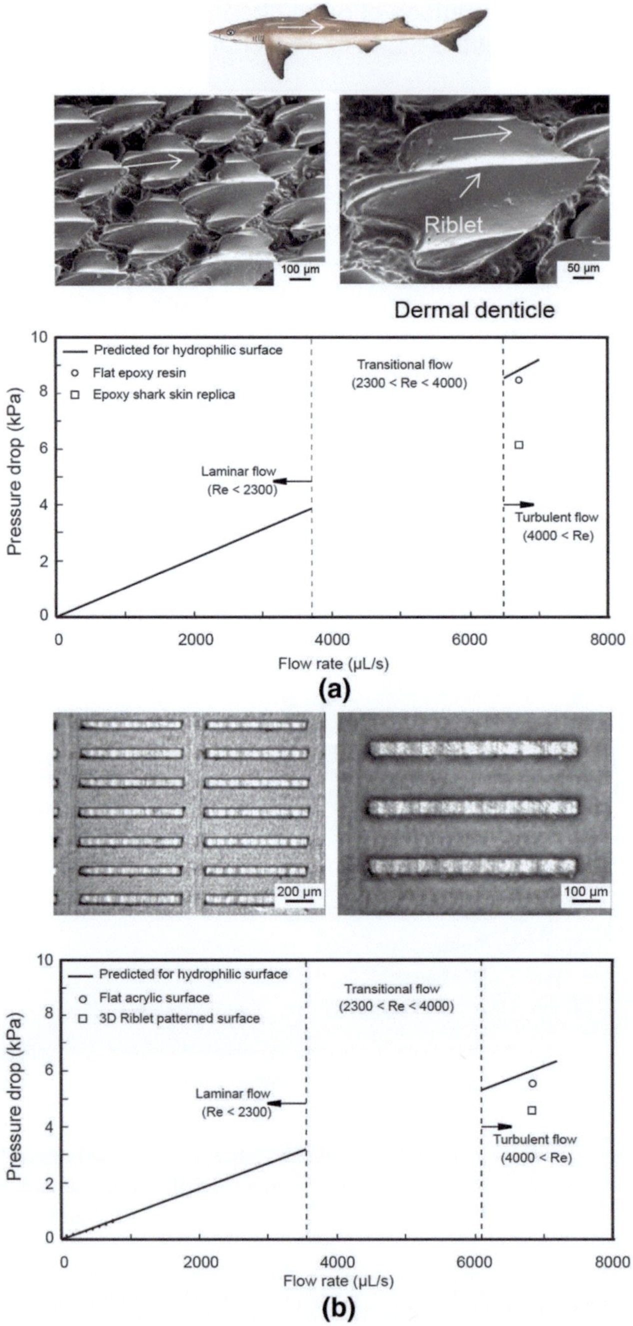
Riblet
100 μm
50 μm
Dermal denticle
Predicted for hydrophilic surface
Flat epoxy resin
Epoxy shark skin replica
Transitional flow
(2300 < Re < 4000)
Laminar flow
(Re < 2300)
Turbulent flow
(4000 < Re)
Pressure drop (kPa)
Flow rate (μL/s)
(a)
200 μm
100 μm
Predicted for hydrophilic surface
Flat acrylic surface
3D Riblet patterned surface
Transitional flow
(2300 < Re < 4000)
Laminar flow
(Re < 2300)
Turbulent flow
(4000 < Re)
Pressure drop (kPa)
Flow rate (μL/s)
(b)

◂**Fig. 11.13** **a** SEM micrographs at two magnifications of replica of spiny dogfish shark skin, and comparison of pressure drop of water flow in a closed rectangular flow channel over flat epoxy surface and shark skin replica surface. **b** Optical micrographs at two magnifications showing segmented blade-type riblets machined from acrylic, and comparison of pressure drop of water flow in a closed rectangular flow channel over flat acrylic surface and segmented blade type riblets. Data in both cases are compared with the predicted pressure drop for a hydrophilic surface (solid lines) (adapted from Jung and Bhushan 2010)

2010). Fabrication limitations and material properties determined the minimum riblet dimensions for the segmented and continuous samples. Blade dimensions were incrementally varied to change the *h/s* and *t/s* values to fully understand their relationship in a closed channel.

A wide range of blade *h/s* and *t/s* values ($0.16 < h/s < 0.62$ and $0.18 < t/s < 0.38$) were studied with *g* and *L* held constant, and *h*, *vw*, and *t* varied individually; see Table 11.4 for dimensions. Samples will be referred to by their "description," in conjunction with their *h/s* and *t/s* values. For example, the segmented blade riblet sample with $h/s = 0.31$ and $t/s = 0.31$ is called "Baseline (0.31, 0.31)" or more simply, "baseline" (Bixler and Bhushan 2013b).

Continuous sawtooth riblets (with $h = t$) were obtained from 3 M Corp. (Minneapolis, Minnesota) and inserted into the flow channel; see Table 11.4 for dimensions

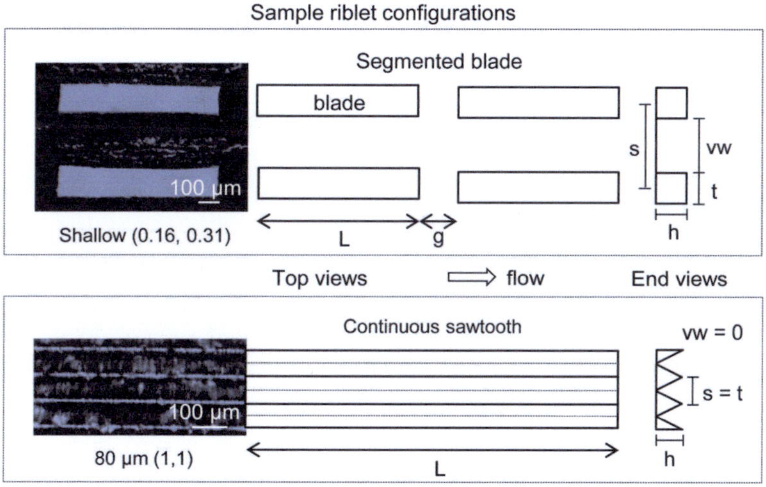

Fig. 11.14 Blade and sawtooth riblet geometries are shown on the top and bottom, respectively. Flow direction in top views is left to right and end view is into the page. Blade riblets are both segmented (top) and continuous (not shown); whereas sawtooth are all continuous (bottom). The riblet height (*h*), thickness (*t*), valley width (*vw*), spacing (*s*), and length (*L*) descript the geometry. Optical images show actual samples (Bixler and Bhushan 2013b)

Table 11.4 Summary of riblet sample dimensions (Bixler and Bhushan 2013b)

Sample set #	Material and fabrication processes	Geometry	Description[a]	Dimensions (μm)	h/s	t/s
1	Laser etched acrylic	Segmentedblade $L = 850$ μm $g = 200$ μm	Effect of dimensions			
			Baseline	$h = 127$ $vw = 280$ $t = 127$	0.31	0.31
			Shallow	½h, vw, t	0.16	0.31
			Deep	2h, vw, t	0.62	0.31
			Narrow	h, ¾vw, t	0.38	0.38
			Thin	h, vw, ½t	0.37	0.18
			Single versus double sided			
			Shallow	½h, vw, t	0.16	0.31
			Effect of coatings			
			Coated	½h, vw, t	0.16	0.31
			Valleys coated	½h, vw, t	0.16	0.31
2		Continuousblade $L = 101$ mm	Baseline	h, vw, t	0.31	0.31
3	Embossed polymer sheets supplied by 3 M Corp.	Continuoussawtooth $L = 101$ mm $vw = 0$	80 μm	$h = s = t =$ 80	1	1
			100 μm	$h = s = t =$ 100	1	1
			150 μm	$h = s = t =$ 150	1	1

[a]Samples are referred to as "Description (h/s, t/s)" in the chapter

(Bixler and Bhushan 2013b). The bottom image of Fig. 11.14 shows the sawtooth pattern configuration.

Fabricated milled sample surfaces were hydrophilic and superoleophilic (Bixler and Bhushan 2013c). A number of studies suggest that superhydrophobic surfaces exhibit lower drag (Ou et al. 2004; Daniello et al. 2009; Martell et al. 2009, 2010; Jung and Bhushan 2010; Bhushan and Jung 2011). In order to make the surface superhydrophobic, a nanostructured coating was selected. The replica was dip-coated with a solution consisting of 50 nm (±15 nm) hydrophobized silica nanoparticles, combined with methylphenyl silicone resin dissolved in tetrahydrofuran and isopropyl alcohol. Coated samples included completely coating the riblets and some included coating only the riblet valleys (Bixler and Bhushan 2013c).

Table 11.5 Summary of the static contact angle, pressure drop in turbulent flow and calculated slip length data for various surfaces. The variation represents ±1 standard deviation (adapted from Jung and Bhushan 2010)

	Static contact angle (deg)	Pressure drop in turbulent flow (%)	Calculated slip length (μm)
(a) *Epoxy resin*			
Hydrophilic (flat epoxy resin)	76 ± 0.9	~0	~0
Hydrophobic (flat with thin lotus wax layer)	119 ± 2.4	~21	~24
Superhydrophobic (hierarchical structure—micropatterned with thin lotus wax layer)	173 ± 0.8	~33	~103
Shark skin replica	89 ± 1.7	~30	~35
(b) *Acrylic resin*			
Flat acrylic resin	82 ± 1.8	~0	–
Riblet patterned surface	146 ± 1.2	~23	–

Closed channel experiments were conducted using the experimental apparatus shown in Fig. 11.7, with varying geometries, configurations, hydrophobicity, oleophilicity, and channel dimensions with laminar through turbulent flow (Bixler and Bhushan 2013b, c). The results presented use the naming convention for samples "sample description (*h/s*, *t/s*)". For example, the 3 M riblet sample with height and spacing equaling 44 μm is written as "44 μm (1, 1)" or more simply "44 μm sample". The baseline blade dimensions were $h = 127$ μm, $vw = 280$ μm, and $t = 127$ μm, with variations presented later in the modeling section. All samples exhibited blade geometry with the aligned-segmented configuration, except the 3 M riblets which exhibited sawtooth geometry and continuous configuration. Also, select riblet samples received a nanostructured coating (e.g. "coated" samples). Such coated samples were either superhydrophobic or superoleophilic, and were believed to improve riblet smoothness. Furthermore, the same nanostructured coating applied to replica shark skin samples indicated that the coating lowered adhesion forces (Bixler and Bhushan 2012b).

Micro-sized closed channel water and air results are presented in Fig. 11.15 and oil results (white paraffin oil with CAS 8012-95-1) in Fig. 11.16. The data are compared to the flat milled sample. Various riblet geometries and dimensions were evaluated in flows ranging from laminar to turbulent regimes. The effect of *h/s* and *t/s*, *h*, hydrophobicity/oleophilicity, and smoothness are shown. The Shallow sample (1/2 height) shows the most pressure drop reduction at 19% in water and 13% in air (Bixler and Bhushan 2013b). Due to their initial performance, these riblets received the nanostructured coating, which then

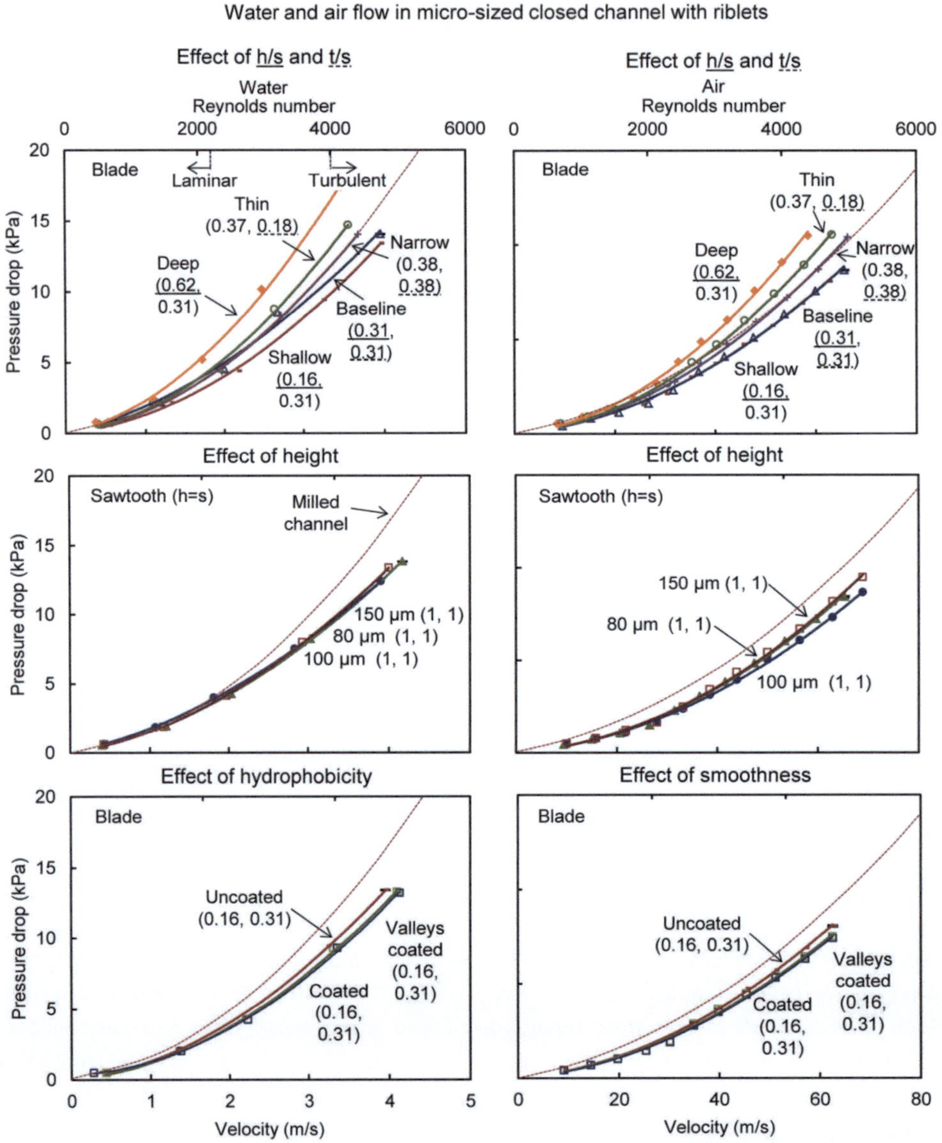

Fig. 11.15 Micro-sized closed channel experimental results with riblets in water and air flow (adapted from Bixler and Bhushan 2013b). Drag reduction is presented as pressure drops between the inlet and outlet of the channel, where lower pressure drop is desirable. Riblets of varying geometries were evaluated in laminar (Re < 2300) through turbulent flow (Re > 4000) conditions. Experiments evaluated the effect of *h/s*, *t/s*, height, hydrophobicity, and smoothness. Riblets were fabricated via laser etching, and a nanostructured coating was applied to achieve the desired hydrophobicity and smoothness. Maximum pressure drop reduction of 34% with water was recorded for hydrophobic riblets in the turbulent regime. Flat milled channel control sample is shown for comparison. Error bars show ±1 standard deviation

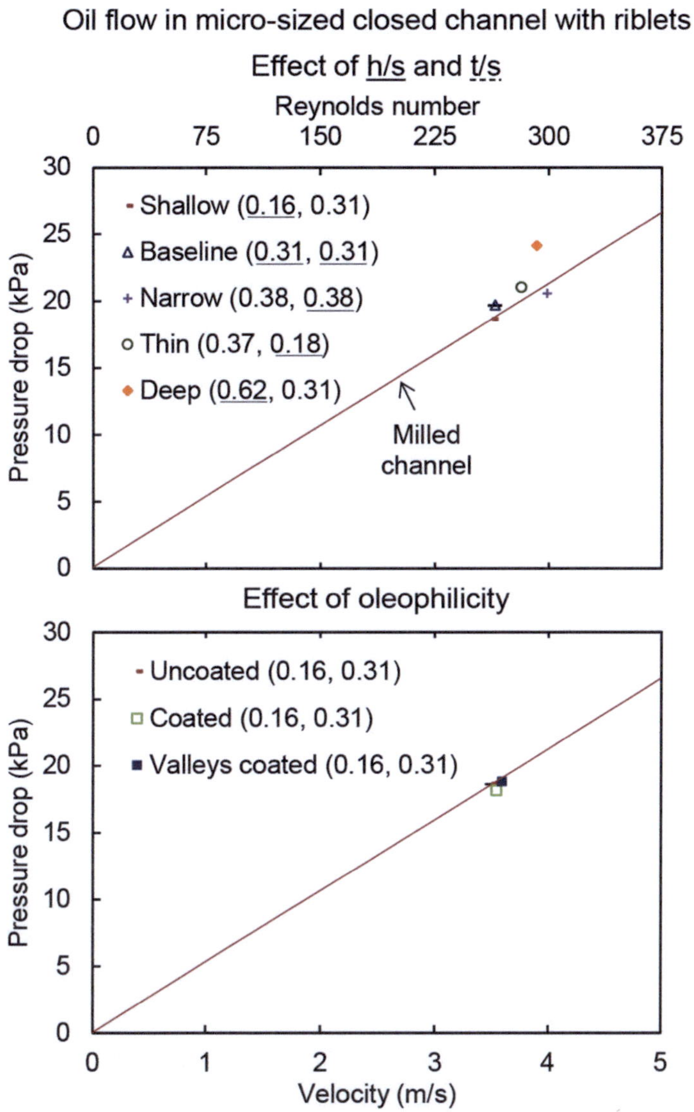

Fig. 11.16 Micro-sized closed channel experimental results with riblets in oil flow (adapted from Bixler and Bhushan 2013c). Drag reduction is presented as pressure drops between the inlet and outlet of the channel, where lower pressure drop is desirable. Riblets of varying geometries were evaluated in laminar ($Re < 375$) flow conditions. Experiments evaluated the effect of *h/s*, *t/s*, and oleophilicity. Riblets were fabricated via laser etching, and a nanostructured coating was applied to achieve the desired oleophilicity. Maximum pressure drop reduction of 9% was recorded for the Narrow riblets. Experiments were conducted using white paraffin oil. Flat milled channel control sample is shown for comparison. Error bars show ±1 standard deviation

produced an improved pressure drop reduction of 34% in water and 24% in air (Bixler and Bhushan 2013b). The 100 μm sample also reduced pressure drop by 30% in water. In oil flow, the Narrow sample (1/4 valley width) shows the most pressure drop reduction at 9% (Bixler and Bhushan 2013c).

From the data, we note that lower *h/s* values appear to provide the greatest drag reduction, although this differs from the open channel results. However, with rectangular micro-sized closed channel experiments, the neighboring wall effects are believed to play a role in drag reduction. In the closed channel experiments, samples with the lowest *h/s* of 0.16 performed the best, presumably due to less interaction between the vortices on the top and bottom or the riblet height remaining in the viscous sublayer (Bixler and Bhushan 2013a).

Understanding the differences in performance between riblet samples in water, oil, and air flow can be explained by considering the Reynolds numbers, as well as the drag reducing mechanisms. It is believed that different mechanisms are at work, mainly due to the effect of viscosity, as well as Reynolds number. For instance, where water and air flow was turbulent, the riblets are believed to lift and pin the vortices, and thus reduce drag. However, oil flow was laminar, benefitting from the so-called thin film effect that reduces skin friction at the solid-liquid interface and thereby reduces drag, to be described later in this section (Bixler and Bhushan 2013c). Additionally, it is believed that turbulent oil flow would benefit from the shark skin effect.

(a) **Continuous versus segmented riblets**

To understand the effects of riblet configurations, namely continuous and aligned-segmented configurations, a series of water, oil, and air flow experiments were conducted (Bixler and Bhushan 2013b, c). Figure 11.17 presents results from the Continuous and Baseline blade samples, representing the continuous and aligned-segmented configurations respectively. Intuitively, the Continuous sample should provide higher drag due to the increased wetted surface area. Experiments with the Continuous sample in water and oil exhibit a higher drag as compared to the Segmented sample. However, there is a decrease in drag with air, for similar Reynolds numbers in water. The continuous blades may reduce cross stream movement with air more efficiently than with segmented blades. With the lower viscosity of air, vortices rotate more easily in the channel, and the gap area between segmented blades perhaps disrupts and unpins the vortices.

(b) **Riblets on top or bottom or both sides**

To understand the effects of riblets on one side versus two sides of the closed channel and with differing micro-sized channel heights, a series of water flow experiments were conducted (Bixler and Bhushan 2013b). Such combinations included experiments with 1× channel height, 2× channel height, riblets on bottom, riblets on top, and riblets on

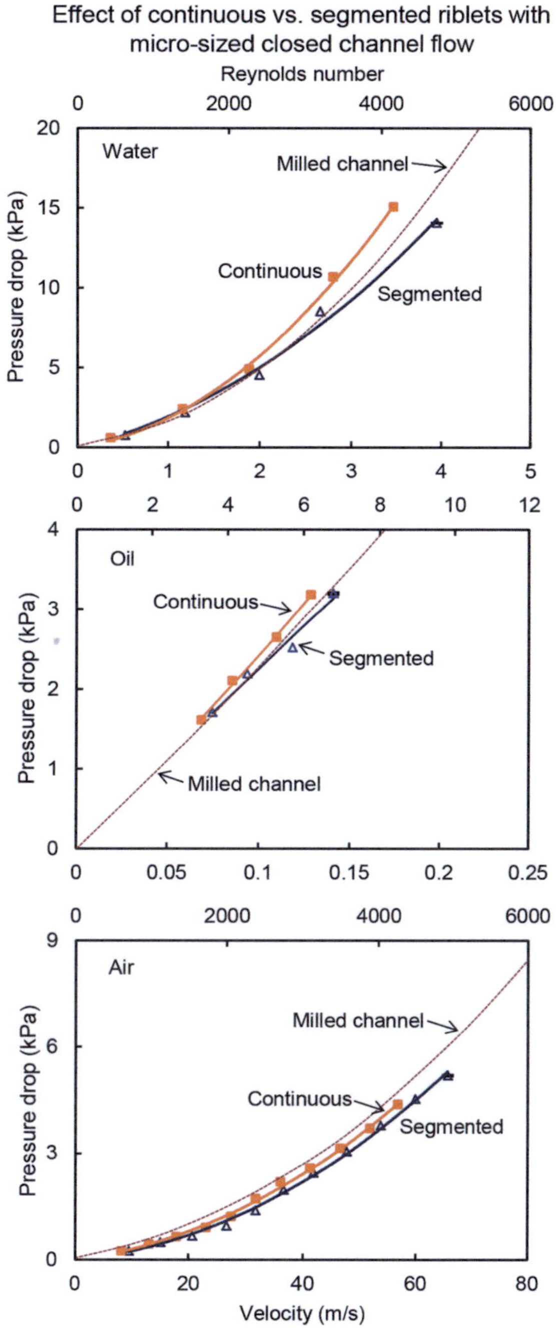

Fig. 11.17 Micro-sized closed channel experimental results showing the effect of continuous versus segmented blade riblets with water, oil, and air flow (adapted from Bixler and Bhushan 2013b, c). Drag reduction is presented as pressure drops between the inlet and outlet of the channel, where lower pressure drop is desirable. Riblets were evaluated in laminar ($Re <$ 2300) through turbulent flow ($Re>4000$) conditions. Results indicate that the continuous riblets increase drag with water and oil, presumably due to the increased wetted surface area, which leads to higher skin friction. Flat milled channel control sample is shown for comparison. Error bars show ± 1 standard deviation

top and bottom; all with Shallow samples. Figure 11.18 (top row) shows results from riblets on the top only and bottom only. Results indicate that riblets located on the bottom perform differently from the same riblets located on the top of the channel. The pressure drop increases by 20% when the riblets are transferred from the bottom to the top (Bixler and Bhushan 2013b). Similar experiments were conducted to study the effect of channel height with riblets on the top or bottom. Results indicated that pressure drop with riblets on the bottom was unaffected by channel height, but pressure drop improved with topside riblets and 2× channel height.

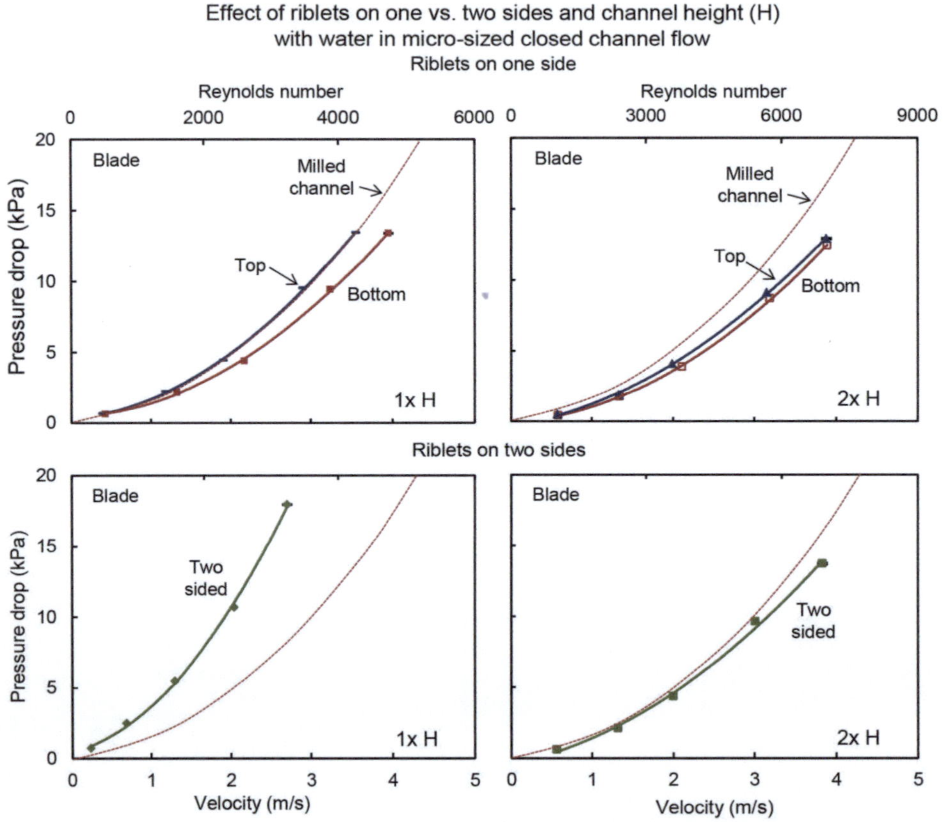

Fig. 11.18 Micro-sized closed channel experimental results showing the effect of channel height and riblets on the top, bottom, or both sides with water flow (adapted from Bixler and Bhushan 2013b). Drag reduction is presented as pressure drops between the inlet and outlet of the channel, where lower pressure drop is desirable. Riblets were evaluated in laminar ($Re < 2300$) through turbulent flow ($Re > 4000$) conditions. Results indicate that riblets on the bottom reduce drag more than when positioned on the top of the channel. Also, 2× channel height reduces drag, presumably due to vortices from the top and bottom moving apart (i.e. neighboring wall effects are negligible). Flat milled channel control sample is shown for comparison. Error bars show ±1 standard deviation

Figure 11.18 (bottom row) shows results from riblets on both the top and bottom of the channel. Neighboring wall effects are believed to be present, since pressure drop increases with the 1× channel height as compared to the flat milled surface. In contrast, the pressure drop decreases with the 2× channel height. It is believed that the vortices from neighboring walls with the 1× channel height interact and increase drag, but the larger gap distance in the 2× channel height reduces this effect. The aforementioned effects are believed to be related to the vortices' interaction (Bixler and Bhushan 2013b).

(c) **The role of riblet wettability**

In the data presented in Figs. 11.15 and 11.16 with and without the nanostructured coating, it is observed that the nanostructured coating provided a drag reducing benefit in all fluids. In water flow, drag reduction improves with superhydrophobicity and low adhesion (Bixler and Bhushan 2012b, 2013b), in air flow with smoothness (Bixler and Bhushan 2013b), and in oil flow with superoleophilicity because of the so-called thin oil film effect (Bixler and Bhushan 2013c). In oil, drag reduction is possible with laminar flow due to this effect. The thin oil film effect can be explained using Fig. 11.19. This model indicates that during flow, oil becomes trapped and holds stationary between the Narrow riblets, and essentially lubricates the solid-liquid interface. The lubricating effect is believed to increase slip length (b), so the flow velocity increases at the channel wall. A similar effect has been observed with the pitcher plant (*Nepenthes genus*) peristome (Koch and Barthlott 2009; Bhushan 2009).

It has been shown that fluid slip has a direct relationship to drag. Jung and Bhushan (2010) measured pressure drop in a flow cell on hydrophilic (CA = 76°), hydrophobic (CA = 119°), and superhydrophobic (CA = 173°) surfaces (used for fabrication of artificial lotus structures) discussed in Chap. 6. Figure 11.20 shows laminar and turbulent pressure drop comparisons for these surfaces in water flow. There is no pressure drop benefit in laminar flow, whereas there is a benefit in turbulent flow. Static contact angle, calculated slip length, and pressure drop data in turbulent flow are summarized in Table 11.5. The greatest decrease in pressure drop of 33% happens with the superhydrophobic surface, followed by 21% with the hydrophobic surface. Reducing the strength of bonding between the fluid molecules and the surface causes a corresponding reduction in fluid drag in both laminar and turbulent regimes.

11.3.6 Summary

Sharks are able to move quickly through water and remain free of fouling due to microstructured riblets covering their skin. Such riblets efficiently reduce skin friction drag by lifting and presumably pinning naturally occurring turbulent vortices. This minimizes the effects of transverse shear stress present in turbulent flow. Lower drag translates

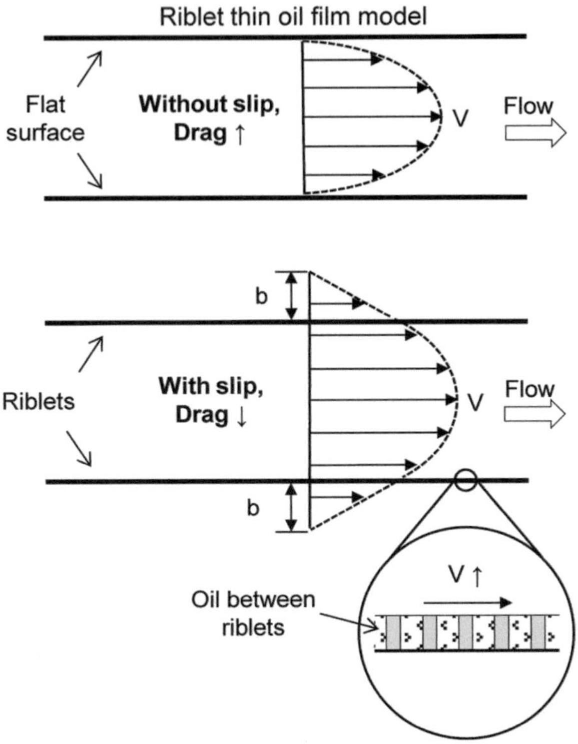

Fig. 11.19 Riblet thin-oil film model. Oil becomes trapped between appropriately sized riblets and essentially lubricates the surface. The lubricating effect is believed to increase slip length (*b*) during laminar oil flow, which translates into lower drag and increased flow rate (Bixler and Bhushan 2013c)

to higher fluid flow velocity at the skin surface, so microorganisms are unable to attach and the skin remains clean. This is further facilitated with specially spaced riblets that do not allow larger microorganisms to colonize. The attempt to understand and mimic such unique shark skin structures has been the focus of many scientific studies. Commercial applications for such low drag and antifouling surfaces are plentiful, ranging from medical devices to marine vessels (Bixler and Bhushan 2013a).

A number of experiments have been performed to understand drag reduction mechanisms in shark skin. These include both open and closed channel flow, using a variety of blade, sawtooth, scalloped, and bullnose riblet geometries in continuous and segmented configurations. These have been studied with water, oil, and air in laminar through turbulent regimes. In open channel flow experiments, drag was measured with flat plates on force balances placed into water, oil, and air flows. Other experiments included airfoils in wind tunnels with drag measured by using wake traverses. In closed channel flow, drag was measured via pressure drop reduction using riblet lined micro-sized closed channels.

Riblets showed an improved drag reduction in turbulent versus laminar flow, presumably due to the lifting and pinning of vortices. Results also indicated that drag reduction improves in most cases with increasing Reynolds number (until expected plateauing). In open channel experiments, the maximum drag reduction up of 10% has been reported

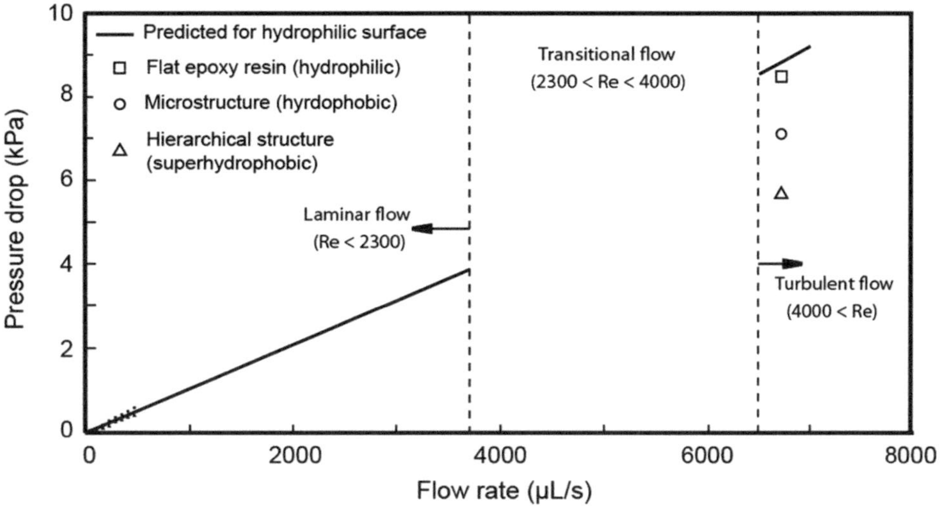

Fig. 11.20 Comparison of pressure drop of water flow in a closed rectangular channel with hydrophilic, hydrophobic, and superhydrophobic surfaces in laminar and turbulent flow regimes. Data are compared with the predicted pressure drop for a hydrophilic surface (solid lines). Decreased pressure drop in turbulent flow corresponds to decreased drag (adapted from Jung and Bhushan 2010)

(Bechert et al. 1997b). In closed channel experiments, maximum pressure drop reduction up to 34% has been reported (Bixler and Bhushan 2013b). Experiments using water, oil, and air also indicated that properly sized riblets combined with an appropriate nanostructured coating further enhances drag reduction. Mechanisms included superhydrophobicity for water, improved smoothness for air, and superoleophilicity for oil flow.

Based on open channel and closed channel data in various fluids, producing optimal and scalable drag reducing riblets for any fluid is believed to require the following design principles: a narrow range of h^+ and s^+, and low t, with $s^+ \approx 15$, $h/s \approx 0.5$, blade geometry, and continuous configuration. In open and closed channel flow, riblets may be scaled up according to the fluid viscosity and flow velocity, where higher viscosity and lower velocity allow for larger riblets (Bixler and Bhushan 2013a).

11.4 Fluid Flow Modeling of Riblets

Various riblet geometries and configurations have been studied through modeling in closed channel flow. By modeling of continuous and sawtooth geometries, Chu and Karniadakis (1993), Choi et al. (1993), and Goldstein et al. (1995) have shown drag reductions of up to 3–6%. A drag reduction is formed by riblets lifting the streamwise vortices formed in turbulent flow that generate transverse shear stresses. Streamwise vortices are referred to as vortices traveling in the predominant flow direction and rotating along an axis line in the streamwise direction. Continuous, blade riblets have been modeled by El-Samni et al. (2007) and Martin and Bhushan (2014, 2016a) both showing a maximum drag reduction of 11% and explained the mechanism as appropriately-sized riblets do not allow vortices to fit within the riblet valleys and bring higher velocity fluid to the wall to increase drag. Martin and Bhushan (2016a) also modeled aligned-segmented and staggered-segmented blade riblets and reported drag reductions of 5 and 3.3%, respectively. Martin and Bhushan (2016b) modeled blade, sawtooth, and scalloped riblet geometries where only the spacing parameter was varied within each case. In addition a new geometry inspired by segmented dermal denticles on a shark was modeled and compared with these riblet geometries, as well as to the experimental data. Based on this study, optimum riblet geometries and dimensions were determined.

In this section, we present fluid flow modeling data for various riblet geometries and configurations in a closed channel based on Martin and Bhushan (2014, 2016a, b). In order to study the effect of riblet geometry as well as the effect of dimensions of vortices on drag reduction, blade riblet structures in continuous, aligned-segmented, and staggered-segmented configurations in a closed channel flow were used. The gaps and offsets in the staggered configurations were independently changed, and the resulting drag and vortex formations were compared to each other as well as to a flat surface and continuous riblet configurations. Sawtooth and scalloped riblet structures with aligned-segment configuration were also studied.

11.4.1 Computational Fluid Dynamic (CFD) Model

To simulate flow over the riblet surfaces, Martin and Bhushan (2014, 2016a, b) used the computational fluid dynamic (CFD) software package ANSYS Fluent 16.0. This software uses the finite volume method to solve the governing equations, which include the momentum Navier-Stokes and pressure-based continuity equations. It contains various turbulence models employed in many studies. It can implement parallel processing, and includes solver improvements, such as solving the momentum and pressure-based continuity equations in a coupled manner (Anonymous 2016a).

In order to balance computational time and accuracy, the large-eddy simulation (LES) model was chosen as the turbulence model (Martin and Bhushan 2014). Because LES filters out small flow field scales and resolves the large field scales using a subgrid-scale (SGS), the computational expense is decreased. This filtering affects the results, but the small scales have less of an effect on the transport of mass and energy than the large scales. The large scales play a greater role on riblet drag. Thus, the mesh resolution and time step were chosen to resolve the large scales. In this research, the Wale SGS, designed to return the correct wall behavior for wall bounded flows, was chosen (Anonymous

2016a). The mesh size and time step were chosen to resolve the large scales with the Smagorinsky-Lilly SGS filtering out the small scales.

Figure 11.21a shows the closed channel computational domain of the riblet geometry where x, y, z axes denote the streamwise, wall-normal, and spanwise directions, respectively (Martin and Bhushan 2014, 2016a, b). The model height, width, and length are H, W, L, respectively. The riblet surface is the bottom wall and the flat surface is the top wall, both prescribed with no-slip boundary conditions. The inlet and outlet surfaces and the left and right surfaces were prescribed with periodic boundary conditions to eliminate wall effects from the results. A closed channel model was chosen to compare drag and vortex structures on flat and riblet surfaces. An open channel model is expected to have similar results, but will require modeling of the flat and riblet surfaces, separately.

In the streamwise direction, a constant instantaneous volume flux boundary condition was applied,

$$\mathcal{Q} = \int V \mathrm{d}A_C = \frac{2}{3} A_c V_l \tag{11.12}$$

where $A_c = HW$ is the cross-sectional area. V is the average velocity in the streamwise direction and V_l is the centerline velocity of a laminar profile with the same volume flux. In (11.12), V_l is 2/3 of V because of the difference between average and maximum velocity in closed channel flows (Munson et al. 2012). For turbulent flow, the Reynolds number was calculated using (11.1) where $D = H/2$, used with periodic boundary conditions. It was selected to be at least 4200 for turbulent flow.

The solution was initialized with a laminar velocity profile with turbulence fluctuations added afterwards. The simulation was advanced until natural turbulence was developed by monitoring wall-shear stresses for quasi-periodic behavior. Next the simulation proceeded for additional time to time average , the simulation had a computational time step of $\Delta T V_l/D = 0.025$, where ΔT is the dimensional time step. The time averaging period was at least 40,000 time steps.

The riblet geometries were blade, sawtooth, and scalloped shown in Fig. 11.21b. The parameters used to describe the geometries were spacing (s), height (h), and thickness (t). To compare vortex structures, y-z planes with vorticity contours and tangential velocity vectors were plotted. From these images, the height, width, and diameter, and distance from the wall for vortices were measured on the surfaces, as described in Fig. 11.21c. To compare drag, percentage drag change was calculated by taking the difference between the time-averaged drag on the flat surface (Martin and Bhushan 2014, 2016a, b).

The computational domain dimensions were chosen to be large enough to include the expected scales of the flow structures and to be greater than the minimal flow unit of 100 nondimensional units in the spanwise direction and 300 nondimensional units in the streamwise direction (Martin and Bhushan 2014, 2016a, b). The length was nondimensionalized similar to (11.3) using the spanwise or streamwise distance as the respective length scale) as determined by Jimenez and Moin (1991). For the continuous models, the

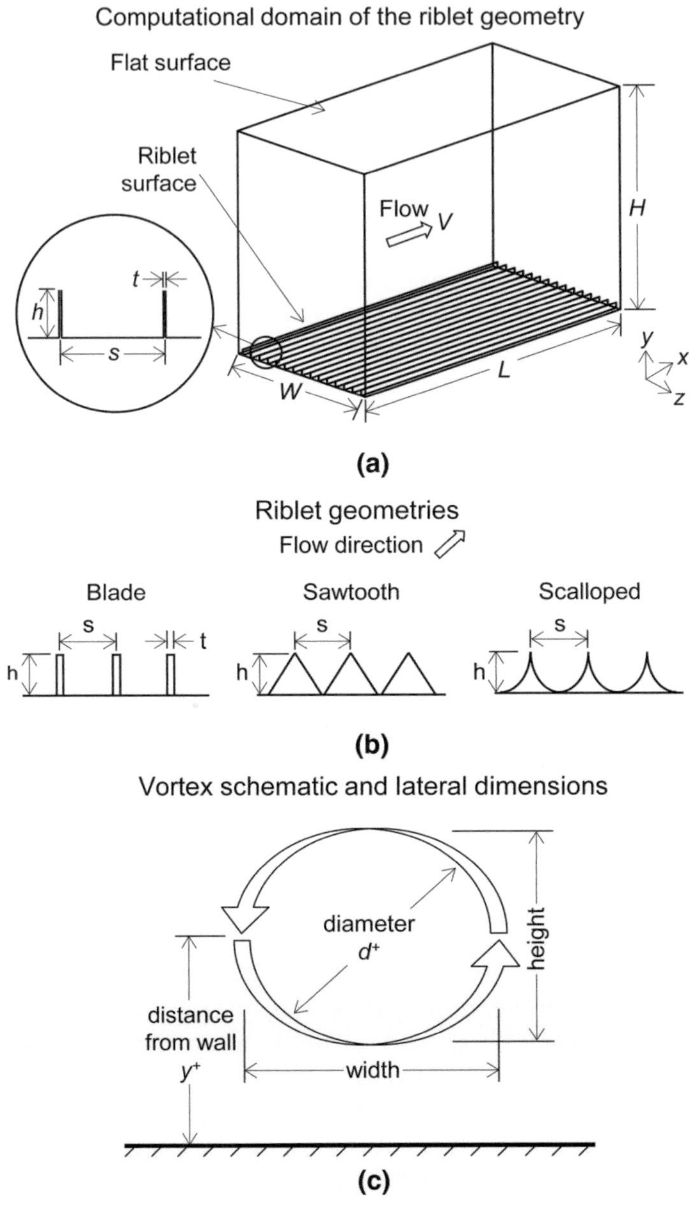

Fig. 11.21 **a** Computational domain of riblet geometry for a blade case showing geometry and riblet dimensions. The flat surface is the top wall, and the riblet surface is the bottom wall. **b** Blade, sawtooth, and scalloped riblet geometries showing spacing, height, and thickness parameters. **c** Vortex schematics showing various dimensions (adapted from Martin and Bhushan 2014, 2016a, b)

total number of hexahedral cells was around one million. An unstructured, tetrahedral mesh was used in order to mesh the various segmented riblet designs. The total number of tetrahedral cells was around four million.

These simulations were run on a supercomputer using four cores with 12 processors each. Each simulation took approximately one week to reach the time-averaging state for fully turbulent flow and approximately one additional week was required to complete the time averaging portion of the simulation due to the transient state.

11.4.2 Modeling of Blade Riblets

The various configurations have been analyzed, shown in Fig. 11.22, which include the continuous; aligned-segmented; and staggered-segmented configurations (adapted from Martin and Bhushan 2016a). In the continuous riblets with a constant length (L), spacing, height, and thickness parameters were varied. In the segmented configurations, the spacing, height, and thickness parameters were kept constant and only the lengths (L_1), gaps (L_2), and offsets (L_3) of the riblet were modified for drag and vortex comparison.

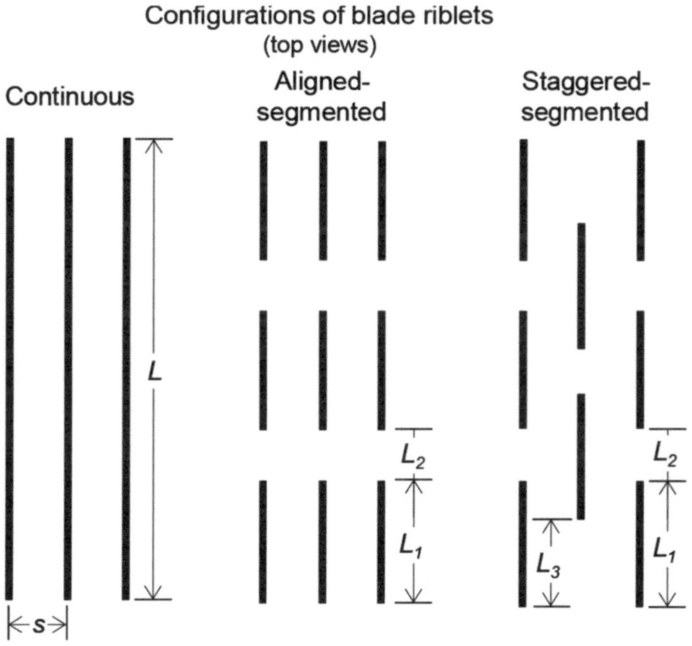

Fig. 11.22 Configurations for the riblet geometry cases showing continuous riblets with various spacing, and aligned-segmented and staggered-segmented styles with the various riblet gap and offset parameters (adapted from Martin and Bhushan 2016a)

The continuous riblet cases had two fluid parameters (velocity V and viscosity ν) and three riblet parameters (spacing s, height h, and thickness t) independently changed. Various test cases are listed in Table 11.6 (Martin and Bhushan 2016a). In the table, the dimensional parameters are listed on the left and the resulting nondimensional parameters are listed on the right. The baseline values shown in bold numbers were chosen for blank cells. The baseline value for spacing was chosen for an s^+ of 16. The baseline values for height and thickness were chosen to keep an h/s ratio of 0.5 and t/s ratio of 0.02 because h/s ratios of about 0.5 and small t/s ratios have been shown to give greater drag reductions.

Table 11.6 Test cases for the simulations of turbulent flow over continuous riblets (Martin and Bhushan 2016a)

	s (mm)	h (mm)	t (μm)	ν/V_l (μm)		s^+	h^+	t^+	Number of riblets
				Varying V_l $\nu = 10^{-6}$ m^2 s^{-1}	Varying ν $V_l =$ 0.42 m s^{-1}				
Baseline	**1.00**	**0.50**	**20.0**	**2.38**	**2.38**	**16.2**	**08.1**	**0.32**	**20**
Effect of V				4.00		10.3	05.1	0.21	20
				2.38		**16.2**	**08.1**	**0.32**	**20**
				1.39		24.5	12.3	0.49	20
				0.98		32.4	16.2	0.65	20
				0.76		46.8	23.4	0.94	20
Effect of ν					4.00	10.7	05.4	0.21	20
					2.38	**16.2**	**08.1**	**0.32**	**20**
					1.39	27.4	13.7	0.55	20
					0.77	46.7	23.3	0.93	20
Effect of s	0.50					08.4	08.4	0.34	40
	1.00					**16.2**	**08.1**	**0.32**	**20**
	1.50					25.3	08.4	0.34	13
	2.50					41.1	08.2	0.33	08
Effect of h		0.25				16.2	04.1	0.32	20
		0.50				**16.2**	**08.1**	**0.32**	**20**
		0.75				16.6	12.5	0.33	20
	0.75	0.22	15.0	1.20		22.9	06.7	0.46	27
	0.75	0.32	15.0	1.20		23.2	09.9	0.46	27
	0.75	0.42	15.0	1.20		23.2	13.0	0.46	27
Effect of t			05.0			16.8	88.4	0.08	20
			10.0			17.0	08.5	0.17	20
			20.0			**16.2**	**08.1**	**0.32**	**20**
			30.0			16.6	08.3	0.50	20

Baseline values shown in bold numbers were chosen for blank cells

For the effect of velocity cases, the viscosity was held constant at $\nu = 10^{-6}$ m^2 s^{-1} (representative of water); for the effect of viscosity cases, the velocity was held constant at $V_l = 0.42$ m s^{-1}. These values gave a baseline Reynolds number of 4180 and a ratio of $\nu/V_l = 2.38$ μm. This Reynolds number is high enough for turbulent flow without becoming too large to increase mesh size and computational expense.

The values of the dimensional parameters were chosen to fulfill several requirements. First, the dimensional parameters were chosen so that the cases would cover a wide nondimensionalized range for the parameter. In addition, the values were chosen so that s^+ of 15–20 would be in the parameter's ranges. This range has been shown for optimal drag reduction. By centering the data on this range, it would allow for test cases on either side of this range to be investigated.

Additionally, the velocity and viscosity values were selected to show the nondimensionalizing method is valid. By modifying the dimensional parameters and getting similar changes in drag for their nondimensional parameters, it shows that only the nondimensional values are important and any parameter in the equation can be modified.

To better understand the relationship between riblet height and drag, two sets of riblet height cases were created. In the first set, only the height was modified with the other parameters matching the baseline case. In the second set, all of the parameters were modified to increase the Reynolds number, center s^+ near 23, and h^+ near 10.

These continuous riblet cases were compared to the segmented riblet cases in which the length of each riblet (L_1), the gap between the riblets (L_2), and the offset between riblets (L_3) were independently changed. These cases are presented in Table 11.7 (Martin and Bhushan 2016a). In the table, the dimensional and nondimensional parameters for the riblet geometry are listed at the top. The dimensional values for the segmented geometries are shown on the left, and the resulting nondimensional values are shown on the right. The lengths were nondimensionalized similar to (11.1) as $L_1^+ = L_1 v_\tau/\nu$, $L_2^+ = L_2 v_\tau/\nu$, and $L_3^+ = L_3 v_\tau/\nu$, respectively.

The dimensional numbers for the various parameters were scaled up from actual riblet dimensions due to limitations with ANSYS Fluent when working with small numbers. Even with double precision, rounding errors could accumulate if actual riblet dimensions were used. Scaling up the numbers does not affect the results since only nondimensional numbers affect the results, and this conclusion will be presented later.

In the segmented riblet cases, the spacing ($s^+ \approx 16$), height ($h^+ \approx 8$), thickness ($t^+ \approx 0.3$), and Reynolds number ($Re = 4180$) were kept constant. These cases were compared to the continuous case with these same riblet parameters. In the aligned-segmented test cases, the length of each riblet (L_1) was varied from 2–6 times the length of the gap between riblets (L_2) and riblet spacing because these ratios can be seen in shark scales. Also in the aligned-segmented test cases, the gap between the riblets (L_2) ranged from 1.0 to 9.4 mm in order to get a wide range of nondimensional riblet gaps of 17–154. This wide range was necessary in order to examine streamwise vortices because they have a

Table 11.7 Test cases for the simulations of turbulent flow over segmented riblets (Martin and Bhushan 2016a)

Riblet geometry	$s = 1.0$ mm, $h = 0.5$ mm, $t = 20$ μm, $\nu/V_l = 2.38$ μm for $Re = 4180$ $s^+ \approx 16$, $h^+ \approx 8$, $t^+ \approx 0.3$						
	L_1 (mm)	L_2 (mm)	L_3 (mm)	L_1^+	L_2^+	L_3^+	Drag change (%)
Continuousblade ($L = 0.035$ m)	∞	0	0	∞	0	0	−9
Aligned-segmented blade	2.0 4.0 6.0	1.0	0	32 67 99	16 17 17	0	−2.1 −3.3 −4.3
	4.0	1.0 2.0 4.0 6.0 9.4	0	67 68 66 68 66	17 34 66 101 154	0	−3.3 −3.6 −3.7 −5.0 +5.2
Staggered-segmented blade	4.0	1.0	0.0 1.0 2.0	67 67 66	17 17 17	0 17 33	−3.3 −3.2 −3.2

great effect on drag in riblets. In the staggered-segmented test cases, three different offsets (L_3) of 0, 1, and 2 mm were chosen in order to compare the effect of offset designs. These offsets were chosen to equally space the offset along the riblet length of 4 mm. The 1 mm offset would be a mirrored geometry of a 3 mm offset and so the 3 mm offset was not modeled.

Next vortex and drag data is presented for the continuous and segmented riblet configurations. This data is used to compare the different configurations and show the differences among drag changes for the different cases.

11.4.2.1 Continuous Riblets

Martin and Bhushan (2014, 2016a) modeled the effect of s^+ by varying s, V, and ν independently, and h^+ and t^+ by varying h and t independently. Nondimensional parameters s^+, h^+, and t^+ are relevant parameters for modeling drag. These parameters can be used in the scaling up of riblet design.

A knowledge of vortex geometry and its location with respect to the wall is important in understanding the drag reduction mechanisms. Table 11.8 presents a summary of typical dimensions for streamwise vortex with cylindrical geometry (Martin and Bhushan 2016b). Figure 11.23 shows representative vortices on a flat surface and two riblet surfaces with different spacings. The vortices hug the flat surface. The location of a vortex changes as the riblet spacing changes (Martin and Bhushan 2016a, b). By choosing appropriately

Table 11.8 Typical values of streamwise vortices data with cylindrical geometry (Martin and Bhushan 2016a, b)

Diameter	Average vortex diameter is between 30 and 40 nondimensional units
Length	Average vortex length is between 100 and 200 nondimensional units
Reynolds number dependence	Average vortex diameter increases with Reynolds number from 20 to 45 nondimensional units for Reynolds numbers from 2500 to 13,000

sized riblets; these vortices can be lifted away from the surface as shown in the figure with $s^+ = 25.3$. However, with too large a spacing, vortices fit within the riblet valleys and lead to a drag increase as shown at $s^+ = 41.2$.

In order to understand the effect of the geometrical parameters on drag change, each parameter listed earlier was individually modified and the resulting drag change was calculated. Selected results are shown in Fig. 11.24 (adapted from Martin and Bhushan 2016a), where the negative drag change values indicate a drag reduction. For the varying spacing models, the maximum drag reduction was ~11% at $s^+ \approx 23$. When the spacing between riblets is small, more riblet peaks contact the vortices, leading to a lower drag reduction. With larger spacing, the vortices can fit within the valleys between the riblet peaks and are not lifted from the surface, leading to a drag increase. When varying only riblet height, maximum drag reduction occurred when $h^+ = 8\text{–}10$. Increasing outside of this range quickly led to a drag increasing state due to the riblets protruding too far into the higher velocity flow, or decreasing below this range led to a less efficient state due to the riblets unable to effectively repel the vortices. When the riblet thickness was modified, little drag change occurred. At much larger thicknesses, the width of the riblet valleys would decrease and the width of the riblet peaks would increase. It is expected that these changes would generate a weaker drag reduction, and eventually a drag increase.

11.4.2.2 Segmented Riblets

In order to compare continuous riblet results to segmented riblet results and better understand the riblet segmentation in sharks, various segmented riblet cases were modelled. The parameters for the length of each riblet (L_1), the gap between the riblets (L_2), and the offset between riblets (L_3) were modified independently.

Figure 11.25a shows the effect of varying the gap length on the location of vortices. Streamwise vortices predominately have a length of 100–200 nondimensional units. When gap lengths are smaller than the length of vortices, they are lifted up away from the surface. Whereas, when the gap lengths are longer, vortices can fall into the gaps between the riblets.

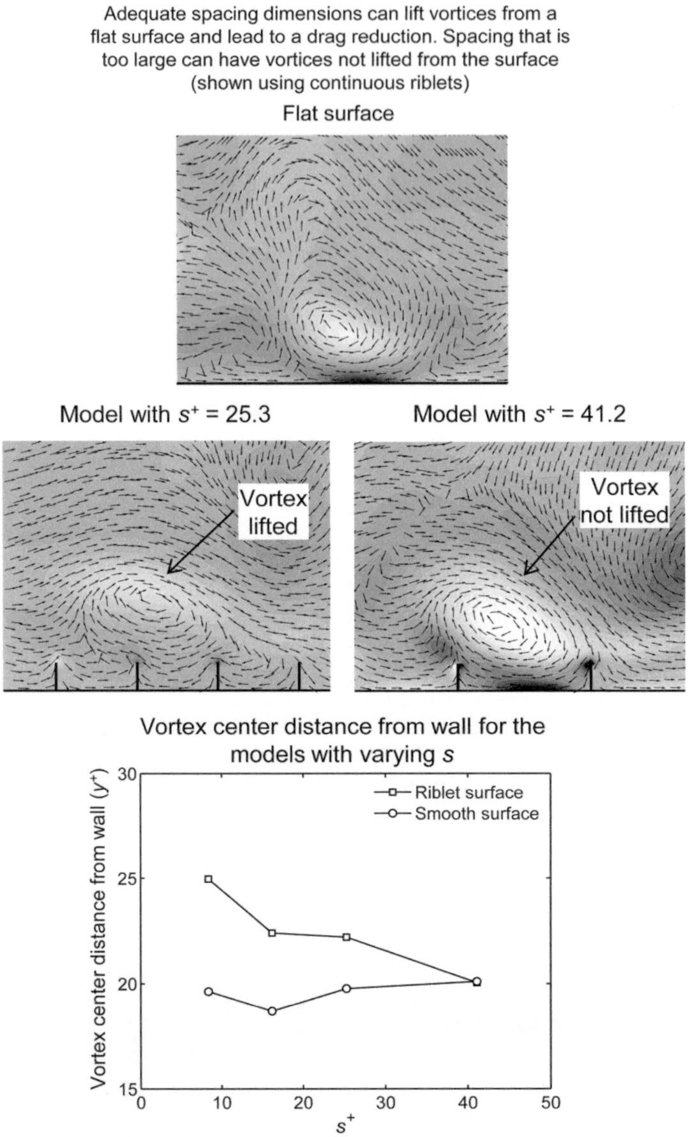

Fig. 11.23 Vortices generated in turbulent flow shown on flat and riblet surfaces. On a flat surface, vortices are near the surface and generate drag. With adequate riblet spacing (such as with $s^+ = 25.3$), vortices are lifted away from the surface reducing drag and creating an overall drag reduction. With too large of a riblet spacing (such as with $s^+ = 41.2$), vortices can fit within the riblet valley leading to a drag increase. Graph shows as spacing increases, average vortex center distance from the riblet wall decreases suggesting vortices falling into the riblet valley (adapted from Martin and Bhushan 2016a, b)

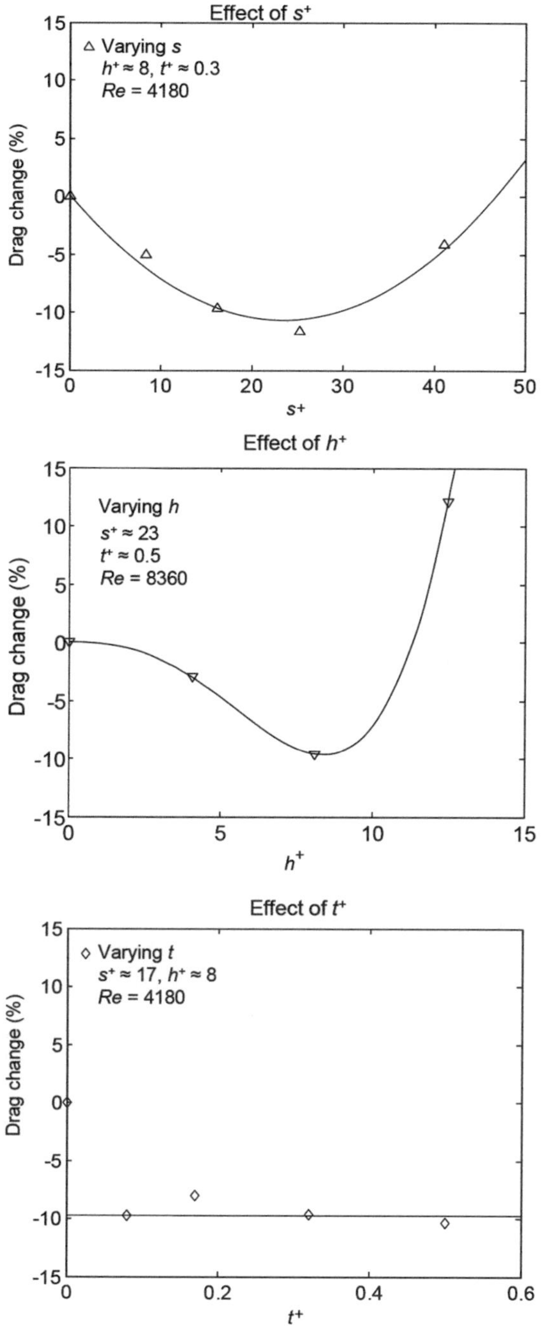

Fig. 11.24 Drag change for the various segmented models plotted against their nondimensionalized parameter—spacing, height, and thickness. From the models with varying height, optimal height was found to be at $s \approx 20$ and $h^+ \approx 8$–10. The range of t^+ did not affect drag notably (adapted from Martin and Bhushan 2016a)

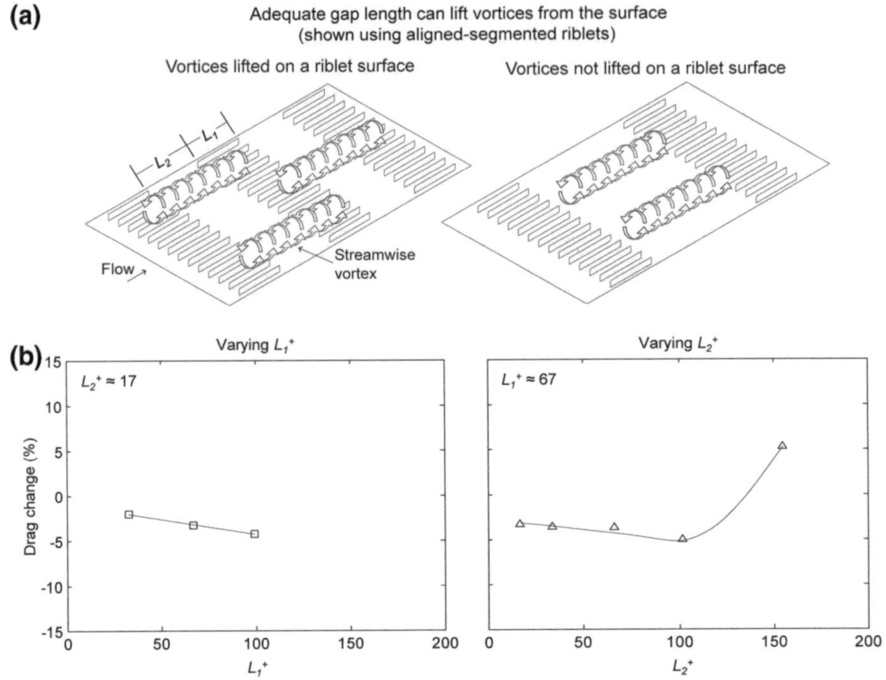

Fig. 11.25 **a** Vortices generated on two riblet surfaces, aligned-segmented with riblet length (L_1), and different gap length (L_2). On the left, the gap is shorter than the length of the streamwise vortex and the vortex is lifted; on the right, the gap is longer and the vortex is not lifted. Streamwise vortices have average lengths of 100–200 nondimensional lengths. **b** Graphs show effect of riblet length and gap length on drag. For increasing riblet length (L_1^*), the geometry approaches the continuous case and the drag reduction increases. As gap length (L_2^*) increases, the drag reduction decreases. Then drag change rapidly increases once the gaps become on the order of the length of the streamwise vortices (adapted from Martin and Bhushan 2016a, b)

The drag change data for the aligned-segmented models with two gap lengths is shown in Fig. 11.25b (Martin and Bhushan 2016a). The aligned-segmented models were plotted for varying L_1^+ with constant L_2^+ and then also with constant L_1^+ with varying L_2^+. In the varying L_1^+ with constant L_2^+ data set, the drag change increase as the lengths of the riblets increase. As the lengths increase, fewer gaps on the riblet surface occur leading to more of a continuous case which showed a greater drag reduction. Fewer gaps lead to less chance of burst and ejection events in the gaps. In the constant L_1^+ with varying L_2^+ data set, the drag change decreases by -5.0% at $L_2^+ = 101$ and then increases after that. With streamwise vortices predominately having nondimensional lengths in the 100–200 range, once the gap becomes larger than 100 nondimensional units, the streamwise vortex is not lifted generating a lot of drag.

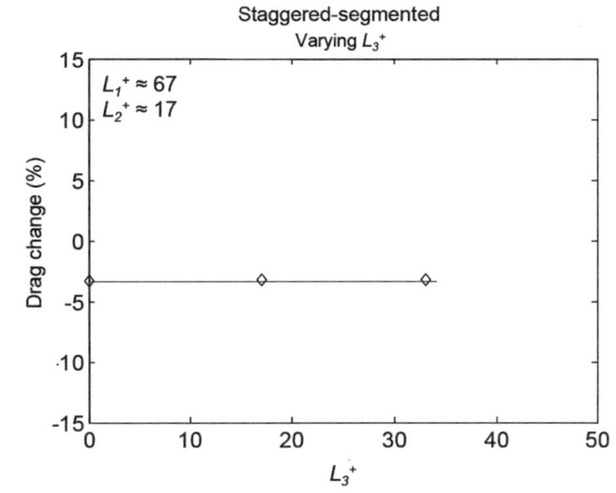

Fig. 11.26 Drag change for the staggered-segmented cases. As L_3^+ increases, no difference in drag change is seen. Staggered-segmented riblets exist to help in ejection of contaminant particles in the riblet valleys (adapted from Martin and Bhushan 2016a)

The staggered-segmented data set is presented in Fig. 11.26. It shows a constant drag reduction of 3.3%. For the values of $L_1^+ = 67$ and $L_2^+ = 17$, varying L_3^+ did not affect the drag change.

Modeling shows that aligned-segmented and staggered-segmented riblets result in drag reduction of 5.0 and 3.3%, respectively, lower than that in the continuous riblet configuration of 11%. Thus it raises the question why do sharks have segmented and/or staggered riblets? It is believed that segmented riblets help the shark to swim by creating a deformable surface and/or in ejection of contaminant particles. Because each scale can individually pivot, when the shark turns, the surface of the shark becomes a deformable surface. Staggered riblets give particles a path to be removed from the surface. Aligned-segmented riblets may have less of a chance for particles to be ejected. Whereas, continuous riblets may have the least chance for contaminants to be removed from the riblet valleys. By discouraging contaminant settlement and eventual growth, the riblet valleys in segmented configurations stay unpolluted and the drag reduction stays effective.

11.4.2.3 Summary

In the continuous configurations, drag reduction of approximately 10% occurs when $s^+ \approx 18$–25 and $h^+ \approx 8$–10 (Martin and Bhushan 2016a), which is similar to $s^+ = 16$ and $h^+ = 8$ from Bechert et al. (1997b) and $s^+ = 18$ and $h^+ = 9$ from El-Samni et al. (2007). With these riblet dimensions, the streamwise vortices were lifted from the surface reducing the drag on the surface. The riblet thickness t^+ should be small.

The aligned-segmented and staggered-segmented riblet configurations had maximum drag reductions of approximately 5 and 3%, respectively (Martin and Bhushan 2016a). The three parameters of riblet length, gap length, and riblet offset each changed the reported drag differently. When the lengths of the riblets increased with constant gap size, the drag reduction increased due to the configuration more closely approximating the continuous configuration. As the gap length approached 100 nondimensional units, the drag reduction increased with the drag increasing due to the increased odds of streamwise vortices falling into the gaps between consecutive riblets. As the riblet offset increased, no differences in drag change were observed.

Table 11.9 summarizes the effect of the continuous and segmented riblet parameters (Martin and Bhushan 2016a). The riblet spacing should be around 15–20 nondimensional units to lift the vortices with the riblet height around 8–10 nondimensional units. If the riblets are too tall, they protrude too far into the flow and increase drag. If they are too short, they do not effectively lift the vortices, which in both cases, increases drag. Thinner riblets are better because the majority of the drag is found at the riblet peaks; however, thin riblets below a certain value do not noticeably change the drag. Riblet gaps should be less than 100 nondimensional units to repel the streamwise vortices that predominately are 100–200 nondimensional units long. Because continuous riblets have a greater drag reduction than aligned-segmented and staggered-segmented riblets, sharks may implement segmentation for two reasons. These configurations may help create a deformable surface for shark movement and/or in ejection of contaminant particles to keep the shark skin surface clean.

11.4.3 Modeling of Blade, Sawtooth and Scalloped Riblets

Various riblet geometries of interest include blade, sawtooth, and scalloped, where only the spacing parameter was varied within each case (Martin and Bhushan 2016b). In addition, a new geometry inspired by the dermal denticles on a Spiny Dogfish shark was modeled with riblets arranged on a scale pattern to compare shark scales to the three riblet geometries. The results were compared to the experimental data, and conclusions were drawn for optimized riblet geometries.

The test cases for the simulations over blade, sawtooth, and scalloped riblets are presented in Table 11.10 (Martin and Bhushan 2016b). In the table, the dimensional parameters are listed on the left and the resulting nondimensional parameters are listed on the right. The viscosity was $\nu = 10^6$ m^2 s^1 (representative of water), and the velocity was $V_l = 0.42$ m s^{-1}. These values gave a ratio of $\nu/V_l = 2.38$ μm and Reynolds number of 4180, which is large enough for turbulent flow (known to be present above 4000) without becoming so large as to increase computational expense.

The riblet height was kept constant for all models at $h = 0.5$ mm in order to get a nondimensional height of $h^+ \approx 8$, which has been shown for optimal drag reduction (Martin and Bhushan 2014). As stated earlier, thickness plays an insignificant role on drag for thin riblets. The thickness dimension was kept at $t = 20$ μm to achieve $t/s = 0.02$ in order to compare to data from Bechert et al. (1997b). The dimensions of riblet spacing were chosen to cover a wide nondimensionalized range. In addition, the riblet

Table 11.9 Summary of effect of riblet geometry on drag (Martin and Bhushan 2016a)

Parameter	Observation	Basis
(a) *Continuous riblet conclusions*		
Spacing (s^+)	Optimal riblet spacing should be around 15–20 nondimensional units	Vortices have a diameter of 30–40 nondimensional units. With optimal riblet spacing, these vortices can be lifted up away from the riblet surface decreasing overall drag
Height (h^+)	Optimal riblet height should be around 8–10 nondimensional units	Riblets that are too tall protrude too far into the flow increasing drag over the entire surface. Riblets that are too short do not effectively lift the vortices away from the surface decreasing sweeps and ejections
Thickness (t^+)	Thinner riblets are better; however, at very small thicknesses, little drag differences are seen	The majority of drag occurs at the riblet peaks and with thinner riblets, there is less drag at the peaks. Most of riblet surface area comes from the spacing and height parameters, and so, thickness plays a smaller overall role in drag change
(b) *Segmented riblet conclusions*		
Riblet gaps	Riblet gaps should be smaller than streamwise vortex lengths to increase chances of lifting vortices. Gaps should have lengths less than 100 nondimensional units	Streamwise vortices predominately have lengths of 100–200 nondimensional units and as gap length increases, the likelihood of a vortex fitting within the gap increases along with increasing drag
Riblet offsets	Riblet offsets by themselves have little effect on drag, but increase the chance for contaminant particles to be ejected from the channels	Riblets with offsets are still able to lift vortices from the surface decreasing sweep and ejection events associated with high drag

Observation	Why segmented?
(c) *Continuous versus segmented configurations*	
Continuous riblets have a drag reduction of approximately 10% whereas aligned-segmented and stagger-segmented have drag reductions of approximately 5 and 3%	Segmented riblets may help sharks to swim by creating a deformable surface since the scales can individually pivot and/or in contaminant ejection to keep the shark clean by providing a path for particles to be removed

Notes Vortex nondimensional diameters increase with *Re*—20 to 45 for *Re* from 2500 to 13,000. Vortex nondimensional lengths predominately range from 100 to 200)

Table 11.10 Test cases for the simulations of turbulent flow over continuous riblets and calculated drag change (Martin and Bhushan 2016b)

Riblet geometry	$v/V_l = 2.38$ μm for $Re = 4180$						
	s (mm)	h (mm)	t (μm)	s^+	h^+	t^+	Drag change (%)
Blade	0.5	0.5	20	08.4	8.4	0.34	−5.0
	1.0			16.2	8.1	0.32	−9.6
	1.5			25.3	8.4	0.34	−11.6
	2.5			41.1	8.2	0.33	−4.2
Sawtooth	0.5	0.5		08.3	8.3		−1.9
	1.0			16.8	8.4		−4.1
	1.5			25.0	8.3		−1.6
	2.0			34.1	8.5		+1.5
Scalloped	1.0	0.5		16.9	8.4		−5.7
	1.5			24.7	8.2		−4.0
	2.0			33.1	8.3		−2.6

V_l is the centerline velocity of a laminar profile with the same volume flux

spacing was chosen so s^+ of 15–20 would be the parameter's ranges. This range has been shown for optimal drag reduction (Bechert et al. 1997b; Martin and Bhushan 2014, 2016a). By including this data range, it would allow for test cases on either side of this range to be investigated.

The shark-inspired scalloped surface shown in Fig. 11.27 was modeled on the dermal denticles of a Spiny Dogfish (Martin and Bhushan 2016b). The approximate scale dimensions were scale length of 450 μm, scale width of 375 μm, gap length of 175 μm, and gap width of 100 μm. The approximate riblet dimensions were spacing of 125 μm and height of 50 μm. These dimensions were roughly scaled to a riblet spacing of 1.0 mm for the model, giving a scale length of 4.0 mm, scale width of 3.0 mm, gap length of 1.0 mm, and a gap width of 0.75 mm. The riblet height was kept at 0.5 mm in order to compare to the test cases presented in Table 11.10. The relevant parameters were: $v/V_l = 2.38$ μm for $Re = 4180$, $s^+ \approx 16$ and $h^+ \approx 8$.

11.4.3.1 Results and Discussion

Images of vortices for the blade, sawtooth, and scalloped geometries are shown in Fig. 11.28 (Martin and Bhushan 2016b). The images on the left show vortices being lifted from the riblet surface where $s^+ = 16.8$–16.9, and the images on the right show vortices not being lifted from the riblet surface where $s^+ = 33.0$–41.2. In all images, vortices remained approximately the same size with a height of 25 and width of 35 nondimensional units. These dimensions result in an approximate diameter of 30 nondimensional units when using the ellipsoidal quadratic mean. When the vortices are lifted from the

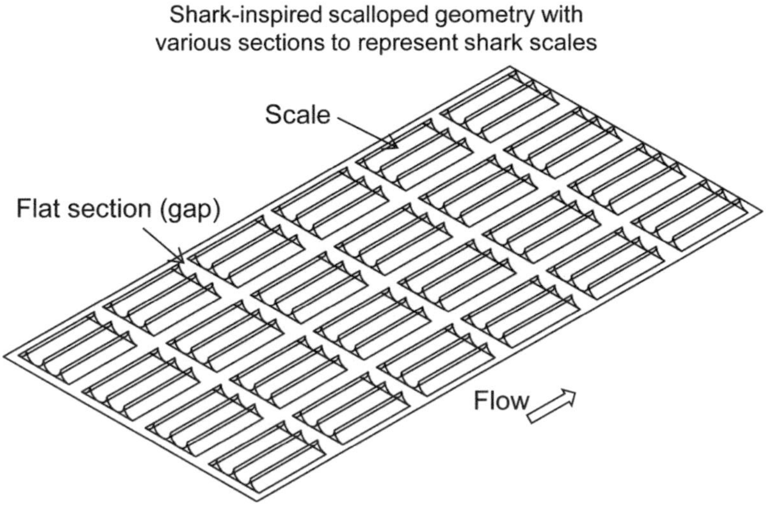

Fig. 11.27 Shark inspired scalloped geometry modeled on the dimensions of dermal denticles on a Spiny Dogfish. Combinations of scales and flat sections (gaps) were used to create a riblet surface (Martin and Bhushan 2016b)

surface, multiple riblets are in contact with the vortex. When the vortex is not lifted from the surface, a vortex is able to fit within the riblet spacing valley. As the vortex rotates in the valley, increased drag results due to the increase of burst and sweep motions. A larger spacing is shown with the blade geometry than with the sawtooth or scalloped geometry to show that as the spacing becomes even larger, the vortex is more able to fit within the valley (Martin and Bhushan 2016b).

Figure 11.29 shows vortices at the scales and at the gap between scales for the shark-inspired scalloped geometry. In both images, the vortices are approximately the same size as each other and the vortex sizes shown in Fig. 11.28. The vortex being lifted on the scales is similar to images of vortices being lifted where the vortex is in contact with multiple riblets. These cases all have a similar riblet spacing of $s^+ = 16$–17. On the gap between scales, the vortex is still lifted from the surface. Vortices would be lifted in the gap length due to the distance of 16 nondimensional units being smaller than average streamwise vortex lengths of 100–200 nondimensional units (Martin and Bhushan 2016b).

The drag change for the sawtooth, scalloped, and blade geometries are shown in Fig. 11.30 and summarized in Table 11.10. Drag change for the shark-inspired scalloped geometry is also shown in Fig. 11.30 with $s^+ = 16$. The blade geometry had the greatest drag reduction of 11.6% at $s^+ = 25.3$, followed by scalloped with a drag reduction of 5.7% at $s^+ = 16.9$, and sawtooth with a drag reduction of 4.1% at $s^+ = 16.8$. For all riblet spacings, the blade geometry had the greatest drag reduction, followed by the scalloped geometry, with the sawtooth geometry coming in last with the least benefit (Martin and Bhushan 2016b).

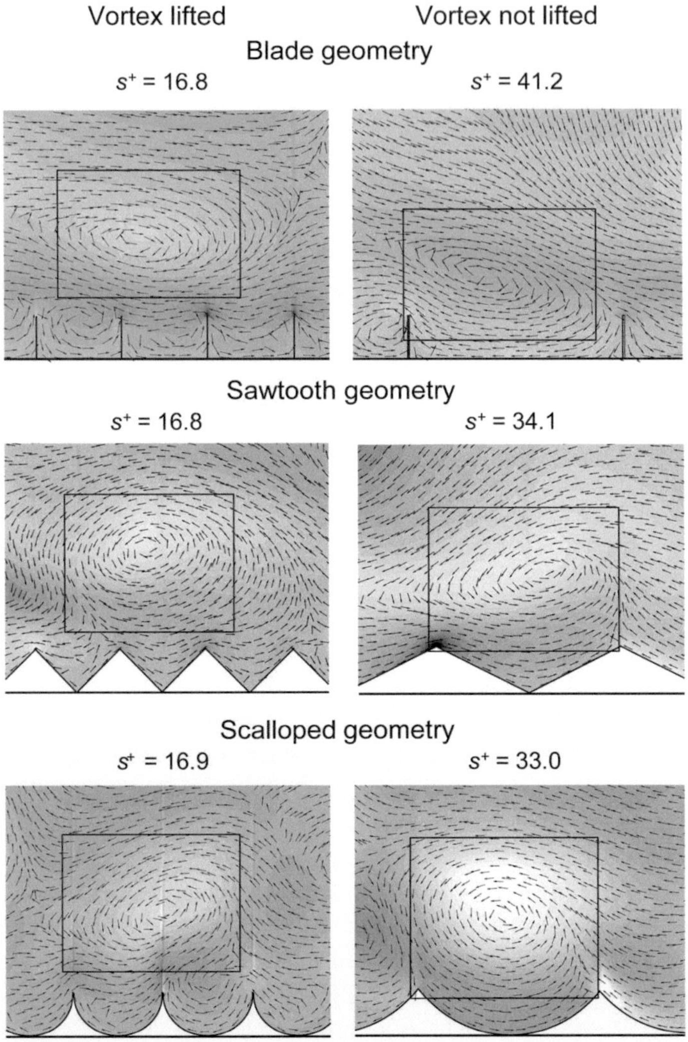

Fig. 11.28 Vortex images on blade, sawtooth, and scalloped geometries showing vortices being lifted on the left and vortices not being lifted on the right. Average vortex diameter was approximately 30 nondimensional units. When the riblet spacing is smaller than the vortex diameter, vortices are lifted; when the riblet spacing is larger than the vortex diameter, vortices can fit within the riblet valley (Martin and Bhushan 2016b)

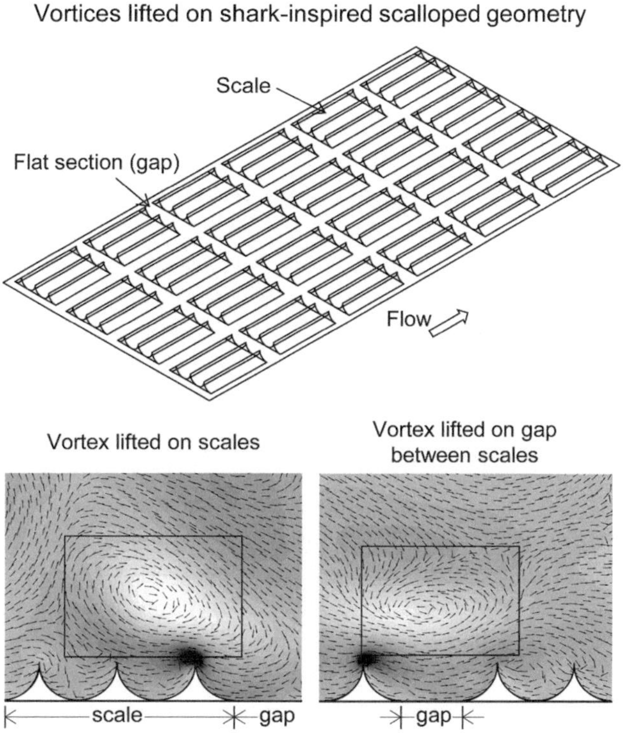

Fig. 11.29 Vortex images on the shark-inspired scalloped geometry. Vortices are being lifted on the scales and on the gaps between scales (Martin and Bhushan 2016b)

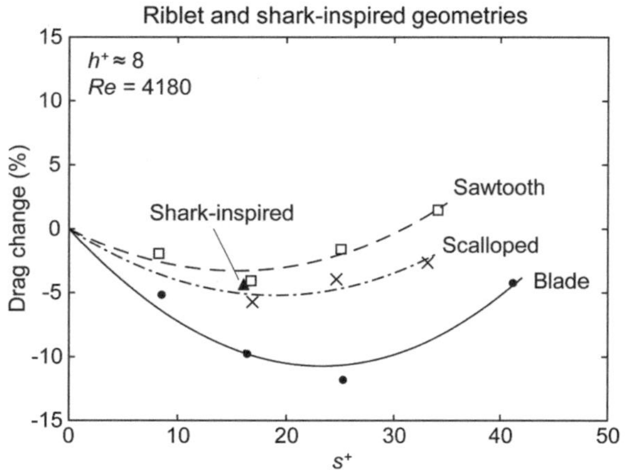

Fig. 11.30 Drag change for three riblet geometries and a shark-inspired scalloped geometry. Blade geometry shows the greatest drag reduction of 11.6% at $s^+ = 25.3$, followed by scalloped at 5.7% at $s^+ = 16.9$, and sawtooth at 4.1% at $s^+ = 16.8$. The shark-inspired geometry shows a drag reduction of 4.5% similar to the continuous scalloped design (Martin and Bhushan 2016b)

The drag change in the experimental data presented in Fig. 11.9b and modeling data in Fig. 11.30 show the same general trends with blade having the greatest drag reduction followed by scalloped and sawtooth. The optimal spacing for drag reduction is approximately the same in both studies at $s^+ = 15$–20. The differences between them likely occur due to the difference in riblet height for the geometries. In Bechert et al. (1997b), a riblet spacing of $s^+ = 28$ yields a riblet height of 14, 20, and 24 nondimensional units for the blade, scalloped, and sawtooth geometries, respectively, due to the constant h/s ratio. However, large riblet heights have a detriment effect on drag reduction. Therefore, in this study, a constant height of $h^+ = 8$ was maintained for all models resulting in drag reduction at riblet spacings of $s^+ > 30$ (Martin and Bhushan 2016b).

Even though the spacing and height for the different geometries were constant, different drag reductions resulted for each geometry (Martin and Bhushan 2016b). The discrepancy occurs due to how the vortex and the riblet surface interact. When a vortex is lifted on a blade surface, the riblet peak is directly in contact with the vortex resulting in a large increase of shear stress over a tiny area. The majority of the surface area on the blade riblets is located at the sides and valley of the riblet. The distance from this area to a lifted vortex is maximized, and the largest drag reduction occurs. In the sawtooth case, the sides of the riblet are closer to the vortex. This decreased distance results in more surface area, with the increased shear stress leading to a weaker drag reduction. The scalloped geometry is a combination of the blade and sawtooth geometry, resulting in a drag reduction between the two geometries.

The drag reduction of the shark-inspired scalloped geometry of 4.5% was similar to the drag reduction for the sawtooth (4.1%) and scalloped (5.7%) geometry with the same riblet spacing (Martin and Bhushan 2016b). The reduction in low-drag efficiency from the continuous scalloped geometry is believed to be due to the gaps in the shark-inspired scalloped geometry. These gaps, both in the spanwise and streamwise direction, are thought to reduce the effectiveness of lifting vortices and lead to pressure drag as an additional source of drag. The shark-inspired surface still has lower drag than the sawtooth geometry due to the sides of the scalloped riblets being further from any vortices than in the sawtooth case. There is not a large difference in drag change between the continuous and shark-inspired geometries due to the small size of the gaps. The gap length is 16 nondimensional units, and the gap width is 12 nondimensional units. Although gaps have an undesirable effect on decreasing drag, sharks have gaps and offsets between dermal denticles. These segmentations likely facilitate in flexion of dermal denticles to help the shark move and in ejection of contaminant particles to keep the shark clean.

11.4.3.2 Summary

Blade, sawtooth, and scalloped riblets were modeled by Martin and Bhushan (2016b) with a constant height of $h^+ = 8$ and varying spacing of $s^+ = 8.3$–41.1. Heights of $h^+ = 8$ have been shown for optimal drag reduction, and this wide range of s^+ would show both drag decreases and increases. Blade riblets were found to have the greatest drag reduction of 11.6%, followed by scalloped at 5.7%, and sawtooth at 4.1%. The optimal riblet spacing for the models were in the range of $s^+ = 16$–25. This range effectively lifted up the vortices, which had an average diameter of 30 nondimensional units in all cases. Smaller riblet spacings have too much surface area and riblet peaks in contact with the vortex. Larger riblet spacings allow the vortex to fall into the riblet valley.

The shark-inspired scalloped geometry had a drag reduction of 4.5% which is approximately the same drag reduction as the sawtooth and scalloped geometries. The differences in drag are due to the gaps between scales, which are expected to reduce the effectiveness of lifting vortices. The gaps between dermal denticles likely facilitate flexion of dermal denticles to help the shark move and in ejection of contaminant particles to keep the shark clean. The roundedness of dermal denticles helps in increased durability (Martin and Bhushan 2016b).

Major observation for riblet optimization are summarized in Fig. 11.31. Blade riblets have the greatest drag reduction on the order of 10% when a riblet spacing of $s^+ =$ 15–20 and riblet height of $h^+ =$ 8–10 are chosen. Therefore, in commercial applications, this geometry with these dimensions should be considered over scalloped and sawtooth designs. Riblets found on the dermal denticles of shark skin are shaped similarly to the scalloped geometry. However, this geometry is not as optimal as the blade geometry for low drag. Sharks have additional considerations of replacing dermal denticles due to damage, flexion for movement, debris ejection, and durability. Thin, rectangular riblets would not be durable and so rounded geometries are commonly found in nature. In commercial

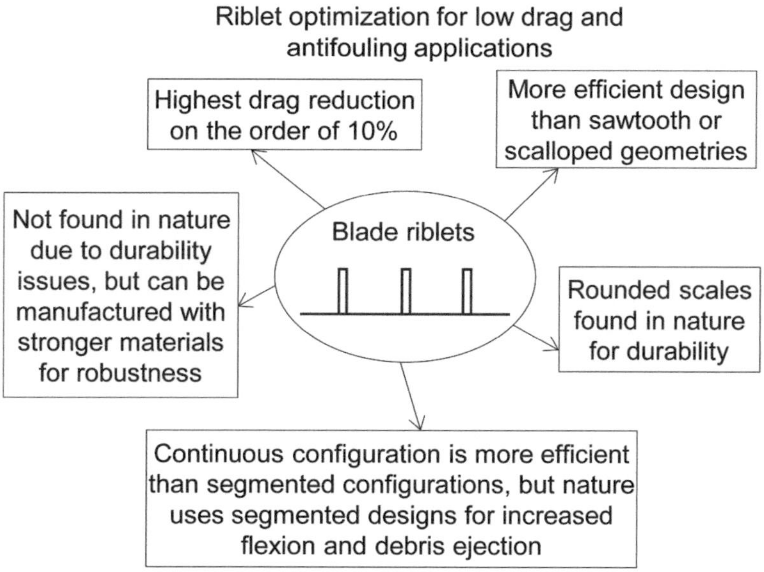

Fig. 11.31 Riblet optimization for low-drag and antifouling applications. For commercial applications, blade riblets should be chosen due to their greatest drag reduction and ability to be manufactured with increased durability using modern industrial techniques (Martin and Bhushan 2016b)

applications, blade riblets could be manufactured with greater durability by using stronger materials (Martin and Bhushan 2016b).

Understanding different riblet configurations helps lead to a better understanding of the riblet mechanism which can be used as drag-reduction design principles for industrial applications in marine, medical, and industrial fields. Riblet dimensions are dependent on the component size and can range from the micro- to nanoscale.

11.5 Application of Riblets for Drag Reduction and Antifouling

11.5.1 Industrial Examples

Riblets provide drag reduction in fluid flow over objects where the dominant form of drag is skin friction and operates in turbulent flow. Major applications for drag reduction include transportation, fluid flow in pipelines, and other flow applications (Marentic and Morris 1992; Kennedy 1993; Davies 2002; Brostow 2008; Bhushan and Jung 2011; Deyuan et al. 2011; Bixler and Bhushan 2012a). About 35–70% of the energy consumed in transportation is estimated to be lost due to drag so drag reduction is an important issue. According to the U.N. International Maritime Organization, about 90% of worldwide goods are carried by sea-faring vessels with over 8.4 billion tons of cargo in 2010 (Anonymous 2012). In 2014, U.S. domestic and international airplanes traveled over 860 billion passenger miles (Anonymous 2016b). In automobiles, energy losses due to drag are also important. However, only some of its areas such as roof and its sides experience high surface drag and for the rest of the shaped surfaces, drag is not so important. High drag reduction is realized in applications operating at a high Reynolds number or high fluid flow velocity divided by fluid kinematic viscosity (V/ν). The kinematic viscosity values of air and water at ambient conditions are about 15.11 and 1.00, respectively.

Therefore, it is more beneficial to use surfaces covered with riblets in water as well as at high velocities in air. Thus, in transportation, water vessels and airplanes flying at high speeds (~10× of automobiles) benefit the most from surfaces covered with riblets.

Another major low-drag application includes pipelines where low pumping energy and large flow rates are required. In 2014, there were over 2.1 million miles of gas distribution pipelines and about 60,000 miles of petroleum pipelines in the U.S. (Anonymous 2016c). In the oil industry, pipeline flow must overcome high drag due to viscous effects and sludge fouling (with Reynolds numbers reaching 1×10^5). Traditionally, drag is lowered using fluid additives or improving the pipeline interior smoothness with corrosion resistant epoxy coatings (Burger et al. 1982; Brostow 2008; Martínez-Palou et al. 2011). Such coatings are most effective when sludge fouling is not present, as found with oil flowing from a well at a high temperature and velocity.

In addition to lowering drag, antifouling surfaces can reduce the spread of infectious diseases by limiting biofouling. Medical biofouling occurs in areas such as prosthetic implants, biosensors, catheters, dental implants, and medical equipment. Problems include implant rejection, malfunction of biosensors, and spread of infectious diseases (Vo-Dinh 2007; Schulz et al. 2009; Shirtliff and Leid 2009; Chan and Wong 2010; Bixler and Bhushan 2012a). Biofilms in the medical field cause close to half of hospital contracted infections. Antibiofouling is also important in food packaging. Biofouling is commonly associated with marine environments where noticeable aquatic growth appears on ships and underwater structures. This increases ship hull drag, corrosion, fuel consumption, and engine stress (Copisarow 1945; Anon 1952; Ray 1959; Melo et al. 1988; Railkin 2004; Bixler and Bhushan 2012a). Industrial fouling occurs in areas such as power plants, water treatment systems, and food/beverage industries (Bixler and Bhushan 2012a). Problems include pipe blockage, decreased membrane flux, contaminated water, and reduced heat exchanger efficiency (Somerscales and Knudsen 1981; Walker et al. 2000; Chan and Wong 2010).

11.5.2 Prototypes and Commercial Applications

In the mid-1980s, vinyl film sawtooth riblets were applied to boat hulls for racing (Choi et al. 1989). Both an Olympic rowing boat and an Americas Cup sailing yacht were covered with riblets during competition. Because skin friction of an airplane accounts for as much as 48% of total drag, vinyl film riblets have also been applied on airplane wings (Han et al. 2002) and fuselage. About 70% of the aircraft was covered with film riblets, and it provided 3% total drag reduction. This 3% drag reduction correlates to a similar 3% savings in fuel costs (Bechert et al. 1997a).

Another large commercial application for riblet technologies is drag reduction in pipe flow. Sawtooth riblet film produced by 3 M has been applied inside pipes (Liu et al. 1990; Rohr et al. 1992; Enyutin et al. 1995; Koury and Virk 1995). Drag reduction tests in pipe flow have been carried out by comparing the pressure drop in a riblet lined pipe with that of a similar pipe with a smooth surface. Drag reduction data for sawtooth riblets for oil flow on a flat plate are shown in Fig. 11.32a (Bechert et al. 1997b), and drag reduction data for water flow in sawtooth riblet lined pipes are shown in Fig. 11.32b (Liu et al. 1990). The cone angle α is described in Fig. 11.9. As riblets are applied to the inside surface of a pipe, the riblet tips are shifted together due to the curve of the pipe wall. Consequently, the optimized s^+ range for riblets in pipe flow is lower than that for the same riblets in flow over a flat surface (Liu et al. 1990). Additionally, the decrease in s^+ suggests that an increased drag reduction in round pipes might be seen with a larger characteristic angle, α, than for flow over riblets on a flat surface.

Machining the surface or applying vinyl film riblets proves difficult for most pipes, and alternatives are being developed. Experimental application of a scratching technique to the inside surface of pipes has created a riblet-like roughness that has provided more than 5% drag reduction benefit (Weiss 1997). Stemming from an old sailors' belief that

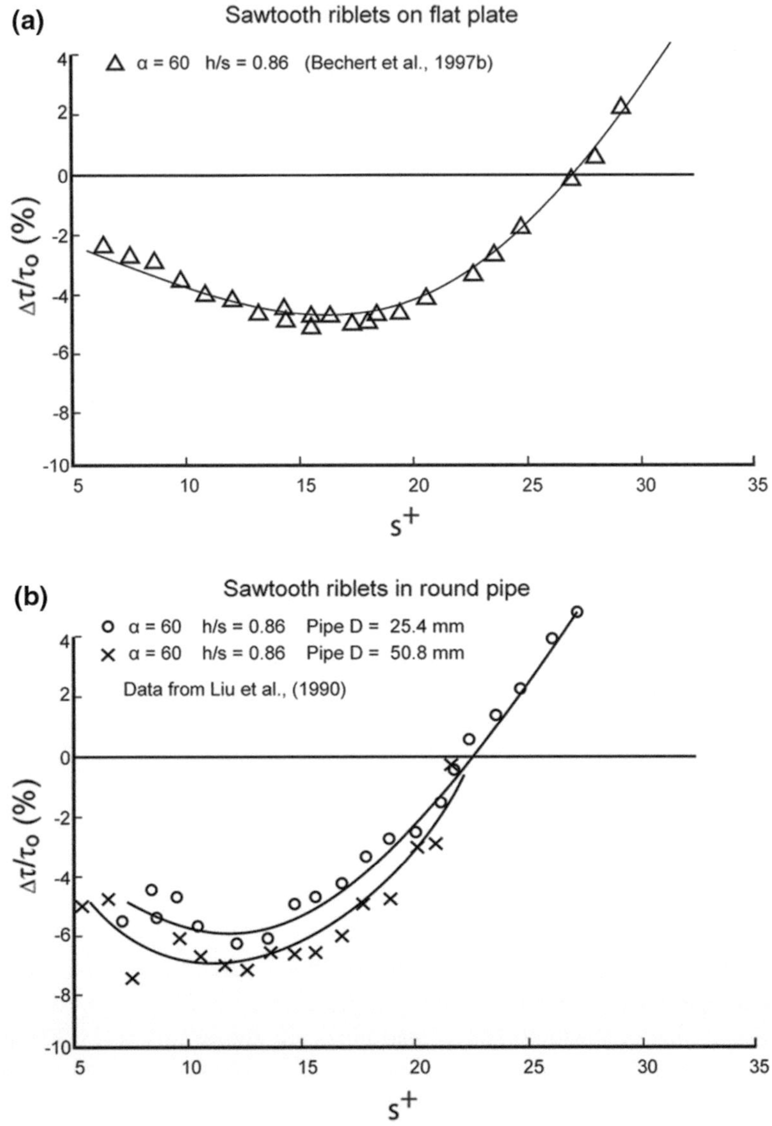

Fig. 11.32 Comparison of riblet performance curves over sawtooth riblets on a **a** flat plate surface and **b** inside round pipes of 25.4 and 50.8 mm diameters. Peak drag reduction for riblets in pipe flow occurs near $s^+ = 12$, which is lower than the flat plate optimal value of $s^+ = 15$. Bechert et al. experiments were performed using oil as the test fluid, and Liu et al. experiments were performed using water as the test fluid in an open channel (adapted from Liu et al. 1990; Bechert et al. 1997b)

ships sail faster when their hulls are sanded in the longitudinal direction, Weiss (1997) fabricated these riblets by using a steel brush moved through the pipeline to create a ridged surface. Studies have shown as much as a 10% reduction in fluid flow with the combined effect of cleaning the pipe and ridging the surface. Tests on a 10 mile long gas pipeline section have confirmed this benefit during commercial operation (Bechert et al. 1997a).

Wind turbines are used for harnessing wind energy which is an important source for renewable energy. Riblet technology can be used on the turbine blades to reduce drag. Riblets can be either fabricated on the blades during production or vinyl film riblets could be applied on the blades.

The only presently available consumer products using riblet technology for drag reduction is competitive swimwear. The FastSkin® suits were introduced by Speedo in 2004 (Krieger 2004). Speedo claimed a drag reduction of several percent in a static test compared to other race suits. However, given the compromises of riblet geometry made during manufacturing, it is hard to believe the full extent of the drag reduction. These suits were worn by American swimmer Michael Phelps in the 2008 Beijing Olympics, where he broke eight Olympic records, seven world records, and won 8 gold medals. Full body swimwear was subsequently banned by the International Swimming Federation (FINA) in 2010 due to an unfair advantage of drag reduction for some swimmers.

The fabric of the FastSkin® swimsuits is knit and the pattern is embossed. The pattern resembles wide blade riblets, shown in Fig. 11.33. Unstretched riblets are tightly packed (Fig. 11.33a). As the fabric stretches, the riblet width and spacing increase (Fig. 11.33b). The associated decrease in h/s ratio depends on the dimensions of each swimmer's body, which is a compromising factor in the design. Riblet thickness is also a factor considered in the design.

Drag reduction is also beneficial in antifouling. Low drag and low adhesion surfaces reduce biofouling. In addition, as indicated in a later chapter, on textured surfaces, microscopic aquatic organisms have difficulty adhering to and colonizing the surface.

Manufacturing techniques for riblets must be chosen specific to their application. Vinyl film riblets are the easiest method, as application of a film to a surface requires less for work small scale applications than other methods. Rolling or grinding methods of riblet application can be used for turbine blades or high volume commercially sold pieces. Laser machining, a two-step molding process using a photolithographically produced master and additive manufacturing can also be used for high volume products. Machining methods are believed to be unfavorable in most instances.

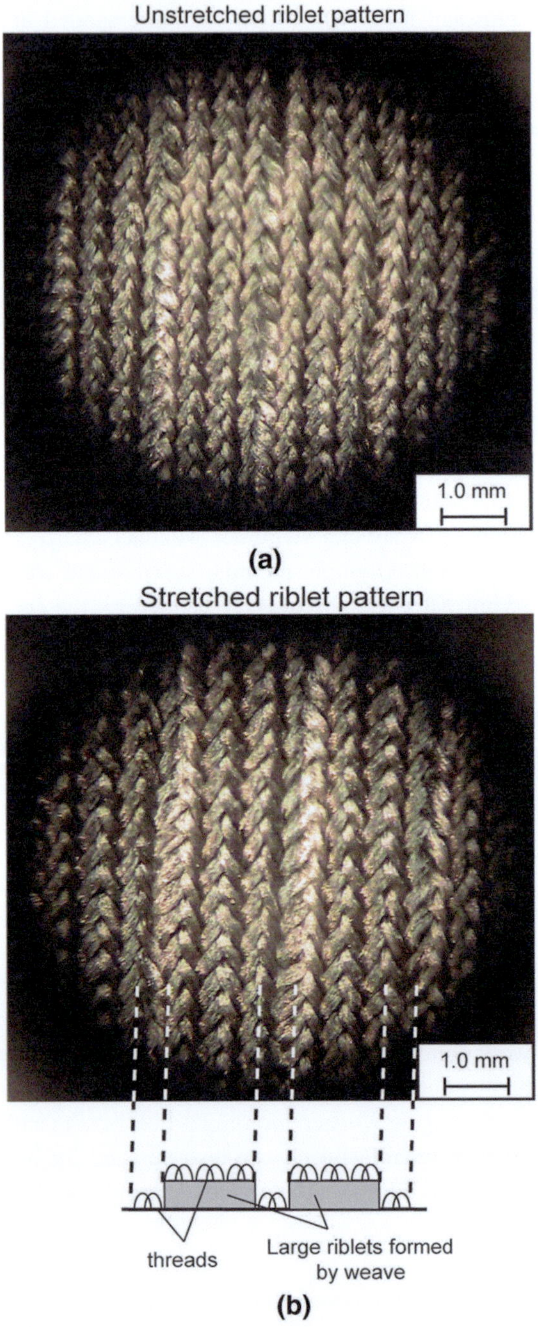

Fig. 11.33 Images of riblet geometries on **a** unstretched and **b** stretched Speedo FastSkin® swimsuit. **c** Schematic showing apparent hierarchical riblet structure formed by threads (Dean and Bhushan 2010)

11.6 Closure

The skin of fast swimming sharks reduces drag and protects against biofouling as they swim through water. Shark skin is covered by very small individual tooth-like scales called dermal denticles (little skin teeth), ribbed with longitudinal grooves (aligned parallel to the real flow direction of water) called riblets. Longitudinal groves encourage anisotropic fluid flow. Fluid drag in the turbulent boundary layer is in large part due to the effects of the streamwise vortices formed in the fluid closest to the surface. Turbulence in the outer boundary layers is in large part due to the translation, ejection, and twisting of these vortices. Additionally, the vortices also cause high velocities at the surface, which create large shear stresses on the surface. Riblets lift the vortices off of the surface and reduce the amount of surface area exposed to the high velocity flow. By modifying the velocity distribution, riblets facilitate a net reduction in shear stress at the surface. Fluid flow experiments and modeling have supported riblet drag reduction theories, Fig. 11.34. To sum up, the microstructured riblets encourage anisotropic flow as well as control vortices on the skin naturally present in turbulent flow which reduces drag. In addition to low drag, the spacing between these dermal denticles is such that microscopic aquatic organisms have difficulty adhering to and colonizing the surface.

Various riblet shapes have been studied experimentally for their drag reducing capabilities, but sawtooth, scalloped, and blade riblets are most common. Optimization studies have been performed by varying flow properties or riblet geometries. Drag reduction by riblet surfaces has been reported to be on the order of 10–30% in water, air, and oil for blade type riblets. A narrow range of h^+ and s^+ and low t^+ with $s^+ \approx 15$ and $h/s \approx 0.5$, blade geometry and continuous configuration has been reported for optimal riblet design.

Fluid flow modeling of continuous and segmented riblet geometries has been carried out. With blade riblets, in the continuous configurations, a drag reduction of approximately 10% occurred when $s^+ \approx 18$–25 and $h^+ \approx 8$–10. With these riblet dimensions, the streamwise vortices were lifted from the surface, reducing the drag on the surface. Furthermore, t^+ should be low for low drag. The aligned-segmented and staggered-segmented riblet configurations had maximum drag reduction of approximately 5 and 3%, respectively. The three parameters of riblet length, gap length, and riblet offset each changed the drag. Longer riblets were better for reducing drag. When gaps between riblets were larger than the length of streamwise vortices, drag increased due to vortices falling between consecutive riblets. Riblet offset did not have much effect. Because continuous riblets have a greater drag reduction than aligned-segmented and staggered-segmented riblets, sharks may implement segmentation for two reasons. These configurations may help create a deformable surface for shark movement and/or in ejection of contaminant particles to keep the shark skin surface clean.

When comparing various riblet geometries, blade riblets were found to have greatest drag reduction of 11% followed by scalloped at 5.7% and sawtooth at 4.1%. The shark inspired scalloped riblet geometry (aligned-segmented) had a drag reduction of

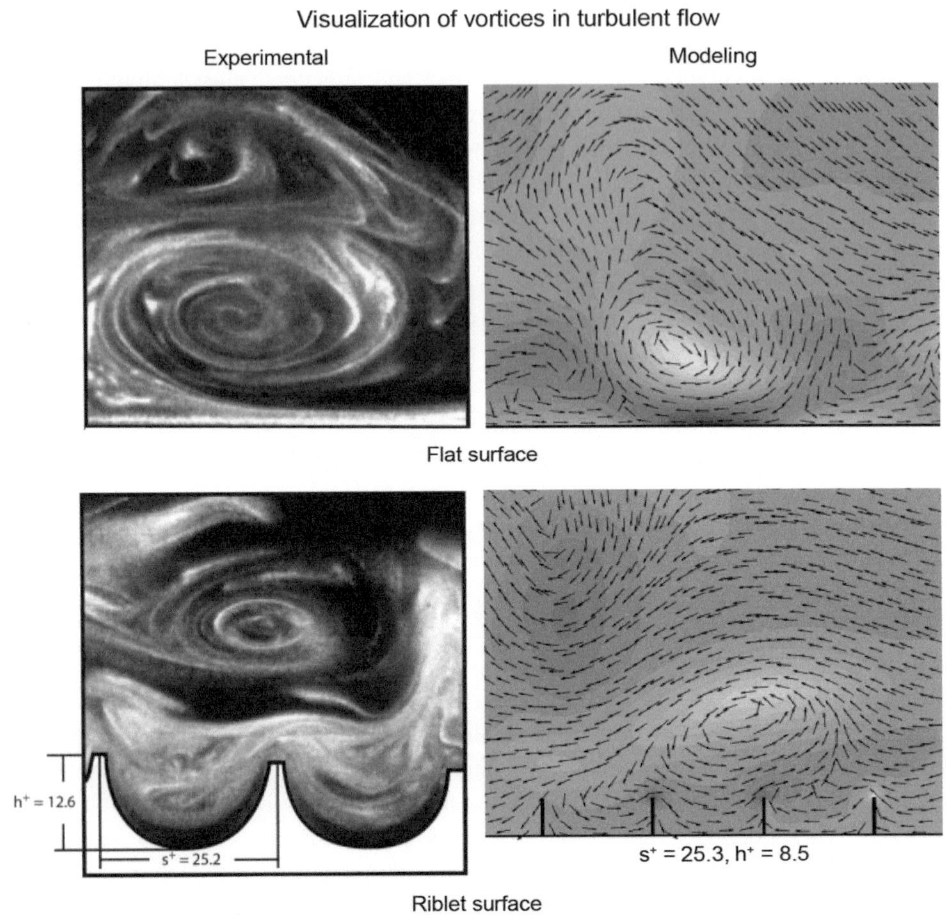

Fig. 11.34 Visualization of streamwise vortices in turbulent flow in a vertical cross-section over flat plate and riblet surfaces in experiments using atomized olive-oil in air (Lee and Lee 2001) and modeling (Martin and Bhushan 2014)

4.5% which is approximately the same drag reduction as the sawtooth and scalloped geometry.

For the greatest drag reduction, blade riblets with an $s^+ = 15$–20 and height of $h^+ = 8$–10 should be chosen on the order of 10%. Nature designs rounded riblets for increased durability. However, with stronger materials, the more effective blade riblets can be utilized. Understanding riblet design allows for low-drag and anti-fouling surfaces in transportation, medical, and industrial applications. Features can range from the micro- to nanoscale dependent upon the scale of the component.

Commercial applications of riblets include competition swimsuits, in which the pattern is embossed, as well as experimental applications to airplanes and boats. Drag reductions in riblet application have been accomplished, and flight applications have seen a fuel savings.

Drag reduction is also helpful in antifouling. Low drag and low adhesion surfaces reduce biofouling. Sharp edges of the dermal denticles make it difficult for microscopic aquatic organisms to adhere to and colonize the surface.

References

Anderson, J.D. (2010), *Fundamentals of Aerodynamics*, fifth ed., McGraw Hill, New York.

Anonymous (1952), *Marine Fouling and its Prevention,* Woods Hole Oceanographic Institute, US Naval Institute, Annapolis, US.

Anonymous (2007), *Pilot's Encyclopedia of Aeronautical Knowledge,* United States Federal Aviation Administration, Skyhorse, New York.

Anonymous (2012), International Maritime Organization, Maritime Knowledge Centre, International Shipping Facts and Figures – Information Resources on Trade, Safety, Security, and Environment Retrieved from <http://www.imo.org/en/KnowledgeCentre/ShipsAndShippingFactsAndFigures/TheRoleandImportanceofIntenationalShipping/Pages/TheRoleandImportanceOfIntenationalShipping.aspx/> 2012

Anonymous (2016a) ANSYS® Fluent, Release 16.0, Help System, Fluent User's Guide, ANSYS, Inc.

Anonymous (2016b) U.S. Department of Transportation, Office of the Assistant Secretary for Research and Technology, Bureau of Transportation Statistics, October 2015 U. S. Airline Traffic Data, Press Release: BTS04-16 Retrieved from <http://www.rita.dot.gov/bts/press_release/bts004_16/> 2016

Anonymous (2016c) U.S. Department of Transportation, Pipeline and Hazardous Materials Safety Administration, Office of Pipeline Safety. Annual Report Summary Retrieved from http://www.phmsa.dot.gov/pipeline/library/data-stats/pipelinemileagefacilities/> 2016.

Barlow, J. B., Rae, W. H., and Pope, A. (1999), *Low-Speed Wind Tunnel Testing,* third ed., Wiley-Interscience, New York.

Batchelor, G. K. (1970), *An Introduction to Fluid Dynamics,* Cambridge University Press, Cambridge, U. K.

Bechert, D. W., Hoppe, G. and Reif, W. E. (1985), "On the Drag Reduction of the Shark Skin," Paper # AIAA-85-0546, presented at AIAA Shear Flow Control Conference, Boulder, CO, AIAA, New York.

Bechert, D. W., Bartenwerfer, M., Hoppe, G., and Reif, W.-E. (1986), "Drag Reduction Mechanisms Derived from Shark Skin," Paper # ICAS-86-1.8.3, Vol. 2 (A86-48-97624-01), pp. 1044-1068, *Proc. 15th ICAS Congress*, AIAA, New York.

Bechert, D. W., Hoppe, G., van der Hoeven, J. G. T. and Makris, R. (1992), "The Berlin Oil Channel for Drag Reduction Research," *Exp. in Fluids* **12**, 251–260.

Bechert, D. W., Bruse, M., Hage, W., and Meyer, R. (1997a), "Biological Surfaces and Their Technological Application – Laboratory and Flight Experiments on Drag Reduction and Separation Control," Paper # AIAA-1997-1960, presented at AIAA 28th Fluid Dynamics Conference, Snowmass Village, CO, AIAA, New York.

Bechert, D. W., Bruse, M., Hage, W., van der Hoeven, J. G. T. and Hoppe, G. (1997b), "Experiments on Drag Reducing Surfaces and their Optimization with an Adjustable Geometry," *J. Fluid Mech.* **338**, 59–87.

Bechert, D. W., Bruse, M. and Hage, W. (2000a), "Experiments with Three-dimensional Riblets as an Idealized Model of Shark Skin," *Exp. Fluids* **28**, 403–412.

Bechert, D. W., Bruse, M., Hage, W. and Meyer, R. (2000b), "Fluid Mechanics of Biological Surfaces and their Technological Application," *Naturwissenschaften* **87**, 157–171.

Bhushan, B. (2009), "Biomimetics: Lessons from Nature – an Overview," *Phil. Trans. R. Soc.* **367**, 1445–1486.

Bhushan, B. (2012), "Bioinspired Structured Surfaces," *Langmuir* **28**, 1698–1714.

Bhushan, B. and Caspers, M. (2017) "An Overview of Additive Manufacturing (3D Printing) for Microfabrication," *Microsyst. Technol.* **23**, 1117–1124.

Bhushan, B. and Jung, Y.C. (2011), "Natural and Biomimetic Artificial Surfaces for Superhydrophobicity, Self-Cleaning, Low Adhesion, and Drag Reduction," *Prog. Mater. Sci.* **56**, 1–108.

Bixler, G. D. and Bhushan, B. (2012a), "Biofouling Lessons from Nature," *Phil. Trans. R. Soc. A,* **370**, 2381–2417.

Bixler, G. D. and Bhushan, B. (2012b), "'Bioinspired Rice Leaf and Butterfly Wing Surface Structures Combining Shark Skin and Lotus Effects," *Soft Matter* **8**, 11271–11284.

Bixler, G. D. and Bhushan, B. (2013a), "Fluid Drag Reduction with Shark-skin Riblet Inspired Microstructured Surfaces," *Adv. Func. Mater.* **23**, 4507–4528.

Bixler, G. D. and Bhushan, B. (2013b), "Shark Skin Inspired Low-drag Microstructured Surfaces in Closed Channel Flow," *J. Colloid Interf. Sci.* **393**, 384–396.

Bixler, G. D. and Bhushan, B. (2013c), "Bioinspired Micro/Nanostructured Surfaces for Oil Drag Reduction in Closed Channel Flow," *Soft Matter* **9**, 1620–1635.

Bixler, G. D. and Bhushan, B. (2013d), "Fluid Drag Reduction and Efficient Self-Cleaning with Rice Leaf and Butterfly Wing Bioinspired Surfaces," *Nanoscale* **5**, 7685–7710.

Blevins, R. D. (1984), *Applied Fluid Dynamics Handbook,* Van Nostrand-Reinhold, New York.

Brennan, A. B., Baney, R. H., Carman, M. I., Estes, T. G., Feinberg, A. W., Wilson, L. H. and Schumacher, J. F.. (2010), "Surface Topographies for Non-Toxic Bioadhesion Control," U. S. Patent no. 7,650,848 B2, Jan. 26.

Brostow, W. (2008), "Drag Reduction in Flow: Review of Applications, Mechanism, and Prediction," *J. Indust and Eng Chem* **14**, 408–416.

Burger, E. D., Munk, W. R. and Wahl, H. A. (1982), "Flow Increase in the Trans Alaska Pipeline Through Use of a Polymeric Drag-Reducing Additive," *J. Petro Tech,* **34**, 377–386.

Buttner, C. C. and Schulz, U. (2011), "Shark Skin Inspired Riblet Structures as Aerodynamically Optimized High Temperature Coatings for Blades of Aeroengines," *Smart Matl. and Struct.* **20**, 1–9.

Caram, J. M. and Ahmed, A. (1991), "Effect of Riblets on Turbulence in the Wake of an Airfoil," *AIAA* **29**, 1769–1770.

Caram, J. M. and Ahmed, A. (1992), "Development of the Wake of an Airfoil with Riblets," *AIAA* **30**, 2817–2818.

Chan, J. and Wong, S. (eds.) (2010), *Biofouling Types, Impact and Anti-Fouling,* Nova Science Publishers, New York.

Choi, K. S., Gadd, G. E., Pearcey, H. H., Savill, A. M., and Svensson, S. (1989), "Tests of Drag-Reducing Polymer Coated on a Riblet Surface," *Appl. Sci. Res.* **46**, 209–216.

Choi, H., Moin, P. and Kim, J. (1993), "Direct Numerical Simulation of Turbulent Flow Over Riblets," *J. Fluid Mech.* **255**, 503–539.

Choi, K. S., Yang, X., Clayton, B. R., Glover, E. J., Altar, M., Semenov, B. N., and Kulik, V. M. (1997), "Turbulent Drag Reduction Using Compliant Surfaces," *Proc. R. Soc A* **453**, 2229–2240.

Chu, D. C. and Karniadakis, G. E. (1993), "A Direct Numerical Simulation of Laminar and Turbulent Flow Over Riblet-mounted Surfaces," *J. Fluid Mech.* **250**, 1–42.

Collins, M. W. and Brebbia, C. A. (eds.) (2004), *Design and Nature II Comparing Design in Nature with Science and Engineering,* WIT Press, Southampton, UK.

Copisarow, M. (1945), "Marine Fouling and its Prevention," *Science* **101**, 406–407.

Coustols, E. (1989), "Behavior of Internal Manipulators: "Riblet" Models in Subsonic and Transonic Flows," Paper # AIAA-89-0963, presented at AIAA Shear Flow Conference, Tempe, AZ, AIAA, New York.

Coustols, E. and Cousteix, J. (1994), "Performances of Riblets in the Supersonic Regime," *AIAA* **32**, 431–433.

Coustols, E. and Schmitt, V. (1990), "Synthesis of Experimental Riblet Studies in Transonic Conditions," in *Turbulence Control by Passive Means* (ed. E. Coustols), pp. 123–140, Kluwer, Dordrecht, The Netherlands.

Daniello, R. J., Waterhouse, N. E. and Rothstein, J. P. (2009), "Drag Reduction in Turbulent Flows over Superhydrophobic Surfaces," *Phys. Fluids* **21**, 085103.

Davies, M. (ed.) (2002), *Standard Handbook for Aeronautical and Astronautical Engineers,* McGraw-Hill, New York.

Dean, B. and Bhushan, B. (2010), "Shark-skin Surfaces for Fluid-Drag Reduction in Turbulent Flow: a Review," *Phil. Trans. R. Soc.* **368**, 4775–4806.

Dean, B. and Bhushan, B. (2012), "The Effect of Riblets in Rectangular Duct Flow," *Appl. Surf. Sci.* **258**, 3936–3947.

Denkena, B., Kohler, J. and Wang, B. (2010), "Manufacturing of Functional Riblet Structures by Profile Grinding," *CIRP J. Manuf. Sci. Tech.* **3**, 14–26.

Deyuan, Z., Yuehao, L., Huawei, C. and Xinggang, J. (2011), "Exploring Drag-Reducing Grooved Internal Coating for Gas Pipelines," *Pipeline & Gas Journal* **238,** 59–60.

Dickinson, P. H. and Proudley, G. M. (1991), "Methods of Manufacture and Surface Treatment Using Laser Radiation," U. S. Patent no. 4,994,639, Feb. 19.

El-Samni, O. A., Chun, H. H. and Yoon, H. S. (2007), "Drag Reduction of Turbulent Flow Over Thin Rectangular Riblets," *Int. J. Eng. Sci.* **45**, 436–454.

Enyutin, G. V., Lashkov, Y. A., and Samoilova, N. V. (1995), "Drag Reduction in Riblet-Lined Pipes," *Fluid Dynamics*, **30**, 45–48.

Frings, B. (1988), "Heterogeneous Drag Reduction in Turbulent Pipe Flows using Various Injection Techniques," *Rheologica Acta* **27**, 92–110.

Gillcrist, M. C. and Reidy, L. W. (1989), "Drag Measurements on Marine Vehicles with a Riblet Surface Coating," in *Drag Reduction in Fluid Flows: Techniques for Friction Control* (eds. R. H. J. Sellin and R.T. Moses), pp. 99–106, Ellis Horwood Publishers, Chichester, England.

Goldstein, D., Handler, R. and Sirovich, L. (1995), "Direct Numerical Simulation of Turbulent Flow Over a Modelled Riblet-covered Surface," *J. Fluid Mech.* **302**, 333–376.

Gruneberger, R. and Hage, W. (2011), "Drag Characteristics of Longitudinal and Transverse Riblets at Low Dimensionless Spacings," *Exp. Fluids,* **50**, 363–373.

Han, M., Huh, J. K., Lee, S. S., and Lee, S. (2002), "Micro-Riblet Film for Drag Reduction," Proceedings of the Pacific Rim Workshop on Transducers and Micro/Nano Technologies, Xiamen, China.

Han, M., Lim, H. C., Jang, Y.-G., Seung, S. L., and Lee, S.-J. (2003), "Fabrication of a Micro-Riblet Film and Drag Reduction Effects on Curved Objects," Paper # 0–7803-7731-1, presented at 12th International Conference on Solid State Sensors, Actuators and Microsystems, Boston, MA.

Hirt, G. and Thome, M. (2007), "Large Area Rolling of Functional Metallic Micro Structures," *Prod. Eng. Res. Devel.* **1**, 351–356.

Hoyt, J. W. (1975), "Hydrodynamic Drag Reduction Due to Fish Slimes," in *Swimming and Flying in Nature* (ed. T. Y. T. Wu), Vol. 2, pp. 653–672.

Jimenez, J. and Moin, P. (1991), "The Minimal Flow Unit in Near-wall Turbulence," *J. Fluid Mech.* **225**, 213–240.

Jung, Y. C. and Bhushan, B. (2010), "Biomimetic Structures for Fluid Drag Reduction in Laminar and Turbulent Flows," *J. Phys.: Condens. Matter* **22**, 1–9.

Kennedy, J. L. (1993), *Oil and Gas Pipeline Fundamentals,* PennWell Books, Tulsa, Oklahoma.

Kesel, A. and Liedert, R. (2007), "Learning from Nature: Non-Toxic Biofouling Control by Shark Skin Effect," *Comp. Biochem. Physiol. A* **146**, S130.

Kline, S. J., Reynolds, W. C., Schraub, F. A. and Runstadler, P. W. (1967), "The Structure of Turbulent Boundary Layers." *J. Fluid Mech.* **30**, 741–773.

Koch, K. and Barthlott, W. (2009), "Superhydrophobic and Superhydrophilic Plant Surfaces: an Inspiration for Biomimetic Materials," *Phil. Trans. R. Soc.* **367**, 1487–1509.

Koury, E. and Virk P. S. (1995), "Drag Reduction by Polymer Solutions in a Riblet-Lined Pipe," *Appl. Sci. Res.* **54**, 323–347.

Krieger, K. (2004), "Do Pool Sharks Really Swim Faster?" *Science* **305**, 636–637.

Lee, S. J. and Lee, S. H. (2001), "Flow Field Analysis of a Turbulent Boundary Layer over a Riblet Surface," *Exp. Fluids* **30**, 153–166.

Lee, J. D., Gregorek, G. M., and Korkan, K. D. (1978), "Testing Techniques and Interference Evaluation in the OSU Transonic Airfoil Facility," Paper # AIAA-1978-1118, presented at 11th Fluid and Plasma Dynamics Conference, Seattle, WA, AIAA, New York.

Liu, K. N., Christodoulou, C., Riccius, O. and Joesph, D. D. (1990), "Drag Reduction in Pipes Lined with Riblets," *AIAA* **28**, 1697–1698.

Lowrey, P. and Harasha, J. (1991), "A Preliminary Assessment of the Feasibility of using Riblets in Internal Flows to Conserve Energy," *Energy* **16**, 631–642.

Luchini, P., Manzo, F., and Pozzi, A. (1991), "Resistance of a Grooved Surface to Parallel Flow and Cross-flow," *J. Fluid Mech.* **228**, 87–109.

Marentic, F. J. and Morris, T. L. (1992), "Drag Reduction Article," U. S. Patent no. 5,133,516, Jul. 28.

Martell, M. B., Perot, J. B., and Rothstein, J. P. (2009), "Direct Numerical Simulations of Turbulent Flows over Drag-reducing Ultrahydrophobic Surfaces," *J. Fluid Mech.* **620**, 31–41.

Martell, M. B., Rothstein, J. P. and Perot, J. B. (2010), "An Analysis of Superhydrophobic Turbulent Drag Reduction Mechanisms using Direct Numerical Simulation," *Phys. Fluids* **22**, 065102.

Martin, S. and Bhushan, B. (2014), "Fluid Flow Analysis of a Shark-Inspired Microstructure," *J. Fluid Mech.* **756**, 5–29.

Martin, S. and Bhushan, B. (2016a), "Fluid Flow Analysis of Continuous and Segmented Riblet Structures," *RSC Adv.* **6**, 10962–10978.

Martin, S. and Bhushan, B. (2016b), "Modeling and Optimization of Shark-inspired Riblet Geometries for Low Drag Application," *J. Colloid Interface Sci.* **474**, 206–215.

Martínez-Palou, R., Mosqueira, M., Zapata-Rendón, B., Mar-Juárez, E., Bernal-Huicochea, C., Clavel-López, J. and Aburto, J. (2011), "Transportation of Heavy and Extra-heavy Crude Oil by Pipeline: A review," *J. Petrol. Sci. Eng,* **75**, 274–282.

Melo, L. F., Bott, T. R. and Bernardo, C. A. (eds.) (1988), *Fouling Science and Technology,* Kluwer Academic Publishers, Dordrecht, The Netherlands.

Munson, B. R., Rothmayer, A. P., Okiishi, T. M., and Huebsch, W. D. (2012), *Fundamentals of Fluid Mechanics,* seventh ed., Wiley, New York.

Neumann, D. and Dinkelacker, A. (1991), "Drag Measurements on V-grooved Surfaces on a Body of Revolution in Axial Flow," *App Sci Res* **48**, 105–114.

Nitschke, P. (1984), "Experimental Investigation of Turbulent Flow in Smooth and Longitudinal Grooved Pipes," *NASA STI/Recon Technical Report* **N88**.

Oeffner, J. and Lauder, G. V. (2012), "The Hydrodynamic Function of Shark Skin and Two Biomimetic Applications," *J. Exp. Biology* **215**, 785–795.

Ou, J., Perot, B. and Rothstein, J. P. (2004), "Laminar Drag Reduction in Microchannels using Ultrahydrophobic Surfaces," *Phys. Fluids* **16**, 4635–4643.

Pritchard, P. J. and Mitchell, J. W. (2015), *Fox and McDonald's Introduction to Fluid Mechanics,* ninth ed., Wiley, New York.

Railkin, A. I. (2004), *Marine Biofouling Colonization Processes and Defenses,* CRC Press, Boca Raton, Florida.

Ralston, E. and Swain, G. (2009), "Bioinspiration – the Solution for Biofouling Control?" *Bioinsp. Biomim.* **4**, 1–9.

Ray, D. L. (ed.) (1959), *Marine Boring and Fouling Organisms,* University of Washington Press, Seattle, WA.

Reidy, L. W. and Anderson, G. W. (1988), "Drag Reduction for External and Internal Boundary Layers Using Riblets and Polymers," Paper # AIAA-1988-0138, presented at 26th AIAA Aerospace Sciences Meeting, Reno, NV, AIAA, New York.

Reif, W. (1985), *Squamation and Ecology of Sharks,* Vol. 78, pp. 1–255, Courier Forschungsinstitut Senckenberg, Frankfurt, Germany.

Rohr, J. J., Andersen, G. W., Reidy, L. W. and Hendricks, E. W. (1992), "A Comparison of the Drag-reducing Benefits of Riblets in Internal and External Flows," *Exp. in Fluids*, **13**, 361–368.

Sareen, A., Deters, R. W., Henry, S. P. and Selig, M. S. (2011), "Drag Reduction Using Riblet Film Applied to Airfoils for Wind Turbines," Paper # AIAA-2011-558, presented at 49th AIAA Aerospace Sciences Meeting, Orlando, FL, AIAA, New York.

Scardino, A. J. (2009), "Surface Modification Approaches to Control Marine Biofouling," in *Advances in Marine Antifouling Coatings and Technologies* (eds. C. Hellio and D. Yebra), pp. 664–692, CRC Press, Boca Raton, Florida.

Schulz, M. J., Shanov, V. N. and Yun, Y. (eds.) (2009), *Nanomedicine Design of Particles, Sensors, Motors, Implants, Robots, and Devices,* Artech House, Boston.

Schumacher, J. F., Aldred, N., Callow, M. E., Finlay, J. A., Callow, J. A., Clare, A. S. and Brennan, A. B. (2007), "Species-Specific Engineered Antifouling Topographies: Correlations between the Settlement of Algal Zoospores and Barnacle Cyprids," *Biofouling* **23,** 307– 317.

Shephard, K. L. (1994), "Functions for Fish Mucus," *Rev. Fish Biology and Fisheries* **4**, 401–429.

Shirtliff, M. and Leid, J. G. (eds.) (2009), *The Role of Biofilms in Device-Related Infections,* Springer-Verlag, Berlin.

Somerscales, E. F. C. and Knudsen, J. G. (eds.) (1981), *Fouling of Heat Transfer Equipment,* Hemisphere Publishing Corporation, Washington, D. C.

Subaschandar, N., Kumar, R., and Sundaram, S. (1999), "Drag Reduction Due to Riblets on a GAW(2) Airfoil," *J. of Aircraft*, **36**, 890–892.

Sundaram, S., Viswanath, P. R., and Rudrakumar, S. (1996), "Viscous Drag Reduction Using Riblets on NACA 0012 Airfoil to Moderate Incidence," *AIAA* **34**, 676–682.

Sundaram, S., Viswanath, P. R., and Subaschandar, N. (1999), "Viscous Drag Reduction Using Riblets on a Swept Wing," *AIAA* **37**, 851–856.

Viswanath, P. R. (1999), "Riblets on Airfoil and Wings: A Review," Paper # AIAA-99-3402, presented at the AIAA 30th Fluid Dynamics Conference, Norfolk, VA, AIAA, New York.

Viswanath, P. R. and Mukund, R. (1995), "Turbulent Drag Reduction Using Riblets on a Supercritical Airfoil at Transonic Speeds," *AIAA* **33**, 945–947.

Viswanath, P. R. (2002), "Aircraft Viscous Drag Reduction using Riblets," *Prog Aerospace Sci*, **38**, 571–600.

Vo-Dinh, T. (ed.) (2007), *Nanotechnology in Biology and Medicine,* CRC Press, Boca Raton, Florida.

Walker, J., Surman, S. and Jass, J. (eds.) (2000), *Industrial Biofouling Detection, Prevention and Control,* Wiley, New York.

Walsh, M. J. (1982), "Turbulent Boundary Layer Drag Reduction using Riblets," Paper # AIAA-1982-0169, presented at AIAA 20th Aerospace Sciences Meeting, Orlando FL, AIAA, New York.

Walsh, M. J. (1990a), "Effect of Detailed Surface Geometry on Riblet Drag Reduction Performance," *J. Aircraft,* **27**, 572–573.

Walsh, M. J. (1990b), "Riblets. Viscous Drag Reduction in Boundary Layers," *Progr. in Astronaut. and Aeronaut.*, **123**, 203–261.

Walsh, M. J. and Lindemann, A. M. (1984), "Optimization and Application of Riblets for Turbulent Drag Reduction," Paper # AIAA-1984-0347, presented at AIAA 22nd Aerospace Sciences Meeting, Reno, NV, AIAA, New York.

Walsh, M. J. and Anders, J. B., Jr. (1989), "Riblet/LEBU Research at NASA Langley," *Appl. Sci. Res.* **46**, 255–262.

Weiss, M. H. (1997), "Implementation of Drag Reduction Techniques in Natural Gas Pipelines," presented at 10th European Drag Reduction Working Meeting, Berlin, Germany, March 19–21.

Wen, L., Weaver, J. C., and Lauder, G. V. (2014), "Biomimetic Shark Skin: Design, Fabrication, and Hydrodynamic Function," *J. Exp. Biology* **217**, 1656–1666.

Wetzel, K. K. and Farokhi, S. (1996), "Interaction of Riblets and Vortex Generators on an Airfoil," Paper # AIAA-1996-2428-CP, presented at the AIAA 14th Applied Aero-dynamics Conference, New Orleans, LA, AIAA, New York.

White, F. (2006), *Viscous Fluid Flow,* third ed., McGraw Hill, New York.

Wilkinson, S. P. and Lazos, B. S. (1988), "Direct Drag and Hot-wire Measurements on Thin-element Riblet Arrays," in *Turbulence Management and Relaminarisation* (eds. H. W. Liepmann and R. Narasimha), pp. 121–131, Springer-Verlag, Berlin, Germany.

12 Gecko Adhesion

12.1 Introduction

The leg attachment pads of several animals are capable of attaching to and detaching from a variety of surfaces and are used for locomotion, even on vertical walls or across the ceiling (Gorb 2001; Bhushan 2007). These include many insects, spiders, and lizards. Biological evolution has led to the optimization of their leg attachment systems. This dynamic attachment ability is referred to as reversible adhesion or smart adhesion (Bhushan et al. 2006). Many insects (e.g., beetles and flies) and spiders have been the subject of investigative interest. However, the attachment pads of geckos have been the most widely studied because they have the highest body mass and exhibit the most versatile and effective adhesive known in nature. Therefore, this chapter will be concerned primarily with gecko adhesion.

There are over 1000 species of geckos (Kluge 2001; Han et al. 2004) that have attachment pads of varying morphology (Ruibal and Ernst 1965). However, the Tokay gecko (*Gekko gecko*) native to south-east Asia, has been the main focus of scientific research (Hiller 1968; Irschick et al. 1996; Autumn et al. 2000; Autumn 2006) because of its size and availability. The Tokay gecko is the second largest gecko species, reaching lengths of approximately 0.3–0.4 m and 0.2–0.3 m for males and females, respectively. They have a distinctive blue or gray body with orange or red spots and can weigh up to 300 g (Tinkle 1992).

Almost 2500 years ago, the ability of the gecko to "run up and down a tree in any way, even with the head downwards," was observed by Aristotle (Aristotle/Thompson 1918, Book IX, Part 9). Even though the adhesive ability of geckos has been known since the time of Aristotle, little was understood about this phenomenon until the late nineteenth century when the microscopic hairs covering the toes of the gecko were first noted. In

B. Bhushan, *Introduction to Biomimetics and Bioinspiration*,
https://doi.org/10.1007/978-3-031-62344-8_12

the 1950s, electron microscopy enabled scientists to view the complex hierarchical morphology that covers the skin on the gecko's toes. Scientific studies have been conducted to determine the factors that allow the gecko to adhere to and detach from surfaces at will, including surface structure (Ruibal and Ernst 1965; Russell 1975, 1986; Williams and Peterson 1982; Schleich and Kästle 1986; Irschick et al. 1996; Autumn and Peattie 2002; Arzt et al. 2003); the mechanisms of adhesion (Wagler 1830; Simmermacher 1884; Schmidt 1904; Hora 1923; Dellit 1934; Ruibal and Ernst 1965; Hiller 1968; Gennaro 1969; Stork 1980; Autumn et al. 2000, 2002; Bergmann and Irschick 2005; Huber et al. 2005b); and adhesion strength (Hiller 1968; Irschick et al. 1996; Autumn et al. 2000; Arzt et al. 2003; Huber et al. 2005a, b). Modeling the gecko attachment system as a system of springs (Bhushan et al. 2006; Kim and Bhushan 2007a, b, c, 2008) has provided valuable insight into adhesion enhancement. van der Waals forces are widely accepted in literature as the dominant adhesion mechanism used by hierarchical attachment systems. Capillary forces created by humidity naturally present in the air can further increase the adhesion force generated by the spatulae (Kim and Bhushan 2008). Both experimental and theoretical work supports these adhesion mechanisms.

There is great interest among the scientific community to study further the characteristics of gecko feet. It is hoped that this information could be applied to the production of micro/nanosurfaces capable of recreating the adhesion forces generated by these lizards (Bhushan 2007). Common man-made adhesives such as tape or glue involve the use of wet adhesives that permanently attach two surfaces. However, replication of the characteristics of gecko feet would enable the development of a superadhesive surface capable of clean, dry adhesion. They have the potential for use in everyday objects such as adhesive tapes, fasteners, and toys and in high technology, such as microelectronic and space applications. These structures also can be used in prosthetics and surgical masks, and can bind to components in microelectronics without the high heat associated with various soldering processes. These structures will never dry out in a vacuum—a common problem in aerospace applications. Replication of the dynamic climbing and peeling ability of geckos could find use in the treads of wall-climbing robots.

This chapter first introduces various hairy attachment systems. Then, it introduces the Tokay Gecko and its construction, and some details on adhesion, peeling, and self-cleaning mechanisms. The next section describes the attachment mechanisms, followed by adhesion measurements and data, adhesion modeling, and adhesion data base of fibrillar structures. The last section presents various fabrication techniques to create structures in the lab.

12.2 Hairy Attachment Systems

There are generally two kinds of attachment pads. The first kind are relatively smooth pads, so-called arolia and euplantulae, which are soft and deformable and are found in some amphibians, such as tree frogs, torrent frogs, cockroaches, grasshoppers, and bugs. They are able to attach to and move over wet or even flooded environments without falling (Federle et al. 2006). Tree frog toe attachment pads consist of a hexagonal array of flat-topped epidermal cells about 10 μm in size separated by approximately 1 μm wide mucus-filled channels. The flattened surface of each cell consists of submicron array of nanopillars or pegs of approximately 100–400 nm diameter. The toe pads are made of an extremely soft, nonhomogeneous material. The epithelium itself has an effective elastic modulus of about 15 MPa, equivalent to silicone rubber. The pads are permanently wetted by mucus secreted from glands that open into the channels between epidermal cells. They attach to mating surfaces by wet adhesion (Federle et al. 2006). They are capable of climbing on wet rocks even when water is flowing over the surface (Barnes et al. 2002). During walking, squeezing of the mucus glands is expected to occur rapidly. Torrent frogs can resist sliding even on flooded surfaces. The surface of their toe pad is similar to that of tree frogs, with some changes in the structure to handle the large flow of water (Ohler 1995).

The second kind of attachment pads are the hairy types, which consist of long, deformable setae, and are found in many insects (e.g., beetles, flies), spiders, and lizards. The microstructures used by beetles, flies, spiders, and geckos have similar structures and can be seen in Fig. 12.1a. As the size (mass) of the creature increases, the radius of the terminal attachment elements decreases. This allows a greater number of setae to be packed into an area, increasing the linear dimension of contact and the adhesion strength. Arzt et al. (2003) determined that the density of the terminal attachment elements, ρ_A, m^{-2} strongly increases with increasing body mass, m, in g. In fact, a master curve can be fitted for all the different species (Fig. 12.1b),

$$\log \rho_A = 13.8 + 0.669 \log m \tag{12.1}$$

The correlation coefficient of the master curve is equal to 0.919. Beetles and flies have the largest attachment pads and the lowest density of terminal attachment elements. Spiders have highly refined attachment elements that cover their legs. Geckos have both the highest body mass and greatest density of terminal elements (spatulae). Spiders and geckos can generate high dry adhesion, whereas beetles and flies increase adhesion by secreting liquid stored generally within a spongy layer of cuticle and delivered at the contacting surface through a system of porous channels (Gorb 2001; Arzt et al. 2003; Kesel et al. 2003). Note that in the smooth attachment system discussed earlier, the secretion is essential for attachment.

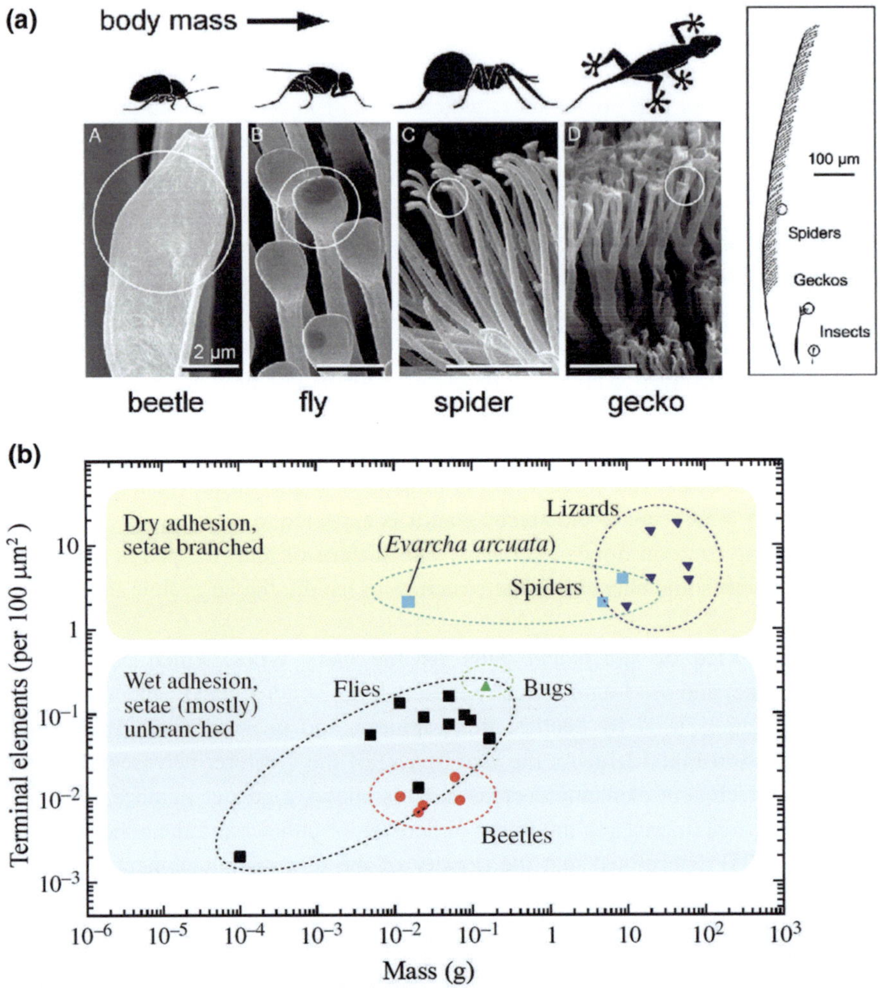

Fig. 12.1 **a** Terminal elements of the hairy attachment pads of a beetle, fly, spider, and gecko shown at two different scales (Arzt et al. 2003), and **b** the dependence of terminal element density on body mass (Federle 2006). Data are from Arzt et al. (2003) and Kesel et al. (2003)

Peattie and Full (2007) have revisited the scaling of terminal attachment elements with the body mass using a phylogenetic approach. In their work, a larger set of species (81) over a wider range of body mass and setal morphology were considered. They found that fiber morphology is better predicted by evolutionary history and adhesion mechanism (dry or wet) than by body mass.

Figure 12.2 shows the scanning electron micrographs (SEM) of the end of the legs of two flies—the fruit fly (drosophila melanogaster) and the syrphid fly. The fruit fly uses setae with flattened tips (spatulae) on the two hairy rods for attachment to smooth surfaces and two front claws for attachment to rough surfaces. The front claws are also used for locomotion. The syrphid fly uses setae on the legs for attachment. In both cases, fluid is secreted at the contacting surface to increase adhesion.

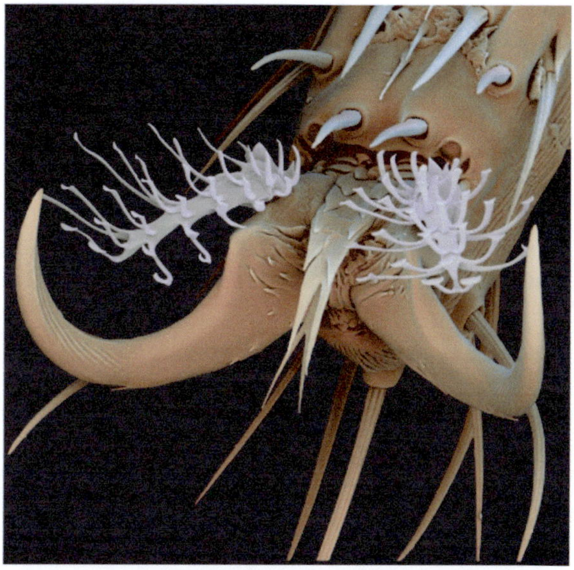

End of the legs of fruit fly (drosophila melanogaster)

End of the leg of syrphid fly

Fig. 12.2 SEM micrographs of the end of the legs of fruit fly (drosophila melanogaster) and syrphid fly (Gorb 2001)

12.3 Tokay Gecko

12.3.1 Construction of Tokay Gecko

The explanation for the adhesion properties of gecko feet can be found in the surface morphology of the skin on the toes of the gecko. The skin is comprised of a complex hierarchical structure of lamellae, setae, branches, and spatulae (Ruibal and Ernst 1965). Figure 12.3 shows various SEM micrographs of a gecko foot, showing the hierarchical structure down to the nanoscale. Figure 12.4 shows a schematic of the structure, and Table 12.1 summarizes the surface characteristics. The gecko attachment system consists of an intricate hierarchy of structures beginning with lamellae, soft ridges 1–2 mm in length (Ruibal and Ernst 1965) that are located on the attachment pads (toes) that compress easily so that contact can be made with rough bumpy surfaces. Tiny curved hairs known as setae extend from the lamellae with a density of approximately 14,000 per square millimeter (Schleich and Kästle 1986). These setae are typically 30–130 μm in length, 5–10 μm in diameter (Ruibal and Ernst 1965; Hiller 1968; Russell 1975; Williams and Peterson 1982), and are composed primarily of β-keratin (Maderson 1964; Russell 1986) with some α-keratin component (Rizzo et al. 2006). Each seta divides into branches that separate into spatulae, with 100 to 1000 total spatulae per seta (Ruibal and Ernst 1965; Hiller 1968). The spatulae are typically 2–5 μm in length with a diameter of 0.1–0.2 μm (Ruibal and Ernst 1965), and they form the points of contact with the surface. The tips of the spatulae are approximately 0.2–0.3 μm in width (Ruibal and Ernst 1965), 0.5 μm in length, and 0.01 μm in thickness (Persson and Gorb 2003) and garner their name from their resemblance to a spatula.

The attachment pads on two feet of the Tokay gecko have an area of about 220 mm^2. About three million setae on their toes that branch off into about three billion spatulae on two feet can produce a clinging ability of about 20 N (vertical force required to pull a lizard down a nearly vertical (85°) surface) (Irschick et al. 1996) and allow them to climb vertical surfaces at speeds over 1 m/s with the capability to attach and detach their toes on the order of 10 ms. In isolated setae, a 2.5 μN preload yielded adhesion of 20–40 μN, and thus the adhesion coefficient, which represents the strength of adhesion as a function of preload, ranges from 8 to 16 (Autumn et al. 2002).

12.3.2 Adhesion Enhancement by Division of Contacts and Multilevel Hierarchical Structure

Typical rough, rigid surfaces are able to make intimate contact with a mating surface only over a very small portion of the perceived apparent area of contact. In fact, the real area of contact is typically two to six orders of magnitude less than the apparent area of contact (Bhushan 2013a, b, 2017a). Autumn et al. (2002) proposed that divided contacts

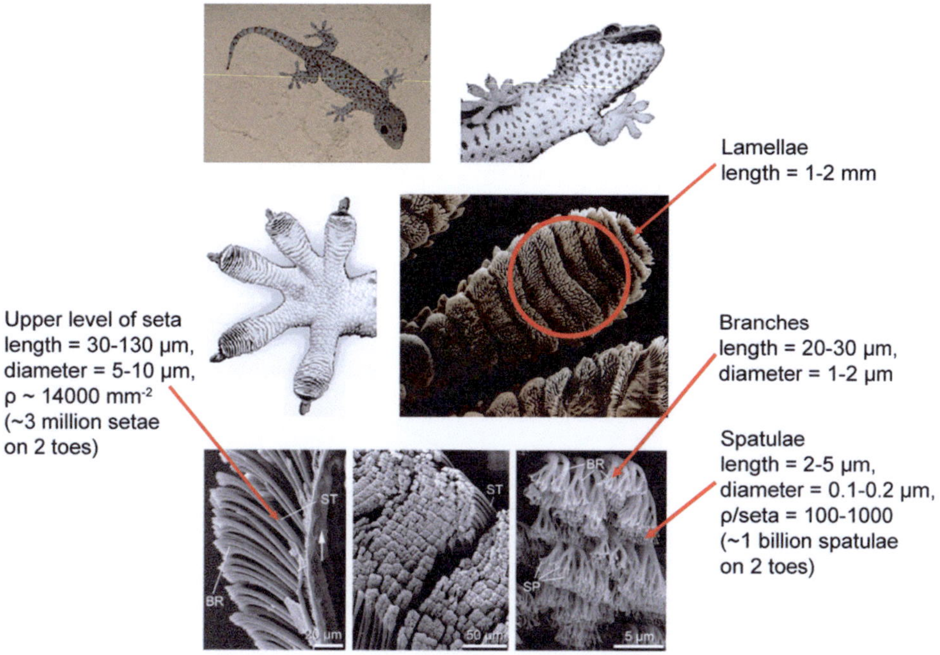

Fig. 12.3 The hierarchical structures of a Tokay gecko foot: a gecko foot (Autumn et al. 2000) and a gecko toe (Autumn 2006). The two feet with an area of 220 mm^2 consist of about three million setae on their toes that branch off into about three billion spatulae. Scanning electron microscope (SEM) micrographs of the setae and the spatulae are shown in the bottom row (Gao et al. 2005). ST seta, SP spatula, BR branch

serve as a means for increasing adhesion. The surface energy approach can be used to calculate adhesion force in a dry environment in order to calculate the effect of division of contacts. If the tip of a spatula is considered as a hemisphere with radius, R, adhesion force of a single contact, F_{ad}, based on the so-called JKR (Johnson-Kendall-Roberts) theory (Johnson et al. 1971) is given as,

$$F_{ad} = \frac{3}{2}\pi W_{ad} R \tag{12.2}$$

where W_{ad} is the work of adhesion (units of energy per unit area). Equation (12.2) shows that the adhesion force of a single contact is proportional to a linear dimension of the contact. For a constant area divided into a large number of contacts or setae, n, the radius of a divided contact, R_1, is given by $R_1 = R/\sqrt{n}$ (self-similar scaling) (Arzt et al. 2003). Therefore, the adhesion force of (12.2) can be modified for multiple contacts such that,

$$F'_{ad} = \frac{3}{2}\pi W_{ad}\left(\frac{R}{\sqrt{n}}\right)n = \sqrt{n}F_{ad} \tag{12.3}$$

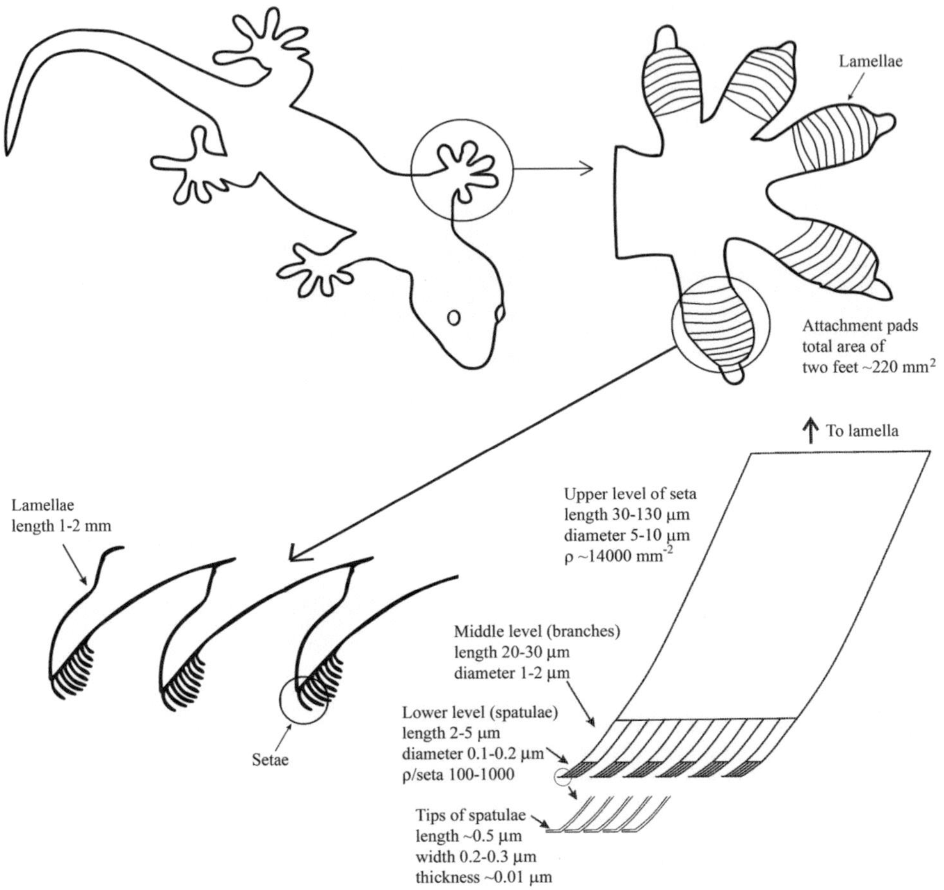

Fig. 12.4 Schematic drawings of a Tokay gecko including the overall body, one foot, a cross-sectional view of the lamellae, and an individual seta. ρ represents number of spatulae

where F'_{ad} is the total adhesion force from the divided contacts. Thus, the total adhesive force is simply the adhesion force of a single contact multiplied by the square root of the number of contacts.

For contact in a humid environment, the meniscus (or capillary) forces further increase the adhesion force (Bhushan 2013a, b, 2017a). The attractive meniscus force (F_m) consists of a contribution by both Laplace pressure and surface tension; discussed later in this Chapter (Orr et al. 1975; Bhushan 2013b). The contribution by Laplace pressure is directly proportional to the meniscus area. The other contribution is from the vertical component of surface tension around the circumference. This force is proportional to the circumference, as is the case for the work of adhesion (Bhushan 2013b). Going through the analysis presented earlier, one can show that the contribution from the component

Table 12.1 Physical dimensions of fibrillar structures on Tokay gecko feet and measured adhesive force data. Young's modulus of surface material, keratin = 1–20 GPa[a,b]

Component	Size	Density	Adhesive force
Seta	30–130[c–f]/5–10[c–f]	~140,00[h,i]	194 μN[j] (in shear)
	Length/diameter (μm)	Setae/mm^2	~20 μN[j] (normal)
Branch	20–30[c]/1–2[c]	–	–
	Length/diameter (μm)		
Spatula	2–5[c]/0.1–0.2[c,g]	100–1000[c,d]	–
	Length/diameter (μm)	Spatulae per seta	
Tip of spatula	~0.5[c,g]/0.2–0.3[c,f]/~0.01[g]	–	11 nN[k]
	Length/width/thickness (μm)		(normal)

[a]Russell (1986); [b]Bertram and Gosline (1987); [c]Ruibal and Ernst (1965); [d]Hiller (1968); [e]Russell (1975); [f]Williams and Peterson (1982); [g]Persson and Gorb (2003); [h]Schleich and Kästle (1986); [i]Autumn and Peattie (2002); [j]Autumn et al. (2000); [k]Huber et al. (2005a)

of surface tension increases with splitting into a larger number of contacts. It increases linearly with the square root of the number of contacts n (self-similar scaling) (Bhushan 2007; Cai and Bhushan 2007, 2008).

$$\left(F'_m\right)_{surface\ tension} = \sqrt{n}(F_m)_{surface\ tension} \qquad (12.4a)$$

where F'_m is the force from the divided contacts, and F_m is the force of an individual contact. This component of meniscus force is significant if the meniscus radius is very small and the contact angles are relatively large.

During separation of two surfaces, a viscous force, F_v, needs to be applied. F_v of the divided contacts is given as (Cai and Bhushan 2007, 2008),

$$F'_v = F_v/n \qquad (12.4b)$$

where F'_v is the force from the divided contacts, and F_v is the force of an individual contact.

The models just presented only consider contact with a flat surface. Multilevel hierarchical structure of the gecko provides compliance and conformability to rough surfaces in order to achieve high adhesion. The flexibility of the body provides conformability at the cm scale. Several toes on the feet provide conformability independently at the several mm scale. Lamellae on the bottom surfaces of the toes provide conformability at the mm scale. The setae on the lamellae provide conformability at the several μm scale. The tips of the setae are divided into spatulae which provide conformability at the few to several hundred nm scale. To summarize, on natural rough surfaces the compliance and adaptability of the hierarchical structure of gecko setae allows for greater contact with a natural

Table 12.2 Young's modulus of gecko skin and other materials for comparison

Material	Young's modulus
β-keratin, mostly present in gecko skin	1–20 GPa
Steel	210 GPa
Cross-linked rubber	1 MPa
Consumer adhesive tape (uncrosslinked rubber)	1 kPa

rough surface than a non-branched attachment system (Sitti and Fearing 2003; Bhushan et al. 2006; Kim and Bhushan 2007a, b, c, 2008). Modeling of the contact between gecko setae and rough surfaces is discussed in detail in Sect. 12.6.

Material properties also play an important role in adhesion. A soft material is able to achieve greater contact with a mating surface than a rigid material. Although gecko skin is primarily comprised of β-keratin, a stiff material with a Young's modulus in the range of 1–20 GPa (Russell 1986; Bertram and Gosline 1987), the effective modulus of the setal arrays on gecko feet is about 100 kPa (Autumn et al. 2006a), which is approximately four orders of magnitude lower than the bulk material. Young's modulus of the gecko skin is compared with that of various materials in Table 12.2. Humans have selected a relatively soft and compliant material for consumer adhesive tape to increase the contact area for high adhesion. Nature has selected a relatively stiff material for the setae to avoid clinging to each other. The non-orthogonal attachment angle of the seta increases the bending stiffness, and the division of contacts and hierarchical structure provide high adhesion. By combining optimal surface structure and material properties, nature has created an evolutionary superadhesive.

12.3.3 Peeling

Although geckos are capable of producing large adhesion forces, they retain the ability to remove their feet from an attachment surface at will by peeling action. The orientation of the spatulae facilitates peeling. Autumn et al. (2000) were the first to experimentally show that the adhesion force of gecko setae is dependent on the three-dimensional orientation as well as the preload applied during attachment (see later in the Chapter). Because of this, geckos have developed a complex foot motion during walking. First, the toes are carefully uncurled during detachment. The maximum adhesion occurs at an attachment angle of 30°—the angle between a seta and mating surface. The gecko is then able to peel its foot from surfaces one row of setae at a time by changing the angle at which its setae contact a surface. At an attachment angle greater than 30°, the gecko will detach from the surface. Two diagonally opposite feet participate in the movement.

Shah and Sitti (2004) determined the theoretical preload required for adhesion as well as the adhesion force generated for setal orientations of 30°, 40°, 50°, and 60°. Consider a solid material (elastic modulus, E, Poisson's ratio, ν) to make contact with the rough surface described by,

$$f(x) = H \sin^2\left(\frac{\pi x}{\lambda}\right) \tag{12.5}$$

where H is the amplitude, and λ is the wavelength of the roughness profile. For a solid adhesive block to achieve intimate contact with the rough surface, neglecting surface forces, it is necessary to apply a compressive stress, σ_c (Jagota and Bennison 2002),

$$\sigma_c = \frac{\pi EH}{2\lambda\left(1 - \nu^2\right)} \tag{12.6}$$

Equation (12.6) can be modified to account for fibers oriented at an angle, θ. The preload required for contact is summarized in Fig. 12.5a. As the orientation angle decreases, so does the required preload. Similarly, adhesion strength is influenced by fiber orientation. As seen in Fig. 12.5b, the greatest adhesion strength occurs at $\theta = 30°$.

Gao et al. (2005) created a finite element model of a single gecko seta in contact with a surface. A tensile force was applied to the seta at various angles, θ, as shown in Fig. 12.5c. Maximum adhesion again occurs at an angle of 30°. For forces applied at an angle of less than 30°, the dominant failure mode was sliding. In comparison, the dominant failure mode for forces applied at angles greater than 30° was detachment. This verifies the results of Autumn et al. (2000) that detachment occurs at attachment angles greater than 30°.

Tian et al. (2006) studied the attachment and detachment process in detail. Schematics of various configurations are shown in Fig. 12.6. In their natural configuration, shown in Fig. 12.6a, the spatula pads are at an angle of about 30° to the seta shaft and are approximately normal to the spatula shafts. This corresponds to a weak adhesion force. When geckos roll their toes inward and grip them, shown in Fig. 12.6b, the pulling angle θ_s of the seta shaft to the substrate will range from 0 to 30°, whereas the pulling angles θ of the spatula shafts to the substrate will range from 0 to 90°. Pulling in of the toes leads to smaller θ and θ_s, with a larger number of spatulae in contact with the substrate. This results in both higher friction and adhesion at the seta shaft. During the rolling out of the toes to detach from a substrate, shown in Fig. 12.6c, the setae move back to their initial free state with $\theta_s \sim 30°$ and $\theta \sim 90°$.

Pesika et al. (2007) developed a tape-peeling model based on the geometry of the peeling zone. They considered geometrical changes within the peel zone. The increased radius of curvature, *R*, increases the length of the peel zone, which in turn increases the peel force, *F*, in the peel direction. Figure 12.7a shows a schematic of the peel zone for two peel regimes: peel angles between cases I and II (constant peel-zone regime) and between cases II and III (variable peel-zone regime). In the constant peel-zone regime, the geometry of the peel zone (i.e., length of the peel zone and the curvature of the tape backing) remains constant while the peel angle, θ, is changed. In contrast, in the variable

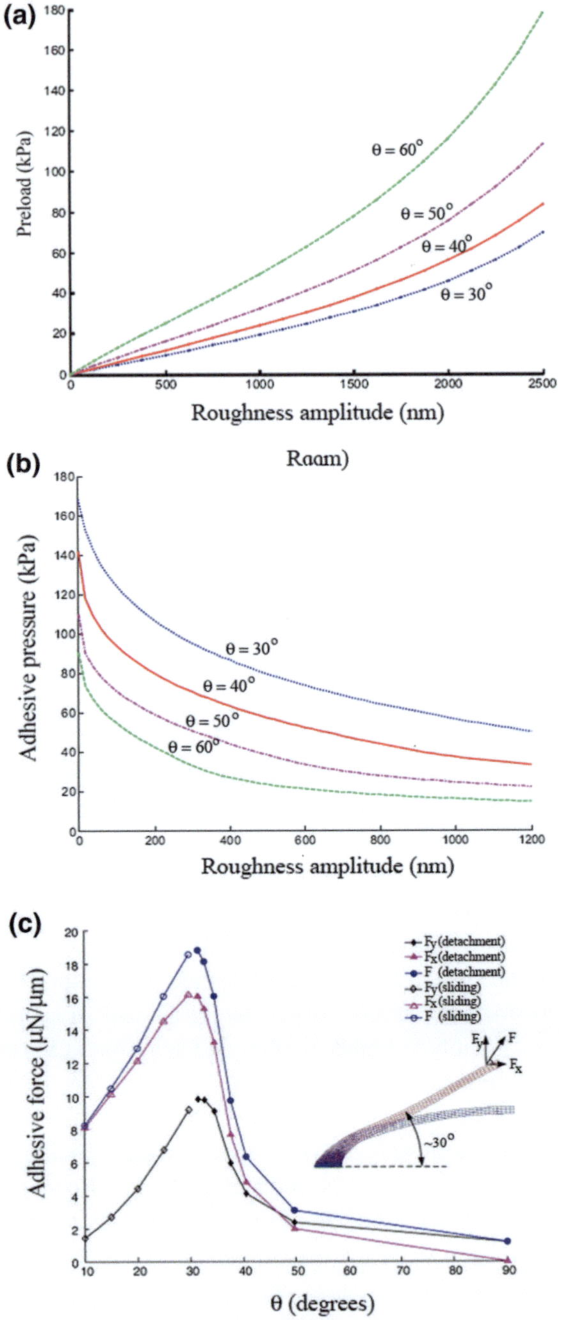

Fig. 12.5 Contact mechanics results for the effect of fiber orientation on **a** preload required for contact, and **b** adhesive pressure for roughness amplitudes ranging from 0 to 2500 nm (Shah and Sitti 2004). **c** Finite element analysis of the adhesive force of a single seta as a function of pull direction (Gao et al. 2005)

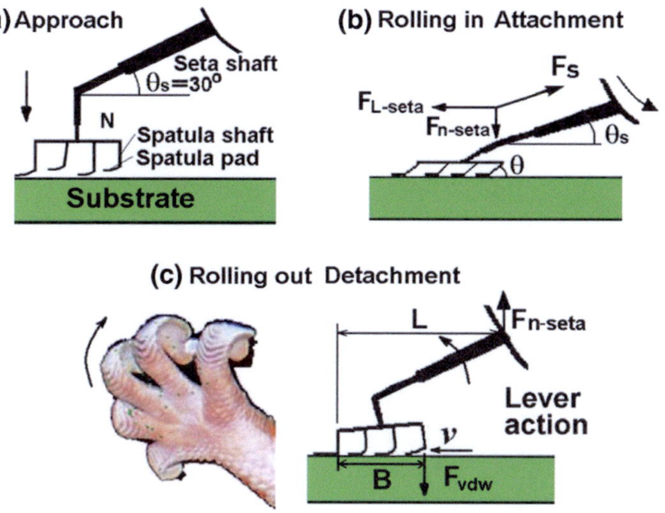

Fig. 12.6 Sketches of attachment and detachment of a single seta by **a** approaching the substrate, **b** rolling (or gripping) in, and **c** rolling (or peeling) out of the toes. The image in **c** left is from the left back foot (Tian et al. 2006)

peel-zone regime, the length of the peel zone and curvature of the tape backing both increase as the peel angle, θ, is changed. Figure 12.7b shows a plot of predicted value of peel force, F, as a function of the peel angle, θ. During detachment, θ increases, resulting in a lower peel force (Fig. 12.5c). The measured peel force for a model tape over glass surface as a function of peel angle is also plotted. The measured data correlate well with the predicted values.

12.3.4 Self-cleaning

Natural contaminants (dirt and dust) and man-made pollutants are unavoidable and have the potential to interfere with the clinging ability of geckos. Particles found in the air are typically less than 10 μm in diameter, while those found on the ground can often be larger (Hinds 1982; Jaenicke 1998). Intuitively, it seems that the great adhesion strength of gecko feet would cause dust and other particles to become trapped in the spatulae, and that they would have no way of being removed without some sort of manual cleaning action on behalf of the gecko. However, geckos are not known to groom their feet like beetles (Stork 1983), nor do they secrete sticky fluids to remove adhering particles like ants (Federle et al. 2002) and tree frogs (Hanna and Barnes 1991), yet they retain adhesive properties. One potential source of cleaning is during the time when the lizards undergo molting, or the shedding of the superficial layer of epidermal cells. However, this process

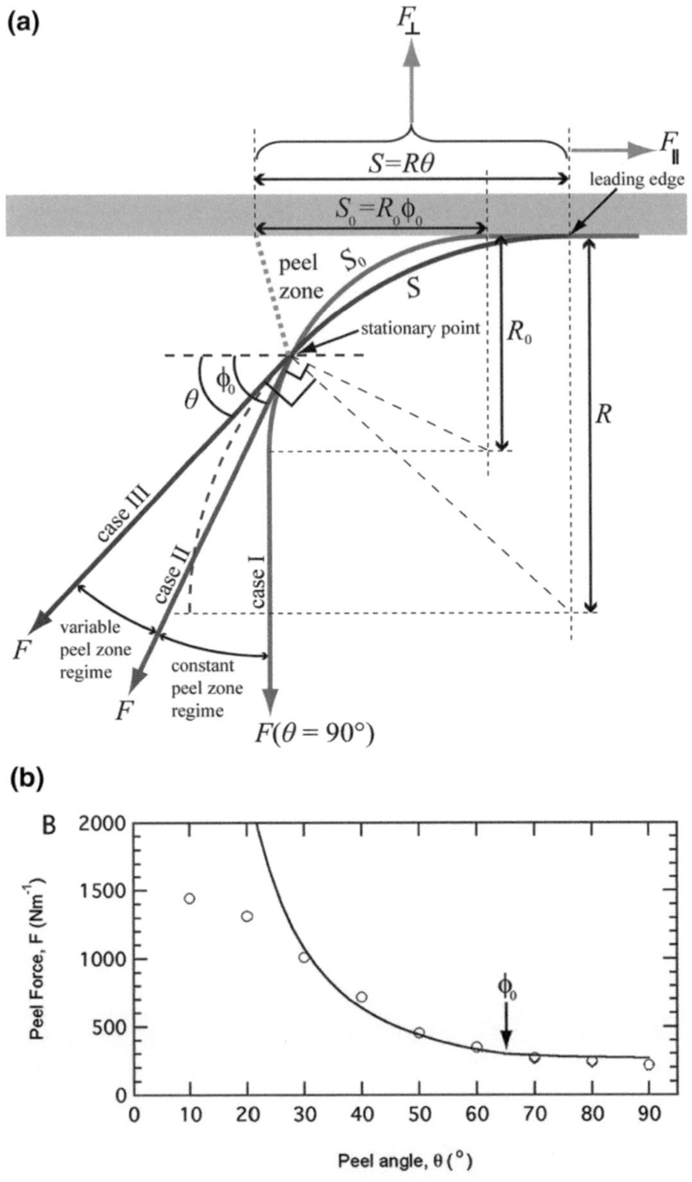

Fig. 12.7 **a** Schematic illustration of the peel zone showing the two peel regimes: constant peel-zone regime and variable peel-zone regimes, and **b** plots of predicted (solid line) and measured (data for a tape) peel force, F, as a function of peel angle, θ (Pesika et al. 2007)

only occurs approximately once per month (Van der Kloot 1992). If molting were the sole source of cleaning, the gecko would rapidly lose its adhesive properties as it was exposed to contaminants in nature (Hansen and Autumn 2005).

Hansen and Autumn (2005) tested the hypothesis that gecko setae become cleaner with repeated use—a phenomenon known as self-cleaning. The cleaning ability of gecko feet was first tested experimentally by applying 2.5 μm radius silica-alumina ceramic microspheres to clean setal arrays. Figure 12.8a shows the setal arrays immediately after dirtying and after five simulated steps. It is noted that a significant fraction of the particles has been removed after five steps. The maximum shear stress that these "dirty" arrays could withstand was measured using a sensor. After each step that the gecko took, the shear stress was once again measured. As seen in Fig. 12.8b, after four steps, the gecko foot was clean enough to withstand its own body weight.

In order to understand this cleaning process, substrate-particle interactions must be examined. The interaction energy between a dust particle, a wall, and spatulae can be modeled as shown in Fig. 12.9 (Hansen and Autumn 2005). The interaction energy between a spherical dust particle and the wall, W_{pw}, can be expressed as (Israelachvili 1992),

$$W_{pw} = \frac{-H_{pw}R_p}{6D_{pw}} \tag{12.7}$$

where p and w refer to the particle and wall, respectively. H is the Hamaker constant, R_p is the radius of the particle, and D_{pw} is the separation distance between the particle and the wall. Similarly, the interaction energy between a spherical dust particle and a spatula,

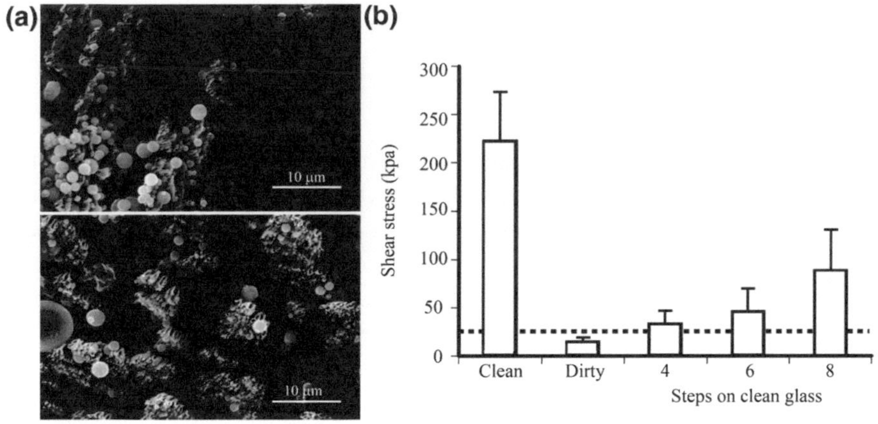

Fig. 12.8 **a** SEM image of spatulae after dirtying with microspheres (top) and after five simulated steps (bottom), **b** mean shear stress exerted by a gecko on a surface after dirtying. The dotted line represents sufficient recovery to support weight by a single toe (Hansen and Autumn 2005)

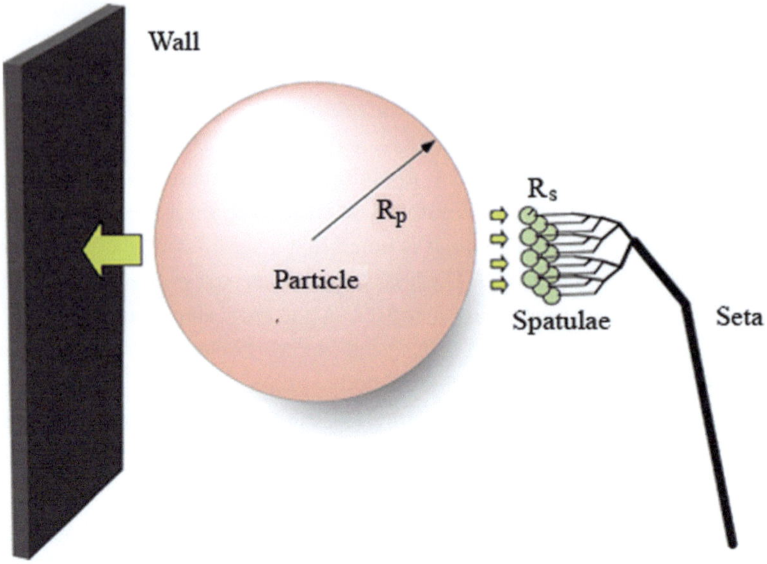

Fig. 12.9 Model of interactions between gecko spatulae of radius R_s, a spherical dirt particle of radius R_p, and a planar wall that enable self-cleaning (Hansen and Autumn 2005)

s, assuming that the spatula tip is spherical is (Israelachvili 1992),

$$W_{ps} = \frac{-H_{ps}R_pR_s}{6D_{ps}(R_p + R_s)} \tag{12.8}$$

The ratio of the two interaction energies, *Z*, can be expressed as

$$Z = \frac{W_{pw}}{W_{ps}} = \left(1 + \frac{R_p}{R_s}\right)\frac{H_{pw}D_{ps}}{H_{ps}D_{pw}} \tag{12.9}$$

When the energy required to separate a particle from the wall is greater than that required to separate it from a spatula ($Z>1$), self-cleaning will occur. For small contaminants ($R_p < 0.5$ μm), there are not enough spatulae available to adhere to the particle. For larger contaminants, the curvature of the particles makes it impossible for enough spatulae to adhere to it. As a result, Hansen and Autumn (2005) concluded that self-cleaning should occur for all spherical spatulae interacting with all spherical particles.

12.4 Attachment Mechanisms

When asperities of two solid surfaces are brought into contact with each other, chemical and/or physical attractions occur. The force developed that holds the two surfaces together is known as adhesion. In a broad sense, adhesion is considered to be either physical or chemical in nature (Bikerman 1961; Zisman 1963; Houwink and Salomon 1967; Israelachvili 1992; Bhushan 1996, 2013a, b, 2017a, b). Chemical interactions such as electrostatic attraction charges (Schmidt 1904), as well as intermolecular forces (Hiller 1968) including van der Waals and capillary forces, have all been proposed as potential adhesion mechanisms in gecko feet. Others have hypothesized that geckos adhere to surfaces through the secretion of sticky fluids (Wagler 1830; Simmermacher 1884), suction (Simmermacher 1884), increased frictional force (Hora 1923), and microinterlocking (Dellit 1934).

Through experimental testing and observations conducted since the mid 1800s, many potential adhesive mechanisms have been eliminated. Observation has shown that geckos lack any glands capable of producing sticky fluids (Wagler 1830; Simmermacher 1884), ruling out the secretion of sticky fluids as a potential adhesive mechanism. Furthermore, geckos are able to create large adhesive forces normal to a surface. Since friction only acts parallel to a surface, the attachment mechanism of increased frictional force has been ruled out. Dellit (1934) experimentally ruled out suction and electrostatic attraction as potential adhesive mechanisms. Experiments carried out in vacuum did not show a difference between the adhesive force at low pressures compared to ambient conditions. Since adhesive forces generated during suction are based on pressure differentials, which are insignificant under vacuum, suction was rejected as an adhesive mechanism (Dellit 1934). Additional testing used X-ray bombardment to create ionized air in which electrostatic attraction charges would be eliminated. It was determined that geckos were still able to adhere to surfaces in these conditions, and therefore, electrostatic charges could not be the sole cause of attraction (Dellit 1934). Autumn et al. (2000) demonstrated the ability of a gecko to generate large adhesive forces when in contact with a molecularly smooth SiO_2 microelectromechanical system (MEMS) semiconductor. Since surface roughness is necessary for microinterlocking to occur, it has been ruled out as a mechanism of adhesion. Two mechanisms, van der Waals forces and capillary forces, remain as the potential sources of gecko adhesion. These attachment mechanisms are described in detail in the following sections.

12.4.1 van der Waals Forces

van der Waals bonds are secondary bonds that are weak in comparison to other physical bonds such as covalent, hydrogen, ionic, and metallic bonds. Unlike other physical bonds, van der Waals forces are always present, regardless of separation. They are effective from very large separations (~50 nm) down to atomic separation (~0.3 nm). The van der Waals force per unit area between two parallel surfaces, f_{vdW}, is given by (Hamaker 1937; Israelachvili and Tabor 1972; Israelachvili 1992),

$$f_{vdW} = \frac{H}{6\pi D^3} \quad \text{for } D < 30 \text{ nm} \tag{12.10}$$

where H is the Hamaker constant, and D is the separation between surfaces.

Hiller (1968) showed experimentally that the surface energy of a substrate is responsible for gecko adhesion. One potential adhesion mechanism would then be van der Waals forces (Stork 1980; Autumn et al. 2000). Assuming van der Waals forces to be the dominant adhesion mechanism used by geckos, the adhesion force of a gecko can be calculated. Typical values of the Hamaker constant range from 4×10^{-20} to 4×10^{-19} J (Israelachvili 1992). In calculation, the Hamaker constant is assumed to be 10^{-19} J, the surface area of a spatula is taken to be 2×10^{-14} m^2 (Ruibal and Ernst 1965; Williams and Peterson 1982; Autumn and Peattie 2002), and the separation distance between the spatula and contact surface is estimated to be 0.6 nm. This equation yields the force of a single spatula to be about 0.5 μN. By applying the surface characteristics of Table 12.1, the maximum adhesion force of a gecko is 150–1500 N for varying spatula density of 100–1000 spatulae per seta. If an average value of 550 spatulae/seta is used, the adhesion force of a single seta is approximately 270 μN, which is in agreement with the experimental value obtained by Autumn et al. (2000), which will be discussed later in this chapter.

Another approach to calculate adhesion force is to assume that spatulae are cylinders that terminate in hemispherical tips. By using (12.2) and assuming that the radius of each spatula is about 100 nm and that the surface energy is expected to be 50 mJ/m^2 (Arzt et al. 2003), the adhesive force of a single spatula is predicted to be 0.02 μN. This result is an order of magnitude lower than the first approach calculated for the higher value of A. For a lower value of 10^{-20} J for the Hamaker constant, the adhesive force of a single spatula is comparable to that obtained using the surface energy approach.

Several experimental results favor van der Waals forces as the dominant adhesive mechanism, including temperature testing (Bergman and Irschick 2005) and adhesion force measurements of a gecko seta with both hydrophilic and hydrophobic surfaces (Autumn et al. 2000). This data will be presented later in this chapter.

12.4.2 Capillary Forces

It has been hypothesized that capillary forces that arise from liquid mediated contact could be a contributing or even the dominant adhesive mechanism used by gecko spatulae (Hiller 1968; Stork 1980). Experimental adhesion measurements (presented later in this Chapter) conducted on surfaces with different hydrophobicities and at various humidities (Huber et al. 2005b) as well as numerical simulations (Kim and Bhushan 2008) support this hypothesis as a contributing mechanism. During contact, any liquid that wets or has a small contact angle on surfaces will condense from vapor in the form of an annular-shaped capillary condensate. Due to the natural humidity present in the air, water vapor will condense to liquid on the surface of bulk materials. During contact this will cause the formation of adhesive bridges (menisci) due to the proximity of the two surfaces and the affinity of the surfaces for condensing liquid. (Zimon 1969; Fan and O'Brien 1975; Phipps and Rice 1979).

In the adhesion model with capillarity by Kim and Bhushan (2008), the tip of the spatula in a single contact was assumed as spherical; see Fig. 12.10. The total adhesion force between a spherical tip and a plane consists of the capillary force and the solid-to-solid interaction. Capillary force can be divided into two components: the Laplace force F_L and the surface tension force F_s, such that the total capillary force F_c,

$$F_c = F_L + F_s \tag{12.11}$$

The Laplace force is caused by the pressure difference across the interface of a curved liquid surface (Fig. 12.10) and depends on the pressure difference multiplied by the meniscus area, which can be expressed as (Orr et al. 1975),

$$F_L = -\pi \kappa \gamma R^2 \sin^2 \phi \tag{12.12}$$

where γ is the surface tension of the liquid, R is the tip radius, ϕ is the filling angle, and κ is the mean curvature of the meniscus. From the Kelvin equation (Israelachvili 1992), which is the thermal equilibrium relation, the mean curvature of the meniscus can be determined as,

$$\kappa = \frac{\Re T}{V\gamma} \ln\left(\frac{p}{p_o}\right) \tag{12.13}$$

where $\Re$ is the universal gas constant, T is the absolute temperature, V is the molecular volume, p_o is the saturated vapor pressure of the liquid at T, and p is the ambient pressure acting outside the curved surface (p/p_o is the relative humidity). Orr et al. (1975) formulated the mean curvature of a meniscus between a sphere and a plane in terms of elliptic integrals. The filling angle ϕ can be calculated from the expression just mentioned and (12.13) using the iteration method. Then, the Laplace force is calculated at a given environment using (12.12).

The surface tension of the liquid results in the formation of a curved liquid-air interface. The surface tension force acting on the sphere is (Orr et al. 1975),

$$F_s = 2\pi R\gamma \sin\phi \sin(\theta_1 + \phi) \tag{12.14}$$

where θ_1 is the contact angle on the sphere.

Hence, the total capillary force on the sphere is

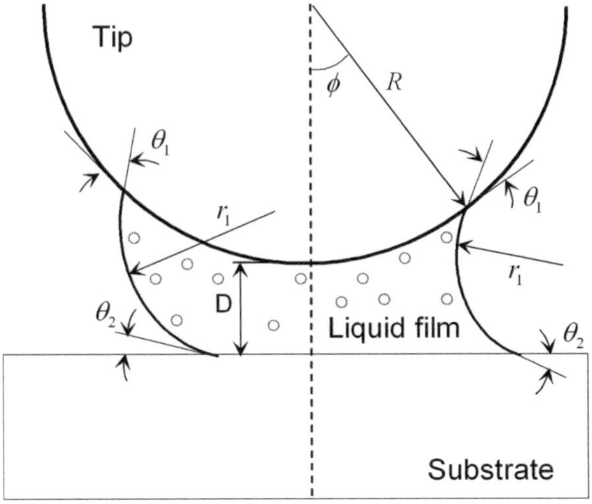

Fig. 12.10 Schematic of a sphere on a plane at distance D with a liquid film in between, forming menisci. In this figure, R is the tip radius, ϕ is the filling angle, θ_1 and θ_2 are contact angles on sphere and plane, respectively, and r_1 and r_2 are the two principal radii of the curved surface in two orthogonal planes (Kim and Bhushan 2008)

$$F_c = \pi R\gamma\left\{2\sin\phi\sin\left(\theta_1+\varphi\right)-\kappa R\sin^2\phi\right\} \tag{12.15}$$

The effect of capillarity on gecko adhesion results will be presented later in this chapter.

12.5 Adhesion Measurements and Data

Experimental measurements of the adhesion force of a single gecko seta (Autumn et al. 2000) and single gecko spatula (Huber et al. 2005a) have been made. The effect of the environment, including temperature (Losos 1990; Bergmann and Irschick 2005) and humidity, (Huber et al. 2005b) has been studied. Some of the data has been used to understand the adhesion mechanism used by the gecko attachment system—van der Waals or capillary forces. The majority of experimental results point towards van der Waals forces as the dominant mechanism of adhesion (Autumn et al. 2000; Bergmann and Irschick 2005). Some research suggests that capillary forces can be a contributing adhesive factor (Huber et al. 2005b; Kim and Bhushan 2008).

12.5.1 Adhesion Under Ambient Conditions

Two feet of a Tokay gecko are capable of producing about 20 N of adhesive force with a pad area of about 220 mm^2 (Irschick et al. 1996). Assuming that there are about 14,000 setae per mm^2, the adhesion force from a single hair should be approximately 7 μN. It is likely that the magnitude is actually greater than this value because it is unlikely that all setae are in contact with the mating surface (Autumn et al. 2000). Setal orientation greatly influences adhesive strength. This dependency was first noted by Autumn et al. (2000). It was determined that the greatest adhesion occurs at 30°. In order to determine the adhesion mechanism(s) used by gecko feet, it is important to know the adhesion force of a single seta. Hence, the adhesion force of gecko foot-hair has been the focus of several investigations (Autumn et al. 2000; Huber et al. 2005a).

12.5.1.1 Adhesion Force of a Single Seta

Autumn et al. (2000) used both a microelectromechanical (MEMS) force sensor and a wire as a force gauge to determine the adhesion force of a single seta. The MEMS force sensor is a dual-axis atomic force microscope (AFM) cantilever with independent piezoresistive sensors which allows simultaneous detection of vertical and lateral forces (Chui et al. 1998). The wire force gage consisted of an aluminum bonding wire that displaced under a perpendicular pull. Autumn et al. (2000) discovered that setal force actually depends on the three-dimensional orientation of the seta, as well as the preloading force applied during initial contact. Setae that were preloaded vertically to the surface exhibited only one-tenth of the adhesive force (0.6 ± 0.7 μN) compared to setae that were pushed vertically and then pulled horizontally to the surface (13.6 ± 2.6 μN). The dependence of the adhesion force of a single gecko spatula on perpendicular preload is illustrated in Fig. 12.11. The adhesion force increases linearly with the preload, as expected (Bhushan 1996, 2013a, b). The maximum adhesion force of a single gecko foot-hair occurred when the seta was first subjected to a normal preload and then slid 5 μm along the contacting surface. Under these conditions, adhesion force measured 194 ± 25 μN (~10 atm adhesive pressure).

12.5.1.2 Adhesive Force of a Single Spatula

Huber et al. (2005a) used atomic force microscopy to determine the adhesion force of an individual gecko spatulae. A seta with four spatulae was glued to an AFM tip. The seta was then brought in contact with a surface and a compressive preload of 90 nN was applied. The force required to pull the seta off of the surface was then measured. As seen in Fig. 12.12, there are two distinct peaks on the graph—one at 10 nN and the other at 20 nN. The first peak corresponds to one of the four spatulae adhering to the contact surface while the peak at 20 nN corresponds to two of the four spatulae adhering to the contact surface. The average adhesion force of a single spatula was found to be 10.8 ± 1 nN. The measured value is in agreement with the measured adhesive strength of an entire gecko (on the order of 10^9 spatulae on a gecko).

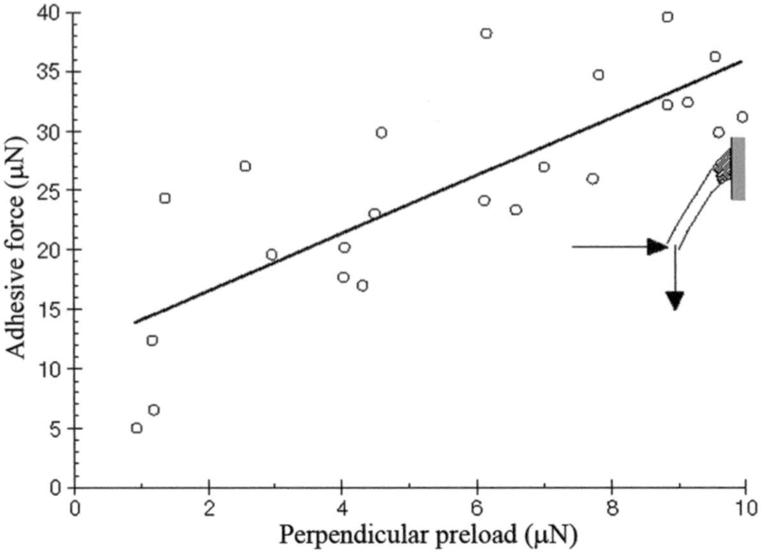

Fig. 12.11 Adhesive force of a single gecko seta as a function of applied preload. The seta was first pushed perpendicularly against the surface and then pulled parallel to the surface (Autumn et al. 2000)

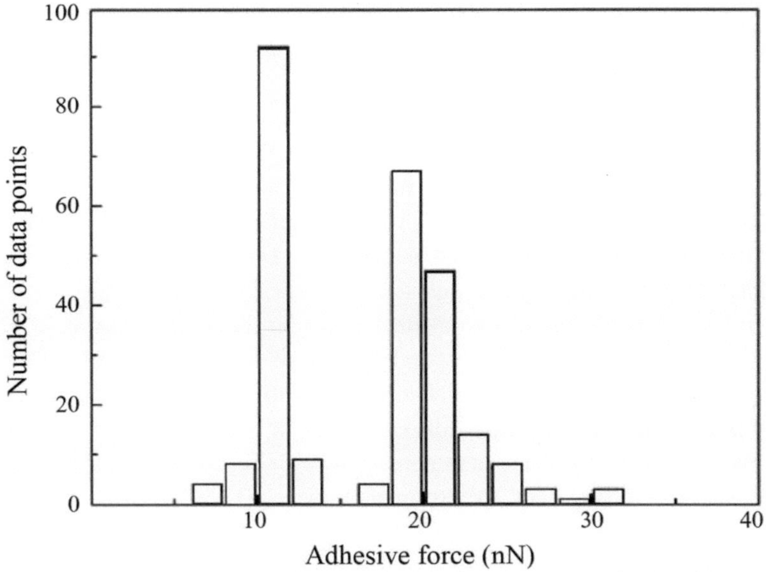

Fig. 12.12 Adhesive force of a single gecko spatula. The peak at 10 nN corresponds to the adhesive force of one spatula and the peak at 20 nN corresponds to the adhesive force of two spatulae (Huber et al. 2005a)

12.5.2 Effects of Temperature

Environmental factors are known to affect several aspects of vertebrate function, including speed of locomotion, digestion rate, and muscle contraction. As a result, several studies have been completed to investigate environmental impact on these functions. Relationships between the environment and other properties such as adhesion are far less studied (Bergmann and Irschick 2005). Only two known studies exist that examine the effect of temperature on the clinging force of the gecko (Losos 1990; Bergmann and Irschick 2005). Losos (1990) examined the adhesion ability of large live geckos at temperatures up to 17 °C. Bergmann and Irschick (2005) expanded upon this research for body temperatures ranging from 15 to 35 °C. The geckos were incubated until their body temperature reached a desired level. The clinging ability of these animals was then determined by measuring the maximum exerted force by the geckos as they were pulled off a custom-built force plate. The clinging force of a gecko for the experimental test range is plotted in Fig. 12.13. It was determined that variation in temperature is not statistically significant in the adhesion force of a gecko. From these results, it was concluded that the temperature independence of adhesion supports the hypothesis of clinging as a passive mechanism (i.e. van der Waals forces). Both studies only measured the overall clinging ability on the macroscale. There have not been any investigations into the effects of temperature on the clinging ability of a single seta on the microscale, and therefore testing in this area would be extremely important.

12.5.3 Effects of Humidity

Huber et al. (2005b) employed similar methods to Huber et al. (2005a) (discussed previously in this Chapter) in order to determine the adhesive force of a single spatula at varying humidity. Measurements were made using an AFM placed in an air-tight chamber. The humidity was adjusted by varying the flow rate of dry nitrogen into the chamber. The air was continuously monitored with a commercially available hygrometer. All tests were conducted at ambient temperature.

As seen in Fig. 12.14, even at low humidity, adhesion force is large. An increase in humidity further increases the overall adhesion force of a gecko spatula. The pull-off force roughly doubled as the humidity was increased from 1.5 to 60%. This humidity effect can be explained in two possible ways: (1) by standard capillarity or (2) by a change of the effective short-range interaction due to absorbed monolayers of water. In other words, the water molecules increase the number of van der Waals bonds that are made. Based on this data, van der Waals forces are the primary adhesion mechanism, and capillary forces are a secondary adhesive mechanism.

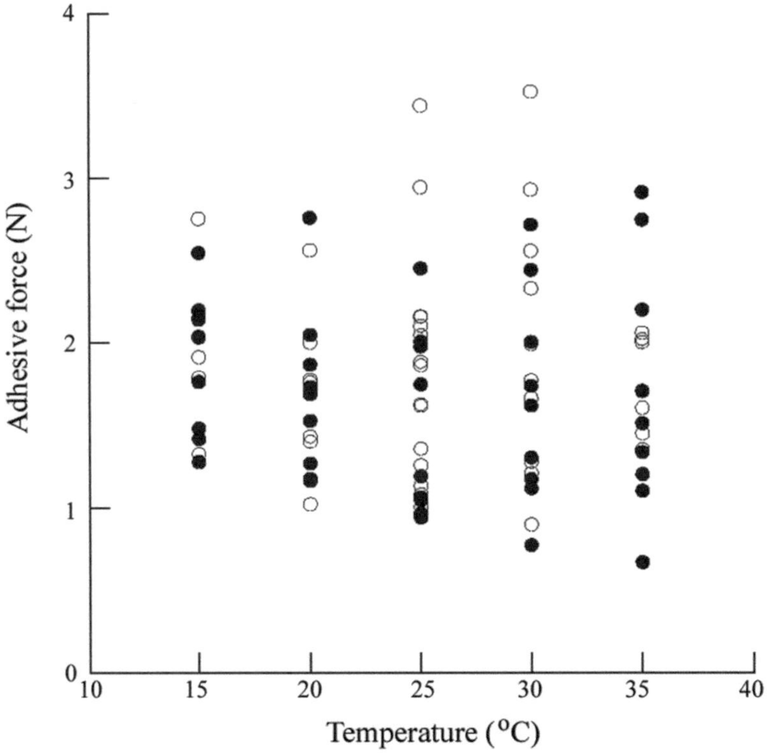

Fig. 12.13 Adhesive force of a gecko as a function of temperature (Bergmann and Irschick 2005)

12.5.4 Effects of Hydrophobicity

To further test the hypothesis the capillary forces play a role in gecko adhesion, the spatular pull-off force was determined for contact with both hydrophilic and hydrophobic surfaces. As seen in Fig. 12.15a, the capillary adhesion theory predicts that a gecko spatula will generate a greater adhesion force when in contact with a hydrophilic surface as compared to a hydrophobic surface. However, the van der Waals adhesion theory predicts that the adhesion force between a gecko spatula and a surface will be the same regardless of the hydrophobicity of the surface (Autumn et al. 2002). Figure 12.15b shows the shear stress of a whole gecko and adhesive force of a single seta on hydrophilic and hydrophobic surfaces. The data shows that the adhesion values are the same on both surfaces. This supports the van der Waals prediction of Fig. 12.15a. Huber et al. (2005b) found that the hydrophobicity of the attachment surface had an effect on the adhesion force of a single gecko spatula as shown in Fig. 12.15c. These results show that adhesion force has a finite value for a superhydrophobic surface, and increases as the surface becomes hydrophilic.

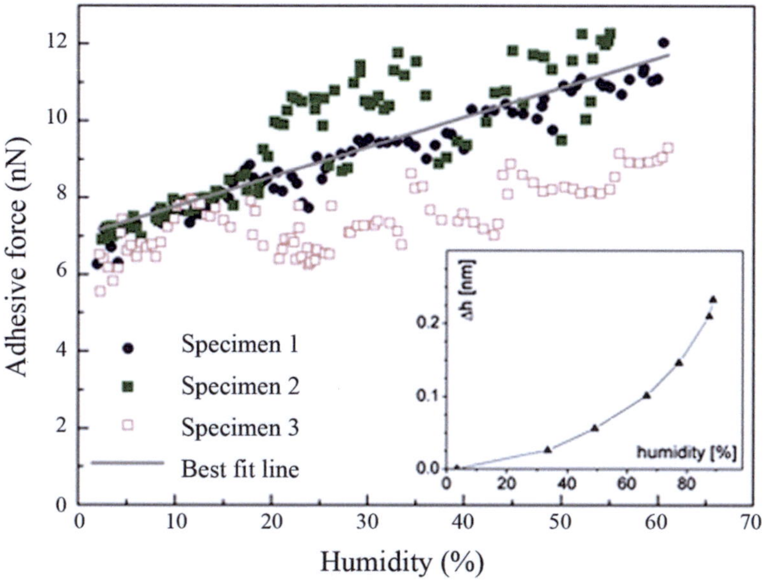

Fig. 12.14 Humidity effects on spatular pull-off force (inset). The increase in water film thickness on a Si wafer with increasing humidity (Huber et al. 2005b)

It is concluded that van der Waals forces are the primary mechanism, and capillary forces further increase the adhesion force generated.

12.6 Adhesion Modeling of Fibrillar Structures

With regard to the natural living conditions of the animals, the mechanics of gecko attachment can be separated into two parts: the mechanics of adhesion of a single contact with a flat surface, and an adaptation of a large number of spatulae to a natural, rough surface. Modeling of the mechanics of adhesion of spatulae to a smooth surface, in the absence of meniscus formation, was developed by Autumn et al. (2002), Jagota and Bennison (2002) and Arzt et al. (2003). As discussed previously in this Chapter, the adhesion force of multiple contacts F'_{ad} can be increased by dividing the contact into a large number (n) of small contacts, while the nominal area of the contact remains the same, $F'_{ad} \sim \sqrt{n} F_{ad}$. However, this model only considers contact with a flat surface. On natural, rough surfaces, the compliance and adaptability of setae are the primary sources of high adhesion. The hierarchical structure of gecko setae allows for a greater contact with a natural, rough surface than a non-branched attachment system (Sitti and Fearing 2003).

Bhushan et al. (2006) and Kim and Bhushan (2007a, b, c, 2008) have approximated a gecko seta in contact with random rough surfaces using a hierarchical spring model. Each

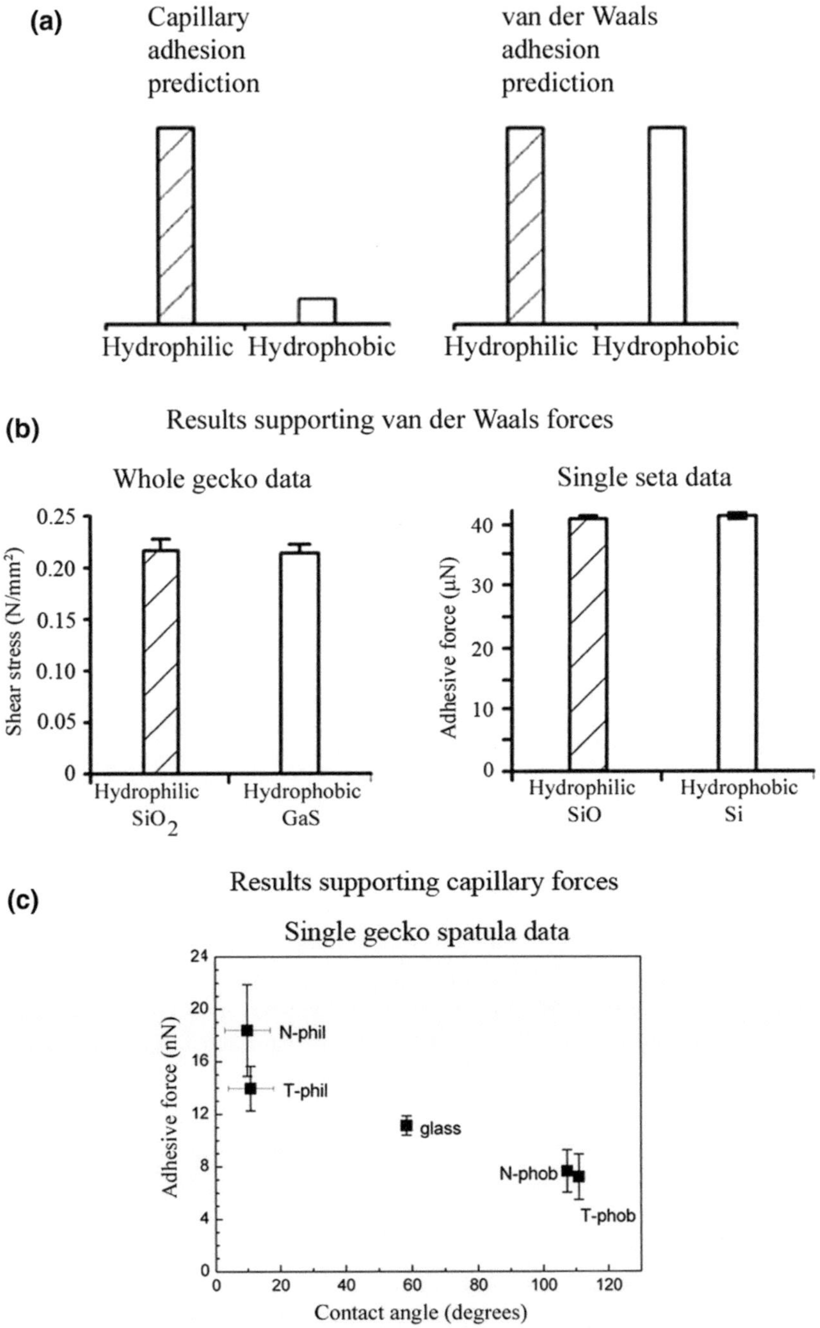
(a)
Capillary adhesion prediction
van der Waals adhesion prediction
Hydrophilic Hydrophobic
Hydrophilic Hydrophobic
(b)
Results supporting van der Waals forces
Whole gecko data
Single seta data
Shear stress (N/mm²)
0.25
0.20
0.15
0.10
0.05
0
Hydrophilic SiO2
Hydrophobic GaS
Adhesive force (μN)
40
30
20
10
0
Hydrophilic SiO
Hydrophobic Si
(c)
Results supporting capillary forces
Single gecko spatula data
Adhesive force (nN)
24
20
16
12
8
4
0
N-phil
T-phil
glass
N-phob
T-phob
0 20 40 60 80 100 120
Contact angle (degrees)

◂**Fig. 12.15** **a** Capillary and van der Waals adhesion predictions for the relative magnitude of the adhesive force of gecko setae with hydrophilic and hydrophobic surfaces (Autumn et al. 2002). **b** Results of adhesion testing for a whole gecko and single seta with hydrophilic and hydrophobic surfaces (Autumn et al. 2002) and **c** results of adhesive force testing of a single gecko spatula with surfaces with different contact angles (Huber et al. 2005b)

Table 12.3 Geometrical size, calculated stiffness, and typical densities of branches of seta for Tokay gecko (Kim and Bhushan 2007a)

Level of seta	Length (μm)	Diameter (μm)	Bending stiffness[a] (N/m)	Typical density (#/mm²)
III upper	75	5	2.908	14×10^3
II middle	25	1	0.126	–
I lower	2.5	0.1	0.0126	$1.4 - 14 \times 10^6$

[a]For elastic modulus of 10 GPa with load applied at 60° to spatula long axis

level of springs in their model corresponds to a level of seta hierarchy. The upper level of springs corresponds to the thicker part of gecko seta, the middle spring level corresponds to the branches, and the lower level of springs corresponds to the spatulae. The upper level is the thickest branch of the seta. It is 75 μm in length and 5 μm in diameter. The middle level, referred to as a branch, has a length of 25 μm and a diameter of 1 μm. The lower level, called a spatula, is the thinnest branch with a length of 2.5 μm and a diameter of about 0.1 μm (Table 12.3). Autumn et al. (2000) showed that the optimal attachment angle between the substrate and a gecko seta is 30° in the single seta pull-off experiment. This finding is supported by the adhesion models of setae as cantilever beams (Shah and Sitti 2004; Gao et al. 2005) (see Sect. 12.3.3 for more details). Therefore, θ was fixed at 30° in the studies by Bhushan et al. (2006) and Kim and Bhushan (2007a, b, c, 2008) presented below.

12.6.1 Single Spring Contact Analysis

In their analysis, Bhushan et al. (2006) and Kim and Bhushan (2007a, b, c, 2008) assumed the tip of the spatula in a single contact to be spherical. The springs on every level of hierarchy have the same stiffness as the bending stiffness of the corresponding branches of seta. If the beam is oriented at an angle θ to the substrate and the contact load F is aligned normal to the substrate, its components along and tangential to the direction of the beam, $F\cos\theta$ and $F\sin\theta$, give rise to bending and compressive deformations, δ_b and δ_c, respectively, as (Young and Budynas 2001),

$$\delta_b = \frac{F\cos\theta\, l_m^3}{3EI}, \quad \delta_c = \frac{F\sin\theta\, l_m}{A_c E} \tag{12.16}$$

where $I = \pi R_m^4/4$ and $A_c = \pi R_m^2$ are the moments of inertia of the beam and the cross-sectional area, respectively, and l_m and R_m are the length and the radius of seta branches, respectively, and m is the level number. The net displacement, $\delta_\perp$ normal to the substrate, is given by,

$$\delta_\perp = \delta_c \sin\theta + \delta_b \cos\theta \tag{12.17}$$

Using (12.16) and (12.17), the stiffness of seta branches, k_m, is calculated as (Glassmaker et al. 2004),

$$k_m = \frac{\pi R_m^2 E}{l_m \sin^2\theta \left(1 + \frac{4 l_m^2 \cot^2\theta}{3 R_m^2}\right)} \tag{12.18}$$

For an assumed elastic modulus, E, of seta material of 10 GPa with a load applied at an angle of 60° to the spatulae long axis, Kim and Bhushan (2007a) calculated the stiffness of every level of seta, as given in Table 12.3.

In the model, both the tips of a spatula and the asperity summits of the rough surface are assumed to be spherical with a constant radius (Bhushan et al. 2006). As a result, a single spatula adhering to a rough surface was modeled as the interaction between two spherical tips. Because β-keratin has a high elastic modulus (Russell 1986; Bertram and Gosline 1987), the adhesion force between two round tips was calculated according to the Derjaguin-Muller-Toporov (DMT) theory (Derjaguin et al. 1975) as,

$$F_{ad} = 2\pi R_c W_{ad} \tag{12.19}$$

where R_c is the reduced radius of contact. It is calculated as $R_c = (1/R_1 + 1/R_2)^{-1}$, where R_1 and R_2 are the radii of contacting surfaces. When $R_1 = R_2$, $R_c = R/2$. The work of adhesion, W_{ad}, is then calculated using the following equation for two flat surfaces separated by a distance, D, (Israelachvili 1992),

$$W_{ad} = -\frac{H}{12\pi D^2} \tag{12.20}$$

where H is the Hamakar constant which depends on the medium between the two surfaces. Typical values of the Hamakar constant for polymers are $H_{air} = 10^{-19}$ J in the air and $H_{water} = 3.7 \times 10^{-20}$ J in the water (Israelachvili 1992). For a gecko seta, which is composed of β-keratin, the value of H is assumed to be 10^{-19} J. The works of adhesion of two surfaces in contact separated by an atomic distance $D \approx 0.2$ nm is approximately equal to 66 mJ/m^2 (Israelachvili 1992). By assuming that the tip radius, R, is 50 nm, using (12.19), the adhesion force of a single contact is calculated as 10 nN (Kim and Bhushan 2007a). This value is identical to the adhesion force of a single spatula measured by Huber et al. (2005a). This adhesion force is used as a critical force in the model for judging whether the contact between the tip and the surface is broken or not during pull-off cycle (Bhushan et al. 2006). If the elastic force of a single spring is less than the adhesion force, the spring is regarded as having been detached.

12.6.2 The Multi-level Hierarchical Spring Analysis

In order to study the effect of the number of hierarchical levels in the attachment system on attachment ability, models with one (Bhushan et al. 2006; Kim and Bhushan 2007a, b), two (Bhushan et al. 2006; Kim and Bhushan 2007a, b) and three (Kim and Bhushan 2007a, b) levels of hierarchy were simulated (Fig. 12.16). The one-level model has springs with length $l_I = 2.5$ μm and stiffness $k_I = 0.0126$ N/m. The length and stiffness of the springs in the two-level model are $l_I = 2.5$ μm, $k_I = 0.0126$ N/m and $l_{II} = 25$ μm, $k_{II} = 0.126$ N/m for levels *I* and *II*, respectively. The three-level model has additional upper level springs with $l_{III} = 75$ μm, $k_{III} = 2.908$ N/m on the springs of the two-level model, which is identical to gecko setae. The base of the springs and the connecting plate between the levels are assumed to be rigid. The distance S_I between the neighboring structures of level *I* is 0.35 μm, obtained from the average value of measured spatula density, 8×10^6 mm^{-2}, calculated by multiplying 14,000 setae/mm^2 by an average of 550 spatula/seta (Schleich and Kästle 1986) (Table 12.3). A 1:10 proportion of the number of springs in the upper level to that in the level below and the same proportion from mid-level to bottom level was assumed (Bhushan et al. 2006). This corresponds to one spring at level *III* which is connected to ten springs on level *II*, and each spring on level *II* is connected with ten springs on level *I*. The number of springs on level *I* considered in the model is calculated by dividing the scan length (2000 μm) with the distance S_I (0.35 μm), which corresponds to 5700.

The spring deflection Δl was calculated as,

$$\Delta l = h - l_0 - z \tag{12.21}$$

where h is the position of the spring base relative to the mean line of the surface; l_O is the total length of a spring structure, which is $l_O = l_I$ for the one-level model, $l_O = l_I + l_{II}$ for the two-level model, and $l_O = l_I + l_{II} + l_{III}$ for the three-level model; and z is the profile height of the rough surface. The elastic force F_{el} arisen in the springs at a distance h from the surface was calculated for the one-level model as (Bhushan et al. 2006),

$$F_{el} = -k_I \sum_{i=1}^{p} \Delta l_i u_i \quad u_i = \begin{cases} 1 \; if \; contact \\ 0 \; if \; no \; contact \end{cases} \tag{12.22}$$

where p is the number of springs in level *I* of the model. For the two-level model, the elastic force was calculated as (Bhushan et al. 2006),

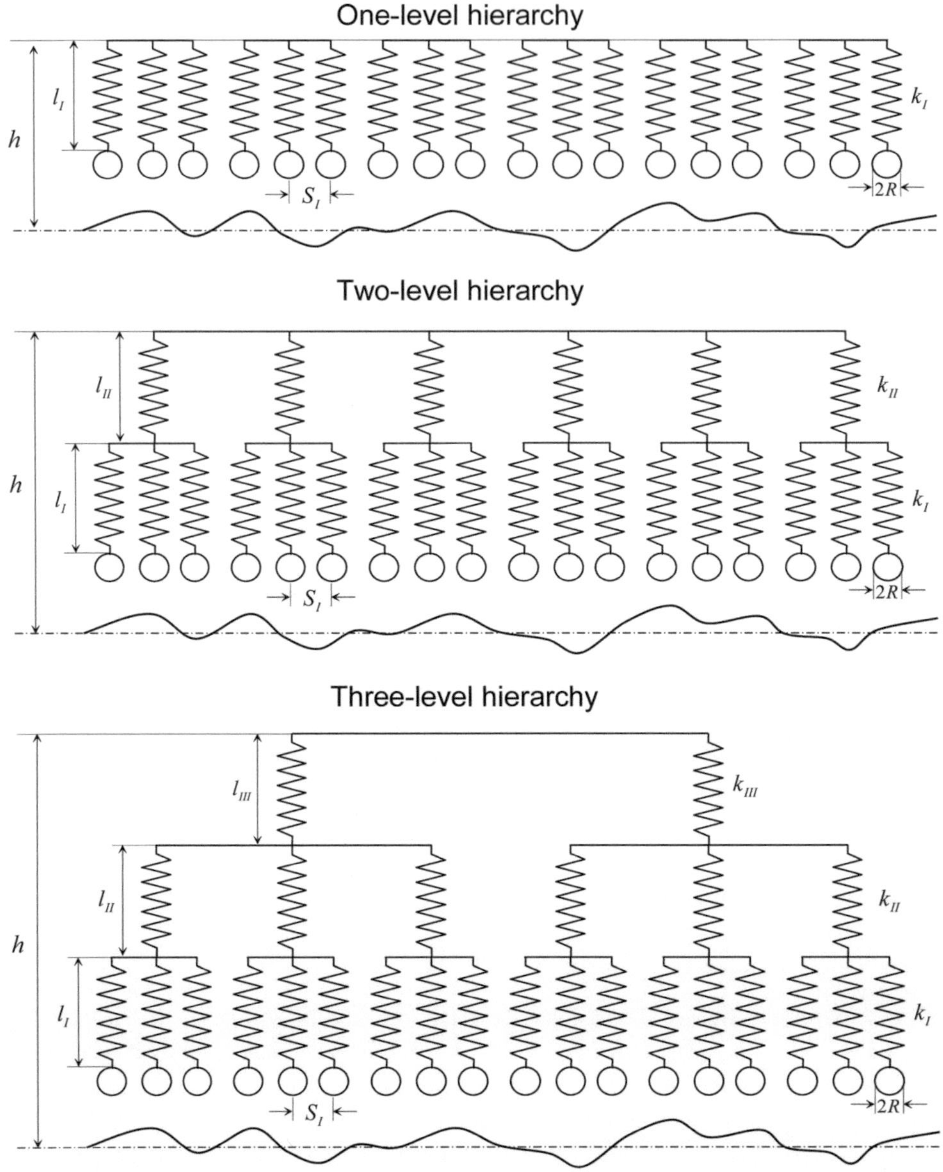

Fig. 12.16 One-, two-, and three-level hierarchical spring models for simulating the effect of hierarchical morphology on interaction of a seta with a rough surface. In this figure, $l_{I,\ II,\ III}$ are lengths of structures, s_I is space between spatulae, $k_{I,\ II,\ III}$ are stiffnesses of structures, *I*, *II*, and *III* are level indexes, *R* is radius of tip, and *h* is distance between upper spring base of each model and mean line of the rough profile (Kim and Bhushan 2007a)

$$F_{el} = -\sum_{j=1}^{q}\sum_{i=1}^{p} k_{ji}\left(\Delta l_{ji} - \Delta l_j\right) u_{ji} \quad u_{ji} = \begin{cases} 1\ \textit{if contact} \\ 0\ \textit{if no contact} \end{cases} \tag{12.23}$$

where q is the number of springs in level II of the model. For the three-level model, the elastic force was calculated as (Kim and Bhushan 2007a),

$$F_{el} = -\sum_{k=1}^{r}\sum_{j=1}^{q}\sum_{i=1}^{p} k_{kji}\left(\Delta l_{kji} - \Delta l_{kj} - \Delta l_j\right) u_{kji} \quad u_{kji} = \begin{cases} 1\ \textit{if contact} \\ 0\ \textit{if no contact} \end{cases} \tag{12.24}$$

where r is the number of springs in level III of the model. The spring force when the springs approach the rough surface is calculated using either (12.22), (12.23), or (12.24) for one-, two- and three-level models, respectively. During pull-off, the same equations are used to calculate the spring force. However, when the applied load is equal to zero, the springs do not detach due to the adhesion attraction, given by (12.19). The springs are pulled apart when pull-off force is equal to adhesion force at the interface. The adhesion force is the lowest value of elastic force F_{el} when the seta has detached from the contacting surface.

The adhesion energy is calculated as,

$$W_{ad} = \int_{\bar{D}}^{\infty} F_{el}(D)\,\mathrm{d}D \tag{12.25}$$

where D is the distance that the spring base moves away from the contacting surface. The lower limit of the distance $\bar{D}$ is the value of D where F_{el} is first zero when the model is pulled away from the contacting surface. Also although the upper limit of the distance is infinity, in practice, the $F_{el}(D)$ curve is integrated to an upper limit where F_{el} increases from a negative value to zero. Figure 12.17 shows the flow chart for the calculation of the adhesion force and the adhesion energy employed by Kim and Bhushan (2007a).

The random rough surfaces used in the simulations were generated on a computer (Bhushan 2013a, b). Several natural (sycamore tree bark and siltstone) and artificial (drywall, wood, laminate, steel, aluminum, and glass) surfaces were chosen to determine the surface parameters of typical rough surfaces that a gecko might encounter. Two-dimensional profiles of surfaces were obtained using a stylus profiler for two scan lengths: 80 μm, which is approximately the size of a single gecko seta, and 2000 μm, which is close to the size of a gecko lamellae (Bhushan et al. 2006). Surface roughness parameters (root mean square (RMS) amplitude σ and correlation length β^*) for scan lengths of 80 and 2000 μm, are presented in Table 12.4. The roughness parameters are scale dependent; therefore, adhesion values also are expected to be scale dependent. As the scan length was increased, the measured values of RMS amplitude and correlation length both increased. At a scale length of 80 μm (size of seta), the roughness amplitude does not

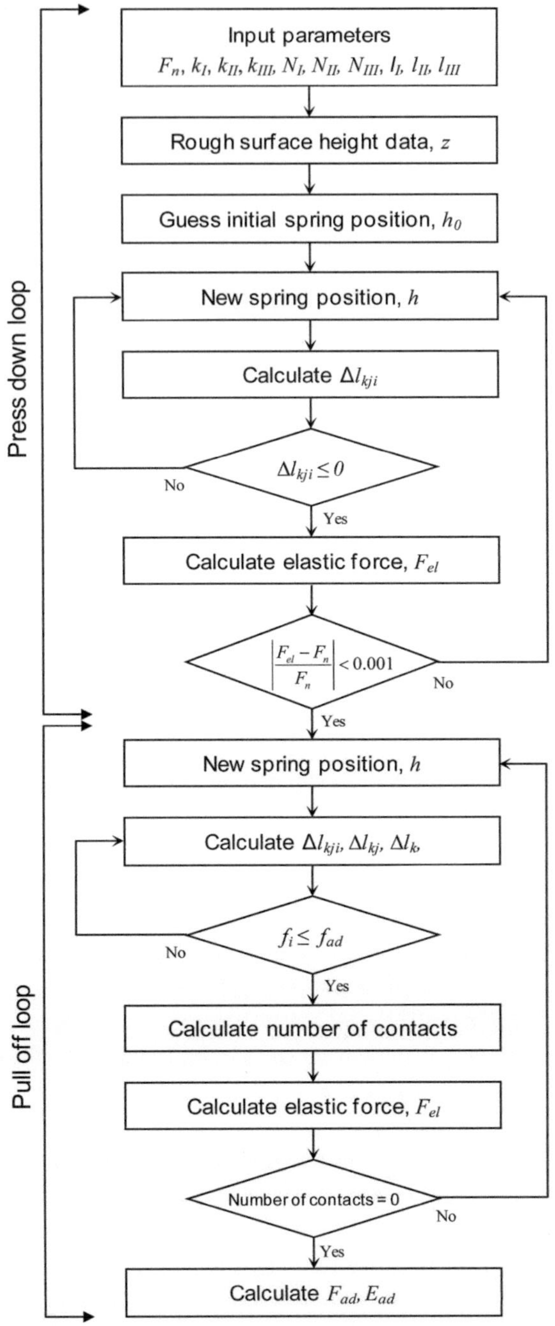

Fig. 12.17 Flow chart for the calculation of the adhesion force (F_{ad}) and the adhesion energy (E_{ad}) for three-level hierarchical spring model. In this figure, F_n is an applied load, $k_{I,\ II,\ III}$ and $l_{I,\ II,\ III}$ are stiffnesses and lengths of structures, Δl_{kji}, Δl_{ki}, and Δl_k are the spring deformations on level *I, II*, and *III*, respectively, *i, j*, and *k* are spring indexes on each level, f_i is the elastic force of a single spring, and f_{ad} is the adhesion force of a single contact (Kim and Bhushan 2007a)

Table 12.4 Scale dependence of surface parameters σ and β* for rough surfaces at scan lengths of 80 and 2000 μm (Bhushan et al. 2006)

Scan length	80 μm		2000 μm	
Surface	σ (μm)	β^* (μm)	σ (μm)	β^* (μm)
Sycamore tree bark	4.4	17	27	251
Siltstone	1.1	4.8	11	268
Painted drywall	1	11	20	93
Wood laminate	0.11	18	3.6	264
Polished steel	0.07	12	0.40	304
Polished 2024 aluminum	0.40	6.5	0.50	222
Glass	0.01	2.2	0.02	152

exceed 5 μm, while at a scale length of 2000 μm (size of lamella), the roughness amplitude is as high as 30 μm. This suggests that setae should adapt to surfaces with roughness on the order of several microns, while lamellae should adapt to roughness on the order of tens of microns. The skin of the gecko should adapt to larger roughness values. The range of values of σ from 0.01 μm to 30 μm and a fixed value of $\beta^* = 200$ μm were used for modeling the contact of a seta with random rough surfaces. The chosen range covers values of roughnesses for relatively smooth, artificial surfaces to natural, rough surfaces.

12.6.3 Adhesion Results of the Multi-level Hierarchical Spring Model

The multi-level hierarchical spring model was developed by Kim and Bhushan (2007a). They obtained various useful results that are presented next. Figure 12.18a shows the calculated spring force–distance curves for the one-, two- and three-level hierarchical models in contact with rough surfaces of different values of RMS amplitude σ, ranging from $\sigma = 0.01$ μm to $\sigma = 30$ μm at an applied load of 1.6 μN, which was derived from the gecko's weight. When the spring model is pressed against the rough surface, contact between the spring and the rough surface occurs at point A. As the spring tip presses into the contacting surface, the force increases up to point B, B′, or B″. During pull off, the spring relaxes, and the spring force passes an equilibrium state (0 N). Tips break free of adhesion forces at point C, C′, or C″ as the spring moves away from the surface. The perpendicular distance from C, C′, or C″ to zero is the adhesion force. The data show that adhesion forces for various levels of hierarchy are comparable when in contact with a smooth surface. However, the adhesion force for the three-level model is the largest when in contact with a rough surface. Thus, multilevel hierarchy facilitates adaptability to various rough surfaces.

Fig. 12.18 **a** Force-distance curves of one-, two-, and three-level models in contact with rough surfaces with different σ values for an applied load of 1.6 μN. **b** The adhesion coefficient, the number of contacts, and the adhesion energy per unit length of profile for one- and multi-level models with an increase of σ value (left figures), and relative increases between multi- and one -level models (right side) for an applied load of 1.6 μN. The value of k_{III} in the analysis is 2.908 N/m (Kim and Bhushan 2007a)

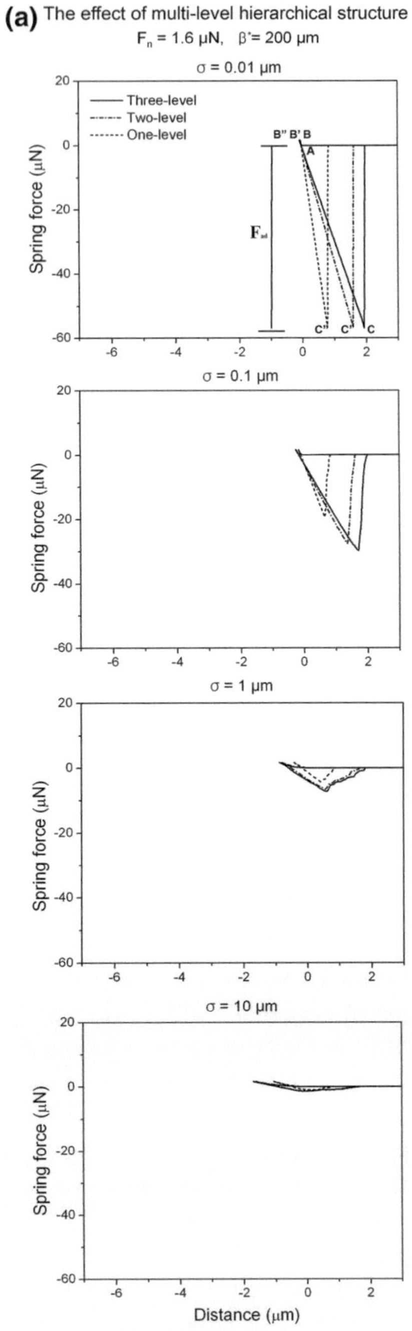

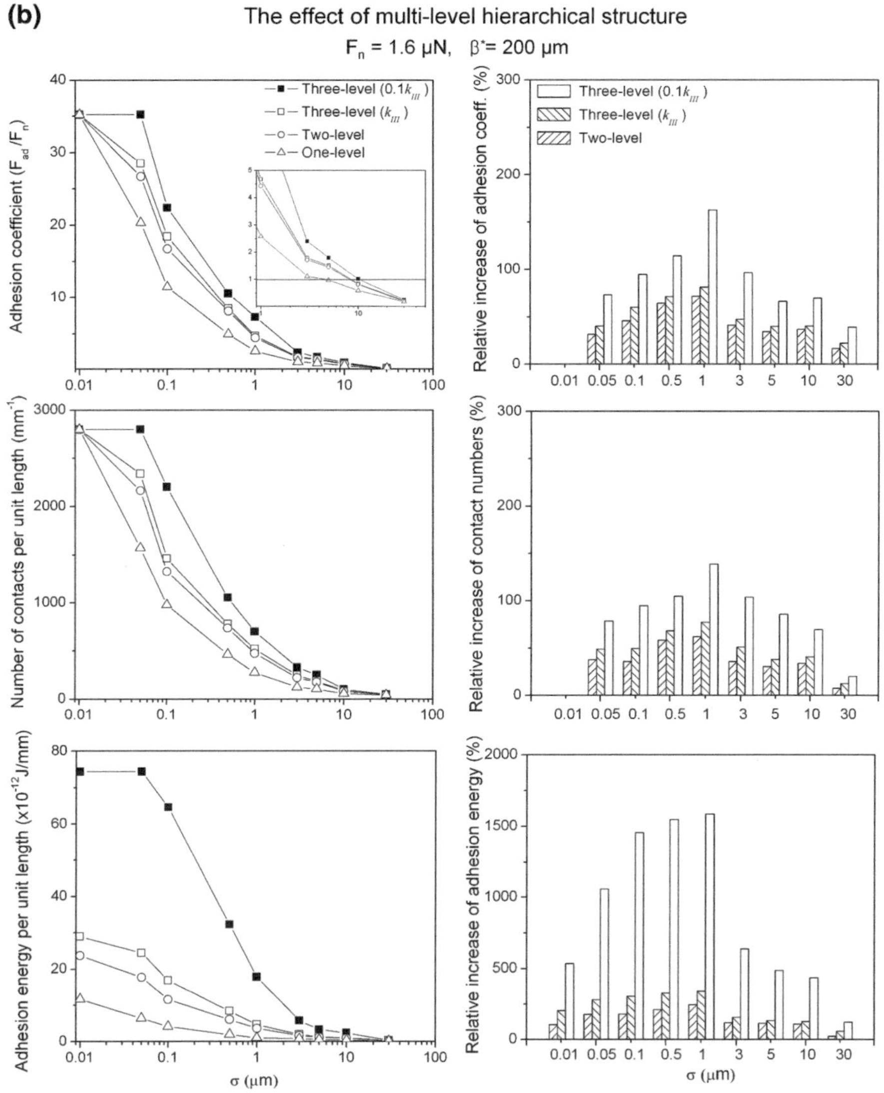

Fig. 12.18 (continued)

Adhesion energy stored during contact can be obtained by calculating the area of the triangle during the unloading part of the curves (12.25). Using the spring force-distance curves, Kim and Bhushan (2007a) calculated the adhesion coefficient, the number of contacts per unit length, and the adhesion energy per unit length of the one-, two-, and three-level models for an applied load of 1.6 μN and a wide range of RMS roughness (σ), as seen in the left graphs of Fig. 12.18b. The adhesion coefficient, defined as the ratio of pull-off force to applied preload, represents the strength of adhesion with respect to the preload. For the applied load of 1.6 μN, which corresponds to the weight of a gecko, the maximum adhesion coefficient is about 36 when σ is smaller than 0.01 μm. This means that a gecko can generate enough adhesion force to support 36 times its bodyweight. However, if σ is increased to 1 μm, the adhesion coefficient for the three-level model is reduced to 4.7. It is noteworthy that the adhesion coefficient falls below 1 when the contacting surface has an RMS roughness σ greater than 10 μm. This implies that the attachment system is no longer capable of supporting the gecko's weight. Autumn et al. (2000, 2002) showed that in isolated gecko setae contacting with the surface of a single crystalline silicon wafer, a 2.5 μN preload yielded adhesion of 20–40 μN and thus a value of adhesion coefficient of 8–16, which supports the simulation results of Kim and Bhushan (2007a).

Figure 12.18b (top left) shows that the adhesion coefficient for the one-level model is lower than that for the three-level model, but that there is only a small difference between the values of the two- and three-level models. In order to show the effect of spring stiffness in the three-level model, another model used springs in level *III* with stiffness that was ten times smaller than the stiffness of the original level *III* springs used. The data are presented in Fig. 12.18b. It can be seen that the three-level model with a third level stiffness of 0.1 k_{III} has a 20–30% higher adhesion coefficient than the three-level model. The results also show that the trends in the number of contacts are similar to that of the adhesive force. The study also investigated the effect of σ on adhesion energy. It was determined that the adhesion energy decreased with an increase of σ. For a smooth surface with $\sigma = 0.01$ μm, the adhesion energies for the two- and three-level hierarchical models are 2 times and 2.4 times larger than that for the one-level model, respectively, but adhesion energy decreases rapidly at surfaces with σ greater than 0.05 μm. Further, in every model adhesion energy finally decreases to zero at surfaces with σ greater than 10 μm. The adhesion energy for the more compliant three-level model is 2–3 times higher than that for the original three-level model.

In order to demonstrate the effect of the hierarchical structure on adhesion enhancement, Kim and Bhushan (2007a) calculated the increases in the adhesion coefficient, the number of contacts, and the adhesion energy of the two- and three-level models, as well as the three-level with 0.1 k_{III} model, relative to the one-level model. These results are shown in the right side of Fig. 12.18b. It was also found that for the two- and three-level models, the adhesion coefficient increases slowly with an increase of σ, and has maximum values of about 70 and 80% at $\sigma = 1$ μm, respectively, and then decreases for surfaces with σ greater than 3 μm. The condition at which a significant enhancement occurs is related to the maximum spring deformation, which is the applied load divided by the spring stiffness. If the maximum spring deformation is greater than 2–3 times larger than the σ value of the surface roughness, a significant adhesion enhancement occurs. The more compliant three-level model shows significant adhesion enhancement. The relative

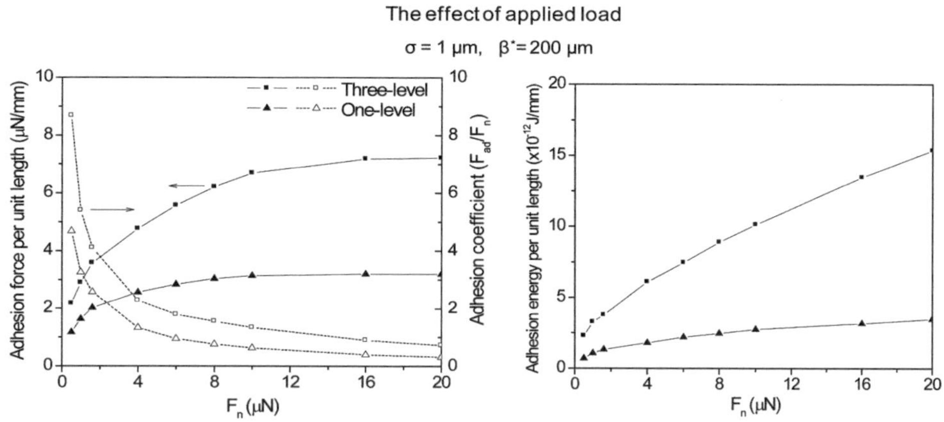

Fig. 12.19 The adhesion force, adhesion coefficient, and adhesion energy as a function of applied loads for both one- and three-level models contacting with the rough surface (Kim and Bhushan 2007a)

increase of the adhesion coefficient and adhesion energy for that model was the highest at $\sigma = 1$ μm.

Figure 12.19 shows the variation of adhesion force and adhesion energy as a function of applied load for both one- and three-level models contacting a surface with $\sigma = 1$ μm. It is shown that as the applied load increases, the adhesion force increases up to a certain applied load and then has a constant value, whereas, adhesion energy continues to increase with an increase in the applied load. The one-level model had a maximum value of adhesion force per unit length of about 3 μN/mm at the applied load of 10 μN, and the three-level model had a maximum value of about 7 μN/mm at the applied load of 16 μN. However, the adhesion coefficient continues to decrease at higher applied loads because adhesion force is constant even if the applied load increases.

The simulation results for the three-level model, which is closest to actual gecko setae, presented in Fig. 12.18 show that roughness reduces the adhesion force. At a surface with σ greater than 10 μm, the ratio of the adhesion force to the gecko weight indicates that it cannot support itself. However, in practice, a gecko can cling or crawl on the surface of ceiling with higher roughness. Kim and Bhushan (2007a) did not consider the effect of lamellae in their study. The authors stated that the lamellae can adapt to the waviness of a surface while the setae allow for the adaptation to micro- or nano- roughness and expect that adding the lamellae of gecko skin to the model would lead to higher adhesion over a wider range of roughness. In addition, their hierarchical model considers deformation normal to the surface and the motion of seta. It should be noted that the measurements of adhesion force of a single gecko seta made by Autumn et al. (2000) demonstrated that a load applied normal to the surface was insufficient for an effective attachment of seta.

Finally, in order to determine the effects of spring stiffness and the number of springs on the adhesion enhancement of a multi-level hierarchical model, one- and three-level models with four different spring stiffnesses and three different numbers of springs were analyzed by Kim and Bhushan (2007b). The stiffness k_I was taken equal to 0.0126 N/m, as before. Other stiffnesses, k_{II} and k_{III}, were normalized with respect to k_I. The three-level model with $k_{III}/k_I = 100$ and $k_{II}/k_I = 10$ has similar stiffness values to gecko's seta presented in Table 12.3 and used in the previous example (Figs. 12.18 and 12.19). The left part in Fig. 12.20 shows the adhesion coefficient, the number of contacts and adhesion energy per unit length for the one- and three-level models with four different spring stiffnesses as a function of σ value for an applied load of 1.6 μN. Trends as a function of σ are the same as observed previously in Fig. 12.18b. For the case of $k_{III} = k_{II} = k_I$, one gets the highest value of adhesion coefficient of 36 on a rough surface, and it remains high up to σ value of about 1 μm, and then it starts to decrease. As the stiffness values of k_{II} and k_{III} increase, the adhesion coefficient starts to decrease at lower values of σ, and it decreases rapidly with an increase in σ. The number of contacts and adhesion energy per unit length as a function of σ have trends similar to that of adhesion coefficient. The right part in Fig. 12.20 shows relative increases between one- and three-level models. The trends are the same as discussed earlier.

For the effect of the number of springs, three different cases of the number of springs in the upper level to that in the lower level were considered. The three-level model with $N_I/N_{II} = 10$ and $N_{II}/N_{III} = 10$ is closest to the actual gecko's setae (discussed earlier). Figure 12.21 shows the adhesion force, adhesion coefficient, and adhesion energy as a function of applied loads for one- and three-level models with different numbers of springs contacting the rough surface. The variation of the number of springs on each level affects the equivalent stiffness of the model. As the number of springs on the lower level increases, the equivalent stiffness decreases. The figure shows that the three-level model with $N_I/N_{II} = 100$ and $N_{II}/N_{III} = 10$ gives the largest adhesion force and adhesion energy among every model, because the equivalent stiffness is lowest.

To sum up, multi-level hierarchical structures with more compliant springs and at an increasing applied load are beneficial to gecko adhesion enhancement. Furthermore, for significant adhesion enhancement, the maximum spring deformation should be two to three times larger than the σ value of the surface roughness.

12.6.4 Capillary Effects

Kim and Bhushan (2008) investigated the effects of capillarity on gecko adhesion by considering capillary force as well as the solid-to-solid interaction. The Laplace and surface tension components of the capillary force are treated the same as earlier in this Chapter. The solid-to-solid adhesive force was calculated by DMT theory according to (12.19) and will be denoted as F_{DMT}.

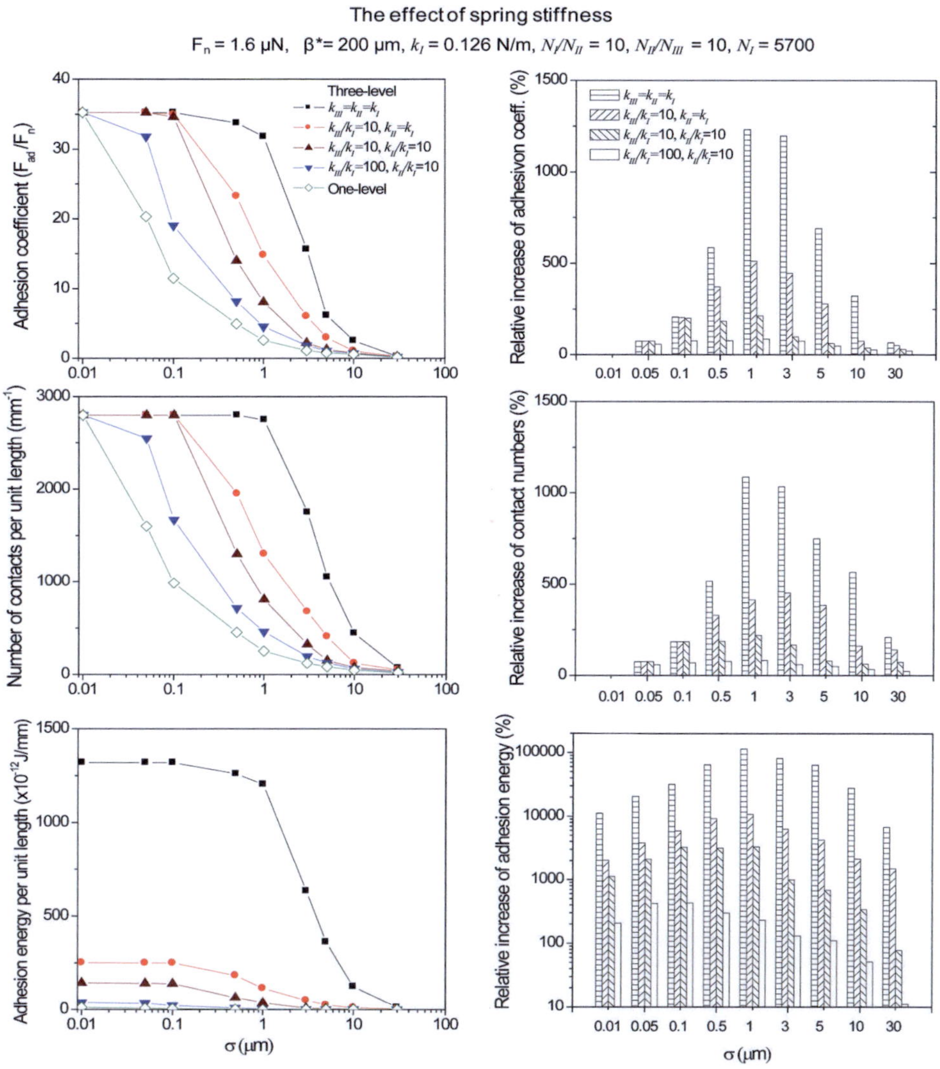

Fig. 12.20 The adhesion coefficient, the number of contacts, and the adhesion energy per unit length of profile for one- and three-level models with different spring stiffnesses as a function of σ value (left figures), and relative increases between one- and three-level models (right figures) for an applied load of 1.6 μN (Kim and Bhushan 2007b)

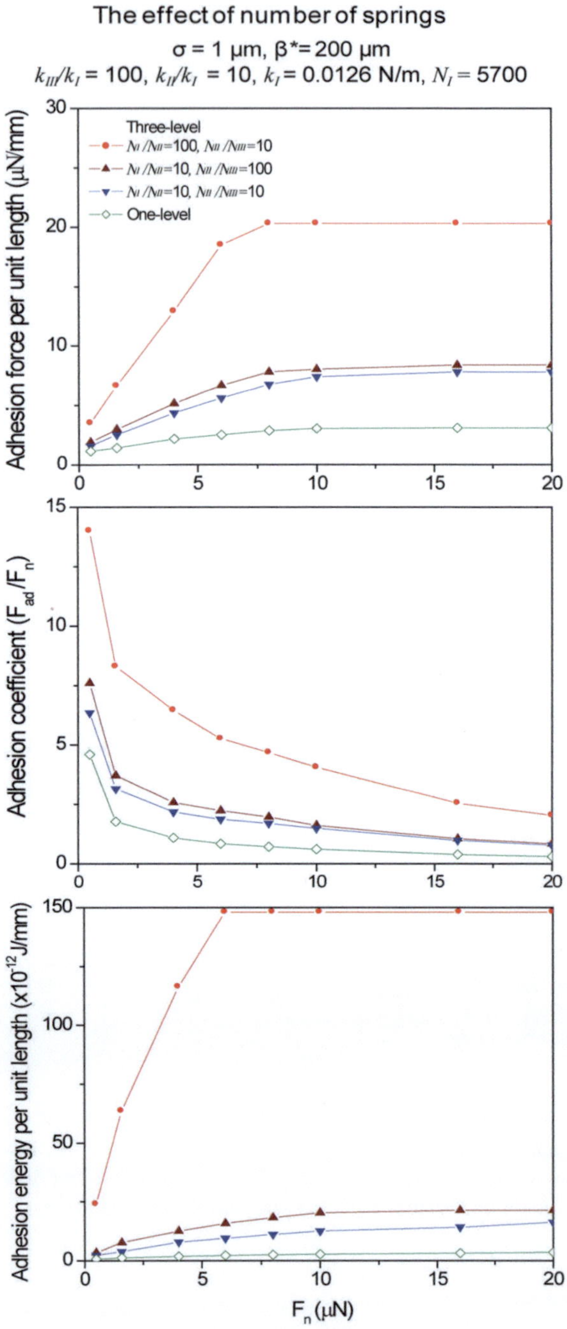

Fig. 12.21 The adhesive force, adhesion coefficient, and adhesion energy as a function of applied loads for one- and three-level models with different number of springs contacting with the rough surface (Kim and Bhushan 2007b)

The work of adhesion was calculated by (12.20). Kim and Bhushan (2008) assumed typical values of the Hamaker constant to be $H_{air} = 10^{-19}$ J in the air and $H_{water} = 6.7 \times 10^{-19}$ J in the water (Israelachvili 1992). The works of adhesion of two surfaces in contact separated by an atomic distance $D \approx 0.2$ nm (Israelachvili 1992) are approximately equal to 66 mJ/m^2 in air and 44 mJ/m^2 in water. Assuming tip radius R is 50 nm, the DMT adhesion forces F_{DMT} of a single contact in air and in water are $F_{DMT}^{air} = 11$ nN and $F_{DMT}^{water} = 7.3$ nN, respectively. As the humidity increases from zero to 100%, the DMT adhesion force will take a value between F_{DMT}^{air} and F_{DMT}^{water}. To calculate the DMT adhesion force for the intermediate humidity, an approximation method by Wan et al. (1992) was used. The work of adhesion W_{ad} for the intermediate humidity can be expressed as,

$$W_{ad} = \int_{D}^{\infty} \frac{H}{6\pi h^3} dh = \int_{D}^{h_f} \frac{H_{water}}{6\pi h^3} dh + \int_{h_f}^{\infty} \frac{H_{air}}{6\pi h^3} dh \tag{12.26}$$

where h is the separation along the plane. The water film thickness at a filling angle ϕ is h_f, which can be calculated as,

$$h_f = D + R(1 - \cos\phi) \tag{12.27}$$

Therefore, using (12.19), (12.26), and (12.27), the DMT adhesion force for the intermediate humidity is given as,

$$F_{DMT} = F_{DMT}^{air} \left\{ 1 - \frac{1}{\left(1 + R(1 - \cos\phi)/D\right)^2} \right\} + F_{DMT}^{water} \left\{ \frac{1}{\left(1 + R(1 - \cos\phi)/D\right)^2} \right\} \tag{12.28}$$

Finally, Kim and Bhushan (2008) calculated the total adhesion force F_{ad} as the sum of (12.15) and (12.28),

$$F_{ad} = F_c + F_{DMT} \tag{12.29}$$

Kim and Bhushan (2008) then used the total adhesion force as a critical force in the three-level hierarchical spring model discussed previously. In the spring model for gecko seta, if the force applied upon spring deformation is greater than the adhesion force, the spring is regarded as having been detached. To simulate the capillary contribution to adhesion force for a gecko spatula, Kim and Bhushan (2008) set the contact angle on a gecko spatula tip θ_1 equal to 128° (Huber et al. 2005b). It was assumed that the spatula tip radius $R = 50$ nm, the ambient temperature $T = 25$ °C, the surface tension of water $\gamma = 73$ mJ/m^2, and molecular volume of water $V = 0.03$ nm^3 (Israelachvili 1992).

Figure 12.22a shows the total adhesion force as a function of relative humidity for a single spatula in contact with surfaces with different contact angles. Total adhesion force decreases with an increase in the contact angle on the substrate, and the difference of total adhesion force among different contact angles is larger in the intermediate humidity regime. As the relative humidity increases, total adhesion force for the surfaces with contact angles less than 60° has a higher value than the DMT adhesion force not considering wet contact, whereas with the value above 60°, total adhesion force has lower values at most relative humidities.

The simulation results of Kim and Bhushan (2008) are compared with the experimental data by Huber et al. (2005b) in Fig. 12.22b. Huber et al. (2005b) measured the pull off force of a single spatula in contact with four different types of Si wafers and glass at a temperature of 25 °C and relative humidity of 52%. According to their description, wafer families "N" and "T" in Fig. 12.22b differ by the thickness of the top amorphous Si oxide layer. The "Phil" type is the cleaned Si oxide surface that is hydrophilic with a water

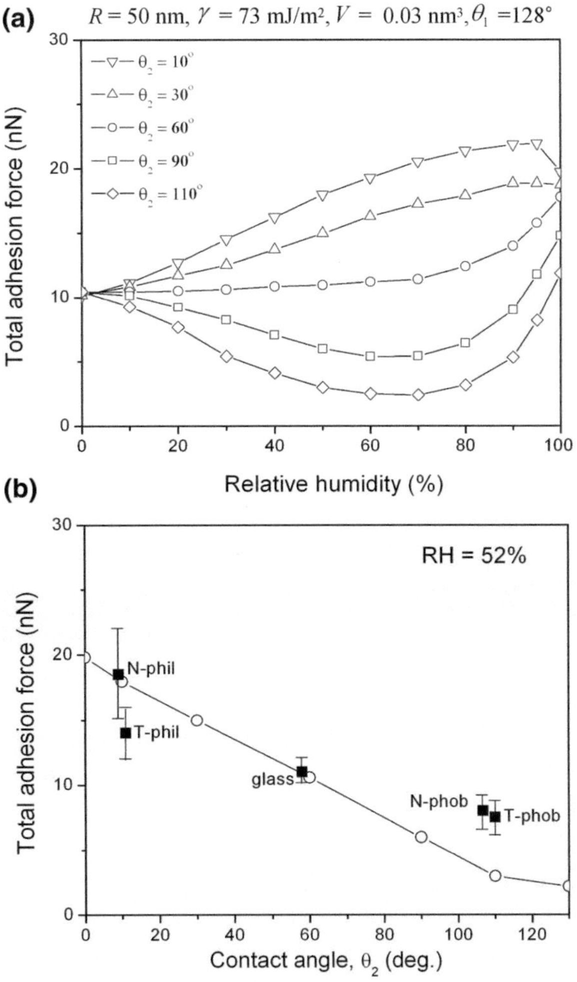

Fig. 12.22 **a** Total adhesion force as a function of relative humidity for a single spatula in contact with surfaces with different contact angles. **b** Comparison of the simulation results of Kim and Bhushan (2008) with the measured data obtained by Huber et al. (2005b) for a single spatula in contact with the hydrophilic and the hydrophobic surfaces (Kim and Bhushan 2008)

contact angle ≈10°, whereas the "Phob" type is covered by a hydrophobic monolayer with a water contact angle>100°. Glass has a water contact angle of 58°. Huber et al. (2005b) showed that the adhesion force of a gecko spatula rises significantly for substrates with increasing hydrophilicity (adhesive force increases by a factor of two as mating surfaces go from hydrophobic to hydrophilic). As shown in Fig. 12.22b, the simulation results of Kim and Bhushan (2008) closely match the experimental data of Huber et al. (2005b).

Kim and Bhushan (2008) carried out adhesion analysis for a three-level hierarchical model for gecko seta. Figure 12.23 shows the adhesion coefficient and number of contacts per unit length for the three-level hierarchical model in contact with rough surfaces with different values of the RMS amplitude σ ranging from $\sigma = 0.01$ μm to $\sigma = 30$ μm for different relative humidities and contact angles of the surface. It can be seen that for a surface with contact angle $\theta_2 = 10°$, the adhesion coefficient is greatly influenced by relative humidity. At 0% relative humidity the maximum adhesion coefficient is about 36 at a value of σ smaller than 0.01 μm, compared to 78 for 90% relative humidity with the same surface roughness. As expected, the effect of relative humidity on increasing the adhesion coefficient decreases as the contact angle becomes larger. For hydrophobic surfaces, relative humidity decreases the adhesion coefficient. Similar trends can be noticed in the number of contacts. Thus, the conclusion can be drawn that hydrophilic surfaces are beneficial to gecko adhesion enhancement.

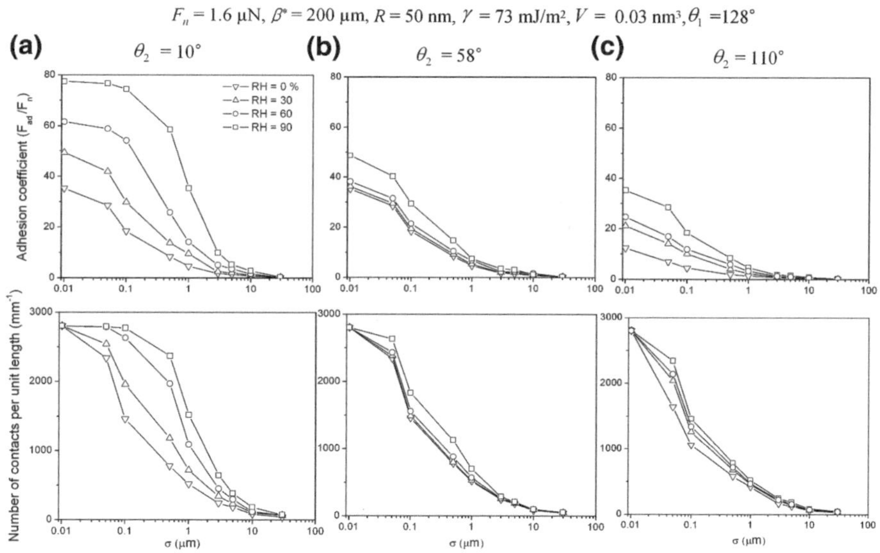

Fig. 12.23 The adhesion coefficient and number of contacts per unit length for three-level hierarchical model in contact with rough surfaces with different values of root mean square amplitudes σ and contact angles for different relative humidities (Kim and Bhushan 2008)

12.7 Adhesion Data Base of Fibrillar Structures

As it relates to the design of biomimetic structures, the mechanics of adhesion between a fibrillar structure and a rough surface has been a topic of investigation by many researchers (Jagota and Bennison 2002; Persson 2003; Sitti and Fearing 2003; Glassmaker et al. 2004, 2005; Gao et al. 2005; Yao and Gao 2006; Kim and Bhushan 2007b, c). Kim and Bhushan (2007c) developed a convenient, general, and useful guideline for understanding biological systems and for improving the attachment of fibrillar structures. This adhesion database was constructed by modeling the fibers as oriented cylindrical cantilever beams with spherical tips. The authors then carried out numerical simulations of the attachment system in contact with random rough surfaces considering three constraint conditions: buckling, fracture, and sticking of the fiber structure. For a given applied load, roughnesses of contacting surface, and fiber material, a procedure to find an optimal fiber radius and aspect ratio for the desired adhesion coefficient was developed.

The model developed by Kim and Bhushan (2007c) is used to find the design parameters for fibers of a single-level attachment system capable of achieving desired properties—high adhesion coefficient and durability. The design variables for an attachment system are as follows: fiber geometry (radius and aspect ratio of fibers, tip radius), fiber material, fiber density, and fiber orientation. The optimal values for the design variables to achieve the desired properties should be selected for fabrication of a biomimetic attachment system.

12.7.1 Fiber Model

The fiber model developed by Kim and Bhushan (2007c) consists of a simple idealized fibrillar structure with a single-level array of micro/nano beams protruding from a backing as shown in Fig. 12.24. The fibers are modeled as oriented cylindrical cantilever beams with spherical tips. In Fig. 12.24, l is the length of fibers, θ is the fiber orientation, R is the fiber radius, R_t is the tip radius, S is the spacing between fibers, and h is the distance between the upper spring base of each model and mean line of the rough profile. The end terminal of the fibers is assumed to be a spherical tip with a constant radius and a constant adhesion force.

12.7.2 Single Fiber Contact Analysis

Kim and Bhushan (2007c) modeled an individual fiber as a beam oriented at an angle θ to the substrate, with the contact load F aligned normal to the substrate. The net displacement normal to the substrate can be calculated according to (12.16) and (12.17). The fiber stiffness ($k = F/F\delta_{\perp}$) is given by (Glassmaker et al. 2004),

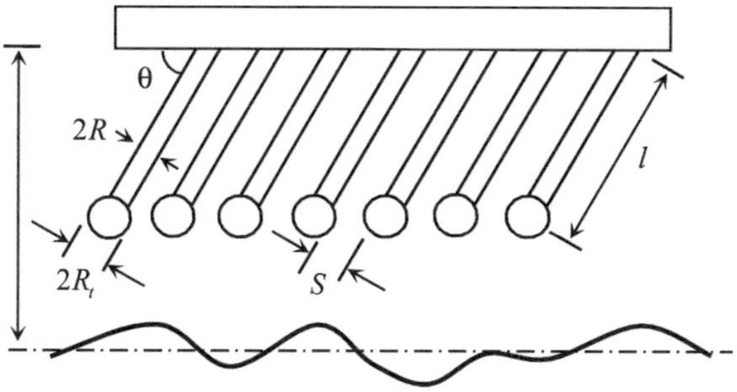

Fig. 12.24 Single-level attachment system with oriented cylindrical cantilever beams with spherical tip. In this figure, l is the length of fibers, θ is the fiber orientation, R is the fiber radius, R_t is the tip radius, S is the spacing between fibers, and h is distance between base of model and mean line of the rough profile (Kim and Bhushan 2007c)

$$k = \frac{\pi R^2 E}{l \sin^2\theta\left(1 + \frac{4l^2 \cot^2\theta}{3R^2}\right)} = \frac{\pi R E}{2\lambda \sin^2\theta\left(1 + \frac{16\lambda^2 \cot^2\theta}{3}\right)} \quad (12.30)$$

where $\lambda = l/2R$ is the aspect ratio of the fiber, and θ is fixed at 30°.

Two alternative models dominate the world of contact mechanics—the Johnson-Kendall-Roberts (JKR) theory (Johnson et al. 1971) for compliant solids, and the Derjaguin-Muller-Toporov (DMT) theory (Derjaguin et al. 1975) for stiff solids. Although gecko setae are composed of β-keratin with a high elastic modulus (Russell 1986; Bertram and Gosline 1987) which is close to the DMT model, in general the JKR theory prevails for biological or artificial attachment systems. Therefore, the JKR theory was applied in the subsequent analysis of Kim and Bhushan (2007c) to compare the materials with wide ranges of elastic modulus. The adhesion force between a spherical tip and a rigid flat surface is thus calculated using the JKR theory as (Johnson et al. 1971),

$$F_{ad} = \frac{3}{2}\pi R_t W_{ad} \quad (12.31)$$

where R_t is the radius of spherical tip, and W_{ad} is the work of adhesion (calculated according to 12.24). Kim and Bhushan (2007c) used this adhesion force as a critical force. If the elastic force of a single spring is less than the adhesion force, they regarded the spring as having been detached.

12.7.3 Constraints

In the design of fibrillar structures, a trade-off exists between the aspect ratio of the fibers and their adaptability to a rough surface. If the aspect ratio of the fibers is too large, they can adhere to each other or even collapse under their own weight as shown in Fig. 12.25a. If the aspect ratio is too small (Fig. 12.25b), the structures will lack the necessary compliance to conform to a rough surface. Spacing between the individual fibers is also important. If the spacing is too small, adjacent fibers can attract each other through intermolecular forces, which will lead to bunching. Therefore, in their analysis, Kim and Bhushan (2007c) considered three necessary constraints to the allowed geometry: buckling, fracture, and sticking of fiber structure.

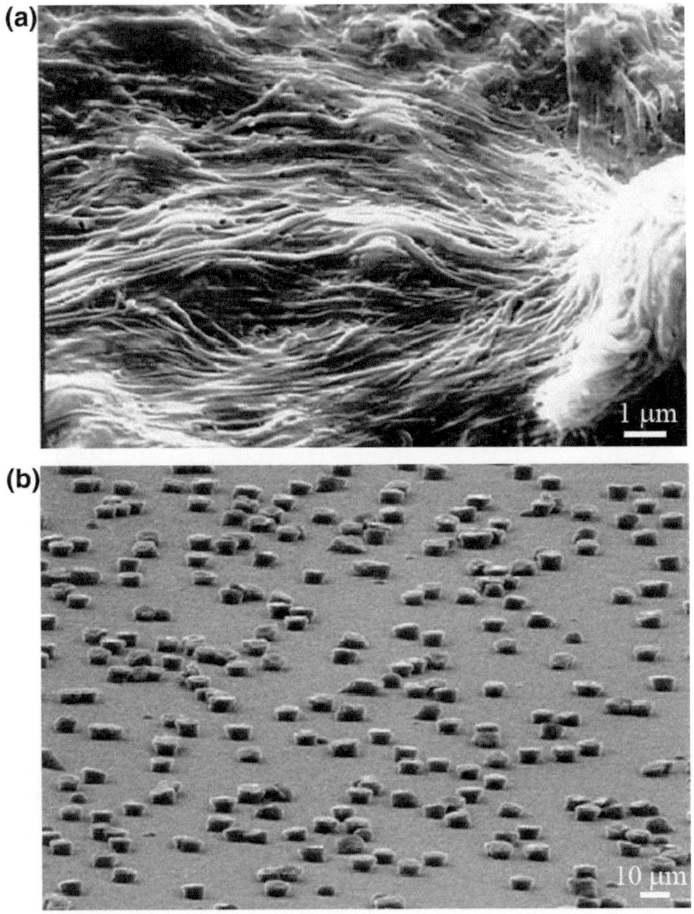

Fig. 12.25 SEM micrographs of **a** high aspect ratio polymer fibrils that have collapsed under their own weight and **b** low aspect ratio polymer fibrils that are incapable of adapting to rough surfaces (Sitti and Fearing 2003)

12.7.3.1 Non-buckling Condition

A fibrillar interface can deliver a compliant response while still employing stiff materials because of bending and micro-buckling of fibers. Based on classical Euler buckling, Glassmaker et al. (2004) established a stress–strain relationship and a critical compressive strain for buckling, ε_{cr}, for the fiber oriented at an angle, θ, to the substrate,

$$\varepsilon_{cr} = -\frac{b_c \pi^2}{3(Al^2/3I)}\left(1 + \frac{A_c l^2}{3I}\cot^2\theta\right) \tag{12.32}$$

where A_c is the cross-sectional area of the fibril, and b_c is a factor that depends on boundary conditions. The factor b_c has a value of 2 for pinned-clamped micro-beams. For fibers having a circular cross section, ε_{cr} is calculated as,

$$\varepsilon_{cr} = -\frac{b_c \pi^2}{3(4l^2/3R^2)}\left(1 + \frac{4l^2}{3R^2}\cot^2\theta\right) = -b_c \pi^2\left(\frac{1}{16\lambda^2} + \frac{\cot^2\theta}{3}\right) \tag{12.33}$$

In (12.33), ε_{cr} depends on both the aspect ratio, λ, and the orientation, θ, of fibers. If $\varepsilon_{cr} = 1$, which means the fiber deforms up to the backing, buckling does not occur. Figure 12.26 plots the critical orientation, θ, as a function of aspect ratio for the case of $\varepsilon_{cr} = 1$. The critical fiber orientation for buckling is 90° at λ less than 1.1. This means that buckling does not occur, regardless of the orientation of the fiber, at λ less than 1.1. For λ greater than 1.1, the critical fiber orientation for buckling decreases with an increase in λ, and has a constant value of 69° at λ greater than 3. Kim and Bhushan (2007c) used a fixed value at 30° for θ, because the maximum adhesive force is achieved at this orientation, and buckling is not expected to occur.

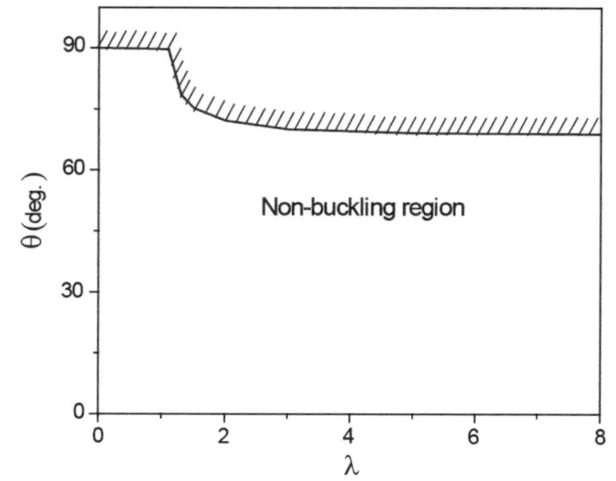

Fig. 12.26 Critical fiber orientation as a function of aspect ratio λ for non-buckling condition for pinned-clamped micro-beams ($b_c = 2$) (Kim and Bhushan 2007c)

12.7.3.2 Non-fiber Fracture Condition

For small contacts, the strength of the system will eventually be determined by fracture of the fibers. Spolenak et al. (2005) suggested the limit of fiber fracture as a function of the adhesion force. The axial stress σ_f in a fiber is limited by its theoretical fracture strength σ_{th}^f as,

$$\sigma_f = \frac{F_{ad}}{R^2\pi} \le \sigma_{th}^f \tag{12.34}$$

Using (12.31), a lower limit for the useful fiber radius R is calculated as,

$$R \ge \sqrt{\frac{3R_t W_{ad}}{2\sigma_{th}^f}} \approx \sqrt{\frac{15R_t W_{ad}}{E}} \tag{12.35}$$

where the theoretical fracture strength is approximated by $E/10$ (Dieter 1988). The lower limit of fiber radius for fiber fracture by the adhesion force depends on elastic modulus. By assuming $W_{ad} = 66$ mJ/m^2 as stated earlier, Kim and Bhushan (2007c) calculated the lower limits of fiber radius for $E = 1$ MPa, 0.1 and 10 GPa to be 0.32, 0.032 and 0.0032 μm, respectively.

The contact stress cannot exceed the ideal contact strength transmitted through the actual contact area at the instant of tensile instability (Spolenak et al. 2005). Kim and Bhushan (2007c) used this condition (12.34) to extract the limit of tip radius, R_t,

$$\sigma_c = \frac{F_{ad}}{a_c^2\pi} \le \sigma_{th} \tag{12.36}$$

where σ_c is the contact stress, a_c is the contact radius, and σ_{th} is the ideal strength of van der Waals bonds, which equals approximately W_{ad}/b, where b is the characteristic length of surface interaction. Based on the JKR theory, for the rigid contacting surface, a_c at the instant of pull-off is calculated as,

$$a_c = \left(\frac{9\pi W_{ad} R_t^2 (1-\upsilon^2)}{8E}\right)^{1/3} \tag{12.37}$$

where ν is Poisson's ratio. The tip radius can then be calculated by combining (12.36) and (12.37) as,

$$R_t \ge \frac{8b^3E^2}{3\pi^2(1-\nu^2)^2 W_{ad}^2} \tag{12.38}$$

The lower limit of tip radius also depends on elastic modulus. Assuming $W_{ad} = 66$ mJ/m^2 and $b = 2 \times 10^{-10}$ m (Dieter 1988), the lower limits of tip radius for $E = 1$ MPa, 0.1 and 10 GPa are calculated as 6×10^{-7}, 6×10^{-3}, and 60 nm, respectively. In this study, Kim and Bhushan (2007c) fixed the tip radius at 100 nm, which satisfies the tip radius condition throughout a wide range of elastic modulus up to 10 GPa.

12.7.3.3 Non-sticking Condition

A high density of fibers is also important for high adhesion. However, if the space S between neighboring fibers is too small, the adhesion forces between them become stronger than the forces required to bend the fibers. Then, fibers might stick to each other and get entangled. Therefore, to prevent fibers from sticking to each other, they must be spaced apart and be stiff enough to prevent sticking or bunching. Several authors (e.g., Sitti and Fearing 2003) have formulated a non-sticking criterion. Kim and Bhushan (2007c) adopted the approach of Sitti and Fearing (2003). Both adhesion and elastic forces will act on bent structures. The adhesion force between two neighboring round tips is calculated as,

$$F_{ad} = \frac{3}{2}\pi R_t' W_{ad} \tag{12.39}$$

where R_t' is the reduced radius of contact, calculated as $R_t' = \left(1/1R_{t_1} + 1/1R_{t_2}\right)^{-1}$; R_{t1} and R_{t2} are the radii of contacting tips. In the case of similar tips, $R_{t1} = R_{t2}$, and thus $R_t' = 2/2R_t$.

The elastic force of a bent structure can be calculated by multiplying the bending stiffness ($k_b = 3\pi R^4 E/4\ l^3$) by a given bending displacement δ as,

$$F_{el} = \frac{3}{4}\frac{\pi R^4 E \delta}{l^3} \tag{12.40}$$

The condition for the prevention of sticking is $F_{el} > F_{ad}$. By combining (12.39) and (12.40), a requirement for the minimum distance S between structures, which will prevent sticking of the structures, is given as (Kim and Bhushan 2007c),

$$S > 2\delta = 2\left(\frac{4}{3}\frac{W_{ad} l^3}{E R^3}\right) = 2\left(\frac{32}{3}\frac{W_{ad}\lambda^3}{E}\right) \tag{12.41}$$

The constant 2 takes into account the two nearest structures. Using distance S, the fiber density, ρ, is calculated as,

$$\rho = \frac{1}{(S + 2R)^2} \tag{12.42}$$

Equation (12.42) was then used to calculate the allowed minimum density of fibers without sticking or bunching. In (12.41), it is shown that the minimum distance, S, depends on both the aspect ratio λ and the elastic modulus E. A smaller aspect ratio and higher elastic modulus allow for greater packing density. However, fibers with a low aspect ratio and high modulus are not desirable for adhering to rough surfaces due to lack of compliance.

12.7.4 Numerical Simulation

The simulation of adhesion of an attachment system in contact with random rough surfaces was carried out numerically. In order to conduct 2D simulations, it is necessary to calculate applied load F_n as a function of applied pressure P_n as an input condition. Using ρ calculated by non-sticking condition, Kim and Bhushan (2007c) calculated F_n as,

$$F_n = \frac{P_n p}{\rho} \tag{12.43}$$

where p is the number of springs in the scan length L, which equals L/(S + 2R).

Fibers of the attachment system are modeled as one-level-hierarchy elastic springs (Fig. 12.16) (Kim and Bhushan 2007c). The deflection of each spring and the elastic force that arises in the springs are calculated according to (12.21) and (12.22), respectively. The adhesion force is the lowest value of elastic force F_{el} when the fiber has detached from the contacting surface. Kim and Bhushan (2007c) used an iterative process to obtain optimal fiber geometry—fiber radius and aspect ratio. If the applied load, the roughness of contacting surface, and the fiber material are given, the procedure for calculating the adhesion force is repeated iteratively until the desired adhesion force is satisfied. In order to simplify the design problem, fiber material is regarded as a known variable. The next step is constructing the design database. Figure 12.27a shows the flow chart for the construction of adhesion design database, and Fig. 12.27b shows the calculation of the adhesion force that is a part of the procedure to construct an adhesion design database.

12.7.5 Results and Discussion

Figure 12.28 shows an example of the adhesion design database for biomimetic attachment systems consisting of single-level cylindrical fibers with an orientation angle of 30° and spherical tips of $R_t = 100$ nm constructed by Kim and Bhushan (2007c). The minimum fiber radius calculated by using non-fiber fracture condition, which plays a role of lower limit of optimized fiber radius, is also added on the plot. The plots in Fig. 12.28 cover all applicable fiber materials from soft elastomer material, such as poly(dimethylsiloxane) (PDMS) , to stiffer polymers, such as polyimide and β keratin. The dashed lines in each plot represent the limits of fiber fracture due to the adhesion force. For a soft material with $E = 1$ MPa in Fig. 12.28a, the range of the desirable fiber radius is more than 0.3 μm and that of the aspect ratio is approximately less than 1. As elastic modulus increases, the feasible range of both fiber radius and aspect ratio also increase as shown in Fig. 12.28b, c. In Fig. 12.28, the fiber radius has a linear relation with the surface roughness on a logarithm scale.

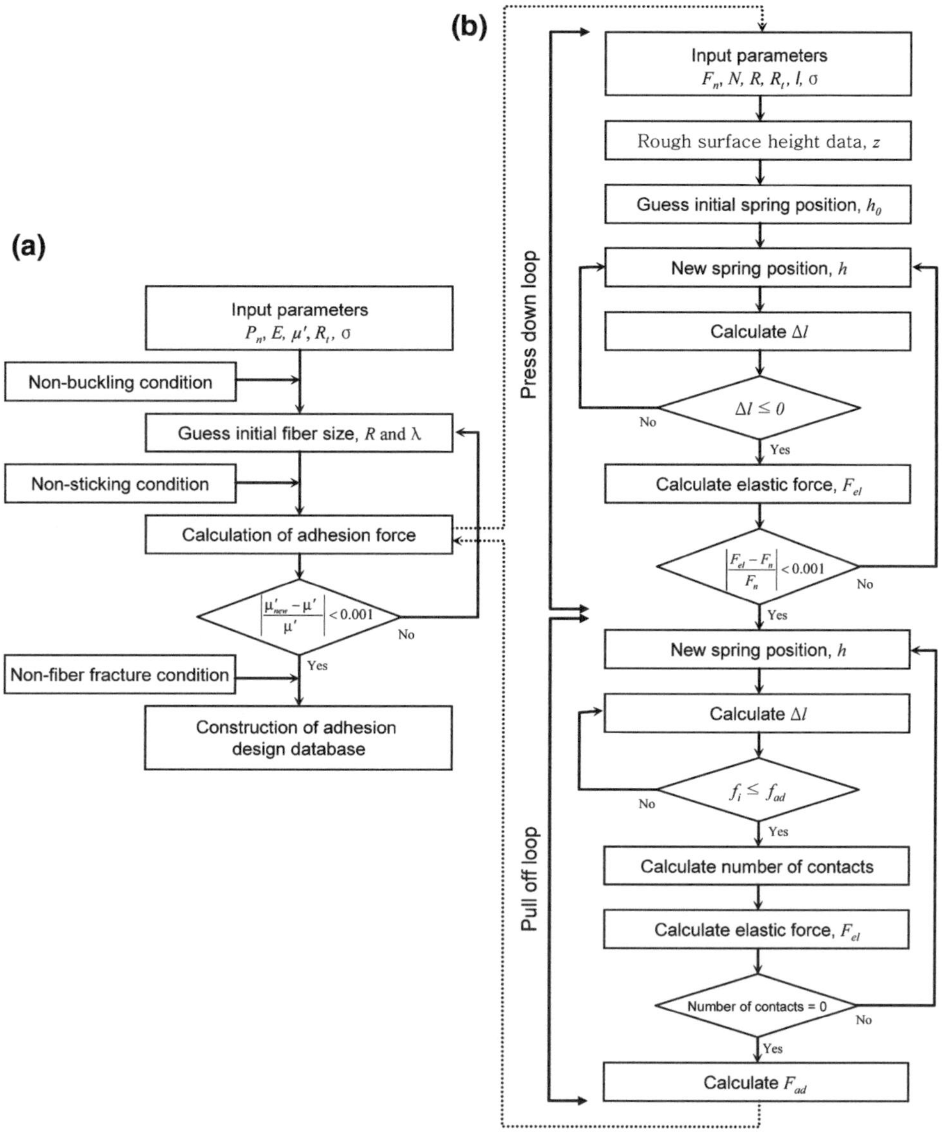

Fig. 12.27 Flow chart for **a** the construction of adhesion design database and **b** the calculation of the adhesion force. In this figure, P_n is the applied pressure, E is the elastic modulus, μ' is the adhesion coefficient, R_t is the tip radius, σ is root mean square amplitude, R is the fiber radius, λ is the aspect ratio of fiber, F_n is the applied load, N is the number of springs, k and l are stiffness and length of structures, Δl is the spring deformation, f_i is the elastic force of a single spring, and f_{ad} is the adhesion force of a single contact (Kim and Bhushan 2007c)

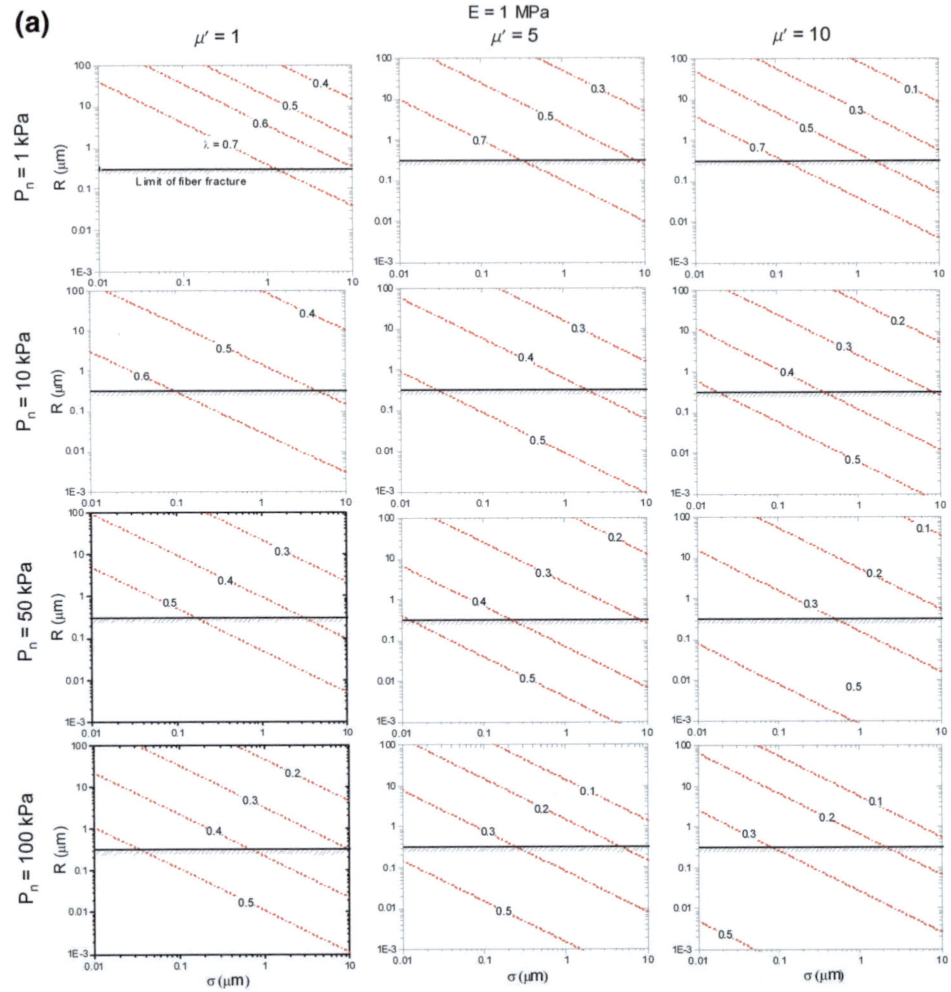

Fig. 12.28 Adhesion design database for biomimetic attachment system consisting of single-level cylindrical fibers with orientation angle of 30° and spherical tips of 100 nm for elastic modulus of **a** 1 MPa, **b** 100 MPa, and **c** 10 GPa (Kim and Bhushan 2007c). The solid lines shown in Figs **b** and **c** correspond to the cases I and II, respectively, which satisfy the specified requirements (Kim and Bhushan 2007c)

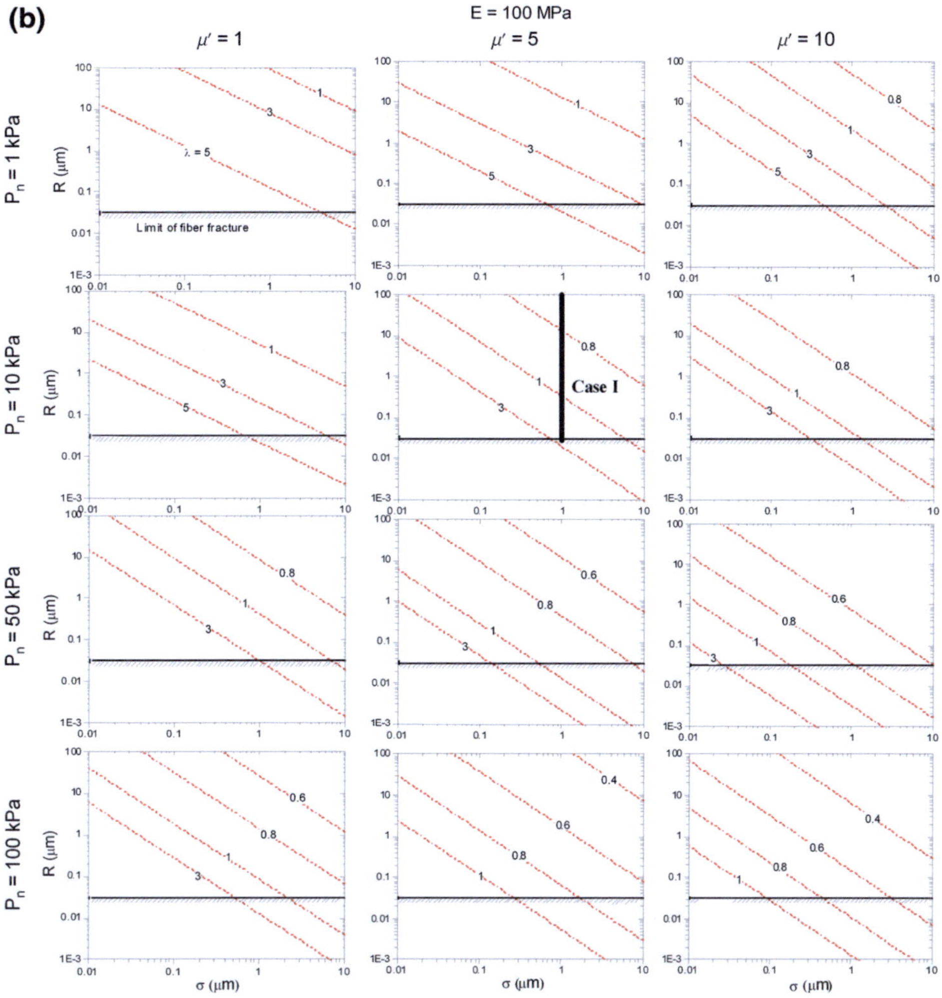

Fig. 12.28 (continued)

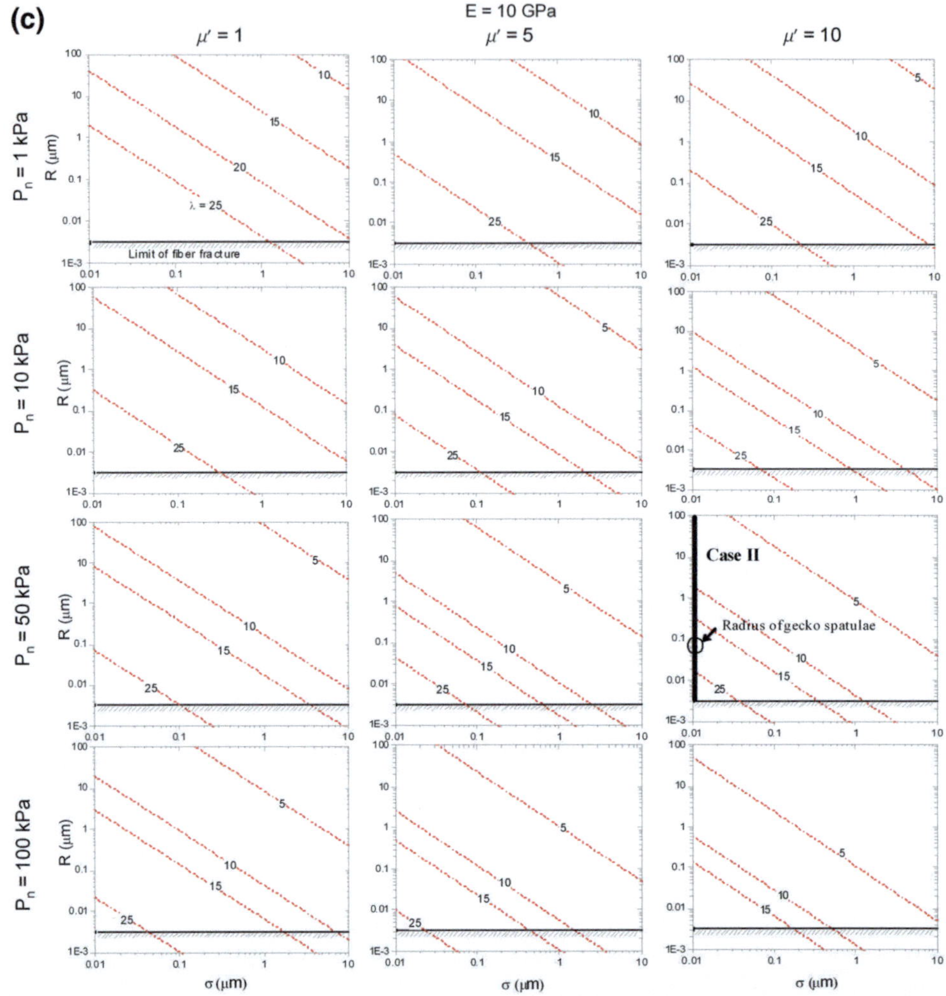

Fig. 12.28 (continued)

If the applied load, the roughness of contacting surface, and the elastic modulus of a fiber material are specified, the optimal fiber radius and aspect ratio for the desired adhesion coefficient can be selected from this design database. Adhesion databases are useful for understanding biological systems and for guiding the fabrication of biomimetic attachment systems. Two case studies (Kim and Bhushan 2007c) are discussed below.

Case study I: Select the optimal size of fibrillar adhesive for a wall climbing robot with the following requirements:

- Material: polymer with $E \approx 100$ MPa
- Applied pressure by weight < 10 kPa
- Adhesion coefficient ≈ 5
- Surface roughness $\sigma < 1$ μm.

The subplot of adhesion database that satisfies the requirement is at the second column and second row in Fig. 12.28b. From this subplot, any values on the marked line can be selected to meet the requirements. For example, fiber radius of 0.4 μm with an aspect ratio of 1 or fiber radius of 10 μm with an aspect ratio of 0.8 satisfies the specified requirements.

Case study II: Compare with adhesion test for a single gecko seta (Autumn et al. 2000, 2002)

- Material: β-keratin with $E \approx 10$ GPa
- Applied pressure = 57 kPa (2.5 μN on an area of 43.6 μm^2)
- Adhesion coefficient = 8–16
- Surface roughness $\sigma < 0.01$ μm.

Autumn et al. (2000, 2002) showed that in isolated gecko setae contacting the surface of a single crystalline silicon wafer, a 2.5 μN preload yielded adhesion of 20–40 μN and thus a value of adhesion coefficient of 8–16. The region that satisfies the above requirements is marked in Fig. 12.28c. The spatulae of gecko setae have an approximate radius of 0.05 μm with an aspect ratio of 25. However, the radius corresponding to $\lambda = 25$ for the marked line is about 0.015 μm. This discrepancy is due to the difference between a simulated fiber model and a real gecko setae model. Gecko setae are composed of a three-level hierarchical structure in practice, so higher adhesion can be generated than in a single-level model (Bhushan et al. 2006; Kim and Bhushan 2007a, b). Given the simplification in the fiber model, this simulation result is very close to the experimental result.

12.8 Fabrication of Gecko Skin-Inspired Structures

Based on the studies reported in the literature, the dominant adhesion mechanism used by gecko attachment systems appears to be van der Waals forces. The hierarchical structure involving complex divisions of the gecko skin (lamellae-setae-branches-spatulae) enable a large number of contacts between the gecko skin and mating surface. As shown in previous calculations, the van der Waals adhesive force for two parallel surfaces is inversely proportional to the cube of the distance between two surfaces. These hierarchical fibrillar microstructured surfaces would be capable of reusable dry adhesion and would have uses in a wide range of applications from everyday objects such as adhesive tapes, fasteners, toys, microelectronic and space applications, and treads of wall-climbing robots. The development of nanofabricated surfaces capable of replicating this adhesion force developed in nature is limited by current fabrication methods. Many different techniques have been used in an attempt to create and characterize bio-inspired adhesive tapes. Attempts are being made to develop climbing robots using gecko inspired structures (Autumn et al. 2006b; Daltorio et al. 2007; Aksak et al. 2008; Cutkosky and Kim 2009).

12.8.1 Single Level Roughness Structures

A soft, compliant fibrillar structure is desirable in order to enable more fibrils to be in close proximity to a mating surface to increase van der Waals forces. Sitti and Fearing (2003) and Cho and Choi (2007) used nanoporous anodic alumina and polycarbonate membranes, as a template to create polymeric nanofibers. Geim et al. (2003) created arrays of polyimide nanofibers using electron-beam lithography and dry etching in oxygen plasma (Fig. 12.29a (left)). By using electron-beam lithography, thermal evaporation of an aluminum film, and lift-off, an array of nanoscale aluminum disks were prepared. These patterns were then transferred in the polyimide film by dry etching in oxygen plasma. A 1 cm^2 sample was able to create 3 N of adhesive force under the new arrangement. This is approximately one-third the adhesive strength of a gecko. They fabricated a Spiderman toy (about 0.4 N) with a hand covered with molded polymer nanohairs, Fig. 12.29b. They demonstrated that it could cling to a glass plate. Bunching of the nanohairs (as described earlier) if they are closely spaced was determined to greatly reduce the both the adhesive strength and durability of the polymer tape. The bunching can be clearly seen in Fig. 12.29a (right). Therefore, an optimal geometry is required.

Gorb et al. (2007) and Bhushan and Sayer (2007) characterized two polyvinylsiloxane (PVS) samples from Gottlieb Binder Inc., Holzgerlingen, Germany: one consisting of mushroom-shaped pillars (Fig. 12.30a), and the other an unstructured control surface (Fig. 12.30b). The structured sample is inspired by the micropatterns found in the attachment systems of male beetles from the family chrysomelidae and is easier to fabricate.

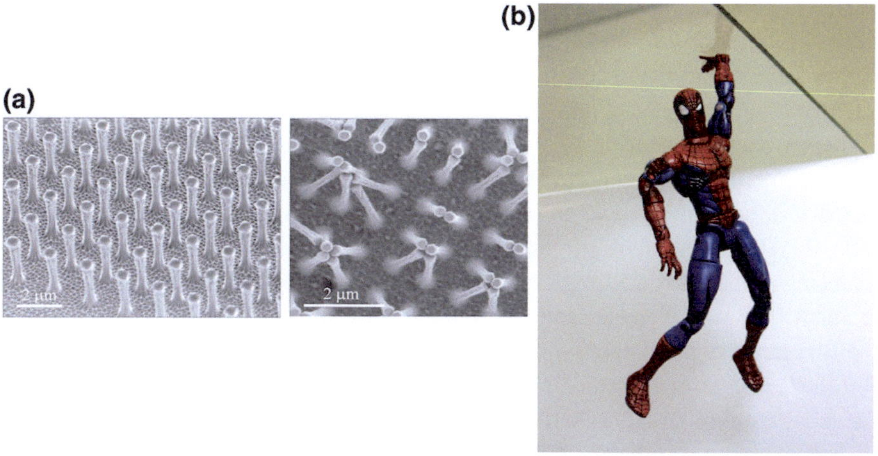

Fig. 12.29 **a** (left) An array of polyimide nanohairs and (right) bunching of the nanohairs, which leads to a reduction in adhesive force. **b** A Spiderman toy (about 0.4 N) with a hand covered with the molded polymer nanohairs, clinging to a glass plate (Geim et al. 2003)

Fig. 12.30 SEM micrographs of the **a** structured and **b** unstructured PVS samples. SH: shaft, NR: neck region, LP: lip (Bhushan and Sayer 2007)

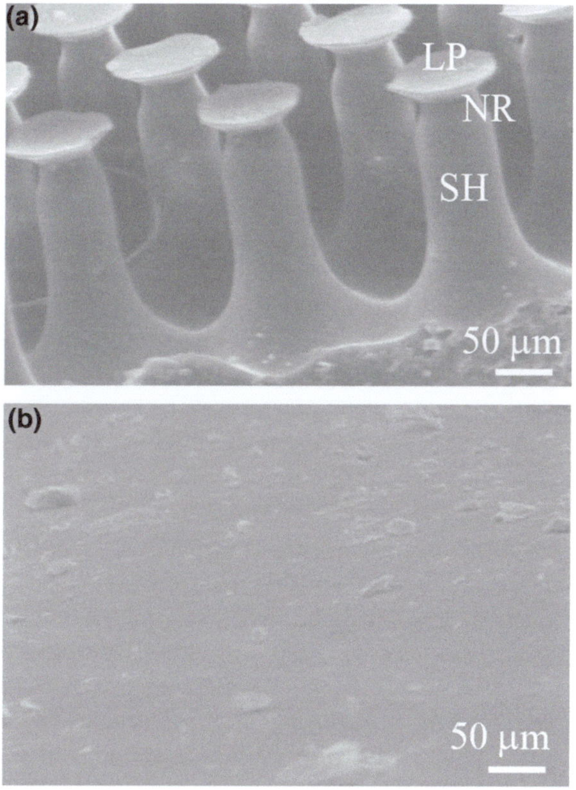

Both sexes possess adhesive hairs on their tarsi. However, males bear hair extremely specialized for adhesion on the smooth surface of female's covering wings during mating. The hairs have broad flattened tips with grooves under the tip to provide flexibility. The mushroom shape provides a larger contact area. The structured samples were produced at room temperature by pouring two-compound polymerizing PVS into the holed template lying on a smooth glass support. The fabricated sample was comprised of pillars that are arranged in a hexagonal order to allow maximum packing density. They were approximately 100 μm in height, 60 μm in base diameter, 35 μm in middle diameter, and 25 μm in diameter at the narrowed region just below the terminal contact plates. These plates were about 40 μm in diameter and 2 μm in thickness at the lip edges. The adhesion force of the two samples in contact with a smooth, flat glass substrate was measured by Gorb et al. (2007) using a microtribometer. Results revealed that the structured specimens featured an adhesion force more than twice that of the unstructured specimens. The adhesion force was also found to be independent of the preload. Moreover, it was found that the adhesive force of the structured sample was more tolerant to contamination compared to the control, and it could be easily cleaned with a soap solution.

Bhushan and Sayer (2007) characterized the surface roughness, friction force, and contact angle of the structured sample and compared the results to an unstructured control. As shown in Fig. 12.31a, the macroscale coefficient of kinetic friction of the structured sample was found to be almost four times greater than the unstructured sample. This increase was determined to be a result of the structured roughness of the sample, and not the random nanoroughness. It is also noteworthy that the static and kinetic coefficients of friction are approximately equal for the structured sample. It is believed that the divided contacts allow the broken contacts of the structured sample to constantly recreate contact. As seen in Fig. 12.31b, the pillars also increased the hydrophobicity of the structured sample, in comparison to the unstructured sample, as expected due to increased surface roughness (Wenzel 1936; Burton and Bhushan 2005; Bhushan and Jung 2011). A large contact angle is important for self-cleaning (Nosonovsky and Bhushan 2008), which agrees with the findings of Gorb et al. (2007) that the structured sample is more tolerant of contamination than the unstructured sample.

To demonstrate the potential of the mushroom-shaped microstructure for real, macroscopic applications, the structure-covered foil was glued with its back to a 20 cm × 20 cm poly(methyl methacrylate) (PMMA) plate with a handle bar. The adhesive performance was tested on a glass plate firmly attached to the ceiling. This device was able to support the weight of a human being, shown in Fig. 12.32.

Directed self-assembly has been proposed as a method to produce regularly spaced fibers (Schäffer et al. 2000; Sitti 2003). In this technique, a thin liquid polymer film is coated on a flat conductive substrate. As demonstrated in Fig. 12.33, a closely spaced metal plate is used to apply a DC electric field to the polymer film. Due to instabilities on the film, pillars will begin to grow until they are touching the upper metal plate. Self-assembly is desirable because the components spontaneously assemble, typically by

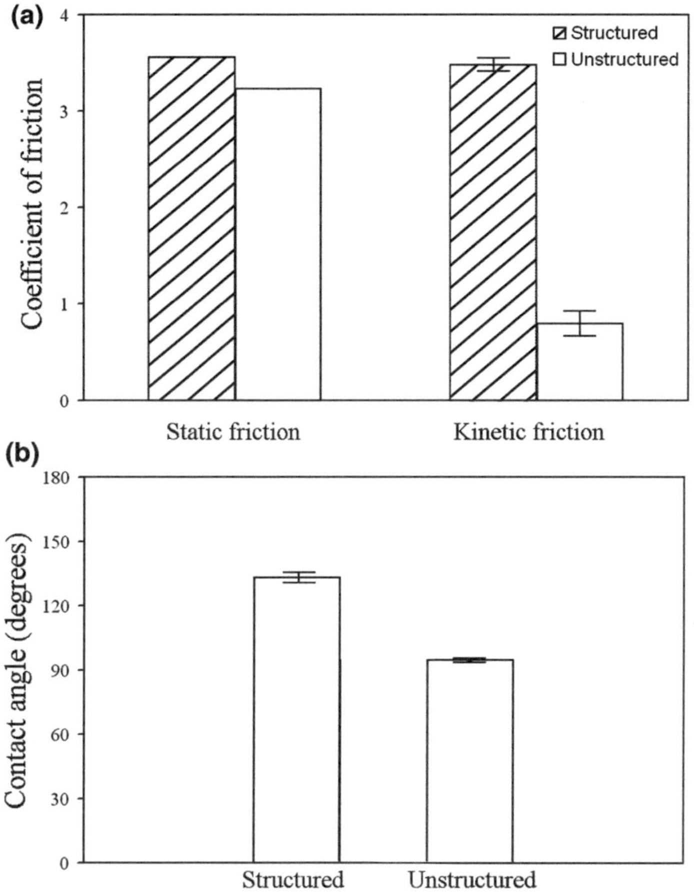

Fig. 12.31 **a** Coefficients of static and kinetic friction for the structured and unstructured samples slid against magnetic tape with a normal load of 130 mN. **b** Water contact angle for the structured and unstructured samples (Bhushan and Sayer 2007)

bouncing around in a solution or gas phase until a stable structure of minimum energy is reached.

Vertically-aligned multiwalled carbon nanotubes (MWCNT) have been used to create nanostructures on polymer surfaces. Yurdumakan et al. (2005) used chemical vapor deposition (CVD) to grow vertically-aligned MWCNT that are 50–100 μm in length on quartz or silicon substrates. The sample with MWCNT site facing up was then dipped in PMMA solution. After the polymerization, PMMA-MWCNT sheets were peeled off from the silicon substrate. The MWCNTs were exposed from the silicon-facing side of the PMMA matrix by etching the top 25 μm with a solvent. SEM images of the MWCNT grown on a silicon substrate as well as transferred into a PMMA matrix and then exposed on

Fig. 12.32 Photograph of a man attached to the glass ceiling by a 20 cm × 20 cm PMMA plate covered by the mushroom-shaped adhesive microstructure (Heepe et al. 2012)

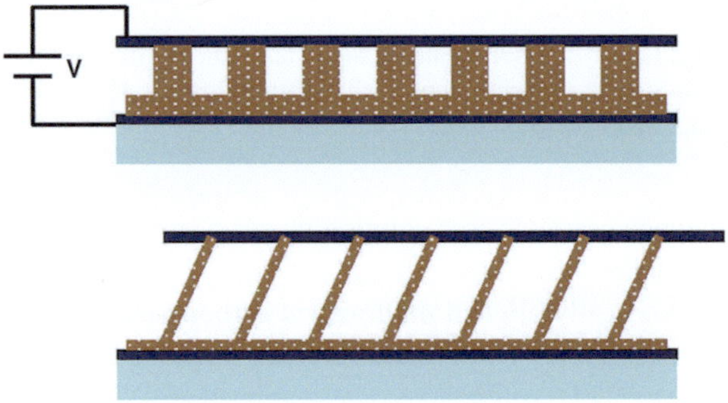

Fig. 12.33 Directed self-assembly based method of producing high aspect ratio micro/nanofibers (Sitti 2003)

the surface can be seen in Fig. 12.34. On a nanoscale, the MWCNT surface was able to achieve adhesive forces two orders of magnitude greater than those of gecko foot-hairs. These structures provided high adhesion on the nanometer level and were not capable of producing high adhesion forces on the macroscale, because they are not compliant. Ge et al. (2007) and others have fabricated nanostructures by transferring micropatterned, vertically-aligned MWCNT arrays onto flexible polymer tape. They reported high adhesion on the macroscale. They also performed peeling experiments. Durability of the adhesive tape is an issue as some of the nanotubes can detach from the substrate with repeated use. Qu et al. (2008) measured adhesion on vertically-aligned MWCNT arrays on Si substrate and reported high adhesion on the nanoscale.

Davies et al. (2008) fabricated mushroom-headed microfibers made of PDMS. In one of the fabrication strategies, a silicon wafer with a thickness that defined the stalk length was obtained, and the masks with mushroom-head features were first used to pattern one side of the silicon wafer with resist. Features were etched to a depth equal to that of the thickness of the mushroom head. Next, the smaller diameter mask was used to pattern the other side of the wafer, which was then etched to produce holes through the entire thickness of the wafer, meeting mushroom headed cavities. This mold was first coated in a fluorocarbon release agent. A PDMS solution was then spun onto this mold and cured to produce mushroom headed microfibers. The resulting casting comprising of stalks and mushroom heads was then pulled through the mold in a single peeling process. To create angled microfiber arrays found in biological attachments using photolithography, Aksak et al. (2007) simply varied the ultraviolet (UV) exposure angle by tilting the wafer during

Fig. 12.34 Multi-walled carbon nanotube structures: (left) grown on silicon by chemical vapor deposition, (right) transferred into a PMMA matrix, and then exposed on the surface after solvent etching (Yurdumakan et al. 2005)

exposure. The fibers were formed at a nonperpendicular angle to the substrate surface (Fig. 12.35a). This master template of angled SU-8 fibers was then used to form many copies of the fiber arrays from curable polyurethanes by molding. They reported that angled fibers exhibited reduced adhesion compared to similar vertical fibers due to a peeling moment. However, angled fibers are favored in biological attachment systems. Murphy et al. (2007) modified angled fiber arrays by adding soft spherical and spatula shaped tips via dipping in a liquid polymer of interest (Fig. 12.35b). To add tips to the fibers, the fiber array sample attached to a micropositioning stage was dipped into a liquid polyurethane layer and retracted, retaining some of the liquid polymer on the tips of the fibers. To form spherical tips, the sample was placed with the fibers facing up and allowed to cure. To form spatula tips, the fiber sample was placed onto a smooth low energy surface and then peeled away after curing. They reported very high adhesion of these fibers with soft tips because of increased contact area.

del Campo et al. (2007a, b) fabricated pillar arrays with controlled 3D tip geometries resembling those found in biological attachments. The fabrication strategy was based on complete or partial soft molding on 2D masters made by lithography with elastomeric precursors followed in some cases by inking and microprinting steps. The patterned master with high-aspect-ratio cylindrical holes was produced by photolithography using SU-8 photoresist films. The SU-8 masters were filled with elastomeric precursors (PDMS supplied as Sylgard 184 by Dow Corning) to produce arrays of cylindrical pillars (Fig. 12.36a). Arrays of pillars with spherical and spatular tips were obtained by inking the Sylgard 184-structured substrates in a thin film of Sylgard 184 precursor. Curing of arrays in upside-down orientation yielded hemispherical tips as a consequence of gravity and surface tension acting on the fluid drop (Fig. 12.36b). Alternatively, the inked stamp can be pressed against a flat substrate and then cured. This leads to pillars with a flat roof (Fig. 12.36c). The roof can be symmetric or asymmetric depending on the tilt of the substrate during curing (Fig. 12.36c, d). They also used silicones used for dental impressions. These materials possess higher initial viscosities and faster crosslinking kinetics than Sylgard 184, which results in incomplete cavity filling. By soft molding these materials after selected delay times after mixing, arrays of tubes and pillars with concave tips (Fig. 12.36e, f) were obtained. They performed adhesion tests on various geometries against a sapphire sphere. They reported that shape of pillar tip affects the contact area and adhesion behavior. Figure 12.37 shows the pull-off strength data as a function of tip radius for various tip geometries. For a given tip radius, pillars with flat punch geometry have significantly higher adhesion than spherical contacts. Pillars with mushroom tips have the highest adhesion.

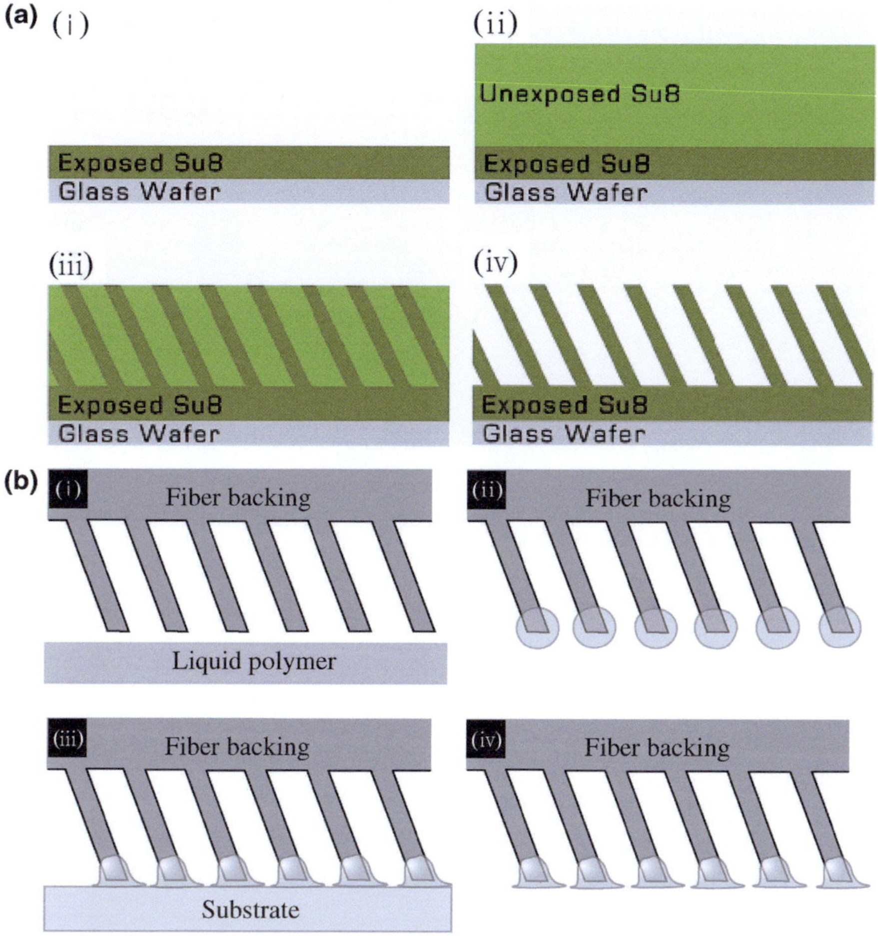

Fig. 12.35 **a** The process steps of the polymer fiber orientation: (i) a thin layer of SU-8 is spun on a glass substrate, then exposed and cured; (ii) a thicker layer of SU-8 is spun that will become the fibers; (iii) the thick layer is patterned with UV exposure by tilting the wafer; and (iv) the SU-8 photoresist is developed, leaving the desired angled fiber array (Aksak et al. 2007). **b** Fiber tip fabrication process: (i) bare fibers are aligned with a layer of liquid polymer; (ii) the fibers are dipped into the liquid and retracted; (iii) the fibers are brought into contact with a substrate; and (iv) the fibers are peeled away from the substrate after curing (Murphy et al. 2007)

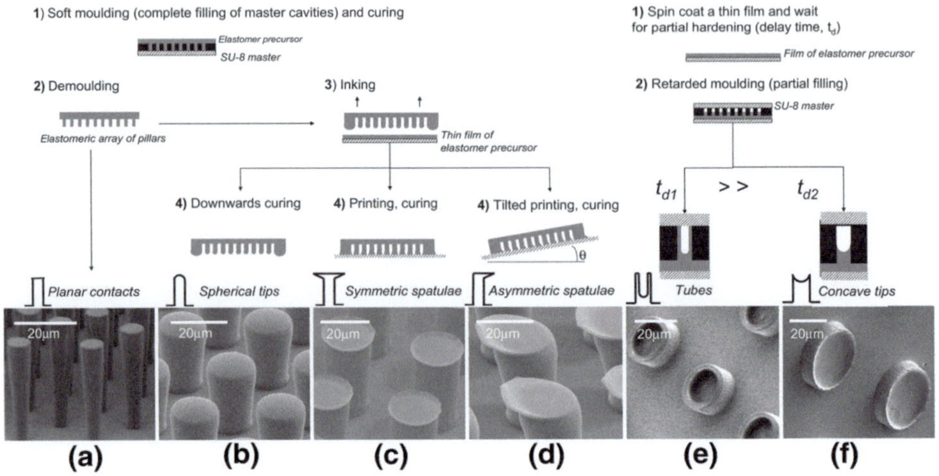

Fig. 12.36 The process steps to fabricate pillar arrays on soft molding and SEM images showing examples of the pillar arrays obtained (del Campo et al. 2007a)

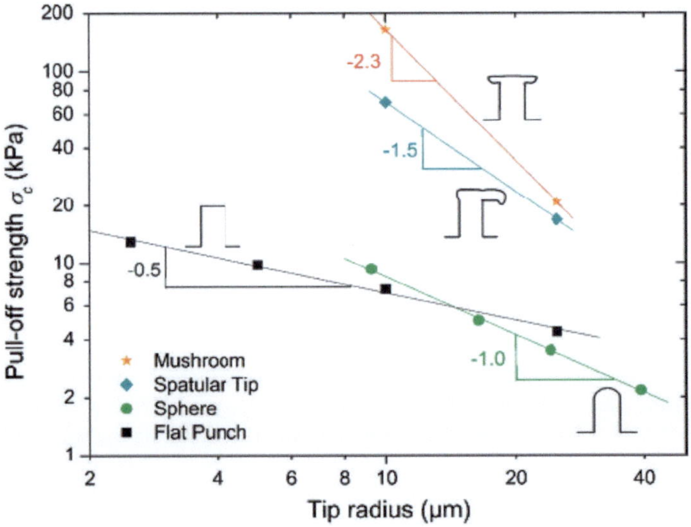

Fig. 12.37 Tip radius dependence of the pull-off force for flat, spherical, spatular, and mushroom-like contacts at a preload of 1 mN. In the case of spherical tips, the radius corresponds to the tip radius. For all other geometries, the pillar radius is used (del Campo et al. 2007b)

12.8.2 Multi-level Hierarchical Structures

The aforementioned fabricated surfaces only have one level of roughness. Although these surfaces are capable of producing high adhesion on the micro/nanoscale, they are not expected to produce large scale adhesion due to a lack of compliance and bunching.

del Campo and Greiner (2007) fabricated a hierarchical structure by two-level photolithography. Figure 12.38 shows a schematic of the process and an example of two-level SU-8 patterns obtained.

Northen and Turner (2005) created a two-level compliant structure by employing a microelectromechanical based approach. The multiscale structures consisted of arrays of organic-looking photoresist nanorods (organorods), ~2 μm tall and 50–200 nm in diameter (comparable in size to gecko spatulae) (Fig. 12.39a); atop photolithographically defined 2 μm thick SiO_2 platforms 100–150 μm on a side (Fig. 12.39b). The platforms of various geometries are supported by single high aspect ratio pillars down to 1 μm in diameter and with heights up to ~50 μm (Fig. 12.39c). The structures were fabricated out of 100 mm single-crystal wafers using standard bulk micromachining techniques. An array of four-fingered platform structures is shown in Fig. 12.39d. Adhesion testing was performed using a nanorod surface on a solid substrate and on the two-level structures. They reported that adhesive pressure of the two-level structures was about four times higher than that of the surfaces with only one level of hierarchy.

Palacio et al. (2013) fabricated two-level hierarchically structured superhydrophobic surfaces made of polyethylene using two-stacked porous polycarbonate membranes as template. Bhushan and Lee (2012) and Lee and Bhushan (2012) fabricated one-level, two-level, and three-level hierarchically structured superhydrophobic surfaces made of polypropylene using one, two, or three stacked porous polycarbonate membranes (Millipore) as a template, respectively, Fig. 12.40. A 1-mm thick polypropylene film and polycarbonate membranes were sandwiched between two PDMS disks. The stack was heated in an oven at 200 °C at above the melting point of polypropylene of 130–170 °C in order for it to flow into pores. The membranes were then etched to realize the polypropylene fibrillar samples.

By changing the density and diameter of the nanofibers, surfaces with either the lotus effect (low adhesion) or gecko effect (high adhesion) could be achieved. Figure 12.41 shows the SEM micrographs of selected micro- and nanostructures as well as a two-level hierarchical structure made from polypropylene (Lee and Bhushan 2012). Figure 12.42 shows wetting properties which include water contact angle (CA) and contact angle hysteresis (CAH) data (Lee and Bhushan 2012). It can be observed that all structure surfaces are superhydrophobic governed by air pocket formation on the surfaces. However, CAHs varies from sample to sample. In surfaces with low CAH, droplets roll off the surface. Whereas they remain stuck if CAH is high. Figure 12.43a shows the adhesive force data obtained using an AFM with 30 μm diameter borosilicate glass tip (Lee and Bhushan

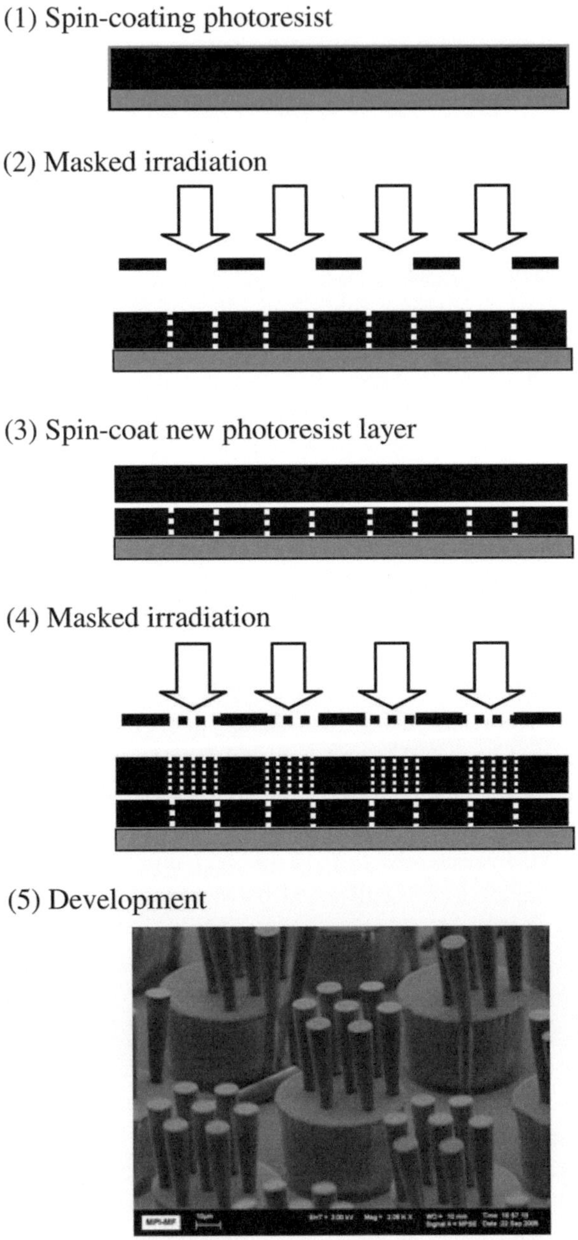

Fig. 12.38 Photolithographic fabrication process and example of two-level hierarchical structure with SU-8. Base pillars have 50 μm diameter and 40 μm height and the top pillars have 9 μm diameter and 35 μm height (del Campo and Greiner 2007)

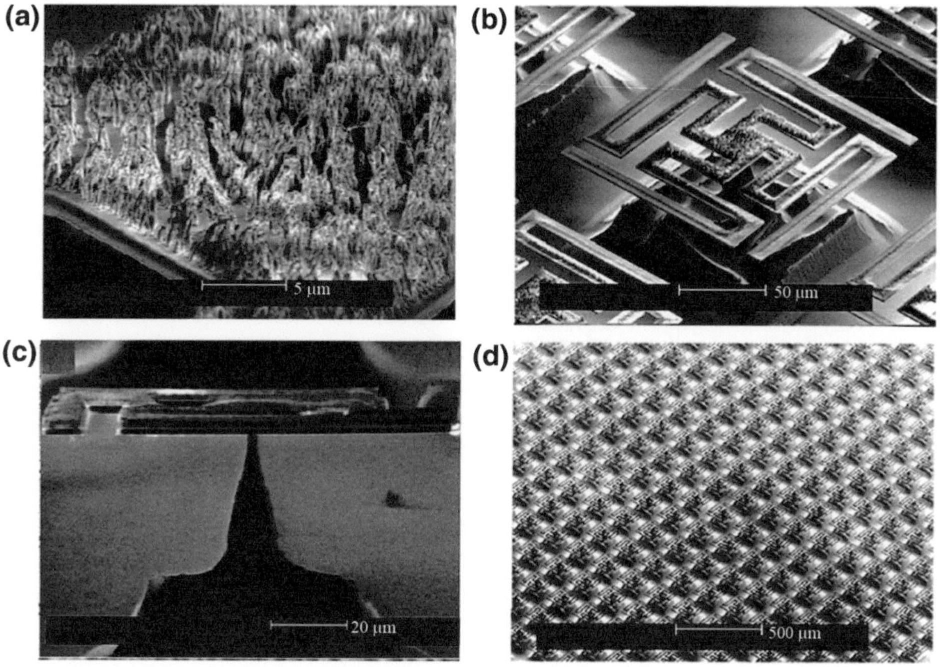

Fig. 12.39 Two-level fabricated adhesive structure composed of **a** organorods atop **b** silicon dioxide platforms. The platforms are supported by **c** support pillars. **d** This structure was repeated multiple times over a silicon wafer (Northen and Turner 2005)

2012). The oriented fibers of 100 and 600 nm diameter and hierarchical structure exhibited the gecko effect (high adhesion) due to their high fiber densities and large contact areas. Conversely, the random fibers of 50 nm and the oriented fibers of 5 and 14 μm diameter exhibited the lotus effect due to their smaller fiber density. To further increase adhesion, a soft, conventional adhesive to coat fiber ends of selected samples were applied. Data for uncoated and coated tips are shown in Fig. 12.43b. As expected, for the coated tips an increase in the adhesive force is observed.

Bhushan and Lee (2012) fabricated one-level, two-level, and three-level hierarchical structured surfaces. The SEM images of fabricated samples are shown in Fig. 12.44. The wetting properties (CA and CAH) of samples are shown in Fig. 12.45. The water CA of the flat polypropylene was 99° which indicates that the surface is hydrophobic having a low surface energy of 30 m J m^{-2}. Fibrous structures increase the CA and decrease the CAH because of air pocket formation. The higher levels have higher effect because of increase in the surface roughness. Adhesion force data for one-level and two-level was measured using AFM with 30 μm diameter glass tip and 30 nm diameter conventional Si_3N_4 tip. Representative force-distance curves are shown in Fig. 12.46. The left column is for data observed using the 30 μm diameter tip, and the right column is for the 30 nm

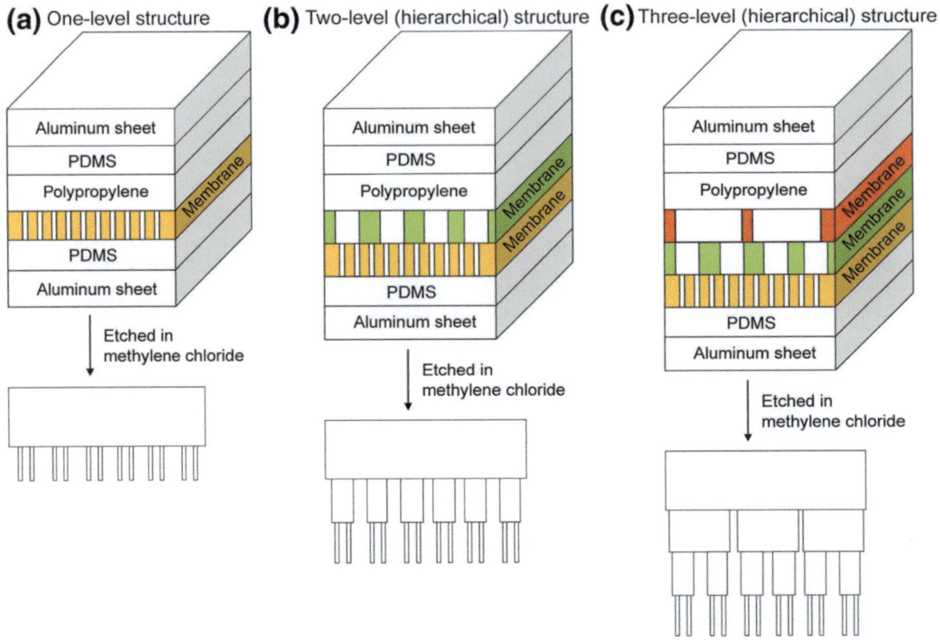

Fig. 12.40 Fabrication processes for **a** the one-level, **b** the two-level, and **c** the three-level fiber structures using one, two, and three polycarbonate membranes in the stack, respectively. A 1-mm thick polypropylene film and polycarbonate membranes are sandwiched between two PDMS disks. After heating the stacks in an oven, the membranes are etched using methylene chloride to obtain the polypropylene fibrillar samples (Bhushan and Lee 2012)

diameter tip. Bending or buckling behavior of the fibers was not observed in the force-distance curves in the left column. In the right column, bending and buckling behavior of one-level and two-level structures are shown while the 30 nm diameter tip travels downward. Figure 12.47 shows the behavior of the two-level single fiber with the AFM tip. Schematics of the fiber behavior at each point as the AFM tip travels are also shown. The tip approaches the fiber (point A to B), and the tip touches the fiber at point B. From point B to C, the tip presses the fiber and is deflected so that the fiber is gradually bent and remains pressed. From point C to D, the fiber buckles. However, the tip keeps pressing the fiber from point D to E, including more compression of the fiber.

From the force-distance curves obtained using the 30 μm diameter tip (Fig. 12.46), the adhesive forces for the samples are calculated. The adhesive force values are plotted in Fig. 12.48. The force on the flat surface is 260 nN, but those on the one- and two-level structures are 290 and 326 nN, respectively. This is attributed to the number of fibers. The flat surface can be considered as a fiber with a diameter larger than the real contact area between the AFM ball tip and the flat surface. As shown in (12.3), with an increase in the number of fibers (n) the adhesive force increases in proportion to the square root of n.

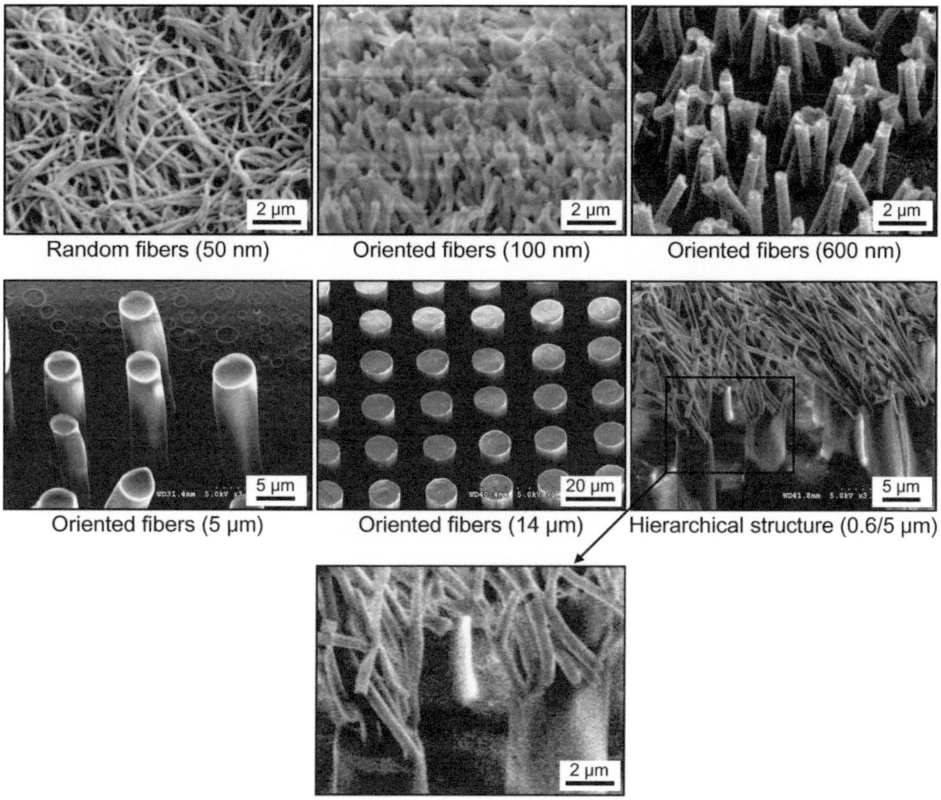

Fig. 12.41 SEM images for various one-level and two-level structures. The one-level structure of 50 nm diameter random nanofiber, 100 and 600 nm oriented nanofibers, 5 and 20 μm diameter oriented microfibers, and the two-level structure of 600 nm diameter with 5 μm diameter oriented fibers are shown (Lee and Bhushan 2012)

Moreover, it is expected that the AFM tip travels deeper into the fiber structures than for the flat sample as shown in the buckling experiment on the fibrillar structures (Fig. 12.47). It results in an additional increase of the adhesive force. The two-level fibrillar structures have an adhesive force slightly larger than the one-level structures. Although the density of the two-level fibrillar structures is lower than that of the one-level, the two-level structure is more compliant which leads to a larger number of fibers coming in contact with the tip, increasing the adhesive force.

To summarize, a simple way to fabricate one-, two-, and three-level fibrillar structures is described. The adhesive forces on various samples have been measured. The adhesive forces of the one- and two-level structures are higher than that of the flat surface due to the increase of the number of fibers on the surfaces. It is shown that the more compliant the fiber structures are, the higher the adhesive forces are (Bhushan and Lee 2012).

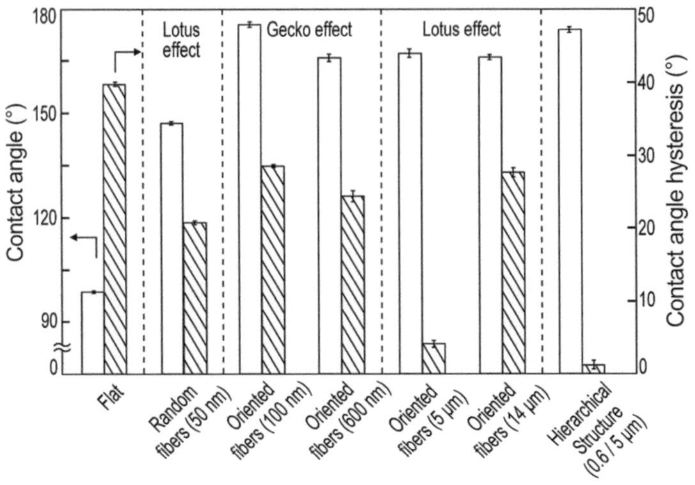

Fig. 12.42 Measured CAs and CAHs on all the one-level and two-level structures (Lee and Bhushan 2012)

12.9 Closure

The adhesive properties of geckos and other creatures, such as flies, beetles and spiders, are due to the hierarchical structures present on each creature's hairy attachment pads. Geckos have developed the most intricate adhesive structures of any of the aforementioned creatures. The attachment system consists of ridges called lamellae that are covered in microscale setae that branch off into nanoscale spatulae, about three billion spatulae on two feet. The so-called division of contacts provide high dry adhesion. Multiple level hierarchically structured surface construction plays an important role in adapting to various rough surfaces, bringing the spatulae in close proximity with the mating surface. These structures, as well as material properties, allow the gecko to obtain a much larger real area of contact between its feet and a mating surface than is possible with a non-fibrillar material. Two feet of a Tokay gecko have about 220 mm^2 of attachment pad area on which the gecko is able to generate approximately 20 N of adhesion force. Although capable of generating high adhesion forces, a gecko is able to detach from a surface at will—an ability known as smart adhesion. Detachment is achieved by a peeling motion of the gecko's feet from a surface.

Experimental results have supported the adhesion theories of intermolecular forces (van der Waals) as a primary adhesion mechanism and capillary forces as a secondary mechanism, and have been used to rule out several other mechanisms of adhesion including the secretion of sticky fluids, suction, and increased frictional forces. Atomic force microscopy has been employed by several investigators to determine the adhesion strength of gecko foot hairs. The measured values of the lateral force required to pull parallel to

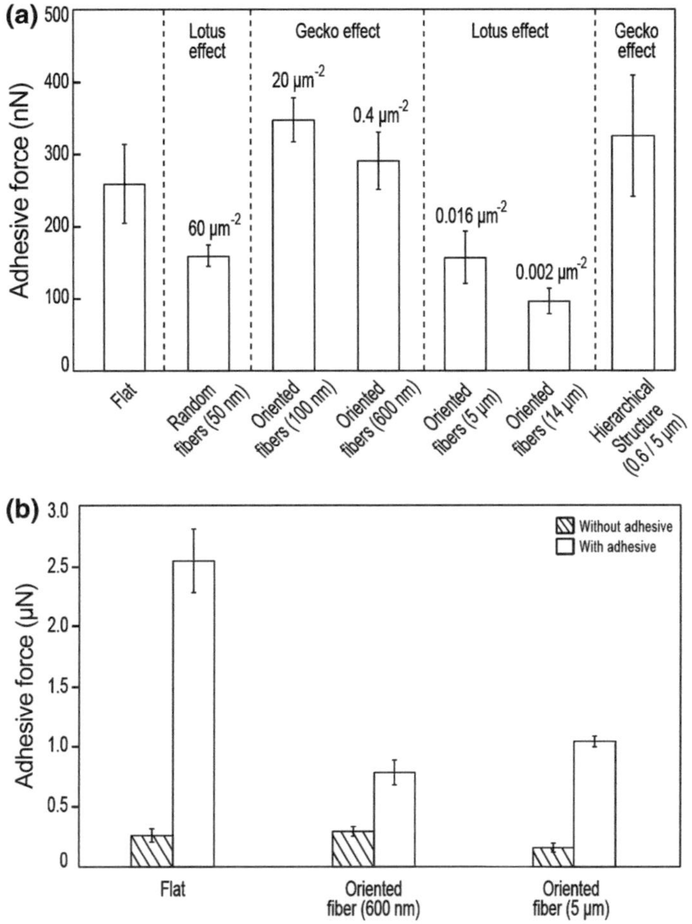

Fig. 12.43 **a** Adhesive force on all the one-level and two-level structures measured using the 30 μm diameter glass tip in an AFM. The values above each bar are the number of fibers in unit area (μm^2). The measurements were performed at a temperature of 21 °C and 45–55% relative humidity, and **b** Adhesive force of the fibers with and without the adhesive applied to the samples (Lee and Bhushan 2012)

the surface for a single seta (194 μN) and adhesive force (normal to the surface) of a single spatula (11 nN) are comparable to the van der Waals prediction of 270 μN and 11 nN for a seta and spatula, respectively. The adhesion force generated by seta increases with preload and reaches a maximum when both perpendicular and parallel preloads are applied. Although gecko feet are strong adhesives, they remain free of contaminant particles through self-cleaning. Spatular size, along with material properties, enables geckos to easily expel any dust particles that come into contact with their feet.

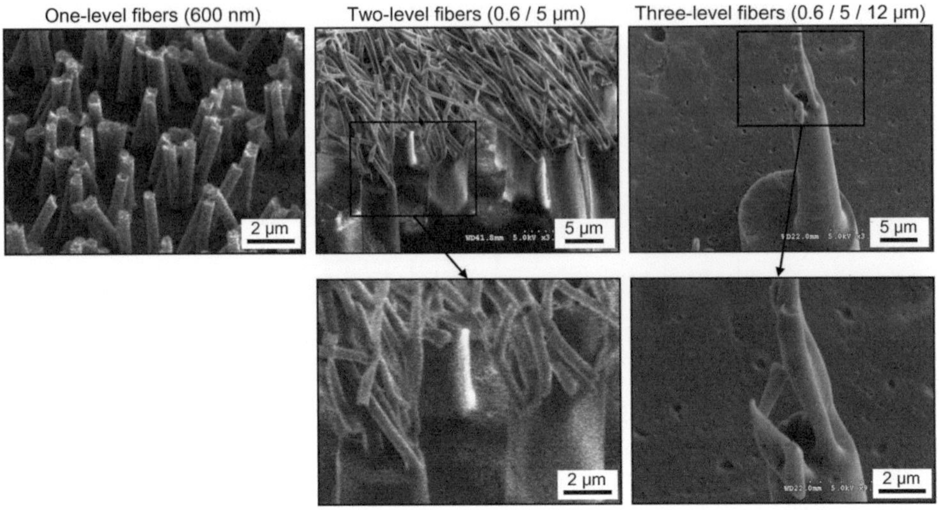

Fig. 12.44 SEM images of the one-level structure of 600 nm diameter fibers, the two-level structure of 600 nm diameter on top of 5 μm diameter fibers, and three-level structure of 600 nm diameter fibers on top of 5 μm diameter fibers on 12 μm diameter fibers (Bhushan and Lee 2012)

Fig. 12.45 The measured water CAs and CAHs on the flat polypropylene sample and the one-, the two-, and the three-level structures (Bhushan and Lee 2012)

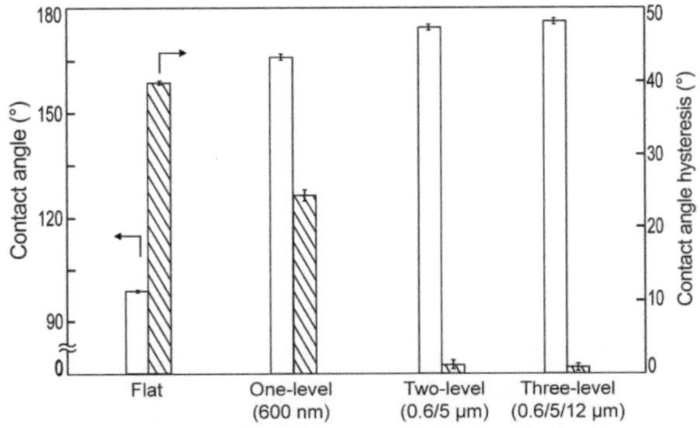

A three-level hierarchical model for a gecko lamella consisting of setae, branches, and spatulae has brought more insight into adhesion of biological attachment systems. One-, two- and three-level hierarchically structured spring models for simulation of a seta contacting with random rough surfaces were considered. The simulation results show that the multi-level hierarchical structure has a higher adhesion force as well as higher adhesion energy than the one-level structure for a given applied load, due to better adaptation and attachment ability. It is concluded that the multi-level hierarchical structure produces adhesion enhancement, and this enhancement increases with an increase in the applied

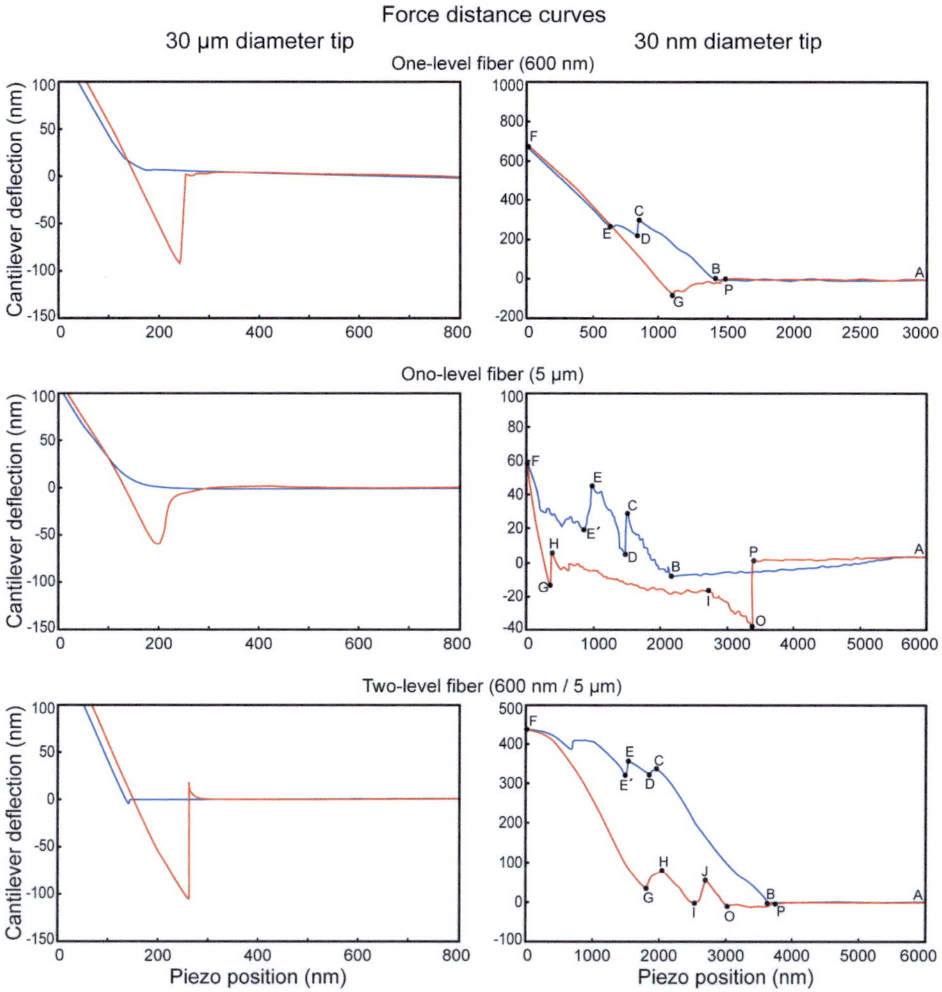

Fig. 12.46 Representative force-distance curves for the samples with the one-level (600 nm and 5 μm) and the two-level single structures (600 nm/5 μm) obtained using the 30 μm diameter AFM tip (left column) and the 30 nm diameter AFM tip (right column) (Bhushan and Lee 2012)

load and a decrease in the stiffness of springs. The condition at which a significant adhesion enhancement occurs appears to be related to the maximum spring deformation. The result shows that significant adhesion enhancement occurs when the maximum spring deformation is two to three times larger than the σ value of surface roughness. For the effect of applied load, as the applied load increases, adhesion force increases up to a certain applied load, and then has a constant value, whereas, adhesion energy continues to increase with an increase in the applied load. For the effect of spring stiffness, the

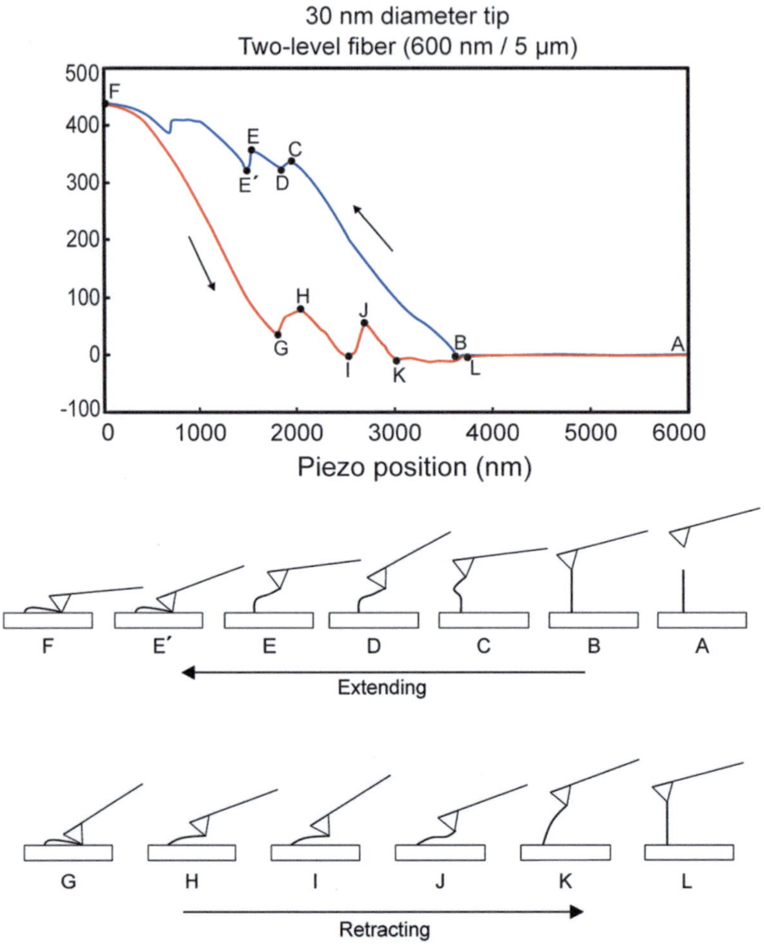

Fig. 12.47 Force distance curve for the two-level single fiber (600 nm/5 μm) obtained using the AFM tip of 30 nm diameter. Schematics of the fiber behavior at each point as the AFM tip travels are shown (Bhushan and Lee 2012)

adhesion coefficient increases with a decrease in the stiffness of springs. The hierarchical model with more compliant springs can generate higher adhesion enhancement in the lower applied load. For the effect of the number of springs, as the number of springs on the lower level increases, the equivalent stiffness decreases. Therefore, the three-level model with larger number of springs on the lowest level gives larger adhesion force and energy. Inclusion of capillary forces in the spring model shows that the total adhesion force decreases as a substrate becomes more hydrophobic (increase in the contact angle of water). The difference of total adhesion force among different contact angles is larger

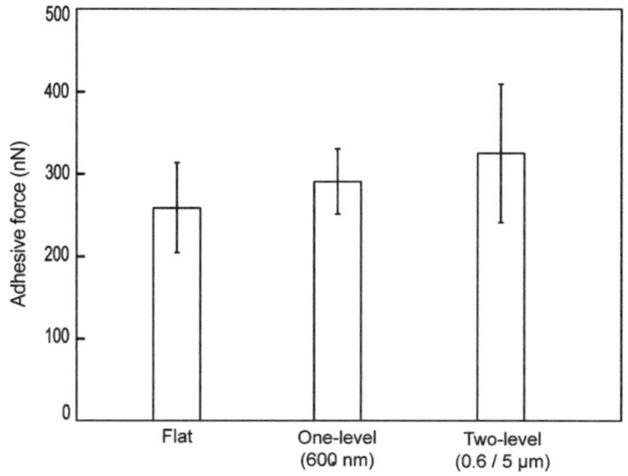

Fig. 12.48 Adhesive force on the flat sample and the one-level and two-level structures measured using the 30 μm diameter tip. The measurements were performed at room temperature (21 °C) and at 45–55% relative humidity (Bhushan and Lee 2012)

in the intermediate humidity regime. In addition, the simulation results match the measured data for a single spatula in contact with both the hydrophilic and the hydrophobic surfaces, which further supports van der Waals forces as the dominant mechanism of adhesion and capillary forces as a secondary mechanism.

There is a great interest among the scientific community to create surfaces that replicate the adhesion strength of gecko feet. These hierarchical fibrillar microstructured surfaces would be capable of reusable dry adhesion, and would have uses in a wide range of applications from everyday objects such as adhesive tapes, fasteners, toys, microelectronics, and space applications, to the treads of wall-climbing robots. In the design of fibrillar structures, it is necessary to ensure that the fibrils are compliant enough to easily deform to the mating surface's roughness profile, yet rigid enough to not collapse under their own weight. Spacing between the individual fibrils is also important. If the spacing is too small, adjacent fibrils can attract each other through intermolecular forces which will lead to bunching. The adhesion design database developed by Kim and Bhushan (2007c) serves as a reference for choosing design parameters.

Nanoindentation, molding, self-assembly, carbon nanotube arrays, lithography, and the use of porous membranes are some of the methods that have been used to create fibrillar structures. The limitations of current machining methods on the micro/nanoscale have resulted in the majority of fabricated surfaces consisting of only one level of hierarchy. Bunching, lack of compliance, and lack of durability are some of the problems that may arise with the aforementioned structures. Multilevel compliant systems have been created using molding, photolithography, a microelectromechanical based approach, and porous membranes. Inspired by work on adding tips to the fibrillar structures, the end of the fibers could be modified to enhance adhesion. For example, a soft adhesive could be used to coat fiber ends to provide added adhesion by conventional adhesives.

Fibrillar structures show great promise in the creation of adhesive structures. Some of the structures have been incorporated into the design of treads of climbing robots.

References

Aksak, B., Murphy, M. P., and Sitti, M. (2007), "Adhesion of Biologically Inspired Vertical and Angled Polymer Microfiber Arrays," *Langmuir* **23**, 3322–3332.

Aksak, B., Murphy, M. P., and Sitti, M. (2008), "Gecko Inspired Micro-Fibrillar Adhesives for Wall Climbing Robots on Micro/Nanoscale Rough Surfaces," *Proc. 2008 IEEE Conf. on Robotics and Automation*, Pasadena, CA, pp. 3058–3063.

Aristotle, *Historia Animalium*, trans. Thompson, D.A.W. (1918), <http://classics.mit.edu/Aristotle/history_anim.html>.

Arzt, E., Gorb, S. and Spolenak, R. (2003), "From Micro to Nano Contacts in Biological Attachment Devices," *Proc. Natl. Acad. Sci. USA* **100**, 10603–10606.

Autumn, K. (2006), "How Gecko Toes Stick," *Am. Scientist*, **94**, 124–132.

Autumn, K. and Peattie, A. M. (2002), "Mechanisms of Adhesion in Geckos," *Integr. Comp. Biol.* **42**, 1081–1090.

Autumn, K., Liang, Y. A., Hsieh, S. T., Zesch, W., Chan, W. P., Kenny, T. W., Fearing, R. and Full, R. J. (2000), "Adhesive Force of a Single Gecko Foot-Hair," *Nature,* **405**, 681–685.

Autumn, K., Sitti, M., Liang, Y. A., Peattie, A. M., Hansen, W. R., Sponberg, S., Kenny, T. W., Fearing, R., Israelachvili, J. N. and Full R. J. (2002), "Evidence for van der Waals Adhesion in Gecko Setae," *Proc. Natl. Acad. Sci. USA* **99**, 12252–12256.

Autumn, K., Majidi, C., Groff, R. E., Dittmore, A., and Fearing, R. (2006a), "Effective Elastic Modulus of Isolated Gecko Setal Arrays," *J. Exp. Biol.* **209**, 3558–3568.

Autumn, K., Dittmore, A., Santos, D., Spenko, M., and Cutkosky, M. (2006b), "Frictional Adhesion, a New Angle on Gecko Attachment," *J. Exp. Biol.* **209**, 3569–3579.

Barnes, W. J. P., Smith, J., Oines, C., and Mundl, R. (2002), "Bionics and Wet Grip," *Tire Technol. Int.* **2002**, 56–60.

Bergmann, P. J. and Irschick, D. J. (2005), "Effects of Temperature on Maximum Clinging Ability in a Diurnal Gecko: Evidence for a Passive Clinging Mechanism?" *J. Exp. Zool.* **303A**, 785–791.

Bertram, J. E. A. and Gosline, J. M. (1987), "Functional Design of Horse Hoof Keratin: The Modulation of Mechanical Properties through Hydration Effects," *J. Exp. Biol.* **130**, 121–136.

Bhushan, B. (1996), *Tribology and Mechanics of Magnetic Storage Devices, second ed.*, Springer-Verlag, New York.

Bhushan, B. (2007), "Adhesion of Multi-Level Hierarchical Attachment Systems in Gecko Feet," *J. Adhesion Sci. Technol.* **21**, 1213–1258.

Bhushan, B. (2013a), *Principles and Applications of Tribology,* second ed., Wiley, New York.

Bhushan, B. (2013b), *Introduction to Tribology,* second ed., Wiley, New York.

Bhushan, B. (2017a) *Nanotribology and Nanomechanics: An Introduction,* fourth ed., Springer International, Cham, Switzerland.

Bhushan, B. (2017b), *Springer Handbook of Nanotechnology*, fourth ed., Springer International, Cham, Switzerland.

Bhushan, B. and Jung, Y. C. (2011), "Natural and Biomimetic Artificial Surfaces for Superhydrophobicity, Self-Cleaning, Low Adhesion, and Drag Reduction," *Prog. Mater. Sci.* **56**, 1–108.

Bhushan, B. and Lee, H. (2012), "Fabrication and Characterization of Multi-level Hierarchical Surfaces," *Faraday Discuss.* **156**, 235–241.

Bhushan, B. and Sayer, R. A. (2007), "Surface Characterization and Friction of a Bio-Inspired Reversible Adhesive Tape," *Microsyst. Technol.* **13**, 71–78.

Bhushan, B., Peressadko, A. G., and Kim, T. W. (2006) "Adhesion Analysis of Two-Level Hierarchical Morphology in Natural Attachment Systems for 'Smart Adhesion'," *J. Adhesion Sci. Technol.* **20**, 1475–1491.

Bikerman, J. J. (1961), *The Science of Adhesive Joints*, Academic Press, New York.

Burton, Z. and Bhushan, B. (2005), "Hydrophobicity, Adhesion, and Friction Properties of Nanopatterned Polymers and Scale Dependence for Micro- and Nanoelectromechanical Systems," *Nano Letters* **5**, 1607–1613.

Cai, S. and Bhushan, B. (2007), "Effects of Symmetric and Asymmetric Contact Angles and Division of Menisci on Meniscus and Viscous Forces during Separation," *Philos. Mag.* **87**, 5505–5522.

Cai, S. and Bhushan, B. (2008), "Meniscus and Viscous Forces During Separation of Hydrophilic and Hydrophobic Surfaces with Liquid Mediated Contacts," *Mater. Sci. Eng. R* **61**, 78–106.

Cho, W. K. and Choi, I. S. (2007), "Fabrication of Hairy Polymeric Films Inspired by Geckos: Wetting and High Adhesion Properties," *Adv. Func. Mater.* **18**, 1089–1096.

Chui, B. W., Kenny, T. W., Mamin, H. J., Terris, B. D. and Rugar, D. (1998), "Independent Detection of Vertical and Lateral Forces with a Sidewall-Implanted Dual-Axis Piezoresistive Cantilever," *Appl. Phys. Lett.* **72**, 1388–1390.

Cutkosky, M. R. and Kim, S. (2009), "Design and Fabrication of Multi-Materials Structures for Bio-Inspired Robots," *Phil. Trans. R. Soc. A* **367**, 1799–1813.

Daltorio, K. A., Gorb, S., Peressadko, A., Horchler, A. D., Ritzmann, R. E., and Quinn, R. D. (2007), "A Robot that Climbs Walls using Micro-Structured Polymer Adhesive," *Proc. 30th Annual Meeting of the Adhesion Society*, pp. 329–331.

Davies, J., Haq, S., Hawke, T., and Sargent, J. P. (2008), "A Practical Approach to the Development of a Synthetic Gecko Tape," *Int. J. Adhesion and Adhesives* **29**, 380–390.

del Campo, A. and Greiner, C. (2007), "SU-8: A Photoresist for High-Aspect-Ratio and 3D Submicron Lithography," *J. Micromech. Microeng.* **17**, R81–R95.

del Campo, A. Greiner, C., Alvares, I., and Arzt, E. (2007a), "Patterned Surfaces with Pillars with Controlled 3D Tip Geometry Mimicking Bioattachment Devices," *Adv. Mater.* **19**, 1973–1977.

del Campo, A., Greiner, C. and Arzt, E. (2007b), "Contact Shape Controls Adhesion of Bioinspired Fibrillar Surfaces," *Langmuir* **23**, 10235–10243.

Dellit, W. D. (1934), "Zur Anatomie und Physiologie der Geckozehe," *Jena. Z. Naturwissen*, **68**, 613–658.

Derjaguin, B. V., Muller, V. M., and Toporov, Y. P. (1975), "Effect of Contact Deformation on the Adhesion of Particles," *J. Colloid Interface Sci.* **53**, 314–326.

Dieter G. E. (1988), *Mechanical Metallurgy*, McGraw Hill, London.

Fan, P. L. and O'Brien, M. J. (1975), "Adhesion in Deformable Isolated Capillaries," *Adhesion Science and Technology*, (ed. L. H. Lee), Vol. 9A, p. 635, Plenum, New York.

Federle, W., (2006) "Why are So Many Adhesive Pads Hairy?" *J. Exp. Biol.* **209**, 2611–2621.

Federle, W., Riehle, M., Curtis, A. S. G. and Full, R. J. (2002), "An Integrative Study of Insect Adhesion: Mechanics of Wet Adhesion of Pretarsal Pads in Ants," *Integr. Comp. Biol.* **42**, 1100–1106.

Federle, W., Barnes, W. J. P., Baumgartner, W., Drechsler, P., and Smith, J. M. (2006), "Wet but Not Slippery: Boundary Friction in Tree Frog Adhesive Toe Pads," *J. R. Soc. Interf.* **3**, 689–697.

Gao, H., Wang, X., Yao, H., Gorb, S. and Arzt, E. (2005), "Mechanics of Hierarchical Adhesion Structures of Geckos," *Mech. Mater.* **37**, 275–285.

Ge, L., Sethi, S., Ci, L., Ajayan, M. and Dhinojwale, A, (2007), "Carbon Nanotube-based Synthetic Gecko Tape," *PNAS* **104**, 10792–10795.

Geim, A. K., Dubonos, S. V., Grigorieva, I. V., Novoselov, K. S., Zhukov, A. A. and Shapoval, S. Y. (2003), "Microfabricated Adhesive Mimicking Gecko Foot-Hair," *Nat. Mater.* **2**, 461–463.

Gennaro, J. G. J. (1969), "The Gecko Grip," *Nat. Hist.* **78**, 36–43.

Glassmaker, N. J., Jagota, A., Hui, C. Y. and Kim, J. (2004), "Design of Biomimetic Fibrillar Interfaces: 1. Making Contact," *J. R. Soc. Interface* **1**, 23–33.

Glassmaker, N. J., Jagota, A. and Hui, C. Y. (2005), "Adhesion Enhancement in a Biomimetic Fibrillar Interface," *Acta Biomaterialia* **1**, 367–375.

Gorb, S. (2001), *Attachment Devices of Insect Cuticles*, Kluwer Academic, Dordrecht, Netherlands.

Gorb, S., Varenberg, M., Peressadko, A. and Tuma, J. (2007), "Biomimetic Mushroom-Shaped Fibrillar Adhesive Microstructures," *J. Royal Soc. Interf.* **4**, 271–275.

Hamaker, H. C. (1937), "London van der Waals Attraction between Spherical Bodies," *Physica*, **4**, 1058.

Han, D., Zhou, K., and Bauer, A. M. (2004), "Phylogenetic Relationships among Gekkotan Lizards Inferred from C-mos Nuclear DNA Sequences and a New Classification of the Gekkota," *Biol. J. Linn. Soc.* **83**, 353–368.

Hanna, G. and Barnes, W. J. P. (1991), "Adhesion and Detachment of the Toe Pads of Tree Frogs," *J. Exper. Biol.* **155**, 103–125.

Hansen, W. R. and Autumn, K. (2005), "Evidence for Self-Cleaning in Gecko Setae," *Proc. Natl. Acad. Sci. USA* **102**, 385–389.

Heepe, L., Kovalev, A. E., Varenberg, M., Tuma, J., and Gorb, S. N. (2012), "First Mushroom-Shaped Adhesive Microstructure: A Review," *Theoretical Appl. Mech. Lett.* **2**, 014008.

Hiller, U. (1968), "Untersuchungen zum Feinbau und zur Funktion der Haftborsten von Reptilien," *Z. Morphol. Tiere*, **62**, 307–362.

Hinds, W. C. (1982), *Aerosol Technology: Properties, Behavior, and Measurement of Airborne Particles,* Wiley, New York.

Hora, S. L. (1923), "The Adhesive Apparatus on the Toes of Certain Geckos and Tree Frogs," *J. Asiat. Soc. Beng.* **9**, 137–145.

Houwink, R. and Salomon, G. (1967), "Effect of Contamination on the Adhesion of Metallic Couples in Ultra High Vacuum," *J. Appl. Phys.* **38**, 1896–1904.

Huber, G., Gorb, S. N., Spolenak, R. and Arzt, E. (2005a), "Resolving the Nanoscale Adhesion of Individual Gecko Spatulae by Atomic Force Microscopy," *Biol. Lett.* **1**, 2–4.

Huber, G., Mantz, H., Spolenak, R., Mecke, K., Jacobs, K., Gorb, S. N. and Arzt, E. (2005b), "Evidence for Capillarity Contributions to Gecko Adhesion from Single Spatula and Nanomechanical Measurements," *Proc. Natl. Acad. Sci. USA* **102**, 16293–16296.

Irschick, D. J., Austin, C. C., Petren, K., Fisher, R. N., Losos, J. B. and Ellers, O. (1996), "A Comparative Analysis of Clinging Ability Among Pad-Bearing Lizards," *Biol. J. Linn. Soc.* **59**, 21–35.

Israelachvili, J. N. (1992), *Intermolecular and Surface Forces*, second ed., Academic Press, San Diego, CA.

Israelachvili, J. N. and Tabor, D. (1972), "The Measurement of van der Waals Dispersion Forces in the Range of 1.5 to 130 nm," *Proc. R. Soc. Lond. A.* **331**, 19–38.

Jaenicke, R. (1998), "Atmospheric Aerosol Size Distribution," *Atmospheric Particles* (eds. R. M. Harrison and R. van Grieken), pp. 1–29, Wiley, New York.
Jagota, A. and Bennison, S. J. (2002), "Mechanics of Adhesion through a Fibrillar Microstructure," *Integr. Comp. Biol.* **42**, 1140–1145.
Johnson, K. L., Kendall, K. and Roberts, A. D. (1971), "Surface Energy and the Contact of Elastic Solids," *Proc. R. Soc. Lond. A.* **324**, 301–313.
Kesel, A. B., Martin, A. and Seidl, T. (2003), "Adhesion Measurements on the Attachment Devices of the Jumping Spider *Evarcha arcuata*," *J. Exp. Biol.* **206**, 2733–2738.
Kim, T. W. and Bhushan, B. (2007a), "The Adhesion Analysis of Multi-Level Hierarchical Attachment System Contacting with a Rough Surface," *J. Adhesion Sci. Technol.* **21**, 1–20.
Kim, T. W. and Bhushan, B. (2007b), "Effect of Stiffness of Multi-Level Hierarchical Attachment System on Adhesion Enhancement," *Ultramicroscopy* **107**, 902–912.
Kim, T. W. and Bhushan, B. (2007c), "Optimization of Biomimetic Attachment System Contacting with a Rough Surface," *J. Vac. Sci. Technol. A* **25** 1003–1012.
Kim, T. W. and Bhushan, B. (2008), "The Adhesion Model Considering Capillarity for Gecko Attachment System," *J. R. Soc. Interf.* **5**, 319–327.
Kluge, A. G. (2001), "Gekkotan Lizard Taxonomy," *Hamadryad* **26**, 1–209.
Lee, H. and Bhushan, B. (2012), "Fabrication and Characterization of Hierarchical Nanostructured Smart Adhesion Surfaces," *J. Colloid Surf. Sci.* **372**, 231–238.
Losos, J. B. (1990), "Thermal Sensitivity of Sprinting and Clinging Performance in the Tokay Gecko (*Gekko gecko*)," *Asiat. Herpetol. Res.* **3**, 54–59.
Maderson, P. F. A. (1964), "Keratinized Epidermal Derivatives as an Aid to Climbing in Gekkonid Lizards," *Nature* **2003**, 780–781.
Murphy, M. P., Aksak, B. and Sitti, M. (2007), "Adhesion and Anisotropic Friction Enhancement of Angled Heterogeneous Micro-fiber Arrays with Spherical and Spatula Tips," *J. Adhes. Sci. Technol.* **21**, 1281–1296.
Northen, M. T. and Turner, K. L., (2005), "A Batch Fabricated Biomimetic Dry Adhesive," *Nanotechnology,* **16**, 1159–1166.
Nosonovsky, M. and Bhushan, B. (2008), *Multiscale Dissipative Mechanisms and Hierarchical Surfaces: Friction, Superhydrophobicity, and Biomimetics*, Springer-Verlag, Heidelberg, Germany.
Ohler, A. (1995), "Digital Pad Morphology in Torrent-living Ranid Frogs" *Asiat. Herpetol. Res.* **6**, 85–96.
Orr, F. M., Scriven, L. E., and Rivas, A. P. (1975), "Pendular Rings between Solids: Meniscus Properties and Capillary Forces," *J. Fluid. Mech.* **67**, 723–742.
Palacio, M. L. B., Bhushan, B., and Schricker, S. R. (2013), "Hierarchical Nanostructured Polymers for Reversible Adhesion in Biomedical Applications," *Materials Letters* **92**, 409–412.
Peattie, A. M. and Full, R. J. (2007), "Phylogenetic Analysis of the Scaling of Wet and Dry Biological Fibrillar Adhesives," *Proc. Natl. Acad. Sci. USA* **104**, 18595–18600.
Persson, B. N. J. (2003), "On the Mechanism of Adhesion in Biological Systems," *J. Chem. Phys.* **118**, 7614–7621.
Persson, B. N. J. and Gorb, S. (2003), "The Effect of Surface Roughness on the Adhesion of Elastic Plates with Application to Biological Systems," *J. Chem. Phys.* **119**, 11437–11444.
Pesika, N. S., Tian, Y., Zhao, B., Rosenberg, K., Zeng, H., McGuiggen, P., Autumn, K., and Israelachvili, J. N. (2007), "Peel-Zone Model of Tape Peeling Based on the Gecko Adhesive System," *J. Adhesion* **83**, 383–401.

Phipps, P. B. P. and Rice, D. W. (1979), "Role of Water in Atmospheric Corrosion," *Corrosion Chemistry, ACS Symposium Series Vol. 89* (ed. by G. R. Brubaker and P. B. P. Phipps), pp. 235–261, American Chemical Society, Washington, D.C.

Qu, L., Dai, L., Stone, M., Xia, Z., and Wang, Z. L. (2008), "Carbon Nanotube Arrays with Strong Shear Binding-On and Easy Normal Lifting-Off," *Science* **322**, 238–242.

Rizzo, N., Gardner, K., Walls, D., Keiper-Hrynko, N., and Hallahan, D. (2006), "Characterization of the Structure and Composition of Gecko Adhesive Setae," *J. Royal Soc. Interf.* **3**, 441–451.

Ruibal, R. and Ernst, V. (1965), "The Structure of the Digital Setae of Lizards," *J. Morphol.* **117**, 271–294.

Russell, A. P. (1975), "A Contribution to the Functional Morphology of the Foot of the Tokay, *Gekko gecko*," *J. Zool. Lond.* **176**, 437–476.

Russell, A. P. (1986), "The Morphological Basis of Weight-Bearing in the Scansors of the Tokay Gecko," *Can. J. Zool.* **64**, 948–955.

Schäffer, E., Thurn-Albrecht, T., Russell, T. P. and Steiner, U. (2000) "Electrically Induced Structure Formation and Pattern Transfer," *Nature*, **403**, 874–877.

Schleich, H. H. and Kästle, W. (1986), "Ultrastrukturen an Gecko-Zehen," *Amphibia Reptilia*, **7**, 141–166.

Schmidt, H. R. (1904), "Zur Anatomie und Physiologie der Geckopfote," *Jena. Z. Naturwissen* **39**, 551.

Shah, G. J. and Sitti, M. (2004), "Modeling and Design of Biomimetic Adhesives Inspired by Gecko Foot-Hairs," *IEEE Int. Conf. on Robotics and Biomimetics*, 873–878.

Simmermacher, G. (1884), "Untersuchungen uber Haftapparate an Tarsalgliedern von Insekten," *Zeitschr. Wissen Zool.* **40**, 481–556.

Sitti, M. (2003), "High Aspect Ratio Polymer Micro/Nano-Structure Manufacturing using Nanoembossing, Nanomolding and Directed Self-Assembly," *Proc. IEEE/ASME Advanced Mechatronics Conf.* **2**, 886–890.

Sitti, M. and Fearing, R. S. (2003) "Synthetic Gecko Foot-Hair for Micro/Nano Structures as Dry Adhesives," *J. Adhesion Sci. Technol.* **17**, 1055–1073.

Spolenak, R., Gorb, S. and Arzt, E., (2005), "Adhesion Design Maps for Bio-Inspired Attachment Systems," *Acta Biomaterialia* **1**, 5–13.

Stork, N. E. (1980), "Experimental Analysis of Adhesion of *Chrysolina polita* on a Variety of Surfaces," *J. Exp. Biol.* **88**, 91–107.

Stork, N. E. (1983), "A Comparison of the Adhesive Setae on the Feet of Lizards and Arthropods," *J. Nat. Hist.*, **17**, 829–835.

Tian, Y., Pesika, N., Zeng, H., Rosenberg, K., Zhao, B., McGuiggan, P., Autumn, K., and Israelachvili, J. (2006), "Adhesion and Friction in Gecko Toe Attachment and Detachment," *Proc. Nat. Acad. Sci. U. S. A.* **103**, 19320–19325.

Tinkle, D. W. (1992), "Gecko," *Encyclopedia Americana*, edited by M. Cummings, Vol. 12, 359, Grolier, U.K.

Van der Kloot, W. G. (1992), "Molting," *Encyclopedia Americana*, edited by M. Cummings, Vol. 19, 336–337, Grolier, U.K.

Wagler, J. G. (1830), *Naturliches System der Amphibien, mit vorangehender Classification der Säugthiere und Vögel. Ein Beitrag zur vergleichenden Zoologie*, J. G. Cotta'schen Buchhandlung, Munich, Germany.

Wan, K. T., Smith, D. T. and Lawn, B. R. (1992), "Fracture and Contact Adhesion Energies of Mica-Mica, Silica-Silica, and Mica-Silica Interfaces in Dry and Moist Atmospheres," *J. Am. Ceram. Soc.* **75**, 667–676.

Wenzel, R. N. (1936), "Resistance of Solid Surfaces to Wetting by Water," *Ind. Eng. Chem.* **28**, 988–994.

Williams, E. E. and Peterson, J. A. (1982), "Convergent and Alternative Designs in the Digital Adhesive Pads of Scincid Lizards," *Science* **215**, 1509–1511.

Yao, H., and Gao, H. (2006), "Mechanics of Robust and Releasable Adhesion in Biology: Bottom-Up Designed Hierarchical Structures of Gecko," *J. Mech. Phys. Solids*, **54**, 1120–1146.

Young, W. C. and Budynas, R. (2001) *Roark's Formulas for Stress and Strain, seventh ed.* McGraw Hill, New York.

Yurdumakan, B., Raravikar, N. R., Ajayan, P. M. and Dhinojwala, A. (2005), "Synthetic Gecko Foot-Hairs from Multiwalled Carbon Nanotubes," *Chem. Comm.* 3799–3801.

Zimon, A. D. (1969), *Adhesion of Dust and Powder*, translated from Russian by M. Corn, Plenum, New York.

Zisman, W. A. (1963), "Influence of Constitution on Adhesion," *Ind. Eng. Chem.* **55** (10), 18–38.

13 Bio- and Inorganic Fouling

13.1 Introduction

Fouling is generally undesirable for most applications (Bhushan 2016, 2017). Fouling includes biological fouling (commonly referred to as biofouling) and inorganic fouling. Biofouling is the accumulation of unwanted biological matter on surfaces, with biofilms created by microorganisms, and macroscale biofouling (simply called macrofouling) created by macroorganisms. Inorganic fouling is fouling that results from the accumulation of non-living particles, and includes deposits of dirt, corrosion, crystals, suspended particles, oil, and ice. The type and extent of fouling depends on the environment, inorganic deposits, and organisms. In biomedical contexts, fouling typically takes the form of biofilms (Stoodley et al. 2002; Vo-Dinh 2007; Schulz et al. 2009; Shirtliff and Leid 2009; Chan and Wong 2010). Whereas marine and industrial biofouling typically include macrofouling (Fingerman et al. 1999; Walker et al. 2000; Railkin 2004; Hellio and Yebra 2009) and inorganic fouling (Somerscales and Knudsen 1981; Bott 1988b; Melo et al. 1988; Pritchard 1988).

Bixler and Bhushan (2012) presented a review of the field. Based on their review, we start with fields susceptible to fouling, followed by biofouling and inorganic fouling formation mechanisms, antifouling strategies from nature, current prevention and cleaning techniques in antifouling, and end with bioinspired rice leaf surfaces for antifouling, based on Bixler et al. (2014).

B. Bhushan, *Introduction to Biomimetics and Bioinspiration*,
https://doi.org/10.1007/978-3-031-62344-8_13

13.2 Fields Susceptible to Fouling

Fouling is prevalent in medical, marine, and industrial fields, causing significant health risks and loss of resources (Walker et al. 2000; Railkin 2004; Vo-Dinh 2007; Hellio and Yebra 2009). Examples of the various types and problems with fouling are shown in Table 13.1, highlighting the many applications susceptible to biofouling and inorganic fouling (Bixler and Bhushan 2012).

Figure 13.1 shows biofouling in the human body can occur in regions such as prosthetic implants, biosensors, catheters, dental implants, and medical equipment (Bixler and Bhushan 2012). Biofouling may lead to implant rejection, malfunction of biosensors, and spread of infectious diseases (Vo-Dinh 2007; Schulz et al. 2009; Shirtliff and Leid 2009; Chan and Wong 2010). Biofilms on medical devices including a ventilation tube, needleless connector, and pacemaker wire are shown in Fig. 13.2 (Bixler and Bhushan 2012). Biofouling also occurs in marine environments where noticeable marine growth appears

Table 13.1 Fields susceptible to biofouling including common examples (adapted from Bixler and Bhushan 2012)

Type	Problems
Medical	
Orthopedic implant	Removal due to infection
Respirator	Ventilator associate pneumonia
Contact lens	Eye infection
Catheter	Urinary tract infections
Hemodialysis	Infectious break-outs
Teeth/dental implant	Periodontal disease, gingivitis
Biosensor	Failure from fibrous encapsulation
Marine	
Ship hull	Increased fuel consumption
Ship engine	Increased stress from extra drag
Marine platform	Increased marine structure load/stress/fatigue
Metal	Increased biocorrosion
Industrial	
Membrane	Reduced flux
Fluid flow	Frictional loss in pipes
Drinking water	Pathogens in potable water
Food, paper, and paint	Food spoilage and worker health risks
Metal cutting fluid	Filter blockage and worker health risks

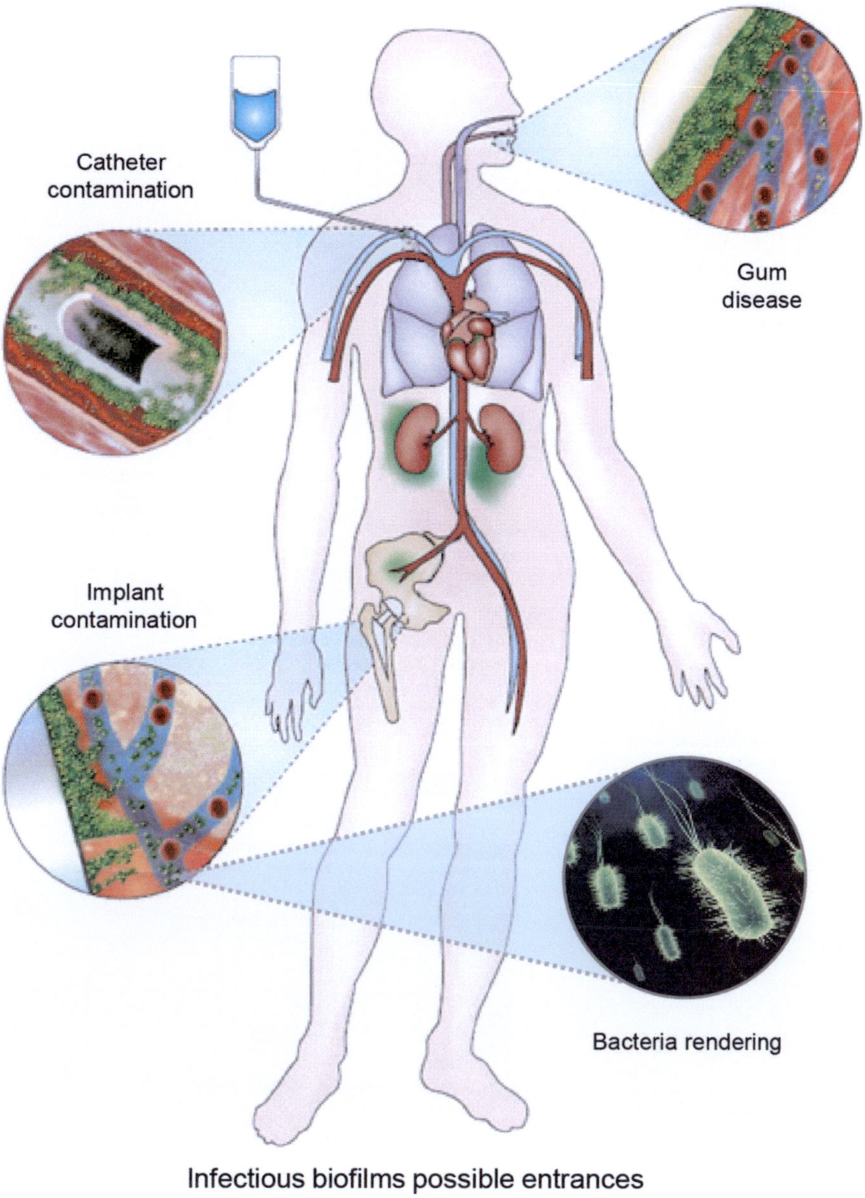

Fig. 13.1 Schematic highlighting areas and entry points of the body susceptible to infectious microorganisms leading to biofilms (Bixler and Bhushan 2012)

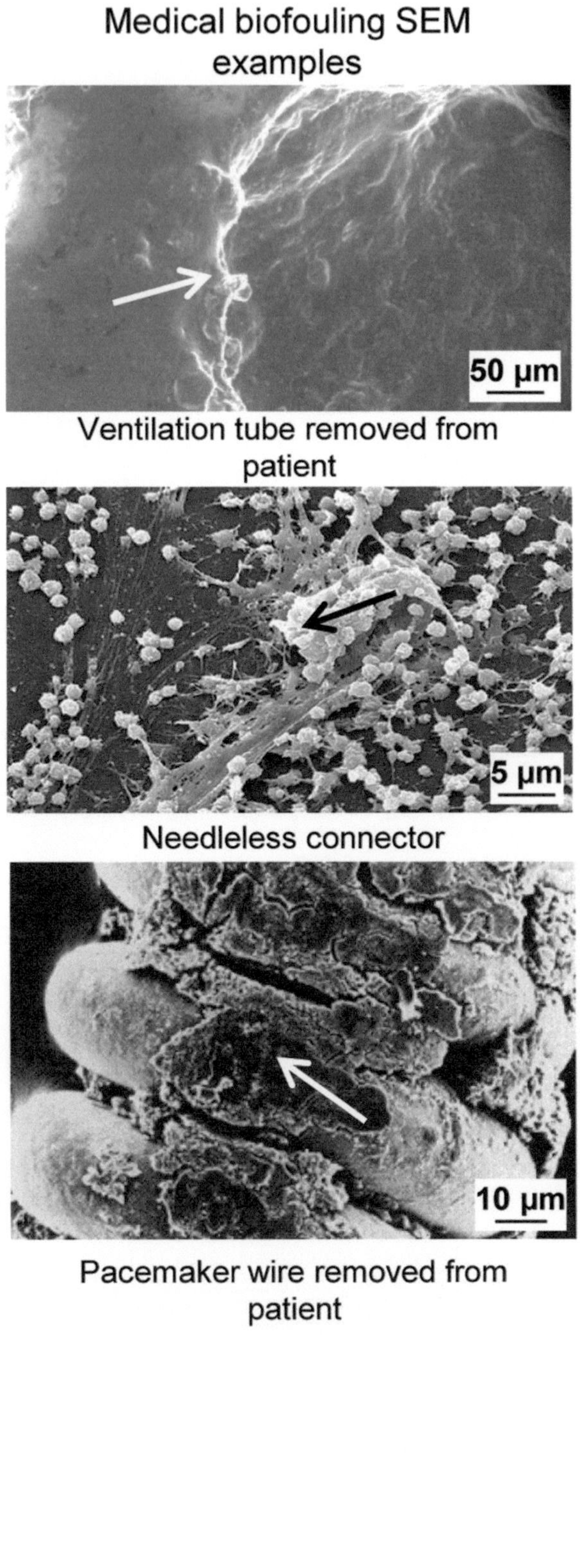

Fig. 13.2 Medical device biofilm examples (Bixler and Bhushan 2012). Images include a ventilation tube (top) (Trinidad et al. 2010), needless connector (middle) (adapted from Janice Carr), and pacemaker wire (bottom) (Marrie and Costerton 1984). Such biofilms may lead to infections that require device removal since antibodies are often ineffective

on ships and underwater structures. This magnifies ship drag, corrosion, fuel consumption, and engine stress (Copisarow 1945; Anonymous 1952; Ray 1959; Melo et al. 1988; Railkin 2004). Industrial biofouling occurs in areas such as power plants, water treatment systems, and food or beverage industries, as shown in Fig. 13.3 (Bixler and Bhushan 2012). Issues include pipe blockage, decreased reverse osmosis (RO) membrane flux, contaminated water, and reduced heat exchanger efficiency (Somerscales and Knudsen 1981; Walker et al. 2000; Chan and Wong 2010). The antimicrobial finishes on fabrics are of interest to protect the consumer against disease or odor (Perera et al. 2013).

Inorganic fouling creates concern in many industries with decreased performance in applications ranging from power production to consumer products. For example, the interior surface of a heat exchanger tube from a nuclear reactor is shown in Fig. 13.4 (Bixler and Bhushan 2012), and corrosion deposits are readily apparent, which leads to inefficient performance (Cunningham et al. 2008; www.hcheattransfer.com). Inorganic fouling also includes undesirable fingerprints on transparent surfaces such as electronic touch screens

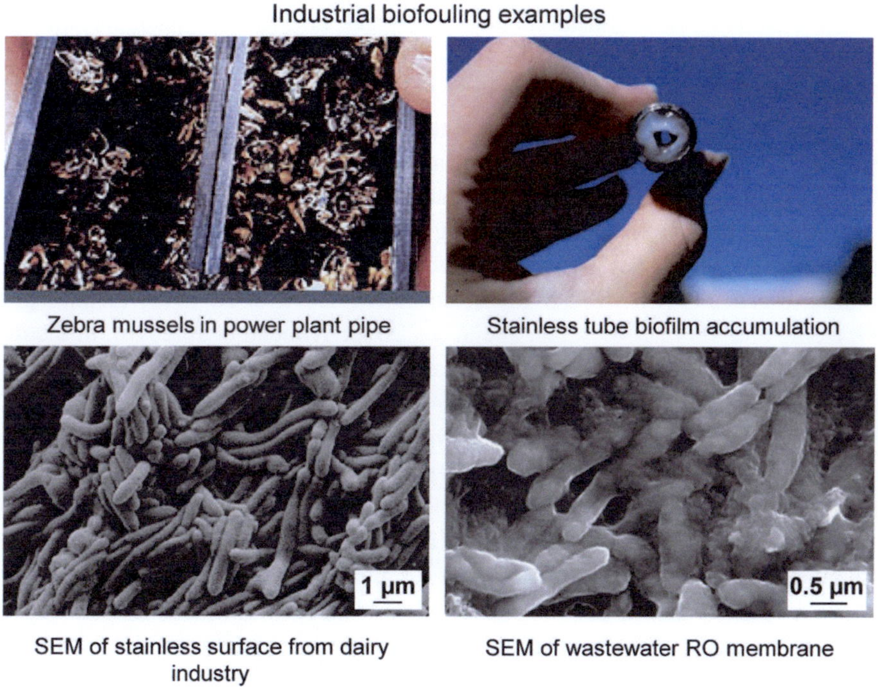

Fig. 13.3 Industrial biofouling examples (Bixler and Bhushan 2012). Zebra mussels (*Dreissena polymorpha*) clogging power plant intake pipe (top left) (photo courtesy of Peter Yates), biofilm accumulation in a stainless steel tube (top right) (Cunningham et al. 2008), biofilm (*B. cereus*) on dairy industry stainless steel plate (bottom left) (Simoes et al. 2010), and biofilm (*P. aeruginosa*) on reverse osmosis (RO) membrane (bottom right) (Herzberg and Elimelech 2007)

(mobile phones, tablets, music players, GPS devices), eyeglasses, computer screens, and window glass (Wang and Bhushan 2015). Furthermore, inorganic fouling includes ice on airplane wings, helicopter blades, oil platforms, power lines, locks and dams, and wind turbines (Cao et al. 2009; Meuler et al. 2010).

13.3 Biofouling and Inorganic Fouling Formation Mechanisms

In this section, we discuss the type, formation, and morphology characteristics of biofouling and inorganic fouling (based on Bixler and Bhushan 2012). Details follow.

13.3.1 Biofouling Formation

Bacterial infection is a major problem in many applications including medical devices, marine surfaces, floors, walls and fabrics and many other surfaces. Hospital acquired infection remains an issue. Bacteria can colonize and proliferate, a phenomenon referred to as "biofouling".

Biofouling begins when microorganisms become stationary on a surface. Microorganisms adhere to one another and to a substrate with an adhesive called the extracellular polymeric substance (EPS). The biofilm continues to grow and become more diverse by attracting additional microorganisms through chemical "messages". Colonization is the process of organisms collecting and growing on a surface, and the morphology is characterized by the thickness, density, structure, composition, bioadhesive strength, and weight of fouling organisms (Marrie and Costerton 1984; Keevil et al. 1999; Flemming et al. 2000; Hellio and Yebra 2009; Chan and Wong 2010; Trinidad et al. 2010). In the marine environment, this usually starts with the formation of a biofilm that attracts larger macrofoulers (Stoodley et al. 2002; Railkin 2004; Costerton 2008; Hellio and Yebra 2009). For instance, tubeworms prefer settling on biofilms, but bryozoans and barnacles do not require a biofilm (Ralston and Swain 2009). Figure 16.5 illustrates the five-stage colonization process, which includes initial attachment, irreversible attachment, initial growth, final growth, and dispersion (adapted from Davies 2003 and Bixler and Bhushan 2012). Initial attachment starts the colonization process, which begins within days to a few weeks in a nutrient rich environment. The biofilm covered surface then attracts other organisms that may have been previously deterred. Initial attachment of microorganisms is reversible, but once they secrete the EPS, the bond becomes irreversible. This permanent attachment allows initial growth, final growth, and dispersion (Stoodley et al. 2002; Railkin 2004; Monroe 2007; Costerton 2008).

Biofouling growth rates depend on the organism, substrate, flow velocity, shear stress, and temperature (Griebe and Flemming 2000; Railkin 2004; Jones 2009). Organism transportation to a surface is either passive from the current or active from the propagule,

Fig. 13.4 Inorganic fouling examples (Bixler and Bhushan 2012). Inside of a clean nuclear power plant heat exchanger tube (top) compared to a fouled tube with corrosion deposits (middle) (Cunningham et al. 2008). Calcium carbonate crystallization fouling on the outside of a heat exchanger (bottom) (photo courtesy of H&C Heat Transfer Solutions, Inc)

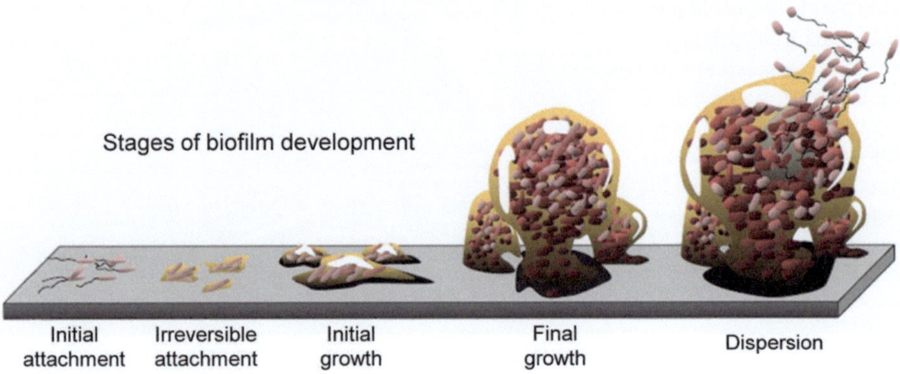

Fig. 13.5 Biofilm development process. Schematic illustrates the five stage colonization process (adapted from Davies 2003 and Bixler and Bhushan 2012)

juvenile, or adult organism. Active mechanisms include electrostatic repulsion, Brownian motion, turbulent pulsations, and cell outgrowths (Railkin 2004; Ralston and Swain 2009). Underwater environments are ideal for biofouling since currents deliver nutrients and carry away wastes, promoting colonization by planktonic and sessile organisms.

Tenacity of the biofouling is dependent upon bioadhesion (Stoodley et al. 2002; Railkin 2004; Monroe 2007; Costerton 2008). Bioadhesion is the adhesion strength of biofouling on a hard surface (Palacio and Bhushan 2012). This depends on the organism type, substrate, and separating fluid (Callow et al. 1986), due to influences of electrostatic forces and surface wettability (Feng and Jiang 2006; Koch and Barthlott 2009; Scardino 2009; Sheng et al. 2010). Biofilm bioadhesion is a two-stage process, starting with the initial attachment and then the irreversible attachment. Initial attachment is controlled by a physical adhesion between the microorganism and the substrate. Also known as adsorption, the initial colonists attach to a surface through weak, reversible van der Waals bonds, which are slightly stronger than electrostatic repulsive force. Irreversible attachment is accomplished with secretion of the EPS, which exhibits a sponge-like matrix. This adhesive permanently bonds the microorganisms to one another and collectively to the surface (Flemming et al. 1999; Stoodley et al. 2002; Railkin 2004).

13.3.2 Inorganic Fouling Formation

Inorganic fouling is composed of non-biological particles that may form in addition to or independently from biofouling. Particles originate from corrosion, crystallization, suspended particles, oil, and ice. For instance, salts from aqueous solutions crystallize and deposit on surfaces. Other deposits may result from minerals found in water such as magnesium, calcium, and barium (Somerscales and Knudsen 1981; Walker et al. 2000; Chan and Wong 2010; Bixler and Bhushan 2012).

Types of inorganic fouling include particulate, freeze, and gas stream particulate. Particulate fouling occurs when suspended solid particles deposit onto a heat transfer surface (Epstein 1988). Deposition of crystals from freeze fouling occurs in locations such as cold region oil pipelines when waxy hydrocarbons contact cold pipe walls. Gas stream particulate fouling occurs in gas lines, reactors, combustion chambers, and heat exchangers. This includes mineral, organic, and inorganic particles, which are common in oil or gas combustion systems (Bott 1988a). Biofouling may initiate inorganic fouling, where biocorrosion causes the formation of corrosion particles. Such fouling is prevalent in boilers, cooling condensers, desalination plants, food processing equipment, geothermal plants, and oil production equipment (Pritchard 1988). Heat exchangers can develop hard deposits called "scale" or porous deposits such as "sludge" (Bott 1988b).

13.3.3 Surface Factors

Biofouling and inorganic fouling depend on the surface factors such as wettability, microtexture, and contours (Fingerman et al. 1999; Railkin 2004; Vadgama 2005; Costerton 2008; Hellio and Yebra 2009; Bixler and Bhushan 2012). For instance, bryozoan and mussel larvae prefer hydrophobic surfaces (Gordon and Mawatari 1992); hydroids, bryozoans, and ascidians prefer microtextured surfaces; larvae, sponges, barnacles, ascidians, and mollusks prefer light colored surfaces; barnacles prefer convex contours; and calcareous sponges prefer concave contours (Railkin 2004).

Surface wettability influences fouler colonization, which ranges from superhydrophobic to superhydrophilic surfaces (Fingerman et al. 1999; Railkin 2004; Vadgama 2005; Costerton 2008; Hellio and Yebra 2009; Bhushan and Jung 2011). Water droplets on a hydrophobic surface will "bead up", whilst droplets on a hydrophilic surface will spread out evenly.

Low adhesive, superhydrophobic surfaces promote contaminant removal with the application of liquid through "self-cleaning". Sheeting action of sliding liquid on a superhydrophilic surface can promote self-cleaning, where the liquid helps collect and remove fouling and disrupt microorganism settlement (Harder and Yee 2009). In general though, microorganisms prefer to colonize hydrophilic surfaces, although some actually prefer hydrophobic surfaces. For example *Ulva linza* prefer hydrophobic surfaces, whereas *Balanus amphitrite* prefer hydrophilic surfaces. Furthermore, microorganisms such as *Ulva linza* attach more easily to hydrophobic surfaces, but experience decreased adhesive strength (Scardino 2009).

Surface microtexture influences organisms such as hydroids, bryozoans, and ascidians that seek shelter against strong currents by settling in grooves, pits, cracks, and crevices

(Fingerman et al. 1999; Railkin 2004). Microorganisms prefer to settle in areas slightly larger than themselves for maximum protection and surface area contact with the substrate. The antifouling occurs when the gap between features is less than the approximate size of fouling microorganisms (Carman et al. 2006; Schumacher et al. 2007a, b; Scardino 2009; Brennan et al. 2010; Bixler and Bhushan 2012). Fewer attachment points between the microorganism and substrate translate into lower bioadhesive strength (Scardino et al. 2008). Bioadhesion is also affected by the effectiveness of EPS flowing into crevices formed by surface roughness, which depends on the adhesive viscosity. Furthermore, when adhesive only contacts the surface asperity peaks or tops of microtexture ridges, applied force to break bonds is significantly reduced (Vladkova 2008).

Several bioassay antifouling experiments have been performed with different microorganisms and microstructured surfaces—using pillars, ribs, and a combination thereof with various diameters, widths, lengths, heights, and pitches (Callow et al. 2002; Hoipkemeier-Wilson et al. 2004; Carman et al. 2006; Schumacher et al. 2007a, b; Scardino et al. 2008; Brennan et al. 2010; Bixler et al. 2014). Laboratory research suggests that samples with appropriately sized topographies prevent colonization by various microorganisms including green algae *Ulva* spores (Carman et al. 2006; Schumacher et al. 2007b), porcine cardiovascular endothelial cells (Carman et al. 2006), and *Balanus amphitrite* cyprids (Schumacher et al. 2007a). Such microorganisms have been exposed to samples and the antifouling effectiveness is determined by counting settled cells using light (Carman et al. 2006) and fluorescence microscopy (Schumacher et al. 2007b). Up to 85% spore settlement reduction is reported with micro-sized ribs when compared to flat control samples (Carman et al. 2006). This reduction is attributed to the micro-sized feature spacing that is slightly smaller than the microorganisms under investigation. The so-called engineered antifouling microtopographies are designed to target specific microorganisms based on their shape and size (Schumacher et al. 2007a, b). Bixler et al. (2014) reported that reduction in biofouling occurs even if the micro-sized featured spacing is larger than the microorganisms. Sharp edges of the features discourage the settling and eventual colonization of microorganisms on the surface.

Figure 13.6 shows surface topography antifouling mechanisms demonstrating wettability and texture properties that influence microorganism settlement and colonization (Bixler and Bhushan 2012). Microorganisms search for an ideal site to settle, which can depend on parameters such as wettability and texture (Feng and Jiang 2006; Scardino 2009). During the search process, microorganisms will settle, move on, or expire in the process. In general, biofouling on superhydrophobic surfaces wash off more easily (self-cleaning) as compared to superhydrophilic surfaces. Images show a variety of examples found in nature, including the hydrophilic elephant ear (*Alocasia odora*) (Koch and Barthlott 2009), microtextured shark skin (*Carcharhinus galapagensis*) (Dean and Bhushan 2010), hydrophobic water fern (*Regnellidium diphyllum*) (Koch and Barthlott 2009), and superhydrophobic nanocolumns on an insect wing (*Cicada orni*) (Lee et al. 2004).

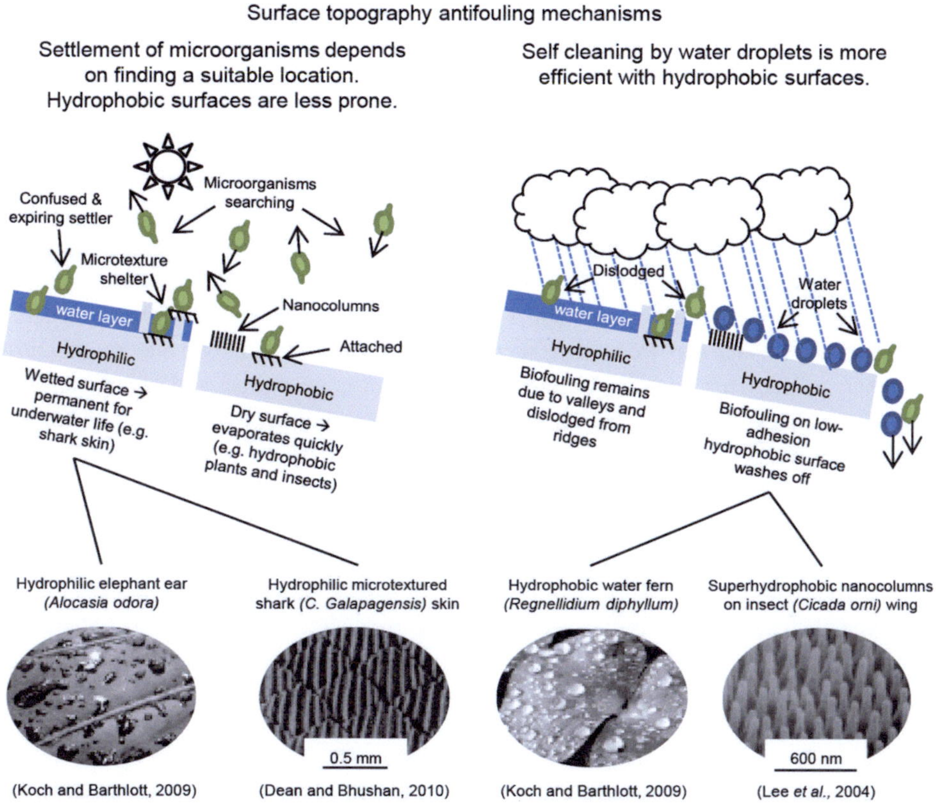

Fig. 13.6 Schematic shows surface topography antifouling mechanisms demonstrating wettability and texture properties that influence microorganism settlement and colonization (adapted from Bixler and Bhushan 2012)

13.4 Antifouling Strategies from Living Nature

In living nature, many plants and animals thrive due to their antifouling and self-cleaning characteristics (Kesel and Liedert 2007; Koch et al. 2008, 2009; Bhushan 2009; Koch and Barthlott 2009; Ralston and Swain 2009; Bixler and Bhushan 2012; Barthlott et al. 2017). Bixler and Bhushan (2012) provide a number of examples of antifouling strategies from living nature, including a variety of both ambient and aquatic species. Common antifouling mechanisms include wettability, low drag, low adhesion, microtexture, grooming/mechanical cleaning, and chemical secretions (Bixler and Bhushan 2012).

Desirable wettability properties include superhydrophobic and superoleophobic low-adhesion self-cleaning surfaces that resist fouling, as liquids wash away contaminant particles (Bixler and Bhushan 2012). Figure 13.7a shows various superhydrophobic examples of lotus leaf, rice leaf, butterfly wings, and hydrophobic pigeon feathers. Figure 13.7b shows examples of superoleophobic-in-water fish scales and shark skin, which exhibit low drag during swimming.

Superhydrophobic lotus leaves exhibit antifouling with self-cleaning properties (Barthlott and Neinhuis 1997; Neinhuis and Barthlott 1997). It is believed that since lotus plants thrive in humid, marshy environments, self-cleaning prevents unwanted fouling, which may inhibit photosynthesis. Lotus leaf hierarchical micropapillae, which are microbumps superimposed with low surface energy waxy nanotubules, provide superior self-cleaning properties (Barthlott and Neinhuis 1997; Bhushan et al. 2009). In nature, rain droplets impact the lotus leaf surface and effectively roll off due to the water repellency effect. This leads to self-cleaning, where droplets collect and remove any contaminant particles, Fig. 13.8a. It should be noted that many other plant species also exhibit self-cleaning properties due to superhydrophobic and low adhesion characteristics.

Fish skin and the skin of fast swimming sharks exhibit low drag, antifouling, and self-cleaning properties (Dean and Bhushan 2010; Bixler and Bhushan 2013a). Fish skin consists of oriented scales with concentric rings overlapping and hinged such that water flow is from head to tail. Sharks exhibit low drag skin in order to swim quickly to catch prey, and that antifouling is necessary in order to maintain the low drag. Shark skin consists of microstructured riblets on dermal denticles that reduce drag and prevent biofouling, as compared to some aquatic species. For instance, whales are often covered with barnacles; however sharks remain clean, see Fig. 13.8b (Bixler and Bhushan 2012, 2013a).

Rice leaves and butterfly wings exploit a combination of antifouling properties from the shark skin and lotus leaf effects (Bixler et al. 2014; Bixler and Bhushan 2015). It is believed that since rice plants thrive in humid, marshy environments, self-cleaning prevents unwanted fouling, which may inhibit photosynthesis. Since butterflies are fragile and unable to clean their wings, these properties are critical to maintain structural coloration and flight control (Wagner et al. 1996). Hierarchical structures consisting of micropapillae superimposed by waxy nanobumps in rice leaves and microgrooves on top of single-like scales in butterfly wings provide superhydrophobicity and low adhesion exhibiting the lotus effect. Sinusoidal grooves in rice leaves and aligned, shingle-like scales in butterfly wings provide the anisotropic flow exhibiting the shark skin effect. A combination of anisotropic flow, superhydrophobicity, and low adhesion leads to drag reduction and antifouling. Due to anisotropic flow characteristics, droplets roll down the blade of the rice leaf and axially away from the butterfly body (Bixler and Bhushan 2013b, 2015). In addition, hierarchical structure reduces coverage and colonization of biofouling products (Bixler et al. 2014).

Pigeon feathers are hydrophobic and remain clean (Bormashenko et al. 2007). Feathers consist of a series of barbs; the barbs themselves are also branched and form the barbules. The hydrophobic properties of the outer feathers (pennae) are due to a combination of this

Fig. 13.7 Antifouling examples from nature using wettability properties. Water promotes self-cleaning on low-adhesion superhydrophobic surfaces, as shown **a** in the lotus leaf, rice leaf, butterfly wings, and pigeon feathers (hydrophobic) examples. **b** Fish scales and shark skin are superoleophobic when submerged in water, and exhibit low drag and antifouling (adapted from Bixler and Bhushan 2012, 2015)

Fig. 13.7 (continued)

complex feather microstructure and the presence of preen oil (preen wax) which is spread mechanically by the bird during the preening procedure. Air pockets are then formed by the feather microstructure that are responsible for hydrophobicity.

Certain microtextured surfaces resist biofouling as organisms seek ideal surface features for settlement, and may be deterred if no suitable surface is found. Various examples are shown in Fig. 13.6.

Grooming/other mechanical cleaning is the physical removal of biofouling from the host, which effectively controls slow- and fast-growing biofouling. Decapods and crustaceans groom other creatures with special brush structures for removing foulers from gills and appendages. Echinoderms and bryozoans use special structures called pedicellaria to groom macroepibionts, while crayfish depend on *Branciobdellid annelids* to feed on foulers in their gills (Bixler and Bhushan 2012).

Any physical action that prevents or removes fouling may be present in conjunction with other antifouling mechanisms, such as low drag, low adhesion and chemical secretions. Figure 13.9a illustrates a mechanical cleaning example, combining wiping with chemical secretions, as demonstrated in the human eye via the lachrymal gland secreting tears, puncta collecting tears, and eyelid wiping the cornea (Menton 2008). Chemical secretion methods range from preventing to removing biofouling. Figure 13.9b shows the red seaweed (*Delisea pulchra*), which uses a halogenated furanone to manipulate colonizing bacteria ‘attraction’ messages (Yen 2010). Other chemical examples include snails that leave a predatory mucous trail, antimicrobial coral egg shells, mucus on shark skin,

Fig. 13.8 Shark skin and lotus leaf effects. **a** The lotus effect is illustrated with a water droplet carrying off contaminants from the lotus leaf surface (Bhushan et al. 2009). **b** Shown is biofouling on the whale and clean shark skin, even though both live in similar environments. The SEM image shows Mako shark skin microstructured riblets that provide low drag and antifouling properties (adapted from Bixler and Bhushan 2013a)

and antifouling chemicals produced by the bacteria *Roseobacter gallaeciensis* (Bixler and Bhushan 2012). Furthermore, blood cell protein adsorption and cell adhesion are inhibited with phosphorylcholine (Bixler and Bhushan 2012).

13.5 Current Prevention and Cleaning Techniques for Antifouling

Reducing or eliminating fouling is accomplished through a variety of methods in medical, marine, and industrial applications (Bixler and Bhushan 2012). To control biofouling, various methods exist which include microorganism resistant coatings and self-cleaning low adhesion surfaces.

Other bioinspired approaches to combat biofouling include the use of nanomaterials (Perera et al. 2013; Bhushan et al. 2014) and nanostructured surfaces (Bixler et al. 2014). Nanomaterials approach prevents bacterial infections whereas nanostructured surfaces approach minimizes adhesion of bacteria to surfaces preventing infection and proliferation leading to biofouling. Details follow.

13.5.1 Current Prevention Techniques

13.5.1.1 Medical Anti-biofouling

Medical biofouling is of great concern for the health of patients (Bixler and Bhushan 2012). Medical devices include catheters, endotracheal tubes, and orthopedic hip implants. Various coatings (Simoes et al. 2010) and hydrophobic polymers with antimicrobial properties (LoVetri et al. 2010; Bixler and Bhushan 2012) have been used. Superhydrophobic and low adhesion "self-cleaning" surfaces are believed to minimize protein bioadhesion, and antimicrobial surfaces reduce ambient microorganisms (Monroe 2007). Surfaces with superhydrophilic polyethylene glycol or polyethylene oxide exhibit less protein adsorption and bioadhesion (Sharma et al. 2007; Vladkova 2008). Other coating examples include self-assembled monolayers (SAMs), which are desirable since they coat surfaces with a molecularly thick biocompatible film. This is especially important for applications such as biomedical micro/nanoelectromechanical systems (BioMEMS/NEMS) (Bhushan et al. 2006; Bhushan 2016, 2017). Figure 13.10 (left) presented by Bixler and Bhushan (2012) shows an orthopedic hip implant covered with a biocompatible coating designed to control biofouling, wear rate, and promote bone growth (Anonymous 1999; Vleugels 2006). Various coated membranes reduce biofouling in medical devices and water purification systems (Cui and Wan 2005). For instance, Advanced Hydro® hydrophilic polymer coatings on water purification and filtration membranes provide a 30–50% reduction in fouling (www.advancedhydro.net). Clean Membranes® utilize a "smart-comb" copolymer

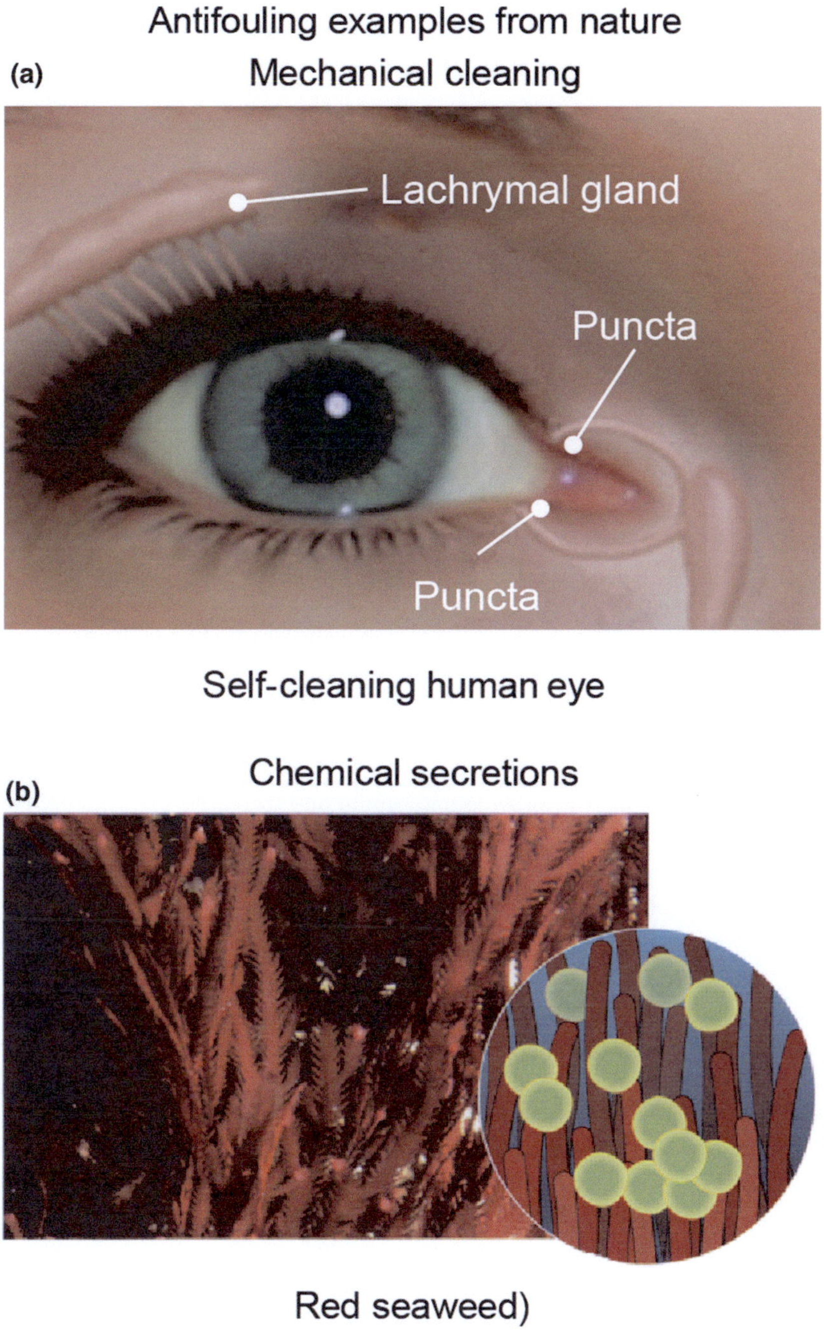

Fig. 13.9 Antifouling examples from nature using mechanical cleaning and chemical secretions. **a** Human eye combines wiping with chemical secretions to clean the cornea. **b** Red seaweed using a halogenated furanone to manipulate colonizing bacteria 'attraction' messages (Bixler and Bhushan 2012)

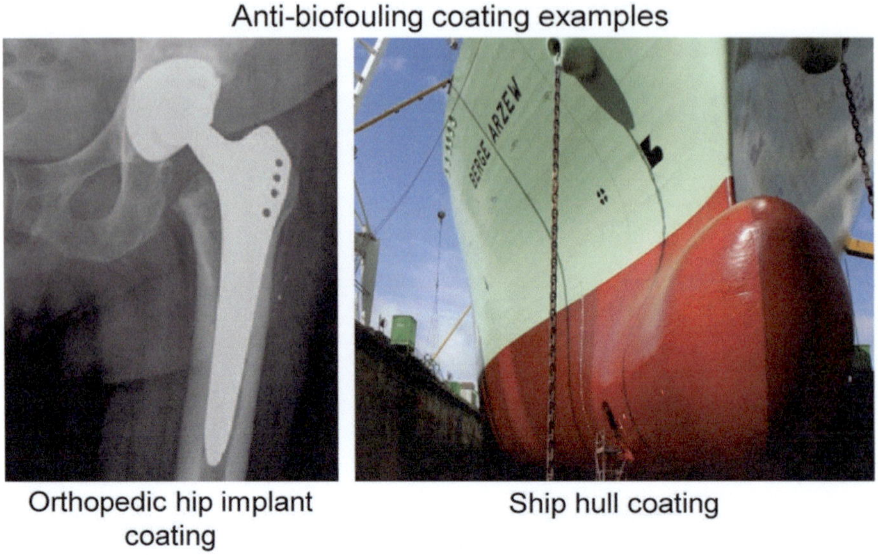

Fig. 13.10 Medical, marine, and industrial antifouling coating examples (Bixler and Bhushan 2012). Orthopedic hip implants are covered with biocompatible coatings designed to reduce biofouling, wear rate, and promote bone growth (left) (photo courtesy of Ir Jef Vleugels). Marine foul-release organic tin-free hull coating called "Green Ocean Coating Heavy Duty" minimizes manual cleaning (right) (photo courtesy of Bayer MaterialScience)

in ultrafiltration membranes to provide antifouling and increased permeability (www.cleanmembranes.com).

Improving biocompatibility can also reduce medical biofouling, since microorganisms will be less likely to target any implanted device.

13.5.1.2 Marine Antifouling

Tributyltin coatings have been efficacious in marine antifouling application, and were commonplace. Unfortunately, toxicity and other properties make tributyltin coatings unsuitable for in vivo use. Because of tributyltin toxicity to marine organisms, antifouling preparations containing it were banned in 2008 by the International Maritime Organization; adding urgency to the development of more environmentally friendly antifouling and foul-release coatings. As a result, many researchers have been exploring environmentally friendly methods such as new nontoxic antifouling paints and foul-release coatings (Callow 1999; Fingerman et al. 1999; Railkin 2004; Hellio and Yebra 2009). Marine foul-release coatings are nontoxic hydrophobic surfaces that promote self-cleaning. Nano/microtextured surfaces control surface energy, charge, conductivity, porosity, roughness, wettability, and friction (Clare and Aldred 2009). Hydrophobic materials such as fluoropolymers and silicone elastomers are nontoxic, adhesion resistant, environmentally friendly, and easy to clean (Clarkson 1999). Siloxane elastomers provide a combination of lower elastic modulus and lower surface energy, creating a hydrophobic, low-stick surface. Silicone foul-release coatings are used for some quick moving boats, but wear away over time (Vladkova 2007, 2008). Figure 13.10 (right) presented by Bixler and Bhushan (2012) shows a foul-release organic tin-free hull coating called "Green Ocean Coating Heavy Duty" by Advanced Marine Coatings, whose smoothness and hardness reduces cleaning cycles (www.bayermaterialscience.com).

13.5.1.3 Industrial Antifouling

Industrial antifouling is accomplished similarly to marine antifouling, with coatings that prevent and remove fouling (Bixler and Bhushan 2012). For instance, industrial heat exchanger applications utilize SAMs to reduce bioadhesive strength between foulers and a heated surface, using low surface energy hydrophobic properties. Materials such as silicon can withstand temperatures up to 200 °C whilst materials such as hexadecyl disulfide can withstand up to 225 °C. SAMs furthermore offer corrosion protection by limiting the oxygen and water diffusion at the substrate surface. This technique could be widely used for heat exchangers but needs long term durability testing (Venkatesan and Murthy 2008).

13.5.2 Self-cleaning Surfaces and Cleaning Techniques

Even though fouling prevention is the goal, in many cases fouling is not completely preventable and a surface must be cleaned. In response, several self-cleaning coatings have been developed for various applications. Bixler and Bhushan (2012) report example self-cleaning products including hydrophobic paints, roof tiles, and fabrics as well as self-cleaning hydrophilic windows and membranes. Examples are Lotusan® paints by Sto (www.stocorp.com), Tegotop® paints by Degussa (www.goldschmidt.com), Erlus Lotus® roof tiles (www.erlus.com), NanoSphere® fabrics by Schoeller Technologies (www.schoeller-textiles.com), and SunClean® glass by PPG Industries (www.ppg.com). Self-cleaning is demonstrated with the hydrophobic Lotusan® roll on and Tegotop® spray paints that create a microstructure surface (www.stocorp.com; www.goldschmidt.com), hydrophobic Erlus Lotus® roof tiles with an etched surface finish (www.erlus.com), and hydrophobic NanoSphere® fabrics with integrated nano particles (www.schoeller-textiles.com). Self-cleaning hydrophilic SunClean® glass with titanium dioxide uses the sheeting action of water to efficiently gather and remove contaminants (www.ppg.com). Figure 13.11 shows the improved cleanliness with Lotusan® hydrophobic self-cleaning paints on concrete and stucco (www.stocorp.com), and better clarity with the SunClean™ hydrophilic self-cleaning glass (www.ppg.com) (Bixler and Bhushan 2012).

Physical cleaning and chemical disinfecting are often considered effective methods to remove fouling (Cologer 1984; Walker et al. 2000; Sharma et al. 2004; Hellio and

Fig. 13.11 Effectiveness of anti-inorganic fouling coatings (Bixler and Bhushan 2012). Demonstrated is the improved cleanliness with Lotusan® hydrophobic self-cleaning paint (top) (www.stocorp.com), and improved clarity with the SunClean™ hydrophilic self-cleaning glass (bottom) (www.ppg.com)

Yebra 2009; Chan and Wong 2010; Bixler and Bhushan 2012). Most employ mechanical methods such as manual scrubbing to disrupt and remove the fouling. Figure 13.12 illustrates common methods of cleaning and disinfecting such as medical sterilization, chemical disinfection, high-pressure water spray, automatic brush scrubbing, and membrane cleaning techniques (Bixler and Bhushan 2012). Included are a steam sterilization medical autoclave, Listerine® antiseptic mouth rinse, water spray on a low-adhesion ship hull, automatic brush washing system, and microfiltration membrane using sonication, chemicals, and water backwash (Lim and Bai 2003; Bixler and Bhushan 2012). Disinfecting is commonly employed in dental applications with antiseptic mouth rinses such as Listerine®. For example, the active ingredient chlorhexidine attacks and reduces dental biofilm to control plaque and gingivitis in the oral cavity (Charles et al. 2004). The biofilm is reduced when the individual bacteria walls rupture (www.jjdentalprofessional.com). Mouth rinses combined with brushing and flossing increase the effectiveness, since such "mechanical methods" disrupt the biofilm (Sharma et al. 2004).

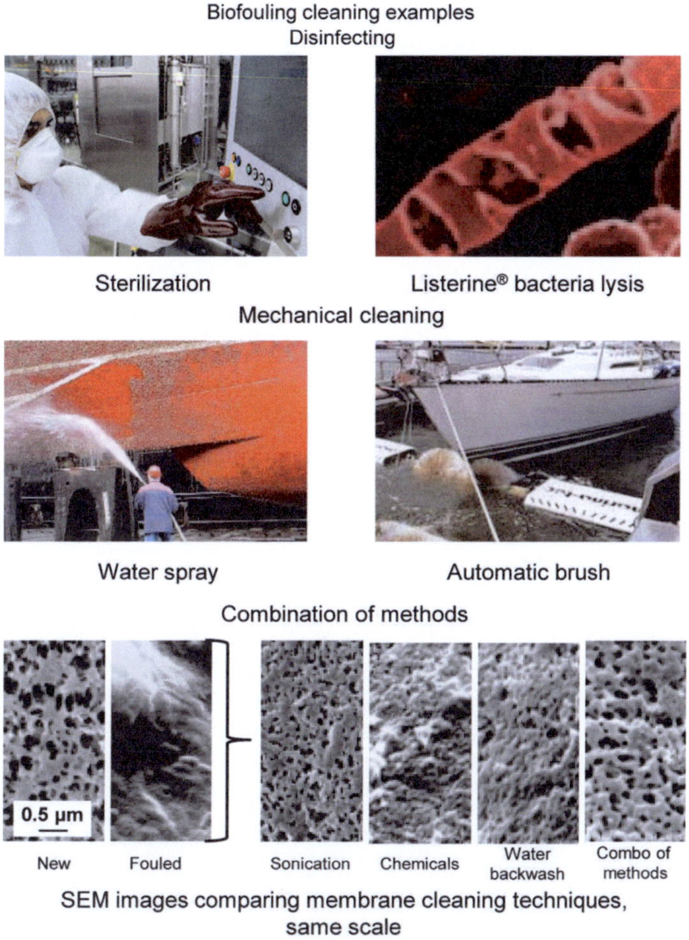

Fig. 13.12 Medical, marine, and industrial biofouling cleaning examples (Bixler and Bhushan 2012). Steam sterilization medical autoclave (top left) (photo courtesy of Life Science Outsourcing), Listerine® effect on bacteria (top right) (adapted from Christoph Schaudinn), water spray on a low-stick ship hull (middle left), automatic brush washing system (middle right) (Willemsen 2005), and microfiltration membrane using sonication, chemicals, and water backwash (bottom) (adapted from Lim and Bai 2003)

13.6 Nanomaterials for Anti-biofouling

Selected nanomaterials have been used for their inherent antimicrobial properties. They prevent bacterial infection by preventing bacterial function. These properties are enhanced or may only exist, when linear dimensions are on nanoscale to increase their surface area. In many cases, antibacterial property increases as particle size decreases. Because of large surface area, nanomaterials can bind easily to bacterial cell walls and antibacterial activity of the material in increased. Antibacterial activity could also be increased by generating reactive oxygen species which kill bacteria (Taylor and Webster 2011).

Various nanomaterials include nanoparticles and nanotubes. Nanoparticles may be made of metals such as silver and gold and metal oxides such as magnesium oxide, titanium dioxide, zinc oxide, and iron oxide. Carbon nanotubes and others have been explored. Colloidal silver has been used for antibacterial purposes for some 50 years. With recent advances in nanoparticles technology, silver nanoparticles are increasingly being explored (Baker et al. 2005; Perera et al. 2013). Antibacterial properties of MgO, ZnO and FeO are found to be size dependent with best bacterial resistance with smallest particles used on the order of 10 nm (e.g. Raghupathi et al. 2011). The use of metal oxide nanoparticles provides high antibacterial activity because of production of reactive oxygen species (Sawai et al. 1996) due to metal ion release (related to surface area) or through their interaction with ultraviolet (UV) light (dependent upon particle size). For example, iron ions generate oxygen free radicals by converting hydrogen peroxide to more reactive hydroxyl radical via the Fenton reaction (Touati 2000). Hydroxyl radicals generated can degrade bacteria.

Bacterial genetic tolerance to antibiotics has been on the rise. Implanted medical devices are susceptible to infections resulting in implant failure. When a device must be removed to eradicate an infection, it is often because of bacteria producing a sticky biofilm matrix on the device surface. Nanoparticles are increasingly used against biofilm-mediated, drug resistant, and device related infections. ZnO nanoparticles have been successfully used for the design of the next generation of therapeutics and biomaterials (Taylor and Webster 2011).

Some self-cleaning coatings are based on UV light-activated photocatalytic TiO_2 thin-films which also exhibit the antimicrobial properties (Sunada et al. 1998). Caballero et al. (2010) reported that by including TiO_2 nanoparticles in paint formulation, bacteria were inactivated in the presence of fluorescent lighting. Antimicrobial paints used indoors (such as in a hospital) have also been formulated through the application of silver nanoparticles. One such technology uses the natural oxidative drying process of vegetable oil to reduce silver into silver nanoparticles creating a nontoxic paint that can be used to coat wood, glass, steel, and various polymers, with strong activity against both gram-positive (*S. aureus*) and gram-negative (*E. coli*) bacteria while releasing silver slowly to increase the longevity of the antimicrobial activity (Puckett et al. 2010). Various processes for the creation of anti-microbial surfaces using nanoparticles are schematically shown in Fig. 13.13 (Taylor and Webster 2011).

Superhydrophilic and antimicrobial nanoparticles in a polyamide thin film coating provide antifouling for reverse osmosis (RO) membranes. Studies show success using TiO_2 nanoparticles as a thin film on reverse osmosis membranes to control biofouling due to the photocatalytic effects that reduce bacteria (Kim et al. 2003). Similarly, ultrafiltration membranes impregnated with antimicrobial and hydrophilic silver nanoparticles (nAg)

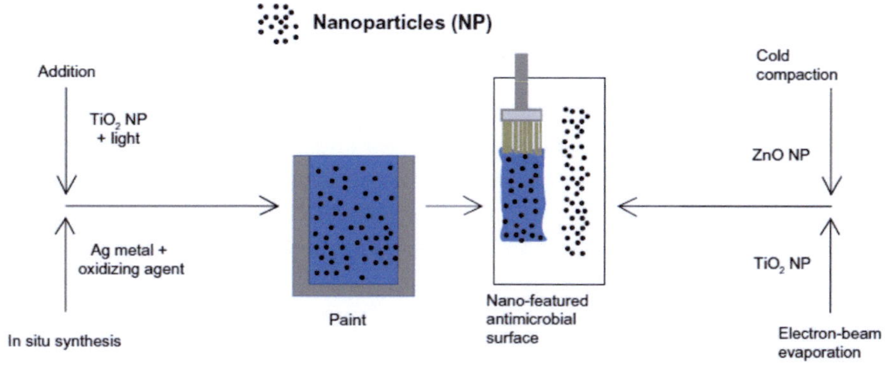

Fig. 13.13 Schematics of processes used for the creation of antimicrobial surfaces: addition, in situ synthesis, cold compaction, or electron-beam evaporation incorporating nanoparticles. During addition, nanoparticles are simply added to the paint and, in one case of TiO_2 nanoparticles, are activated using a light source such as fluorescent light. In situ synthesis uses the properties of paint to reduce silver (Ag) metal into nanoparticles, for example using metal-catalyzed free-radical-mediated oxidation process during the production of silver nanoparticles in vegetable oil-based paints. During cold compaction, nanoparticles are pressurized into a nano-featured surface. Electron-beam evaporation concentrates a large amount of heat produced by high-energy electron-beam bombardment on the source material to be deposited, in this case pure Ti pellets, whereby heating and vaporizing occur (Taylor and Webster 2011)

provide antifouling benefit. Figure 13.14 compares such treated and untreated polysulfone ultrafiltration membranes, showing fewer *E. coli* (*Escherichia coli*) attached to the nAg membrane (Zodrow et al. 2009).

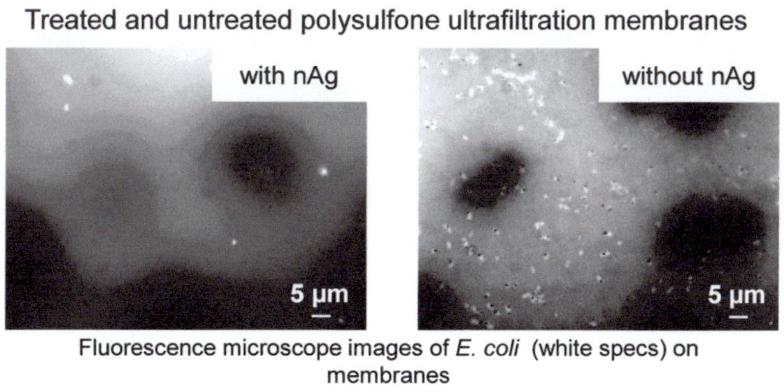

Fig. 13.14 Polysulfone ultrafiltration membrane treated with silver nanoparticles (nAg) shows antifouling benefit with fewer *E. coli* attached (Zodrow et al. 2009)

The antimicrobial finishes on fabrics are of interest to protect the consumer against disease or odor (Perera et al. 2013). It is of interest in healthcare industry, schools, and hotels where protection from disease is a growing concern (Hipler and Elsner 2006; Mather et al. 2011; Windler et al. 2013). It is of particular interest in military applications because of potential exposure to deadly chemicals and various microorganisms. Fabrics treated with nanostructured materials with inorganic active agents such as silver, gold, titanium dioxide and zinc oxide, are used for antimicrobial activity (Dastjerdi and Montazar 2010). Silver nanoparticles are preferred over the other organic and inorganic nanostructures as an antimicrobial agent. Silver in different forms, such as silver nitrate, silver sulphadiazine, and metallic silver, has been used in human health care for a long time. Silver nanoparticles have a proven healing activity against various microorganisms containing bacteria, viruses and fungi (Hipler and Elsner 2006). Silver nanoparticles are highly active in extremely small quantities due to the greater ion release from larger surface area per unit of mass compared to bulk silver. The silver ions released from the silver nanoparticles may interact with microbial enzymes affecting DNA replication and cell division leading to microbial cell death. The antimicrobial mechanism of silver nanoparticles is believed to be related to the generation of free radicals and reactive oxygen species which may cause cell membrane damage of microorganisms. The membrane damage caused by the silver nanoparticles may reduce the ability of microorganisms to extrude silver ions out of cells to limit their effect (Perera et al. 2013).

Chemical treatment of fabric using silver nanoparticles can be employed ex situ or in situ. Ex situ approach, which includes blending of prepared silver nanoparticles with fabric, is employed commercially in the production of odorless socks, surgical gowns, dressing bandages, bed sheets, pillow cases, carpets and pet clothing. However, the ex situ treatment lack durability of textiles. Durability directly relates to toxicity concerns that occurs from washing, rubbing and abrasion which shortens shelf life of nanosilver treated products. The in situ approach involves the application of nanoparticles by the in situ reduction of metal salts. This involves the use of a metal ion precursor, a reducing agent and a stabilizing agent to prevent aggregation of nanoparticles on the fabric. The porous cellulose fibers have been successfully treated with terminal aldehyde or carboxylic groups as a reducing agent and a stabilizer. Perera et al. (2013) reported that the in situ approach could retain the properties close to the original fabric after subjected to various screening tests. Detail follow.

13.6.1 Surface Treatment of Cotton Fabrics

Perera et al. (2013) applied silver nanoparticles on woven cotton fabrics using in situ method. 1.5×10^{-3} mol/l $AgNO_3$ solution was used as the precursor to produce silver nanoparticles. Cotton fabric samples were first immersed in $AgNO_3$ solutions, padded and dried in the oven at 50 °C for 10 min. The silver ions in the $AgNO_3$ solution were diffused and absorbed into the cotton fibers with the help of the polar hydroxyl groups within the fiber structure. Later, the treated samples were irradiated using a UV lamp (Philips, G36T8) at a distance of 100 mm for 1 h to synthesize silver nanoparticles in situ on the cotton surface. For comparison purpose, untreated fabrics were also padded, dried and cured using DI water (blank process).

13.6.2 Morphology and Contact Angle

The SEM images for untreated cotton fibers with a typically smooth, grooved surface are shown in Fig. 13.15. In these images, swelling of the fiber is visible, which was associated with wetting of the fibers after applying the blank process treatment to untreated cotton fabrics with DI water. Images representing the $AgNO_3$ + UV method show a uniform distribution of silver nanoparticles on the cotton fibers. This was likely due to Ag^+ easily penetrating the yarns and fiber, or by trapping in the fibril network of the cotton fiber structures with the help of the ion-dipole attraction between positive silver ions and negative hydroxyl groups in the cellulose structure (Perera et al. 2013).

The wettability of all fabric samples with water was assessed by making water contact angle measurements. The droplet images are presented in Fig. 13.16 (Perera et al. 2013). Uncoated fabric is hydrophobic soon after the application of the water droplets. However, the contact angle decreases to below 90° after 5 min of application of water droplet exhibiting the hydrophilic nature of the cotton. The silver treated samples were superhydrophobic. The contact angle decreased somewhat after 5 min of application of water droplet. The data are summarized in Table 13.2.

The presence of cotton fibers protruding from the cotton yarns create air pockets under the water droplet responsible for initial hydrophobicity. Higher value of contact angle of silver treated samples possibly occurred due to an increased number of air pockets due to the presence of nanoparticles. After application of water droplet for 5 min, the contact angle decreased due to the hydrophilic OH groups present in the cotton fibers which start to absorb water decreasing the contact angle.

13.6.3 Durability of the Treatment After Wash

To understand the stability of the silver treatment on fabric samples, washing experiments were done at 50 °C in the presence of non-ionic detergent Ultravon CN Ciba for 45 min according to AATCC 61 (2A) method, which usually is used to determine the color fastness to home and commercial laundering (Perera et al. 2013). This procedure is similar to five home washing cycles in ambient conditions at 38 ± 3 °C. The concentration of silver in the effluent after 5 and 10 washing cycles was quantified using atomic absorption

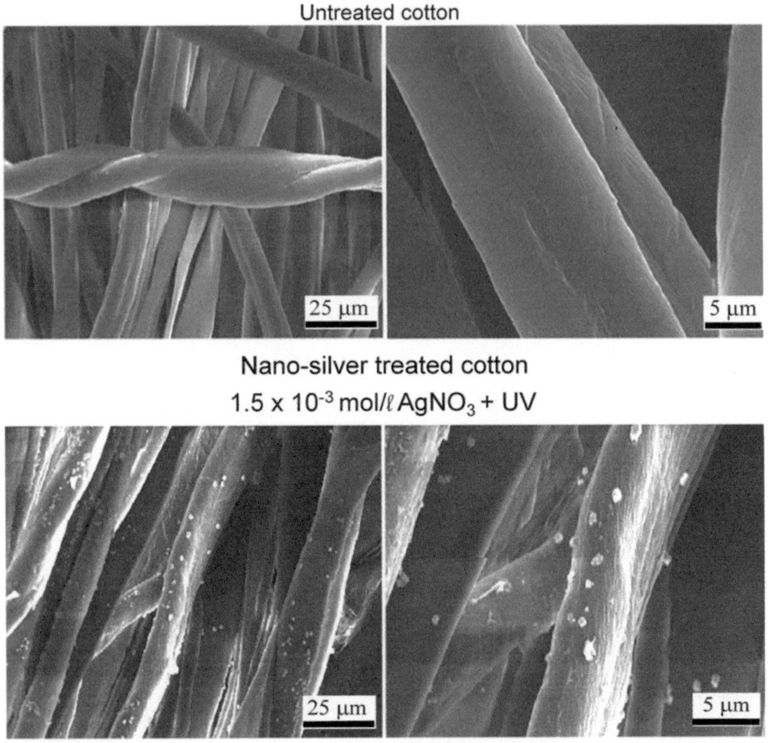

Fig. 13.15 Scanning electron microscope image of untreated cotton fibers, and silver nanoparticles on fibers of cotton fabrics treated with 1.5×10^{-3} mol/l $AgNO_3$ + UV (adapted from Perera et al. 2013)

spectroscopy. The stability of the silver treatments was compared with silver concentration in the freshly treated fabrics.

The atomic absorption results of the silver nanoparticle treated fresh samples and effluent after 5 and 10 washes are shown in Fig. 13.17. The data shows that Ag concentration decreases with wash cycles. It is believed that the repeated washing reduces the durability of the antimicrobial property.

13.6.4 Antimicrobial Properties

Qualitative antimicrobial efficiency of silver nanoparticle treated fresh cotton fabric samples and washed samples was evaluated using the disk diffusion method against strains of *S. aureus* (ATCC 6538) and *E. coli* (ATCC 10536) (Perera et al. 2013). The plates were individually inoculated with each of the bacterial isolates. Silver nanoparticles treated

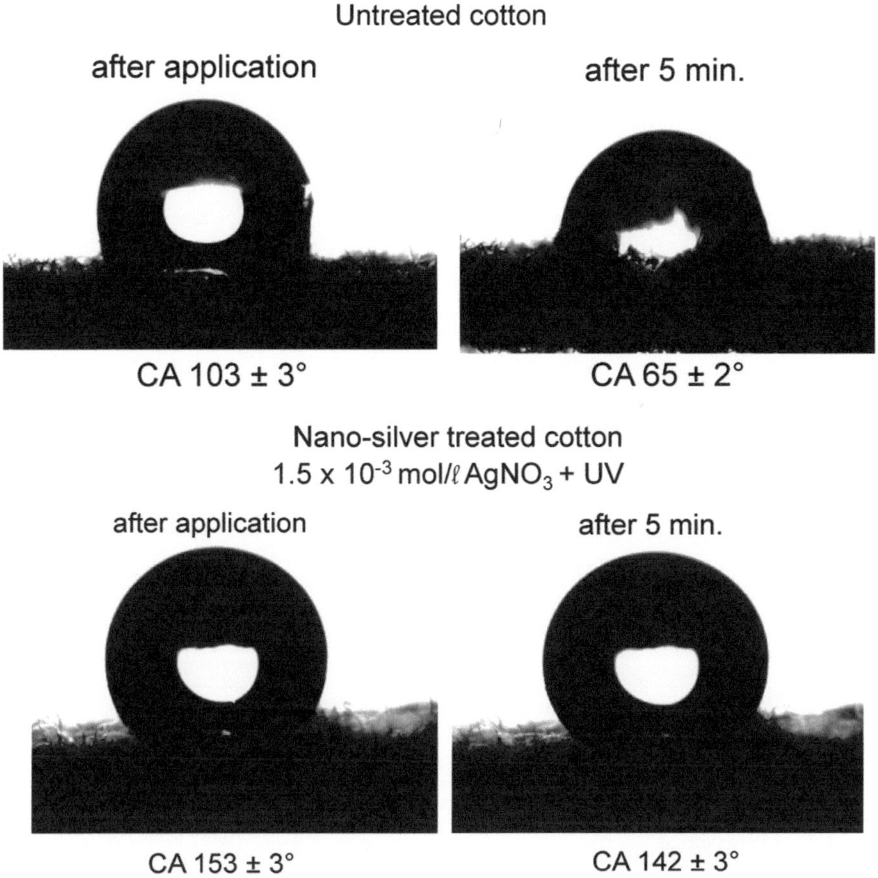

Fig. 13.16 Images of water droplets on untreated cotton fabric, and silver nanotreated cotton fabrics with 1.5×10^{-3} mol/l $AgNO_3$ + UV (adapted from Perera et al. 2013)

sterile fabric samples were dispensed on *S. aureus* and *E. coli* inoculated plates. All plates were then incubated at 37 °C for 24 h. Antibacterial activity of the silver treated samples was assessed by examining the clear inhibition zone around the samples according to ISO 20645 protocol. The experiments were carried out in triplicate under sterile conditions. Antimicrobial activity was evaluated using a method described by Pollini et al. (2009). A rating from "Good" to "Poor" was given according to the width of the inhibition zone around the samples in the agar medium. Studies were done on fresh, 5 and 10 times wash samples.

The data are summarized in Table 13.2. All untreated cotton fabric samples grew bacteria, demonstrating a typical bacterial attack on the cotton in a nutrient medium. Treated

Table 13.2 Measured contact angles and antimicrobial properties of nanoparticle silver treated cotton fabrics (adapted from Perera et al. 2013)

		Untreated cotton	Nanoparticle-silver treated cotton
			1.5×10^{-3} mol/l $AgNO_3$ + UV
Contact angle (°) (5 μl drop)	After application	103 ± 3	153 ± 3
	After 5 min	65 ± 2	142 ± 3
Antimicrobial properties			
Against *S. aureus* bacteria			
	Before wash	Poor	Good
	After 5 wash	Poor	Good
	After 10 wash	Poor	Good
Against *E. coli* bacteria			
	Before wash	Poor	Fairly good
	After 5 wash	Poor	Sufficient
	After 10 wash	Poor	Sufficient

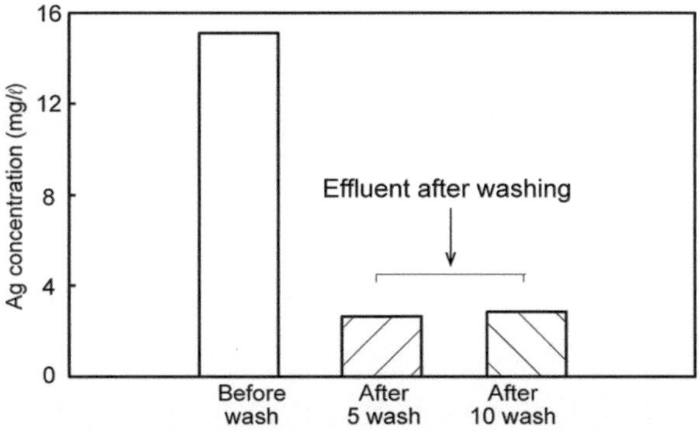

Fig. 13.17 Silver concentration on treated cotton fabrics after 5 and 10 washing cycles. The data are also presented for before wash (adapted from Perera et al. 2013)

samples demonstrated "Good" susceptibility against *S. aureus* and fairly good against *E. coli*. The protection was maintained after wash.

13.7 Nanostructured Surfaces for Antifouling

Nanostructured surfaces reduce adhesion and coverage of infection and further proliferation (Schumacher et al. 2007a, b; Scardino et al. 2008; Bixler et al. 2014). Microorganisms slightly larger than the gap between features are less likely to colonize the surface.

For microorganisms smaller than the gap between features, the sharp edges of the asperites present on structured surfaces discourage the setting and eventual colonization of microorganisms on the surface.

In this section, we present data on an anti-biofouling study carried out by Bixler et al. (2014) on rice leaf inspired surfaces. The rice leaf is particularly intriguing due to its relatively simple micropapillae morphology. Bioinspired rice leaf surfaces using micro-sized pillar patterns can be produced with silicon master patterns fabricated by photolithography (Bixler and Bhushan 2014). Sample geometrical dimensions are based on actual rice leaf morphology. Actual rice leaf surfaces are covered by a sinusoidal pattern of micropapillae with height 2–4 μm, diameter 2–4 μm, pitch 5–10 μm, and peak radius 0.5–1 μm (Bixler and Bhushan 2013b, 2015).

13.7.1 Fabrication of Micropatterned Samples

The bioinspired rice leaf surface shown in Fig. 13.18 was created to replicate the rice and butterfly wing effect, and was designed to include anisotropic flow and superhydrophobic characteristics. Samples were created using silicon master patterns and the soft-lithography technique (Bixler and Bhushan 2014). Silicon master patterns with pits (negative vs. positive features) were fabricated in order to emboss polymer films and to accurately produce master patterns. Polydimethylsiloxane (PDMS) replicas were produced from silicon master patterns using a three-step soft lithography procedure (Bixler and Bhushan 2014). PDMS (184 Silicone Elastomer by Dow Corning) was chosen due to its low surface energy property which leads to high contact angle. A nanostructured superhydrophobic coating with 50 nm hydrophobized silica nanoparticles and methylphenyl silicone resin was applied to make them superhydrophobic.

Micropatterned plastic sheets were also produced using a commercial hot embossing procedure in a clean room environment (Bixler and Bhushan 2014). The hot embosser produced single-sided embossed plastic sheets from a silicon master pattern using a combination of heat and pressure. Poly(methyl methacrylate) (PMMA) plastic sheets were selected, which were optically clear and 175 μm in thickness. PMMA molds relatively easily due to its softening temperature.

13.7.2 Anti-biofouling Measurements

For studies to evaluate anti-biofouling benefits, Bixler et al. (2014) considered several foulers primarily based on their size and availability. These include zoospores (*Enteromorpha*) (Callow et al. 2002) and (*Ulva linza*) (Finlay et al. 2002), proteins (Banerjee et al. 2011), *E. coli* (Ma et al. 2000), and MRSA (*Methicillin-resistant Staphylococcus*

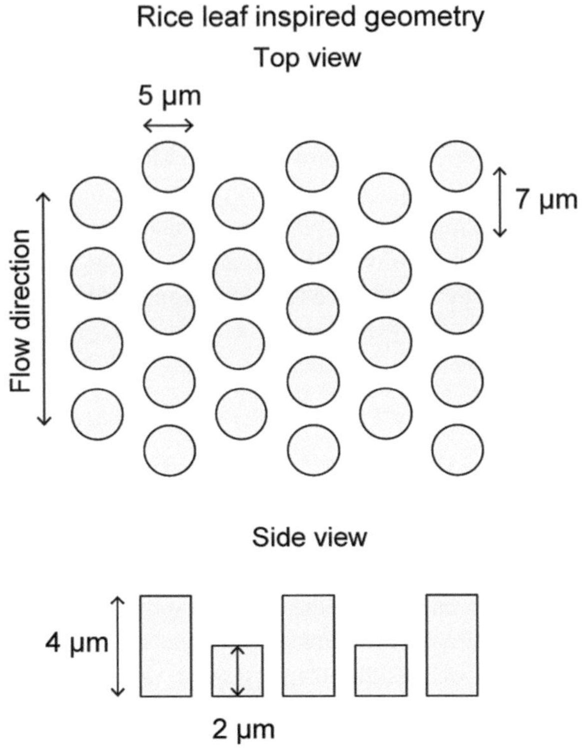

Fig. 13.18 Rice leaf inspired geometry for antifouling. Shown are micropatterned surfaces with dual height pillars. Pillars are arranged in a hexagonal array in order to maintain consistent gaps to promote antifouling (adapted from Bixler and Bhushan 2014)

aureus) (Fletcher et al. 2007). They selected the microorganism *E. coli* due to its prevalence in the medical industry, availability, and access to facilities for safe handling. The multi-step anti-biofouling measurement process is shown in Fig. 13.19 with each process from the *E. coli* broth creation to the final counting (Bixler et al. 2014). The dimensions of single wild type *E. coli* cells were approximately 0.5–0.75 μm wide and 2 μm long. When actively growing and dividing, the cells stretch to ~4 μm before dividing into two cells.

The bioassay procedure began with stock solution of *E. coli* (HB2151 *E. coli* clones from The Medical Research Council, Cambridge, England) streaked onto TYE (Tryptone and Yeast Extract) dehydrated agar plates for bacterial colony growth. The plates were then placed in an incubator (Labnet, model 311 DS) at 37 °C overnight, and then stored at 4 °C until use. To prepare *E. coli* broth, a single colony was picked from streak plates and inoculated into 50 ml sterile LB (Luria Broth, Lennox from Fisher Scientific, Pittsburgh, PA) in a beveled flask. *E. coli* was grown for 16 h with shaking at 250 rpm at a temperature of 37 °C. After incubation, *E. coli* solution was diluted to 1/100 and 1/10 into sterile LB broth and placed on ice until use.

Bixler et al. (2014) placed the flat control sample (PDMS), a replica sample (PDMS) and an embossed sample (PMMA) in the bottom of a sterile, polystyrene culture dish and

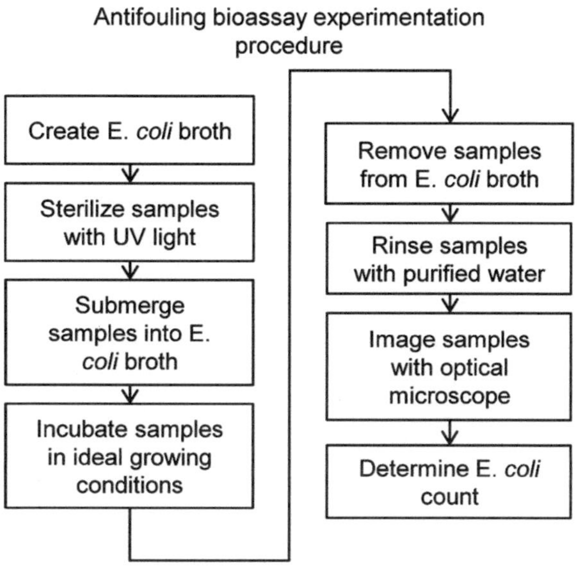

Fig. 13.19 Antifouling bioassay experimental procedure with *E. coli* cells in the laboratory. As shown, samples are sterilized, submerged into broth, placed in incubator, rinsed to remove loose cells, and imaged with oil immersion light microscope to determine cell counts (Bixler et al. 2014)

sterilized for 10 min via ultraviolet exposure (30W T8 Hot Cathode Germicidal Sterilamp fluorescent light bulb, Philips TUV30W G30T8 360164). *E. coli* broth was added to the culture dish until all samples were covered and then incubated at 37 °C for either 2 or 4 h. The aforementioned concentrations and incubation times were determined from trial and error with samples to achieve an appropriate level of *E. coli* growth. The goal was to visually determine a difference at 1000× magnification between flat control and other samples under investigation.

After incubation, samples were removed from *E. coli* broth and rinsed at a 45° angle. Rinsing removed any unattached *E. coli* from the surface prior to imaging. An auto-pipette (Pipet-Aid from Drummond Scientific, Broomall, PA) was clamped and held on a stand over the rinsing station with a liquid fall height (distance between pipette tip end and sample) of 7 cm. Samples were rinsed with 10 ml of sterile PBS at a flow rate of approximately 2 ml/s, and then immediately placed onto a glass microscope slide with a glass cover slip on top. Using a light microscope and CCD camera (Nikon, Diagnostics Instruments, Inc.), a 140 μm by 140 μm area of each sample was imaged and flat samples analyzed with image processing software (SPIP 5.1.11, Image Metrology A/S, Horshølm, Denmark). Images were collected with a 100× oil immersion objective lens for a total magnification of 1000×. For flat control samples, image processing software (SPIP 5.1.11, Image Metrology A/S, Horshølm, Denmark) was utilized to determine total coverage area of *E. coli*. For the samples, the *E. coli* was manually counted five times in circular areas 35 μm in diameter using two samples within the same bioassay.

13.7.3 Anti-inorganic Fouling Measurements

To understand the anti-inorganic fouling effectiveness of the samples, self-cleaning experiments were conducted (Bixler and Bhushan 2014). These experiments included contaminating samples, employing a wash technique, and determining the percentage of inorganic particles removed. Depositing contaminate particles on the tilted (45°) replicas involved a glass contamination chamber (0.3 m in diameter and 0.6 m high). A tray containing 0.2 g of hydrophilic silicon carbide (SiC) contaminants (400 mesh particle size by Aldrich, with sizes ranging from 10 to 15 μm) was placed in the top chamber with an air hose directed in the center. These particles were chosen because of their similar properties to natural dirt (shape, size, and hydrophilicity). Contaminants were blown with laboratory air for 10 s at 300 kPa, and then allowed to settle for 30 s before the separator panel was removed.

After 30 min, the samples were removed and subjected to pre-wash experiment particle analysis. Using a light microscope and CCD camera, a 280 μm by 210 μm area of each sample was imaged and analyzed with SPIP image processing software. This process was performed before and after each wash experiment. Wash experiments consisted of exposing the tilted (45°) replicas to water droplets falling from a specified heights and drip rates (total duration of 2 min using 10 mL water). Droplet velocity was approximately 5.6 m/s at a height of 0.4 m. This translates into pressures of 4000 Pa.

13.7.4 Results and Discussion

13.7.4.1 Anti-biofouling

To understand the antifouling effectiveness of the various samples, Bixler et al. (2014) conducted a series of bioassay experiments. Figure 13.20 shows light microscope images of the *E. coli* colonized on a flat sample at two magnifications (Bixler et al. 2014). As shown, the *E. coli* are long cylindrically shaped microorganisms and tend to cluster in groups and align themselves from head to tail.

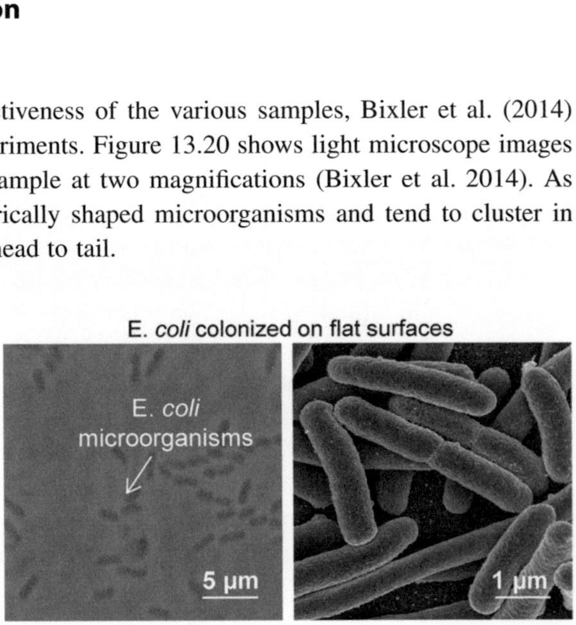

Fig. 13.20 Images of *E. coli* on flat surfaces after the bioassay procedure (Bixler et al. 2014) and at a higher magnification (adapted from www.universityofcalifornia.edu)

Figure 13.21a shows light microscope images of flat PDMS samples at the two concentrations and two incubation times (with *E. coli* highlighted by the SPIP software) (Bixler et al. 2014). Figure 13.21b shows the tabulated results of percentage coverage area of the *E. coli* on the surfaces for 1/100 and 1/10 concentrations at 2 and 4 h incubation time. The greatest coverage occurs at the higher incubation time as *E. coli* growth occurs exponentially over time. Furthermore, as expected, the lower concentration does not provide as much coverage area, however the differences are within 10% for both incubation times.

Figure 13.22 shows light microscopic images of the flat PDMS control sample, replica sample, and the embossed sample. Antifouling is determined by counting the total number of *E. coli* on the flat control samples and comparing that to the other samples, with results reported in percentage *E. coli* reduced coverage. Figure 13.23 shows the percentage reduced coverage compared to the flat control samples. As anticipated due to the discontinuities of the various surfaces, patterned samples provided antifouling benefits. In general, the antifouling benefit is greater with the 2 h incubation duration as compared to the 4 h duration. It is believed that the longer duration leads to more significant coverage, and the features provide less overall benefit. It should be noted that the replica sample and embossed sample performed similarly, indicating minimal differences between samples fabricated from PDMS or PMMA.

Bixler et al. (2014) presented conceptual models to describe anti-biofouling mechanisms. Figure 13.24 shows schematic illustrations of colonization of microorganisms exposed to flat and patterned samples. Microorganisms are unable to effectively colonize patterned morphology as well as they can on the flat control sample. If the length of the microorganism is larger than the pillar spacing, they cannot invade the spacing and instead land on top of the pillars. If the length of the microorganism is smaller than the spacing, as is the case with *E. coli*, the sharp edges of the pillars are believed to discourage the settling and eventual colonization of microorganisms on the surface. The edges provide an inconvenient landing zone for settlement, and thus the microorganism must continue the search for an ideal location to colonize. It is believed during this process that many *E. coli* that do not attach remain metabolically active throughout the experiment or others may expire. In many cases, the *E. coli* may be able to settle in clusters and around the bases of the pillars, which provide multiple attachment points of contact.

When the microorganism concentrations are low and the incubation times are short, only a few microorganisms are able to colonize the sample surface. However, when the concentrations are high and the incubation times are long, the microorganisms eventually will cover over the pillars and create a biofilm. Nevertheless, the discontinuities found with the sharp edged micro sized pillars are demonstrated to provide antifouling benefit and extend the time at which a particular surface becomes colonized.

To summarize, there are a number of examples of micropatterned surfaces in the literature that show that, if the micro-sized feature spacing is slightly smaller than the microorganisms, they exhibit anti-biofouling. The data presented in this section shows that even if the spacing is larger, they exhibit anti-biofouling.

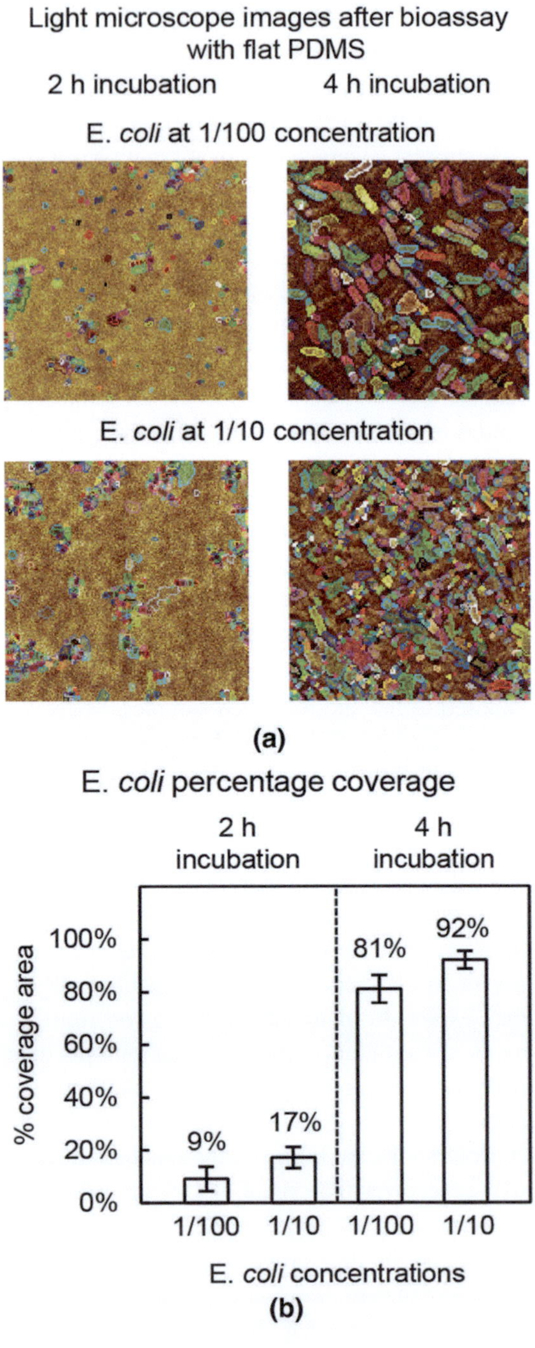

Fig. 13.21 **a** Light microscope images of flat PDMS control samples and **b** percent coverage area after *E. coli* bioassay with two concentrations and two incubation times. *E. coli* covers the majority of the surface after 4 h in ideal growing conditions. This baseline serves to compare the replica sample and embossed sample when calculating percentage reduced coverage (adapted from Bixler et al. 2014)

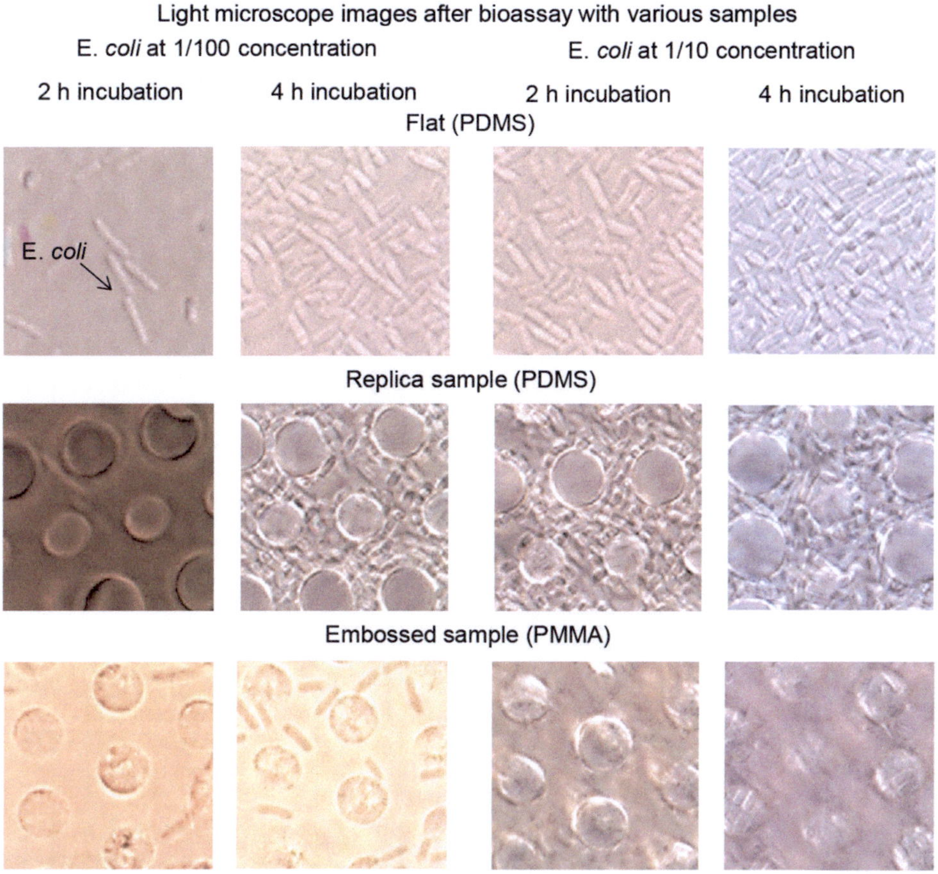

Fig. 13.22 Light microscope images using oil immersion techniques at 1000× magnification of samples after bioassay experiments using *E. coli*. Images of flat PDMS, the replica sample, and the embossed sample are shown with two *E. coli* concentrations and two incubation times. Long and narrow cylindrically shaped objects are the *E. coli*. Antifouling analysis was conducted using the images and particle analysis software to quantify effectiveness (adapted from Bixler et al. 2014)

13.7.4.2 Anti-inorganic Fouling

To understand the anti-inorganic fouling effectiveness of the various samples, Bixler and Bhushan (2014) conducted a series of self-cleaning wash experiments. Figure 13.25 presents the before and after images for flat PDMS and PMMA control samples, the replica sample, and the embossed sample (Bixler and Bhushan 2014). These images were analyzed using SPIP imaging software to quantify the percentage of particles removed. Figure 13.26 shows self-cleaning data for wash experiments. Self-cleaning is more evident with a patterned sample and an embossed sample; as compared to flat samples (Bixler and Bhushan 2014). To understand wettability of samples, Bixler and Bhushan (2014)

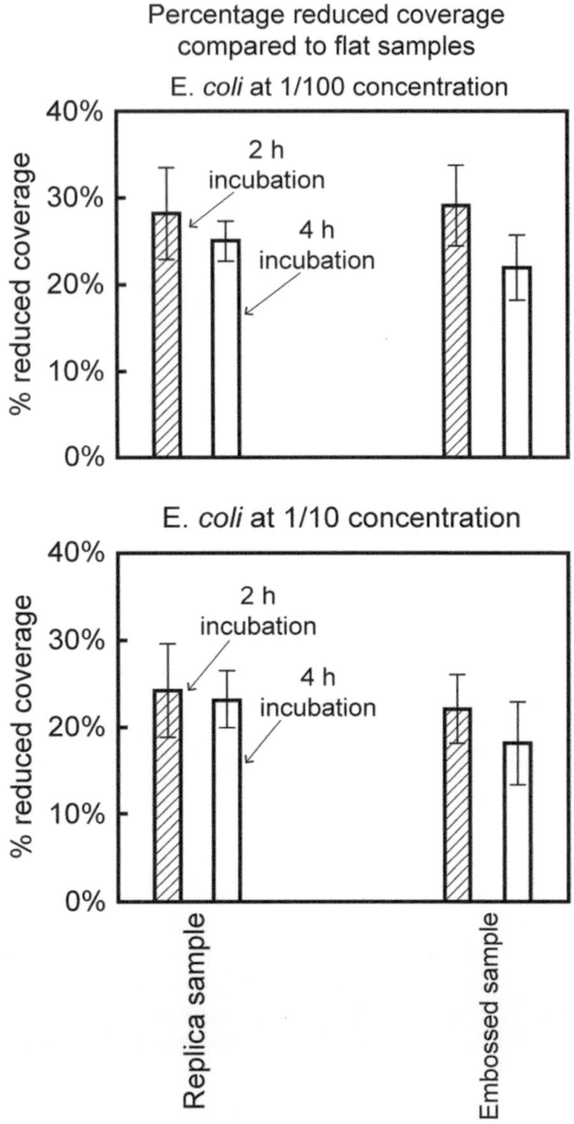

Fig. 13.23 Anti-biofouling results from *E. coli* bioassay experiments. Results are presented for the replica sample, and the embossed sample. The percentage reduced coverage is calculated from flat PDMS control samples. It is believed that antifouling efficiency improves with high CA and low CAH. Error bars show ±1 standard deviation (adapted from Bixler et al. 2014)

made contact angle measurements. They measured contact angle (CA) and contact angle hysteresis (CAH) values. The replica sample was found to be superhydrophobic (~150°) with low CAH, whereas the embossed sample was hydrophobic (~101°). High CA and low CAH is believed to amplify the self-cleaning abilities of the samples, as it is believed that the droplets are able to roll and collect the particles after impact.

An illustration of anti-inorganic fouling mechanisms is shown in Fig. 13.27 (Bixler and Bhushan 2014). The schematic shows that foulers may be removed from a surface

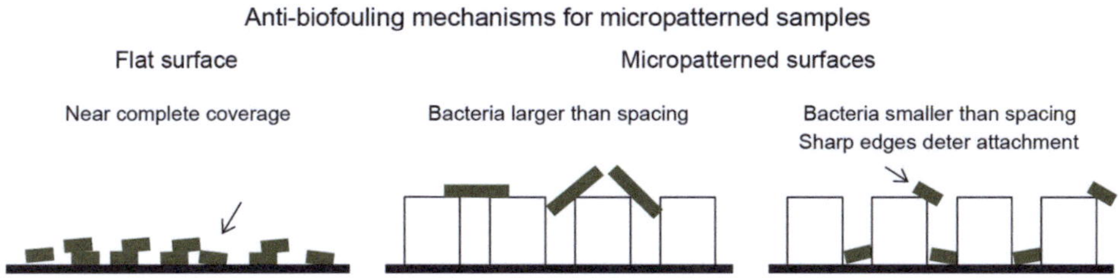

Fig. 13.24 Conceptual models of anti-biofouling mechanisms with micropatterned samples. Shown are colonization on flat and micropatterned surfaces. If the length of the microorganism is larger than the pillar spacing, it cannot invade the spacing and instead it land on top of the pillars. If the length of the microorganism is smaller than the spacing, microstructured discontinuities and sharp edges are believed to discourage colonization—where they may expire in the process of searching for an ideal landing site

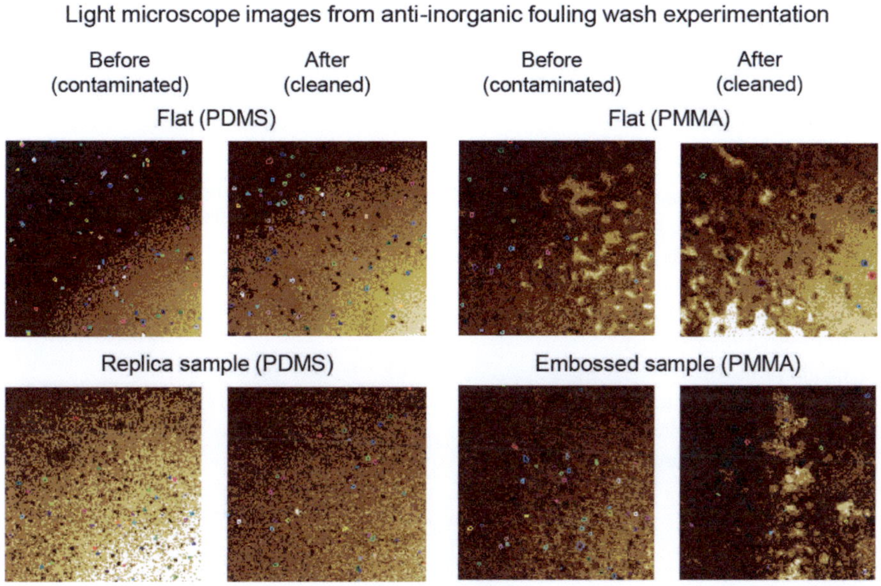

Fig. 13.25 Light microscope images of samples before and after anti-inorganic fouling wash experimentation, with specs representing silicon carbide (SiC) particles (Bixler and Bhushan 2014). Results are presented for flat PDMS and PMMA, the replica sample, and the embossed sample. Self-cleaning analysis was conducted using the images and particle analysis software to quantify particle removal

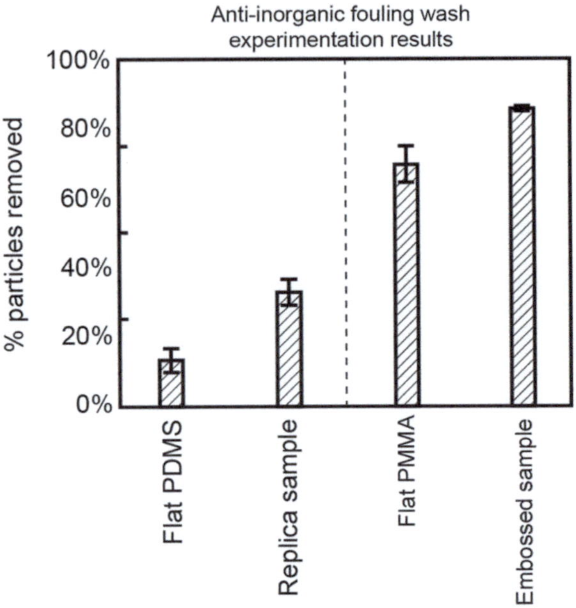

Fig. 13.26 Anti-inorganic fouling wash experimentation results (adapted from Bixler and Bhushan 2014). Results are presented for flat PDMS, replica sample, flat PMMA sample, and embossed sample. The embossed sample shows the greatest self-cleaning efficiency followed by the replica sample. Micropatterned samples are superior in self-cleaning efficiency compared to their respective flat samples. Error bars show ±1 standard deviation

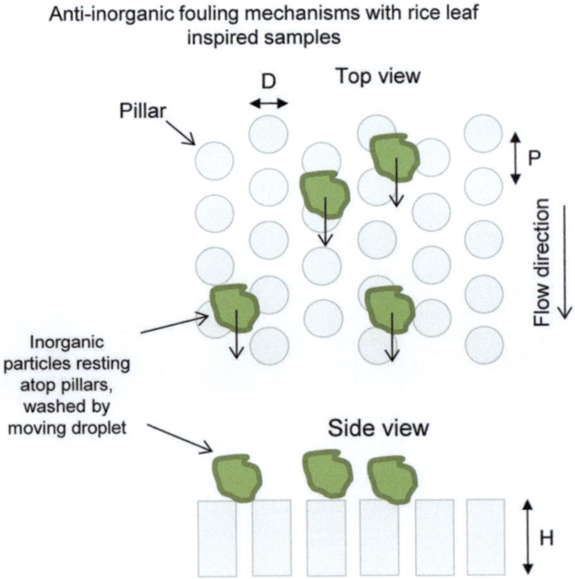

Fig. 13.27 Anti-inorganic fouling mechanisms with rice leaf inspired samples (adapted from Bixler and Bhushan 2014). Model shown utilizes uniformly spaced micro-sized single height pillars in a hexagonal array. Inorganic fouling may be washed away in fluid flow via self-cleaning

by water droplets or fluid flow when the rice and butterfly wing effect is present. Such a pattern is believed to encourage self-cleaning and directional fluid control, which is believed to provide antifouling benefit.

13.8 Closure

In this chapter, biofouling and inorganic fouling are introduced. Next, fields susceptible to fouling are summarized, followed by biofouling and inorganic fouling formation mechanism, antifouling strategies from living nature, current prevention and cleaning techniques for antifouling, and ends with demonstrations of the use of silver nanoparticles and bioinspired rice leaf surfaces for anti-fouling. The data shows that surfaces treated with nanoparticles and micropatterned surfaces provide anti-biofouling.

References

Anonymous (1952), *Marine Fouling and its Prevention*, Woods Hole Oceanographic Institute, US Naval Institute, Annapolis, Maryland.

Anonymous (1999), *Why Is Titanium the Metal of Choice for Medical Applications from Head to Toe?* International Titanium Association, Boulder, Colorado.

Baker, C., Pradhan, A., Pakstis, L., Pochan, D. J., and Shah, S. I. (2005), "Synthesis and Antibacterial Properties of Silver Nanoparticles," *J. Nanosci. Nanotechnol.* **5**, 244–249.

Banerjee, I., Pangule, R. C., and Kane R. S. (2011), "Antifouling Coatings: Recent Developments in the Design of Surfaces That Prevent Fouling by Proteins, Bacteria, and Marine Organisms," *Adv. Mater.* **23**, 690–718.

Barthlott, W. and Neinhuis, C. (1997), "Purity of the Sacred Lotus, or Escape from Contamination in Biological Surfaces," *Planta* **202**, 1–8.

Barthlott, W., Mail, M., Bhushan, B., and Koch, K. (2017), "Plant Surfaces: Structures and Functions for Biomimetic Innovations," *Nano-Micro Lett.* **9**, 23.

Bhushan, B. (2009), "Biomimetics: Lessons from Nature—An Overview," *Phil. Trans. R. Soc. A.* **367**, 1445–1486.

Bhushan, B. (2016), *Encyclopedia of Nanotechnology*, second ed., Vol. 1–6, Springer International, Cham, Switzerland.

Bhushan, B. (2017), *Springer Handbook of Nanotechnology*, fourth ed., Springer International, Cham, Switzerland.

Bhushan, B. and Jung, Y. C. (2011), "Natural and Biomimetic Artificial Surfaces for Superhydrophobicity, Self-Cleaning, Low Adhesion, and Drag Reduction," *Prog. Mater. Sci.* **56**, 1–108.

Bhushan, B., Luo, D. Schricker, S. R., Sigmund, W., and Zauscher, S. (2014), *Handbook of Nanomaterials Properties*, Volumes 1–2, Springer, Heidelberg, Germany.

Bhushan, B., Hansford, D., and Lee, K. K. (2006), "Surface Modification of Silicon and Polydimethylsiloxane Surfaces with Vapor-phase-deposited Ultrathin Fluorosilane Films for Biomedical Nanodevices," *J. Vac. Sci. Technol. A* **24**, 1197–1202.

Bhushan, B., Jung, Y. C., and Koch, K. (2009), "Micro-, Nano- and Hierarchical Structures for Superhydrophobicity, Self-Cleaning and Low Adhesion," *Phil. Trans. R. Soc. A* **367**, 1631–1672.

Bixler, G. D. and Bhushan, B. (2012), "Biofouling Lessons from Nature," *Phil. Trans. R. Soc. A* **370**, 2381–2417.

Bixler, G. D. and Bhushan, B. (2013a), "Fluid Drag Reduction with Shark-Skin Riblet Inspired Microstructured Surfaces," *Adv. Funct. Mater.* **23**, 4507–4528.

Bixler, G. D. and Bhushan, B. (2013b), "Fluid Drag Reduction and Efficient Self-Cleaning with Rice Leaf and Butterfly Wing Bioinspired Surfaces," *Nanoscale* **5**, 7685–7710.

Bixler, G. D. and Bhushan, B. (2014), "Rice- and Butterfly-Wing Effect Inspired Self-Cleaning and Low Drag Micro/nanopatterned Surfaces in Water, Oil, and Air Flow," *Nanoscale* **6**, 76–96.

Bixler, G. D. and Bhushan, B. (2015), "Rice- and Butterfly-Wing Effect Inspired Low Drag and Antifouling Surfaces: A Review," *Crit. Rev. Solid State Mater. Sci.* **40**, 1–37.

Bixler, G. D., Theiss, A., Bhushan, B., and Lee, S. C. (2014), "Anti-fouling Properties of Microstructured Surfaces Bio-inspired by Rice Leaves and Butterfly Wings," *J. Colloid Interface Sci.* **419**, 114–133.

Bormashenko, E., Bormashenko, Y., Stein, T., Whyman, G., and Bormashenko, E. (2007), "Why Do Pigeon Feathers Repel Water? Hydrophobicity of Pennae, Cassie-Baxter Wetting Hypothesis, and Cassie-Wenzel Capillary-induced Wetting Transition," *J. Colloid Interface Sci.* **311**, 212–216.

Bott, T. R. (1988a), "Crystallization of Organic Materials," in *Fouling Science and Technology* (eds. L. F. Melo, T. R. Bott, and C. A. Bernardo), pp. 275–280, Kluwer Academic Publishers, Dordrecht, The Netherlands.

Bott, T. R. (1988b), "Crystallization Fouling—Basic Science and Models," in *Fouling Science and Technology* (eds. L. F. Melo, T. R. Bott, and C. A. Bernardo), pp. 251–260, Kluwer Academic Publishers, Dordrecht, The Netherlands.

Brennan, A. B., Baney, R. H., Carman, M. I., Estes, T. G., Feinberg, A. W., Wilson, L. H., and Schumacher, J. F. (2010), "Surface Topographies for Non-Toxic Bioadhesion Control," United States Patent no. 7 650 848.

Caballero, L., Whitehead, K. A., Allen, N. S., and Verran, J. (2010), "Photoinactivation of *Escherichia coli* on Acrylic Paint Formulations Using Fluorescent Light," *Dyes Pigm.* **86**, 56–62.

Callow, M. E. (1999), "The Status and Future of Biocides in Marine Biofouling Prevention," in *Recent Advances in Marine Biotechnology* (eds. M. Fingerman, R. Nagabhushanam, and M. F. Thompson), pp. 109–126, Science Publishers, Enfield, New Hampshire.

Callow, M. E., Pitchers, R. A., and Milne A. (1986), "The Control of Fouling by Non-Biocidal Systems," in *Algal Biofouling* (eds. L. V. Evans and K. D. Hoagland), pp. 145–158, Elsevier Science Publishers, Amsterdam.

Callow, M. E., Jennings, A. R., Brennan, A. B., Seegert, C. E., Gibson, A., Wilson, L., Feinberg, A., Baney, R., and Callow, J. A. (2002), "Microtopographic Cues for Settlement of Zoospores of the Green Fouling Alga Enteromorpha," *Biofouling* **18**, 237–245.

Cao, L., Jones, A. K., Sikka, V. K., Wu, J., and Gao, D. (2009), "Anti-Icing Superhydrophobic Coatings," *Langmuir* **25**, 12444–12448.

Carman, M. L., Estes, T. G., Feinburg, A. W., Schumacher, J. F., Wilkerson, W., Wilson, L. H., Callow, M. E., Callow, J. A., and Brennan, A. B. (2006), "Engineered Antifouling Microtopographies—Correlating Wettability with Cell Attachment," *Biofouling* **22**, 11–21.

Chan, J. and Wong, S. (eds.) (2010), *Biofouling Types, Impact and Anti-fouling*, Nova Science Publishers, New York.

Charles, C. H., Mostler, K. M, Bartels, L. L., and Mankodi, S. M. (2004), "Comparative Antiplaque and Antigingivitis Effectiveness of and Chlorhexidine and an Essential Oil Mouth Rinse: 6-Month Clinical Trial," *J. Clin. Periodontal.* **31**, 878–884.

Clare, A. S. and Aldred, N. (2009), "Surface Colonization by Marine Organisms and its Impact on Antifouling Research," in *Advances in Marine Antifouling Coatings and Technologies* (eds. C. Hellio and D. Yebra), pp. 46–79, CRC Press, Boca Raton, Florida.

Clarkson, N. (1999), "The Antifouling Potential of Silicone Elastomer Polymers," in *Recent Advances in Marine Biotechnology* (eds. M. Fingerman, R. Nagabhushanam, and M. F. Thompson), pp. 87–108, Science Publishers, Enfield, New Hampshire.

Cologer, C. P. (1984), "Six Year Interaction of Underwater Cleaning with Copper Based Antifouling Paints on Navy Surface Ships," *Nav. Eng. J.* **96**, 200–208.

Copisarow, M. (1945), "Marine Fouling and its Prevention," *Science* **101**, 406–407.

Costerton, J. W. (ed.) (2008), *Springer Series on Biofilms*, Springer-Verlag, Berlin.

Cui, Z. and Wan, Y. (2005), "Biofouling in Membrane Separation Systems," in *Surfaces and Interfaces for Biomaterials* (ed. P. Vadgama), pp. 493–544, CRC Press, Boca Raton, Florida.

Cunningham, A. B., Lennox, J. E., and Ross, R. J. (2008), Biofilms in Industrial Environments. Retrieved June 27, 2011, from http://biofilmbook.hypertextbookshop.com/public_version/contents/contents.html.

Dastjerdi, R, and Montazar, M. (2010), "A Review on the Application of Inorganic Nano-structured Materials in the Modification of Textiles: Focus on Anti-microbial Properties," *Colloids. Surf. B* **79**, 5–18.

Davies, D. (2003), "Understanding Biofilm Resistance to Antibacterial Agents," *Nat. Rev.* **2**, 114–122.

Dean, B. and Bhushan, B. (2010), "Shark-Skin Surfaces for Fluid-Drag Reduction in Turbulent Flow: A Review," *Phil. Trans. R. Soc. A.* **368**, 4775–4806.

Epstein, N. (1988), "Particulate Fouling of Heat Transfer Surfaces: Mechanisms and Models," in *Fouling Science and Technology* (eds. L. F. Melo, T. R. Bott, and C. A. Bernardo), pp. 143–164, Kluwer Academic Publishers, Dordrecht, The Netherlands.

Feng, X. and Jiang, L. (2006), "Design and Creation of Superwetting/Antiwetting Surfaces," *Adv. Mater.* **18**, 3063–3078.

Fingerman, M., Nagabhushanam, R., and Thompson, M. F. (eds.) (1999), *Recent Advances in Marine Biotechnology*, Science Publishers, Inc., Enfield, New Hampshire.

Finlay, J. A., Callow, M. E., Schultz, M. P., Swain, G. W., and Callow, J. A. (2002), "Adhesion Strength of Settled Spores of the Green Alga Enteromorpha," *Biofouling* **18**, 251–256.

Flemming, H. C., Wingender, J., Moritz, R., Borchard, W., and Mayer, C. (1999), "Physio-chemical Properties of Biofilms—A Short Review," in *Biofilms in the Aquatic Environment* (eds. C. W. Keevil, A. Godfree, D. Holt, and C. Dow), pp. 1–12, The Royal Society of Chemistry, Cambridge, UK.

Flemming, H. C., Szewzyk, U., and Griebe, T. (eds.) (2000), *Biofilms Investigative Methods and Applications*, Technomic Publishing Co., Lancaster, Pennsylvania.

Fletcher, J. T., Finlay, J. A., Callow, M. E., Callow, J. A., and Ghadiri, M. R. (2007), "A Combinatorial Approach to the Discovery of Biocid al Six-Residue Cyclic D, L-α-Peptides Against the Bacteria Methicillin-Resistant Staphylococcusaureus (MRSA) and *E. coli* and the Biofouling Algae Ulva linza and Navicula perminuta," *Chem. Eur. J.* **13**, 4008–4013.

Gordon, D. P. and Mawatari, S. F. (1992), *Atlas of Marine-fouling Bryozoa of New Zealand Ports and Harbours*, New Zealand Oceanographic Institute, Wellington.

Griebe, T. and Flemming, H. C. (2000), "Rotating Annular Reactors for Controlled Growth of Biofilms," in *Biofilms Investigative Methods and Applications* (eds. H. C. Flemming, U. Szewzyk, and T. Griebe), pp. 1–22, Technomic Publishing Co, Inc., Lancaster, Pennsylvania.

Harder, T. and Yee, L. H. (2009), "Bacterial Adhesion and Marine Fouling," in *Advances in Marine Antifouling Coatings and Technologies* (eds. C. Hellio and D. Yebra), pp. 113–131, CRC Press, Boca Raton, Florida.

Hellio, C. and Yebra, D. (eds.) (2009), *Advances in Marine Antifouling Coatings and Technologies*, CRC Press, Boca Raton, Florida.

Herzberg, M. and Elimelech, M. (2007), "Biofouling of Reverse Osmosis Membranes: Role of Biofilm-enhanced Osmotic Pressure," *J. Membr. Sci.* **295**, 11–20.

Hipler, U. -C. and Elsner, P. (eds.) (2006), *Biofunctional Textiles and the Skin*, Karger, Basel, Switzerland.

Hoipkemeier-Wilson, L., Schumacher, J. F., Carman, M. L., Gibson, A. L., Feinberg, A. W., Callow, M. E., Finlay, J. A., Callow, J. A., and Brennan, A. B. (2004), "Antifouling Potential of Lubricious, Micro-engineered, PDMS Elastomers against Zoospores of the Green Fouling Alga Ulva (*Enteromorpha*)," *Biofouling* **20**, 53–63.

Jones, G. (2009), "The Battle Against Marine Biofouling: A Historical Review," in *Advances in Marine Antifouling Coatings and Technologies* (eds. C. Hellio and D. Yebra), pp. 19–45, CRC Press, Boca Raton, Florida.

Keevil, C. W., Godfree, A., Holt, D., and Dow, C. (eds.) (1999), *Biofilms in the Aquatic Environment*, The Royal Society of Chemistry, Cambridge, UK.

Kesel, A. and Liedert, R. (2007), "Learning from Nature: Non-Toxic Biofouling Control by Shark Skin Effect," *Comp. Biochem. Physiol. A* **146**, S130.

Kim, S. H., Kwak, S. Y., Sohn, B. H., and Park, T. H. (2003), "Design of TiO_2 Nanoparticle Self-Assembled Aromatic Polyamide Thin-film-composite (TFC) Membrane as an Approach to Solve Biofouling Problem," *J. Membr. Sci.* **211**, 157–165.

Koch, K. and Barthlott, W. (2009), "Superhydrophobic and Superhydrophilic Plant Surfaces: An Inspiration for Biomimetic Materials," *Phil. Trans. R. Soc.* **367**, 1487–1509.

Koch, K., Bhushan, B., and Barthlott, W. (2008), "Diversity of Structure, Morphology, and Wetting of Plant Surfaces," *Soft Matter* **4**, 1943–1963.

Koch, K., Bhushan, B., and Barthlott, W. (2009), "Multifunctional Surface Structures of Plants: An Inspiration for Biomimetics," *Prog. Mater. Sci.* **54**, 137–178.

Lee, W., Jin, M. K., Yoo, W. C., and Less, J. K. (2004), "Nanostructuring of a Polymeric Substrate with Well-Defined Nanometer-Scale Topography and Tailored Surface Wettability," *Langmuir* **20**, 7665–7669.

Lim, A. L. and Bai, R. (2003), "Membrane Fouling and Cleaning in Microfiltration of Activated Sludge Wastewater," *J. Membr. Sci.* **216**, 279–290.

LoVetri, K., Gawande, P. V., Yakandawala, N., and Madhyastha S. (2010), "Biofouling and Antifouling of Medical Devices," in *Biofouling Types, Impact and Anti-fouling* (eds. J. Chan and S. Wong), pp. 105–128, Nova Science Publishers, New York.

Ma, H., Bowman, C. N., and Davis, R. H. (2000), "Membrane Fouling Reduction by Backpulsing and Surface Modification," *J. Membr. Sci.* **173**, 191–200.

Marrie, T. J. and Costerton, J. W. (1984), "Morphology of Bacterial Attachment to Cardiac Pacemaker Leads and Power Packs," *J. Clin. Microbiol.* **19**, 911–914.

Mather R. R. and Wardman R. H. (eds.) (2011), *The Chemistry of Textile Fibers*, Royal Society of Chemistry, Cambridge.

Melo, L. F., Bott, T. R., and Bernardo, C. A. (eds.) (1988), *Fouling Science and Technology*, Kluwer Academic Publishers, Dordrecht, The Netherlands.

Menton, D. (2008), "The Seeing Eye," *AnswersMagazine.com*, July–Sept. pp. 76–79.

Meuler, A. J., Smith, J. D., Varanasi, K. K., Mabry, J. M., McKinley, G. H., and Cohen, R. E. (2010), "Relationships Between Water Wettability and Ice Adhesion," *ACS Appl. Mater. Interfaces* **2**, 3100–3110.

Monroe, D. (2007), "Looking for Chinks in the Armor of Bacterial Biofilms," *PLoS Biol.* **5**, 2458–2461.

Neinhuis, C. and Barthlott, W. (1997), "Characterization and Distribution of Water-Repellent, Self-Cleaning Plant Surfaces," *Ann. Bot.* **79**, 667–677.

Palacio, M. L. B. and Bhushan, B. (2012), "Bioadhesion: A Review of Concepts and Applications," *Phil. Trans. R. Soc. A* **370**, 2321–2347.

Perera, S., Bhushan, B., Bandara, R., Rajapakse, G., Rajapakse, S., and Bandara, C. (2013), "Morphological, Antimicrobial, Durability, and Physical Properties of Untreated and Treated Textiles Using Silver-Nanoparticles," *Colloid Surf. A* **436**, 975–989.

Pollini M., Russo M., Licciulli E. A., Sannino A., and Maffezzoli E. A. (2009), "Characterization of Antibacterial Silver Coated Yarns," *J. Mater. Sci: Mater. Med.* **20**, 2361–2366.

Pritchard, A. M. (1988), "Deposition of Hardness Salts," in *Fouling Science and Technology* (eds. L. F. Melo, T. R. Bott, and C. A. Bernardo), pp. 261–274, Kluwer Academic Publishers, Dordrecht, The Netherlands.

Puckett, S. D., Lee, P. P., Ciombor, D. M., Aaron, R. K., and Webster, T. J. (2010), "Nano-textured Titanium Surfaces for Enhancing Skin Growth on Transcutaneous Osseointegrated Devices," *Acta Biomater.* **6**, 2352–2362.

Raghupathi, K. R., Koodali, R. T., and Manna, A. C. (2011), "Size-Dependent Bacterial Growth Inhibition and Maechanisms of Antibacterial Activity of Zinc Oxide nanoparticles," *Langmuir* **27**, 4020–4028.

Railkin, A. I. (2004), *Marine Biofouling Colonization Processes and Defenses*, CRC Press, Boca Raton, Florida.

Ralston, E. and Swain, G. (2009), "Bioinspiration—The Solution for Biofouling Control?" *Bioinsp. Biomim.* **4**, 1–9.

Ray, D. L. (ed.) (1959), *Marine Boring and Fouling Organisms*, University of Washington Press, Seattle.

Sawai, J., Kawada, E., Kanou, F, Igarashi, H., Hashimoto, A, Kokugan, T., and Shimizu, M. (1996), "Detection of Active Oxygen Generated from Ceramic Powders having Antibacterial Activity," *J. Chem. Eng. Jpn.* **29**, 627–633.

Scardino, A. J. (2009), "Surface Modification Approaches to Control Marine Biofouling," in *Advances in Marine Antifouling Coatings and Technologies* (eds. C. Hellio and D. Yebra), pp. 664–692, CRC Press, Boca Raton, Florida.

Scardino, A. J., Guenther, J., and de Nys, R. (2008), "Attachment Point Theory Revisited: The Fouling Response to a Microtextured Matrix," *Biofouling* **24**, 45–53.

Schulz, M. J., Shanov, V. N., and Yun, Y. (eds.) (2009), *Nanomedicine Design of Particles, Sensors, Motors, Implants, Robots, and Devices*, Artech House, Boston.

Schumacher, J. F., Aldred, N., Callow, M. E., Finlay, J. A., Callow, J. A., Clare, A. S., and Brennan, A. B. (2007a), "Species-Specific Engineered Antifouling Topographies: Correlations between the Settlement of Algal Zoospores and Barnacle Cyprids," *Biofouling* **23**, 307–317.

Schumacher, J. F., Carman, M. L., Estes, T. G., Feinberg, A. W., Wilson, L. H., Callow, M. E., Callow, J. A., Finlay, J. A., and Brennan, A. B. (2007b), "Engineered Antifouling Microtopographies—Effect of Feature Size, Geometry, and Roughness on Settlement of Zoospore of the Green Alga *Ulva*," *Biofouling* **23**, 55–62.

Sharma, N., Charles, C. H., Lynch, M. C., Qaqish, J., McGuire, J. A., Galustians, J. G., and Kumar, L. D. (2004), "Adjunctive Benefit of an Essential Oil-Containing Mouth Rinse in Reducing Plaque and Gingivitis in Patients Who Brush and Floss Regularly," *J Am. Dent. Assoc.* **135**, 496–504.

Sharma, S., Popat, K. C., and Desai, T. A. (2007), "Design and Biological Applications of Nanostructured Poly(Ethylene Glycol) Films," in *Nanotechnology in Biology and Medicine Methods, Devices, and Applications* (ed. T. Vo-Dinh), pp. 39–1, CRC Press, Boca Raton, Florida.

Sheng, X., Ting, Y. P., and Pehkonen, S. O. (2010), "Force Measurements of Bacterial Adhesion to Metals Using Cell Probe in Atomic Force Microscopy," in *Biofouling Types, Impact and Antifouling* (eds. J. Chan and S. Wong), pp. 129–154, Nova Science Publishers, New York.

Shirtliff, M. and Leid, J. G. (eds.) (2009), *The Role of Biofilms in Device-Related Infections*, Springer-Verlag, Berlin.

Simoes, M., Simoes, L. C., and Vieira, M. J. (2010), "A Review of Current and Emergent Biofilm Control Strategies," *LWT Food Sci. Technol.* **43**, 573–583.

Somerscales, E. F. C. and Knudsen, J. G. (eds.) (1981), *Fouling of Heat Transfer Equipment*, Hemisphere Publishing Corporation, Washington.

Stoodley, P., Sauer, K., Davies, D. G., and Costerton, J. W. (2002), "Biofilms as Complex Differentiated Communities," *Annu. Rev. Microbiol.* **56**, 187–209.

Sunada, K., Kikuchi, Y., Hashimoto, K., and Fujishima, A. (1998), "Bactericidal and Detoxification Effects of TiO_2 Thin Film Photocatalysts," *Environ. Sci. Technol.* **32**, 726–728.

Taylor, E. and Webster, T. J. (2011), "Reducing Infections through Nanotechnology and Nanoparticles," *Int. J. Nanomed.* **6**, 1463–1473.

Touati, D. (2000), "Iron and Oxidative Stress in Bacteria," *Arch. Biochem. Biophys.* **373**, 1–6.

Trinidad, A., Ibanez, A., Gomez, D., Garcia-Berrocal, J. R., and Ramierz-Camacho, R. (2010), "Application of Environmental Scanning Electron Microscopy for Study of Biofilms in Medical Devices," *Microsc.: Sci. Technol. Appl. Educ.* **1**, 204–210.

Vadgama, P. (ed.) (2005), *Surfaces and Interfaces for Biomaterials*, CRC Press, Boca Raton, Florida.

Venkatesan, R. and Murthy, P. S. (2008), "Macrofouling Control in Power Plants," in *Springer Series on Biofilms* (ed. J. W. Costerton), pp. 265–290, Springer-Verlag, Berlin.

Vladkova, T. (2007), "Surface Engineering for Non-Toxic Biofouling Control," *J. Univ. Chem. Technol. Metall.* **43**, 239–256.

Vladkova, T. (2008), "Surface Modification Approach to Control Biofouling," in *Springer Series on Biofilms* (ed. J. W. Costerton), pp. 135–162, Springer-Verlag, Berlin.

Vleugels, I. R. (2006), "Multi-Functional Coating Solution," *Euprofile Public Serv. Rev.: Eur. Union* **18**, 1–2.

Vo-Dinh, T. (ed.) (2007), *Nanotechnology in Biology and Medicine*, CRC Press, Boca Raton, Florida.

Wagner, T., Neinhuis, C., and Barthlott, W. (1996), "Wettability and Contaminability of Insect Wings as a Function of Their Surface Sculptures," *Acta Zool.* **77**, 213–225.

Walker, J., Surman, S., and Jass, J. (eds.) (2000), *Industrial Biofouling Detection, Prevention and Control*, Wiley, New York.

Wang, Y. and Bhushan, B. (2015), "Wear-Resistant and Antismudge Superoleophobic Coating on Polyethylene Terephthalate Substrate Using SiO_2 Nanoparticles," *ACS Appl. Mater. Interfaces* **7**, 743–755.

Willemsen, P. (2005), "Biofouling in European Aquaculture: Is There an Easy Solution?" *Eur. Aquac. Soc. Spec. Publ.* **35**, 101–118.

Windler L., Height M., and Nowack B. (2013), "Comparative Evaluation of Antimicrobials for Textile Applications," *Environ. Int.* **53**, 62–73.

Yen, J. (2010), "Marine Dynamics," in *Bulletproof Feathers How Science Uses Nature's Secrets to Design Cutting-Edge Technology* (ed. R. Allen), pp. 22–43, University of Chicago Press, Chicago.

Zodrow, K., Brunet, L., Mahendra, S., Li, D., and Zhang, A. (2009), "Polysulfone Ultrafiltration Membranes Impregnated with Silver Nanoparticles Show Improved Biofouling Resistance and Virus Removal," *Water Res.* **43**, 715–723.

Bioinspired Strategies for Water Harvesting from Fog and Condensation 14

The Human Right to Water: UN General Assembly Resolution 64/292 (2010), "Recognizes the right to safe and clean drinking water and sanitation as a human right that is essential for the full enjoyment of life and all human rights."

11th Annual Cisco Visual Networking Index Global Mobile Data Traffic Forecast, "By 2021, more members of the global population will be using mobile phones (5.5 billion) than running water (5.3 billion)."

14.1 Introduction

Access to a safe water supply is a human right (United Nations 2015). Fresh water sustains human life and is vital for human health. Water availability depends on both its physical abundance and its safety for human consumption. Some of the arid regions of the world lack adequate safe drinking water. Figure 14.1 shows the proportion of the population with access to safe drinking water in 2015 (WHO/UNICEF 2017; WWAP 2019; Bhushan 2018, 2019, 2020a, b). In a total of 181 countries, more than 75% of the population had different degrees of accessibility (Bhushan 2020a, b). Due to bad economies or poor infrastructure in some parts of the world, water accessibility continues to worsen, particularly in these regions.

It is estimated that about 800 million people worldwide lack basic access to drinking water (WHO/UNICEF 2017; Bhushan 2018, 2019, 2020a, b). This means that they cannot reach a protected source of drinking water within a total walking distance of 30 min. About 2.2 billion people, representing nearly one-third of the world's population, do not have access to a safe water supply. That means no drinking water on the property that is

B. Bhushan, *Introduction to Biomimetics and Bioinspiration*,
https://doi.org/10.1007/978-3-031-62344-8_14

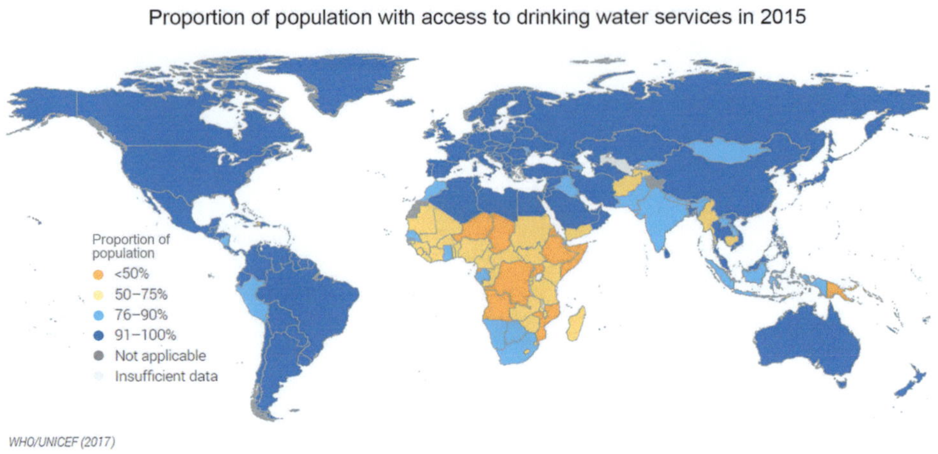

Fig. 14.1 Proportion of population with access to basic drinking water services in 2015. 181 countries had a coverage of over 75% (Adapted from WHO/UNICEF 2017; WWAP 2019, Bhushan 2018, 2019, 2020a, b)

available at all times and free of contamination. Rural areas are particularly affected by the lack of drinking water, with 80% of those lacking access residing in such regions. Almost half of people drinking water from unprotected sources live in sub-Saharan Africa. In addition, women and girls facing discrimination in accessing safe drinking water, factors such as ethnicity, disability, age, and health status contribute to inequalities.

Because of global warming and an increased number of drought periods, water supply continues to shrink. Figure 14.2 shows the level of physical water stress around the world, showing the percent of total fresh water withdrawn annually (UN 2018; Bhushan 2018, 2019, 2020a, b). Over 2 billion people, representing nearly one third of the world population, live in countries experiencing high water stress, meaning that water resources consumed are not regenerated to the necessary extent by rain or the return of the purified water (UN 2018). About 4 billion people experience severe water scarcity during at least one month of the year (WWAP 2019). Stress levels will continue to increase as water demand grows, and the effects of climate change intensify. A majority of people affected by water stress reside in North Africa, the Middle East, and South Asia, populations in North and South America, including the southwestern United States, have also experienced heightened impacts. (Bhushan 2018, 2019, 2020a, b).

Figure 14.3a shows a child drinking contaminated water from the river (Adapted from image provided by Water Crisis in Africa). In many parts of the world, people need to transport water from far, which is often not so clean (Fig. 14.3b (Adapted from, left image provided by Picsio, Creative Commons, and right image provided by Reuters). It is often children and women who do most of the water transporting (Bhushan 2020a, b).

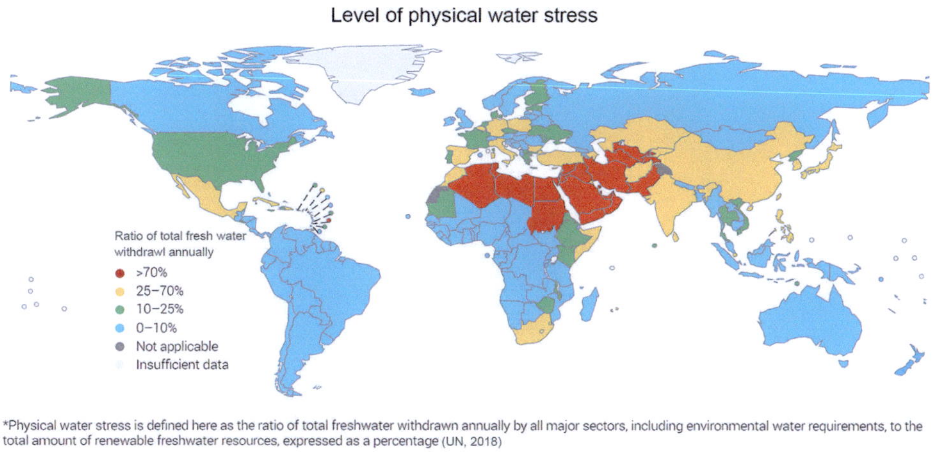

Fig. 14.2 Level of physical water stress showing percent of total fresh water withdrawn annually (Adapted from UN 2018; Bhushan 2018, 2019, 2020a, b)

14.1.1 Water on Earth

Roughly, 70% of the Earth's surface is covered by water, however, the vast majority of water is contained in the oceans with freshwater accounting for only about 2.5% of all water, as shown in Fig. 14.4a (Shiklomanov 1993; Brown and Bhushan 2016; Bhushan 2018, 2019, 2020a, b). Water continuously moves in a cycle due to evaporation, condensation, precipitation, surface and channel runoff, and subsurface flow, as shown in Fig. 14.4b (Brown and Bhushan 2016; Brown and Bhushan 2016; Bhushan 2018, 2019, 2020a, b). The evaporative phase can help purify water by separating it from contaminants picked up in other phases of the cycle, including salt in the oceans. Of the 2.5% fresh water, the majority is trapped as ice in glaciers and snow (1.74% in total). Only 0.79% (11 quadrillion m^3) of all water is found as surface water in lakes and rivers (0.03% or 400 trillion m^3), and as groundwater (0.76%) (Shiklomanov 1993). The distribution of this water is not uniform across the world, with around 20% of the world's surface fresh water found in the North American Great Lakes (EPA 1995).

Given that about 97.5% of water is saline water, desalination has become increasingly important in all parts of the world. Ocean water is contaminated by salt, as well as by bacteria and particulates. For human consumption, ocean water must be desalinated and purified. However, desalination remains an energy intensive process and is prohibitively expensive (Elimelech and Phillip 2011). Desalination may consume as much as 30 times more energy than other water purification approaches. Furthermore, for every 50 L of seawater, desalination produces about 1 L of clean water as well as 2 L of brine solution which poses serious environmental issues. Globally, desalination provides a small fraction of purified clean water supply. In 2020, one fourth of all global desalination of water

Fig. 14.3 **a** A child drinking contaminated water from the river (Adapted from image provided by Water Crisis in Africa), **b** people transporting water from far (Adapted from, left image provided by Picsio, Creative Commons, and right image provided by Reuters) (Adapted from Bhushan 2020a, b)

was carried out in Saudi Arabia and U.A.E. In the United States, Carlsbad Desalination Plant in California is the largest desalination plant. It delivers about 200,000 L of fresh desalinated water to San Diego County, enough to serve about 400,000 people (Bhushan 2018, 2019, 2020a, b).

In addition to all water on Earth, the Earth's atmosphere contains about 0.001% of global water (on the order of 13 trillion m^3 or 1.3×10^{16} L). It is found in the form of clouds, fog, mist, and water vapor. It is a small amount in global terms, but it is about 3.3% of water in rivers and lakes which can be a major source of drinking water. In arid deserts, many plants and animals survive by harvesting water from the atmosphere. Water is a recyclable natural resource with potential to water the arid regions of the world (Gleick 1993; Bhushan 2018, 2019, 2020a, b). Scientists and engineers have been working to develop bioinspired water harvesters (Bhushan 2019, 2020a, b).

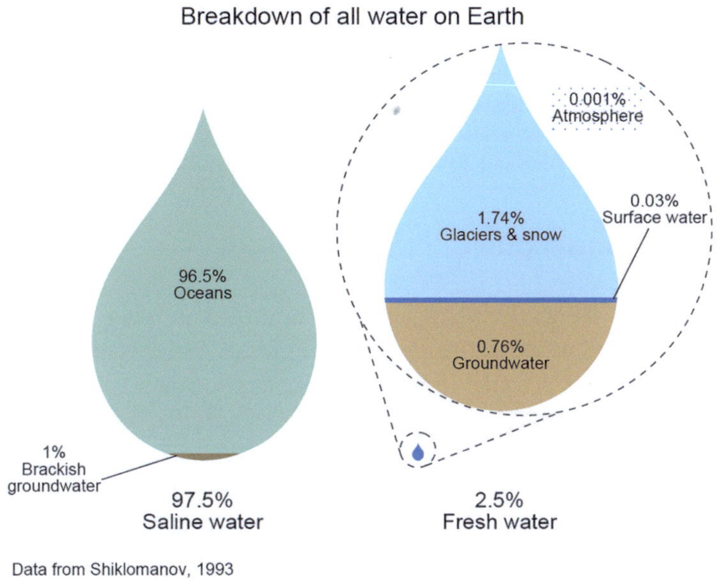
Breakdown of all water on Earth
0.001%
Atmosphere
0.03%
Surface water
1.74%
Glaciers & snow
96.5%
Oceans
0.76%
Groundwater
1%
Brackish
groundwater
97.5%
Saline water
2.5%
Fresh water
Data from Shiklomanov, 1993

(a)

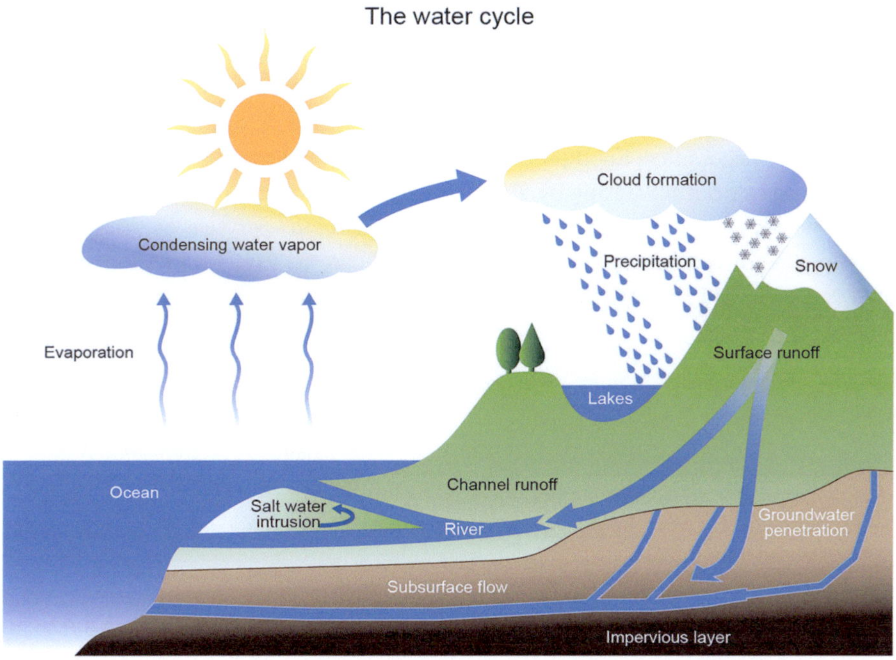
The water cycle
Cloud formation
Condensing water vapor
Precipitation
Snow
Evaporation
Surface runoff
Lakes
Ocean
Channel runoff
Salt water
intrusion
River
Groundwater
penetration
Subsurface flow
Impervious layer

(b)

◂**Fig. 14.4** **a** Percentage breakdown of all water on Earth. 2.5% is fresh water, with surface water in rivers and lakes only accounting for 0.03% of all water (400 trillion m^3) and groundwater only 0.79% of all water (11 quadrillion m^3) (Adapted from Shiklomanov 1993; Brown and Bhushan 2016; Bhushan 2018, 2019,2020a, b), and (b) the water cycle where of all water the evaporative phase helps purify water via separation from contaminants picked up in other phases of the cycle (Adapted from Brown and Bhushan 2016; Bhushan 2018, 2019, 2020a, b)

14.1.2 Water Consumption

About 60% of body weight in adults and 75% in infants is water. The body constantly loses water. It needs to be replenished by drinking water. Water is needed for food preparation and sanitation and personal hygiene. A summary of water consumption by humans is presented in Table 14.1 (Day et al. 2019; Bhushan 2020a, b). The amount of water required to meet basic human water needs is about 7.5–15 L day^{-1}, with 7.5 L day^{-1} being the absolute minimum of the individual's water needs (Day et al. 2019; Bhushan 2020a, b). In the U.S.A., about 600 L of water per capita and per day is consumed; whereas, in Europe it is about 300 L per capita and in the rest of the world is much less. Typical water requirements per capita are on the order of 50 L day^{-1}. It is noted that these numbers do not include water requirements for food production and other consumer needs, which can be much higher than water consumption by humans.

Water consumption continues to grow worldwide, driven by a combination of population growth, socio-economic development, and rising demand in the industrial and domestic sectors. With an ever-growing population, there is an increasing demand for food and other necessities. Food production requires a significant amount of water. Water consumption has been increasing by about 1–2% per year since the 1980s (WWAP 2019). Global water consumption is expected to reach 6 trillion m^3 per year by 2030 as the population moves toward a projected 8.2 billion, as shown in Fig. 14.5 (Anonymous 2009; United Nations 2015; Brown and Bhushan 2016; Bhushan 2018, 2019, 2020a, b).

Table 14.1 Water consumption by various processes by humans (Adapted from Day et al. 2019; Bhushan 2020a, b)

Process	Amount of water lost/used (L day^{-1})
The body constantly releases water through evaporation from the skin, through urine production and through the release of water from the lungs with the air we breathe. It needs to be replenished by drinking water	~2.4
Food preparation	3–6
Sanitation and personal hygiene practices	2–6
Total basic needs	**7.5–15**
Typical needs worldwide	**50**

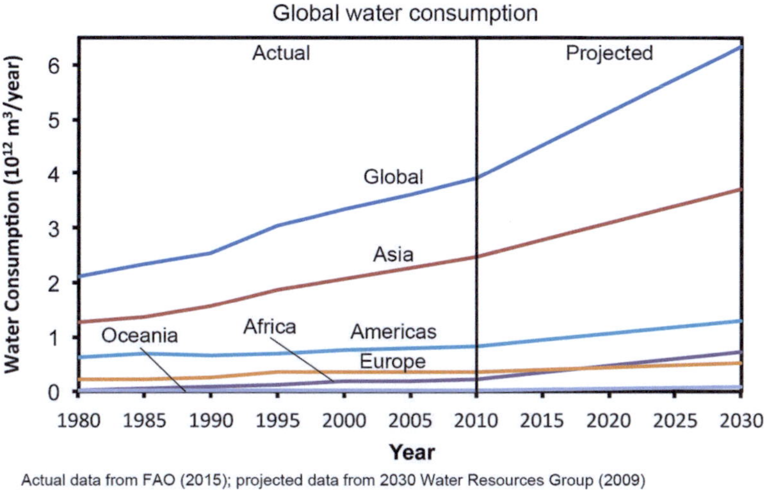

Fig. 14.5 Global water consumption based on actual (1980–2010) data (Adapted from United Nations 2015) and projected (2030) data (Adapted from Anonymous 2009; Brown and Bhushan 2016; Bhushan 2018, 2019, 2020b)

14.1.3 Water Contamination

Water contamination due to human activity, unsafe industrial practices, and population growth continues to increase. Water contamination is one of the major environmental and natural resource concerns in the twenty-first century (Clark et al. 1986; Rao et al. 2012; Brown and Bhushan 2016; Fingas 2017; Bhushan 2018, 2020b). Contamination by oil can occur during operation of machinery, oil exploration and oil transportation, and due to operating environment. Oil is commonly used as a lubricant in machinery and can leak and produce contamination. Oil spills occasionally occur during oil exploration and transportation. They will continue to occur as long as society is dependent upon oil (Brown and Bhushan 2016; Bhushan 2018). The Deepwater Horizon oil spill in the Gulf of Mexico in 2010 was the worst accidental offshore oil spill in history. It spilled about 200 million U.S. gallons of oil (Fig. 14.6). The partial cleanup by British Petroleum took several years and full cleanup was never accomplished. Oil spills cause damage to the environment and contaminate sources of potable water. Another source of contamination stems from the emergence of fracking in the U.S., where water-based fluids (containing sand and chemicals) are injected under high pressure to fracture rocks to release oil and gas, which lead to oil-contaminated wastewater (Bhushan 2018, 2020a, b).

Water contamination with a variety of chemicals is another major concern with growing world population and industrialization with unsafe practices of waste disposal. Various organic contaminants present in contaminated water pose health risks.

Fig. 14.6 Photograph of oil-contaminated water after the 2010 Deepwater Horizon oil spill in the Gulf of Mexico (Adapted from Bhushan 2018, 2020b)

Access to "clean" and safe water is a global problem. While the situation is particularly severe for large number of countries in Africa, Asia, and the Middle East, Latin America, Caribbean, and some countries in Eastern Europe, many people in Western and Coastal Europe as well as in North America also suffer from lack of inequitable access to safe drinking water (WWAP 2019; Bhushan 2018, 2019, 2020a, b).

14.1.4 Lessons from Nature for Water Harvesting to Supplement Water Supply

It is apparent that the current supply of fresh water needs to be supplemented to meet future needs. To find new sources of water supply, living nature may provide solutions. In living nature, after some 3.8 billion years of evolution, many plant and animal species in arid regions exhibit efficient solutions for water harvesting from fog and condensation of water vapor. These solutions typically involve species possessing unique surface structures and chemistry on or within their bodies that help to direct the movement of water to where it is either consumed or stored, before it is evaporated (Brown and Bhushan 2016; Bhushan 2018, 2019, 2020a, b; Gurera and Bhushan 2020b).

By studying the surface structures and chemistries involved, bioinspired fog harvesters are being developed (Bhushan 2018, 2019, 2020a, b). The ambient temperature during desert nights is low, as low as 0° – 4°C. This temperature can be lower than the dew point, which would lead to water condensation of water vapor from ambient. Bioinspired water condensation can also be used for water harvesting (Bhushan 2018, 2019, 2020a, b).

Bioinspired water harvesters can be used to provide a supplemental water source for communities in the arid regions where fog and condensation of water vapor is common, such as the coastal regions of Africa, the Southwestern coast of South America, and the Southwestern United States (Brown and Bhushan 2016; Bhushan 2018, 2019, 2020a, b). In addition, these harvesters can be used in various emergency and defense applications (Bhushan 2019, 2020a, b). Emergency applications, such as natural disasters, could benefit for short periods from portable units which could be dropped from air. The cost of clean water in military bases in a combat zone can be high, and water harvesters become attractive. Various designs for large and portable 3D towers have been developed.

Bioinspired approaches are attractive to develop materials and surfaces in an environmentally friendly and sustainable manner to supplement water supply and remove contamination.

14.1.5 Scope of the Chapter

In this chapter, an overview of bioinspired water harvesting methods from fog and water condensation of water vapor from ambient, to supplement water supply is presented. First, an overview of arid desert conditions and plants and animals is presented. Next, various examples of water harvesting from living nature are presented with a discussion of the surface structures and chemistries used. Then, water collection data for various designs of bioinspired surfaces, both from fog and condensation from ambient are presented, which can be used to develop optimal designs. Various consumer, emergency and defense applications are discussed. Various designs of water harvesting towers and projections for water collection are presented (Brown and Bhushan 2016; Bhushan 2018, 2019, 2020a, b).

14.2 Overview of Arid Desert Conditions, Water Sources, and Desert Plants and Animals

A desert is a barren landscape where little precipitation occurs (Walker 1992; Costa 1995; Mares 1999; Harris 2003; Allaby 2006; Laity 2008; Greenberger 2009; Gurera and Bhushan 2020b; Bhushan, 2018, 2019, 2020a, b). A desert is referred to as a land area that receives no more than 250 mm of precipitation per year. The amount of evaporation in a desert often greatly exceeds the annual rainfall, which results in a moisture deficit over the course of a year. All deserts are arid or dry. There is little water available for any living nature to survive. Consequently, living conditions are hostile for living nature including plant and animal life.

Deserts cover about one-third of the Earth's land surface area or about 10% of the total Earth surface. They are home to around 1 billion people– one-sixth of the Earth's population. A map of the world's deserts with non-polar and arid land, with relative degrees of aridness is shown in Fig. 14.7 (Gurera and Bhushan 2020b; Bhushan 2019, 2020a, b). The largest (non-polar) hot desert in the world is the Sahara in North Africa. It has one of the harshest environments on Earth, covering some 5.8 million sq. km, nearly a third of the African continent, about the size of the United States (including Alaska and Hawaii). It is the third largest after the two polar deserts. The Sahara Desert is also the hottest desert in the world. The driest desert in the world is the Atacama Desert in South America. The Atacama Desert had no rain for 401 years, between 1570 and 1971. The Namib Desert in southwest Africa is one of the most arid regions in the world with average annual rainfall of only 18 mm, and it is not uncommon to experience consecutive years with no rainfall at all. Despite the low rainfall, prevailing south-westerly winds form fog along the coast from 60–200 days per year, which can be blown for up to 50 km inland (Shanyengana et al. 2002). In addition, during cold nights, condensation of water vapor occurs (Bhushan 2019, 2020a, b; Gurera and Bhushan 2020b).

Very few species survive in the desert. The species surviving in the deserts need water to survive. Water appears in deserts from various sources including rivers, lakes, oases, and groundwater which are used by roots in the case of plants, metabolic water in the case of animals, rainwater, snow, fog, mist, and water condensation. In deserts, fog, mist and condensation as reliable water sources, whereas rainfall is unpredictable (Bhushan 2019, 2020a, b; Gurera and Bhushan 2020b).

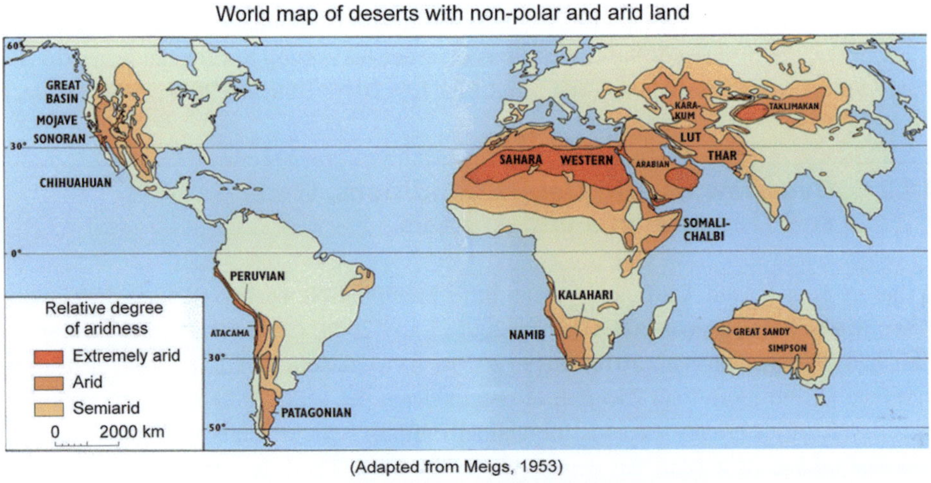

Fig. 14.7 World map of deserts with non-polar and arid land with relative degree of aridness. Arid deserts exist in parts of Africa, the Middle East, Southwestern South America and Southwestern United States (Adapted from Meigs 1953; Bhushan 2019, 2020a, b)

In foggy regions during the cold nights, particularly in the coastal desert regions, fog, mist and water vapor exist in the atmosphere. Fog and mist consist of water droplets suspended in air which can be coaxed and coalesced. In humid climates, air consists of a significant amount of water vapor which can be nucleated into droplets (dew droplets) on cold surfaces and coalesced. It is referred to as condensation of water vapor (Bhushan 2019, 2020a, b).

14.2.1 Water Source

Water moves continuously above and below the Earth's surface. Bodies of water, clouds, evaporation and condensation all are part of the water cycle. Fog is composed of micron-sized water droplets that form when air becomes saturated with water vapor. Fog is a thick cloud that remains suspended in the atmosphere. Dew is the deposit of water droplets formed on cold surfaces, with temperature lower than the dew point, by condensation of water vapor in the air or fog. In many plants and animals, living nature uses fog and condensation as a vital source of water, particularly in arid areas that receive little rainfall (Brown and Bhushan 2016; Bhushan 2019, 2020a, b). Fog and dew always exist when the temperature decreases late at night and in the early morning. There is evidence that over 5000 years ago, hunter–gatherer groups were able to populate arid areas along the southern coast of Peru by utilizing fresh water from fog and condensation, though the collection method is unknown (Beresford-Jones et al. 2015; Bhushan 2019, 2020a, b).

Fog and Mist

Fog and mist occur when the water vapor concentration in the atmosphere reaches saturation. They are a visible aerosol consisting of tiny water droplets suspended in air, which is found close to the land surface in cold nights. It is more frequent near lakes or oceans. In coastal regions, fog cover can extend inland up to 75 km, even during days in the high desert regions such as in the Mohave desert. Fog is formed when warm moist air above the desert land encounters the cooler surface of land or blows over the cold water in the ocean, which changes the invisible gas to tiny visible droplets. The formation is aided by the presence of nucleation sites on which the suspended water phase can coagulate. As an example, small particulates can enhance the fog or mist formation. There is less coalescence of water droplets in mist which makes it less dense and quicker to dissipate. The difference between fog and mist is in their location and density. Mist is referred to as cloud cover, suspended in air, and normally found above mountains. Whereas, fog is a cloud that reaches the ground level, normally found above a water surface. In addition fog is denser and lasts for a longer period and mist is thinner and one can see more clearly through it. Visibility in fog is less than about 1 km; whereas in mist, it is 1–2 km (Bhushan 2019, 2020a, b).

Fog consists of small water droplets whose size, typically, are on the order of 10–50 microns, with a concentration on the order of 10–100 microdroplets per cm^3 (Mares 1999; Anonymous 1999). Air in fog has a relative humidity, generally above 95%. Density of water in fog is on the order of 0.05–0.5 g m^{-3}. For reference, density of water is 1000 kg m^{-3}. This source of water is clean from impurities (Basu et al. 2018). This is a great benefit of water collected from fog harvesting (Bhushan 2019, 2020a, b).

The number of fog days and amount of fog depends upon the geographical position of the desert as well as to offshore ocean currents. Thick fog is formed near coastal deserts (Meigs 1966; Mares 1999). As an example, in the Namibian coastal region, fog occurs on 60–200 days per year along the coast and extends for some distance inland (Shanyengana et al. 2002). In arid and semi-arid regions, fog dominates as a water source. During low rainfall months, many species survive on fog (Bhushan 2019, 2020a, b).

Fog deposition occurs primarily during the summer months of June through October (Hiatt et al. 2012). The highest fog collection occurs when the temperature of the land is hot, during the summer months of June and July, during nights and early mornings. As rainfall decreases in the summer months, fog deposition increases, and as the region receives rainfall in fall, fog deposition decreases. Figure 14.8 shows the rate of fog water collected in different regions and the number of days fog occurs annually (Fessehaye et al. 2014). The rate of fog water collected ranges from 1.5 to 12 L m^{-2} day^{-1}. Avearge fog water collection is on the order of 2 L m^{-2} day^{-1} or on the order of 0.1 L m^{-2} h^{-1}, dependent upon the location. The rate of water collection depends on various environmental factors including the difference between air temperature and dew point temperature, wind speed, wind direction, and land temperature (Hiatt et al. 2012; Bhushan 2019, 2020a, b).

As a reference, a consumer humidifier produces fog with water content on the order of 50 –1000 L m^{-2} h^{-1} with nozzle diameter on the order of a couple of cm. The fog emitted by the humidifier gets dissipated over a large area (Bhushan 2019, 2020a, b).

Condensation of Water Vapor

Dew occurs as a result of phase transition, in which water vapor is transformed into liquid when it comes in contact with a surface. The primary condition for the formation of dew is that the temperature of the surface on which condensation takes place may be lower than or equal to the dew point temperature. During phase transition or condensation, nucleation of the liquid phase occurs. Nucleation is the formation of the nanoscale water droplets that are thermodynamically stable (Sigsbee 1969; Pruppacher and Klett 2010). The nucleation rate depends upon the degree of subcooling, relative humidity, and the wetting properties of the surface. Once a water droplet has nucleated on the surface, it begins to grow due to the formation of concentration gradient of water molecules around the droplet. This is followed by droplet coalescence (Bhushan 2020a, b).

The atmosphere contains a large quantity of moisture in the form of water vapor. Dew point is the temperature to which air must be cooled to become saturated with water vapor.

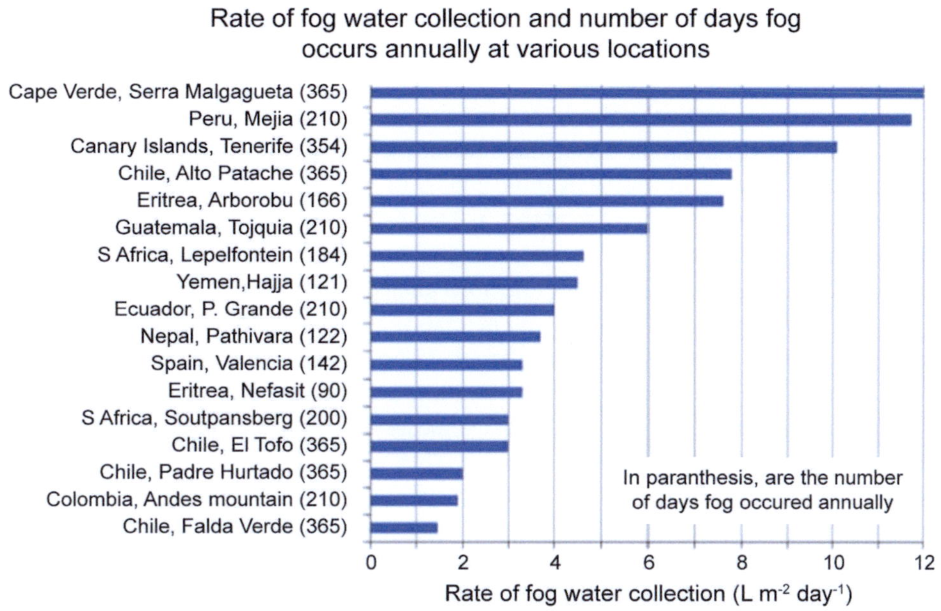

Fig. 14.8 Rate of fog water collection and number of days fog occurs annually at various desert locations (Adapted from Fessehaye et al. 2014; Bhushan 2019, 2020a, b)

When further cooled, the airborne water vapor will condense to form water droplets or dew. At a temperature lower than the dew point, the saturated vapor pressure of water in the ambient air decreases (Alduchov and Eskridge 1996), which leads to a higher amount of water condensation (Pruppacher and Klett 2010; Moran et al. 2018). It should be noted that the dew point temperature should be above the freezing point (0 ºC) as frost forms at or below 0 ºC (Bhushan 2020a, b).

The ambient temperature in the desert during nights can be low, as low as about 0ºC. If the temperature is lower than the dew point, it leads to water condensation from ambient; referred to as dew. Dew may also freeze as the temperature descends below 0ºC and is called white dew. The moisture from dew is generally available throughout the year with a larger amount available during dry summer months when living species are under highest stress when rainfall is small (Malek et al. 1999; Mares 1999; Bhushan, 2019, 2020a, b).

Malek et al. (1999) measured moisture contributed by dew in northeastern Nevada and central Utah in the U.S. Based on data collected for precipitation and dew moisture contribution for northeastern Nevada with area of about 1113 km^2, between June 10, 1993 and Sept. 30, 1994, they reported that annual precipitation was about 131 mm $year^{-1}$ and dew contribution was about 14 mm $year^{-1}$. Mares (1999) reported that dew contribution can range from as much as 10 L m^{-2} (10 mm) $year^{-1}$ in cold climates to as much as 75 L m^{-2} (75 mm) $year^{-1}$ in subhumid warm areas. In some desert areas, especially semiarid and Mediterranean habitats, dew may contribute a significant percentage of the total precipitation that is available in the area during the year for plants and animals (Bhushan 2020a, b).

Dew can be a significant source of water for plants and animals in many of the world's deserts, particularly during the hot summer months when rainfall is small (Monteith 1963; Baier 1966; Kidron et al. 2002; Kidron 2005). In Namibia and Chile, dew is among the primary sources of water for plants and animals. Various organisms have special adaptations to capture dew and use it to supply their water needs. In some cases, dew contribution may be a large fraction of the total annual water input on any given day. Dew is a source of water for insects and small animals. Examples are the *Diacamma rugosum*, a common ant in India that acquires a substantial fraction of water requirements from dew, and the snail *Trichoidea seetzenii* which has been found to use dew as a source of water (Agam and Berliner 2006). Hill et al. (2015) measured the role of dew in several plants which are native to the Negev desert in Israel. They reported that these plants received on the order of half of their water from dew. In certain conditions, dew formation may play an important role as a factor influencing the weathering of rocks (Bhushan 2020a, b).

It should be noted that water collection from condensation on an annual basis is believed to be lower than that collected from the fog for plants and animals (Bhushan 2020a, b).

14.2.2 Desert Plants and Animals

Many plants and animals can survive in deserts (Cloudsley-Thompson and Chadwick 1964; Axelrod 1979; Costa 1995; Anonymous 1996; Mares 1999; Van Rheede van Oudtshoorn and Van Rooyen 1999; Harris 2003; Bredeson 2009; Murphy 2012; Brown and Bhushan 2016; Bhushan 2018, 2019, 2020a, b; Gurera and Bhushan 2020b). Despite the harsh conditions of hot and arid regions like the Sahara, Namib, and Atacama deserts, a diverse array of plants and animals thrive, adapting to survive with minimal water. Based on estimates by World Wildlife Fund, some 500 species of plants, some 70 known mammalian species, some 90 avian species, some 100 reptilian species, many species of spiders, scorpions and other small arthropods are known to exist. Desert plants can be broadly classified into four classes—wildflowers, cacti and other succulents, trees and shrubs, and grasses, mosses, and lichens (Gurera and Bhushan 2020b). Desert animals include vertebrates and invertebrates (Gurera and Bhushan 2020b). Vertebrates can be broadly categorized into five classes: mammals, birds, reptiles, fishes, and amphibians. Examples of invertebrates include insects and crustaceans. Figure 14.9 presents a montage of selected plants from each class (Anonymous 1996; Bhushan 2020b; Gurera and Bhushan 2020b). Figures 14.10 and 14.11 present montages of selected animals—vertebrates and invertebrates, respectively (Bhushan 2020b; Gurera and Bhushan 2020b).

Fig. 14.9 Montage of selected desert plants with their class for each class (Adapted from Bhushan 2020b; Gurera and Bhushan 2020b)

Desert plants and animals have special adaptations. Unique structures and morphologies found in desert plants and animals assist them in water collection from ambient. Some plants may survive by growing roots that reach deep underground to find buried water sources and have leaves that are shaped into spines the minimize water loss. Plants such as the cactus have thick, fleshy stems that help them store water. Creosote bushes and mesquite trees have small leaves which lose less water in dry desert air than larger leaves would.

Both plants and animals have developed mechanisms to collect water on their leaves or bodies, which then travels to be consumed or stored.

Fig. 14.10 Montage of selected desert animals (vertebrates) for each class (Adapted from Bhushan 2020b; Gurera and Bhushan 2020b)

14.3 Water Harvesting—Lessons from Living Nature

In living nature, many plants and animals (e. g., Figs. 14.9, 14.10 and 14.11) in arid regions have evolved surface structures and chemistries that enable them to harvest water from fog and/or condensation of water vapor (Brown and Bhushan 2016; Bhushan 2018, 2019, 2020a, b; Gurera and Bhushan 2020b). External structures feature areas where droplets from fog or condensation get deposited and grow before eventually being transported to where it is consumed or stored. A key design feature is that collected water needs to be transported to where it is stored or consumed before it is evaporated. For example, in the case of some plants, the collected water is transported towards the roots

Fig. 14.11 Montage of selected desert animals (invertebrates) (Adapted from Bhushan 2020b; Gurera and Bhushan 2020b)

or trunk for storage. In the case of desert beetles and lizards, collected water is transported towards their mouths for consumption. High droplet velocity is needed for water collection purposes to minimize water loss to evaporation (Bhushan 2020a, b).

Figures 14.12 and 14.13 show water harvesting mechanisms for cacti, grasses, moss, bushes, Namib desert beetles, lizards, rattlesnakes, and spider webs (Brown and Bhushan 2016; Bhushan 2018, 2019, 2020a, b; Gurera and Bhushan 2020b). Details follow.

14.3.1 Cactus

Cacti commonly survive in arid regions. They are mostly succulents, a type of plant with thick, fleshy, and swollen parts that are adapted to store water and minimize water loss (Ogburn and Edwards 2009). Mooney et al. (1977) reported that cacti species use fog as a supplemental source of water. *Copiapoa haseltoniana*, a specie native to the Atacama Desert, utilizes the runoff from nightly fog events to survive in arid deserts (Vesilind 2003). *Opuntia microdasys* is a specie of cacti endemic to Mexico, which features small barbs atop conical spines that help in the collection of water (Ju et al. 2012). *Ferocactus latispinus* is another specie which features conical spines with tiny barbs (Bhushan 2020a, b; Gurera and Bhushan 2020b). *Gymnocalycium baldianum*, a specie in neighboring Argentina, collects and transports water through microcapillaries (Liu et al. 2015).

Figure 14.12 shows a schematic of spines with microscopic conical barbs on *Opuntia microdasys* specie. Water droplets collect on the conical spine and move towards the base due to curvature gradient providing Laplace pressure gradient. Droplets may also

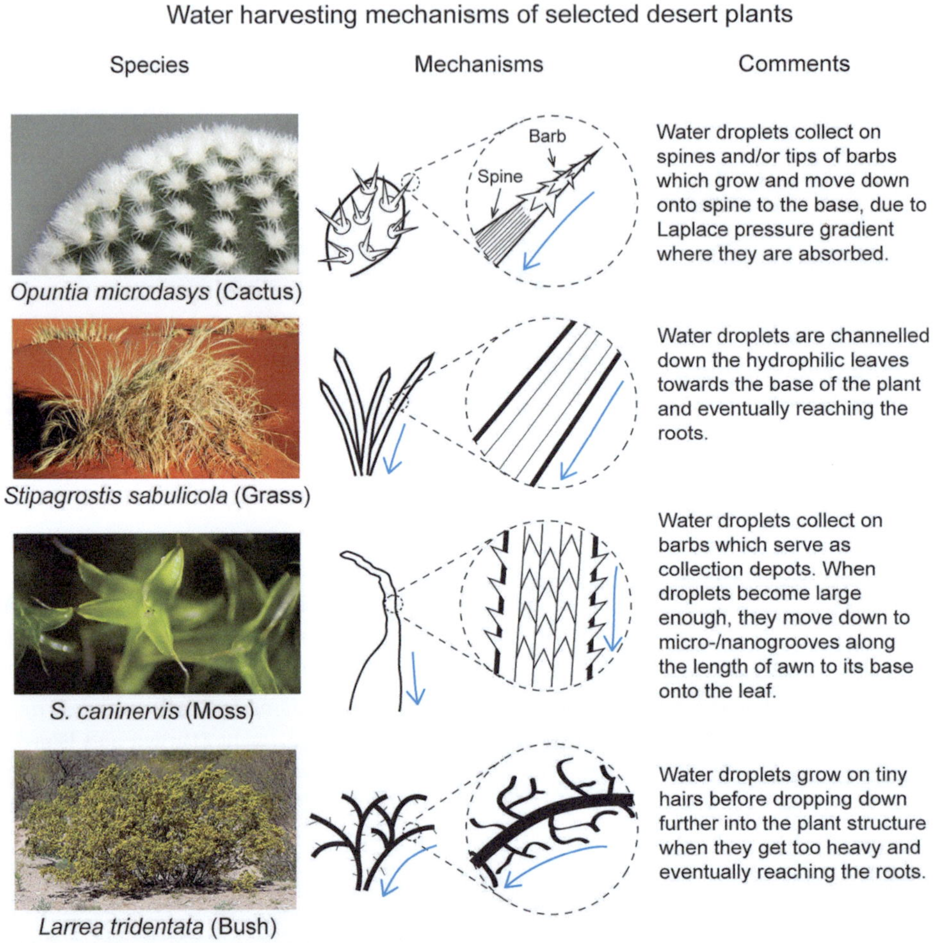

Fig. 14.12 Summary of desert plants found to collect water from fog and condensation of water vapor. A combination of surface structure and chemistry results in interception of water from fog and transport to the roots or another area where it can be consumed or stored (Adapted from Bhushan 2018, 2020a, b; Gurera and Bhushan 2020b)

collect on the tips of the small barbs and once they reach critical size, they move onto the conical spine. On the spine, the droplets move towards the base due to curvature gradient, providing Laplace pressure gradient (Appendix 14.A). Once at the base of the spine, the plant absorbs the water (Brown and Bhushan 2016; Bhushan 2018, 2019, 2020a, b). Laplace pressure gradient is large enough that water droplets can defy gravity and climb upward, as shown in Fig. 14.14 (Bhushan 2018, 2019, 2020b).

In addition to the presence of Laplace pressure gradient on a conical object, the surface slope results in gravitational forces acting on the surface. Therefore, for a conical object,

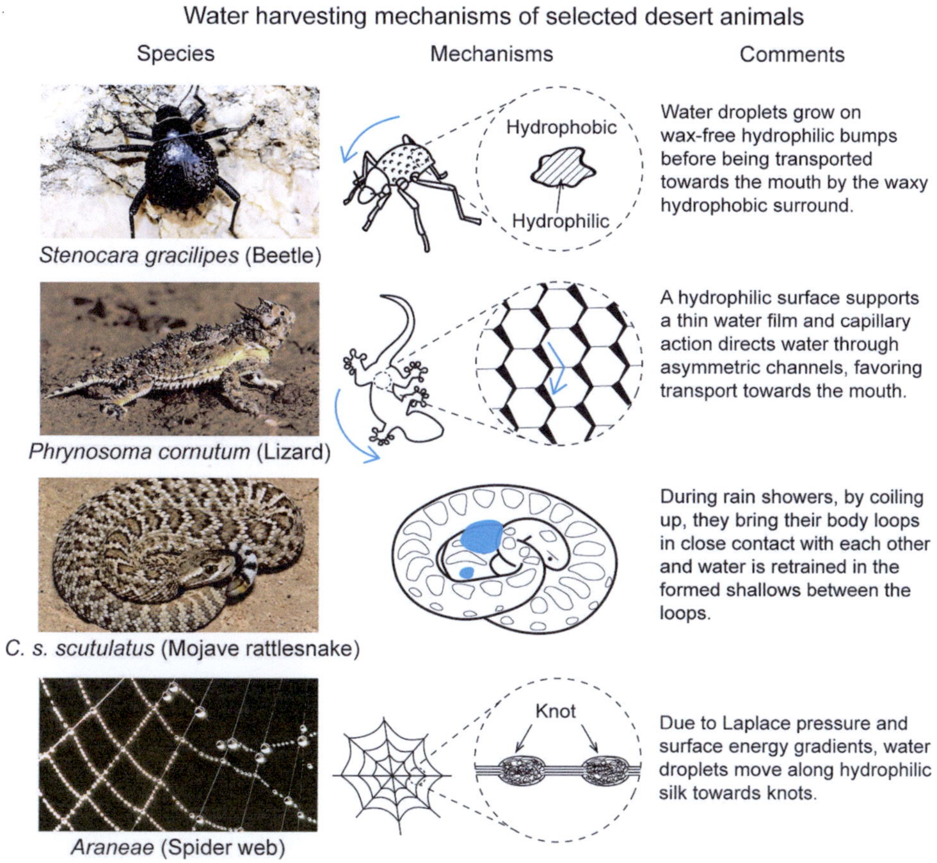

Fig. 14.13 Summary of desert animals found to collect water from fog and condensation of water vapor. A combination of surface structure and chemistry results in interception of water from fog and transport to the mouth or another area where it can be consumed or stored (Adapted from Bhushan 2018, 2020a, b; Gurera and Bhushan 2020b)

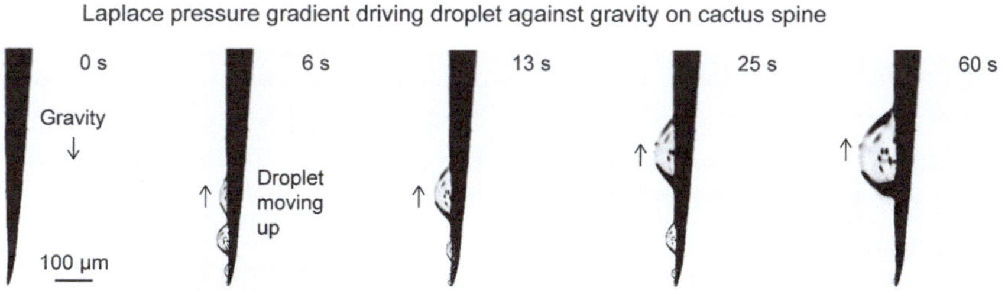

Fig. 14.14 Photographs of water droplets initially collected at the barb tip and/or spines of *Opuntia microdasy*, climbed up over the cactus spine due to Laplace pressure gradient. Droplets defy gravity (Adapted from Bhushan 2019, 2020a, b)

with tip pointed outward, and inclined with 0° or higher with respect to the horizontal axis, droplet movement is facilitated both by Laplace pressure gradient and gravitational forces (Bhushan 2018, 2019, 2020b).

Several water collection experiments have been performed on conical surfaces and on triangular geometry on flat surfaces to demonstrate that Laplace pressure gradient assists in droplet transport which increases the water collection rate (Bhushan 2018, 2019, 2020a, b).

14.3.2 Grass

Other plant species adapted to collect water from fog include *Stipagrostis sabulicola*, a grass endemic to the Namib desert (Brown and Bhushan 2016). Water droplets collect on the leaf before coalescing and running down towards the base of the plant (Louw and Seely, 1980; Ebner et al. 2011; Roth-Nebelsick et al. 2012; Bhushan 2019, 2020a, b). The leaves feature longitudinal ridges, which dictate this fluid flow, shown in Fig. 14.12 (Bhushan 2018, 2019, 2020a, b). Another type of grass, *Setaria viridis*, is found to collect water with a similar structure and mechanism to that found on the *Opuntia microdasys* cacti (Xue et al. 2014).

14.3.3 Desert Moss

Desert moss can survive in extremely arid regions. *Syntrichia caninervis* are common desert mosses in the desert. This plant is unique because its leaf surfaces must be wet for photosynthesis to occur, and its root-like structures do not collect water from the soil. The 0.5–2 mm long hair-like structures, referred to as awns or trichomes, exist at the tip of each leaf of the *S. carninervis*, which are used to collect water from fog, dew, and rain (Fig. 14.12) (Koch et al. 2008; Pan et al. 2016; Bhushan 2018, 2019, 2020a, b).

Awns are hydrophobic and have multiscale structure which facilitate water collection and transport. The awn is covered with nano- and micro- scale grooves and larger scale elongated conical barbs. Barbs serve as collection depots where collected water forms small droplets. When droplets become large enough, they move downward due to Laplace pressure gradient to the micro-/nano-groves present along the length of the awn. Droplets then move towards its base on the leaf surface. The awns provide water collection, droplet formation, and rapid transportation to the leaf to keep the moss alive (Bhushan 2018, 2019, 2020a, b).

14.3.4 Bushes

Larrea tridentata bush can survive in arid regions and is known to collect water from fog. Fine hairs on the leaves of the plant intercept fog droplets where they coalesce and grow before dropping down into the plant structure when they become too heavy, and eventually reach to the roots, as shown in Fig. 14.12 (Bhushan 2018, 2019, 2020a, b).

14.3.5 Namib Desert Beetles

The Namib desert in southern Africa is one of the most arid regions in the world with average annual rainfall of only 1.8 cm, and it is not uncommon to experience consecutive years with no rainfall at all (Shanyengana et al. 2002). *Stenocara gracilipes* and *Onymacris unguicularis* are beetles which are native to this region. The beetles survive as a result of collection of water from fog. The first observation of fog harvesting in the Namib desert was made by Hamilton and Seely (1976), with beetles emerging during nocturnal fogs and lowering their heads while oriented into the wind. The water was found to trickle down the body of the beetle and into the mouth (Bhushan 2018, 2019, 2020a, b).

The back of the beetle is comprised of a random array of about 0.5 mm diameter bumps spaced 0.5–1.5 mm apart, shown in Fig. 14.13 (Bhushan 2020a, b). The bumps are smooth, while the surrounding area is covered in microstructured wax (Parker and Lawrence 2001). The bumps are hydrophilic while the background wax is hydrophobic. Water from the fog lands on the bumps and droplets begin to grow. The droplet continues to grow (up to 4–5 mm) until the weight of the droplet overcomes the capillary force and the droplet detaches and gets transported to the neighboring hydrophobic region and eventually rolls down the tilted beetle's back to its mouth (Bhushan 2018, 2019, 2020a, b).

Gurera and Bhushan (2019a) reported that superhydrophobic surfaces with superhydrophilic spots increase the fog water collection as compared to the flat surfaces with various homogeneous wettability. These experiments verified that an array of hydrophilic bumps on a hydrophobic background is efficient for fog harvesting.

14.3.6 Lizards

Moloch horridus is a species of lizard native to arid regions in western and southern Australia, shown in Fig. 14.13 (Bhushan 2018, 2019, 2020a, b). Bentley and Blumer (1962) reported that water droplets spread out over the skin before reaching the mouth. Water movement occurs due to capillary action along open channels within the skin. Similarly, in the case of *Phrynocephalus helioscopus*, a species of lizard native to arid regions in Asia, the lizard adopted a posture where the head was depressed and the hindquarters were elevated which helped guide the water to the mouth (Schwenk and Greene 1987).

Comanns et al. (2015) reported that in the case of *Phrynosoma cornutum* or the Texas horned lizard, water placed on the skin flowed preferentially towards the mouth, with capillary forces dominating over gravitational and viscous forces. A network of capillary channels exists with a narrowing of individual channels in the direction of the mouth (longitudinal), shown in Fig. 14.13 (Bhushan 2020a, b). An abrupt widening from one channel to the next, in addition to an interconnecting narrower channel running laterally, would follow. The narrowing of the channel results in favorable water transport in that direction due to the curvature of the liquid–air interface. The lateral interconnecting channels overcome the effects of the abrupt widening and help maintain an advancing liquid front (Comanns et al. 2015). In the backward direction, liquid flow is stopped as the channel widens and pressure would need to be applied to force the liquid in that direction (Brown and Bhushan 2016; Bhushan 2018, 2019, 2020a, b).

14.3.7 Rattlesnakes

Many moisture-harvesting reptiles collect water from fog. For example, desert snakes emerge from their dens or any potential roofing during rain showers and coil up in the open. By coiling up, they bring their body loops in close contact with each other, and water is retained in the formed shallows between the loops, shown in Fig. 14.13 (Cardwell 2006; Gorb and Gorb 2017; Bhushan 2020a, b). Some species also exhibit a dorso-ventral flattening of the body, probably to increase exposed surface area. With the snout continuously in contact with the shallows and the head moving contrary to scales' overlap, the collected water is ingested.

14.3.8 Spider Webs

Spider webs are known to collect water, as evidenced by capturing a dew-glistened web, shown in Fig. 14.13 (Bhushan 2018, 2019, 2020a, b). Dry webs undergo moisture-induced structural reorganization. First, the hygroscopic nature of the proteins contained within the silk results in the condensation of water droplets and the swelling of the cylindrical silk thread. This cylinder is then broken up due to Rayleigh instability, where a cylinder of fluid will break up into smaller drops to lower its surface area, resulting in the formation of a "beads on a string" structure with a series of knots periodically spaced along the thread (Edmonds and Vollrath 1992; Brown and Bhushan 2016). It is believed that this rebuilding of the web structure and subsequent water capture is beneficial for the spider, as not only does it provide a source of drinking water, but it also results in improved capture of prey due to the water-swollen knots possessing enhanced adhesive properties (Edmonds and Vollrath 1992; Bhushan 2018, 2019, 2020a, b).

14.4 Bioinspired Flat and Conical Surfaces for Water Harvesting

As mentioned earlier, beetles use heterogeneously wettable surfaces, cactus spines use conical geometry, and blades of grass use longitudinal grooves to drive water droplets for water transport and storage/use before they evaporate. Bioinspired surfaces for water harvesting from fog have been inspired by the beetle (Garrod et al. 2007; Bai et al. 2014; Gurera and Bhushan 2019a), cactus (Ju et al. 2013; Gurera and Bhushan 2019a, b, c, d; Schriner and Bhushan 2019), and grass (Azad et al. 2015; Gurera and Bhushan 2019a, 2019b). For water harvesting from condensation of water vapor, cactus-inspired conical surfaces have also been used (Gurera and Bhushan 2020a).

Cactus-inspired triangular patterns also have been used for water transport (Sect. 14.5). The crucial thing for a droplet to move inside a triangular pattern is that it should be touching the borders. This is not the case in cones. Laplace pressure gradient always acts on a droplet sitting on a cone. This presents a clear advantage of cones over flat triangular patterns.

Water collection studies from fog on flat and conical surfaces were carried out by Gurera and Bhushan (2019a, b, c, d). Heterogeneous wettability, Laplace pressure gradient, gravitational effects, and grooved surfaces were investigated on flat and conical water collectors. The flat surfaces with different wettability and surface roughness, as well as heterogeneous wettability, were characterized for their water collection abilities. Conical surfaces with different geometry, ungrooved and grooved, and different wettability were characterized for droplet movement and their water collection abilities at different inclination angles. Cones with nonlinear surface profiles, to maximize the effects of Laplace pressure gradient and gravitational effects, were also studied.

Water collection studies from condensation of water vapor on surfaces was carried out by Gurera and Bhushan, (2020a). Metallic cones of different tip angle, cone length and surface area were used. The effect of inclination angle and array was also studied. The results were compared with the water collection from fog.

In this section, an overview of studies of water harvesting from fog using flat and conical surfaces carried out by Gurera and Bhushan (2019a, b, c, d) is presented. Based on the data, design guidelines for water harvesting towers with heterogeneous wettability and/or consisting of conical arrays are presented (Bhushan 2018, 2019, 2020a, 2020b).

14.4.1 Flat Surfaces with Homogeneous and Heterogeneous Wettability

Flat surfaces with homogeneous wettability included surfaces with superhydrophobicity, hydrophobicity, hydrophilicity, and superhydrophilicity (Bhushan 2017, 2018). Beetle-inspired surfaces consisted of superhydrophilic spots over a superhydrophobic

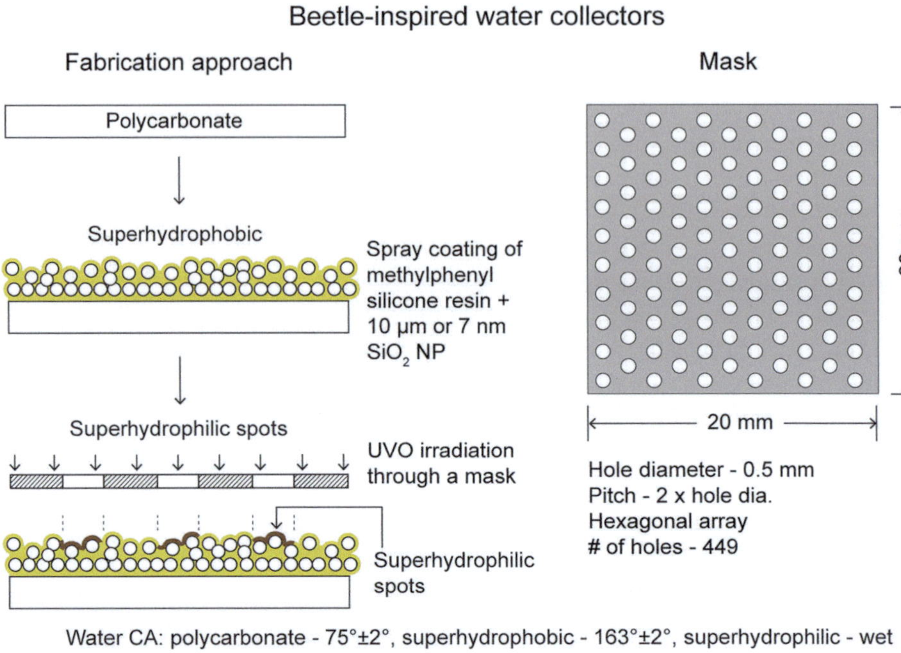

Fig. 14.15 Fabrication of surfaces with heterogeneous wettability for beetle-inspired water collectors (Adapted from Bhushan 2019, 2020a, b; Gurera and Bhushan 2019a)

background. A spot size of 0.5 mm diameter and a pitch of 1 mm was chosen (Fig. 14.15) (Gurera and Bhushan 2019a).

14.4.2 Conical Surfaces with and Without Grooves and Homogeneous and Heterogeneous Wettability

Cactus inspired conical surfaces were fabricated with and without grooves, inspired by grass, and some with heterogeneous wettability. Heterogeneous wettability in grass and cactus does not exist in nature. It was incorporated with inspiration coming from the desert beetle. These were fabricated using additive manufacturing or 3D printing (Objet30 Prime, Stratasys) that allows flexibility in designing and scalability (Gurera and Bhushan 2019a, 2019b). Various designs are presented in Fig. 14.16 (Gurera and Bhushan 2019a).

Cylinder Versus Cone

Conical and cylindrical surfaces were fabricated to study the role of conical geometry. The design was based on the dimensions of natural species. A base diameter of the cylinder of 3 mm was chosen to mimic the grass's diameter. The length was chosen to be 35 mm.

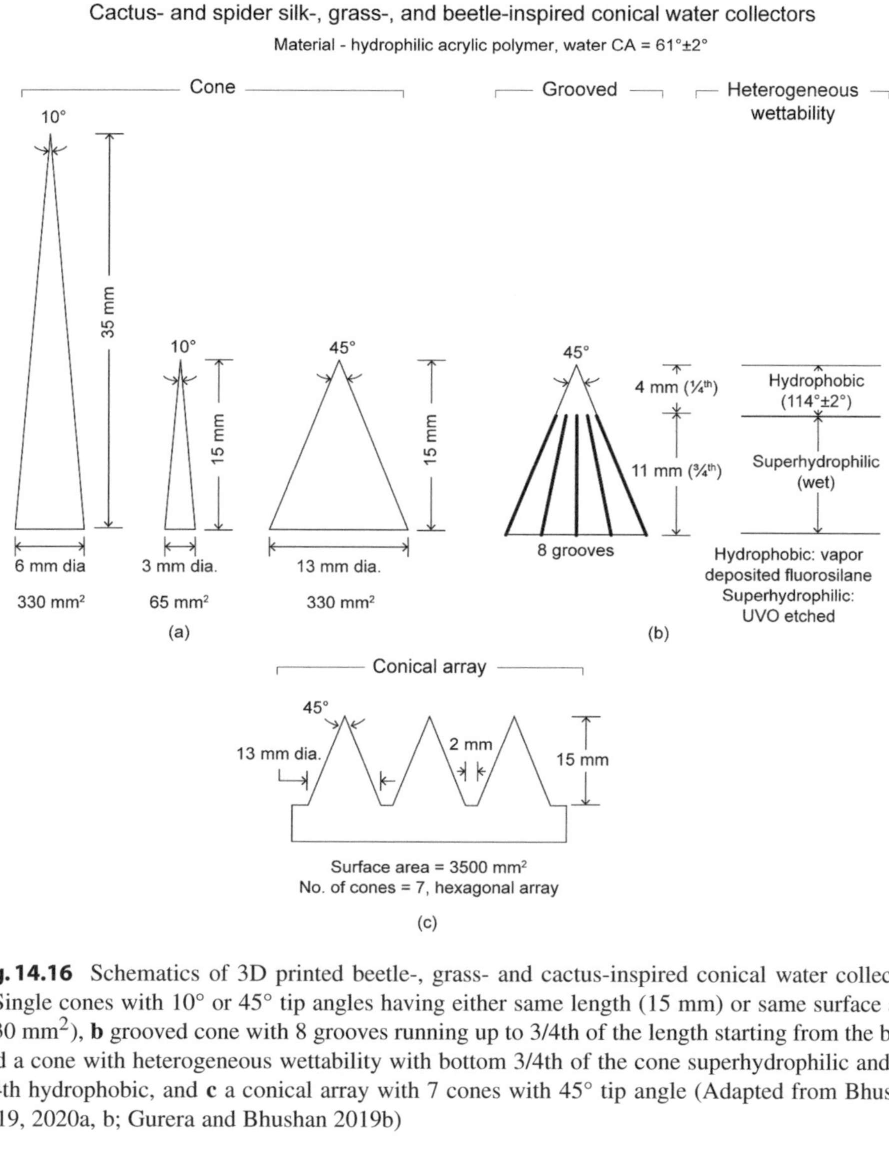

Fig. 14.16 Schematics of 3D printed beetle-, grass- and cactus-inspired conical water collectors **a** Single cones with 10° or 45° tip angles having either same length (15 mm) or same surface area (330 mm^2), **b** grooved cone with 8 grooves running up to 3/4th of the length starting from the base, and a cone with heterogeneous wettability with bottom 3/4th of the cone superhydrophilic and top 1/4th hydrophobic, and **c** a conical array with 7 cones with 45° tip angle (Adapted from Bhushan 2019, 2020a, b; Gurera and Bhushan 2019b)

To keep the surface area and the length constant, the base diameter of the cone, 6 mm, was doubled to that of cylinder, and the tip angle was kept at 10°, as seen in cactus spines (Gurera and Bhushan 2019a).

Single Cone

(a) Geometry

To study the effect of conical geometry, two tip angles were chosen: 10° and 45°. Cacti conical spines have a tip angle of about 10°. For comparison, a larger tip angle of 45° was chosen (Gurera and Bhushan 2019b). When considering the collection of water from fog, a comparison between cones of different tip angles and the same surface area is needed because fog intercepts the entire surface, and droplets are formed across the whole surface. Therefore, having a similar number of water nucleation sites presents a fair comparison. It is known that a droplet moves from the tip to the base under a fog flow. Therefore, comparing tip angles with same lengths, presents another fair comparison. This results in the traveling distance for droplets to be the same on each cone (Gurera and Bhushan 2019b).

Three cones with two tip angles were fabricated—one pair with the same surface area (330 mm^2) and another pair with the same length (15 mm), as shown in Fig. 14.16 (Gurera and Bhushan 2019b). The surface area, A, is given by $\pi l^2(\tan(\theta/2)/\cos(\theta/2))$, where l is the cone length, and θ is the tip angle.

(b) Grooves

A representative cone of 45° tip angle and 15 mm length, was chosen to investigate the effect of grooves inspired by desert grass and cactus, as shown in Fig. 14.16 (Gurera and Bhushan 2019b). The number of grooves selected were 8; they were spaced equidistantly and ran up to 3/4th of the cone's length, as inspired from cactus. The thin grooves found on a cactus spine could not be fabricated using the 3D printer. Therefore, grooves' cross-sectional dimension, 0.4 mm by 0.4 mm, inspired from grass was used.

(c) Heterogeneity

A representative cone of 45° tip angle and 15 mm length, was chosen to investigate the effect of heterogeneity, as shown in Fig. 14.16 (Gurera and Bhushan 2019b). Heterogeneous wettability in grass and cactus does not exist in nature. It was incorporated with inspiration coming from the desert beetle. Heterogeneous wettability in the conical surfaces included a hydrophobic tip with a superhydrophilic base. A hydrophobic tip was selected as it can collect more water, and a superhydrophilic base because it can transport the collected water quickly to the base. Superhydrophobic tips are not recommended because droplets will not stick to the surface and will fly away instead because of the low tilt angle.

For fabrication, the entire surface of the cones were made superhydrophilic using a UVO lamp. The bottom 3/4th was then covered with Teflon tape and the entire object was subjected to fluorination; afterwards the tape was removed. The fluorination was achieved via vapor deposition of fluorosilane (448,931, Sigma Aldrich). The samples were placed upside down and a droplet of fluorosilane was placed 1 cm below the tip in an enclosure, and was left for 30 min (Brown and Bhushan 2015).

Conical Array

To design arrays, representative cones with 45° tip angle and 15 mm length, were chosen. Seven cones were selected in the array, with end-to-end spacing between two adjacent cones of about 2 mm, as shown in Fig. 14.16 (Gurera and Bhushan 2019b).

14.4.3 Experimental Apparatuses for Water Collection from Fog

Figure 14.17a shows a schematic of the water collection apparatus used to collect water from fog (Gurera and Bhushan 2019a, b). The conical samples were mounted to a base. The sample base was inclined at either 0º or 45º from vertical axis (θ). A commercial humidifier was used to produce a stream of fog onto a surface. The falling water droplets were collected in a container and were measured using an analytical balance underneath. The minimum mass the balance could measure was 1 mg. The humidifier which emits fog at about 10 cm/s was kept about 20 cm away from the surface. The flow speed was calculated by measuring the volume of water lost over time, and by knowing the diameter of the pipe through which the fog was blown out.

When the fog is turned on, droplets form all across the intercepted surface area. There are two forces which drive these droplets to the base—gravity and force due to the Laplace pressure gradient. To characterize the effect of both forces individually, two inclination angles were chosen—0° and 45°; here 0° means the cone axis is parallel to the fog flow (horizontal). At 0° inclination angle, the only driving force is the Laplace pressure gradient. At 45° inclination angle, the force due to both gravity and the Laplace pressure gradient act on the droplets. In the experiments, mass range of each collected drop was about 30 to 60 mg.

A representative data of amount of water collection as a function of time for two surfaces is presented in Fig. 14.17b. The two surfaces were cones with a 10° tip angle and a flat surface. Both were hydrophilic, had the same surface area of about 330 mm^2, and were inclined at a 45° angle. It is observed that water collection rate for cone is larger than that of the flat surface.

14.4.4 Results and Discussion

Water collection data from fog for flat, cylindrical, and conical surfaces are presented in this section (Gurera and Bhushan 2019a, b). First, water collection data by flat surfaces of various wettability and beetle-inspired surfaces at two inclination angles is presented. To study the role of conical geometry, a cylinder vs. cone with the same surface area

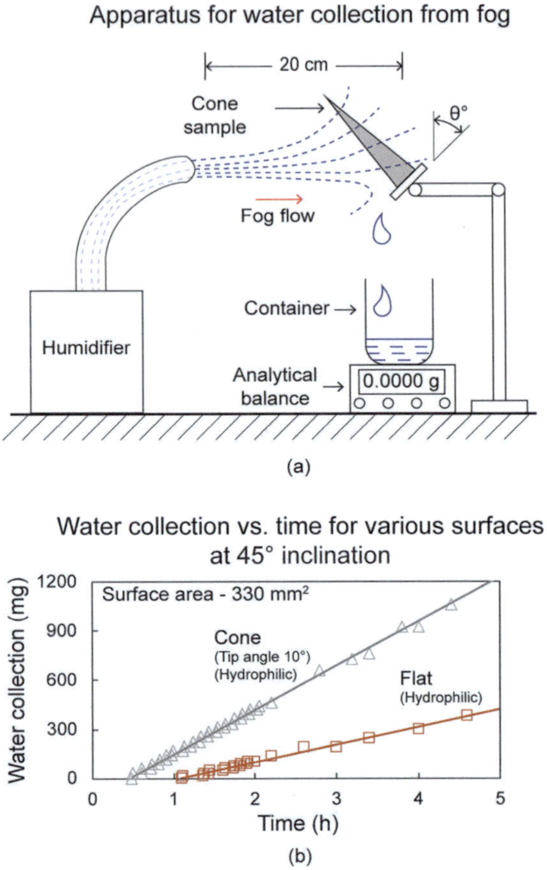

Fig. 14.17 **a** Schematic of experimental apparatus for water collection from fog. A commercial humidifier was used to throw a stream of fog on a cone sample. The sample base was inclined at either 0° or 45° from vertical axis (θ). Inclination angle of 0° means that the cone axis is parallel to the fog flow (horizontal). The collected water was measured by an analytical balance underneath. **b** A representative water collection (mg) versus time (h) plot for hydrophilic flat surface, and hydrophilic cone surface of 10° tip angle and 35 mm length, both at 45° inclination (Adapted from Bhushan 2019, 2020b; Gurera and Bhushan 2019a, b)

and two inclination angles is presented. Next water collection data from fog on cones with various geometries and wettability at two inclination angles, and conical arrays are presented. Finally, design guidelines for water collection tower are presented (Bhushan 2020a, b).

Flat Surfaces with Various Wettability and Beetle-Inspired Surfaces at 45°and 0° Inclination Angles

To study the effect of various wettability and heterogeneity on flat surfaces, water collection rate per unit area was measured at a 45° inclination angle. The data are shown in Fig. 14.18 (Gurera and Bhushan 2019a). The data shows that the higher the repellency (higher contact angle) on the flat surface, the higher the water collection. The decreasing order of the water collection rate per unit area on flat surfaces with homogeneous wettability is superhydrophobic, hydrophobic, hydrophilic, and superhydrophilic. This is attributed to the fact that the higher the repellency, the more spherical the droplet shape

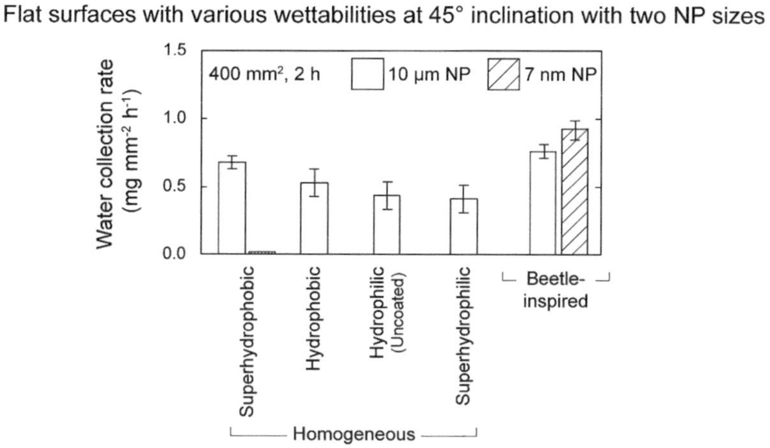

Fig. 14.18 Water collection rates per unit surface area for flat surfaces with various wettability and beetle-inspired surfaces, at 45° inclination. The wettability includes superhydrophobic, hydrophobic, hydrophilic (uncoated), and superhydrophilic surfaces, and a beetle-inspired surface (which includes 0.5 mm diameter circular superhydrophilic spots surrounded by a superhydrophobic surface (Adapted from Bhushan 2019, 2020a, b; Gurera and Bhushan 2019a)

will be, which results in a lower contact area between the droplet and the surface. Therefore, less heat will be transferred to the droplet and less evaporation will be observed. Large spherical droplets are formed on the superhydrophobic surface which roll down. In the case of hydrophilic and superhydrophilic surfaces, the droplets spread across the surface leading to higher evaporation (Gurera and Bhushan 2019a).

A beetle-inspired surface was found to collect more water than flat surfaces with various homogeneous wettability. In beetle-inspired surface, because of the heterogeneity, droplets can slide/roll at a faster rate, and maintain a spherical droplet shape. The faster rate also leads to less evaporation (Gurera and Bhushan 2019a).

Gurera and Bhushan (2019a) also performed experiments on various surfaces at 0° inclination angle. Water collection rates were lower at 0° inclination angle because of the differences in fog interception. At 0° inclination angle, fog forms a whirl, and some droplets get lost to the wind. Whereas at 45° inclination angle, fog intercepts and slides tangentially leading to deposition of larger number of droplets.

To summarize, it was found that an increase in water repellency on flat surfaces resulted in higher water collection rates. Water collection rates at 45° inclination angle (with respect to the vertical axis) was larger than at 0° inclination angle. Surfaces with heterogeneous wettability had higher water collection rate than surfaces with homogeneous wettability.

Cylinder Versus Cone at 0° Inclination Angle

To study the role of conical geometry, water collection rates on a cylinder and a cone were measured. Figure 14.19a shows the water collection data for single cylindrical and conical surfaces, with same surface area and length, at 0° inclination angle (Gurera and Bhushan 2019a). The 0° angle was chosen to minimize the effect of gravity on the movement of the droplets along the length. The left half of the bar chart shows water collection for the complete object. However, in nature, the collected water that matters is that which reaches the base, therefore, the right half of the bar chart presents the water collection rate per unit area at base half-length. The complete water collection rate per unit area by cylinder and cone was found to be comparable. However, it was negligible at the base half-length for cylinder. Figure 14.19b shows comparison of mass and frequency of droplet from a cylinder and a cone in complete collection. Cones collect heavier droplets with a lower frequency.

Figure 14.19c shows optical images of droplets growing and moving on a cylinder and a cone at three-time steps (Gurera and Bhushan 2019a). On the cylinder, water droplets nucleate at the tip, grow, attain a critical size, then fall. Droplets mostly nucleate at the tip because they intercept the fog flow perpendicularly. On the rest of the curved surface, little to no droplets were observed since the surface is parallel to the fog flow. The smaller droplets became visible much later; at about the 4-h mark. Since the droplets are small and sit on the surface for a long time, a significant amount of their volume is lost to evaporation. Therefore, the droplets nucleating at the cylinder's tip are mostly responsible for the water collection rate per unit surface area (Gurera and Bhushan 2019a)

The droplets growing at the cylinder's tip do not move along the length, since there is no force acting in that direction. However, this is not the case in cones. Droplets nucleate at the tip, grow and move along its length. For a droplet resting on a conical surface, the changing radius results in a pressure difference inside the droplet. Laplace pressure gradient causes droplets to move on a cone from regions of low radius to regions of higher radius.

Irrespective of the motion of the droplet, both a cylinder and a cone have similar water collection rate per unit area. As mentioned earlier, in nature, the water collection that matters is at the base. A cylinder does not achieve any water collection at the base, whereas a cone achieves the same water collection at the base that was observed in complete collection. As a result, cones have an advantage over cylinders in water collection at the base.

To illustrate that Laplace pressure gradient defies gravity, Fig. 14.20 shows video stills of droplets on a cone with 10° tip angle pointed downward and a cylinder. It can be seen that droplets grew larger but did not move on a cylinder. However, droplets grew, coalesced and climbed upward on a cone, defying gravity.

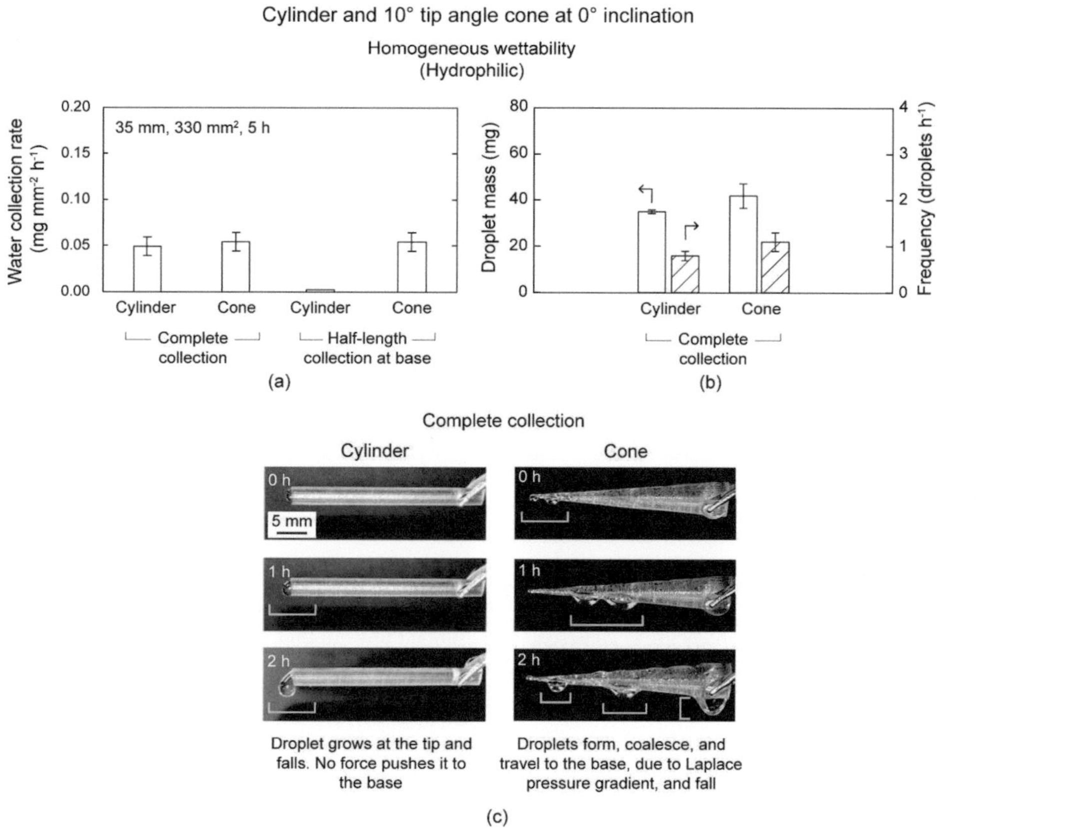

Fig. 14.19 **a** Water collection rates for a cylinder and a cone at 0° inclination. The left half of the bar chart shows the data when water was collected under the entire surface (complete collection). The right half shows the data when the water was collected under the half-length near the base (as in nature), **b** mass and frequency of falling droplets from a cylinder and a cone in complete collection, and **c** The optical images showing movement of the water droplets on a cylinder and a cone in complete collection (Adapted from Bhushan 2019,2020a, b; Gurera and Bhushan2019a)

Cones at 45° Inclination Angle and Comparison with Flat Surfaces with Homogeneous and Heterogeneous Wettability

To study the effect of inclination angle, water collection rates on a cone at 45° inclination angle were measured at the base. Data are shown in Fig. 14.21 (Gurera and Bhushan 2019a). The water collection rate at 45° inclination is about an order of magnitude higher that that at 0° inclination (Fig. 14.19). At 45° inclination angle, the entire surfaces intercept the fog flow, not just its tip. Droplets nucleate, grow, slide along the length, and drop at the base. Droplets fall at the base because there is gravity (g sin θ), in addition to Laplace pressure gradient, aiding the movement of the droplets towards the base. It was reported by Gurera and Bhushan (2019a) that at 45° inclination angle, the frequency

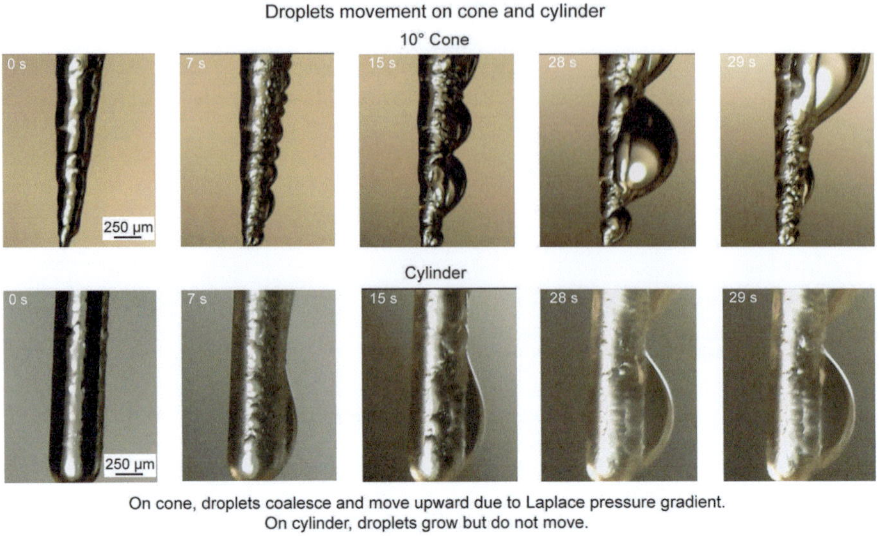

Fig. 14.20 Video stills of water droplets on a cone with 10° tip angle, pointed downward, and a cylinder, both inclined in the vertical direction. Water droplets grew larger but did not move on a cylinder; however, droplets grew, coalesced and climbed upward on a cone due to Laplace pressure gradient, defying gravity (Bhushan 2019, 2020a, b)

of the droplets falling was about 6–8 times higher as compared to the 0° inclination. However, aid by gravity did not increase the mass of the droplets falling. To summarize, higher inclination increased the water collection rate.

The cone data is also compared with the data for flat and beetle-inspired surfaces in Fig. 14.21. Cones and beetle-inspired surfaces collect about twice as much water as that for a hydrophilic flat surface. Water collection for cones and beetle-inspired surfaces is

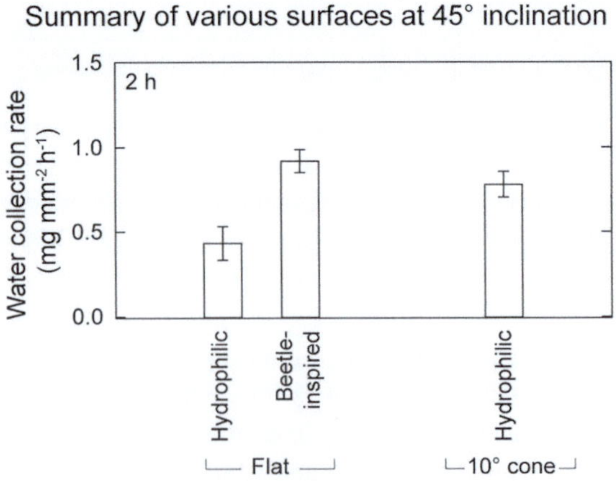

Fig. 14.21 Summary of water collection rates for flat hydrophilic and beetle-inspired surfaces and a hydrophilic cone at 45° inclination (Adapted from Bhushan 2019, 2020a, b; Gurera and Bhushan 2019a)

comparable. Water is transported faster on cones and beetle-inspired surfaces than on the flat surfaces which also reduces evaporation. Finally, in the case of cones, a surface covered with cones (conical array) would increase the surface area to increase collection by several fold, as much as factor of 10 or even more.

Cones—Effect of Geometry, Inclination, Grooves, and Heterogeneous Wettability

Two sets of cones were tested to evaluate the effect of tip angles—one with the same surface area and another with the same length at two inclination angles. Cones with grooves and heterogeneous wettability were also studied. Next, a conical array was tested (Gurera and Bhushan 2019b).

(a) Two Tip Angles with Same Surface Area or with Same Length

Gurera and Bhushan (2019b) measured water collection rate for cones with two tip angles and with the same surface areas at 0° inclination (Data not shown). They reported that the cone with higher tip angle had higher water collection rate. This is because cone length of the larger tip angle cone is shorter and droplets take lesser time to reach to the base. They also reported that although the mass of the falling droplets was similar, the cone with a larger tip angle had a higher frequency of falling droplets.

Gurera and Bhushan (2019b) also measured water collection rates for cones with two tip angles of the same length at 0° inclination. The cones of same length were found to have similar water collection rate. Laplace pressure gradient was found to be more effective in a smaller tip angle cone. However, due to the difference in surface areas, more water droplets are formed on the surface of the 45° tip angle cone. This makes the water collection rate for both the cones similar. Droplet mass and frequency were reported to be similar for the two cones.

(b) Inclination Angle

Figure 14.22 shows water collection rates for cones with two tip angles at two inclination angles (Gurera and Bhushan 2019b). The cones of 10° and 45° tip angles, having either the same surface area or same length were compared at two inclination angles of 0° and 45°.

As it has been mentioned before, there are two forces driving droplets on the conical surfaces due to Laplace pressure gradient and gravity. In the previous section, it was established that in the absence of driving gravitational force, at 0° inclination angle, the cone length dictates the water collection rate. The shorter the cone, the higher the water collection rate, irrespective of the tip angle. This is because the distance travelled by the droplets from the tip to the base is shorter in shorter cones.

At a 45° inclination angle, gravity also drives droplets. In addition to role of Laplace pressure gradient and gravity, distribution of droplets on the cone surface also plays a role. At 0° inclination angle, as indicated earlier, larger droplets were observed near the

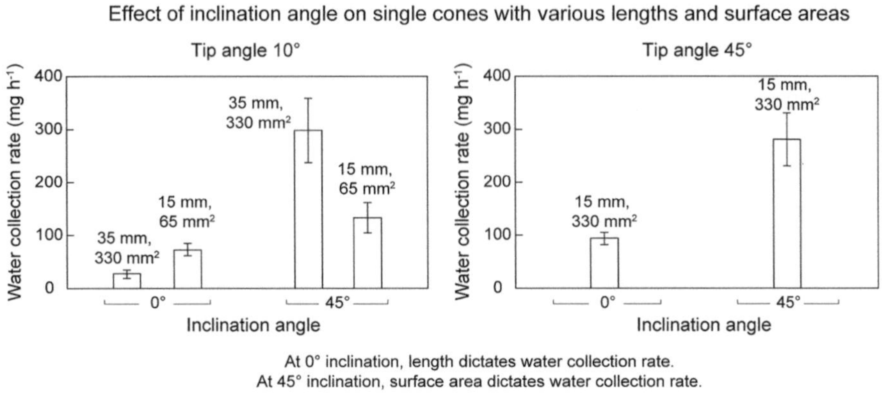

Fig. 14.22 Summary of water collection rates for cones of tip angles of 10° and 45°, with either the same length or the same surface area, and at two inclination angles of 0° and 45° (Adapted from Bhushan 2019,2020a, b ; Gurera and Bhushan 2019a)

tip because of higher fog interception. These droplets grow further and move towards the base which takes time. At 45° inclination angle, water droplets were observed on the entire surface area. More uniform distribution will increase the water collection rate. Therefore, increasing the inclination angle increases the water collection rate, irrespective of the cone tip angle. The increase is more for cones with a larger surface area. At a higher inclination angle it is the number of droplets formed on the surface that dictates the water collection rate, not the length.

(c) Grooves

To study the effect of grooves, a representative cone with a tip angle of 45° and a length of about 15 mm was selected. Figure 14.23a shows the water collection rates (mg h^{-1}) of ungrooved and grooved cones at two inclination angles, 0° and 45° (Gurera and Bhushan 2019b). At a 0° inclination angle, grooves help in increasing the water collection rate. This is because grooves help in channeling the water. It is believed that they also create an increased Laplace pressure gradient via the gradient in grooves spacing. However, at a 45° inclination angle, the improvement due to grooves is not observed. That is probably because the gravitational force is overpowering the advantage of grooves.

To study the differences in the channeling of a water droplet on the ungrooved and grooved cones, a fixed volume of water droplet, 5 μL, was placed on the cones at similar locations (Gurera and Bhushan 2019b). The optical images are shown in Fig. 14.23b. The droplets between the grooves appears to be more channeled towards the base. This elongation of the droplet is the reason for the increase in the water collection rate in the grooved cone.

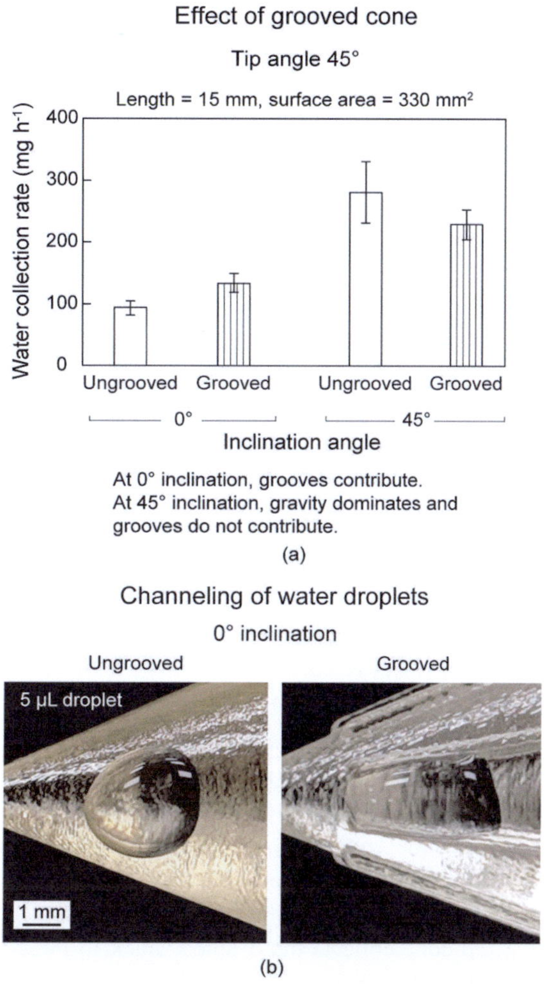

Fig. 14.23 **a** Water collection rates for ungrooved and grooved cones with tip angle of 45°, at 0° and 45° inclination angles. **b** A water droplet of volume of 5 μL was placed on ungrooved and grooved cones to demonstrate the channeling in the grooved cone (Adapted from Bhushan, 2019, 2020a, b; Gurera and Bhushan, 2019b)

(d) Heterogeneous Wettability

To study the effect of heterogeneous wettability, a cone with a tip angle of 45° and a length of about 15 mm was selected. Figure 14.24 shows the effect of heterogeneity on water collection rate (mg h^{-1}) at two inclination angles of 0° and 45° (Gurera and Bhushan 2019b). The water collection rates are compared with the cones with heterogeneous wettability. At 0° inclination angle, the heterogeneity increases the water collection rate. This is because the heterogeneity assists in transporting the water quickly. However, at a 45° inclination angle, an improvement due to heterogeneity is not observed. This is probably because the gravitational force dominates the transport and not the heterogeneity.

Multistep wettability gradient studies are reported by Gurera and Bhushan (2019d) and not included in the chapter.

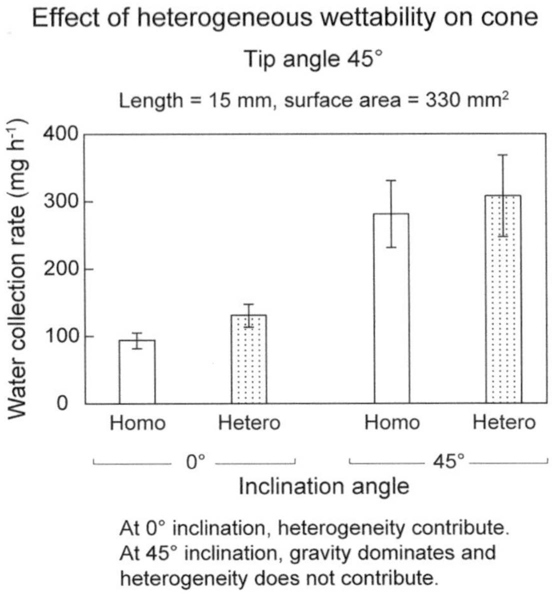

Fig. 14.24 Water collection rates for cones with a tip angle of 45° and homogeneous and heterogeneous wettability, and at 0° and 45° inclination angles (Adapted from Bhushan 2019,2020a, b; Gurera and Bhushan 2019b)

(e) Conical Array

An array with cones with a tip angle of 45° and a length of about 15 mm were selected to evaluate effects of arrays at two inclination angles of 0° and 45°. Figure 14.25a shows a water collection rate (mg h^{-1}) for an array (Gurera and Bhushan 2019b). The array data is presented as water collection rate per number of cones. The data are compared with a single cone with a tip angle of 45°. At a 0° inclination angle, having an array increased the water collection rate per number of cones. It is believed that this increase is due to a cascading effect on a falling droplet. This is because a droplet falling from the top cone collects the droplets stuck to the cone underneath increasing the net water collection rate per cone, as shown in Fig. 14.25b (Gurera and Bhushan 2019b). At a 45° inclination angle, the data for a single cone and the array per cone is comparable. Water collection rates at 45° inclination angle is larger than at 0° inclination angles. As suggested earlier, this occurs because the gravitational forces dominate the water transport at a higher inclination.

(f) Summary

For water collection on conical surfaces from fog, water droplets are driven by the Laplace pressure gradient and gravity. During travel, droplets coalesce and travel farther because of larger mass. The Laplace pressure gradient dominates at 0° inclination angle, which results in higher water collection rate for shorter length. Gravity dominates the water collection rate at 45° inclination angle, which results in higher water collection rate for larger surface area. The water collection rate remains independent of the tip angle for same

Fig. 14.25 **a** Water collection rate of a single cone and an array at 0° and 45° inclination angles. The array data is presented as water collection rate per cone. **b** Optical images of falling droplets on array at 0° inclination angle showing coalescence (Adapted from Bhushan 2019, 2020a, b; Gurera and Bhushan 2019b)

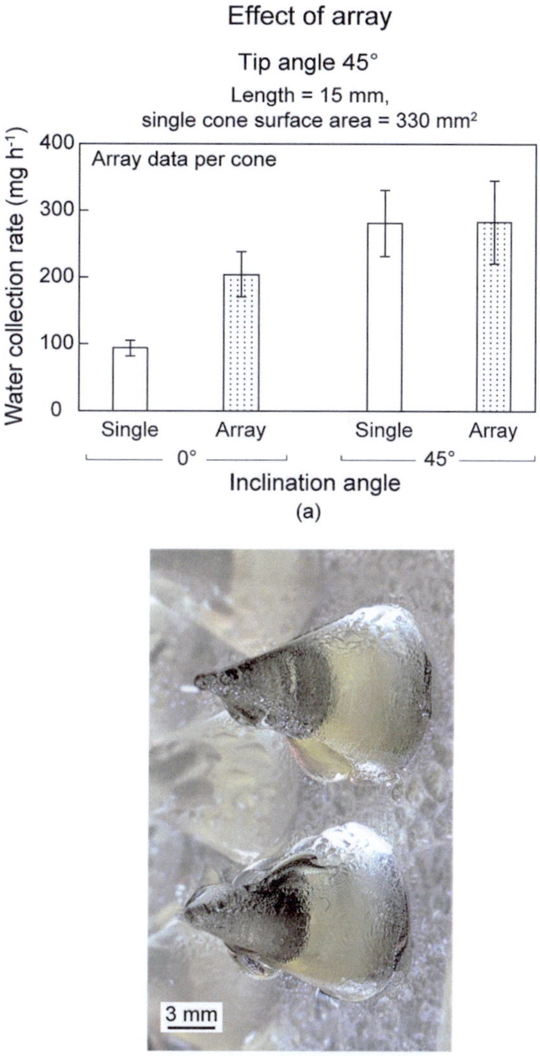

Droplets falling from the top cone coalesce with droplets stuck to cone underneath.
(b)

length but larger at larger tip angle for same surface area, irrespective of the inclination angle. The water collection rate always increases with an increase in inclination angle, regardless of the cone, because of gravity effects. Grooves and heterogeneous wettability also increase the water collection rate. In conical arrays, water collection per cone is higher at 0° inclination angle than a single cone. However, it is similar at 45° inclination (Gurera and Bhushan 2019b).

Projection for Water Collection Rates

In fog deposition at 0° inclination angle (samples placed horizontally), for flat surfaces with any wettability, water will not be collected in the reservoir. In the case of hydrophilic cones, water collection occurs in the reservoir (near their base), data shown in Figs. 14.19 and 14.22. It is also expected, in the case of beetle-inspired surfaces, some water collection would occur.

At 45° inclination angle, based on Figs. 14.19, 14.21 and 14.22, water collection rates for a hydrophilic cone at 45° inclination angle is more than factor of 10 larger than at 0° inclination angle. Water collection rate is larger than that at 0° inclination angle because of the effect of gravity.

At 45° inclination angle, based on Fig. 14.18, for beetle-inspired surfaces and hydrophilic cones, water collection rates per unit surface area is about 1 mg mm^{-2} h^{-1}, which is about twice as much water as that for a hydrophilic flat surface. Based on Figs. 14.23, 14.24, 14.25, cones with grooves and heterogeneous wettability and conical arrays provide slightly larger water collection. Furthermore, collected water is transported at a faster rate on a beetle-inspired and cone surfaces than that on a flat surface which reduces the evaporation of water droplets before reaching storage/use location. Finally, based on Fig. 14.22, at 45° inclination angle, water collection rates are larger for larger surface area for both cone angles. For the data shown, an increase of 3 to 6 times can be achieved. Therefore, cones of larger surface area for the same base area should be used.

Based on these observations, for a conical array with grooves and heterogeneity, water collection rates per unit base area as compared to that of flat hydrophilic surface can be increased by about a factor of 10 or more. In addition, water is transported faster which reduces evaporation. A summary of projected collection rates is presented in Table 14.2 (Bhushan 2020a, b).

For completeness, we compare typical water collection rates from fog in the desert with that obtained in the lab. Typical water collection rate in the desert, based on Fig. 14.8 is on the order of 2 L m^{-2} day^{-1} (or 2 mg mm^{-2} day^{-1}). Hourly collection rate is about 1 mg mm^{-2} h^{-1}. In the lab, based on Fig. 14.21, water collected on a flat hydrophilic surface is about 0.5 mg mm^{-2} h^{-1}. Therefore, for the experimental conditions used in the lab, water collection rate is about five times that in the desert.

14.4.5 Design Guidelines for Water Harvesting Systems

Condensation and fog often exist simultaneously in deserts. Fog mostly exists in the nights, and condensation is found during early morning hours. Therefore, to design water harvesting systems both sources should be considered (Bhushan 2020a, b).

In cones, both the Laplace pressure gradient and gravity are the important factors in driving water droplets towards the base. The Laplace pressure gradient dominates at 0° inclination angle (with samples placed horizontally), and shorter cone length will provide

Table 14.2 Summary of projected water collection rates from fog (Bhushan 2020a, b)

Surfaces	Increase in projected water collection rates
0° inclination angle (horizontal)	
Flat surfaces with any wettability	No collection in the reservoir
Beetle-inspired surfaces and hydrophilic cones	Finite collection
45° inclination angle	
Beetle-inspired surfaces and hydrophilic cones relative to hydrophilic flat surface	2x
Increase in surface area of cones for the same base area	> 3x
Cones with grooves and/or heterogeneous wettability	> 1x
Conical array per cone relative to a single cone	> 1x
Total	~ 10x

The water collection rates at 45° inclination angle are about an order of magnitude larger than that at 0° inclination angle

higher water collection rate at the base. The gravity dominates at 45° inclination angle, and higher surface area will provide higher water collection rate at the base. Higher inclination angle provides a higher water collection rate, as compared to a lower inclination angle because of the contributions by gravity. Therefore, for a high water collection rate, a higher surface area, inclined at higher inclination angle is desired. One can increase the surface area which would increase water collection by several fold, an order of magnitude or more. The surface area can be maximized by having a larger number of cones per unit base area and longer cones. The dimensions of the cones are limited by their structural integrity. Grooves, heterogeneous wettability, and nonlinear shapes can also increase water collection rates (Bhushan 2019, 2020a, b).

It was found that, from fog, both beetle-inspired and conical surfaces provide about twice the water collection rates per unit surface area than that of homogeneous wettable, flat surfaces. For conical arrays with grooves and heterogeneous wettability, the water collection rates per unit base area can be increased by about a factor of 10. In addition, water is transported faster which reduces evaporation (Bhushan 2019, 2020a, b).

For scaleup, nets or towers can be designed from the beetle-inspired or beetle-, grass- and cactus-inspired designs. For beetle-inspired, the junctions of the nets or towers should have superhydrophilic spots. For beetle-, grass- and cactus-inspired, the junctions of the nets or towers should have heterogeneous wettable cones with grooves and inclined at 45°. The heterogeneity includes hydrophobic tip and superhydrophilic base (Bhushan 2019, 2020a, b).

14.5 Bioinspired Triangular Patterns on Flat Surfaces for Water Harvesting

As mentioned earlier, cactus spines take advantage of the conical geometry to drive water droplets by Laplace pressure gradient for water transport and storage/use before they get evaporated. There have been some attempts to produce bioinspired surfaces with conical geometry for water harvesting from fog and/or condensation of water vapor (Bhushan 2018, 2019, 2020a, b). A triangular geometry on flat surfaces, inspired by cactus spines from fog and/or condensation, has been pioneered by Song and Bhushan (2019a, b, c, d).

The triangular patterns with various geometry and wettability surrounded with a region of less wettability to constrain condensed water droplets inside the patterns have been studied. The wettability of the triangular pattern affects the water collection process and the hydrophilic pattern has been shown to be desirable for water transport compared to the superhydrophilic and hydrophobic patterns (Song and Bhushan 2019a). Hydrophilic triangular patterns of various geometry have been investigated for high water collection rates.

An important consideration in efficient water collection is to transport water as rapidly as possible. Droplets can be driven on surfaces with wettability gradient (Brochard 1989; Chaudhury and Whitesides 1992; Gurera and Bhushan 2019b; Feng and Bhushan 2020). When a droplet sits on a surface with heterogeneous wettability, wettability gradient provides an unbalanced force on both sides of the droplet as a driving force for directional transport of the droplet. Multistep wettability gradient has been used on triangular patterns for rapid droplet transport (Feng and Bhushan 2020). The effect of wettability gradient and triangular geometry on water condensation and transport have been investigated.

One of the limitations of using triangular patterns is that the magnitude of the Laplace pressure gradient decreases along the length of the pattern and the droplet volume required for transport becomes larger. To overcome these limitations, Bhushan and Feng (2020) developed nested triangular patterns. Based on water condensation studies on the nested triangular patterns, they reported that the nested pattern increases the droplet speed.

Water collection studies from fog on triangular patterns have been carried out by Song and Bhushan (2019c) and from combined fog and condensation by Song and Bhushan (2019d).

In this section, an overview of systematic studies of water harvesting from condensation of water vapor using a triangular geometry carried out by Song and Bhushan (2019a, b, d) is presented. The triangular patterns of various geometry with hydrophilic surfaces were used. Surrounding regions with less wettability than the triangular patterns were selected to constrain collected water in the pattern. For water condensation studies, a low temperature of 5 °C was used to decrease saturated vapor pressure to promote water condensation.

14.5.1 Samples with Triangular Patterns

Two types of samples were fabricated. One type of sample contained a single triangular pattern, which was used to investigate the movement of the condensed droplets, as shown in Fig. 14.26 (Song and Bhushan 2019a, b). The single triangular pattern with a 20 mm length, with region A being hydrophilic, was surrounded by a superhydrophobic rim (0.5 mm wide). The superhydrophobic rim was produced to serve as a dam to the condensed liquid in the triangular pattern. If any droplets are transferred in the superhydrophobic dam region, they will slide to the outer less wettable region and will be prevented from sliding back into the pattern. Three included angles ($\alpha = 5°$, 9° and 17°) were selected to investigate the effect of α on the droplet condensation and transport process (Bhushan 2019, 2020a, b).

Song and Bhushan (2019a) studied the effect of wettability of triangular patterns surrounded regions. The hydrophilic triangular patterns surrounded by superhydrophobic regions were found to be more desirable for water collection than hydrophobic triangular patterns (Bhushan 2019, 2020a, b).

The other type of sample contained an array of triangular patterns that were located on both sides of a rectangular reservoir, to increase the amount of the collected water for better measurement accuracy, as shown in Fig. 14.26. An array with an included angle of 9° and length of 10 mm, was used. To study the effect of included angle, arrays with four included angles of 9°, 17°, 22° and 30°, with a length of 10 mm were selected. To study the effect of length, arrays with four lengths of 5, 10, 20 and 30 mm with an included angle of 9° were selected (Bhushan 2019, 2020a, b).

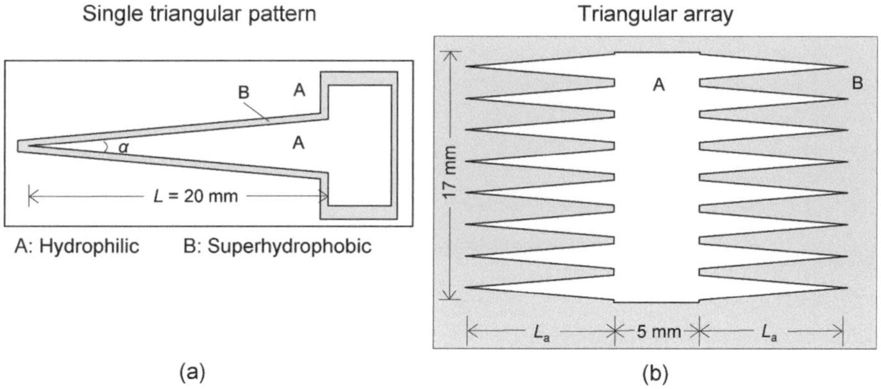

Fig. 14.26 A single triangular pattern, and a rectangular reservoir with an array of triangular patterns (Adapted from Bhushan 2019, 2020a, b; Song and Bhushan 2019b)

14.5.2 Experimental Apparatuses for Water Collection from Condensation

Figure 14.27 shows a schematic of the apparatus for water collection from condensation (Song and Bhushan 2019a, b). The samples were placed on top of an aluminum block on the horizontal table and cooled by a thermoelectric Peltier cooler (GeekTeches) down to about 5 ± 1 °C. A digital camera was used to record the water collection process.

A property of interest in water condensation is dew point temperature. It is the temperature to which air must be cooled to become saturated with water vapor. When further cooled below the dew point, airborne water vapor will condense on the surface to form water droplets (dew). At a temperature lower than dew point the saturated vapor pressure of water in the ambient air decreases which leads to water condensation (Alduchov and Eskridge 1996). The dew point temperature, T_d, can be calculated using the following equation (Lawrence 2005; Pruppacher and Klett 2010; Moran et al. 2018).

$$T_d = T - ((100 - RH)/5) \tag{14.1}$$

where RH is relative humidity of air–water mixture defined as partial pressure of water vapor in the mixture divided by saturation vapor pressure. Using Eq. 14.1, the dew point temperature at the ambient (22 °C and 50% RH) is about 12 °C, higher than the sample temperature of 5 °C.

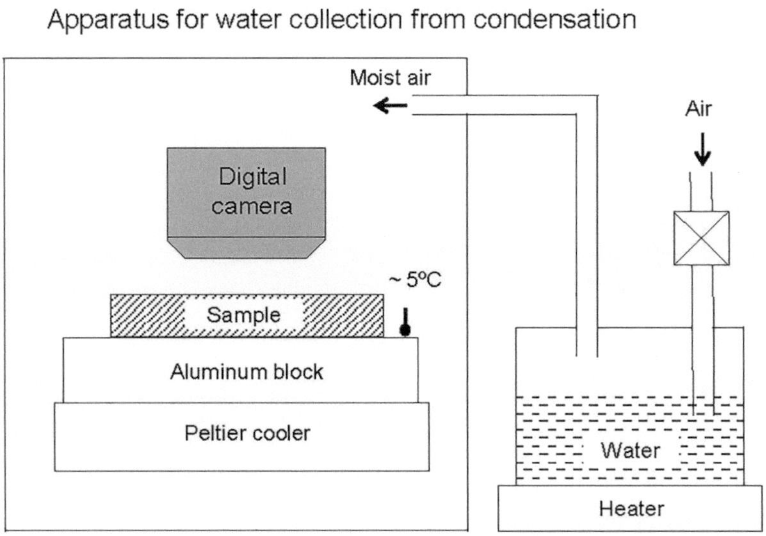

Fig. 14.27 Schematic of apparatus for water collection from condensation (Adapted from Bhushan 2019, 2020a, b; Song and Bhushan 2019a, b)

The samples were housed in an acrylic chamber (1 m × 0.5 m × 0.8 m). The air temperature in the chamber was 22 ± 1 ºC. Relative humidity (RH) was controlled by injection of humid air. The humid air was produced by an air stream that passed through a tank of hot water. By changing the temperature of the hot water and the flow rate of the air stream, RH in the chamber could be controlled between 30 and 95%. A digital camera was used to capture the condensation process.

Experiments were conducted at a relative humidity of 85 ± 5%, except those performed to study the effect of relative humidity. When the sample was cooled down to 5 °C at RH > 50%, which is lower than the dew point of the ambient air, water vapor continuously condensed on the triangular region.

14.5.3 Results and Discussion

Water collection data from condensation for single and array of triangular patterns are presented in this section (Song and Bhushan 2019a, b).

Triangular Patterns

Figure 14.28a shows the water condensation, coalescence and transport process on a hydrophilic triangular pattern surrounded by superhydrophobic region (Song and Bhushan 2019a). Droplets spread and touched the pattern boundaries to facilitate droplet movement. It took about 25 min for the self-transportation of condensed droplets to occur from the tip to a distance of about 5.5 mm, (Song and Bhushan 2019a)

To clearly illustrate water condensation and transport process as a function of time on a hydrophilic pattern, Fig. 14.28b presents schematics showing droplets nucleation and movement on the hydrophilic triangular pattern surrounded by superhydrophobic region (Song and Bhushan, 2019a). The droplets nucleate and coalesce. Once they touch the two sides, they start to move because of Laplace pressure gradient (Song and Bhushan 2019a; Bhushan 2019, 2020a, b).

Array of Hydrophilic Triangular Patterns

To increase the water collection rate, samples containing a reservoir with an array of triangular patterns were used. Figure 14.29a shows the optical image of the reservoir with triangular patterns with 10 mm length and an included angle of 9°, after being exposed to condensation for 450 min (Song and Bhushan 2019b). The condensed water on the triangular area was transported to the reservoir whose mass was measured via a paper tissue. To evaluate the additional mass of water condensed at the reservoir area, a rectangular hydrophilic area was placed beside the array and the mass of the condensed water was measured as well. When calculating the condensation rate on the triangular patterns, the mass of the water on the rectangular region was deducted from the mass of the water on the reservoir with array (Song and Bhushan 2019b).

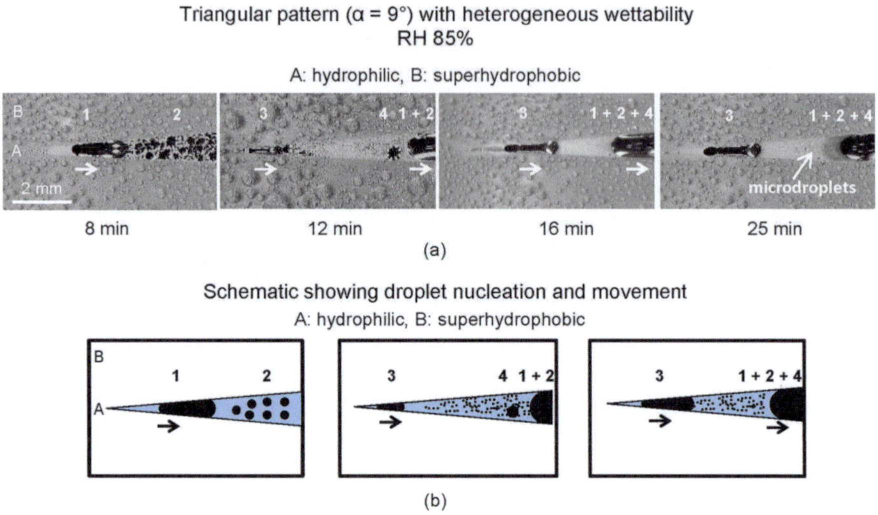

Fig. 14.28 **a** Selected optical images of condensation on a hydrophilic triangular pattern surrounded by superhydrophobic region. The condensed droplets on the triangular patterns move in the direction with larger triangular width when tiny droplets coalesce into one droplet with a size larger than the local triangular width. Arrows shown below some droplets are based on the movement of the center of the droplet observed in the videos, and **b** Schematic illustration of droplet nucleation and movement on a hydrophilic triangular pattern surrounded by a superhydrophobic region (Adapted from Bhushan 2019, 2020a, b; Song and Bhushan 2019a)

The effect of relative humidity on the condensation rate is shown in Fig. 14.29b (Song and Bhushan 2019b). The condensation rate increased linearly with relative humidity in the measured range of 50 and 85%.

Next, for optimization of water collection designs, the effect of the included angle, α, and the length of the triangular patterns, L_a, were studied. As shown in Fig. 14.29c, the included angle did not affect the condensation rate (Song and Bhushan 2019b). Even though the droplet transported slowly on a triangle with a larger included angle, the size of the droplet was larger which may provide similar condensation rates. Figure 14.29c also shows the condensation rate as a function of the length of the triangular patterns. The condensation rate decreased when the length increased. Since a shorter distance requires less time to transport the condensed droplets and though droplets being removed are expected to be smaller, the removal rate increased the condensation rate (Bhushan 2019, 2020a, b).

Summary

The water condensation and transport process of triangular patterns was investigated. The hydrophilic triangular patterns surrounded by superhydrophobic regions were found to be desirable for water collection. For droplet transport, when the droplets were constrained

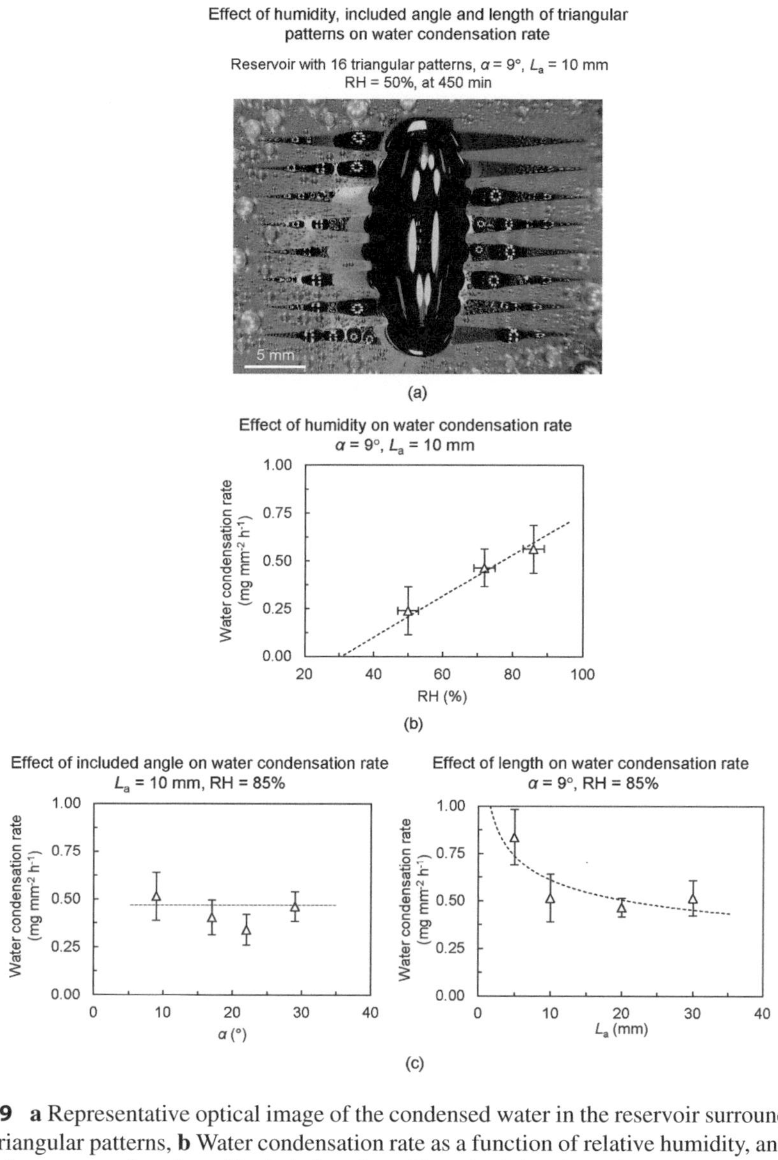

Fig. 14.29 **a** Representative optical image of the condensed water in the reservoir surrounded by an array of triangular patterns, **b** Water condensation rate as a function of relative humidity, and **c** Water condensation rate as a function of included angle and length of the triangular patterns (Adapted from Bhushan 2019, 2020a, b; Song and Bhushan 2019b)

within the triangular patterns, they started to move after coalescence, and they reached a critical size and touched the two sides. A water collection reservoir was fabricated with an array of triangular patterns to measure the condensation rate. The relative humidity increased the condensation rate. The included angle did not affect the condensation rate. A decrease in the length of the patterns increased the condensation rate (Song and Bhushan 2019a, b).

To design a condensation water collection tower, a hydrophilic pattern surrounded by a superhydrophobic region should be used. A larger number of triangular patterns with smaller included angles and shorter lengths can be used for a higher condensation rate. Relative humidity of the ambient air increased the condensation rate; therefore, high relative humidity is desirable. However, it may not be possible to control it in practical applications (Song and Bhushan 2019a, b; Bhushan 2019, 2020a, b).

14.6 Commercial Applications and Various Water Harvesting Tower Designs

14.6.1 Commercial Applications

Bioinspired large water harvesting towers and portable water harvesting towers are of commercial interest. Large water harvesting towers can be used to supply water to a community. Portable units can be used to supply a home or a camper (Fig. 14.30). Unlike in desalination, these units can be operated inland. Manufacturing and maintenance costs and energy consumption will be an important consideration for economic viability in these applications.

In addition to consumer applications, these towers can be used for various emergency and defense applications. Emergency applications, such as natural disasters, could benefit for short periods from portable units which could be dropped from air (Fig. 14.31). In emergency applications, human life is at stake, and cost is not an issue. Defense applications include combat and military bases in combat zones. The cost of clean water in Forward Operating Bases (FOBs) in a combat zone such as in Afghanistan can be as much as US $350/gallon because of the dangers posed in transportation. Large towers can be installed at military bases located in deserts which can supply safe drinking water at a minimal cost, possible couple of orders of magnitude lower than that of transported water (Fig. 14.32) (Bhushan 2020a, b).

14.6.2 Projection of Water Collection Rates in Water Harvesting

Water collection rates for bioinspired water harvesters are projected and compared with the collection rates on flat surfaces and living species in deserts. In arid deserts, based on

Fig. 14.30 Schematic of a water harvesting tower placed in a home for personal use (Image provided by Getty Images) (Adapted from Bhushan 2020a, b)

some data presented in Sect. 14.2.1, water collection rates on flat surfaces are on the order of 2 L m^{-2} day^{-1} (2 mg mm^{-2} day^{-1}). The source of water is fog and condensation of water vapor mostly in nights. The bioinspired surfaces covered with conical arrays, based on data presented in Sect. 14.4 (Table 14.2), can provide higher collection rates about an order of magnitude larger than that of a flat hydrophilic surface. This means that bioinspired surfaces can collect water on the order of 20 L m^{-2} day^{-1}. For an example of water collection of 20 L m^{-2} day^{-1}, a medium size tower covered with bioinspired surface with a surface area of 200 m^2, water collected would be about 4,000 L day^{-1}. If the water consumption per capita is 50 L day^{-1}, based on Table 14.1 (Sect. 14.1), a tower can provide sufficient water for about twenty families with 4 people in each family. The data are summarized in Table 14.3.

14.6.3 Design of Water Harvesters

Since the earlier attempts by Carlos Espinosa in Chile in 1957, passive net-based fog harvesters have been built in several countries (Klemm et al. 2012). The net is directly exposed to the atmosphere, and the foggy air is pushed through the net by the wind. Fog droplets are deposited on the net which combine to form large droplets and travel by

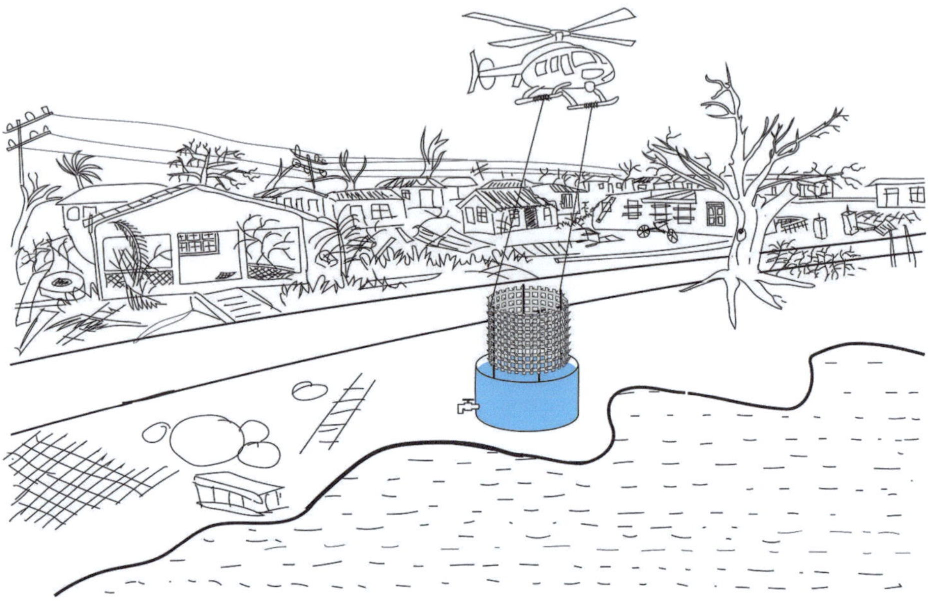

Fig. 14.31 Schematic of a portable water harvesting tower being dropped by a helicopter in an emergency zone on an island (Adapted from Bhushan 2020a, b)

gravity to a collection dish at the bottom. In North America, many organizations such as FogQuest use 2D nets for fog interception and harvesting that can provide a supplemental source of water in arid regions, Fig. 14.33a (Brown and Bhushan 2016; Bhushan 2018, 2019, 2020a, b). The yearly water collection rates from 3 to 10 L m^{-2} per day have been reported. To increase water collection rates, newer designs were developed, such as a tower by the charity Warka Water project, Fig. 14.33b (Brown and Bhushan 2016; Bhushan 2018, 2019, 2020a, b). However, the nets in these devices are typically made of material and structures that have not been optimized based on bioinspiration for water collection.

Based on the research reported in this chapter, large bioinspired 3D towers using either triangular patterns or cones can be designed for water collection higher than that of the designs so far. Bioinspiration provides lessons for high water collection. Slotted tower surfaces were selected to reduce swirl by intercepting wind. Figures 14.34 and 14.35 show schematics of water harvesting towers covered with triangular patterns or cones, respectively (Bhushan 2019, 2020a, b). In the case of cones, heterogeneous wettability and grooves are used and are mounted at 45° to benefit from gravitational forces. Grooves on the ribs are provided to facilitate water flow down to the reservoir. Water droplets falling on the cones underneath provide cascading effect to accelerate fluid flow. In the

Fig. 14.32 Schematic of a water harvesting tower placed on a military base in the combat zone in the desert (Adapted from image provided by Jeffrey McGovern/U. S. Air Force; Bhushan 2020a, b)

Table 14.3 Water collection rates from fog and condensation of water vapor (Bhushan 2020a, b)

Type of surface	Approximate water collection rate (L m^{-2} day)
Flat surface in arid desert (Sect. 14.2)	2
Bioinspired surfaces covered with cones or triangular patterns (Sects. 14.4 and 14.5)	20[a] (× 10 increase)

[a] A medium size bioinspired tower with a surface area of 200 m^2 can collect 4000 L day^{-1}

case of triangular patterns, patterns are placed with tip pointed outward on the shims oriented horizontally to the vertical ribs. The collected water flows down the grooved ribs to the reservoir. In addition, water droplets will be deposited on the slot edges and will eventually fall down to the reservoir (Bhushan 2020a, b).

For condensation of water vapor, the cooling process of the cones or shims with triangular patterns in the towers is provided by the low temperatures in desert nights. If additional cooling is required, one could take advantage of the renewable daytime solar energy. In the field, the water harvesting systems will get covered by dust. They will require filters to keep the active collection area clean. Filters may require high pressure air, which again can be provided by solar energy. The towers could be operated with near zero energy (Bhushan 2020a, b).

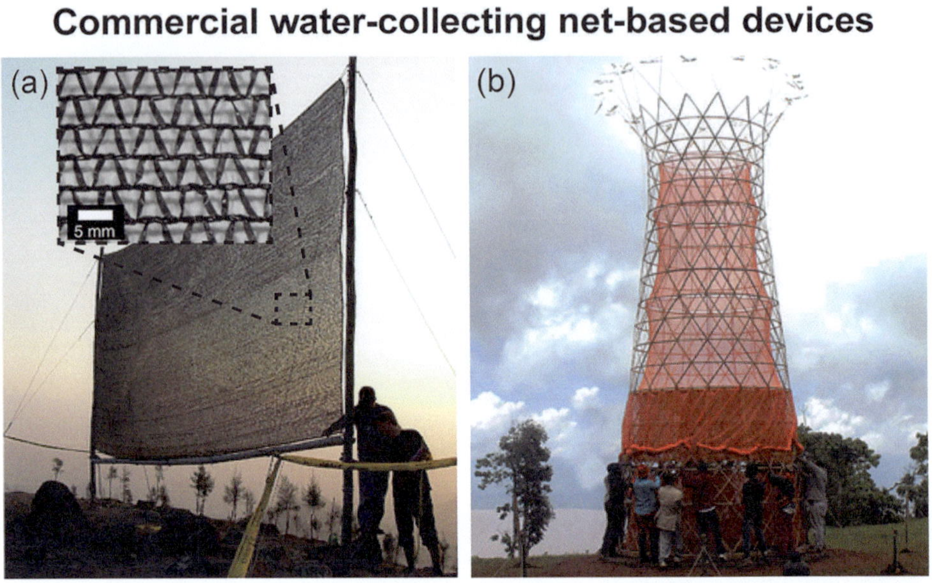

Fig. 14.33 Net-based water collectors currently available from **a** FogQuest (photograph by Anne Lummerich, inset adapted from Park et al. 2013), and **b** Warka Water (photograph by Architecture and Vision) (Adapted from Brown and Bhushan 2016; Bhushan 2019, 2020a, b)

14.6.4 Operational and Maintenance Cost

The operational and maintenance cost needs to be minimized for field use, particularly for consumer applications. As a reference, the cost of bottled water can range from US $0.20 to $1.50 in different countries. One liter of desalinated water costs about 1 cent US. The target for bioinspired designs for consumer applications is less than 2 cents US (data summarized in Table 14.4). Given that operating cost for water harvesting tower is negligible and only maintenance cost exists, the target cost is achievable (Bhushan 2020a, b).

It should be noted that cost sensitivity is dependent upon the application. For emergency and defense applications, the cost is not an issue. Bioinspired approaches become very attractive (Bhushan 2020a, b).

14.6.5 Scaleup and Commercialization Issues

Significant investments are needed for scale up designs, technology transfer and product launch. Major innovation prizes have a long record of spurring innovation in private, public and philanthropic sectors (Bhushan 2015). They have potential to facilitate quantum

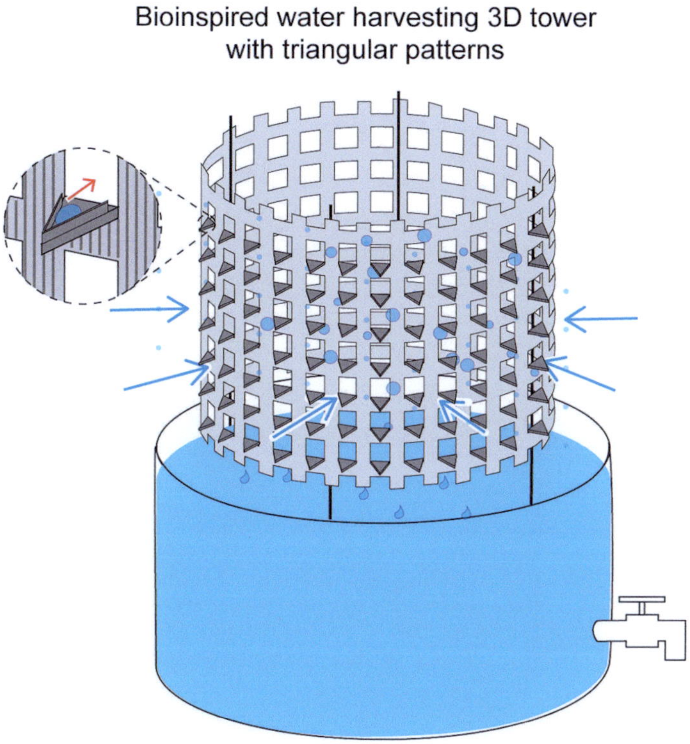

Fig. 14.34 Schematic of a grass- and cactus-inspired 3D water harvesting tower covered with triangular patterns (Adapted from Bhushan 2020a, b)

jump in technologies by attracting investors across the spectrum including those out-of-discipline. Although total investments by various competitive teams may be several folds they also only pay for success. In 2016, X-Prize Foundation announced US $1.75 M Water Abundance XPrize, a two year competition to develop energy-efficient technologies that harvest water from thin air. The goal was to develop a water harvesting device that can deliver 2000 L of water per day at a cost of no more than 2 cents US using only renewable energy. In Oct 2018, the Skysource/Skywater Alliance was awarded a grand prize of US $1.5 M (Bhushan 2020a, b).

14.7 Closure

Fresh water sustains human life and is vital for human health. It is estimated that about 800 million people worldwide lack basic access to drinking water. About 2.2 billion people (nearly a third of the global population) do not have access to a safe water supply, free

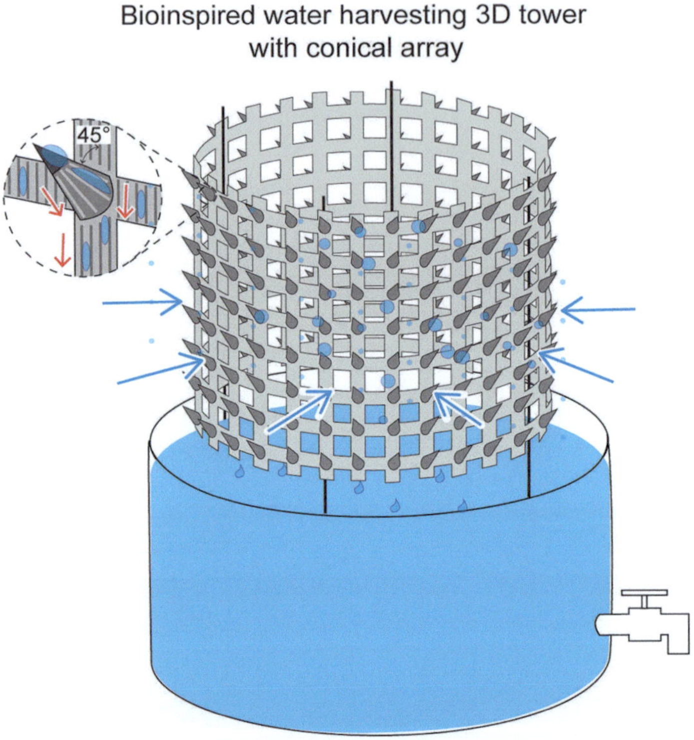

Fig. 14.35 Schematic of a grass- and cactus-inspired 3D water harvesting tower covered with cones (Adapted from Brown and Bhushan 2016; Bhushan 2019, 2020a, b)

Table 14.4 Water collection costs for various approaches (Bhushan 2020a, b)

Water collection approach	Approximate cost (L^{-1})
Bottled water (for reference)	US \$0.20 to \$1.50
Desalination	1 cent US
Bioinspired design target	2 cents US (achievable)

of contamination. Also, over 2 billion people live in countries experiencing high water stress. The current supply of fresh water needs to be supplemented to meet future needs. Living nature has evolved species which can survive in the most arid regions of the world by water collection from fog and condensation in the night. Collected water evaporates as soon as the sun comes out in the early mornings. Before it evaporates, species have mechanisms to transport water to a location where either it is stored or consumed. These species possess unique chemistry and structures on or within the body for collection and transport of water. One needs to understand the mechanisms used in nature. Then, one can fabricate bioinspired surfaces to supplement safe water supply. This technology would be of particular interest in arid regions to develop means to harvest water from fog and condensation.

In this chapter, an overview of arid desert conditions, water sources and plants and animals, lessons from nature for water harvesting, and data on bioinspired surfaces from fog and condensation are presented. Consumer, emergency and military applications are discussed. Various designs of water harvesting towers and projections for water collection are presented.

Appendix 14.A: Laplace Pressure Gradient on a Conical Surface

For a spherical droplet sitting on a surface, capillary pressure or Laplace pressure in the liquid p_L is proportional to the surface tension of the liquid in air (γ_{LA}) divided by the local radius, R (Adamson and Gast 1997; Bhushan 2013a, 2013b, 2018, 2019, 2020b)

$$p_L = \frac{\gamma_{LA}}{R} \tag{14.2}$$

The Laplace pressure can be attractive or repulsive depending on whether the surface is hydrophilic or hydrophobic, respectively. The p_L remains constant on a flat surface.

Next, we consider a liquid droplet sitting on a conical object. We consider two adjacent locations A and B with the local radii of the cone, as R_A and R_B, respectively, Fig. 14.36 (Bhushan 2018, 2019, 2020b). The surface curvature gradient results in the Laplace pressure difference between the two opposite ends of the droplet along the surface. The Laplace pressure difference is given as,

$$\Delta p_L = \gamma_{LA}\left(\frac{1}{R'_A} - \frac{1}{R'_B}\right) \tag{14.3}$$

where R'_A and R'_B are the radii of curvature at the front and rear contact lines of the droplet, respectively. The curvature gradient leading to Laplace pressure difference that acts on the contact area A, produces the Laplace force F_L,

$$F_L = \iint_A \Delta p_L \mathrm{dA} \tag{14.4}$$

where the contact area is approximately equal to volume of the droplet, $V_{droplet}$ divided by the length L of the droplet,

$$A \sim \frac{\sin\alpha}{(R_B - R_A)} V_{droplet} \tag{14.5}$$

where α is the half apex angle of the cone. Combining Eqs.14.3–14.5, we get,

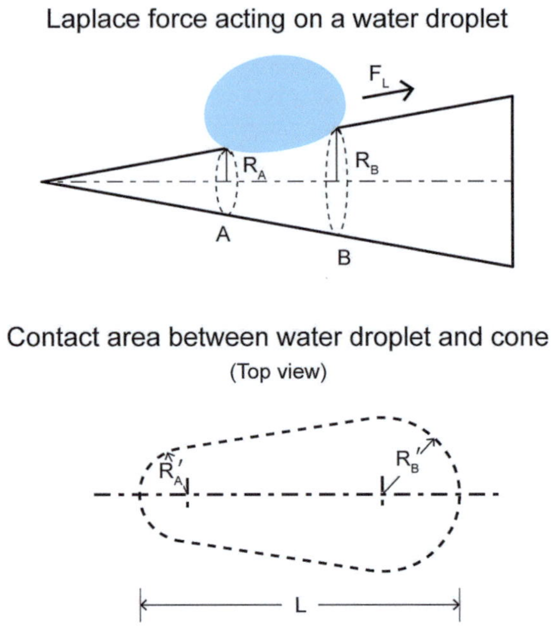

Fig. 14.36 Schematic of a liquid droplet on a conical object with a cone angle of 2α and local radii R_A and R_B at two locations of A and B, respectively. Shown is the Laplace force (F_L) which helps in driving the droplet towards a larger radius. 2020b (Adapted from Bhushan 2018, 2019,)

$$F_L \sim \gamma_{LA}\left(\frac{1}{R'_A} - \frac{1}{R'_B}\right)\frac{\sin\alpha}{R_B - R_A}V_{droplet} \tag{14.6}$$

The Laplace force acting on a conical object drives the droplet from regions of lower radius to larger radius as long as the Laplace force is larger than adhesion force. During this droplet movement, new droplets may be deposited in the path, which coalesce resulting in a large liquid volume and provide the additional movement.

References

Anonymous (2009) "Charting Our Water Future: Economic Frameworks to Inform Decision-making," The 2030 Water Resources Group Report, see http://www.2030wrg.org/wp-content/uploads/2014/07/Charting-Our-Water-Future-Final.pdf

Adamson, A. V. and Gast, A. P. (1997), *Physical Chemistry of Surfaces*, sixth ed., Wiley, New York.

Agam, N. and Berliner, P. R. (2006), "Dew Formation and Water Vapor Adsorption in Semi-Arid Environments– A Review," J. Arid Environ. **65**, 572-590.

Alduchov, O. A. and Eskridge, R. E. (1996), "Improved Magnus Form Approximation of Saturation Vapor Pressure," J. Appl. Meteorol. **35**, 601-609.

Allaby, M. (2006), *Deserts*, Chelsea House Publishers, New York, New York.

Anonymous (1996), "DesertUSA," https://www.desertusa.com/

Anonymous (1999), *Sourcebook of Alternative Technologies for Freshwater Augmentation in Some Countries in Asia*, United Nations Environment Programme, Stationery Office Books, Norwich, U.K. http://www.unep.or.jp/ietc/Publications/TechPublications/TechPub-8e/fog.asp

Axelrod, D. I. (1979), *Age and Origin of Sonoran Desert Vegetation*, California Academy of Sciences, San Francisco, California.

Azad, M. A. K., Ellerbrok, D., Barthlott, W., and Koch, K. (2015), "Fog Collecting Biomimetic Surfaces: Influence of Microstructure and Wettability," Bioinspir. Biomim. **10**, 016004.

Bai, H., Wang, L., Ju, J., Sun, R., Zheng, Y., and Jiang, L. (2014), "Efficient Water Collection on Integrative Bioinspired Surfaces with Star-Shaped Wettability Patterns," *Adv. Mater.* **26**, 5025–5030.

Baier, W. (1996), "Studies on Dew Formation under Semi-arid Conditions," *Agric. Meteorol.* **3**, 103-112.

Basu, S., Agarwal, A. K., Mukhopadhyay, A., and Patel, C. (Eds.) (2018), *Droplet and Spray Transport: Paradigms and Applications*, Springer, New York.

Bentley, P. J. and Blumer, W. F. C. (1962), "Uptake of Water by the Lizard, *Moloch horridus*," *Nature* **194**, 699–700.

Beresford-Jones, D., Pullen, A. G., Whaley, O. Q., Moat, J., Chauca, G., Cadwallader, L., Arce, S., Orellana, A., Alarcón, C., Gorriti, M., Maita, P. K., Sturt, F., Dupeyron, A., Huaman, O., Lane, K. J., and French, C. (2015), "Re-evaluating the Resource Potential of Lomas Fog Oasis Environments for Preceramic Hunter–gatherers under Past ENSO Modes on the South Coast of Peru," *Quat. Sci. Rev.* **129**, 196–215.

Bhushan, B. (2013a), *Introduction to Tribology,* second ed., Wiley, New York.

Bhushan, B. (2013b), *Principles and Applications of Tribology,* second ed., Wiley, New York.

Bhushan, B. (2015), "Perspective: Science and Technology Policy – What is at Stake and Why Should Scientists Participate?" *Sci. Publ. Policy* **42**, 887-900.

Bhushan, B. (2017), *Springer Handbook of Nanotechnology*, fourth ed., Springer International, Cham, Switzerland.

Bhushan, B. (2018), *Biomimetics: Bioinspired Hierarchical-Structured Surfaces for Green Science and Technology*, third ed., Springer International, Cham, Switzerland.

Bhushan, B. (2019), "Bioinspired Water Collection Methods to Supplement Water Supply," *Phil. Trans. R. Soc. A* **377**, 20190119.

Bhushan, B. (2020a), Design of Water Harvesting Towers and Projections for Water Collection from Fog and Condensation," *Phil. Trans. R. Soc. A* **378**, 20190440.

Bhushan, B. (2020b), *Bioinspired Water Harvesting, Purification, and Oil-Water Sep*aration, Springer International, Cham, Switzerland.

Bhushan, B. and Feng, W. (2020), "Water Collection and Transport in Bioinspired Nested Triangular Patterns," *Phil. Trans. R. Soc. A* **378**, 20190441.

Bredeson, C. (2009), *Baby Animals of the Desert*, Enslow Publishers, Berkeley Heights, New Jersey.

Brochard, F. (1989), "Motions of Droplets on Solid Surfaces Induced by Chemical or Thermal Gradients," *Langmuir* **5**, 432-438.

Brown, P. S., and Bhushan, B. (2015), "Bioinspired, Roughness-Induced, Water and Oil Super-Philic and Super-Phobic Coatings Prepared by Adaptable Layer-by-Layer Technique," *Sci. Rep.* **5**, 14030.

Brown, P. S. and Bhushan, B. (2016), "Bioinspired Materials for Water Supply and Management: Water Collection, Water Purification and Separation of Water from Oil," *Phil. Trans. R. Soc. A* **374**, 20160135.

Cardwell, M. D. (2006), "Rain Harvesting in a Wild Population of *Crotalus s. scutulatus* (Serpentes: Viperidae)," *Herpetol. Rev.* **37**, 142-144.

Chaudhury, M. K. and Whitesides, G. M. (1992), "How to Make Water Run Uphill," *Science* **256**, 1539–1541

Cloudsley-Thompson, J. L. and Chadwick, M. J. (1964), *Life in Deserts*, Dufour Editions, Philadelphia, Pennsylvania.

Comanns, P., Buchberger, G., Buchbaum, A., Baumgartner, R., Kogler, A., Bauer, S., and Baumgartner, W. (2015), "Directional, Passive Liquid Transport: the Texas Horned Lizard as a Model for a Biomimetic 'Liquid Diode'," *J. R. Soc. Interface* **12**, 20150415 pp 1–8.

Costa, G. (1995), *Behavioral Adaptations of Desert Animals*, Springer-Verlag, Berlin, Germany.

Day, S. J., Forster, T., Himmelsbach, J., Korte, L., Mucke, P., Radtke, K., Theilbörger, P., and Weller, D. (2019), *World Risk Report 2019 – Focus: Water Supply*, Bündnis Entwicklung Hilft, Berlin, Germany.

Ebner, M., Miranda, T., and Roth-Nebelsick, A. (2011), "Efficient Fog Harvesting by *Stipagrostis sabulicola* (Namib Dune Bushman Grass)," *J. Arid Environ.* **75**, 524–531.

Edmonds, D. T. and Vollrath, F. (1992), "The Contribution of Atmospheric Water Vapour to the Formation and Efficiency of a Spider's Capture Web," *Proc. R. Soc. Lond. B* **248**, 145–148.

Elimelech, M. and Phillip, W. A. (2011), "The Future of Seawater Desalination: Energy, Technology, and the Environment," *Science* **333**, 712–717.

Environmental Protection Agency (1995), *The Great Lakes: An Environmental Atlas and Resource Book 3rd Edition,* EPA, Chicago, IL.

Feng, W. and Bhushan, B. (2020), "Multistep Wettability Gradient in Bioinspired Triangular Patterns for Water Condensation and Transport," *J. Colloid Interface Sci.* **560**, 866-873.

Fessehaye, M., Abdul-Wahab, S. A., Savage, M. J., Kohler, T., Gherezghiher, T., and Hurni, H. (2014), "Fog-Water Collection for Community Use," *Renew. Sust. Energ. Rev.* **29**, 52–62.

Garrod, R. P., Harris, L. G., Schofield, W. C. E., McGettrick, J., Ward, L. J., Teare, D. O. H., and Badyal, J. P. S. (2007), "Mimicking a Stenocara Beetle's Back for Microcondensation Using Plasmachemical Patterned Superhydrophobic–Superhydrophilic Surfaces," *Langmuir* **23**, 689–693.

Gorb, S. N. and Gorb, E. V. (eds.) (2017), *Functional Surfaces in Biology III – Diversity of the Physical Phenomena*, Springer International, Cham, Switzerland.

Greenberger, R. (2009), *Deserts–The Living Landscape*, The Rosen Publishing Group, New York, New York.

Gurera, D. and Bhushan, B. (2019a), "Designing Bioinspired Surfaces for Water Collection from Fog," *Phil. Trans. R. Soc. A* **377**, 20180269.

Gurera, D. and Bhushan, B. (2019b), "Optimization of Bioinspired Conical Surfaces for Water Collection from Fog," *J. Colloid Interface Sci.* **551**, 26-38.

Gurera, D. and Bhushan, B. (2019c), "Bioinspired Conical Design for Efficient Water Collection from Fog," *Phil. Trans. R. Soc. A* **377**, 20190125.

Gurera, D. and Bhushan, B. (2019d), "Multistep Wettability Gradient on Bioinspired Conical Surfaces for Water Collection from Fog," *Langmuir* **35**, 16944-16947.

Gurera, D. and Bhushan, B. (2020a), "Designing Bioinspired Conical Surfaces for Water Collection from Condensation," *J. Colloid Interface Sci.* **560**, 138-148.

Gurera, D. and Bhushan, B. (2020b), "Passive Water Harvesting by Desert Plants and Animals: Lessons from Nature," *Phil Trans. R. Soc. A* **378**, 20190444.

Hamilton, W. J. and Seely, M. K. (1976), "Fog Basking by the Namib Desert Beetle, *Onymacris unguicularis*," *Nature* **262**, 284–285.

Harris, N. (2003), *Atlas of the World's Deserts*, Taylor and Francis Group, New York, NY.

Hiatt, C., Fernandez, D., and Potter, C. (2012), "Measurements of Fog Water Deposition on the California Central Coast," *Atmospheric Clim. Sci.* **02**, 525.

Hill, A. J., Dawson, T. E., Shelef, O., and Rachmilevitch, S. (2015)." The Role of Dew in Negev Desert Plants," *Oecologia*, **178**, 317–327.

Ju, J., Bai, H., Zheng, Y., Zhao, T., Fang, R., and Jiang, L. (2012), "A Multi-structural and Multi-functional Integrated Fog Collection System in Cactus," *Nat. Commun.* **3**, Art. 1247.

Ju, J., Xiao, K., Yao, X., Bai, H., and Jiang, L. (2013), "Bioinspired Conical Copper Wire with Gradient Wettability for Continuous and Efficient Fog Collection," *Adv. Mater.* **25**, 5937–5942.

Kidron, G. J. (2005), "Angle and Aspect Dependent Dew and Fog Precipitation in the Negev Desert," *J. Hydrol.* **301**, 66–74.

Kidron, G. J., Herrnstadt, I., Barzilay, E. (2002), "The Role of Dew as a Moisture Source for Sand Microbiotic Crust in the Negev Desert Israel," *J Arid Environ* **52**, 517–533.

Klemm, O., Schemenauer, R. S., Lummerich, A., Cereceda, P., Marzol, V., and Corell, D., et al. (2012), "Fog as a Fresh-Water Resource: Overview and Perspectives," *Ambio.* **41**, 221–234.

Koch, K., Bhushan, B., and Barthlott, W. (2008), "Diversity of Structure, Morphology and Wetting of Plant Surfaces," *Soft Matter* **4**, 1943-1963.

Laity, J. (2008), *Deserts and Desert Environments*, Wiley–Blackwell Publishing, Hoboken, New Jersey.

Lawrence, M. G. (2005), "The Relationship between Relative Humidity and the Dewpoint Temperature in Moist Air," *BAMS* **100**, 225-233.

Liu, C., Xue, Y., Chen, Y., and Zheng, Y. (2015), "Effective Directional Self-gathering of Drops on Spine of Cactus with Splayed Capillary Arrays," *Sci. Rep.* **5**, 17757 pp 1–8.

Louw, G. N. and Seely, M. K. (1980), "Exploitation of Fog Water by a Perennial Namib Dune Grass, *Stipagrotis sabulicola*," *S. Afr. J. Sci.* **76**, 38–39.

Malek, E., McCurdy, G., and Giles, B. (1999). "Dew Contribution to the Annual Water Balances in Semi-arid Desert Valleys," *J. Arid Environ.* **42**, 71-80.

Mares, M. A. (Ed.) (1999), *Encyclopedia of Deserts*, University of Oklahoma Press, Norman, Oklahoma.

Meigs, P. (1953), "World Distribution of Arid and Semi-Arid Honoclimates, In *Reviews of Research on Arid Zone Hydrology*, Arid Zone Programme-1, pp. 203-209, UNESCO, Paris, France.

Meigs, P. (1966), *Geography of Coastal Deserts*, Report No. NS.64/III.33/A, United Nations Educational, Scientific and Cultural Organization (UNESCO), Paris, France.

Monteith, J. L. (1963), Dew: Facts and Fallacies. In *The Water Relations of Plants.* Rutter, A. J., Whitehead, F. H. (eds.), *Q. J. Roy. Meteor. Soc.*, pp. 37–56, Wiley, New York.

Mooney, H. A., Weisser, P. J., and Gulmon, S. L. (1977), "Environmental Adaptations of the Atacaman Desert Cactus *Oopiapoa haseltoniana,*" *Flora, Bd.* **166**, S. 117-124.

Moran, M. J., Shapiro, H. N., Boettner, D. D., and Bailey, M. B. (2018), *Fundamentals of Engineering Thermodynamics*, Ninth ed., Wiley, New York.

Murphy, J. A., (2012), *Desert Animal Adaptations*, Capstone Press, North Mankata, Minnesota

Ogburn, R. M. and Edwards, E. J. (2009), "Anatomical Variation in Cactaceae and Relatives: Trait Lability and Evolutionary Innovation," *Am. J. Bot.* **96**, 391–408.

Pan, Z., Pitt, W. G., Zhang, Y., Wu, N., Tao, Y. and Truscott, T. T. (2016), "The Upside-down Water Collection System of *Syntrichia caninervis*," *Nat. Plants* **2**, 16076.

Park, K.-C., Chhatre, S. S., Srinivasan, S., Cohen, R. E., and McKinley, G. H. (2013) "Optimal Design of Permeable Fiber Network Structures for Fog Harvesting," *Langmuir* **29**, 13269–13277.

Parker, A. R. and Lawrence, C. R. (2001), "Water Capture by a Desert Beetle," *Nature* **414**, 33–34.

Pruppacher, H. R. and Klett, J. D. (2010), *Microphysics of Clouds and Precipitation*, second ed., Springer, New York.

Roth-Nebelsick, A., Ebner, M., Miranda, T., Gottschalk, V., Voigt, D., Gorb, S., Stegmaier, T., Sarsour, J., Linke, M., and Konrad, W. (2012), "Leaf Surface Structures Enable the Endemic Namib Desert Grass *Stipagrostis sabulicola* to Irrigate Itself with Fog Water," *J. R. Soc. Interface* **9**, 1965–1974.

Schriner, C. T. and Bhushan, B. (2019), "Water Droplet Dynamics on Bioinspired Conical Surfaces," *Phil. Trans. R. Soc. A* **377**, 20190118.

Schwenk, K. and Greene, H. W. (1987), "Water Collection and Drinking in *Phrynocephalus helioscopus*: A Possible Condensation Mechanism," *J. Herpetol.* **21**, 134–139.

Shanyengana, E. S., Henschel, J. R., Seely, M. K., and Sanderson, R. D. (2002), "Exploring Fog as a supplementary water source in Namibia," *Atmospheric Res.* **64**, 251-259.

Shiklomanov, I. A. (1993), "World Fresh Water Resources." In *Water in Crisis: A Guide to the World's Fresh Water Resources* (ed. P. H. Gleick), p 13. New York: Oxford University Press.

Sigsbee, R. A. (1969), Nucleation, Marcel Dekker, New York.

Song, D. and Bhushan, B. (2019a), "Water Condensation and Transport on Bioinspired Triangular Patterns with Heterogeneous Wettability at a Low Temperature," *Phil. Trans. R. Soc. A* **377**, 20180335.

Song, D. and Bhushan, B. (2019b), "Optimization of Bioinspired Triangular Patterns for Water Condensation and Transport," *Phil. Trans. R. Soc. A* **377**, 20190127.

Song, D. and Bhushan, B. (2019c). "Bioinspired Triangular Patterns for Water Collection from Fog," *Phil. Trans. R. Soc. A* **377**, 20190128.

Song, D. and Bhushan, B. (2019d), "Enhancement of Water Collection and Transport in Bioinspired Triangular Patterns from Combined Fog and Condensation," *J. Colloid Interface Sci.* **557**, 528-538.

United Nations (2002) "General Comment No. 15: The Right to Water (arts. 11 and 12 of the International Covenant on Economic, Social and Cultural Rights)," see http://www.unhcr.org/49d095742.html

United Nations (2018), *Sustainable Development Goal 6: Synthesis Report 2018 on Water Sanitation*, United Nations, New York. www.unwater.org/app/uploads/2018/07/SDG6_SR2018_web_v5.pdf.

Van Rheede van Oudtshoorn, K. and Van Rooyen, M. W. (1999), *Dispersal Biology of Desert Plants*, Springer-Verlag, Berlin, Germany.

Vesilind, P. J. (2003), "Atacama Desert," *National Geographic* (August, 2003) see http://ngm.nationalgeographic.com/features/world/south-america/chile/atacama-text

Walker, A. S. (1992), *Desert: Geology and Resources*, U. S. Geological Survey, Denver Colorado.

WHO/UNICEF (2017), *Progress on Drinking Water, Sanitation and Hygiene: 2017 Update and SDG Baselines*. Geneva, WHO/UNICEF.washdata.org/sites/default/files/documents/reports/2018–01/JMP-2017-report-final.pdf

WWAP (2019), *The United Nations World Water Development Report 2019: Leaving No One Behind*, WWAP (UNESCO World Water Assessment Programme), UNESCO, Paris, France.

Xue, Y., Wang, T., Shi, W., Sun, L., and Zheng, Y. (2014), "Water Collection Abilities of Green Bristlegrass Bristle," *RSC Adv.* **4**, 40837–40840.

15 Mosquitoes' Locomotion and Painless Piercing

15.1 Introduction

Some 10 million different species of animals and plants are known to exist on earth (Stork 2009; Mora et al. 2011; Barthlott et al. 2016). Of which, arthropods make up the largest category with roughly 5 million species (McGavin 1993; Chapman et al. 2013). The word, arthropods, is derived from two Greek words: *arthron* (joint) + *podos* (foot), meaning jointed feet. These are spineless animals with segmented bodies, paired and jointed legs, exoskeletons, and bilateral symmetry. Arthropods, being one of the major class of animals, create medical problems (Service 1978; Durden and Mullen 2009; Goddard 2013). They attack other organisms, cause allergic reactions, release toxins and venoms, and may even cause death (Bhushan 2018).

Figure 15.1 shows arthropods of medical-veterinary interest (Phylum, Arthropoda), which are classified on the basis of the well established ranking system (Parker 1997; Durden and Mullen 2009; Schmidt 2009; Stork 2009; Goddard 2013). Decreasing order of the ranking system is Phylum, Subphylum, Class, Order, and common names of the species. Starting from the top, Phylum *Arthropoda* is divided into four Subphyla *Hexapoda*, *Chelicerata*, *Myriapoda*, and *Crustacea*. The Subphyla are further divided into six Classes and seventeen Orders. Each of the Order lists examples of the arthropods. The class *Insecta* includes all of bugs and are 6-legged. This class is remarkably large and diverse (Bhushan 2018).

Even nature's most disgusting creatures—arthropods have important secrets. Although one wants to keep away from them, they have important lessons that can lead to bioinspired optimum designs for various applications. Some arthropods simply attack their prey with stingers and others feed on their host on plant nectar, blood, and other organisms. A few medically important arthropods are classified based on their attacking mechanism in Fig. 15.2 (Dickerson and Lavoipierre 1959; Durden and Mullen 2009; Chapman et al.

B. Bhushan, *Introduction to Biomimetics and Bioinspiration*,
https://doi.org/10.1007/978-3-031-62344-8_15

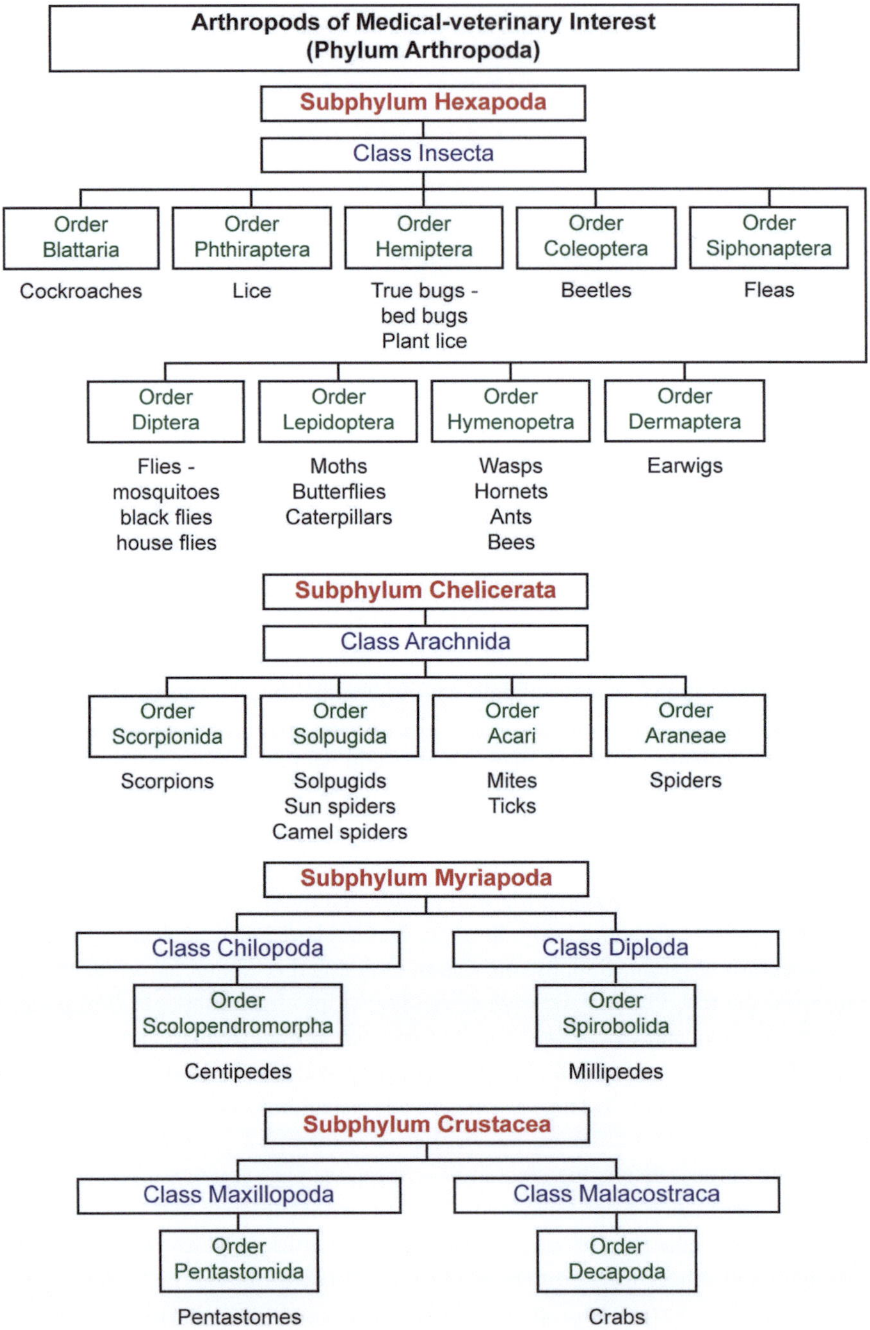

Fig. 15.1 Arthropods of medical-veterinary interest (*Adapted from* Parker 1997; Durden and Mullen 2009; Stork 2009; Goddard 2013)

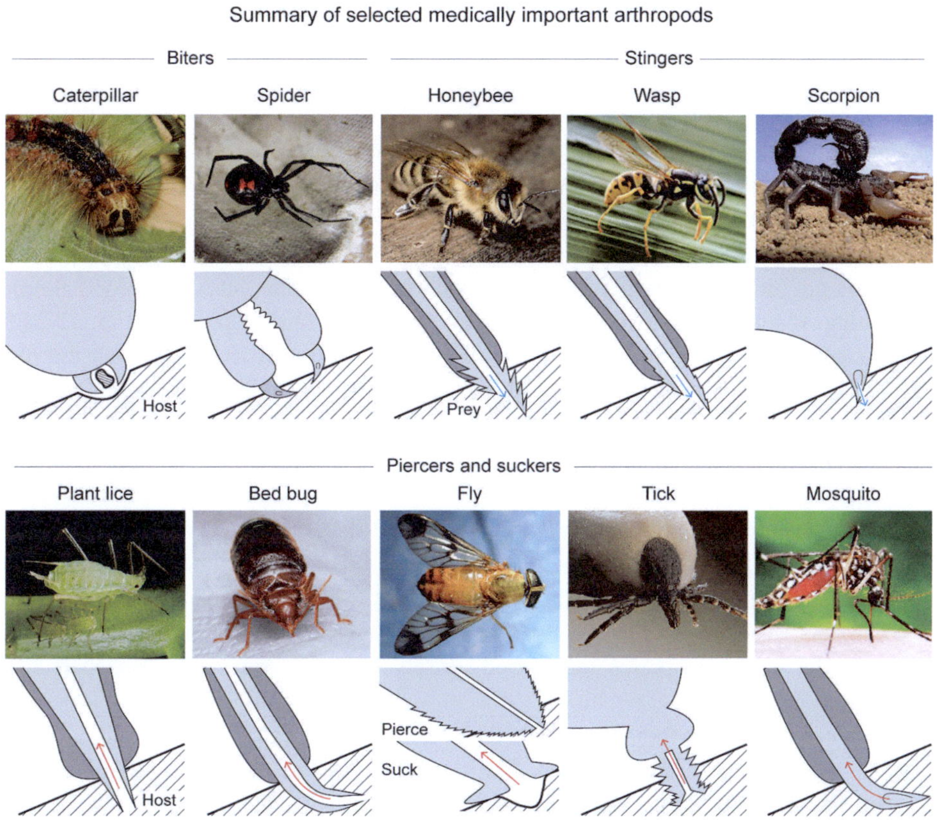

Fig. 15.2 Selected medically important arthropods classified based on their attacking mechanisms. The arthropods which bite their host with their mouthparts are classified as biters. The arthropods which sting with stingers located on their rear end and release venom into their prey are classified as stingers. The arthropods which pierce and suck bodily fluids out of their hosts are classified with their mouthparts as piercers and suckers (*Adapted from* Bhushan 2018; Gurera et al. 2018)

2013; Gurera et al. 2018). They are classified as biters, stingers, piercers, and suckers. The arthropods which bite with their mouthparts are classified as biters. Examples of bitters include caterpillars and spiders. Caterpillars use a pair of jaw-like-structures (mandibles) to bite onto their hosts. Spiders use a pair of sharp jaw-like-structures (spider fang) to bite onto their hosts, and later its serrated structure (chelicera teeth) to chew on them (Bhushan 2018).

Next, the arthropods which sting into their prey with stingers located on their rear end and release venom, are classified as stingers. Examples of the stingers include honeybees, wasps, and scorpions (Das et al. 2018). Among these wasps are known to cause most pain (Akre et al. 1981; Schmidt 2009). Honeybees sting using a pair of barbed stylets covered in a blunt sheath which moves back and forth alternatively while stinging the prey. The stylet is a medical name for slender, piercing parts. To release venom, honeybees use openings found between the stylets. Wasps sting and release venom in a similar fashion to honeybees. However, as compared to the honeybees, wasps have smoother barbed stylets (Akre et al. 1981). Scorpions use a sharp tail (aculeus) to sting their prey. The tail has an opening that releases venom to paralyze their prey (Bhushan 2018).

Lastly, the arthropods that pierce and suck bodily fluids out of their hosts, with their mouthparts, are classified as piercers and suckers. Examples of piercers and suckers include plant lice that suck plant sap, and bed bugs, flies, ticks, and mosquitoes that suck blood. Blood sucking arthropods find a blood vessel and then suck the blood through an opening. Plant lice use two pairs of stylets (only one pair is shown in Fig. 15.2) for piercing into their hosts, and use the opening formed between stylets to suck the plant sap. These stylets are covered in a sheath. Bed bugs pierce into their host using two pairs of stylets (only one pair is shown) and suck blood through the opening formed by one of the pair. Flies pierce the host using a pair of barbed stylets and suck the blood using the opening formed between the other mouthparts. Ticks use a pair of barbed stylets (chelicerae) by repeatedly flexing and relaxing the tips of the stylets. After reaching a certain distance, the pair of stylets along with another mouthpart (hypostome, not shown) suck blood through the opening formed between the three. Mosquitoes use a vibrating bundle of six stylets (fascicle) to pierce into hosts. Then they use one of the stylets (labrium shown in the figure) to pierce into blood vessels and suck the blood (Bhushan 2018; Gurera et al. 2018).

Biters only bite into their host with their teeth. A bite can introduce saliva as well as additional pathogens and diseases. The stingers penetrate the epidermis of the prey skin with a sharp tip with a hollow, narrow, canal through which the venom is injected for several minutes. It should be noted that all stings are not venomous. Piercers and suckers suck blood from the host and may infect with virus leading to malaria or other diseases. Stinging arthropods including honeybees, wasps, and scorpions, stinging other animals including humans mostly in self-defense. They create problems ranging from minor pain including local swelling for a couple of hours, to a much severe pain, including cardiovascular compromise, anaphylaxis and even death due to respiratory block. Wasps and bees do not transfer human diseases or viruses from one human to another through stinging. Bees lose their stingers once they sting. Wasp stingers do not carry enough blood following a wasp sting to transmit a virus. However, since wasps and bees are scavengers, they are known to carry bacteria that can lead to infections during stinging. As a result, they are also known to spread diseases (Bhushan 2018).

The mosquito's saliva can carry viruses and protozoa that it received from hosts that it had attacked previously. It is these viruses that pass on the diseases from host to host, such as malaria, yellow fever, West Nile virus, dengue fever, and Zika virus and can cause death. It is considered to be the deadliest animal in the world. Mosquitoes, specifically by malaria alone, have been reported to cause deaths on the order of 1–3 million every year (Ito et al. 2002; WHO 2016). In 2015, mosquitoes piercing was reported to be

Table 15.1 Summary of number of deaths caused by various animals in 2015 (Adapted from Gates III 2016)

Animal	Number of deaths
Mosquito	830,000 (~54%)
Human	580,000 (~38%)
Snake	60,000 (~4%)
Sandfly	24,200 (~1.5%)
Dog	17,400 (~1%)
Kissing bug	8000 (~0.5%)
Others (scorpion, lion, etc.)	15,370 (~1%)

responsible for about half of the total deaths caused by animals, as shown in Table 15.1 (Gates III 2016). The rest of the deaths, in decreasing order, were caused by humans, snakes, sandflies, dogs, kissing bugs, scorpions, lions, and so on.

Mosquitos have unusual locomotory attributes. A study of their locomotion leading to bioinspiration is also of interest.

Wasp anatomy and stinging process have been studied by Das et al. (2018). An overview of wasp's structure and nanomechanical properties data, modeling of the penetration process, and stinger-inspired microneedle design guidelines have been provided by Bhushan (2018). This chapter provides an overview of lessons from mosquitoes' locomotion and their painless piercing. Based on the understanding, conceptual schematics of bioinspired microneedle design is presented for biomedical applications, such as microsyringe needle system for drug delivery, hard tissue biopsy needle, and surgical suture. First, based on Bhushan (2018), mosquitoes locomotory attributes, which are used by them to hunt a host, are described. Next, based on Gurera et al. (2018), microanatomy, feeding process, and nanomechanical properties data of mosquitoes' labrum are presented. A hypothesis behind the painless piercing is presented followed by mosquito-inspired microneedle design guidelines.

15.2 Mosquitoes' Locomotion

A mosquito, irrespective of its sex, can be visually described as shown in Fig. 15.3 (top). It has a part to feed itself, three parts to sense, and two parts to move (Clements 1999; Chapman et al. 2013). The part to feed itself is known as the proboscis. Proboscis is used to pierce into a host and to suck fluids out of it. The proboscis has a length of about 2 mm and its diameter of about 80 μm. Next, three parts to sense are antennas, maxillary palps, and eyes, which helps mosquitoes to hear, taste, and see, respectively. The antennas in mosquitoes are about 1 mm long, and mosquitoes hear using them when vibrations are induced in the antennas by its surroundings (Johnston 1855). The maxillary palps taste various components of air, such as carbon dioxide (Kellogg 1970; Jones et al. 2007). As always, eyes are used by mosquitoes to see (Gao et al. 2007). The three sensing parts, antennas, maxillary palps, and eyes, exist in pairs. Next, two parts to move are wings and legs; however, the former is used more. Mosquitoes have a pair of wings and three pairs of legs. The wings are about 3–6 mm long and about 1 mm wide (Christophers 1960). The legs are about 7–10 mm long. Moreover, a mosquito weighs about 1–4 mg.

The features mentioned, however, do not appear the same for both sexes of mosquitoes, as shown in Fig. 15.3 (bottom) (Gordon and Lavoipierre 1962; Gurera et al. 2018). Visually, male mosquitoes are smaller in size, have longer maxillary palps, and have more bushy antennas (Foster and Walker 2009). There is also a difference in their hearing frequency. Best frequency of the female antennas is about 230 Hz and the male's is about 380 Hz. The males' best frequency happens to be the fundamental frequency of females flying,

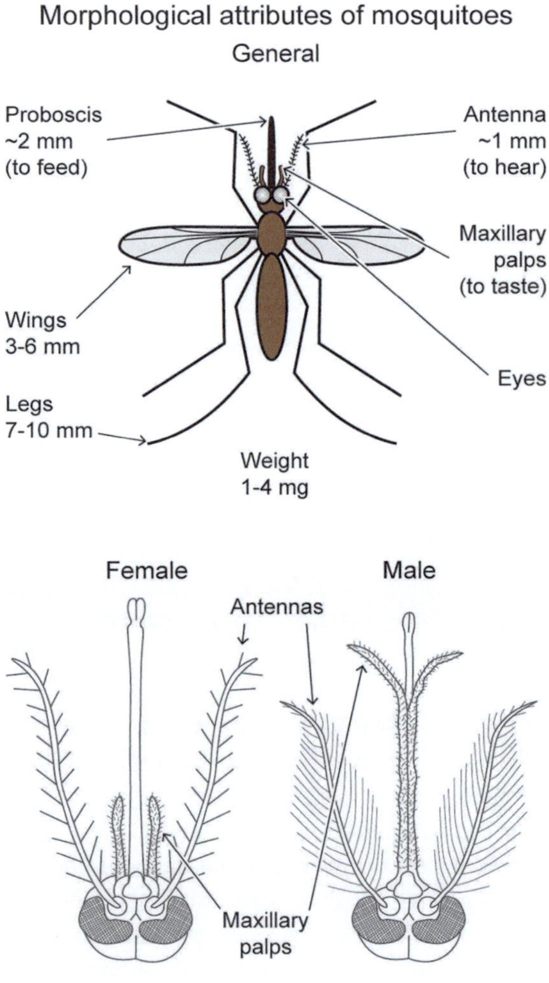

Fig. 15.3 (Top) A mosquito, generally, irrespective of its sex, can be morphologically described as shown. (Bottom) Moreover, there are visual differences between a female and a male mosquito. The male mosquitoes have longer maxillary palps and bushy antennas (*Adapted from* Bhushan 2018; Gurera et al. 2018)

so that they may need it to find their mates (Clements 1999). Length of the female antenna is about 2 mm and male's mosquito is about 1.5 mm. Male and female mosquitoes also have different dietary preferences. Male mosquitoes feed on plant nectar instead of blood. Since the males do not have to pierce the skin, the piercing subparts of the proboscis are not used, and they become nonfunctional.

Mosquitoes have three unusual locomotory attributes: standing on water, sticking to any surface, and flying in rain, Fig. 15.4 (Dickerson et al. 2012; Kong et al. 2015; Gurera 2018). These attributes give them an edge while hunting for a host.

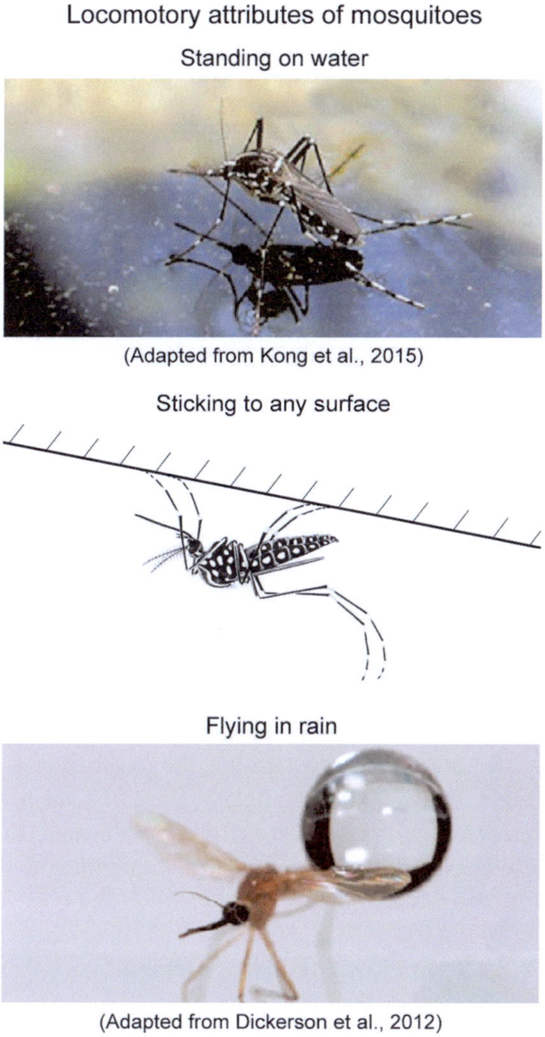

Fig. 15.4 Mosquitoes have three locomotory attributes, which are standing on water, sticking to any surface, and flying in rain (*Adapted from* Bhushan 2018)

15.2.1 Standing on Water

As discussed before, mosquitoes have six legs, which are about 7–10 mm long, and its diameter ranges from about 200 μm at the base, to about 80 μm at the tip (Wu et al. 2007). The leg consists of three parts: femur, tibia, and tarsus. The femur is attached directly to the body and tarsus is the farthest part of the leg from its body. The tarsus is about 6 mm long whereas the other two parts are about 2.5 mm long. Being the thinnest part, the tarsus is more flexible. These parts, in turn, are attached together via joints (Bhushan 2018).

As shown in Fig. 15.5 (top), the legs are covered with hierarchically microstructured scales (Wu et al. 2007; Gurera 2018). The microstructures of the scales consist of longitudinally running ridges and cross running ribs. This, along with the orientation of the scales, along with the leg, provide superhydrophobicity with water contact angle > 150° on the surface, as shown in Fig. 15.5 (bottom). This water repellency gives mosquitoes the ability to stand on water (Bhushan 2018).

When standing on water, only the distal part of the leg, known as tarsus, is in contact with water (Kong et al. 2015). It has been reported that a tarsus can take load up to 20 times the weight of the mosquito, and the whole leg can take up to 23 times the weight of a mosquito. There are two key factors playing role in making tarsus such an efficient weight bearer: flexibility and angle of stepping in the water (Bhushan 2018).

Since the tarsus is flexible, it means it can bend easily at a load. When a tarsus is pressed against water, it will continue to deform above the surface of the water. If it would not have been flexible, for the same load, the contact area of the tarsus and the water would have been lesser, and the tarsus would have pierced the water easily. However, in the case of actual tarsus, the contact area will keep on increasing and hence total pressure on water will decrease. Therefore, it will be difficult for tarsus to pierce water easily (Bhushan 2018).

The tarsus is at an inclination when it comes in contact with water. The inclination angle also affects the standing abilities of mosquitoes. Similar theory, which was just discussed, can be applied here too—when the angles are low, they result in more contact area, hence lesser pressure is applied. This low pressure can account for the high-bearing loads. It has been reported that the largest forces appear in a range of 0°–20° inclination angle. At an inclination angle of about 70°, the bearing load appears to diminish, since the leg pierces water as soon as the contact begins (Kong et al. 2015). In practice, mosquitoes can change this angle by flexing or extending their legs accordingly (Bhushan 2018).

15.2.2 Sticking to Any Surface

As shown in Fig. 15.6 (top), mosquitoes feet are a combination of claws and setae that help them attach (Wu et al. 2007; Chapman et al. 2013; Gurera 2018). The distal part

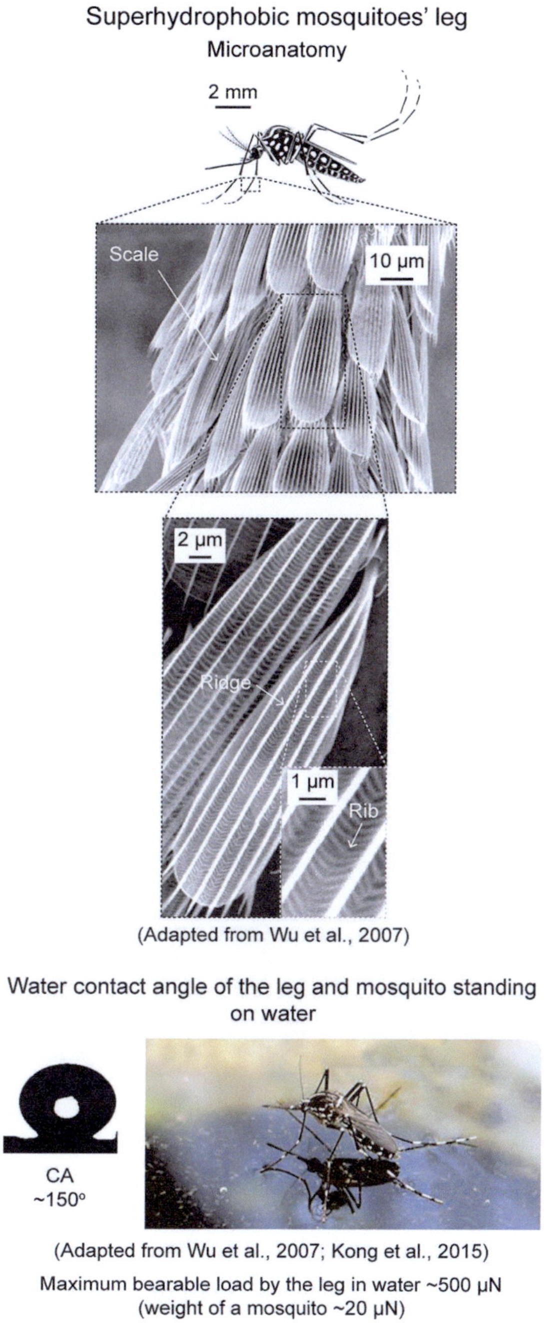

Fig. 15.5 Mosquitoes' legs are covered in micro- and nanostructures (top), which give them the ability to stand in water. (Bottom) A water contact angle measurement on the leg and a mosquito standing on water (*Adapted from* Bhushan 2018)

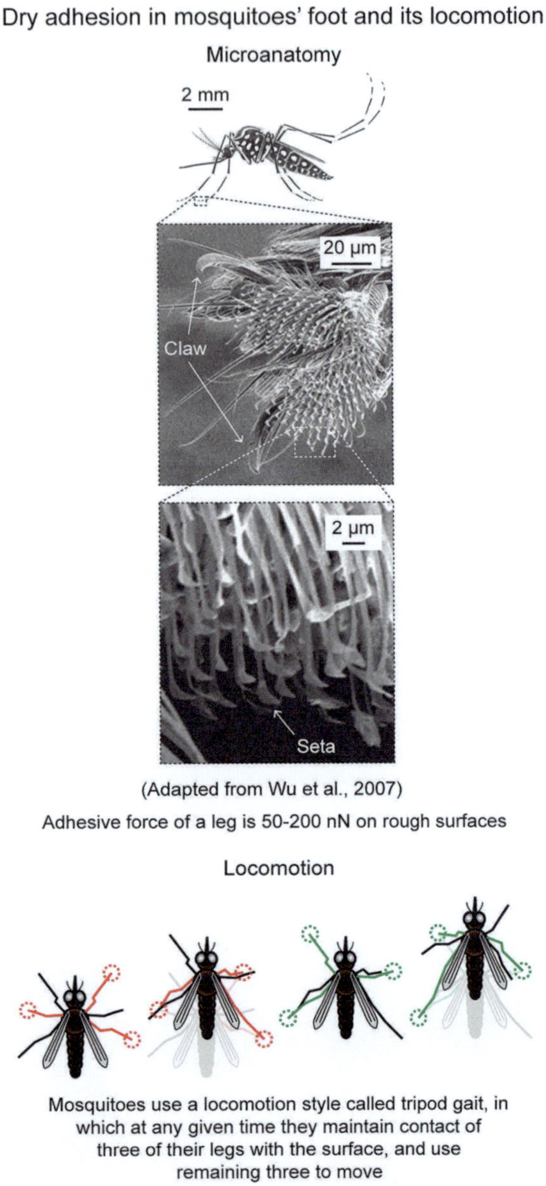

Fig. 15.6 Mosquitoes' feet have a combination of claws and setae (top). Claws help them attach to rough surfaces and setae help them attach to smooth surfaces. Mosquitoes use tripod gait to move (bottom). The dotted circles represent foot adhered to the surface (*Adapted from* Bhushan 2018)

of the tarsus is referred to as the foot. The claws are used to attach to rough surfaces and the setae are used to attach to smoother surfaces. The setae at their end have a pad with a contact area of about 2 μm^2. These pads attach to a smoother surface using van der Waals force, in a manner similar to geckos. The mechanism can generate 50–200 nN force per leg (Pashazanusi et al. 2017). However, these forces are on the higher side if the surface is rough. The wettability of the surface, to which mosquitoes attach, also changes the mechanism of attaching. If the surface is hydrophobic, van der Waals forces are predominant. If the surface is hydrophilic, capillary forces also played a role in increasing the adhesive force (Bhushan 2018).

After attaching to a surface, it uses tripod gait, which is common in most six-legged arthropods, as shown in Fig. 15.6 (bottom). In the gait, at any given time it maintains contact of three of their legs with the surface, and use remaining three to move (Bhushan 2018).

15.2.3 Flying in Air and Rain

Leading edge vortex is a commonly used mechanism for flying in nature (Dudley 2002). In the mechanism, flying species get an extra lift, by developing a vortex formed at the front edge of wings. The vortex decreases air pressure at the top of the wing and gives them a lift. However, in the case of mosquitoes, the trailing edge vortex plays a crucial role in generating the lift, which is exclusive to mosquitoes (Bomphrey et al. 2017). During flapping, as the back edge of the wing flips around, it captures the swirling wave of the previous flap, allowing it to capture some of that energy. Another exclusive feature of mosquitoes is its high flapping frequency, of about 700 Hz, and low flapping amplitude of about 40°. The flapping amplitude is less than half of any other flying animal studied. The authors are unsure of any advantage of this mechanism (Bhushan 2018).

When flying in rain, microscales and nanohairs on the wing, as shown in Fig. 15.7a, are beneficial (Sudo et al. 2000; Gurera 2018). The structures help the wings remain hydrophobic, water contact angle on the wing is about 95° (Pal 1950). Only the veins of the wing are covered with microscales (Chapman et al. 2013), and the length of a scale is about 90 μm. The wing is also covered with hairs, which measure about 35 μm in length. These features are beneficial to mosquitoes during flying in rain (Bhushan 2018).

Given the fact that the mass ratio of a raindrop to mosquito can be 2–50, a raindrop can produce a substantially larger force than the weight of a mosquito. Legs and wings are the most probable area of an impact since they are the longest body parts of mosquitoes. The insect may be killed if it is flying to close to the ground. After getting an impact it takes the insect about 15–60 mm distance to recover (Dickerson et al. 2012). Also, after recovering from an impact of a raindrop, especially, on its wing, it rotates up to an angle about 50° (Bhushan 2018).

Flying in a variety of conditions, like fog, may adversely affect mosquito wings. In fog, as compared to rain, drops are smaller; a raindrop is on the order of 1000–5000 μm (NASA 2016), and a fog droplet is on the order of 1–50 μm in diameter (United Nations Environment Programme 1998). So, when a water drop is smaller than the distance between micro-asperities, it will follow the Wenzel regime, and stick to the surface. After an accumulation of drops, they all will coalesce to form a big drop that is attached

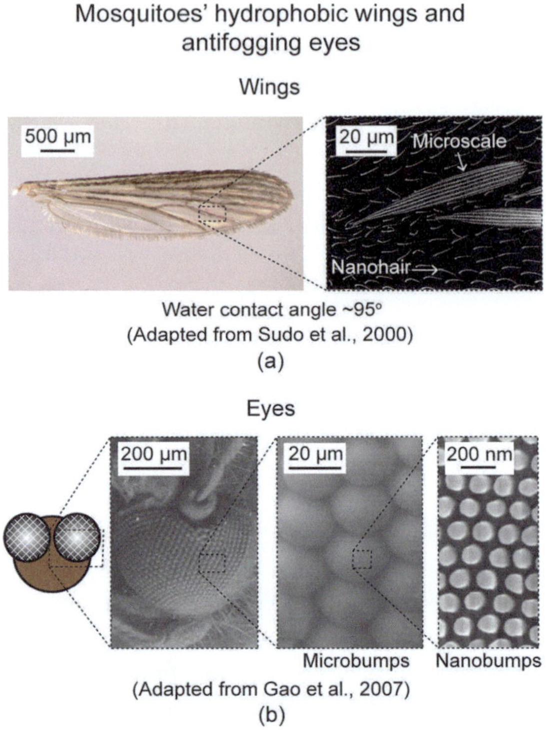

Fig. 15.7 **a** Mosquitoes wings have a combination of microscales and nanohairs, which makes the wing hydrophobic, and gives mosquitoes ability to fly in rain. **b** Mosquitoes' hemispherical eyes consist of hexagonally arranged micro- and nanobumps. The hierarchically structured eyes repel water and give mosquitoes the ability to see in the fog (*Adapted from* Bhushan 2018)

to the wing. It has been reported that the big drop's water tension will be enough to overcome the bending stiffness of the wing (Dickerson et al. 2015). Most of the stiffness of wings are provided by its veins. However, they run in a longitudinal direction, hence the drop gets folded in the transverse direction. The authors of the study report that when a wing thickness is less than 50 μm, the wing folds; and thickness of mosquitoes' wing is on the order of 10 μm. In practice, mosquitoes attempt to remove the excess water by increasing flapping frequency, hard landing, and/or taking off at different angles from the ground (Bhushan 2018).

Mosquitoes also have antifogging eyes, which help them to keep their vision clear (Gao et al. 2007). During fog, the wetting issue that existed on the wings does not exist in the case of the eyes. On the wing, the space between the asperities is more than the fog drop diameter. In the case of the eyes, it is not so. As shown in Fig. 15.7b, the eyes consist of hexagonally, closely packed microbumps with a diameter of about 25 μm, and loosely packed nanobumps with a diameter of about 100 nm and a spacing of about 50 nm (Gao et al. 2007). Since the spacing is less than the fog drop diameter, which is 10–50 μm in diameter, the drops follow Cassie-Baxter regime. The regime states that there is an air pocket formation between the drop and the surface; hence, the drop can roll-off easily.

In addition, since the eyes, on a macroscale, are hemispherical, the rolling of drops may take place in real time (Bhushan 2018).

15.2.4 Summary

Mosquitoes can stand on water, stick to any surface, and fly in rain. Mosquitoes are able to stand on water because of their hydrophobic legs, which are covered with hierarchically structured scales. They are able to stick to any surface due to dry adhesion in their feet, which is due to a combination of claws and setae. The foot can apply an adhesive force of one to two orders of nN on a rough surface. Lastly, mosquitoes are able to fly in rain because of their hydrophobic wings covered with microscales and nanohairs and antifogging eyes covered with micro- and nanobumps (Bhushan 2018).

15.3 Mosquitoes' Painless Piercing

15.3.1 Microanatomy

Mosquitoes feed on a host using their mouthpart, known as the proboscis. A detailed schematic of the proboscis (left column) and its various scanning electron micrographs (SEM) (right column) are shown in Fig. 15.8 (Jones 1978; Gurera et al. 2018). The proboscis is a collective term used for mosquito's fascicle and its retractable, thick, outer cover, known as labium, as shown in the top of the figure. The labium being blunt does not pierce the host and retracts during the piercing. However, the fascicle, being sharper, pierces the host (Bhushan 2018; Gurera et al. 2018).

A fascicle is a term for the bundle of stylets, shown in Fig. 15.8. The largest of them is a pointed and hollow stylet, the labrum, which sucks the blood. The others include a pair each of sharp mandibles and serrated-shaped maxilla, which sit on both sides of the labrum. These provide a major contribution to piercing since these are sharp and serrated. The maxilla, having serrated edges, also helps anchor the fascicle in place once inside the host. The final stylet of the fascicle is the hypopharynx which releases saliva during the initial piercing (Choumet et al. 2012; Bhushan 2018; Gurera et al. 2018).

The mosquitoes' saliva serves several purposes including numbing pain during piercing and stopping the blood clotting during sucking. Mosquitoes' salivary glands contain 0.3 μg of protein with more than 1000 compositions which are responsible for various functions (Ribeiro et al. 2001). As an example for anopheline mosquitoes, anosphensin family of protein, which is responsible for numbing, inhibits growth of a pain-inducing peptide knows as bradykinin (Ribeiro et al. 2010). For the same species of mosquitoes,

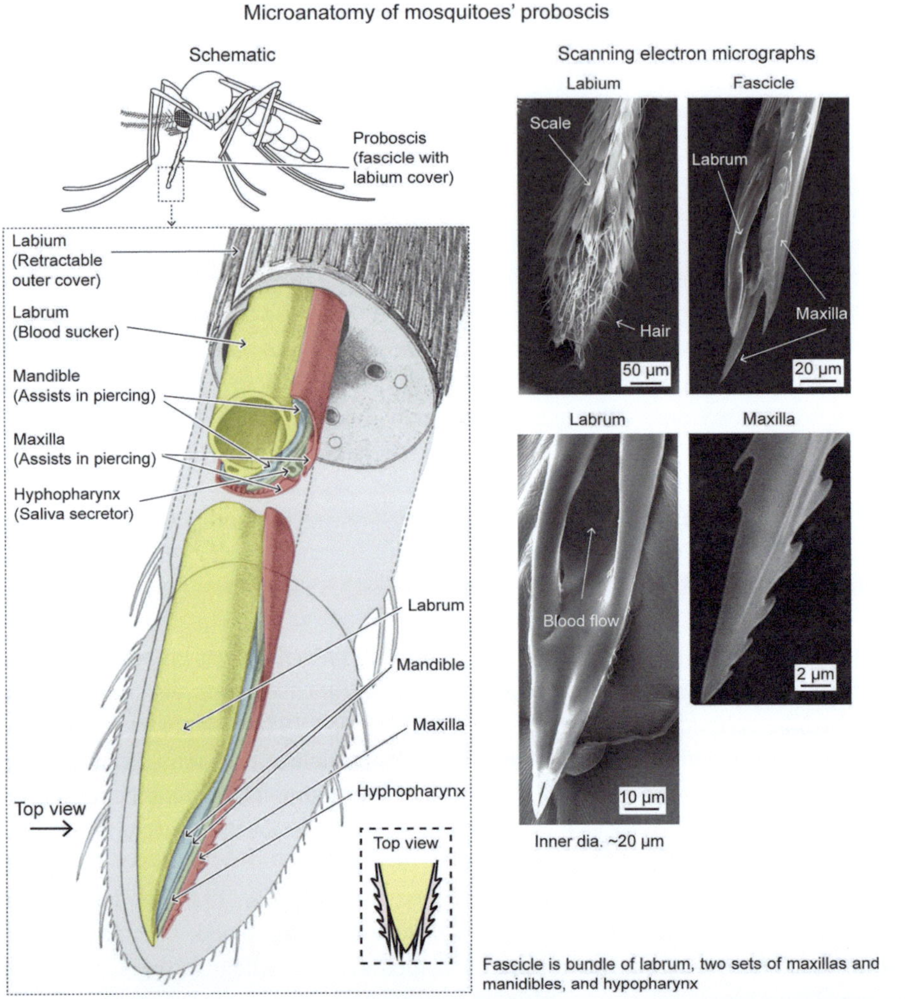

Fig. 15.8 Schematics showing microanatomy of the mosquitoes' proboscis in the left column, and SEMs of the stylets of the proboscis in the right column (*Adapted from* Bhushan 2018; Gurera et al. 2018)

anophelin family of protein, which is responsible for anti-blood-clotting, inhibits processing of fibrinogen to fibrin protein that is known to clot the blood. Incidentally, the malaria parasite is also transferred into the host via this saliva (Bhushan 2018; Gurera et al. 2018).

SEM micrographs of labium, fascicle, labrum, and maxilla are shown in the right column of Fig. 15.8. The labium is covered with scales all around it and has hairs at its tip. The hairs help in sensing a suitable piercing spot on the host (Foster and Walker 2009). The fascicle is shown with the labrum and a pair of maxillae bundled together.

The labrum is shown which has an outer diameter of about 30 μm and the inner diameter of about 20 μm. The maxilla is shown with the serrated edges (Bhushan 2018; Gurera et al. 2018).

15.3.2 Feeding

A detailed schematic of the feeding process of mosquitoes is shown in Fig. 15.9 (Jones 1978; Gurera et al. 2018). First, mosquitoes search for a suitable spot on the surface of the host, using the tip of its labium. After locating a spot, the fascicle consisting of a serrated design (with sharp teeth) is inserted inside the host at about 15 Hz with vibratory motion of about 40–90 μm (Kong and Wu 2010). The saliva, which acts as a numbing agent, is released under the skin by the hypopharynx (Bhushan 2018; Gurera et al. 2018).

The insertion force used by mosquitoes is mostly in the range of 10–20 μN and is about three times lower than the lowest reported insertion force reported by an artificial needle. When insertion forces are lowered, lesser tissue area will be deformed, which reduces nerve sensation. In other words, less pain or sensation will be felt by the host. Therefore, this vibratory motion is believed to help in reducing pain in the host. It is known that vibratory actuation in needles helps to reduce the insertion force. It should be noted that the labium is a retractable outer cover and it does not pierce but provides lateral support to the fascicle (Ramasubramanian et al. 2008). Without the labium, the fascicle will not be successful in piercing and will buckle instead (Bhushan 2018; Gurera et al. 2018).

Next, once inside the host, the labrum constantly searches for a blood vessel to pierce in. The labrum being flexible, is able to bend at about 90°, and also can be maneuvered in any direction (Gordon and Lumsden 1939). It uses specialized sensors at the tip of the labrum which sense blood vessels (Hudson 1970; Foster and Walker 2009). It moves fascicle around in the skin tissue searching for an ideal place to tap for the blood source. After locating a blood vessel, a vibratory motion is used again to pierce in. However, this time only the labrum pierces and at a lowered frequency of about 5 Hz (Gurera et al. 2018). It uses a specialized pumping system for suction through its feeding tube, labrum. Mosquitoes use about 8 kPa pressure to suck blood out of their host (Chapman et al. 2013). During the blood sucking, the injected saliva, in addition to acting as a numbing agent mentioned earlier, prevents blood clotting and inflammation to minimize discomfort, and keeps the blood flowing (Bhushan 2018).

There are reasons for the lowered frequency at end of the piercing process (Gurera et al. 2018). First, the material to be pierced, the blood vessel, is a softer material as compared to the skin. Therefore, the less vigorous motion will do the work. Second, there is an increase in friction between the labrum and the host's tissues. As the labrum keeps going inside the host, there will be an increase in contact area between labrum and

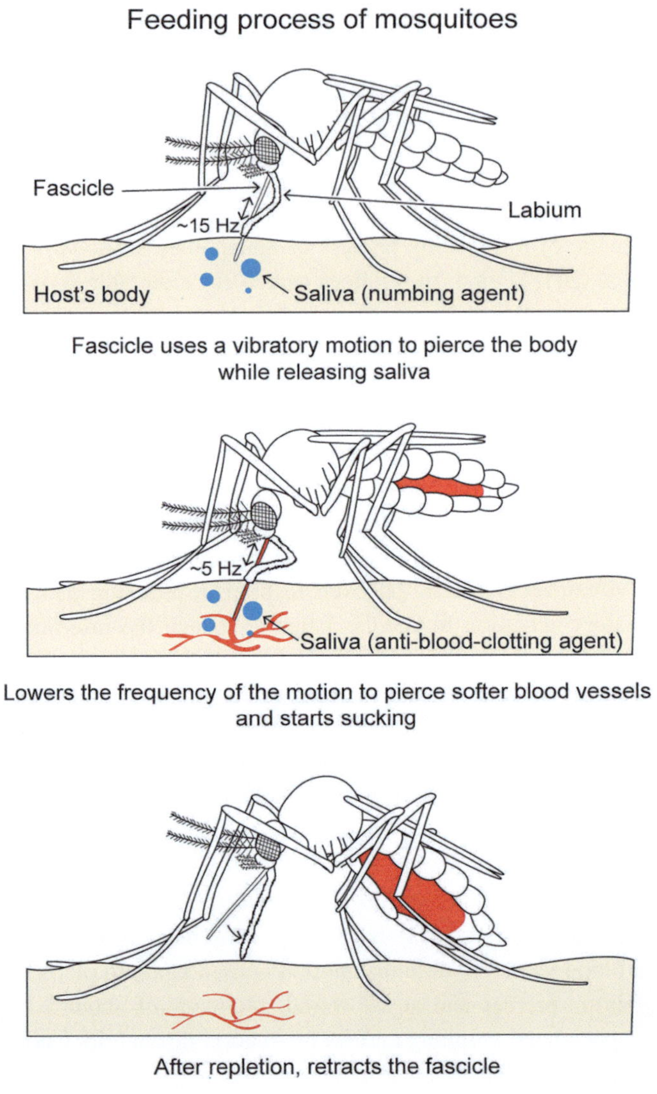

Fig. 15.9 Schematics of various steps in the feeding process used by mosquitoes (*Adapted from* Bhushan 2018; Gurera et al. 2018)

the host. That may increase the frictional forces; hence mosquitoes are unable to keep up the same frequency (Bhushan 2018).

Finally, after the mosquito is repleted, the fascicle is retracted in the labium cover.

Finding a suitable location on the host can take 5–30 s depending on the age of a mosquito. Once inside the host, finding a blood vessel can take on the order of 150–400 s. The blood feeding can take on the order of 5 min. The average amount of sucked blood by mosquitoes is on the order of few milligrams, which is comparable to their bodyweight (Nayar and Sauerman 1975). Mosquito piercing generally causes a localized inflammatory response. The itchy bump one gets from mosquito bite is caused by dilating of blood vessels and host body's immune response to mosquito saliva, which is an allergen. Itching stops several hours after piercing (Bhushan 2018).

To sum up, in the painless piercing process, the tip of the labium is used to search for a suitable location on the host. Once the location is found, the fascicle consisting of a serrated design (with sharp teeth) pierces in and the saliva, which acts as a numbing agent, is released by the hypopharynx. A vibratory motion is used to pierce the sharp mandible and serrated maxilla inside the host. The serration in the maxilla also helps in keeping the parts anchored to the host's tissues. The wider labrum helps the entire fascicle pierce the skin. Once the labrum is in, mosquitoes lower their vibration frequency and start looking for a blood vessel using the sensor located at tip of the labrum. After locating a blood vessel, the labrum pierces the blood vessels and starts sucking in blood, until repletion. During the blood sucking, the injected saliva, in addition to acting as a numbing agent, prevents blood clotting and inflammation to minimize discomfort, and keeps the blood flowing. Finally, the fascicle retracts, and it goes back inside the labium cover (Bhushan 2018; Gurera et al. 2018).

15.3.3 Nanomechanical Property Measurements of Labium

To further understand the piercing process, material properties of mosquitoes' mouthparts are needed. During piercing, the thin mandible and maxilla initiate the piercing, but it is the labrum, which is about 35 times wider, which helps in the insertion of the entire fascicle into a host. Therefore, the labrum's tip has been chosen for nanomechanical characterization (Gurera et al. 2018).

Mosquito Species Selection

The type of mosquito to be used in this study were selected based on its dietary preference and availability (Gurera et al. 2018). Females were selected because, as discussed earlier, only they feed on humans, whereas the males feed on plants. Among the females, a specie named *Aedes vexans* was chosen because they are aggressive piercers, high in population, and are found near residential areas (DeGroote et al. 2007). Moreover, it is one of the common species with an epidemiological significance (Lemine et al. 2017) as it was among the responsible species for spreading West Nile Virus across North America in 1999.

Nanomechanical Characterization Technique

To measure local properties as well as gradients in mechanical properties over small regions of labrum with a thickness of about 10 μm, nanomechanical property measurements are needed with depth of penetration of few hundred nm. A depth-sensing nanoindentation technique was used to measure elastic modulus and hardness under static and dynamic conditions (Bhushan et al. 1996; Bhushan 2017). A TI 950 TriboIndenter® (Hysitron, Inc., Minneapolis, Minnesota, USA), with a Berkovich tip (tip radius is on the order of 100 nm), calibrated with fused silica was used. In the technique, a ramp-up load is applied to the sample using a nanoprobe. As the load is being applied, the load and the probe's penetration depth are continuously measured. The load versus penetration depth data was used to calculate elastic modulus and nanohardness (Bhushan 2018; Gurera et al. 2018).

The labrum samples were prepared by dissecting the labrum from a mosquito and pasting it on a glass substrate using double sided tape. To characterize the labrum's tip, seven indentation locations distributed uniformly across the tip were chosen as shown in Fig. 15.10 (Gurera et al. 2018). The seven locations are defined by the intersection of three non-parallel longitudinal lines, L1–L3, with three parallel transverse lines, T1–T3. A nomenclature of the locations has also been chosen in the study. For example, the location formed by the intersection of L2 and T1 will be called "L2, T1".

Biological materials have viscoelastic material properties which are loading rate dependent. Since mosquitoes vibrate their labrum to pierce into their host, their properties were measured in dynamic conditions, in addition to static conditions. As mentioned before, mosquitoes use the insertion load of 10–20 μN while piercing a prey and the frequency of 5–15 Hz. Therefore, the measurements were made at a static load of 10 μN and at an oscillating load of the frequency ranging from 5–200 Hz, to measure static and dynamic effects, respectively (Bhushan 2018; Gurera et al. 2018).

For static properties, a trapezoidal shaped load function was used in which the sample is loaded to 10 μN in six seconds, the load is held for two seconds, and then it is unloaded to zero-load in the last six seconds, shown in Fig. 15.11 (Gurera et al. 2018). The load cycle was based on the standard protocol used in nanoindentation studies (Hysitron 2018). For dynamic properties, a small dynamic load of 0.2 μN amplitude and a frequency

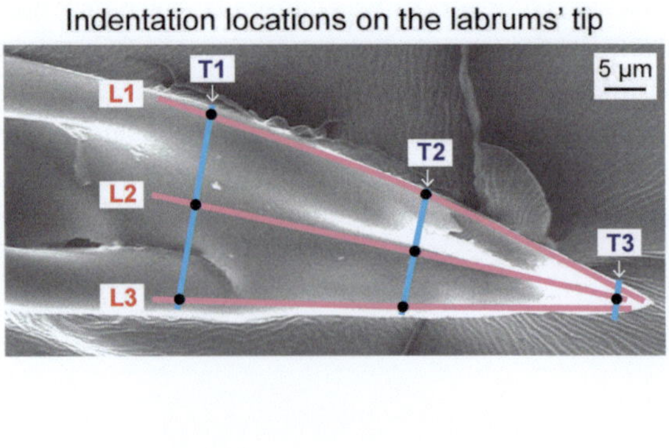

Fig. 15.10 The labrum's tip was characterized at presented seven locations. The locations were formed by the intersection of three longitudinal lines, L1–L3, with three transverse lines, T1–T3 (*Adapted from* Bhushan 2018; Gurera et al. 2018)

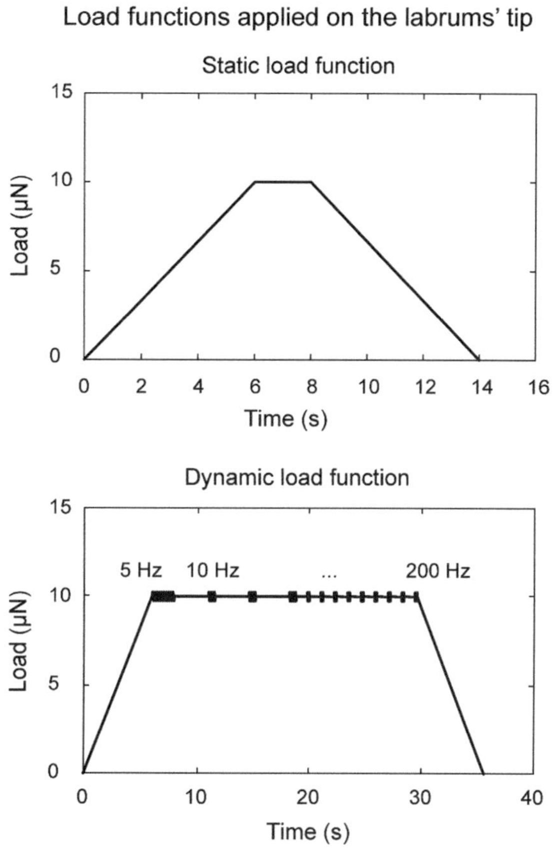

Fig. 15.11 Labrum's tip was characterized using two loading functions: static (top) and dynamic (bottom). The static load function has a peak load of 10 μN. The dynamic load function has a quasistatic load of 10 μN and dynamic load of 0.3 μN. The tested frequencies were 5, 10, 15, 20, 40, 60, 80, 100, 120, 140, 160, 180, and 200 Hz (*Adapted from* Bhushan 2018; Gurera et al. 2018)

sweeping from 5 to 200 Hz were superimposed on a static load of 10 μN, shown in Fig. 15.11. The frequencies used were 5, 10, 15, 20, 40, 60, 80, 100, 120, 140, 160, 180, and 200 Hz.

Measurements were made on five different labrum samples for each type of measurement, and their average values are reported. For each sample, all of the seven locations were tested right after the sample preparation. This was done because it is known that desiccation affects mechanical properties of a biological material. The penetration depths of all the presented data were measured to be less than 10% of the labrum's thickness—10 μm, except for the location — "L2, T3", at end of the tip. This implies that there was no effect on the substrate on which the labrums were mounted. The data were taken in a temperature- and humidity-controlled environment, in which the temperature and relative humidity ranged from 21°–23°C and 40%–50%, respectively (Bhushan 2018; Gurera et al. 2018).

Experimental Data

Static properties data of the labrum's tip is presented in Fig. 15.12 (Gurera et al. 2018). Figure 15.12a shows representative load–displacement data on two extreme locations. Note that the material creeps when the load is held for two seconds. Of the two locations, "L2, T1" and "L2, T3", higher displacement was observed in the latter one.

Trends observed in the mechanical properties of the labrum's tip are shown as line plots in Fig. 15.12b and as maps in Fig. 15.12c. To understand the trends, discussion on the plots follows. In Fig. 15.12b, the left and the right columns show the trends along L1–L3, and T1–T3, respectively. The solid lines shown are the average values and the error bars shown are one standard deviation from the average. It is observed that the elastic modulus decreases along L1–L3. Elastic modulus is a measure of elastic deformation and is also referred as stiffness. This implies that the end of the tip is compliant. It is also observed that the modulus along L2 is lower than along the adjacent L1 and L3. This implies that the tip is stiffer on sides as compared to its middle (Bhushan 2018; Gurera et al. 2018).

Similar trends were observed in the case of nanohardness (Gurera et al. 2018). Hardness is a measure of resistance to plastic deformation. It is observed that the nanohardness along L2 is lower than along the adjacent L1 and L3. This implies that the tip is softer at the end of the tip. It should be noted that the labrum tip is one order less hard than another piercing part, the human tooth (He and Swain 2011).

Dynamic properties data of the labrum's tip is shown in Fig. 15.13 (Gurera et al. 2018). Figure 15.13a shows trends in mechanical properties as a function of the frequency in a form of line plots and as maps in Fig. 15.13b. To facilitate the understanding of the trends, a discussion of the line plots follows. In Fig. 15.13a, the left and the right column shows the mechanical properties at all locations along L1, and at all locations along T1, respectively. The mechanical properties presented are storage modulus and tan δ. The solid lines shown are curve fitted lines and the error bars are deviations of the average values from the fitted lines (Bhushan 2018; Gurera et al. 2018).

Trends for storage modulus are presented, which are similar to those of elastic modulus. Storage modulus is a measure of elastic properties of the material. The storage modulus was found to be decreasing along L1–L3, however, only the trend along L1 is shown. As before, the modulus along L2 was also found to be lower, as compared to the sides L1 and L3. In addition, it was found that the magnitude of the storage modulus increases with frequency, which implies that labrum's tip gets stiffer with increase in the frequency (Bhushan 2018; Gurera et al. 2018).

Figure 15.13 also presents trends in tan δ versus frequency (Gurera et al. 2018). Tan δ is a ratio of viscous properties to elastic properties. It is a measure of the amount of viscosity in a material as compared to its elasticity. It is the parameter used for comparing the dynamic nature of different materials. Given the scatter in the data, there appear to be no trends in the tan δ values. However, the data is of the same order as that of dragonflies' wing (Sun and Bhushan 2012) and beetles' cuticle (Sun et al. 2016).

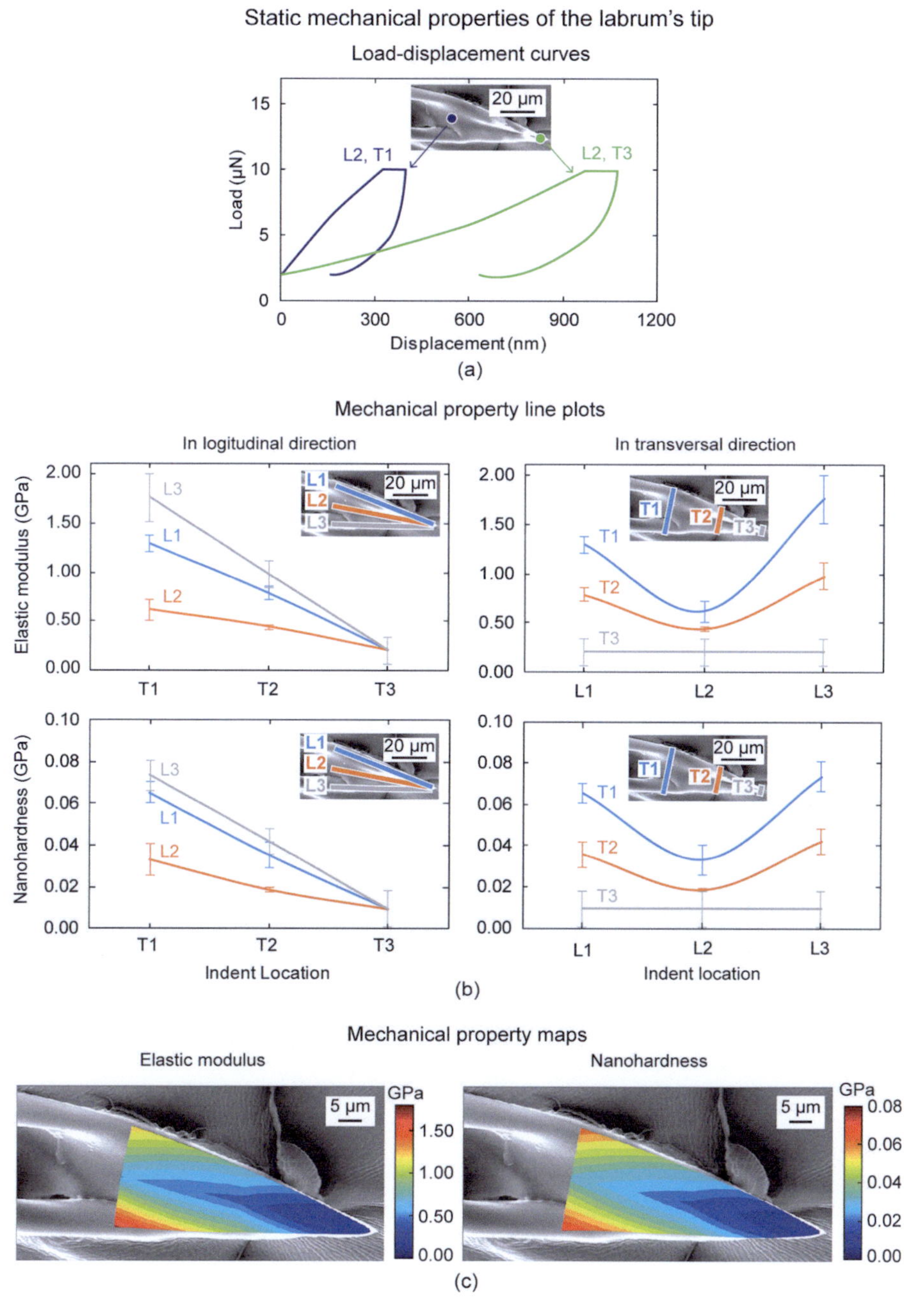

Fig. 15.12 **a** Representative load–displacement curves at two locations—"L2, T1" and "L2, T3". **b** Mechanical property line plots and **c** maps of elastic modulus and nanohardness. The tip has a gradient in mechanical properties, which decreases along L1–L3, and has a dip along T1–T3. The stiffness and hardness decrease along the tip and is lower in the middle across the tip (*Adapted from* Bhushan 2018; Gurera et al. 2018)

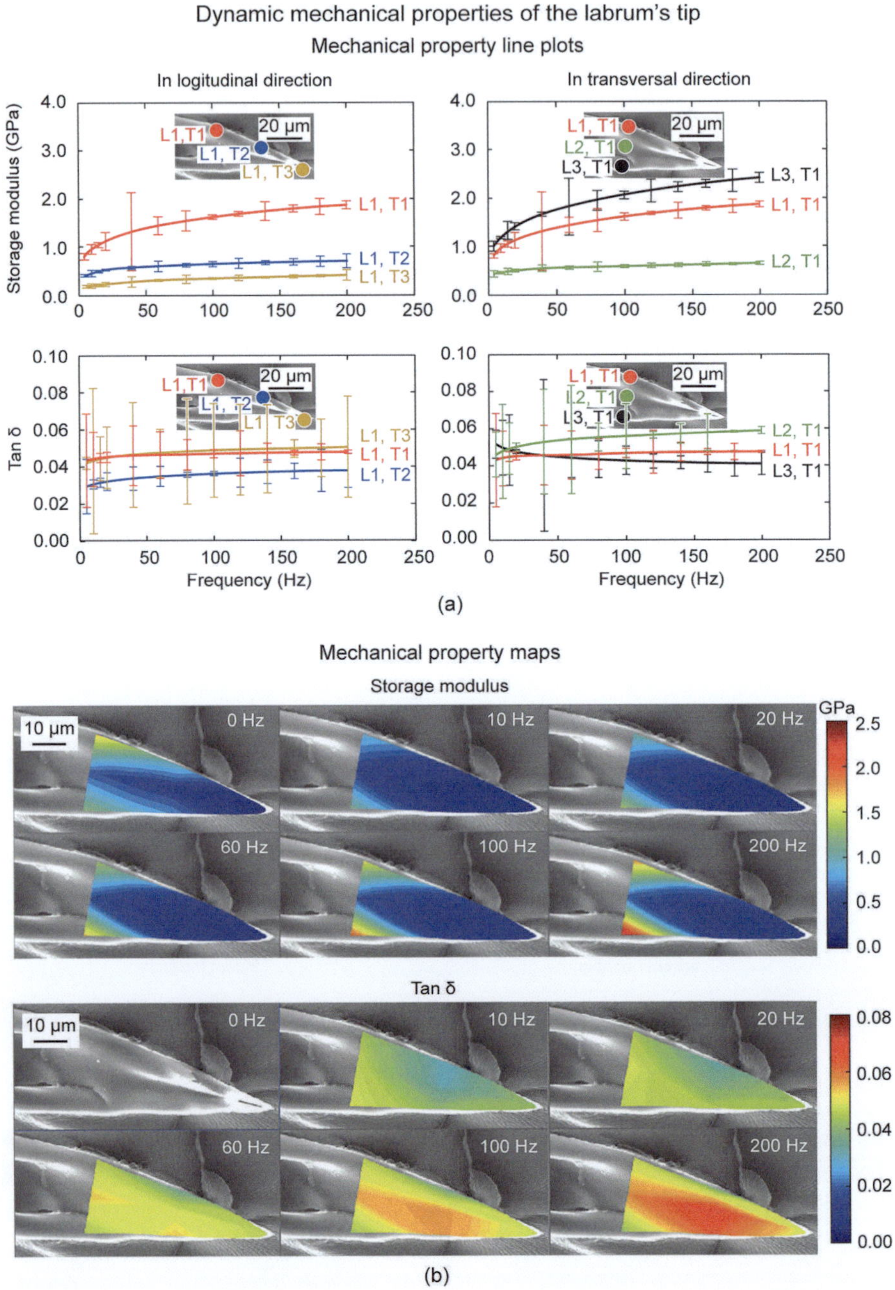

Fig. 15.13 **a** Line plots and **b** maps of storage modulus and tan δ as a function of frequency. Storage modulus increases with the increase in frequency, however the tan δ does not change. Irrespective of the frequency, the stiffness decreases along the tip and has lower stiffness in the middle across the tip (*Adapted from* Bhushan 2018; Gurera et al. 2018)

Relevance of Nanomechanical Properties in Piercing Mechanism
It is noted that elastic modulus and nanohardness decrease along the labrum, with end of the tip being most compliant and softest. Across the labrum, the properties are lowest in the middle. It is also noted that the labrum gets stiffer at higher frequencies. It is expected that the more compliant and softer end of the tip may cause lesser pain while piercing the host, because it will deform the skin less. The lesser deformation sensed by the nerve endings in the skin will cause lesser pain. The end of the tip may also be compliant and soft because it is used for sensing blood vessels inside the host. Furthermore, the stiffer and harder sides of the tip will also help the labrum to resist bending/buckling during piercing. Finally, increase in stiffness at the high frequency may maintain lower deformation at higher stresses (Bhushan 2018; Gurera et al. 2018).

15.3.4 Lessons from Mosquito Piercing and Conceptual Schematic of a Painless Mosquito-Inspired Microneedle

A mosquito can insert a needle-like fascicle into the host skin and draws blood for several minutes without host even noticing. The mechanism behind the painless piercing includes the combination of the frequency dependent graded mechanical properties, numbing agent secretion, serrated design, and vibratory actuation (Gurera et al. 2018). As discussed earlier, the most compliant and softest end of the labrum's tip as compared to the rest of it, reduces the pain felt by the host. However, the compliant and soft end of the tip will increase the insertion force. To compensate for that increase, the serrated design and the vibratory actuation assist in easy insertion. The serrated design increases the number of high-stressed points on the skin, which means easy insertion, and vibratory actuation has also been known to lower the insertion force. In addition, vibratory motion has also been known to increase stiffness of the labrum, which will limit any excessive deformation in the labrum's tip during the vibratory piercing. The pain will also be further reduced by the numbing agent secretion (Bhushan 2018; Gurera et al. 2018).

There are conical-shaped microneedles commercially available that are made up of a medical grade polymer, of length on the order of 500–1500 μm (3M 2015; Brady and Donnelly 2016; NanoPass Technologies Ltd 2017). However, they induce pain when inserted. One can learn from the mosquito's piercing mechanism and this understanding could be used as a starting point to create a painless microneedle for biomedical purposes. A conceptual schematic of a painless, bioinspired microneedle has been proposed, shown in Fig. 15.14 (Bhushan 2018; Gurera et al. 2018).

In Fig. 15.14 (top), the microneedle has a channel for fluid passage and serrated design, similar to the labrum and the maxilla, respectively (Gurera et al. 2018). The microneedle is also accompanied by a numbing agent secretor, which together are enclosed within a thick covering, similar to the hypopharynx and the labium, respectively. The tip of the microneedle shown in the cross-sectional detailed view is based on the combination of

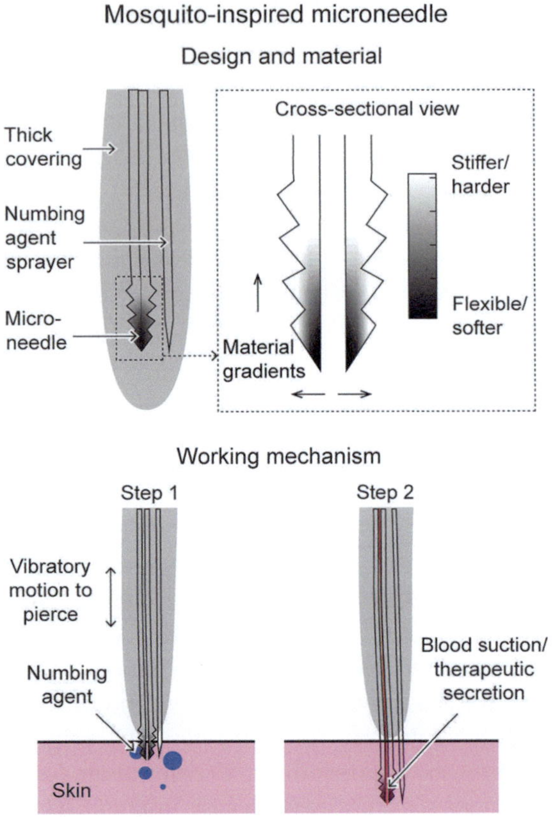

Fig. 15.14 Conceptual schematic of a painless mosquito-inspired microneedle, along with its material and design (top) and its working mechanism (bottom) (*Adapted from* Bhushan 2018; Gurera et al. 2018)

the mechanical properties discussed earlier. The microneedle is made up of a viscoelastic material having gradients in mechanical properties in both longitudinal and transverse directions similar to that of the labrum.

The microneedle also has a working mechanism similar to that of mosquitoes, as shown in Fig. 15.14 (bottom) (Gurera et al. 2018). The microneedle would be inserted using a vibratory actuation, which along with the serrated design to assist in piercing. At the start of insertion, a numbing agent would be secreted. The factors responsible for easy insertion, along with the numbing agent secretion and the longitudinal gradient in mechanical properties are believed to reduce the pain. The lateral material gradient along with the increasing stiffness with the frequency will also help in a successful piercing, by keeping the microneedle away from bending/buckling and having higher deformations at higher stresses, respectively. In addition, the outer covering will prevent the entire microneedle from buckling. After successful insertion, either blood can be sucked or a therapeutic agent can be secreted, based on the need (Bhushan 2018; Gurera et al. 2018).

15.3.5 Summary

Among various medically important arthropods which feed on living hosts, mosquitoes' piercing spreads viruses which causes the highest number of deaths annually. The primary cause of the deaths is malaria, which is spread by infected mosquitoes' piercing. Lessons from mosquito locomotion and painless piercing are presented. To survive, mosquitoes need to hunt for a host for blood. It is relatively easy for mosquitoes since they possess unusual locomotory attributes: standing on water, sticking to any surface, and flying in rain, because of their superhydrophobic legs, dry adhesion in the foot, and hydrophobic wings and antifogging eyes, respectively. Mosquitoes pierce using their fascicle, which is a bundle of coherently functioning six stylets. Mosquitoes painlessly pierce using a combination of the numbing by releasing saliva to act as a numbing agent, the fascicle's serrated design, the vibratory actuation, and the graded and frequency-dependent mechanical properties of the labrum with a soft tip. Once they pierce the blood vessel, they suck the blood until repletion. After the repletion, the fascicle is retracted. Based on this understanding, conceptual schematic of a painless mosquito-inspired microneedle has been proposed (Bhushan 2018; Gurera et al. 2018).

15.4 Closure

Arthropods are the largest group of living organisms. They attack other organisms by biting, stinging, or piercing and sucking. In this chapter, the piercing and sucking mechanism in mosquitoes is described. Based on lessons learned from mosquitoes, conceptual schematic of bioinspired painless microneedle designs is presented.

It is expected that this study will provide an impetus for fabrication of painless bioinspired microneedles requiring low insertion forces for biomedical applications such as micro-syringe needle system for drug delivery, hard tissue biopsy needle, and surgical suture.

References

3M (2015), "3M^{TM} Hollow Microstructured Transdermal System," http://multimedia.3m.com/mws/media/996631O/hollow-microstructured-transdermal-system-hmts-flyer.pdf

Akre, R. D., Greene, A., MacDonald, J. F., Landolt, P. J., and Davis, H. G. (1981), *The Yellowjackets of America North of Mexico*, Agricultural Handbook No. 552, U.S. Department of Agriculture, Washington, D.C.

Barthlott, W., Rafiqpoor, M. D., and Erdelen, W. R. (2016), "Bionics and Biodiversity—Bio-Inspired Technical Innovation for a Sustainable Future," In *Biomimetic Research for Architecture and Building Construction: Biological Design and Integrative Structures*, (eds. J. Knippers, K. G. Nickel, and T. Speck), first ed., pp. 11–55, Springer International, Cham, Switzerland.

Bhushan, B. (2017), "Depth-Sensing Nanoindentation Measurement Techniques and Applications," *Microsyst. Technol.* **23**, 1595–1649.

Bhushan, B. (2018), *Biomimetics: Bioinspired Hierarchical-Structured Surfaces for Green Science and Technology*, third ed., Springer International, Cham, Switzerland.

Bhushan, B., Kulkarni, A. V., Bonin, W., and Wyrobek, J. T. (1996), "Nanoindentation and Picoindentation Measurements Using a Capacitive Transducer System in Atomic Force Microscopy," *Philos. Mag. A* **74**, 1117–1128.

Bomphrey, R. J., Nakata, T., Phillips, N., and Walker, S. M. (2017), "Smart Wing Rotation and Trailing-Edge Vortices Enable High Frequency Mosquito Flight," *Nature* **544**, 92–95.

Brady, A. J., and Donnelly, R. F. (2016), "Clinical Impact and Patient Safety: The Potential of Microneedles in Changing the Form and Perception of Transdermal Drug Delivery," *Nanoscience in Dermatology* (eds. M. R. Hamblin, P. Avci, and T. W. Prow), pp. 47–56, Academic Press, London, U. K.

Chapman, R. F., Simpson, S. J., and Douglas, A. E. (2013), *The Insects: Structure and Function*, fifth ed., Cambridge University Press, Cambridge, U. K.

Choumet, V., Attout, T., Chartier, L., Khun, H., Sautereau, J., Robbe-Vincent, A., Brey, P., Huerre, M., and Bain, O. (2012), "Visualizing Non Infectious and Infectious Anopheles Gambiae Blood Feedings in Naive and Saliva-Immunized Mice," *PLoS ONE* **7**, e50464.

Christophers, S. R. (1960), *Aëdes Aegypti (L.) the Yellow Fever Mosquito: Its Life History, Bionomics and Structure*, Cambridge University Press, London, U. K.

Clements, A. N. (1999), *The Biology of Mosquitoes, Volume 2, Sensory Reception and Behaviour*, CABI Publishing, Oxon, U. K.

Das, R., Yadav, R., Sihota, P., Uniyal, P., Kumar, N., and Bhushan, B. (2018), "Biomechanical Evaluation of Wasp and Honeybee Stingers," *Sci. Rep.—Nature* **8**, 14945.

DeGroote, J., Mercer, D. R., Fisher, J., and Sugumaran, R. (2007), "Spatiotemporal Investigation of Adult Mosquito (Diptera: Culicidae) Populations in an Eastern Iowa County, USA," *J. Med. Entomol.* **44**, 1139–1150.

Dickerson, G. and Lavoipierre, M. M. J. (1959), "Studies on the Methods of Feeding of Bloodsucking Arthropods," *Trop. Med. Parasitol.* **53**, 347–357.

Dickerson, A. K., Liu, X., Zhu, T., and Hu, D. L. (2015), "Fog Spontaneously Folds Mosquito Wings," *Phys. Fluids* **27**, 21901–21901.

Dickerson, A. K., Shankles, P. G., Madhavan, N. M., and Hu, D. L. (2012), "Mosquitoes Survive Raindrop Collisions by Virtue of Their Low Mass," *Proc. Natl. Acad. Sci. U.S.A.* **109**, 9822–9827.

Dudley, R. (2002), *The Biomechanics of Insect Flight: Form, Function, Evolution*, Princeton University Press, Princeton, New Jersey.

Durden, L. A., and Mullen, G. R. (2009), "Introduction," In *Medical and Veterinary Entomology* (eds. G. R. Mullen, and L. A. Durden), second ed., pp. 1–12, Academic Press, Burlington, MA.

Foster, W. A., and Walker, E. D. (2009), "Mosquitoes," In *Medical and Veterinary Entomology* (eds. G. R. Mullen, and L. A. Durden), second ed., pp. 207–260, Academic Press, Burlington, MA.

Gao, X., Yan, X., Yao, X., Xu, L., Zhang, K., Zhang, J., Yang, B., and Jiang, L. (2007), "The Dry-Style Antifogging Properties of Mosquito Compound Eyes and Artificial Analogues Prepared by Soft Lithography," *Adv. Mater.* **19**, 2213–2217.

Gates III, W. H. (2016), "Why I'd Rather Cuddle with a Shark than a Kissing Bug," https://www.gatesnotes.com/Health/Most-Deadly-Animal-Mosquito-Week-2016

Goddard, J. (2013), *Physician's Guide to Arthropods of Medical Importance*, sixth ed., CRC Press, Boca Raton, Florida.

Gordon, R. M., and Lavoipierre, M. M. J. (1962), *Entomology for Students of Medicine*, Blackwell Scientific, Oxford, U. K.

Gordon, R. M., and Lumsden, W. H. R. (1939), "A Study of the Behaviour of the Mouth-parts of Mosquitoes When Taking Up Blood from Living Tissue; Together with Some Observations on the Ingestion of Microfilariae," *Ann. Trop. Med. Parasitol.* **33**, 259–278.

Gurera, D. (2018), "Lessons from Nature and Bioinspired Fabrication: Mosquito Bite and Lotus Leaf Inspired Superliquiphobic Leather," M. S. Thesis, Department of Mechanical and Aerospace Engineering, The Ohio State University, Columbus, Ohio.

Gurera, D., Bhushan, B., and Kumar, N. (2018), "Lessons from Mosquitoes' Painless Piercing," *J. Mech. Behav. Biomed.* **84**, 178-187.

He, L.-H., and Swain, M. V. (2011), "Nanoindentation of Tooth Tissues," M. L. Oyen (Ed.), *Handbook of Nanoindentation: With Biological Applications*, pp. 239–278, Pan Stanford, Singapore.

Hudson, A. (1970), "Notes on the Piercing Mouthparts of Three Species of Mosquitoes (Diptera: Culicidae) Viewed with the Scanning Electron Microscope," *Can. Entomol.* **102**, 501–509.

Hysitron (2018), "Probe Calibration," https://www.hysitron.com/media/1683/t-003-v30-probe-calibration.pdf

Ito, J., Ghosh, A., Moreira, L. A., Wimmer, E. A., and Jacobs-Lorena, M. (2002), "Transgenic Anopheline Mosquitoes Impaired in Transmission of a Malaria Parasite," *Nature* **417**, 452–455.

Johnston, C. (1855), "Original Communications: Auditory Apparatus of the Culex Mosquito," *J. Cell. Sci.* **s1–3**, 97–102.

Jones, J. C. (1978), "The Feeding Behavior of Mosquitoes," *Sci. Am.* **238**, 138–148.

Jones, W. D., Cayirlioglu, P., Kadow, I. G., and Vosshall, L. B. (2007), "Two Chemosensory Receptors Together Mediate Carbon Dioxide Detection in Drosophila," *Nature* **445**, 86–90.

Kellogg, F. E. (1970), "Water Vapour and Carbon Dioxide Receptors in Aedes Aegypti," *J. Insect Physiol.* **16**, 99–108.

Kong, X. Q., and Wu, C. W. (2010), "Mosquito Proboscis: An Elegant Biomicroelectromechanical System," *Phys. Rev. E* **82**, 011910.

Kong, X. Q., Liu, J. L., Zhang, W. J., and Qu, Y. D. (2015), "Load-Bearing Ability of the Mosquito Tarsus on Water Surfaces Arising from Its Flexibility," *AIP Adv.* **5**, 037101.

Lemine, A. M. M., Lemrabott, M. A. O., Ebou, M. H., Lekweiry, K. M., Salem, M. S. O. A., Brahim, K. O., Moukah, M. O., Bouraya, I. N. O., Brengues, C., Trape, J.-F., Basco, L., Bogreau, H., Simard, F., Faye, O., and Boukhary, A. O. M. S. (2017), "Mosquitoes (Diptera: Culicidae) in Mauritania: A Review of Their Biodiversity, Distribution and Medical Importance," *Parasit. Vectors* **10**, 1–13.

McGavin, G. (1993), *Bugs of the World*, Facts on File, New York.

Mora, C., Tittensor, D. P., Adl, S., Simpson, A. G. B., and Worm, B. (2011), "How Many Species Are There on Earth and in the Ocean?," *PLoS Biol.* **9**, e1001127.

NanoPass Technologies Ltd (2017), "MicronJet600 Advanced Intradermal Solution," http://www.nanopass.com/product/

NASA (2016), "Size Matters: NASA Measures Raindrop Sizes from Space to Understand Storms," https://www.nasa.gov/feature/goddard/2016/size-matters-nasa-measures-raindrop-sizes-from-space-to-understand-storms

Nayar, J. K., and Sauerman, J. D. M. (1975), "The Effects of Nutrition on Survival and Fecundity in Florida Mosquitoes: Part 2. Utilization of a Blood Meal for Survival," *J. Med. Entomol.* **12**, 99–103.

Pal, R. (1950), "The Wetting of Insect Cuticle," *Bull. Entomol. Res.* **41**, 121–139.

Parker, S. P. (Ed.) (1997), *McGraw-Hill Encyclopedia of Science & Technology*, eighth ed., McGraw-Hill, New York.

Pashazanusi, L., Lwoya, B., Oak, S., Khosla, T., Albert, J. N. L., Tian, Y., Bansal, G., Kumar, N., and Pesika, N. S. (2017), "Enhanced Adhesion of Mosquitoes to Rough Surfaces," *ACS Appl. Mater. Interfaces* **9**, 24373–24380.

Ramasubramanian, M. K., Barham, O. M., and Swaminathan, V. (2008), "Mechanics of a Mosquito Bite with Applications to Microneedle Design.," *Bioinspir. Biomim.* **3**, 1–10.
Ribeiro, J. M. C., Charlab, R., and Valenzuela, J. G. (2001), "The Salivary Adenosine Deaminase Activity of the Mosquitoes Culex Quinquefasciatus and Aedes Aegypti," *J. Exp. Biol.* **204**, 2001–2010.
Ribeiro, J. M. C., Mans, B. J., and Arcà, B. (2010), "An Insight into the Sialome of Blood-Feeding Nematocera," *Insect Biochem. Mol. Biol.* **40**, 767–784.
Schmidt, J. O. (2009), "Wasps," In *Encyclopedia of Insects* (eds. V. H. Resh and R. T. Carde), second ed., pp. 1049–1052, Elsevier Science & Technology, Oxford, U. K.
Service, M. W. (1978), "A Short History of Early Medical Entomology," *J. Med. Entomol.* **14**, 603–626.
Stork, N. E. (2009), "Biodiversity," In *Encyclopedia of Insects,* V. H. Resh and R. T. Carde (Eds.), second ed., pp. 75–80, Elsevier Science and Technology, Oxford, U. K.
Sudo, S., Tsuyuki, K., and Tani, J. (2000), "Wing Morphology of Some Insects," *JSME Int. J. Ser. C* **43**, 895–900.
Sun, J., and Bhushan, B. (2012), "The Structure and Mechanical Properties of Dragonfly Wings and Their Role on Flyability," *Comptes Rendus Mécanique* **340**, 3–17.
Sun, J., Wu, W., Ling, M., Bhushan, B., and Tong, J. (2016), "A Dynamic Nanoindentation Technique to Investigate the Nanomechanical Properties of a Colored Beetle," *RSC Adv.* **6**, 79106–79113.
United Nations Environment Programme (1998), "Sourcebook of Alternative Technologies for Freshwater Augumentation in Africa," http://www.unep.or.jp/ietc/publications/techpublications/techpub-8a/fog.asp
WHO (2016), "World Malaria Report 2015," http://www.who.int/malaria/publications/world-malaria-report-2015/report/en/
Wu, C. W., Kong, X. Q., and Wu, D. (2007), "Micronanostructures of the Scales on a Mosquito's Legs and Their Role in Weight Support," *Phys. Rev. E* **76**, 17301.

Index

B. Bhushan, *Introduction to Biomimetics and Bioinspiration*,
https://doi.org/10.1007/978-3-031-62344-8

D

E

N

R

T

MIX
Papier aus verantwortungsvollen Quellen
Paper from responsible sources
FSC® C105338

If you have any concerns about our products,
you can contact us on
ProductSafety@springernature.com

In case Publisher is established outside the EU,
the EU authorized representative is:
Springer Nature Customer Service Center GmbH
Europaplatz 3, 69115 Heidelberg, Germany

Printed by Libri Plureos GmbH
in Hamburg, Germany